Available with MyStatLab™ for Your Business Statistics Courses

MyStatLab is the market-leading online learning management program for learning and teaching business statistics.

Statistical Software Support

Built-in tutorial videos and functionality make using the most popular software solutions seamless and intuitive. Tutorial videos, study cards, and manuals (for select titles) are available within MyStatLab and accessible at the point of use. Easily launch exercise and eText data sets into Excel or StatCrunch, or copy and paste into any other software program.

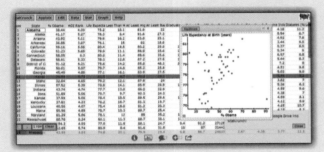

Leverage the Power of StatCrunch

MyStatLab leverages the power of StatCrunch —powerful, web-based statistical software. In addition, access to the full online community allows users to take advantage of a wide variety of resources and applications at www.statcrunch.com.

Bring Statistics to Life

Virtually flip coins, roll dice, draw cards, and interact with animations on your mobile device with the extensive menu of experiments and applets in StatCrunch. Offering a number of ways to practice resampling procedures, such as permutation tests and bootstrap confidence intervals, StatCrunch is a complete and modern solution.

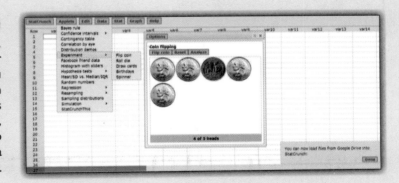

www.mystatlab.com

Business Statistics
A First Course

Business Statistics
A First Course
3rd Edition

Norean R. Sharpe
Georgetown University

Richard D. De Veaux
Williams College

Paul F. Velleman
Cornell University
With Contributions by David Bock

PEARSON

Boston Columbus Indianapolis New York San Francisco Amsterdam Cape Town
Dubai London Madrid Milan Munich Paris Montrèal Toronto Delhi Mexico City
São Paulo Sydney Hong Kong Seoul Singapore Taipei Tokyo

Editor in Chief: Deirdre Lynch
Acquisitions Editor: Patrick Barbera
Editorial Assistant: Justin Billing
Program Manager: Chere Bemelmans
Project Manager: Sherry Berg
Product Marketing Manager: Tiffany Bitzel
Field Marketing Manager: Adam Goldstein
Marketing Assistant: Jennifer Myers
Program Management Team Lead: Karen Wernholm
Project Management Team Lead: Peter Silvia
Senior Author Support/Technology Specialist: Joe Vetere
Associate Director of Design: Andrea Nix

Program Design Lead/Cover Designer: Beth Paquin
Text Design: Studio Montage
Manager, Rights Management: Gina M. Cheselka
Media Producer: Aimee Thorne
MathXL Senior Project Manager: Robert Carroll
QA Manager, Assessment Content: Marty Wright
Senior Procurement Specialist: Carol Melville
Full-Service Project Management, Composition, and
    Illustrations: Lumina Datamatics, Inc.
Cover Photo: Silhouette of business women in a skyscraper,
    ©Yvan Dubé/Getty Images

Credits and acknowledgments borrowed from other sources and reproduced, with permission, in this textbook appear on the appropriate page within text or in Appendix C, which is hereby made part of this copyright page.

The student edition of this book has been cataloged by the Library of Congress as follows:

Library of Congress Cataloging-in-Publication Data
Sharpe, Norean Radke.
   Business Statistics: A First Course / Norean R. Sharpe, Richard D. De Veaux, Paul F. Velleman,
   with contributions by David Bock. —Third edition.
       pages cm
   Includes bibliographical references and index.
   ISBN 0-13-418244-8 (hardcover)
   1. Commercial statistics. I. De Veaux, Richard D. II. Velleman, Paul F., 1949– III. Title.
   HF1017.S468 2017
   519.5—dc23
                                                    2015016942

1 2 3 4 5 6 7 8 9 10—RRD—19 18 17 16

www.pearsonhighered.com

ISBN 10: 0-13-418244-8
ISBN 13: 978-0-13-418244-5

To my husband, Peter, for his patience and support
—*Norean*

To my family
—*Dick*

To my father, who taught me about ethical business practice by his constant example as a small businessman and parent
—*Paul*

Meet the Authors

Norean Radke Sharpe (Ph.D. University of Virginia) has developed an international reputation as an educator, administrator, and consultant on assessment and accreditation. She is a professor at the McDonough School of Business at Georgetown University, where she is also Senior Associate Dean and Director of Undergraduate Programs. Prior to joining Georgetown, Norean taught business statistics and operations research courses to both undergraduate and MBA students for fourteen years at Babson College. Before moving into business education, she taught statistics for several years at Bowdoin College and conducted research at Yale University. Norean is coauthor of the recent text, *A Casebook for Business Statistics: Laboratories for Decision Making*, and she has authored more than 30 articles—primarily in the areas of statistics education and women in science. Norean currently serves as Associate Editor for the journal *Cases in Business, Industry, and Government Statistics*. Her scholarship focuses on business forecasting, statistics education, and student learning. She is co-founder of the DOME Foundation, a nonprofit foundation that works to increase Diversity and Outreach in Mathematics and Engineering, and she currently serves on two other nonprofit boards in the Washington, D.C. area. Norean has been active in increasing the participation of women and underrepresented students in science and mathematics for several years and has two children of her own.

Richard D. De Veaux (Ph.D. Stanford University) is an internationally known educator, consultant, and lecturer. Dick has taught statistics at a business school (Wharton), an engineering school (Princeton), and a liberal arts college (Williams). While at Princeton, he won a Lifetime Award for Dedication and Excellence in Teaching. Since 1994, he has taught at Williams College, although he returned to Princeton for the academic year 2006–2007 as the William R. Kenan Jr. Visiting Professor of Distinguished Teaching. He is currently the C. Carlisle and Margaret Tippit Professor of Statistics at Williams College. Dick holds degrees from Princeton University in Civil Engineering and Mathematics and from Stanford University in Dance Education and Statistics, where he studied with Persi Diaconis. His research focuses on the analysis of large data sets and data mining in science and industry. Dick has won both the Wilcoxon and Shewell awards from the American Society for Quality. He is an elected member of the International Statistics Institute (ISI) and a Fellow of the American Statistical Association (ASA). He currently serves on the Board of Directors of the ASA. Dick is also well known in industry, having consulted for such *Fortune* 500 companies as American Express, Hewlett-Packard, Alcoa, DuPont, Pillsbury, General Electric, and Chemical Bank. He was named the "Statistician of the Year" for 2008 by the Boston Chapter of the American Statistical Association for his contributions to teaching, research, and consulting. In his spare time he is an avid cyclist and swimmer. He also is the founder and bass for the doo-wop group the Diminished Faculty and is a frequent singer and soloist with various local choirs, including the Choeur Vittoria of Paris, France. Dick is the father of four children.

Paul F. Velleman (Ph.D. Princeton University) has an international reputation for innovative statistics education. He designed the Data Desk® software package and is also the author and designer of the award-winning ActivStats® multimedia software, for which he received the EDUCOM Medal for innovative uses of computers in teaching statistics and the ICTCM Award for Innovation in Using Technology in College Mathematics. He is the founder and CEO of Data Description, Inc. (www.datadesk.com), which supports both of these programs. He also developed the Internet site *Data and Story Library* (DASL; lib.stat.cmu.edu/DASL/), which provides data sets for teaching Statistics. Paul coauthored (with David Hoaglin) the book *ABCs of Exploratory Data Analysis*. Paul teaches Statistics at Cornell University in the Department of Statistical Sciences and in the School of Industrial and Labor Relations, for which he has been awarded the MacIntyre Prize for Exemplary Teaching. His research often focuses on statistical graphics and data analysis methods. Paul is a Fellow of the American Statistical Association and of the American Association for the Advancement of Science. Paul's experience as a professor, entrepreneur, and business leader brings a unique perspective to the book.

Richard De Veaux and Paul Velleman have authored successful books in the introductory college and AP High School market with David Bock, including *Intro Stats*, Fourth Edition (Pearson, 2014); *Stats: Modeling the World*, Fourth Edition (Pearson, 2015); and *Stats: Data and Models*, Fourth Edition (Pearson, 2016).

Contents

Preface

The question that should motivate a business student's study of Statistics should be "How can I make better decisions?"[1] As entrepreneurs and consultants, we know that in today's data-rich environment, knowledge of Statistics is essential to survive and thrive in the business world. But, as educators, we've seen a disconnect between the way business statistics is traditionally taught and the way it should be used in making business decisions. In *Business Statistics: A First Course*, we try to narrow the gap between theory and practice by presenting relevant statistical methods that will empower business students to make effective, data-informed decisions.

Of course, students should come away from their statistics course knowing how to think statistically and how to apply statistics methods with modern technology. But they must also be able to communicate their analyses effectively to others. When asked about statistics education, a group of CEOs from *Fortune* 500 companies recently said that although they were satisfied with the technical competence of students who had studied statistics, they found the students' ability to communicate their findings to be woefully inadequate.

Our Plan, Do, Report rubric provides a structure for solving business problems that mimics the correct application of statistics to solving real business problems. We emphasize the often neglected thinking (Plan) and communication (Report) steps in problem solving in addition to the methodology (Do). This approach requires up-to-date, real-world examples and data. So we constantly strive to illustrate our lessons with current business issues and examples.

What's New in This Edition?

We've been delighted with the reaction to previous editions of *Business Statistics: A First Course*. We've streamlined the third edition further to help students focus on the central material. And, of course, we continue to update examples and exercises so that the story we tell is always tied to the ways Statistics informs modern business practice.

- **Recent data.** We teach with real data whenever possible, so we've updated data throughout the book. New examples reflect current stories in the news and recent economic and business events. The Brief Cases have been updated with new data and new contexts.

- **Improved organization.** We have retained our "data first" presentation of topics because we find that it provides students with both motivation and a foundation in real business decisions on which to build an understanding.

  - Chapters 1–4 have been streamlined to cover collecting, displaying, summarizing, and understanding data in four chapters. We find that this provides students with a solid foundation to launch their study of probability and statistics.
  - Chapters 5–9 introduce students to randomness and probability models. They then apply these new concepts to sampling. This provides a gateway to the core material on statistical inference. We've moved the discussion of probability trees and Bayes' rule into these chapters.
  - Chapters 10–13 cover inference for both proportions and means. We introduce inference by discussing proportions because most students are better acquainted with proportions reported in surveys and news stories. However, this edition ties in the discussion of means immediately so students can appreciate that the reasoning of inference is the same in a variety of contexts.
  - Chapters 14 and 15 cover regression-based models for decision making.
  - Chapter 16 introduces Data Mining and points the way forward for further study.

- **Streamlined design.** Our goal has always been an accessible text. This edition sports a new design that clarifies the purpose of each text element. The major theme of each

---

[1]Unfortunately, not the question most students are asking themselves on the first day of the course.

chapter is more linear and easier to follow without distraction. Supporting material is clearly boxed and shaded, so students know where to focus their study efforts.

- **Enhanced Technology Help with expanded Excel 2013 coverage.** We've updated Technology Help with instructions for R and XLStat and added detailed instructions for Excel 2013 to almost every chapter.

- **Updated Ethics in Action features.** We've updated more than half of our Ethics in Action features. Ethically and statistically sound alternative approaches to the questions raised in these features and a link to the American Statistical Association's Ethical Guidelines are now presented in the Instructor's Solutions Manual, making the Ethics features suitable for assignment or class discussion.

- **Updated examples to reflect the changing world.** The time since our last revision has seen marked changes in the U.S. and world economies. Our examples and exercises have been updated to keep pace.

- **Increased focus on core material.** Statistics in practice means making smart decisions based on data. Students need to know the methods, how to apply them, and the assumptions and conditions that make them work. We've tightened our discussions to get students there as quickly as possible, focusing increasingly on the central ideas and core material.

- **Used MyStatLab performance data to improve exercises.** The authors analyzed aggregated student usage and performance data from MyStatLab for the previous edition of this text. The results of this analysis helped improve the quality and quantity of exercises that matter the most to instructors and students.

Our Approach

Statistical Thinking

For all of our improvements, examples, and updates in this edition of *Business Statistics: A First Course* we haven't lost sight of our original mission—writing a modern business statistics text that addresses the importance of *statistical thinking* in making business decisions and that acknowledges how Statistics is actually used in business.

Statistics is practiced with technology, and this insight informs everything from our choice of forms for equations (favoring intuitive forms over calculation forms) to our extensive use of real data. But most important, understanding the value of technology allows us to focus on teaching statistical thinking rather than calculation. The questions that motivate each of our hundreds of examples are not "How do you find the answer?" but "How do you think about the answer?", "How does it help you make a better decision?", and "How can you best communicate your decision?"

Our focus on statistical thinking ties the chapters of the book together. An introductory Business Statistics course covers an overwhelming number of new terms, concepts, and methods, and it is vital that students see their central core: how we can understand more about the world and make better decisions by understanding what the data tell us. From this perspective, it is easy to see that the patterns we look for in graphs are the same as those we think about when we prepare to make inferences. We can see that the many ways to draw inferences from data are several applications of the same core concepts. And it follows naturally that when we extend these basic ideas into more complex (and even more realistic) situations, the same basic reasoning is still at the core of our analyses.

Our Goal: Read This Book!

The best textbook in the world is of little value if it isn't read. Here are some of the ways we made *Business Statistics: A First Course* more approachable:

- *Readability.* We strive for a conversational, approachable style, and we introduce anecdotes to maintain interest. Instructors report (to their amazement) that their students read ahead of their assignments voluntarily. Students tell us (to *their* amazement) that they actually enjoy the book. In this edition, we've tightened our discussions to be more focused on the central ideas we want to convey.

- ***Focus on assumptions and conditions.*** More than any other textbook, *Business Statistics: A First Course* emphasizes the need to verify assumptions when using statistical procedures. We reiterate this focus throughout the examples and exercises. We make every effort to provide templates that reinforce the practice of checking these assumptions and conditions, rather than rushing through the computations. Business decisions have consequences. Blind calculations open the door to errors that could easily be avoided by taking the time to graph the data, check assumptions and conditions, and then check again that the results and residuals make sense.

- ***Emphasis on graphing and exploring data.*** We consistently emphasize the importance of displaying data. Examples often illustrate the value of examining data graphically, and the Exercises reinforce this. Good graphics reveal structures, patterns, and occasional anomalies that could otherwise go unnoticed. These patterns often raise new questions and inform both the path of a resulting statistical analysis and the business decisions. Hundreds of new graphics found throughout the book demonstrate that the simple structures that underlie even the most sophisticated statistical inferences are the same ones we look for in the simplest examples. This helps tie the concepts of the book together to tell a coherent story.

- ***Consistency.*** We work hard to avoid the "do what we say, not what we do" trap. Having taught the importance of plotting data and checking assumptions and conditions, we are careful to model that behavior throughout the book. (Check the Exercises in the chapter on multiple regression and you'll find us still requiring and demonstrating the plots and checks that were introduced in the early chapters.) This consistency helps reinforce these fundamental principles and provides a familiar foundation for the more sophisticated topics.

- ***The need to read.*** In this book, important concepts, definitions, and sample solutions are not always set aside in boxes. The book needs to be read, so we've tried to make the reading experience enjoyable. The common approach of skimming for definitions or starting with the exercises and looking up examples just won't work here. (It never did work as a way to learn about and understand Statistics.)

Coverage

The topics covered in a Business Statistics course are generally mandated by our students' needs in their studies and in their future professions. But the *order* of these topics and the relative emphasis given to each is not well established. *Business Statistics: A First Course* presents some topics sooner or later than other texts. Although many chapters can be taught in a different order, we urge you to consider the order we have chosen.

We've been guided in the order of topics by the fundamental goal of designing a coherent course in which concepts and methods fit together to provide a new understanding of how reasoning with data can uncover new and important truths. Each new topic should fit into the growing structure of understanding that students develop throughout the course. For example, we teach inference concepts with proportions first and then with means. Most people have a wider experience with proportions, seeing them in polls and advertising. And by starting with proportions, we can teach inference with the Normal model and then introduce inference for means with the Student's *t* distribution.

We introduce the concepts of association, correlation, and regression early. Our experience in the classroom shows that introducing these fundamental ideas early makes Statistics useful and relevant even at the beginning of the course. By Chapter 4, students can discuss relationships among variables in a meaningful way. Later in the semester, when we discuss inference, it is natural and relatively easy to build on the fundamental concepts learned earlier and enhance them with inferential methods.

GAISE Report

We've been guided in our choice of what to emphasize by the GAISE (Guidelines for Assessment and Instruction in Statistics Education) Report, which emerged from extensive studies of how students best learn Statistics (**http://www.amstat. org/education/gaise/GaiseCollege_full.pdf**). Those recommendations, now officially adopted and recommended by the American Statistical Association, urge (among other detailed suggestions) that Statistics education should:

1. Emphasize statistical literacy and develop statistical thinking.
2. Use real data.
3. Stress conceptual understanding rather than mere knowledge of procedures.
4. Foster active learning.
5. Use technology for developing conceptual understanding and analyzing data.
6. Make assessment a part of the learning process.

In this sense, this book is thoroughly modern.

Continuing Features

A textbook isn't just words on a page. A textbook is many elements that come together to form a big picture. The features in *Business Statistics: A First Course* provide a real-world context for concepts, help students apply these concepts, promote problem solving, and integrate technology—all of which help students understand and see the big picture of Business Statistics.

Providing Real-World Context

Motivating Vignettes. Each chapter opens with a motivating vignette, often taken from the authors' consulting experiences. Companies featured include Amazon.com, Zillow.com, and Keen Inc. We analyze data from or about the companies in the motivating vignettes throughout the chapter.

Brief Cases. Each chapter includes one or more Brief Cases that use real data and ask students to investigate a question or make a decision. Students define the objective, plan the process, complete the analysis, and report a conclusion. Data for the Brief Cases are available online, formatted for various technologies.

Case Studies. Each of the parts of the book ends with a Case Study. Students are given realistically large data sets (online) and challenged to respond to open-ended business questions using the data. Students can bring together methods they have learned throughout the book to address the issues raised. Students will have to use a computer to work with the large data sets that accompany these Case Studies.

What Can Go Wrong? In each chapter, What Can Go Wrong? highlights the most common statistical errors and the misconceptions about Statistics. The most common mistakes for the new user of Statistics often involve misusing a method—not miscalculating a statistic. One of our goals is to arm students with the tools to detect statistical errors and to offer practice in debunking misuses of Statistics, whether intentional or not.

Applying Concepts

For Examples. Almost every section of every chapter includes a focused example that illustrates and applies the concepts or methods of that section to a real-world business context.

Step-by-Step Guided Examples. The answer to a statistical question is almost never just a number. Statistics is about understanding the world and making better decisions with data. Guided Examples model a thorough solution in the right column with commentary in the left column. The overall analysis follows our innovative **Plan, Do, Report** template. Each analysis begins with a clear question about a business decision and an examination of the data (**Plan**), moves to calculating the selected statistics (**Do**), and finally concludes with a **Report** that specifically addresses the question. To emphasize that our goal is to address the motivating question, we present the **Report** step as a business memo that summarizes the results in the context of the example and states a recommendation if the data are able to support one. To preserve the realism of the example, whenever it is appropriate, we include limitations of the analysis or models in the concluding memo, as one should in making such a report.

By Hand. Even though we encourage the use of technology to calculate statistical quantities, we recognize the pedagogical benefits of occasionally doing a calculation by hand. The By Hand boxes break apart the calculation of some of the simpler formulas and help the student through the calculation of a worked example.

Reality Check. We regularly offer reminders that Statistics is about understanding the world and making decisions with data. Results that make no sense are probably wrong, no matter how carefully we think we did the calculations. Mistakes are often easy to spot with a little thought, so we ask students to stop for a reality check before interpreting results.

Notation Alert. Throughout this book, we emphasize the importance of clear communication. Proper notation is part of the vocabulary of Statistics, but it can be daunting. We've found that it helps students when we are clear about the letters and symbols statisticians use to mean very specific things, so we've included Notation Alerts whenever we introduce a special notation that students will see again.

Math Boxes. In many chapters, we present the mathematical underpinnings of the statistical methods and concepts. We set proofs, derivations, and justifications apart from the narrative, so the underlying mathematics is there for those who want greater depth, but the text itself presents the logical development of the topic at hand without distractions.

What Have We Learned? Each chapter ends with a What Have We Learned? summary that includes learning objectives and definitions of terms introduced in the chapter. Students can think of these as study guides.

Promoting Problem Solving

Just Checking. Throughout each chapter we pose short questions to help students check their understanding. The answers are at the end of the exercise sets in each chapter to make them easy to check. The questions can also be used to motivate class discussion.

Ethics in Action. Statistics is not just plugging numbers into formulas; most statistical analyses require a fair amount of judgment. Ethics in Action vignettes—updated for this edition—in each chapter provide a context for some of the judgments needed in statistical analyses. Possible errors, a link to the American Statistical Association's Ethical Guidelines, and ethically and statistically sound alternative approaches are presented in the Instructor's Solutions Manual.

Section Exercises. The exercises for each chapter begin with straightforward exercises targeted at the topics in each section. These are designed to check understanding of specific topics. Because they are labeled by section, it is easy to turn back to the chapter to clarify a concept or review a method.

Chapter Exercises. These exercises are designed to be more realistic than Section Exercises and to lead to conclusions about the real world. They may combine concepts and methods from different sections, and they contain relevant, modern, and real-world questions. Many come from news stories; some come from recent research articles. The exercises marked with a **T** indicate that the data are provided online (at the book's companion website, **www.pearsonhighered.com/sharpe**) in a variety of formats. We pair the exercises so that each odd-numbered exercise (with answer in the back of the book) is followed by an even-numbered exercise on the same Statistics topic. Exercises are roughly ordered within each chapter by both topic and by level of difficulty.

Integrating Technology

Data and Sources. Most of the data used in examples and exercises are from real-world sources and whenever we can, we include URLs for Internet data sources. The data we use are on the companion website, **www.pearsonhighered.com/sharpe**.

Videos with Optional Captioning. Videos, featuring the *Business Statistics: A First Course* authors, review the high points of each chapter. The presentations feature the same student-friendly style and emphasis on critical thinking as the textbook. In addition, 10 *Business Insight Videos* feature Deckers, Southwest Airlines, Starwood, and other companies and focus on statistical concepts as they pertain to the real world. Videos are available with captioning and can be viewed from within the online MyStatLab course.

Technology Help. In business, Statistics is practiced with computers using a variety of statistics packages. In Business-school Statistics classes, however, Excel is the software most often used. In Technology Help at the end of each chapter, we summarize what students can find in the most common software, often with annotated output. Updated for this edition, we offer extended guidance for Excel 2013, and start-up pointers for Minitab, SPSS, XLStat, R, and JMP, formatted in easy-to-read bulleted lists. This advice is not intended to replace the documentation for any of the software, but rather to point the way and provide start-up assistance.

Resources for Success

MyStatLab® Online Course *Business Statistics:*
A First Course, 3rd edition, Sharpe/De Veaux/Velleman
(access code required)

MyStatLab is available to accompany Pearson's market leading text offerings. To give students a consistent tone, voice, and teaching method each text's flavor and approach is tightly integrated throughout the accompanying MyStatLab course, making learning the material as seamless as possible.

New! Launch Exercise Data in Excel

Students are now able to quickly and seamlessly launch data sets from exercises within MyStatLab into a Microsoft Excel spreadsheet for easy analysis. As always, students may also copy and paste exercise data sets into most other software programs.

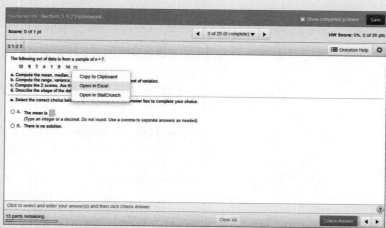

Business Insight Videos

Ten engaging videos showing managers at top companies using statistics in their everyday work. Assignable questions built in to your MyStatLab course encourage discussion.

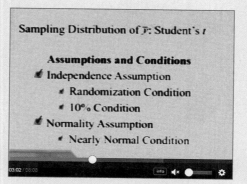

Chapter Overview Videos

Featuring each of the book's authors, 5-10 minute chapter overview videos walk students through key concepts covered in each chapter. These videos are accessible through the multimedia library and are assignable.

Resources for Success

Instructor Resources

Additional resources can be downloaded from **www.pearsonhighered.com** or MyStatLab; hardcopy resources can be ordered from your sales representative.

Instructor's Edition

Instructor's Edition contains answers to all exercises.

Instructor's Resource Guide

Instructor's Resource Guide (download only), written by the authors, contains chapter-by-chapter comments on the major concepts, tips on presenting topics (and what to avoid), teaching examples, suggested assignments, basic exercises, and web links and lists of other resources.

Online Test Bank

Online Test Bank (download only), includes chapter quizzes and part level tests.

Instructor's Solutions Manual

Instructor's Solutions Manual (download only) contains detailed solutions to all of the exercises.

TestGen®

TestGen® (www.pearsoned.com/testgen) enables instructors to build, edit, print, and administer tests using a computerized bank of questions developed to cover all the objectives of the text. TestGen is algorithmically based, so instructors can create multiple but equivalent versions of the same question or test with the click of a button. Instructors can also modify test bank questions or add new questions.

PowerPoint® Lecture Slides

PowerPoint Lecture Slides provide an outline to use in a lecture setting, presenting definitions, key concepts, and figures from the text.

Learning Catalytics

Learning Catalytics is a web-based engagement and assessment tool. As a "bring-your-own-device" direct response system, Learning Catalytics offers a diverse library of dynamic question types that allow students to interact with and think critically about statistical concepts. As a real-time resource, instructors can take advantage of critical teaching moments both in the classroom or through assignable and gradable homework.

XLStat for Pearson

XLStat for Pearson is an Excel® add-in that offers a wide variety of functions to enhance the analytical capabilities of Microsoft Excel, making it the ideal tool for your everyday data analysis and statistics requirements. Developed in 1993, XLStat is used by leading businesses and universities around the world. XLStat is compatible with all Excel versions from version 97 to version 2013 (except Mac 2008) including PowerPC and Intel-based Mac systems.

Student Resources

Additional resources for student success.

Student's Solutions Manual

Student's Solutions Manual provides detailed, worked-out solutions to odd-numbered exercises.

Study Cards for Business Statistics Software

This series of study cards, available for Excel® 2013 with XLSTAT, Excel 2013 with Data Analysis Toolpak, Minitab, JMP, SPSS, and StatCrunch provide students with easy step-by-step guides to the most common business statistics software.

Acknowledgments

This book would not have been possible without many contributions from David Bock, our coauthor on several other texts. Many of the explanations and exercises in this book benefit from Dave's pedagogical flair and expertise. We are honored to have him as a colleague and friend.

Many people have contributed to this book from the first day of its conception to its publication. *Business Statistics: A First Course* would have never seen the light of day without the assistance of the incredible team at Pearson. Our Editor in Chief, Deirdre Lynch, was central to the development and realization of the book from day one. We thank Acquisitions Editor Patrick Barbera for his support throughout this edition. Sherry Berg, Project Manager, and Nancy Kincade, Project Manager at Lumina Datamatics, kept us on task as much as humanly possible. We are indebted to them. Chere Bemelmans, Program Manager; Justin Billing, Editorial Assistant; Tiffany Bitzel, Product Marketing Manager; Jennifer Myers, Marketing Assistant; and Dona Kenly, Senior Market Development Manager, were essential in managing all of the behind-the-scenes work that needed to be done. Aimee Thorne, Media Producer, put together a top-notch media package for this book. Senior Procurement Specialist Carol Melville worked miracles to get this book in your hands.

We'd also like to thank our accuracy checker whose monumental task was to make sure we said what we thought we were saying: Dirk Tempelaar, Maastricht University.

We also thank those who provided feedback through focus groups, class tests, and reviews:

Hope M. Baker, Kennesaw State University

John F. Beyers, University of Maryland—University College

Scott Callan, Bentley College

Laurel Chiappetta, University of Pittsburgh

Anne Davey, Northeastern State University

Joan Donohue, The University of South Carolina

Robert Emrich, Pepperdine University

Michael Ernst, St. Cloud State

Mark Gebert, University of Kentucky

Kim Gilbert, University of Georgia

Nicholas Gorgievski, Nichols College

Clifford Hawley, West Virginia University

Kathleen Iacocca, University of Scranton

Chun Jin, Central Connecticut State University

Austin Lampros, Colorado State University

Roger Lee, Salt Lake Community College

Monnie McGee, Southern Methodist University

Richard McGowan, Boston College

Mihail Motzev, Walla Walla University

Robert Potter, University of Central Florida

Eugene Round, Embry-Riddle Aeronautical University

Sunil Sapra, California State University—Los Angeles

Dmitry Shishkin, Georgia Gwinnett College

Courtenay Stone, Ball State University

Gordon Stringer, University of Colorado—Colorado Springs

Arnold J. Stromberg, University of Kentucky

Joe H. Sullivan, Mississippi State University

Timothy Sullivan, Towson University
Minghe Sun, University of Texas—San Antonio
Patrick Thompson, University of Florida
Jackie Wroughton, Northern Kentucky University
Ye Zhang, Indiana University—Purdue Indianapolis

Finally, we want to thank our families. This has been a long project, and it has required many nights and weekends. Our families have sacrificed so that we could write the book we envisioned.

Norean Sharpe
Richard De Veaux
Paul Velleman

Index of **Applications**

BE = Boxed Example; E = Exercises; EIA = Ethics in Action; GE = Guided Example; IE = In-Text Example; JC = Just Checking; P = Project;
TH = Technology Help

Accounting
Administrative and Training Costs (E), 44, 403–404
Annual Reports (E), 42
Audits and Tax Returns (E), 172, 296, 368
Bookkeeping (E), 262
Budgets (E), 365, 366
Company Assets, Profit, and Revenue (BE), 121; (E), 41, 43–44, 201, 480, 483, 524, 526–527; (IE), 2, 7, 97, 266, 372
Cost Cutting (E), 448, 450
Expenses (IE), 8
Financial Close Process (E), 408
Probability Calculations and Plots (TH), 227–228
Purchase Records (E), 49; (IE), 4
Random numbers, generating (TH), 167
Random Variables and Probability Models (TH), 199
Spreadsheets (IE), 4

Advertising
Ads (E), 329, 332–334, 409–410, 522–523
Advertising in Business (BE), 304; (E), 43–44, 47–48, 409, 415–416, 522–523; (GE), 154–156; (IE), 2, 6
Branding (E), 409
Free Products (IE), 306, 355
International Advertising (E), 175
Predicting Sales (E), 138–139
Product Claims (BE), 373; (E), 233, 411, 414, 416, 446; (EIA), 124–125
Target Audience (E), 175, 204, 406; (EIA), 546; (JC), 343
Truth in Advertising (E), 332

Agriculture
Agricultural Discharge (EIA), 253
Beef and Livestock (E), 364
Drought and Crop Losses (E), 412
Farmers' Markets (E), 203
Lobster Fishing Industry (E), 525–526
Seeds (E), 294

Banking
Annual Percentage Rate (P), 200
ATMs (E), 168; (IE), 371
Certificates of Deposit (CDs) (P), 200
Credit Card Bank (P), 39
Credit Card Charges (E), 83, 296–297, 365, 487; (GE), 64–65, 315–316, 390–393; (IE), 266
Credit Card Companies (BE), 282; (E), 83, 291, 296–297, 327, 365, 392; (GE), 9, 104–105, 146, 265–266, 282, 315–316, 371–373, 377–382, 531–533; (JC), 315; (P), 14
Credit Card Customers (BE), 282; (E), 204, 296–297, 327, 365, 449; (GE), 64–65, 315–316, 377–379, 390–393; (IE), 265–266, 268, 282, 371–372
Credit Card Debt (E), 410
Credit Card Offers (BE), 282; (E), 296–297; (GE), 315–316, 377–382; (IE), 9, 146–147, 266, 282, 372–373; (P), 14, 548

Credit Scores (IE), 145–146
Credit Unions (E), 327; (EIA), 285
Interest Rates (E), 133, 170; (IE), 265, 266; (P), 200
Maryland Bank National Association (IE), 265–266
Mortgages (E), 17, 133; (GE), 270–271
Subprime Loans (IE), 145
World Bank (E), 15, 94, 136

Business (General)
Attracting New Business (E), 367
Best Places to Work (E), 451, 484
Bossnapping (E), 290; (GE), 278–280; (JC), 280
Business Planning (IE), 97, 317
Chief Executives (E), 93, 177, 233, 364–365, 448, 449; (IE), 72–73, 347–348
Company Case Reports and Lawyers (GE), 270–271
Company Databases (IE), 7, 9
Contract Bids (E), 173, 202
Elder Care Business (EIA), 471
Enterprise Resource Planning (E), 408–409, 451
Entrepreneurial Skills (E), 449
Forbes 500 Companies (E), 95, 364–365
Fortune 500 Companies (BE), 74; (E), 258, 290, 451, 480, 483, 484; (IE), 347–348
Franchises (EIA), 124–125, 471
Industry Sector (E), 450
International Business (E), 40, 48, 258–259, 295; (IE), 238; (P), 256–257
Job Growth (E), 452, 484
Organisation for Economic Cooperation and Development (OECD) (E), 88
Outside Consultants (IE), 35
Outsourcing (E), 450
Research and Development (E), 44; (IE), 97–98; (JC), 390
Small Business (E), 42, 132, 135, 173, 202, 366, 450, 523; (IE), 2
Start-Up Companies (E), 16, 89, 297, 333; (EIA), 165
Trade Secrets (IE), 455
Women-Led Businesses (E), 201, 332–333

Company Names
Adair Vineyards (E), 83
AIG (GE), 66–67; (IE), 49–50, 52, 58
Alpine Medical Systems, Inc. (EIA), 514
Amazon.com (IE), 2, 4, 97–98, 101–103, 108–111, 114
American Express (IE), 371
Arby's (E), 15
Bank of America (IE), 265, 371
BMW (E), 139
Buick (E), 135
Burger King (E), 528
Capital One (IE), 9, 3
Chevy (E), 410
Circuit City (E), 362
Cisco Systems (E), 42
Coca-Cola (E), 41

CompUSA (E), 362
Cypress (JC), 104
Desert Inn Resort (E), 171
Diners Club (IE), 371
eBay (E), 204
Expedia.com (IE), 489
Fair Isaac Corporation (IE), 145–146
Fisher-Price (E), 42
Ford (E), 135, 410; (IE), 249
General Electric (IE), 299
GfK Roper (E), 43, 258, 295, 448; (GE), 31–32; (IE), 25, 238, 242, 431; (P), 256–257
Google (E), 43, 44, 451; (IE), 20–24, 188, 189–191
Guinness & Co. (BE), 194; (IE), 335–336
Holes-R-Us (E), 93
Honda (E), 134, 135
Hostess (IE), 241
Intel (JC), 104
Jeep (E), 176–177
KEEN (IE), 19–20
Kelly's BlueBook (E), 176
Kraft Foods, Inc. (P), 474
L.L. Bean (E), 16
Lycos (E), 258
Mattel (E), 42
Metropolitan Life (MetLife) (IE), 179–180
Microsoft (E), 42, 85; (IE), 489
M&M/Mars (E), 172, 292, 330; (GE), 154–156
Nambé Mills, Inc. (GE), 462–464; (IE), 455–456, 466–468
National Beverage (E), 41
Nissan (IE), 218
PepsiCo (E), 41, 171, 328
Pew Research (E), 42, 47, 168–169, 260, 443, 444, 447–449; (IE), 9, 25, 150, 240, 426
Pontiac (E), 135
Roper Worldwide (JC), 195
Starbucks (IE), 8
Suzuki (E), 528
Systemax (E), 362
Time-Warner (BE), 269, 270
Toyota (E), 135, 480
UPS (IE), 537
Visa (IE), 371–372
Wal-Mart (E), 415–416, 524, 526–527
WinCo Foods (E), 415–416
Yahoo (IE), 22–23
Zenna's Café (EIA), 77
Zillow.com (IE), 61, 489–490

Consumers
Categorizing Consumers (E), 446, 449; (IE), 6–7, 242–243
Consumer Confidence Index (IE), 272
Consumer Groups (E), 332, 368, 410
Consumer Loyalty (E), 325; (IE), 2; (JC), 378; (P), 326, 442

1

Data and Decisions

E-Commerce

E-Commerce and mobile commerce have dramatically changed the way the world shops. Online shoppers can buy clothes, food, even cars with the click of a mouse and a digital swipe of their credit card—24 hours a day, 7 days a week. Companies now reach their customers in ways no one could even imagine just a generation ago. Online sales in some sectors, such as clothing and electronics, already account for over 15% of total sales, which is about double what it was five years ago. U.S. adults, on average, currently spend about $1200 a year online, but some projections put that number at nearly $2000 a year by 2016.

The trend in online shopping is worldwide. The amount Australians spend online is expected to grow by $10B in the next five years. The research firm Forrester estimates that global digital retailing is headed toward 15 to 20% of total sales worldwide in the near future.

A few generations ago, many store owners knew their customers well. With that knowledge, they could personalize their suggestions, guessing which items that particular customer might like. Online marketers rely on similar information about customers and potential customers to make decisions. But in today's digital age retailers never meet their customers, so, that information has to be obtained in other ways. How do today's companies know which ads to place on your browser or what order to list the websites from your search? How do marketers know what to advertise and to whom?
The answer is …
Data.

1.1   What Are Data?

Q: What is Statistics?

A: Statistics is a way of reasoning, along with a collection of tools and methods, designed to help us understand the world.

Q: What are statistics?

A: Statistics (plural) are quantities calculated from data.

Q: So what is data?

A: You mean, "what *are* data?" Data is the plural form. The singular is datum.

Q: So, what are data?

A: Data are values along with their context.

Businesses have always relied on data for planning and to improve efficiency and quality. Now, more than ever before, businesses rely on the information in data to compete in the global marketplace. Every time you make an online purchase, much more information is actually captured than just the details of the purchase itself. What pages did you search in order to get to your purchase? How much time did you spend looking at each? Companies use this information to make decisions about virtually all phases of their business, from inventory to advertising to website design. These data are recorded and stored electronically, in vast digital repositories called **data warehouses**.

In the past few decades these data warehouses have grown enormously in size, but with the use of powerful computers, the information contained in them is accessible and used to help make decisions. The huge capacity of these warehouses has given rise to the term **Big Data** to describe data sets so large that traditional methods of storage and analysis are inadequate. Even though the data amounts are huge, some decisions can be made quickly. When you pay with your credit card, for example, the information about the transaction is transmitted to a central computer where it is processed and analyzed. A decision whether to approve or deny your purchase is made and transmitted back to the point of sale, all within a few seconds. But data alone can't help you make better business decisions. You must be able to summarize, model, and understand what the data can tell you. That collection of tools and its associated reasoning is what we call "Statistics."

Statistics plays a role in making sense of our complex world in an astonishing number of ways. Statisticians assess the risk of genetically engineered foods or of a new drug being considered by the Food and Drug Administration (FDA). Statisticians predict the number of new cases of AIDS by regions of the country or the number of customers likely to respond to a sale at the supermarket. And statisticians help scientists, social scientists, and business leaders understand how unemployment is related to environmental controls, whether enriched early education affects the later performance of school children, and whether vitamin C really prevents illness. Whenever you have data and a need to understand the world or make an informed decision, you need Statistics.

An instructor who wanted to analyze student perceptions of business ethics (a question we'll come back to in a later chapter), couldn't administer a survey to every single university student in the United States. That wouldn't be practical or cost-effective. Instead, she could survey a smaller, representative group of students. Statistics can help us make the leap from a smaller sample of data we have at hand to an understanding of the world at large. Chapter 8 discusses sampling. The theme of generalizing from the specific to the general is one that we revisit throughout this book. We hope this text will empower *you* to draw conclusions from data and make valid business decisions in response to such questions as:

- Will the new design of your website increase click-through rates and result in more sales?
- What is the effect of advertising on sales?
- Do aggressive, "high-growth" mutual funds really have higher returns than more conservative funds?
- Is there a seasonal cycle in your firm's revenues and profits?
- What is the relationship between shelf location and cereal sales?
- Do students around the world perceive issues in business ethics differently?
- Are there common characteristics about your customers and why they choose your products?—and, more importantly, are those characteristics the same among those who aren't your customers?

The Essence of Statistics

Our ability to answer questions such as these and make sound business decisions with data depends largely on our ability to understand *variation*. That may not be the term you expected to find at the end of that sentence, but it is the essence of Statistics. The key to learning from data is understanding the variation that is all around us.

Data vary. People are different. So are economic conditions from month to month. We can't see everything, let alone measure it all. And even what we do measure, we measure imperfectly. So the data we base our decisions on provide, at best, an imperfect picture of the world. Variation lies at the heart of what Statistics is all about. How to make sense of it is the central challenge of Statistics.

Companies use data to make decisions about nearly every aspect of their business. By studying the past behavior of customers and predicting their responses, they hope to better serve their customers and to compete more effectively. This process of using data, especially of **transactional data** (data collected for recording the companies' transactions), to make decisions and predictions is sometimes called **data mining** or *predictive analytics*. The more general term **business analytics** (or sometimes simply analytics) describes *any* use of data and statistical analysis to drive business decisions from data whether the purpose is predictive or simply descriptive.

Leading companies are embracing business analytics. Reed Hastings, a former computer science major, is the founder and CEO of Netflix. Netflix uses analytics on customer information both to recommend new movies and to adapt the website that customers see to individual tastes. Netflix offered a $1 million prize to anyone who could improve on the accuracy of their recommendations by more than 10%. That prize was won in 2009 by a team of statisticians and computer scientists using data-mining techniques. eBay used analytics to examine its own use of computer resources. Although not obvious to their own technical people, once they crunched the data they found huge inefficiencies. According to Forbes, they were able to "save millions in capital expenditures within the first year."

To begin to make sense of **data**, we first need to understand its context. Newspaper journalists know that the lead paragraph of a good story should establish the "Five W's": *who, what, when, where*, and (if possible) *why*. Often, we add *how* to the list as well. Answering these questions can provide a **context** for data values and make them meaningful. The answers to the first two questions are essential. If you can't answer *who* and *what*, you don't have data, and you don't have any useful information.

> THE W'S:
> WHO
> WHAT
> WHEN
> WHERE
> WHY

Table 1.1 shows purchase records from an online music retailer. A table like this is called a **data table**. Each row represents a purchase of a music album. In general, rows of a data table correspond to individual **cases** about which we've recorded some characteristics called **variables**.

Order Number	Name	State/Country	Price	Area Code	Album Download	Gift?	Stock ID	Artist
105-2686834-3759466	Katherine H.	Ohio	5.99	440	Identity	N	B0000OI5Y6	James Fortune & Flya
105-9318443-4200264	Samuel P.	Illinois	9.99	312	Port of Morrow	Y	B000002BK9	The Shins
105-1872500-0198646	Chris G.	Massachusetts	9.99	413	Up All Night	N	B000068ZVQ	Syco Music UK
103-2628345-9238664	Monique D.	Canada	10.99	902	Fallen Empires	N	B0000010AA	Snow Patrol
002-1663369-6638649	Katherine H.	Ohio	11.99	440	Sees the Light	N	B002MXA7Q0	La Sera

Table 1.1 Example of a data table. The variable names are in the top row. Typically, the *Who* of the table are found in the leftmost column.

Cases go by a variety of names. Individuals who answer a survey are referred to as **respondents**. People on whom we experiment are **subjects** or (in an attempt to acknowledge the importance of their role in the experiment) **participants**, but animals, plants, websites, and other inanimate subjects are often called **experimental units**. Often we call cases just what they are: for example, *customers, economic quarters*, or *companies*. In a database, rows are called **records**—in this example, purchase records. Perhaps the most generic term is cases. In Table 1.1, the cases are the individual orders.

The column titles (variable names) tell *what* has been recorded. What does a row of Table 1.1 represent? Be careful. Even if people are involved, the cases may not correspond to people. For example, in Table 1.1, each row represents a different order and not the customer who made the purchases (notice that the same person made two different orders). A common place to find the *who* of the table is the leftmost column. It's often an identifying variable for the cases, in this example, the order number.

If you collect the data yourself, you'll know what the cases are and how the variables are defined. But, often, you'll be looking at data that someone else collected. The information about the data, called the metadata, might have to come from the company's database administrator or from the *information technology* department of a company. **Metadata** typically contains information about *how*, *when*, and *where* (and possibly *why*) the data were collected, *who* each case represents, and the definitions of all the variables.

A data table like the one shown in Table 1.1 is sometimes called a **spreadsheet**. Although spreadsheets were designed for accounting, it is common to keep modest-size datasets in a spreadsheet even if no accounting is involved. It is usually easy to move a data table from a spreadsheet program to a program designed for statistical graphics and analysis, either directly or by copying the data table and pasting it into the statistics program.

Spreadsheets are not convenient for really large amounts of data. Amazon has tens of millions of customers and millions of products. But very few customers have purchased more than a few dozen items, so almost all the entries in a spreadsheet of customers by items would be blank—not a very efficient way to store information. For that reason large organizations use relational databases.

In a **relational database**, two or more separate data tables are linked together so that information can be merged across them. Each data table is a *relation* because it is about a specific set of cases with information about each of these cases for all (or at least most) of the variables ("fields" in database terminology). For example, a table of customers, along with demographic information on each, is such a relation. A data table of all the items sold by the company, including information on price, inventory, and past history, is another relation. Transactions may be held in a third "relation" that references each of the other two relations. Table 1.2 shows a small example.

In statistics, analyses are typically performed on a single relation because all variables must refer to the same cases. But often the data must be retrieved from a relational database, which may require expertise with that software. In the rest of the book, we'll assume that the data have been retrieved and placed in a data table or spreadsheet with variables as columns and cases as the rows.

Customers

Customer Number	Name	City	State	Zip Code	Customer since	Gold Member?
473859	R. De Veaux	Williamstown	MA	01267	2007	No
127389	N. Sharpe	Washington	DC	20052	2000	Yes
335682	P. Velleman	Ithaca	NY	14580	2003	No
…						

Items

Product ID	Name	Price	Currently in Stock?
SC5662	Silver Cane	43.50	Yes
TH2839	Top Hat	29.99	No
RS3883	Red Sequined Shoes	35.00	Yes
…			

Transactions

Transaction Number	Date	Customer Number	Product ID	Quantity	Shipping Method	Free Ship?
T23478923	9/15/13	473859	SC5662	1	UPS 2nd Day	N
T23478924	9/15/13	473859	TH2839	1	UPS 2nd Day	N
T63928934	10/20/13	335682	TH2839	3	UPS Ground	N
T72348299	12/22/13	127389	RS3883	1	Fed Ex Ovnt	Y

Table 1.2   A relational database shows all the relevant information for three separate relations linked together by customer and product numbers.

FOR EXAMPLE   Identifying variables and the W's

Carly, a marketing manager at a credit card bank, wants to know if an offer mailed 3 months ago has affected customers' use of their cards. To answer that, she asks the information technology department to assemble the following information for each customer: total spending on the card during the 3 months before the offer (*Pre Spending*); total spending for 3 months after the offer (*Post Spending*); the customer's *Age* (by category); what kind of expenditure they made (*Segment*); if customers are enrolled in the website (*Enroll?*); what offer they were sent (*Offer*); and the amount each customer spent on the card in their segment (*Segment Spend*). She gets a spreadsheet whose first six rows look like this:

Account ID	Pre Spending	Post Spending	Age	Segment	Enroll?	Offer	Segment Spend
393371	$2,698.12	$6,261.40	25–34	Travel/Ent	NO	None	$887.36
462715	$2,707.92	$3,397.22	45–54	Retail	NO	Gift Card	$5,062.55
433469	$800.51	$4,196.77	65+	Retail	NO	None	$673.80
462716	$3,459.52	$3,335.00	25–34	Services	YES	Double Miles	$800.75
420605	$2,106.48	$5,576.83	35–44	Leisure	YES	Double Miles	$3,064.81
473703	$2,603.92	$7,397.50	<25	Travel/Ent	YES	Double Miles	$491.29

(continued)

> **QUESTION** Identify the cases and the variables. Describe as many of the W's as you can for this data set.
>
> **ANSWER** The cases are individual customers of the credit card bank. The data are from the internal records of the credit card bank for the past 6 months (3 months before and 3 months after an offer was sent to the customers). The variables include the account ID of the customer (*Account ID*) and the amount charged on the card before (*Pre Spending*) and after (*Post Spending*) the offer was sent out. Also included are the customer's *Age*, marketing *Segment*, whether they enrolled on the website (*Enroll?*), what offer they were sent (*Offer*), and how much they charged on the card in their marketing segment (*Segment Spend*).

1.2   Variable Types

When the values of a variable are the names of categories, we call it a **categorical**, or **qualitative, variable**. When the values of a variable are measured numerical quantities with **units**, we call it a **quantitative variable**.

Descriptive responses to questions are often categories. For example, the responses to the questions "What type of mutual fund do you invest in?" or "What kind of advertising does your firm use?" yield categorical values. An important special case of categorical variables is one that has only two possible responses (usually "yes" or "no"), which arise naturally from questions like "Do you invest in the stock market?" or "Do you make online purchases from this website?"

Categorical, or Quantitative?

When area codes were first introduced all phones had dials. To reduce wear and tear on the dials and to speed calls, the lowest-digit codes (the fastest to dial—those for which the dial spun the least) were assigned to the largest cities. So, New York City was given 212, Chicago 312, LA 213, and Philadelphia 215, but rural upstate New York was 607, Joliet was 815, and San Diego 619. Back then, the numerical value of an area code could be used to guess something about the population of its region. But after dials gave way to push buttons, new area codes were assigned without regard to population and area codes are now just categories.

Question	Categories or Responses
Do you invest in the stock market?	__ Yes __ No
What kind of advertising do you use?	__ Newspapers __ Internet __ Direct mailings
What is your class at school?	__ Freshman __ Sophomore __ Junior __ Senior
I would recommend this course to another student.	__ Strongly Disagree __ Slightly Disagree __ Slightly Agree __ Strongly Agree
How satisfied are you with this product?	__ Very Unsatisfied __ Unsatisfied __ Satisfied __ Very Satisfied

Table 1.3   Some examples of categorical variables.

In a purchase record, price, quantity, and time spent on the website are all quantitative values with units (dollars, count, and seconds). For quantitative variables, the units tell how each value has been measured. Even more important, units such as yen, cubits, carats, angstroms, nanoseconds, miles per hour, or degrees Celsius tell us the *scale* of measurement, so we know how far apart two values are. Without units, the values of a measured variable have no meaning. It does little good to be promised a raise of 5000 a year if you don't know whether it will be paid in euros, dollars, yen, or Swazi lilangeni. An essential part of a quantitative variable is its units.

The distinction between categorical and quantitative variables seems clear, but there are reasons to be careful. First, some variables can be considered as either categorical or quantitative, depending on the kind of questions we ask about them. For example, the variable *Age* would be considered quantitative if the responses were numerical and they had units. A doctor would certainly consider *Age* to be quantitative. The units could be years, or for infants, the doctor would want even more precise units, like months, or days. On the other hand, a retailer might lump

together the values into categories like "Child (12 years or less)," "Teen (13 to 19)," "Adult (20 to 64)," or "Senior (65 or over)." For many purposes, like knowing which song download coupon to send you, that might be all the information needed. Then *Age* would be a categorical variable.

How to classify some variables as categorical or quantitative may seem obvious. But be careful. Area codes may look quantitative, but are really categories. What about ZIP codes? They are categories too, but the numbers do contain information. If you look at a map of the United States with ZIP codes, you'll see that as you move West, the first digit of ZIP codes increases, so treating them as quantitative might make sense for some questions.

Data mining and other statistical analysis programs often must guess whether a variable is categorical or quantitative. When the variable contains symbols other than numbers, the software will correctly treat the variable as categorical, but just because a variable has numbers doesn't mean it is quantitative. Data miners spend much of their time going back through data sets to correctly identify variables as categorical or quantitative. Chapter 2 discusses summaries and displays of categorical variables more fully. Chapter 3 discusses quantitative variables.

Identifiers

Identifier variables are categorical variables that assign a unique identifier code to each individual in the data set. Your student ID number, social security number, and mobile phone number are all identifiers.

Identifier variables are crucial for Big Data because, they make it possible to combine data from different sources, protect confidentiality, and provide unique labels. Your school's grade transcripts are likely in a different relation than your bursar bill records. Your student ID is what links them. Most companies keep such relational databases. The identifiers in Table 1.2 are the *Customer Number*, *Product ID*, and *Transaction Number*.

Other Data Types

Many companies follow up with customers after a service call or sale with an online questionnaire. They might ask:

"How satisfied were you with the service you received?"

1) Not satisfied; 2) Somewhat satisfied; 3) Moderately satisfied; or 4) Extremely satisfied.

Is this variable categorical or quantitative? There is certainly an *order* of perceived worth; higher numbers indicate higher perceived worth. An employee whose customer responses average around 4 seems to be doing a better job than one whose average is around 2, but are they *twice* as good? When the values of a categorical variable have an intrinsic order, we can say that the variable is **ordinal**. By contrast, a categorical variable with unordered categories is sometimes called **nominal**. Values can be individually ordered (e.g., the ranks of employees based on the number of days they've worked for the company) or ordered in classes (e.g., Freshman, Sophomore, Junior, Senior). Ordering is not absolute; how the values are ordered depends on the purpose of the ordering. For example, are the categories Infant, Youth, Teen, Adult, and Senior ordinal? Well, if we are ordering on age, they surely are. But if we are ordering on purchase volume, it is likely that either Teen or Adult will be the top group.[1]

---

[1] Some people differentiate quantitative variables according to whether their measured values have a defined value for zero. This is a technical distinction and usually not one we'll need to make. (For example, it isn't correct to say that a temperature of 80°F is twice as hot as 40°F because 0°F is an arbitrary value. On the Celsius scale those temperatures are 26.67°C and 4.44°C—a ratio of 6.) The term *interval scale* is sometimes applied to quantitative variables that lack a defined zero, and the term *ratio scale* is applied to measurements for which such ratios are appropriate.

Year	Total Revenue (in $M)
2002	3288.9
2003	4075.5
2004	5294.2
2005	6369.3
2006	7786.9
2007	9441.5
2008	10,383.0
2009	9774.6
2010	10,707
2011	11,700
2012	13,300

Table 1.4 Starbucks's total revenue (in $M) for the years 2002 to 2012.

Cross-Sectional and Time Series Data

The quantitative variable *Total Revenue* in Table 1.4 is an example of a time series. A **time series** is an ordered sequence of values of a single quantitative variable measured at regular intervals over time. Time series are common in business. Typical measuring points are months, quarters, or years, but virtually any consistently spaced time interval is possible.

By contrast, most of the methods in this book are better suited for **cross-sectional data**, where several variables are measured at the same time point. If we collect data on sales revenue, number of customers, and expenses for last month at *each* Starbucks (more than 21,000 locations as of 2014) at one point in time, this would be cross-sectional data. Cross-sectional data may contain some time information (such as dates), but it isn't a time series because it isn't measured at regular intervals. Because different methods are used to analyze these different types of data, it is important to be able to identify both time series and cross-sectional data sets.

FOR EXAMPLE Identifying the types of variables

QUESTION Before she can continue with her analysis, Carly (from the example on page 5) must classify each variable as being quantitative or categorical (or possibly both), and whether the data are a time series or cross-sectional. For quantitative variables, what are the units? For categorical variables, are they nominal or ordinal?

ANSWER

Account ID – categorical (nominal, identifier)

Pre Spending – quantitative (units $)

Post Spending – quantitative (units $)

Age – categorical (ordinal). Could be quantitative if we had more precise information

Segment – categorical (nominal)

Enroll? – categorical (nominal)

Offer – categorical (nominal)

Segment Spend – quantitative (units $)

The data are cross-sectional. We do not have successive values over time.

1.3   Data Sources: Where, How, and When

In addition to knowing the *who* and *what* of data, we'd like to know the *where*, *how*, and *when* of data as well. Values recorded in 1947 may mean something different than similar values recorded last year. Values measured in Abu Dhabi may differ in meaning from similar measurements made in Mexico.

How the data are collected can make the difference between insight and nonsense. As we'll see later, data that come from a voluntary survey on the Internet are almost always worthless. Chapter 8 discusses sound methods for *designing* a *survey* or poll to help ensure that the inferences you make are valid.

Another way to collect valid data is by performing an experiment in which you actively manipulate variables (called factors) to see what happens. Most of the "junk mail" credit card offers that you receive are actually experiments done by marketing groups in those companies. They may make different versions of an offer to selected groups of customers to see which one works best before rolling out the winning idea to the entire customer base.

Data can be found in many places. Companies analyze data from their own databases. Some organizations may charge you a fee for accessing or downloading their data. The U.S. government collects information on nearly every aspect of life in the United States, both social and economic (see for example www.census.gov, or more generally, www.usa.gov), as the European Union does for Europe (see ec.europa.eu/eurostat). International organizations such as the World Health Organization (www.who.org) and polling agencies such as Pew Research (www.pewresearch.org) offer information on a variety of current social and demographic trends. Data like these are usually collected for different purposes than to answer your particular business question, so you should be cautious when generalizing from data like these. Unless the data were collected in a way that ensures that they are representative of the population in which you are interested, you may be misled. Chapter 16 discusses data mining, which attempts to use Big Data to make hypotheses and draw insights.

There's a World of Data on the Internet

These days, one of the richest sources of data is the Internet. With a bit of practice, you can learn to find data on almost any subject. We found many of the data sets used in this book by searching on the Internet. The Internet has both advantages and disadvantages as a source of data. Among the advantages are the fact that often you'll be able to find even more current data than we present. One disadvantage is that references to Internet addresses can "break" as sites evolve, move, and die. Another disadvantage is that important metadata—information about the collection, quality, and intent of the data—may be missing.

Our solution to these challenges is to offer the best advice we can to help you search for the data, wherever they may be residing. We usually point you to a website. We'll sometimes suggest search terms and offer other guidance.

Some words of caution, though: Data found on Internet sites may not be formatted in the best way for use in statistics software. Although you may see a data table in standard form, an attempt to copy the data may leave you with a single column of values. You may have to work in your favorite statistics or spreadsheet program to reformat the data into variables. You will also probably want to remove commas from large numbers and such extra symbols as money indicators ($, ¥, £, €); few statistics packages can handle these.

Throughout this book, we often provide a margin note for a new dataset listing some of the W's of the data. When we can, we also offer a reference for the source of the data. It's a habit we recommend. The first step of any data analysis is to know why you are examining the data (what you want to know), whom each row of your data table refers to, and what the variables (the columns of the table) record. These are the *Why*, the *Who*, and the *What*. Identifying them is a key part of the *Plan* step of any analysis. Make sure you know all three before you spend time analyzing the data.

FOR EXAMPLE Identifying data sources

On the basis of her initial analysis, Carly asks her colleague Ying Mei to e-mail a sample of customers from the Travel and Entertainment segment and ask about their card use and household demographics. Carly asks another colleague, Gregg, to design a study about their double miles offer. In this study, a random sample of customers receives one of three offers: the standard double miles offer; a double miles offer good on any airline; or no offer.

(continued)

QUESTION For each of the three data sets—Carly's original data set and Ying Mei's and Gregg's sets—state whether they come from a designed survey or a designed experiment or are collected in another way.

ANSWER Carly's data set was derived from transactional data, not part of a survey or experiment. Ying Mei's data come from a designed survey, and Gregg's data come from a designed experiment.

JUST CHECKING

An insurance company that specializes in commercial property insurance has a separate database for their policies that involve churches and schools. Here is a small portion of that database.

Policy Number	Years Claim Free	Net Property Premium ($)	Net Liability Premium ($)	Total Property Value ($000)	Median Age in ZIP Code	School?	Territory	Coverage
4000174699	1	3107	503	1036	40	FALSE	AL580	BLANKET
8000571997	2	1036	261	748	42	FALSE	PA192	SPECIFIC
8000623296	1	438	353	344	30	FALSE	ID60	BLANKET
3000495296	1	582	339	270	35	TRUE	NC340	BLANKET
5000291199	4	993	357	218	43	FALSE	OK590	BLANKET
8000470297	2	433	622	108	31	FALSE	NV140	BLANKET
1000042399	4	2461	1016	1544	41	TRUE	NJ20	BLANKET
4000554596	0	7340	1782	5121	44	FALSE	FL530	BLANKET
3000260397	0	1458	261	1037	42	FALSE	NC560	BLANKET
8000333297	2	392	351	177	40	FALSE	OR190	BLANKET
4000174699	1	3107	503	1036	40	FALSE	AL580	BLANKET

1 List as many of the W's as you can for this data set.

2 Classify each variable as to whether you think it should be treated as categorical or quantitative (or both); if quantitative, identify the units.

WHAT CAN GO WRONG?

- **Don't label a variable as categorical or quantitative without thinking about the data and what they represent.** The same variable can sometimes take on different roles.

- **Don't assume that a variable is quantitative just because its values are numbers.** Categories are often given numerical labels. Don't let that fool you into thinking they have quantitative meaning. Look at the context.

- **Always be skeptical.** One reason to analyze data is to discover the truth. Even when you are told a context for the data, it may turn out that the truth is a bit (or even a lot) different. The context colors our interpretation of the data, so those who want to influence what you think may slant the context. A survey that seems to be about all students may in fact report just the opinions of those who visited a fan website. The question that respondents answered may be posed in a way that influences responses.

ETHICS IN ACTION

Sarah Potterman, a doctoral student in educational psychology, is researching the effectiveness of various interventions recommended to help children with learning disabilities improve their reading skills. One particularly intriguing approach is an interactive software system that uses analogy-based phonics.

Sarah contacted the company that developed this software, RSPT Inc., to obtain the system free of charge for use in her research. RSPT Inc. expressed interest in having her compare its product with other intervention strategies and was quite confident that its approach would be the most effective. Not only did the company provide Sarah with free software, but RSPT Inc. also generously offered to fund her research with a grant to cover her data collection and analysis costs.

- Identify the ethical dilemma in this scenario.

- What are the undesirable consequences?

- Propose an ethical solution that considers the welfare of all stakeholders.

Jim Hopler is operations manager for a local office of a top-ranked full-service brokerage firm. With increasing competition from both discount and online brokers, Jim's firm has redirected attention to attaining exceptional customer service through its client-facing staff, namely brokers. In particular, management wished to emphasize the excellent advisory services provided by its brokers.

Results from surveying clients about the advice received from brokers at the local office revealed that 20% rated it *poor*, 5% rated it *below average*, 15% rated it *average*, 10% rated it *above average*, and 50% rated it *outstanding*. With corporate approval, Jim and his management team instituted several changes in an effort to provide the best possible advisory services at the local office. Their goal was to increase the percentage of clients who viewed their advisory services as *outstanding*.

Surveys conducted after the changes were implemented showed the following results: 5% *poor*, 5% *below average*, 20% *average*, 40% *above average*, and 30% *outstanding*. In discussing these results, the management team expressed concern that the percentage of clients who considered their advisory services *outstanding* fell from 50% to 30%.

One member of the team suggested an alternative way of summarizing the data. By coding the categories on a scale from 1 = poor to 5 = outstanding and computing the average, they found that the average rating increased from 3.65 to 3.85 as a result of the changes implemented. Jim was delighted to see that their changes were successful in improving the level of advisory services offered at the local office. In his report to corporate, he only included average ratings for the client surveys.

- Identify the ethical dilemma in this scenario.

- What are the undesirable consequences?

- Propose an ethical solution that considers the welfare of all stakeholders.

WHAT HAVE WE LEARNED?

Learning Objectives

Understand that data are values, whether numerical or labels, together with their context.

- *Who, what, why, where, when* (and *how*)—the W's—help nail down the context of the data.
- We must know *who, what,* and *why* to be able to say anything useful based on the data. The *who* are the cases. The *what* are the variables. A variable gives information about each of the cases. The *why* helps us decide which way to treat the variables.
- Stop and identify the W's whenever you have data, and be sure you can identify the cases and the variables.

Identify whether a variable is being used as categorical or quantitative.

- Categorical variables identify a category for each case. Usually we think about the counts of cases that fall in each category. (An exception is an identifier variable that just names each case.)

- Quantitative variables record measurements or amounts of something; they must have units.
- Sometimes we may treat the same variable as categorical or quantitative depending on what we want to learn from it, which means some variables can't be pigeonholed as one type or the other.

Consider the source of your data and the reasons the data were collected. That can help you understand what you might be able to learn from the data.

Terms

Big Data	The collection and analysis of data sets so large and complex that traditional methods typically brought to bear on the problem would be overwhelmed.
Business analytics	The process of using statistical analysis and modeling to drive business decisions.
Case	A case is an individual about whom or which we have data.
Categorical (or qualitative) variable	A variable that names categories (whether with words or numerals) is called categorical or qualitative.
Context	The context ideally tells *who* was measured, *what* was measured, *how* the data were collected, *where* the data were collected, and *when* and *why* the study was performed.
Cross-sectional data	Data taken from situations that vary over time but measured at a single time instant is said to be a cross-section of the time series.
Data	Recorded values whether numbers or labels, together with their context.
Data mining	The process of using a variety of statistical tools to analyze large data bases or data warehouses.
Data table	An arrangement of data in which each row represents a case and each column represents a variable.
Data warehouse	A large data base of information collected by a company or other organization usually to record transactions that the organization makes, but also used for analysis via data mining.
Experimental unit	An individual in a study for which or for whom data values are recorded. Human experimental units are usually called subjects or participants.
Identifier variable	A categorical variable that records a unique value for each case, used to name or identify it.
Metadata	Auxiliary information about variables in a database, typically including *how*, *when*, and *where* (and possibly *why*) the data were collected; *who* each case represents; and the definitions of all the variables.
Nominal variable	The term "nominal" can be applied to a variable whose values are used only to name categories.
Ordinal variable	The term "ordinal" can be applied to a variable whose categorical values possess some kind of order.
Participant	A human experimental unit. Also called a subject.
Quantitative variable	A variable in which the numbers are values of measured quantities with units.
Record	Information about an individual in a database.
Relational database	A relational database stores and retrieves information. Within the database, information is kept in data tables that can be "related" to each other.
Respondent	Someone who answers, or responds to, a survey.
Spreadsheet	A spreadsheet is layout designed for accounting that is often used to store and manage data tables. Excel is a common example of a spreadsheet program.
Subject	A human experimental unit. Also called a participant.
Time series	Data measured over time. Usually the time intervals are equally spaced or regularly spaced (e.g., every week, every quarter, or every year).
Transactional data	Data collected to record the individual transactions of a company or organization.
Units	A quantity or amount adopted as a standard of measurement, such as dollars, hours, or grams.
Variable	A variable holds information about the same characteristic for many cases.

TECHNOLOGY HELP: Data

Most often we find statistics on a computer using a program, or *package*, designed for that purpose. There are many different statistics packages, but they all do essentially the same things.

If you understand what the computer needs to know to do what you want and what it needs to show you in return, you can figure out the specific details of most packages pretty easily.

For example, to get your data into a computer statistics package, you need to tell the computer:

- Where to find the data. This usually means directing the computer to a file stored on your computer's disk or to data on a database. Or it might just mean that you have copied the data from a spreadsheet program or Internet site and it is currently on your computer's clipboard. Usually, the data should be in the form of a data table. Most computer statistics packages prefer the *delimiter* that marks the division between elements of a data table to be a tab character and the delimiter that marks the end of a case to be a *return* character.
- Where to put the data. (Usually this is handled automatically.)
- What to call the variables. Some data tables have variable names as the first row of the data, and often statistics packages can take the variable names from the first row automatically.

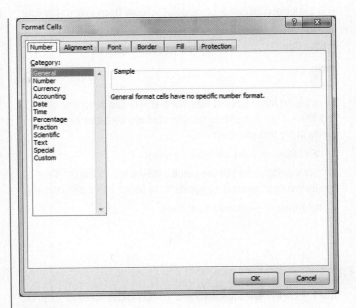

EXCEL

To open a file containing data in Excel:

- Choose **File > Open**.
- Browse to find the file to open. Data files provided with this text are tab-delimited text files (.txt) or comma-delimited text files (.csv).Excel supports many other file formats.
- You can also copy tables of data from other sources, such as Internet sites, and paste them into an Excel spreadsheet. Excel can recognize the format of many tables copied this way, but this method may not work for some tables.
- When opening a data file, Excel may not recognize the format of the data. If data include dates or other special formats ($, €, ¥, etc.), identify the desired format. Select the cells or columns to reformat and choose **Format > Cell**. Often, the General format is the best option.

JMP

To import a text file, choose:

- **File > Open** and select the file from the dialog. At the bottom of the dialog screen you'll see **Open As:**—be sure to change to **Data (Using Preview)**. This will allow you to specify the delimiter and make sure the variable names are correct. (**JMP** also allows various formats to be imported directly, including .xls files.)

You can also paste a data set in directly (with or without variable names) by selecting:

- **File > New > New Data Table** and then **Edit > Paste (or Paste with Column Names** if you copied the names of the variables as well).

Finally, you can import a data set from a URL directly by selecting:

- **File > Internet Open** and pasting in the address of the web site. JMP will attempt to find data on the page. It may take a few tries and some edits to get the data set in correctly.

MINITAB

To import a text or Excel file, choose:

- **File > Open Worksheet**. From **Files of type**, choose **Text (*.txt)** or **Excel (*.xls; *xlsx).**

- Browse to find and select the file.
- In the lower right corner of the dialog, choose **Open** to open the data file alone, or **Merge** to add the data to an existing worksheet.
- Click **Open**.

R can import many types of files, but text files (tab or comma delimited) are easiest. If the file is tab delimited and contains the variable names in the first row, then:

> mydata = read.delim(file.choose())

will give a dialog where you can pick the file you want to import. It will then be in a data frame called mydata. If the file is comma delimited, use:

> mydata = read.csv(file.choose())

Comments

(RStudio provides an interactive dialog that may be easier to use). For other options including the case that the file does not contain variable names, consult **R** help.

SPSS

To import a text file, choose:

- **File > Open > Data**. Under "Files of type", choose **Text (*.txt,*.dat)**. Select the file you want to import. Click **Open.**
- A window will open called **Text Import Wizard**. Follow the steps, depending on the type of file you want to import.

Brief **Case**

Credit Card Bank

Like all credit and charge card companies, this company makes money on each of its cardholders' transactions. Thus, its profitability is directly linked to card usage. To increase customer spending on its cards, the company sends many different offers to its cardholders, and market researchers analyze the results to see which offers yield the largest increases in the average amount charged.

On your disk (in the file **Credit Card Bank**) is part of a database like the one used by the researchers. For each customer, it contains several variables in a spreadsheet.

Examine the data in the data file. List as many of the W's as you can for these data and classify each variable as categorical or quantitative. If quantitative, identify the units.

EXERCISES

SECTION 1.1

1. A real estate major collected information on some recent local home sales. The first 6 lines of the database appear below. The columns correspond to the house identification number, the community name, the ZIP code, the number of acres of the property, the year the house was built, the market value, and the size of the living area (in square feet).

a) What does a row correspond to in this data table? How would you best describe its role: as a participant, subject, case, respondent, or experimental unit?
b) How many variables are measured on each row?

House_ID	Neighborhood	Mail_ZIP	Acres	Yr_Built	Full_Market_Value	Size
41340053	Greenfield Manor	12859	1.00	1967	$1,00,400	960
4128001474	Fort Amherst	12801	0.09	1961	$1,32,500	906
412800344	Dublin	12309	1.65	1993	$1,40,000	1620
4128001552	Granite Springs	10598	0.33	1969	$67,100	900
412800352	Arcady	10562	2.29	1955	$1,90,000	1224
413400322	Ormsbee	12859	9.13	1997	$1,26,900	1056

2. A local bookstore is keeping a database of its customers to find out more about their spending habits so that the store can start to make personal recommendations based on past purchases. Here are the first five rows of their database:

a) What does a row correspond to in this data table? How would you best describe its role: as a participant, subject, case, respondent, or experimental unit?
b) How many variables are measured on each row?

Transaction ID	Customer ID	Date	ISBN Number of Purchase	Price	Coupon?	Gift?	Quantity
29784320912	4J438	11/12/2009	345-23-2355	$29.95	N	N	1
26483589001	3K729	9/30/2009	983-83-2739	$16.99	N	N	1
26483589002	3K729	9/30/2009	102-65-2332	$9.95	Y	N	1
36429489305	3U034	12/5/2009	295-39-5884	$35.00	N	Y	1
36429489306	3U034	12/5/2009	183-38-2957	$79.95	N	Y	1

SECTION 1.2

3. Referring to the real estate data table of Exercise 1,

a) For each variable, would you describe it as primarily categorical, or quantitative? If quantitative, what are the units? If categorical, is it ordinal or simply nominal?
b) Are these data a time series, or are these cross-sectional? Explain briefly.

4. Referring to the bookstore data table of Exercise 2,

a) For each variable, would you describe it as primarily categorical, or quantitative? If quantitative, what are the units? If categorical, is it ordinal or simply nominal?
b) Are these data a time series, or are these cross-sectional? Explain briefly.

SECTION 1.3

5. For the real estate data of Exercise 1, do the data appear to have come from a designed survey or experiment? What concerns might you have about drawing conclusions from this data set?

6. A student finds data on an Internet site that contains financial information about selected companies. He plans to analyze the data and use the results to develop a stock investment strategy. What kind of data source is he using? What concerns might you have about drawing conclusions from this data set?

CHAPTER EXERCISES

For each description of data in Exercises 7 to 26, identify the W's, name the variables, specify for each variable whether its use indicates it should be treated as categorical or quantitative, and for any quantitative variable identify the units in which it was measured (if they are not provided, give some possible units in which they might be measured). Specify whether the data come from a designed survey or experiment. Are the variables time series or cross-sectional? Report any concerns you have as well.

7. The news. Find a newspaper or magazine article in which some data are reported (e.g., see *The Wall Street*

Journal, Financial Times, Business Week, or *Fortune*). For the data discussed in the article, answer the questions above. Include a copy of the article with your report.

8. The Internet. Find an Internet site on which some data are reported. For the data found on the site, answer as many of the questions above as you can. Include a copy of the URL with your report.

9. Survey. An automobile manufacturer wants to know what college students think about electric vehicles. They ask you to conduct a survey that asks students, "Do you think there will be more electric or gasoline powered vehicles on the road in 2025?" and "How likely are you to buy an electric vehicle in the next 10 years?" (scale of 1 = not at all likely to 5 = very likely).

10. Your survey. Think of a question that you'd like to know the answer to that might be answered with a survey. What are the questions? Identify the variables and answer the questions above.

11. World databank. The World Bank provides economic data on most of the world's countries at their website (databank.worldbank.org/data/home.aspx). Select 5 indicators that they provide and answer the questions above for these variables.

12. Arby's menu. A listing posted by the Arby's restaurant chain gives, for each of the sandwiches it sells, the type of meat in the sandwich, number of calories, and serving size in ounces. The data might be used to assess the nutritional value of the different sandwiches.

13. MBA admissions. A school in the northeastern United States is concerned with the recent drop in female students in its MBA program. It decides to collect data from the admissions office on each applicant, including: sex of each applicant, age of each applicant, whether or not they were accepted, whether or not they attended, and the reason for not attending (if they did not attend). The school hopes to find commonalities among the female accepted students who have decided not to attend the business program.

14. MBA admissions II. An internationally recognized MBA program outside of Paris intends to also track the GPA of the MBA students and compares MBA performance to standardized test scores over a six-year period (2009–2014).

15. Pharmaceutical firm. Scientists at a major pharmaceutical firm conducted an experiment to study the effectiveness of an herbal compound to treat the common cold. They exposed volunteers to a cold virus, then gave them either the herbal compound or a sugar solution known to have no effect on colds. Several days later they assessed each patient's condition using a cold severity scale ranging from 0–5. They found no evidence of the benefits of the compound.

16. Start-up company. A start-up company is building a database of customers and sales information. For each customer, it records name, ID number, region of the country (1 = East, 2 = South, 3 = Midwest, 4 = West), date of last purchase, amount of purchase, and item purchased.

17. Vineyards. Business analysts hoping to provide information helpful to grape growers sent out a questionnaire to a sample of growers requesting these data about vineyards: size, number of years in existence, state, varieties of grapes grown, average case price, gross sales, and percent profit.

18. Spectrem Group polls. Spectrem Group (www. spectrem .com) provides services for the affluent and retirement markets. In a recent survey, they found that millionaires tend to prefer dogs to cats from a question asking them to list the pets they own. They also found that senior executives are more likely to buy treats and toys for their pet than regular investors by asking, "What services and products do you buy for your pet?"

19. EPA. The Environmental Protection Agency (EPA) tracks fuel economy of automobiles. Among the data EPA analysts collect from the manufacturer are the manufacturer (Ford, Toyota, etc.), vehicle type (car, SUV, etc.), weight, horsepower, and gas mileage (mpg) for city and highway driving.

20. Consumer Reports. In 2013, Consumer Reports published an article comparing smart phones. It listed 46 phones, giving brand, price, display size, operating system (Android, iOS, or Windows Phone), camera image size (megapixels), and whether it had a memory card slot.

21. Zagat. Zagat.com provides ratings from customer experiences on restaurants. For each restaurant, the percentage of customers that liked it, the average cost and ratings of the food, decor, and service (all on a 30-point scale) are reported.

22. L.L. Bean. L.L. Bean is a large U.S. retailer that depends heavily on its catalog sales. It collects data internally and tracks the number of catalogs mailed out, the number of square inches in each catalog, and the sales ($ thousands) in the four weeks following each mailing. The company is interested in learning more about the relationship (if any) among the timing and space of their catalogs and their sales.

23. Stock market. An online survey of students in a large MBA Statistics class at a business school in the northeastern United States asked them to report their total personal investment in the stock market ($), total number of different stocks currently held, total invested in mutual funds ($), and the name of each mutual fund in which they have invested. The data were used in the aggregate for classroom illustrations.

24. Theme park sites. A study on the potential for developing theme parks in various locations throughout Europe in 2013 collects the following information: the country where the proposed site is located, estimated cost to acquire site, size of population within a one-hour drive of the site, size of the site, and availability of mass transportation within five minutes of the site. The data will be used to present to prospective developers.

T 25. Indy 2014. The 2.5-mile Indianapolis Motor Speedway has been the home to a race on Memorial Day nearly every year since 1911. Even during the first race there were controversies. Ralph Mulford was given the checkered flag first but took three extra laps just to make sure he'd completed 500 miles. When he finished, another driver, Ray Harroun, was being presented with the winner's trophy, and Mulford's protests were ignored. Harroun averaged 74.6 mph for the 500 miles. Here are the data for the first few and six recent Indianapolis 500 races.

Year	Winner	Chassis	Engine	Time (hrs)	Speed (mph)	Car #
1911	Ray Harroun	Marmon	Marmon	6.7022	74.602	32
1912	Joe Dawson	National	National	6.3517	78.719	8
1913	Jules Goux	Peugeot	Peugeot	6.5848	75.933	16
...		...				
...		...				
2009	Hélio Castroneves	Dallara	Honda	3.3262	150.318	3
2010	Dario Franchitti	Dallara	Honda	3.0936	161.623	10
2011	Dan Wheldon	Dallara	Chevrolet	2.9366	170.265	98
2012	Dario Franchitti	Dallara	Honda	2.9809	167.734	50
2013	Tony Kanaan	Dallara	Chevrolet	2.6676	187.433	12
2014	Ryan Hunter-Reay	Dallara	Honda	2.6801	186.563	19

T **26. Kentucky Derby 2014.** The Kentucky Derby is a horse race that has been run every year since 1875 at Churchill Downs, Louisville, Kentucky. The race started as a 1.5-mile race, but in 1896 it was shortened to 1.25 miles because experts felt that three-year-old horses shouldn't run such a long race that early in the season. (It has been run in May every year but one—1901—when it took place on April 29.) The table at the bottom of the page shows the data for the first few and a few recent races. http://www.kentuckyderby.ag/kentuckyderby-results.php and http://horseracing.about.com/od/history/l/blderbywin.htm

Year	Winner	Jockey	Duration (seconds)	Track Condition
1875	Aristides	O. Lewis	157.75	Fast
1876	Vagrant	B. Swim	158.25	Fast
1877	Baden-Baden	W. Walker	158	Fast
1878	Day Star	J. Carter	157.25	Dusty
1879	Lord Murphy	C. Shauer	157	Fast
...				
2010	Super Saver	Calvin Borel	124.45	Fast
2011	Animal Kingdom	John R. Velazquez	122.04	Fast
2012	I'll Have Another	Mario Gutierrez	121.83	Fast
2013	Orb	J. Rosario	122.89	Sloppy
2014	California Chrome	Victor Espinoza	123.66	Fast

When you organize data in a spreadsheet, it is important to lay it out as a data table. For each of these examples in Exercises 27 to 30, show how you would lay out these data. Indicate the headings of columns and what would be found in each row.

27. Mortgages. For a study of mortgage loan performance: amount of the loan, the name of the borrower.

28. Employee performance. Data collected to determine performance-based bonuses: employee ID, average contract closed (in $), supervisor's rating (1–10), years with the company.

29. Company performance. Data collected for financial planning: weekly sales, week (week number of the year), sales predicted by last year's plan, difference between predicted sales and realized sales.

30. Command performance. Data collected on investments in Broadway shows: number of investors, total invested, name of the show, profit/loss after one year.

For the following examples in Exercises 31 to 34, indicate whether the data are time-series or cross-sectional.

31. Car sales. Number of cars sold by each salesperson in a dealership in September.

32. Motorcycle sales. Number of motorcycles sold by a dealership in each month of 2014.

33. Cross sections. Average diameter of trees brought to a sawmill in each week of a year.

34. Series. Attendance at the third World Series game recording the age of each fan.

JUST CHECKING ANSWERS

1 Who—policies on churches and schools

   What—policy number, years claim free, net property premium ($), net liability premium ($), total property value ($000), median age in ZIP code, school?, territory, coverage

   How—company records

   When—not given

2 Policy number: identifier (categorical)

   Years claim free: quantitative

   Net property premium: quantitative ($)

   Net liability premium: quantitative ($)

   Total property value: quantitative ($)

   Median age in ZIP code: quantitative

   School?: categorical (true/false)

   Territory: categorical

   Coverage: categorical

2

Displaying and Describing Categorical Data

KEEN, Inc.

KEEN, Inc. was started to create a sandal designed for a variety of water activities. The sandals quickly became popular due to their unique patented toe protection—a black bumper to protect the toes when adventuring out on rivers and trails. Today, the KEEN brand offers over 300 different outdoor performance and outdoor inspired casual footwear styles as well as bags and socks.

Few companies experience the kind of growth that KEEN has in its first nine years. Amazingly, they've done this with relatively little advertising and by selling primarily to specialty footwear and outdoor stores in addition to online outlets.

After the 2004 tsunami disaster in Japan, KEEN cut its advertising budget almost completely and donated over $1 million to help the victims and establish the KEEN Foundation to support environmental and social causes. Philanthropy and community projects continue to play an integral part of the KEEN brand values. In fact, KEEN has established a giving program with a philanthropic effort devoted to helping the environment, conservation, and social movements involving the outdoors.

WHO	Visits to the KEEN, Inc. website
WHAT	Source (search engine or other) that led to KEEN's website
WHEN	February 2013
WHERE	Worldwide
HOW	Data compiled by KEEN
WHY	To understand customer use of the website and how they got there

KEEN, Inc., like most companies, collects data on visits to its website. Each visit to the site and each subsequent action the visitor takes (changing the page, entering data, etc.) is recorded in a file called a usage, or access weblog. These logs contain a lot of potentially worthwhile information, but they are not easy to use. Here's one line from a log:

```
245.240.221.71 -- [1/Apr/2013:13:15:08-0800]" GET http://
www.keenfootwear.com/us/en/product/shoes/men/cnx/clear
water%20cnx/forest%20night!rust "http://www.google.com/"
"Mozilla/5.0WebTV/1.2 (compatible; MSIE 2.0)"
```

KEEN, like many other small and mid-sized companies, uses *Google Analytics* to collect and summarize its log data.

In February 2013, there were 226,925 visits to the KEEN site. We wouldn't be able to see any patterns in a data table with 226,925 rows. And seeing is exactly what we want to do. We need ways to show the data so that we can see patterns, relationships, trends, and exceptions.

2.1   Summarizing a Categorical Variable

KEEN might want to know how people reach their website. They might use the information to allocate their advertising revenue to various search engines, putting ads where they'll be seen by the most potential customers. The variable *Source* records, for each visit to KEEN's website, where the visitor came from. The categories are all the search engines used, plus the label "Direct," which indicates that the customer typed in KEEN's web address (or URL) directly into the browser. To make sense of the 226,925 visits for which they have data, they'd like to summarize the variable and display the information in a way that can easily communicate the results to others.

Frequency Tables

A **frequency table** records the counts for each of the categories of the variable. Some tables report percentages, and many report both. For example, Table 2.1 shows the ways that customers found their way to the KEEN website.

Source	Visits	Visits by %
Google	130,158	57.36
Direct	52,969	23.34
E-mail	16,084	7.09
Bing	9,581	4.22
Yahoo	7,439	3.28
Facebook	2,253	0.99
Mobile	1,701	0.75
Other	6,740	2.97
Total	**226,925**	**100.00**

Table 2.1   A frequency table of the *Source* used by visitors to the KEEN, Inc. website. Notice the label "Other." When the number of categories gets too large, we often lump together values of the variable into "Other." When to do that is a judgment call, but it's a good idea to have fewer than about a dozen categories. (Source: KEEN, Inc., personal communication.)

FOR EXAMPLE   Making frequency and relative frequency tables

The Super Bowl, the championship game of the National Football League of the United States, is an important annual social event for Americans, with tens of millions of viewers. The ads that air during the game are expensive: a 30-second ad during the 2013 Super Bowl cost about $4M. The high price of these commercials makes them high-profile and much anticipated, and so the advertisers feel pressure to be innovative, entertaining, and often humorous. Some people, in fact, watch the Super Bowl mainly for the commercials. Polls often ask whether respondents are more interested in the game or the commercials. Here are 40 responses from one such poll. (NA/Don't Know = No Answer or Don't Know):

Won't Watch	Game	Commercials	Won't Watch	Game
Game	Won't Watch	Commercials	Game	Game
Commercials	Commercials	Game	Won't Watch	Commercials
Game	NA/Don't Know	Commercials	Game	Game
Won't Watch	Game	Game	Won't Watch	Game
Game	Won't Watch	Won't Watch	Game	Won't Watch
Won't Watch	Commercials	Commercials	Game	Won't Watch
NA/Don't Know	Won't Watch	Game	Game	Game

QUESTION Make a frequency table for this variable. Include the percentages to display both a frequency and relative frequency table at the same time.

ANSWER There were four different responses to the question about watching the Super Bowl. Counting the number of participants who responded to each of these gives the following table:

Response	Counts	Percentage
Commercials	8	20.0
Game	18	45.0
Won't Watch	12	30.0
No Answer/Don't Know	2	5.0
Total	**40**	**100.0**

100.01%?

Sometimes if you carefully add the percentages of all categories, you will notice the total isn't exactly 100.00% even though we know that that's what the total has to be. The discrepancy is due to individual percentages being rounded. You'll often see this in tables of percents, sometimes with explanatory footnotes.

2.2 Displaying a Categorical Variable
The Three Rules of Data Analysis

There are three things you should always do with data:

1. **Make a picture.** A display of your data will reveal things you are not likely to see in a table of numbers and will help you to *plan* your approach to the analysis and think clearly about the patterns and relationships that may be hiding in your data.
2. **Make a picture.** A well-designed display will *do* much of the work of analyzing your data. It can show the important features and patterns. A picture will also reveal things you did not expect to see: extraordinary (possibly wrong) data values or unexpected patterns.
3. **Make a picture.** The best way to *report* to others what you find in your data is with a well-chosen picture.

These are the three rules of data analysis. These days, technology makes drawing pictures of data easy, so there is no reason not to follow the three rules.

Some displays communicate information better than others. We'll discuss some general principles for displaying information honestly in this chapter.

Data visualization has become a special discipline in its own right. A well-designed display can show features of even a large, complex data set. Figure 2.1 on the next page is a specially designed visualization showing the connections between two categorical variables, *College major* and *Career choice*, for 15,600 alumni of Williams College. Innovative visualizations such as this one—many of them interactive or animated—are becoming more common as Big Data is mined for unanticipated patterns and relationships.

Figure 2.1 Visualization of the link between major in college and career of Williams College alumni. Each individual is graphed as an arc connecting his or her major on the left with a career area on the right. Each major is assigned a color: Humanities in the blue range, Social Sciences in the reds and oranges, and Sciences in greens. It is easy to see the expected large arc connecting Biology and Health/Medicine and the spread of Math majors to many careers. Possibly less expected is that Economics majors choose a wide range of careers. Banking/Finance draws many from Economics, but also quite a few from History, Political Science, and the Humanities. (This image was created by Satyan Devadoss, Hayley Brooks, and Kaison Tanabe using the CIRCOS software; an interactive version of this graph can be found at http://cereusdata. com.)

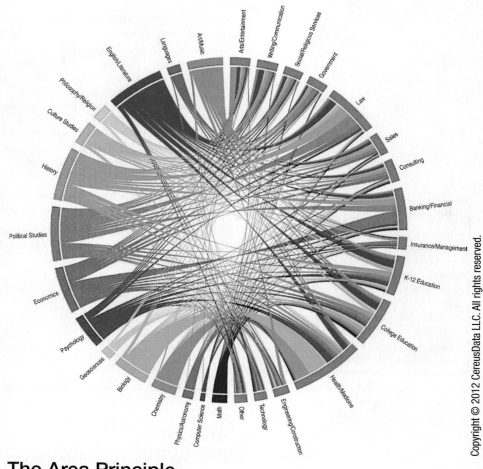

The Area Principle

We can't make just any display; a bad picture can distort our understanding rather than help it. For example, Figure 2.2 is a graph of the frequencies of Table 2.1. What impression do you get of the relative frequencies of visits from each source? You can easily see from both the table and the figure that the most popular source was

Figure 2.2 Although the length of each sandal corresponds to the correct number, the impression we get is all wrong because we perceive the entire area of the sandal. In fact, only about 57% of all visitors used Google to get to the website.

Google. But the impression given by Figure 2.2 doesn't seem to correspond well to the numbers in the table.

Although it's true that the majority of people came to KEEN's website from Google, in Figure 2.2 it looks like nearly all did. That doesn't seem right. What's wrong? The lengths of the sandals *do* match the frequencies in the table. But our eyes tend to be more impressed by the *area* (or perhaps even the *volume*) than by other aspects of each sandal image, and it's that aspect of the image that we notice. Since there were about two and a half as many people who came from Google as those who typed in the URL directly, the sandal depicting the number of Google visitors is about two and a half times longer than the sandal below it, but it occupies more than six times the area. As you can see from the frequency table, that just isn't a correct impression.

The best data displays observe a fundamental principle of graphing data called the **area principle**, which says that the area occupied by a part of the graph should correspond to the magnitude of the value it represents.

Bar Charts

Figure 2.3 gives us a chart that obeys the area principle. It's not as visually entertaining as the sandals, but it does give a more *accurate* visual impression of the distribution. The height of each bar shows the count for its category. The bars are the same width, so their heights determine their areas, and the areas are proportional to the counts in each class. Now it's easy to see that nearly half the site hits came from places other than Google. We can also see that there were about two and a half times as many visits that originated with a Google search as there were visits that came directly. Bar charts make these kinds of comparisons easy and natural.

Figure 2.3 Visits to the KEEN, Inc. website by *Source*. With the area principle satisfied, the true distribution is clear.

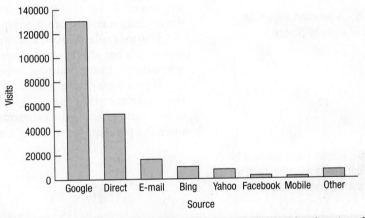

A **bar chart** displays the **distribution** of a categorical variable, showing the counts for each category next to each other for easy comparison. Bar charts should have small spaces between the bars to indicate that these are freestanding bars that could be rearranged into any order. The bars are lined up along a common base with labels for each category. The variable name is often used as a subtitle for the horizontal axis.

Bar charts are usually drawn vertically in columns,

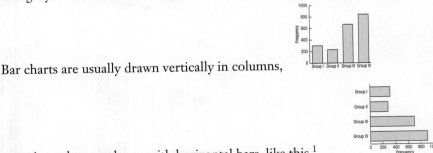

but sometimes they are drawn with horizontal bars, like this.[1]

---

[1] Excel refers to this display as a column chart when the bars are vertical and a bar chart when they are horizontal, but that's not standard statistics terminology.

Figure 2.4 The relative frequency bar chart looks the same as the bar chart (Figure 2.3) but shows the proportion of visits in each category rather than the counts.

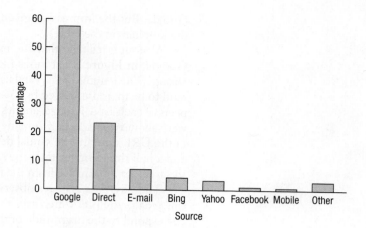

If we want to draw attention to the relative *proportion* of visits from each *Source*, we could replace the counts with percentages and use a **relative frequency bar chart**, like the one shown in Figure 2.4.

Pie Charts

A **pie chart** shows how a whole group breaks into several categories. Pie charts show all the cases as a circle sliced into pieces whose areas are proportional to the fraction of cases in each category.

Because we're used to cutting up pies into 2, 4, or 8 pieces, pie charts are good for seeing relative frequencies near 1/2, 1/4, or 1/8. For example, in Figure 2.5, you can easily see that the slice representing Google is just a bit more than half the total. Unfortunately, other comparisons are harder to make with pie charts.

For example, Figure 2.6 shows three pie charts that look pretty much alike along with bar charts of the same data. The bar charts show three distinctly different patterns, but it is almost impossible to see those in the pie charts.

If you want to make a pie chart or relative frequency bar chart, you'll need to also make sure that the categories don't overlap, so that no individual is counted in two categories. If the categories do overlap, it's misleading to make a pie chart, since the percentages won't add up to 100%.

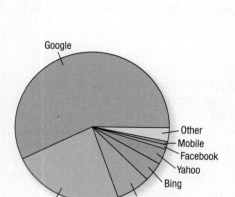

Figure 2.5 A pie chart shows the proportion of visits by *Source*.

Figure 2.6 Patterns that are easy to see in the bar charts are often hard to see in the corresponding pie charts.

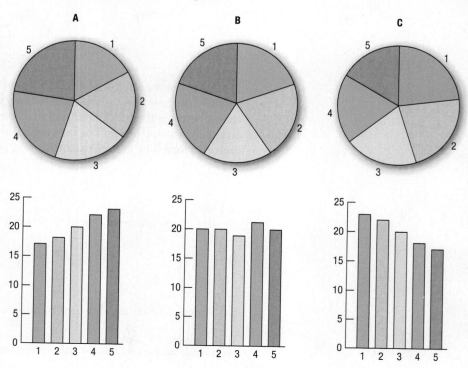

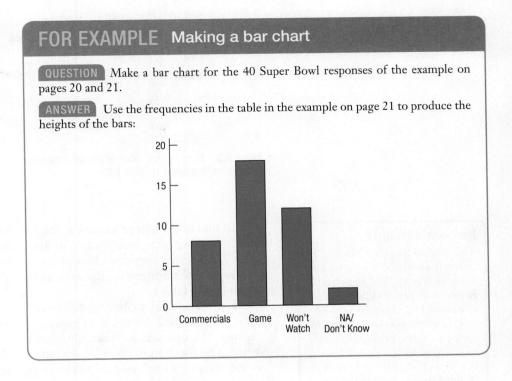

FOR EXAMPLE   Making a bar chart

QUESTION   Make a bar chart for the 40 Super Bowl responses of the example on pages 20 and 21.

ANSWER   Use the frequencies in the table in the example on page 21 to produce the heights of the bars:

2.3   Exploring Two Categorical Variables: Contingency Tables

WHO	Respondents in the Pew Research Worldwide Survey
WHAT	Responses to question about social networking
WHEN	2012
WHERE	Worldwide
HOW	Data collected by Pew Research using a multistage design. For details see www.pewglobal.org/2012/12/12/survey-methods-43/
WHY	To understand penetration of social networking worldwide

In December 2012 Pew Research conducted surveys in countries across the world (www.pewglobal.org/2012/12/12/social-networking-popular-across-globe/). One question of interest to business decision makers is how common it is for citizens of different countries to use social networking and whether they have it available to them. Table 2.2 gives a table of responses for several of the surveyed countries. Note that N/A means "not available" because respondents lacked internet access—a situation that marketers planning for the future might expect to see change.

The pie chart (Figure 2.7) shows clearly that fewer than half of respondents said that they had access to social networking and used it.

But if we want to target our online customer relations with social networks differently in different countries, wouldn't it be more interesting to know how social networking use varies from country to country?

Social Networking	Count	Relative frequency
No	1249	24.787
Yes	2175	43.163
N/A	1615	32.050

Table 2.2   A combined frequency and relative frequency table for the responses from 5 countries (Britain, Egypt, Germany, Russia, and the U.S.) to the question "Do you use social networking sites?" N/A means "Not Available."

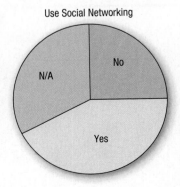

Figure 2.7   Responses to the question "Do you use social networking sites?" N/A means "No Internet Available."

	Britain	Egypt	Germany	Russia	U.S.	Total
No	336	70	460	90	293	**1249**
Yes	529	300	340	500	506	**2175**
N/A	153	630	200	420	212	**1615**
Total	**1018**	**1000**	**1000**	**1010**	**1011**	**5039**

Table 2.3   Contingency table of *Social Networking* and *Country*. The right margin "Totals" are the values that were in Table 2.2.

Percent of What?

The English language can be tricky. If asked, "What percent of those answering 'Yes' were from Russia?" it's pretty clear that you should focus only on the *Yes* row. The question itself seems to restrict the who in the question to that row, so you should look at the number of those in each country among the 2175 people who replied "Yes." You'd find that in the row percentages.

But if you're asked, "What percent were Russians who replied 'yes'?" you'd have a different question. Be careful. That question really means "what percent of the entire sample were both from Russia and replying 'Yes'?", so the *who* is all respondents. The denominator should be 5039, and the answer is the table percent.

Finally, if you're asked, "What percent of the Russians replied 'yes'?" you'd have a third question. Now the *who* is Russians. So the denominator is the 1010 Russians, and the answer is the column percent.

To find out, we need to look at the two categorical variables *Social Networking* and *Country* together, which we do by arranging the data in a two-way table such as Table 2.3. Because they show how individuals are distributed along each variable depending on, or *contingent on*, the value of the other variable, tables like this are called **contingency tables**.

The margins of a contingency table give totals. The totals in the right-hand column of Table 2.3 show the frequency distribution of the variable *Social Networking*. We can see, for example, that Internet access is certainly not yet universal. The totals in the bottom row of the table show the frequency distribution of the variable *Country*—how many respondents Pew obtained in each country. When presented like this, at the margins of a contingency table, the frequency distribution of either one of the variables is called its **marginal distribution**. The marginal distribution for a variable in a contingency table is the same as its frequency distribution.

Each **cell** of a contingency table (any intersection of a row and column of the table) gives the count for a combination of values of the two variables. For example, in Table 2.3 we can see that 153 respondents did not have internet access in Britain. Looking across the *Yes* row, you can see that the largest number of responses in that row (529) is from Britain. Are Egyptians less likely to use social media than Britons? Questions like this are more naturally addressed using percentages.

We know that 300 Egyptians report that they use social networking. We could display this count as a percentage, but as a percentage of what? The total number of people in the survey? (300 is 5.95% of the total.) The number of Egyptians surveyed? (300 is 30% of the 1000 Egyptians surveyed.) The number of respondents who use social networking? (300 is 13.8% of social networking users.) Most statistics programs offer a choice of **total percent**, **row percent**, or **column percent** for contingency tables. Unfortunately, they often put them all together with several numbers in each cell of the table. The resulting table (Table 2.4) holds lots of information but is hard to understand.

Conditional Distributions

The more interesting questions are contingent on something. We'd like to know, for example, whether these countries are similar in use and availability of social networking. That's the kind of information that could inform a business decision. Table 2.5 shows the distribution of social networking conditional on country.

By comparing the frequencies conditional on *Country*, we can see interesting patterns. For example, Germany stands out as the country in which the largest percentage (46%) have Internet access but don't use social networking ("No"). Russia and Egypt may have more respondents with no Internet access, but those who have

	Britain	Egypt	Germany	Russia	U.S.	Total
No	336	70	460	90	293	**1249**
	26.9	5.6	36.8	7.2	23.5	**100**
	33.0	7.0	46.0	8.9	29.0	**24.8**
	6.7	1.4	9.1	1.8	5.8	**24.8**
Yes	529	300	340	500	506	**2175**
	24.3	13.8	15.6	23.0	23.3	**100**
	52.0	30.0	34.0	49.5	50.0	**43.2**
	10.5	6.0	6.8	9.9	10.0	**43.2**
N/A	153	630	200	420	212	**1615**
	9.5	39.0	12.4	26.0	13.1	**100**
	15.0	63.0	20.0	41.6	21.0	**32.1**
	3.0	12.5	4.0	8.3	4.2	**32.1**
Total	**1018**	**1000**	**1000**	**1010**	**1011**	**5039**
	20.2	**19.8**	**19.8**	**20.0**	**20.1**	**100**
	100	**100**	**100**	**100**	**100**	**100**
	20.2	**19.8**	**19.8**	**20.0**	**20.1**	**100**

Table contents:
Count
Percent of Row Total
Percent of Column Total
Percent of Table Total

Table 2.4 Another contingency table of *Social Networking* and *Country* showing the counts and the percentages these counts represent. For each count, there are three choices for the percentage: by row, by column, and by table total. There's probably too much information here for this table to be useful.

		Britain	Egypt	Germany	Russia	U.S.	Total
Social Networking	**No**	336	70	460	90	293	**1249**
		33.0	7.0	46.0	8.9	29.0	**24.8**
	Yes	529	300	340	500	506	**2175**
		52.0	30.0	34.0	49.5	50.0	**43.2**
	N/A	153	630	200	420	212	**1615**
		15.0	63.0	20.0	41.6	21.0	**32.1**
	Total	**1018**	**1000**	**1000**	**1010**	**1011**	**5039**
		100	**100**	**100**	**100**	**100**	**100**

Table 2.5 The conditional distribution of *Social Networking* conditioned on 5 values of *Country*. This table shows the column percentages.

access are very likely to use social networking. A distribution like this is called a **conditional distribution** because it shows the distribution of one variable for just those cases that satisfy a condition on another. In a contingency table, when the distribution of one variable is the same for all categories of another variable, we say that the two variables are **independent**. That tells us there's no association between these variables. We'll see a way to check for independence formally later in the book. For now, we'll just compare the distributions.

FOR EXAMPLE Contingency tables and side-by-side bar charts

Here is a contingency table of the responses for 1008 adult U.S. respondents to the question about watching the Super Bowl discussed in the previous For Example.

	Sex		
	Female	**Male**	**Total**
Game	198	277	**475**
Commercials	154	79	**233**
Won't Watch	160	132	**292**
NA/Don't Know	4	4	**8**
Total	**516**	**492**	**1008**

QUESTION Does it seem that there is an association between what viewers are interested in watching and their sex?

ANSWER First, find the conditional distributions of the four responses for each sex:

For Men:

Game = 277/492 = 56.3%

Commercials = 79/492 = 16.1%

Won't Watch = 132/492 = 26.8%

NA/Don't Know = 4/492 = 0.8%

For Women:

Game = 198/516 = 38.4%

Commercials = 154/516 = 29.8%

Won't Watch = 160/516 = 31.0%

NA/Don't Know = 4/516 = 0.8%

Now display the two distributions with side-by-side bar charts:

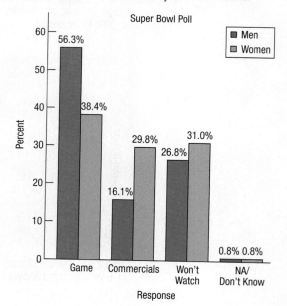

Based on this poll it appears that women were only slightly less interested than men in watching the Super Bowl telecast: 31% of the women said they didn't plan to watch, compared to just under 27% of men. Among those who planned to watch, however, there appears to be an association between the viewer's sex and what the viewer is most looking forward to. While more women are interested in the game (38%) than the commercials (30%), the margin among men is much wider: 56% of men said they were looking forward to seeing the game, compared to only 16% who cited the commercials.

JUST CHECKING

So that they can balance their inventory, an optometry shop collects the following data for customers in the shop.

		Eye Condition			
		Nearsighted	**Farsighted**	**Need Bifocals**	**Total**
Sex	**Males**	6	20	6	**32**
	Females	4	16	12	**32**
	Total	**10**	**36**	**18**	**64**

1 What percent of females are farsighted?

2 What percent of nearsighted customers are female?

3 What percent of all customers are farsighted females?

4 What's the distribution of *Eye Condition*?

5 What's the conditional distribution of *Eye Condition* for males?

6 Compare the percent who are female among nearsighted customers to the percent of all customers who are female.

7 Does it seem that *Eye Condition* and *Sex* might be dependent? Explain.

2.4 Segmented Bar Charts and Mosaic Plots

Everyone knows what happened in the North Atlantic on the night of April 14, 1912 as the *Titanic*, thought by many to be unsinkable, sank, leaving almost 1500 passengers and crew members on board to meet their icy fate. Women and children first was the rule for those commanding the lifeboats, but how did the class of ticket held enter into the order?

Here is a contingency table of the 2201 people on board, categorized by *Survival* and *Ticket Class*.

			Class				
			First	**Second**	**Third**	**Crew**	**Total**
Survival	**Alive**	**Count**	203	118	178	212	**711**
		% of Column	62.5%	41.4%	25.2%	24.0%	**32.3%**
	Dead	**Count**	122	167	528	673	**1490**
		% of Column	37.5%	58.6%	74.8%	76.0%	**67.7%**
	Total	**Count**	**325**	**285**	**706**	**885**	**2201**
			100%	**100%**	**100%**	**100%**	**100%**

Table 2.6 A contingency table of *Class* by *Survival* with only counts and column percentages. Each column represents the conditional distribution of *Survival* for a given category of ticket *Class*.

Looking at how the percentages change across each row, it sure looks like ticket class mattered in whether a passenger survived. To make it more vivid, we could display the percentages for surviving and not for each *Class* in a side-by-side bar chart such as in Fig 2.8 on the next page.

Now it's easy to compare the risks. Among first-class passengers, 37.5% perished, compared to 58.6% for second-class ticket holders, 74.8% for those in third class, and 76.0% for crew members.

Figure 2.8 *Side-by-side bar chart* showing the conditional distribution of *Survival* for each category of ticket *Class*.

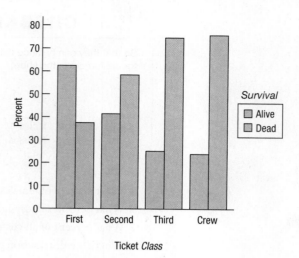

We could also display the *Titanic* information by dividing up bars rather than circles (as we did for pie charts). The resulting **segmented (or stacked) bar chart** treats each bar as the "whole" and divides it proportionally into segments corresponding to the percentage in each group. We can clearly see that the distributions of ticket *Class* are different, indicating again that survival was not independent of ticket *Class*.

Figure 2.9 A segmented bar chart for *Class* by *Survival*. Notice that although the totals for survivors and nonsurvivors are quite different, the bars are the same height because we have converted the numbers to *percentages*. Compare this display with the bar chart in Figure 2.8.

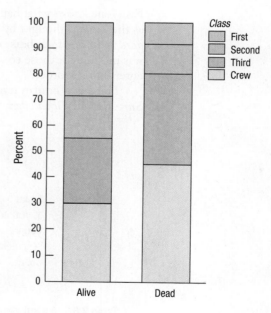

A variant of the segmented bar chart, a **mosaic plot** (Figure 2.10), looks like a segmented bar chart, but obeys the area principle better by making the bars proportional to the sizes of the groups. Now, each rectangle is proportional to the number of cases in the data set. Mosaic plots are increasingly popular for displaying contingency tables and are found in many software packages.

Figure 2.10 A mosaic plot for *Class by Survival*. The plot is just like the segmented bar chart in Figure 2.9 except that the space has been taken out between the categories on the *x*-axis and the rectangles are proportional to the number of cases of the *x* variable as well. We can easily see that the number of survivors was far less than the nonsurvivors, something that we can't in the bar charts.

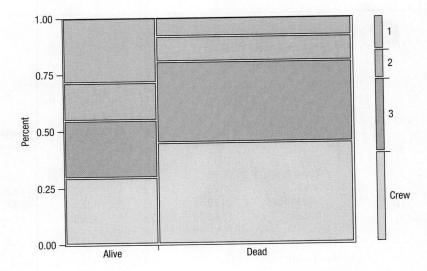

GUIDED EXAMPLE   Food Safety

Food storage and food safety are major issues for multinational food companies. A client wants to know if people of all age groups have the same degree of concern so GfK Roper Consulting asked 1500 people in five countries whether they agree with the following statement: "I worry about how safe the food I buy is." We would want to report to the client how concerns about food safety are related to age.

PLAN	Setup

- State the objectives and goals of the study.
- Identify and define the variables.
- Provide the time frame of the data collection process.

Determine the appropriate analysis for data type.

The client wants to examine the distribution of responses to the food safety question and see whether they are related to the age of the respondent. GfK Roper Consulting collected data on this question in the fall of 2005 for their 2006 Worldwide report. We will use the data from that study.

The variable is *Food Safety*. The responses are in nonoverlapping categories of agreement, from Agree Completely to Disagree Completely (and Don't Know). There were originally 12 age groups, which we can combine into five:

Teen	13–19
Young Adult	20–29
Adult	30–39
Middle Aged	40–49
Mature	50 and older

Both variables, *Food Safety* and *Age*, are ordered categorical variables. To examine any differences in responses across age groups, it is appropriate to create a contingency table and a side-by-side bar chart. Here is a contingency table of "Food Safety" by "Age".

(continued)

Mechanics For a large data set like this, we rely on technology to make tables and displays.

		Food Safety					
		Agree Completely	Agree Somewhat	Neither Disagree Nor Agree	Disagree Somewhat	Disagree Completely	Don't Know	Total
Age	Teen	16.19	27.50	24.32	19.30	10.58	2.12	**100%**
	Young Adult	20.55	32.68	23.81	14.94	6.98	1.04	**100%**
	Adult	22.23	34.89	23.28	12.26	6.75	0.59	**100%**
	Middle Aged	24.79	35.31	22.02	12.43	5.06	0.39	**100%**
	Mature	26.60	33.85	21.21	11.89	5.82	0.63	**100%**

A side-by-side bar chart is particularly helpful when comparing multiple groups.

A side-by-side bar chart shows the percent of each response to the question by age group.

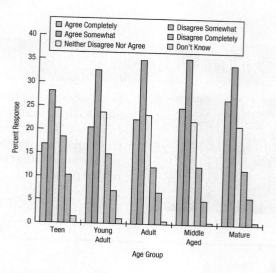

 REPORT **Summary and Conclusions**

Summarize the charts and analysis in context. Make recommendations if possible and discuss further analysis that is needed.

MEMO

Re: Food safety concerns by age

Our analysis of the GfK Roper Reports™ Worldwide survey data shows a weak pattern of concern about food safety that generally increases from youngest to oldest.

Our analysis thus far has not considered whether this trend is consistent across countries. If it were of interest to your group, we could perform a similar analysis for each of the countries.

The enclosed tables and plots provide support for these conclusions.

2.5 Simpson's Paradox

Here's an example showing that combining percentages across very different values or groups can give confusing results. Suppose there are two sales representatives, Peter and Katrina. Peter argues that he's the better salesperson, since he managed to close 83% of his last 120 prospects compared with Katrina's 78%. But let's look at the data a little more closely. Here (Table 2.7) are the results for each of their last 120 sales calls, broken down by the product they were selling.

Springfield Industries

Founded	1983
Employees	8536
Stock price	12.625
Average	3510.54

		Product		
		Printer Paper	**USB Flash Drive**	**Overall**
Sales Rep	**Peter**	90 out of 100 90%	10 out of 20 50%	100 out of 120 83%
	Katrina	19 out of 20 95%	75 out of 100 75%	94 out of 120 78%

Table 2.7 Look at the percentages within each Product category. Who has a better success rate closing sales of paper? Who has the better success rate closing sales of Flash Drives? Who has the better performance overall?

Look at the sales of the two products separately. For printer paper sales, Katrina had a 95% success rate, and Peter only had a 90% rate. When selling flash drives, Katrina closed her sales 75% of the time, but Peter only 50%. So Peter has better "overall" performance, but Katrina is better selling each product. How can this be?

This problem is known as **Simpson's Paradox**, named for the statistician who described it in the 1960s. Although it is rare, there have been a few well-publicized cases of it. As we can see from the example, the problem results from inappropriately combining percentages of different groups. Katrina concentrates on selling flash drives, which is more difficult, so her *overall* percentage is heavily influenced by her flash drive average. Peter sells more printer paper, which appears to be easier to sell. With their different patterns of selling, taking an overall percentage is misleading. Their manager should be careful not to conclude rashly that Peter is the better salesperson.

The lesson of Simpson's Paradox is to be sure to combine only comparable measurements for comparable individuals. Be especially careful when combining across different levels of a second variable. It's usually better to compare percentages *within* each level, rather than across levels.

Discrimination?

One famous example of Simpson's Paradox arose during an investigation of admission rates for men and women at the University of California at Berkeley's graduate schools. As reported in an article in *Science*, about 45% of male applicants were admitted, but only about 30% of female applicants got in. It looked like a clear case of discrimination. However, when the data were broken down by school (Engineering, Law, Medicine, etc.), it turned out that within each school, the women were admitted at nearly the same or, in some cases, much *higher* rates than the men. How could this be? Women applied in large numbers to schools with very low admission rates. (Law and Medicine, for example, admitted fewer than 10%.) Men tended to apply to Engineering and Science. Those schools have admission rates above 50%. When the total applicant pool was combined and the percentages were computed, the women had a much lower *overall* rate, but the combined percentage didn't really make sense.

WHAT CAN GO WRONG?

- **Don't violate the area principle.** This is probably the most common mistake in a graphical display. Violations of the area principle are often made for the sake of artistic presentation. Consider this pie chart of ways that respondents said they commute to work.

3% — 1% — 0% — 1% — 21%

Drive Alone
Shared
Taxi
Commuter Walk
Commuter Drive
Other Walk
Other Drive

Would it surprise you to learn that the fraction who "Shared" rides to work is 33%, while the fraction who "Drive Alone" is 41%? This pie chart was made in Excel, but overuse of features that make it look interesting has hurt its ability to convey accurate information.

- **Keep it honest.** Here's a pie chart that displays data on the percentage of high school students who engage in specified dangerous behaviors as reported by the Centers for Disease Control. What's wrong with this plot?

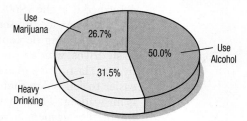

Use Marijuana 26.7%
Use Alcohol 50.0%
Heavy Drinking 31.5%

Try adding up the percentages. Or look at the 50% slice. Does it look right? Then think: What are these percentages of? Is there a "whole" that has been sliced up? In a pie chart, the proportions shown by each slice of the pie must add up to 100%, and each individual must fall into only one category. Of course, showing the pie on a slant makes it even harder to detect the error.

Here's another one. This chart shows the average number of texts sent by American cell phone customers in the period 1999 to 2014.

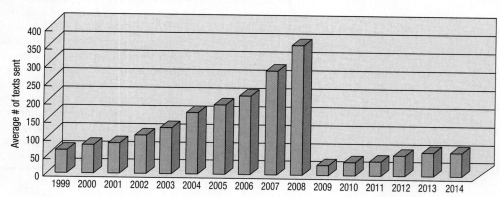

It may look as though text messaging decreased suddenly sometime around 2009, which probably doesn't seem right to you. In fact, this chart has several problems. First, it's not a bar chart. Bar charts display counts of categories. This bar chart is a plot of a quantitative variable (average number of texts) against year. Of course, the real problem is that starting in 2009, they reported the data for texts sent per day, not per month. Mistakes like this in graphics are more common than you think.

- **Don't confuse percentages.** Many percentages based on conditional and joint distributions sound similar, but are different (see Table 2.4):

  - The percentage of Russians who answered "Yes": This is 500/1010 or 49.5%.
  - The percentage of those who answered "Yes" who were Russian: This is 500/2175 or 23%.
  - The percentage of those who were Russian *and* answered "Yes": This is 500/5039 or 9.92%.

  In each instance, pay attention to the wording that makes a restriction to a smaller group (those who are Russian, those who answered "Yes," and all respondents, respectively) before a percentage is found. This restricts the who of the problem and the associated denominator for the percentage. Your discussion of results must make these differences clear.

- **Don't forget to look at the variables separately, too.** When you make a contingency table or display a conditional distribution, be sure to also examine the marginal distributions. It's important to know how many cases are in each category.

- **Be sure to use enough individuals.** When you consider percentages, take care that they are based on a large enough number of individuals (or cases). Take care not to make a report such as this one:

  We found that 66.67% of the companies surveyed improved their performance by hiring outside consultants. The other company went bankrupt.

- **Don't overstate your case.** Independence is an important concept, but it is rare for two variables to be *entirely* independent. We can't conclude that one variable has no effect whatsoever on another. Usually, all we know is that little effect was observed in our study. Other studies of other groups under other circumstances could find different results.

ETHICS IN ACTION

Mount Ashland Promotions Inc. is organizing one of its most popular events, the ZenNaturals Annual Trade Fest. At this trade show, producers, manufacturers, and distributors in the natural foods market display the latest trends in organic foods, herbal supplements, and natural body care products. The Trade Fest attracts a wide variety of participants, from large distributors who display a wide range of products to small, independent companies.

As in previous years, Nina Li and her team at Mount Ashland are in charge of managing the event, which includes all advertising and publicity as well as arranging spots for exhibitors. The success of this event depends on Nina's ability to attract large numbers of small independent retailers in the natural foods market who are looking to expand their product lines. She knows that these small retailers tend to be zealously committed to the principles of healthful lifestyle. Moreover, many are members of the Organic Trade Federation (OTF), an organization that advocates ethical consumerism.

The OTF has been known to boycott trade shows that include too many products with controversial ingredients such as ginkgo biloba, hemp, or kava kava. Nina is aware that some herbal diet teas have been receiving lots of negative attention lately in trade publications and the popular press. These teas claim to be "thermogenic" or fat burning, and typically contain ma huang (or ephedra). Ephedra is particularly controversial, not only because it can be unsafe for people with certain existing health conditions, but because this fast-acting stimulant commonly found in diet and energy products is contrary to the OTF's principles and values.

Worried that too many products at the ZenNaturals Trade Fest may be thermogenic teas, Nina decides to take a closer look at vendors already committed to participate in the event. Based on the data that her team pulled together, she finds that more than 33% of them do indeed include teas in their product lines. She was quite surprised to find that this percentage is so high. She decides to categorize the vendors into four groups: (1) those selling herbal supplements only; (2) those selling organic foods and herbal supplements; (3) those selling organic foods, herbal supplements, and natural body care products; and (4) all others. She finds that only 2% of groups 1, 2, and 4 include tea in their product lines, while 34% of the third group do. Even though group 3 contains most of the vendors, Nina instructs her team to use the average percentage 10% in its communications, especially with the OTF, about the upcoming ZenNaturals Annual Trade Fest.

- Identify the ethical dilemma in this scenario.

- What are the undesirable consequences?

- Propose an ethical solution that considers the welfare of all stakeholders.

WHAT HAVE WE LEARNED?

Learning Objectives	
	Make and interpret a frequency table for a categorical variable. • We can summarize categorical data by counting the number of cases in each category, sometimes expressing the resulting distribution as percentages.
	Make and interpret a bar chart or pie chart. • We display categorical data using the area principle in either a **bar chart** or a **pie chart**.
	Make and interpret a contingency table. • When we want to see how two categorical variables are related, we put the counts (and/or percentages) in a two-way table called a **contingency table**.
	Make and interpret bar charts and pie charts of marginal distributions. • We look at the **marginal distribution** of each variable (found in the margins of the table). We also look at the **conditional distribution** of a variable within each category of the other variable. • Comparing conditional distributions of one variable across categories of another tells us about the association between variables. If the conditional distributions of one variable are (roughly) the same for every category of the other, the variables are **independent**.
Terms	
Area principle	In a statistical display, each data value is represented by the same amount of area.
Bar chart (relative frequency bar chart)	A chart that represents the count (or percentage) of each category in a categorical variable as a bar, allowing easy visual comparisons across categories.

Cell	Each location in a contingency table, representing the values of two categorical variables, is called a cell.
Column percent	The proportion of each column contained in the cell of a frequency table.
Conditional distribution	The distribution of a variable restricting the *who* to consider only a smaller group of individuals.
Contingency table	A table displaying the frequencies (sometimes percentages) for each combination of two or more variables.
Distribution	The distribution of a variable is a list of:
	• all the possible values of the variable
	• the relative frequency of each value
Frequency table (relative frequency table)	A table that lists the categories in a categorical variable and gives the number (the percentage) of observations for each category.
Independent variables	Variables for which the conditional distribution of one variable is the same for each category of the other.
Marginal distribution	In a contingency table, the distribution of either variable alone. The counts or percentages are the totals found in the margins (usually the right-most column or bottom row) of the table.
Mosaic plot	A mosaic plot is a graphical representation of a (usually two-way) contingency table. The plot is divided into rectangles so that the area of each rectangle is proportional to the number of cases in the corresponding cell.
Pie chart	Pie charts show how a "whole" divides into categories by showing a wedge of a circle whose area corresponds to the proportion in each category.
Row percent	The proportion of each row contained in the cell of a frequency table.
Segmented bar chart	A segmented bar chart displays the conditional distribution of a categorical variable within each category of another variable.
Simpson's paradox	A phenomenon that arises when averages, or percentages, are taken across different groups, and these group averages appear to contradict the overall averages.
Total percent	The proportion of the total contained in the cell of a frequency table.

TECHNOLOGY HELP: Displaying Categorical Data

Although every package makes a slightly different bar chart, they all have similar features:

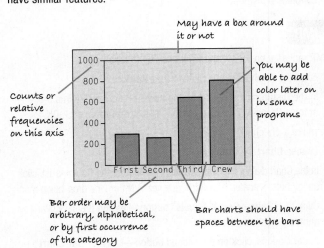

May have a box around it or not

You may be able to add color later on in some programs

Counts or relative frequencies on this axis

Bar order may be arbitrary, alphabetical, or by first occurrence of the category

Bar charts should have spaces between the bars

Sometimes the count or a percentage is printed above or on top of each bar to give some additional information. You may find that your statistics package sorts category names in annoying orders by default. For example, many packages sort categories alphabetically or by the order the categories are seen in the data set. Often, neither of these is the best choice.

EXCEL

Excel offers a versatile and powerful tool it calls a *PivotTable*. A pivot table can summarize, organize, and present data from an Excel spreadsheet. Pivot tables can be used to create frequency distributions and contingency tables. They provide a starting point for several kinds of displays. Pivot tables are linked to data in your Excel spreadsheet so they will update when you make changes to your data. They can also be linked directly to a *PivotChart* to display the data graphically.

In a pivot table, all types of data are summarized into a row-by-column table format. Pivot table cells can hold counts, percentages, and descriptive statistics.

To create a pivot table:

- Open a data file in Excel. At least one of the variables in the dataset should be categorical.

- Choose **Insert > PivotTable** or **Data > PivotTable** (Mac). If you are using a PC, choose to put the pivot table in a new worksheet. Macintosh users should choose the option to create a custom pivot table.

- The PivotTable builder has five boxes:
  - *Field List* (top): variables from the data set linked to the *PivotTable*. (The *PivotTable* tool calls the variables "fields.") Fields can be selected using the checkbox or dragged and dropped into one of the areas below in the *PivotTable* builder.
  - *Report Filter* (middle left): Variables placed here filter the data in the pivot table. When selected, the filter variable name appears above the pivot table. Use the drop-down list to the right of the variable name to choose values to display.
  - *Row Labels* (bottom left): Values of variables placed here become row labels in the pivot table.
  - *Column Labels* (middle right): Values of variables placed here become column labels in the pivot table.
  - *Values* (bottom right): Variables placed here are summarized in the cells of the table. Change settings to display count, sum, minimum, maximum, average, and more or to display percentages and ranks.

To create a frequency distribution pivot table:

- Drag a categorical variable from the Field List into **Row Labels**.

- Choose another variable from the data set and drag it into **Values**.

- To change what fact or statistics about the **Values** variable is displayed, click the arrow next to the variable in the **Values** box and open the **Value Field Settings**. For a frequency distribution, select **count of [VARIABLE]**. When changing **Value Field Settings**, note the tab **Show Values As,** which provides other display options (e.g., % of row, % of column).

The result will be a frequency table with a column for count.

To create a contingency table using *PivotTable:*

- Drag a categorical variable from the Field List into **Row Labels**.

- Drag a second categorical variable from the Field List into **Column Labels**.

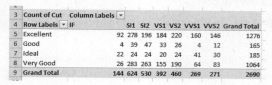

- Choose another variable from the dataset and drag it into **Values**. The resulting pivot table is a row-by-column contingency table.

3	Count of Cut	Column Labels							
4	Row Labels	IF	SI1	SI2	VS1	VS2	VVS1	VVS2	Grand Total
5	Excellent	92	278	196	184	220	160	146	1276
6	Good	4	39	47	33	26	4	12	165
7	Ideal	22	24	24	20	24	41	30	185
8	Very Good	26	283	263	155	190	64	83	1064
9	Grand Total	144	624	530	392	460	269	271	2690

NOTE: As with the frequency distribution, you can use the **Value Field Settings** to change the type of summary.

To create a chart from a pivot table frequency distribution or contingency table:

- Place the cursor anywhere on the pivot table.

- Click **PivotTable Tools > PivotChart**.

- Choose the type of chart: options include pie chart, bar chart, and segmented bar graph.

- Move the chart to a new worksheet by right-clicking the chart and selecting **Move chart.**

- In a bar chart created from a contingency table, by default, rows display on the x-axis and the columns are separate bars. To change this, place cursor in chart and choose **PivotChart Tools > Design > Switch Row/Column**.

- On Macs, choose the **Charts** tab and select your chart from the ribbon or choose a chart type from the **Chart** menu.

XLSTAT

To create a contingency table from unsummarized data:

- On the XLStat tab, choose **Preparing data**.

- From the menu, choose **Create a contingency table**.

- In the dialog box, enter your data range on the General tab. Your data should be in two columns, one of which is the row variable and the other is the column variable.

- On the Outputs tab, check the box next to **Contingency table** and optionally choose **Percentages/Row or Column** to see the conditional distributions.

JMP

JMP makes a bar chart and frequency table together.

- From the **Analyze** menu, choose **Distribution**.

- In the Distribution dialog, drag the name of the variable into the empty variable window beside the label **Y, Columns**; click **OK**.

To make a pie chart,

- Choose **Chart > Graph** menu.

- In the Chart dialog, select the variable name from the Columns list, click on the button labeled **Statistics**, and select **N** from the drop-down menu.

- Click the "**Categories, X, Levels**" button to assign the same variable name to the x-axis.

- Under Options, click on the second button—labeled "**Bar Chart**"—and select **Pie Chart** from the drop-down menu.

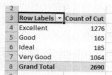

3	Row Labels	Count of Cut
4	Excellent	1276
5	Good	165
6	Ideal	185
7	Very Good	1064
8	Grand Total	2690

MINITAB

To make a bar chart,

- Choose **Bar Chart** from the **Graph** menu.
- Then select a Simple, Cluster, or Stack chart from the options and click **OK**.
- To make a **Simple** bar chart, enter the name of the variable to graph in the dialog box.
- To make a relative frequency chart, click **Chart Options**, and choose **Show Y as Percent**.
- In the Chart dialog, enter the name of the variable that you wish to display in the box labeled "Categorical variables."
- Click **OK**.

R

To make a bar chart or pie chart in **R**, you first need to create the frequency table for the desired variable:

- **table(X)** will give a frequency table for a single variable X.
- **barplot(table(X))** will give a bar chart for X.
- Similarly **pie(table(X))** will give a pie chart.

Comments

Stacked bar charts of two variables, X and Y, can be made using **barplot(xtabs($\sim$X + Y))** or directly from a two-way table of counts or percentages. Legends and other options are available for all charts using various functions.

SPSS

To make a bar chart,

- Open the **Chart Builder** from the **Graphs** menu.
- Click the **Gallery** tab.
- Choose **Bar Chart** from the list of chart types.
- Drag the appropriate bar chart onto the canvas.
- Drag a categorical variable onto the x-axis drop zone.
- Click **OK**.

Comments

A similar path makes a pie chart by choosing **Pie chart** from the list of chart types.

Brief **Case**

Credit Card Bank

In Chapter 1, you identified the W's for the data in the file **Credit Card Bank**. For the categorical variables in the data set, create frequency tables, bar charts, and pie charts using your software. What might the bank want to know about these variables? Which of the tables and charts do you find most useful for communicating information about the bank's customers? Write a brief case report summarizing your analysis and results.

EXERCISES

SECTION 2.1

1. As part of the human resource group of your company you are asked to summarize the educational levels of the 512 employees in your division. From company records, you find that 164 have no college degree (None), 42 have an associate's degree (AA), 225 have a bachelor's degree (BA), 52 have a master's degree (MA), and 29 have PhDs. For the educational level of your division:

a) Make a frequency table.
b) Make a relative frequency table.

2. As part of the marketing group at Pixar, you are asked to find out the age distribution of the audience of Pixar's latest film. With the help of 10 of your colleagues, you conduct exit interviews by randomly selecting people to question at 20 different movie theaters. You ask them to tell you if they are younger than 6 years old, 6 to 9 years old, 10 to 14 years old, 15 to 21 years old, or older than 21. From 470 responses, you find out that 45 are younger than 6, 83 are 6 to 9 years old, 154 are 10 to 14, 18 are 15 to 21, and 170 are older than 21. For the age distribution:

a) Make a frequency table.
b) Make a relative frequency table.

SECTION 2.2

3. From the educational level data described in Exercise 1:

a) Make a bar chart using counts on the *y*-axis.
b) Make a relative frequency bar chart using percentages on the *y*-axis.
c) Make a pie chart.

4. From the age distribution data described in Exercise 2:

a) Make a bar chart using counts on the *y*-axis.
b) Make a relative frequency bar chart using percentages on the *y*-axis.
c) Make a pie chart.

5. For the educational levels described in Exercise 1:

a) Write two to four sentences summarizing the distribution.
b) What conclusions, if any, could you make about the educational level at other companies?

6. For the ages described in Exercise 2:

a) Write two to four sentences summarizing the distribution.
b) What possible problems do you see in concluding that the age distribution from these surveys accurately represents the ages of the national audience for this film?

SECTION 2.3

7. From Exercise 1, we also have data on how long each person has been with the company (tenure) categorized into three levels: less than 1 year, between 1 and 5 years, and more than 5 years. A table of the two variables together looks like:

	None	AA	BA	MA	PhD
<1 Year	10	3	50	20	12
1–5 Years	42	9	112	27	15
More Than 5 Years	112	30	63	5	2

a) Find the marginal distribution of the tenure. (*Hint*: Find the row totals.)
b) Verify that the marginal distribution of the education level is the same as that given in Exercise 1.

8. In addition to their age levels, the movie audiences in Exercise 2 were also asked if they had seen the movie before (Never, Once, More than Once). Here is a table showing the responses by age group:

	Under 6	6–9	10–14	15–21	Over 21
Never	39	60	84	16	151
Once	3	20	38	2	15
More Than Once	3	3	32	0	4

a) Find the marginal distribution of their previous viewing of the movie. (*Hint:* Find the row totals.)
b) Verify that the marginal distribution of the ages is the same as that given in Exercise 2.

SECTION 2.4

9. For the table in Exercise 7:

a) Find the column percentages.
b) Looking at the column percentages in part a, does the *tenure* distribution (how long the employee has been with the company) for each educational level look the same? Comment briefly.
c) Make a segmented or stacked bar chart showing the *tenure* distribution for each educational level.
d) Is it easier to see the differences in the distributions using the column percentages or the segmented bar chart?
e) How would a mosaic plot help to accurately display these data?

10. For the table in Exercise 8:

a) Find the column percentages.
b) Looking at the column percentages in part a, does the distribution of how many times someone has seen the movie look the same for each age group? Comment briefly.
c) Make a segmented bar chart, showing the distribution of viewings for each age level.
d) Is it easier to see the differences in the distributions using the column percentages or the segmented bar chart?
e) How would a mosaic plot represent these data more appropriately?

CHAPTER EXERCISES

11. Graphs in the news. Find a bar graph of categorical data from a business publication (*Bloomberg Businessweek*, *Fortune*, *The Wall Street Journal*, etc.).

a) Is the graph clearly labeled?
b) Does it violate the area principle?
c) Does the accompanying article tell the W's of the variable?
d) Do you think the article correctly interprets the data? Explain.

12. Graphs in the news, part 2. Find a pie chart of categorical data from a business publication (*Bloomberg Businessweek*, *Fortune*, *The Wall Street Journal*, etc.).

a) Is the graph clearly labeled?
b) Does it violate the area principle?
c) Does the accompanying article tell the W's of the variable?
d) Do you think the article correctly interprets the data? Explain.

13. Tables in the news. Find a frequency table of categorical data from a business publication (*Bloomberg Businessweek*, *Fortune*, *The Wall Street Journal*, etc.).

a) Is it clearly labeled?
b) Does it display percentages or counts?
c) Does the accompanying article tell the W's of the variable?
d) Do you think the article correctly interprets the data? Explain.

14. Tables in the news, part 2. Find a contingency table of categorical data from a business publication (*Bloomberg Businessweek, Fortune, The Wall Street Journal*, etc.).

a) Is it clearly labeled?
b) Does it display percentages or counts?
c) Does the accompanying article tell the W's of the variable?
d) Do you think the article correctly interprets the data? Explain.

15. U.S. market share. An article in *The Wall Street Journal* (March 18, 2011) reported the 2010 U.S. market share of leading sellers of carbonated soft drinks, summarized in the following pie chart:

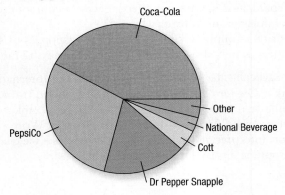

a) Is this an appropriate display for these data? Explain.
b) Which company had the largest share of the market?

16. World market share. *The Wall Street Journal* article described in Exercise 15 also indicated the market share of the leading *brands* of carbonated beverages. The following bar chart displays the values:

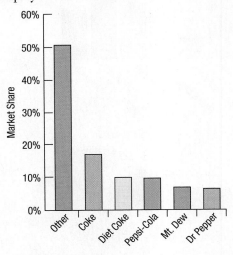

a) Is this an appropriate display for these data? Explain.
b) Which brand had the largest share of the beverage market?
c) Which brand had the larger market share—Mountain Dew or Dr Pepper?

17. Market share again. Here's a bar chart of the data in Exercise 15.

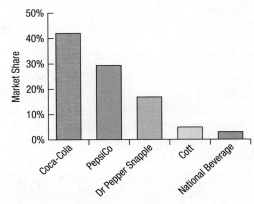

a) Compared to the pie chart in Exercise 15, which is better for displaying the relative portions of market share? Explain.
b) What is missing from this display that might make it somewhat misleading?

18. World market share again. Here's a pie chart of the data in Exercise 16.

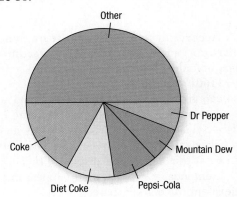

a) Which display of these data is best for comparing the market shares of these brands? Explain.
b) Does Mountain Dew or Dr Pepper have a bigger market share? Is that comparison easier to make with the pie chart or the bar chart of Exercise 16?

19. Insurance company. An insurance company is updating its payouts and cost structure for their insurance policies. Of particular interest to them is the risk analysis for customers currently on heart or blood pressure medication. The Centers for Disease Control and Prevention (www.cdc.gov) lists causes of death in the United States during one year as follows.

Cause of Death	Percent
Heart disease	30.3
Cancer	23.0
Circulatory diseases and stroke	8.4
Respiratory diseases	7.9
Accidents	4.1

a) Is it reasonable to conclude that heart or respiratory diseases were the cause of approximately 38% of U.S. deaths during this year?

b) What percent of deaths were from causes not listed here?

c) Create an appropriate display for these data.

20. College value? In March and April of 2011, the Pew Research Center asked 2142 U.S. adults and 1055 college presidents whether they would "rate the job the higher education system is doing in providing value for the money spent by students and their families" as Excellent, Good, Only Fair, or Poor.

	Poor	Only Fair	Good	Excellent	No Answer/ Don't Know
U.S. Adults	321	900	750	107	64
College Presidents	32	222	622	179	0

a) Compare the distribution of opinions between U.S. adults and college presidents on the value of higher education.

b) Is it reasonable to conclude that 5.00% of *all* U.S. adults think that the higher education system provides an excellent value?

21. SaaS. According to a 2013 report by Synergy Research Group (www.informationweek.com/telecom /unified-communications/cisco-rules-saas-uc-conferencing -market/231601562) the market for desktop conferencing apps has been growing rapidly, spurred by a 33% jump in software-as-a-service-delivered conferencing apps revenue. Cisco Systems has a 58% share of this market, followed by Citrix with a 13% share and Microsoft with 11%. Create an appropriate graphical display of this information and write a sentence or two that might appear in a newspaper article about the market share.

22. Mattel. In their 2013 annual report, Mattel Inc. reported that their domestic market sales were broken down as follows: 49.6% Mattel Girls and Boys brand, 36.1% Fisher-Price brand, and the rest of their over $3.5 billion revenues were due to their American Girl brand. Create an appropriate graphical display of this information and write a sentence or two that might appear in a newspaper article about their revenue breakdown.

23. Small business financing. The Wells Fargo/Gallup Small Business Index asked 604 small business owners in October 2011 "how difficult or easy do you think it will be for your company to obtain credit when you need it?" 22% said "Very difficult," 21% "Somewhat difficult," 28% "About Average," 11% "Somewhat easy," and 11% "Very Easy."

a) What do you notice about the percentages listed? How could this be?

b) Make a bar chart to display the results and label it clearly.

c) Would a pie chart be an effective way of communicating this information? Why or why not?

d) Write a couple of sentences on how small businesses felt about the difficulty of obtaining credit in late 2011.

24. Small business cash flow. The Wells Fargo/Gallup Small Business Index survey from Exercise 23 also asked 604 small businesses about their cash flow over the next 12 months. 13% responded "Very Good," 37% "Somewhat good," 21% "Neither good nor poor," 20% "Somewhat poor," and 7% "Very poor."

a) What do you notice about the percentages listed?

b) Make a bar chart to display the results and label it clearly.

c) Would a pie chart be an effective way of communicating this information? Why or why not?

d) Write a couple of sentences on the responses of small business owners about their cash flow in the next 12 months.

25. Environmental hazard 2014. Data from the International Tanker Owners Pollution Federation Limited (www.itopf .com) give the cause of spillage for 459 large oil tanker accidents from 1970 to 2014. Here are the displays. Write a brief report interpreting what the displays show. Is a pie chart an appropriate display for these data? Why or why not?

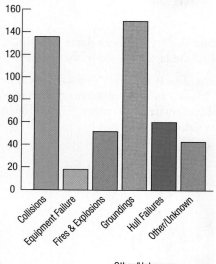

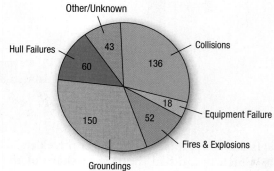

26. Olympic medals. In the history of the modern Olympics, the United States has won more medals than any other country. But the United States has a large population.

Perhaps a better measure of success is the number of medals won *per capita*—that is the number of medals divided by the population. By that measure, the leading countries are Liechtenstein (255.42 medals/cap), Norway (95.271), Finland (86.514), and Sweden (66.455). The following table summarizes the medals/capita counts for the 100 countries with the most medals.

a) Try to make a display of these data. What problems do you encounter?
b) Can you find a way to organize the data so that the graph is more successful?

Medals/capita	# Countries	Medals/capita	# Countries
0	72	130	0
10	13	140	0
20	3	150	0
30	3	160	0
40	2	170	0
50	0	180	0
60	1	190	0
70	0	200	0
80	1	210	0
90	1	220	0
100	0	230	0
110	0	240	0
120	0	250	1

27. Importance of wealth. GfK Roper Reports Worldwide surveyed people, asking them "How important is acquiring wealth to you?" The percent who responded that it was of more than average importance were: 71.9% China, 59.6% France, 76.1% India, 45.5% U.K., and 45.3% U.S. There were about 1500 respondents per country. A report showed the following bar chart of these percentages.

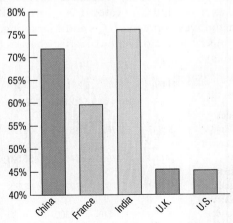

a) How much larger is the proportion of those who said acquiring wealth was important in India than in the United States?
b) Is that the impression given by the display? Explain.
c) How would you improve this display?
d) Make an appropriate display for the percentages.
e) Write a few sentences describing what you have learned about attitudes toward acquiring wealth.

28. Importance of power. In the same survey as that discussed in Exercise 27, GfK Roper Consulting also asked "How important is having control over people and resources to you?" The percent who responded that it was of more than average importance are given in the following table:

China	49.1%
France	44.1%
India	74.2%
U.K.	27.8%
U.S.	36.0%

Here's a pie chart of the data:

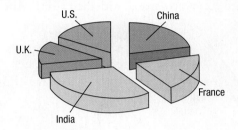

a) List the errors you see in this display.
b) Make an appropriate display for the percentages.
c) Write a few sentences describing what you have learned about attitudes toward acquiring power.

29. Google financials. Google Inc. derives revenue from three major sources: advertising revenue from their websites, advertising revenue from the thousands of third-party websites that comprise the Google Network, and licensing and miscellaneous revenue. The following table shows the percentage of all revenue derived from these sources for the period from 2009 to 2013.

	2009	2010	2011	2012	2013
Google Websites	67%	66%	69%	68%	67%
Google Network Members' Websites	30%	30%	28%	27%	24%
Other Revenues	3%	4%	3%	5%	9%

a) Are these row or column percentages?
b) Make an appropriate display of these data.
c) Write a brief summary of this information.

30. Real estate pricing. A study of a sample of 1057 houses in upstate New York reports the following percentages of houses falling into different Price and Size categories.

		Price			
		Low	Med Low	Med High	High
Size	Small	61.5%	35.2%	5.2%	2.4%
	Med Small	30.4%	45.3%	26.4%	4.7%
	Med Large	5.4%	17.6%	47.6%	21.7%
	Large	2.7%	1.9%	20.8%	71.2%

a) Are these column, row, or total percentages? How do you know?

b) What percent of the highest priced houses were small?

c) From this table, can you determine what percent of all houses were in the low price category?

d) Among the lowest priced houses, what percent were small or medium small?

e) Write a few sentences describing the association between *Price* and *Size*.

31. Stock performance. The following table displays information for 470 of the S&P 500 stocks, on how their one-day change on October 24, 2011 (a day on which the S&P 500 index gained 1.23%) compared with their year to date change.

		Year to Date	
		Positive Change	Negative Change
October 24, 2011	Positive Change	164	233
	Negative Change	48	25

a) What percent of the companies reported a positive change in their stock price over the year to date?

b) What percent of the companies reported a positive change in their stock price over both time periods?

c) What percent of the companies reported a negative change in their stock price over both time periods?

d) What percent of the companies reported a positive change in their stock price over one period and a negative change in the other period?

e) Among those companies reporting a positive change in their stock price on October 24 over the prior day what percentage also reported a positive change over the year to date?

f) Among those companies reporting a negative change in their stock price on October 24 over the prior day what percentage reported a positive change over the year to date?

g) What relationship, if any, do you see between the performance of a stock on a single day and its year-to-date performance?

32. New product. A company started and managed by business students is selling campus calendars. The students have conducted a market survey with the various campus constituents to determine sales potential and identify which market segments should be targeted. (Should they advertise in the Alumni Magazine and/or the local newspaper?) The following table shows the results of the market survey.

		Buying Likelihood			
		Unlikely	Moderately Likely	Very Likely	Total
Campus Group	Students	197	388	320	905
	Faculty/Staff	103	137	98	338
	Alumni	20	18	18	56
	Town Residents	13	58	45	116
	Total	333	601	481	1415

a) What percent of all these respondents are alumni?

b) What percent of these respondents are very likely to buy the calendar?

c) What percent of the respondents who are very likely to buy the calendar are alumni?

d) Of the alumni, what percent are very likely to buy the calendar?

e) What is the marginal distribution of the campus constituents?

f) What is the conditional distribution of the campus constituents among those very likely to buy the calendar?

g) Does this study present any evidence that this company should focus on selling to certain campus constituents?

33. Real estate. The *Greenville, South Carolina Real Estate Hub* keeps track of home sales in their area. They reported that sales were down in 2010 by about 3.7% from the previous year. Here are the number of homes sold in Greenville for the last 5 months of 2009 and 2010:

	August	September	October	November	December
2010	475	466	502	423	495
2009	607	597	596	581	447

a) What percent of all homes in these ten months were sold in October of 2009?

b) What percent of all homes in these 10 months were sold in 2010?

c) What percent of all homes in these 10 months were sold in December?

d) How did the percent of homes sold in November change from 2009 to 2010?

34. Google financials, part 2. Google Inc. divides their total costs and expenses into five categories: Costs of Revenues, Research and Development, Sales and Marketing, General and Administrative, and Dept of Justice charges (amounts in $Millions).

	2009	2010	2011	2012	2013
Cost of Revenues	$8844	$10,417	$13,188	$20,634	$21,993
Research and Development	$2843	$3762	$5162	$6793	$7137
Sales and Marketing	$1984	$2799	$4589	$6143	$6554
General and Administrative	$1667	$1962	$2724	$3845	$4432
Dept of Justice	$0	$0	$500	$0	$0
Total Costs and Expenses	$15,338	$18,940	$26,163	$37,415	$40,116

a) What percent of total costs and expenses were sales and marketing in 2009? In 2013?

b) What percent of total costs and expenses were due to research and development in 2009? In 2013?

c) Have general and administrative costs grown as a percentage of total costs and expenses over this time period?

35. Movie ratings 2014. The movie ratings system is a voluntary system operated jointly by the Motion Picture Association of America (MPAA) and the National Association of Theatre Owners (NATO). The ratings themselves are given by a board of parents who are members of the Classification and Ratings Administration (CARA). The board was created in response to outcries from parents in the 1960s for some kind of regulation of film content, and the first ratings were introduced in 1968. Here is information on the ratings of 279 movies that came out in 2014, also classified by their genre.

		Rating				
		G	PG	PG-13	R or NC-17	Total
Genre	Action/Adventure	2	14	26	17	59
	Comedy	1	6	19	53	79
	Drama	0	14	41	62	117
	Thriller/Suspense	0	0	14	44	58
	Others*	0	4	2	2	8
	Total	3	38	102	178	321

*Others include Westerns, musicals etc.

a) Find the conditional distribution (in percentages) of movie ratings for action/adventure films.
b) Find the conditional distribution (in percentages) of Genres for PG-13 rated films.
c) Create a graph comparing the ratings for the four genres.
d) Are *Genre* and *Rating* independent? Write a brief summary of what these data show about movie ratings and the relationship to the genre of the film.

36. CyberShopping. It has become more common for shoppers to "comparison shop" using the Internet. Respondents to a Pew survey in 2013 who owned cell phones were asked whether they had, in the past 30 days, looked up the price of a product while they were in a store to see if they could get a better price somewhere else. Here is a table of their responses by income level.

	<$30K	$30K–$49.9K	$50K–$74.9K	>$75K
Yes	207	115	134	204
No	625	406	260	417

(Source: www.pewinternet.org/Reports/2012/In-store-mobile-commerce.aspx)

a) Find the conditional distribution (in percentages) of income distribution for those who do not compare prices on the Internet.
b) Find the conditional distribution (in percentages) of income distribution for shoppers who do compare prices.
c) Create a graph comparing the income distributions of those who compare prices with those who don't.
d) Do you see any differences between the conditional distributions? Write a brief summary of what these data show about Internet use and its relationship to income.

37. MBAs. A survey of the entering MBA students at a university in the United States classified the country of origin of the students, as seen in the table.

		MBA Program		
		Two-Year MBA	Evening MBA	Total
Origin	Asia/Pacific Rim	31	33	64
	Europe	5	0	5
	Latin America	20	1	21
	Middle East/Africa	5	5	10
	North America	103	65	168
	Total	164	104	268

a) What percent of all MBA students were from North America?
b) What percent of the Two-Year MBAs were from North America?
c) What percent of the Evening MBAs were from North America?
d) What is the marginal distribution of origin?
e) Obtain the column percentages and show the conditional distributions of origin by MBA Program.
f) Do you think that origin of the MBA student is independent of the MBA program? Explain.

38. MBAs, part 2. The same university as in Exercise 37 reported the following data on the gender of their students in their two MBA programs.

		Type		
		Two-Year	Evening	Total
Sex	Men	116	66	182
	Women	48	38	86
	Total	164	104	268

a) What percent of all MBA students are women?
b) What percent of Two-Year MBAs are women?
c) What percent of Evening MBAs are women?
d) Do you see evidence of an association between the *Type* of MBA program and the percentage of women students? If so, why do you believe this might be true?

T **39. Top-producing movies, 2014.** The following table shows the Motion Picture Association of America (MPAA; www.mpaa.org) ratings for the top 20 grossing films in the United States for each of the 10 years from 2005 to 2014. (Data are number of films.)

	Rating				
Year	G	PG	PG-13	R/NC-17	Total
2014	0	4	13	3	20
2013	1	4	11	4	20
2012	0	6	12	2	20
2011	1	4	11	4	20
2010	1	9	9	1	20
2009	0	7	12	1	20
2008	2	4	10	4	20
2007	1	5	11	3	20
2006	1	4	13	2	20
2005	1	4	13	2	20
Total	8	51	115	26	200

a) What percent of all these top 20 films are G rated?

b) What percent of all top 20 films in 2005 were G rated?

c) What percent of all top 20 films were PG-13 and came out in 2010?

d) What percent of all top 20 films produced in 2010 or later were PG-13?

e) What percent of all top 20 films produced from 2005 to 2009 were rated PG-13 or R/NC-17?

f) Compare the conditional distributions of the ratings for films produced in 2010 or later to those produced from 2005 to 2009. Write a couple of sentences summarizing what you see.

T 40. Movie admissions 2013. The following table shows attendance data collected by the Motion Picture Association of America during the period 2009 to 2013. Figures are the number (in millions) of frequent moviegoers in each age group.

	Age						
	2–11	12–17	18–24	25–39	40–49	50–59	60 +
2013	4.3	5.5	7.2	8.2	3.2	4.2	3.8
2012	2.8	6.3	8.7	9.9	5.8	3.3	4.6
2011	2.5	5.7	6.6	9.7	3.3	3.1	4.1
2010	3.1	6.1	7.4	7.7	3.5	3.0	4.3
2009	2.8	5.7	6.3	6.3	4.5	2.9	3.4

a) What percent of all frequent moviegoers during this period were people between the ages of 12 and 24?

b) What percent of the frequent moviegoers in 2011 were people between the ages of 12 and 39?

c) What percent of *all* frequent moviegoers during this period were people between the ages of 18 and 24 who went to the movies in 2009?

d) What percent of frequent moviegoers in 2010 were people 60 years old and older?

e) What percent of *all* frequent moviegoers in this period were people 60 years old and older who went to the movies in 2010?

f) Compare the conditional distributions of the age groups across years. Write a couple of sentences summarizing what you see.

41. Tattoos. A study by the University of Texas Southwestern Medical Center examined 626 people to see if there was an increased risk of contracting hepatitis C associated with having a tattoo. If the subject had a tattoo, researchers asked whether it had been done in a commercial tattoo parlor or elsewhere. Write a brief description of the association between tattooing and hepatitis C, including an appropriate graphical display.

	Tatto Done in Commercial Parlor	Tattoo Done Elsewhere	No Tattoo
Has Hepatitis C	17	8	18
No Hepatitis C	35	53	495

42. Poverty and region 2013. In 2013, the following data were reported by the U.S. Census Bureau. The data show the number of people (in thousands) living above and below the poverty line in each of the four regions of the United States. Based on these data do you think there is an association between region and poverty? Explain.

	Below Poverty Level	Above Poverty Level
Northeast	7046	48,432
Midwest	8590	58,195
South	18,870	98,091
West	10,812	62,930

43. Being successful. In a random sample of U.S. adults surveyed in December 2011, Pew research asked how important it is "to you personally" to be successful in a high-paying career or profession. Here is a table reporting the responses. (Percentages may not add to 100% due to rounding.) (Data from www.pewsocialtrends.org/files/2012/04/Women-in-the-Workplace.pdf)

	Women		Men	
Age	18–34	35–64	18–34	35–64
One of the most important things	18%	7%	11%	9%
Very important, but not the most	48%	35%	47%	34%
Somewhat important	26%	34%	31%	37%
Not important	8%	24%	10%	20%
	100%	100%	100%	100%

a) What percent of young women consider it very important or one of the most important things for them personally to be successful?

b) How does that compare with young men?

c) From this table, can you determine what percent of all women responding felt this way? Explain.

d) Write a few sentences describing the association between the sex of young respondents and their attitudes toward the importance of financial or professional success.

T 44. Minimum wage workers 2013. The U.S. Department of Labor (www.bls.gov) collects data on the number of U.S.

workers who are employed at or below the minimum wage. Here is a table showing the number of hourly workers by *Age* and *Sex* and the number who were paid at or below the prevailing minimum wage in 2013:

Age	Hourly Workers (in thousands)		At or Below Minimum Wage (in thousands)	
	Men	Women	Men	Women
16–24	7558	7552	655	1007
25–34	9281	8326	286	418
35–44	7112	7082	116	239
45–54	7181	7916	83	231
55–64	4915	5798	55	99

a) What percent of all women were in the 16–24 *Age* group?
b) Using side-by-side bar graphs, compare the proportions of men and women who worked at or below minimum wage at each *Age* group. Use the total number of workers at or below minimum wage as the denominator. Write a couple of sentences summarizing what you see.

45. Moviegoers and ethnicity. The Motion Picture Association of America studies the ethnicity of moviegoers to understand changes in the demographics of moviegoers over time. Here are the numbers of moviegoers (in millions) classified as to whether they were Hispanic, African-American, Caucasian, and Other for the year 2010. Also included are the numbers for the general U.S. population and the number of tickets sold.

	Caucasian	Hispanic	African-American	Other	Total
Population	204.6	49.6	37.2	18.6	310
Moviegoers	88.8	26.8	16.9	8.5	141
Tickets	728	338	143	91	1300
Total	1021.4	414.4	197.1	118.1	1751

a) Compare the conditional distribution of *Ethnicity* for all three groups: the entire population, moviegoers, and ticket holders.
b) Write a brief description of the association between population groups and *Ethnicity*.

46. Department store. A department store is planning its next advertising campaign. Since different publications are read by different market segments, they would like to know if they should be targeting specific age segments. The results of a marketing survey are summarized in the following table by *Age* and *Shopping Frequency* at their store.

	Shopping	Under 30	30–49	50 and Over	Total
Frequency	Low	27	37	31	95
	Moderate	48	91	93	232
	High	23	51	73	147
	Total	98	179	197	474

(header spanning Under 30, 30–49, 50 and Over is "Age")

a) Find the marginal distribution of *Shopping Frequency*.
b) Find the conditional distribution of *Shopping Frequency* within each age group.
c) Compare these distributions with a segmented bar graph.
d) Write a brief description of the association between *Age* and *Shopping Frequency* among these respondents.
e) Does this prove that customers ages 50 and over are more likely to shop at this department store? Explain.

47. Success II. Look back at the table in exercise 43 concerning desires for success and a high-paying career. That table presented only the percentages, but Pew Research reported the numbers of respondents in the major categories:

	Women		Men	
Age	18–34	35–64	18–34	35–64
Count	610	571	703	605

With this additional information you should be able to answer these questions. (Note: Percentages were rounded to whole numbers, so estimated cell counts will have fractions. You need not round estimated cell counts to whole numbers for the purpose of answering these questions.)

a) What percentage of 18- to 34-year-olds (both male and female) reported that being successful in a high-paying career or profession was "one of the most important things" to them personally?
b) What percentage of 18- to 34-year-olds who said that such success was "one of the most important things" were women?
c) Write a few sentences describing how the opinions of young women differ from those of older female respondents.

48. Advertising. A company that distributes a variety of pet foods is planning their next advertising campaign. Since different publications are read by different market segments, they would like to know how pet ownership is distributed across different income segments. The U.S. Census Bureau (www.allcountries.org/uscensus/424_household_pet_ownership_and_by_selected.html) reports the number of households owning various types of pets. Specifically, they keep track of dogs, cats, birds, and horses.

a) Do you think the income distributions of the households who own these different animals would be roughly the same? Why or why not?

Percent Distribution of Households Owning Pets

		Pets			
		Dog	**Cat**	**Bird**	**Horse**
Income Range	**Under $12,500**	12.7	13.9	17.3	9.5
	$12,500 to $24,999	19.1	19.7	20.9	20.3
	$25,000 to $39,999	21.6	21.5	22.0	21.8
	$40,000 to $59,999	21.5	21.2	17.5	23.1
	$60,000 and over	25.2	23.7	22.3	25.4

b) The table shows the percentages of income levels for each type of animal owned. Are these row percentages, column percentages, or total percentages?
c) Do the data support that the pet food company should not target specific market segments based on household income? Explain.

49. Insurance company, part 2. An insurance company that provides medical insurance is concerned with recent data. They suspect that patients who undergo surgery at large hospitals have their discharges delayed for various reasons—which results in increased medical costs to the insurance company. The recent data for area hospitals and two types of surgery (major and minor) are shown in the following table.

		Discharge Delayed	
		Large Hospital	**Small Hospital**
Procedure	**Major Surgery**	120 of 800	10 of 50
	Minor Surgery	10 of 200	20 of 250

a) Overall, for what percent of patients was discharge delayed?
b) Were the percentages different for major and minor surgery?
c) Overall, what were the discharge delay rates at each hospital?
d) What were the delay rates at each hospital for each kind of surgery?
e) The insurance company is considering advising their clients to use large hospitals for surgery to avoid postsurgical complications. Do you think they should do this?
f) Explain, in your own words, why this confusion occurs.

50. Delivery service. A company must decide which of two delivery services they will contract with. During a recent trial period, they shipped numerous packages with each service and have kept track of how often deliveries did not arrive on time. Here are the data.

Delivery Service	Type of Service	Number of Deliveries	Number of Late Packages
Pack Rats	Regular	400	12
	Overnight	100	16
Boxes R Us	Regular	100	2
	Overnight	400	28

a) Compare the two services' overall percentage of late deliveries.
b) Based on the results in part a, the company has decided to hire Pack Rats. Do you agree they deliver on time more often? Why or why not? Be specific.
c) The results here are an instance of what phenomenon?

51. Graduate admissions. A 1975 article in the magazine *Science* examined the graduate admissions process at Berkeley for evidence of gender bias. The following table shows the number of applicants accepted to each of four graduate programs.

Program	Males Accepted (of Applicants)	Females Accepted (of Applicants)
1	511 of 825	89 of 108
2	352 of 560	17 of 25
3	137 of 407	132 of 375
4	22 of 373	24 of 341
Total	**1022 of 2165**	**262 of 849**

a) What percent of total applicants were admitted?
b) Overall, were a higher percentage of males or females admitted?
c) Compare the percentage of males and females admitted in each program.
d) Which of the comparisons you made do you consider to be the most valid? Why?

52. Simpson's Paradox. Develop your own table of data that is a business example of Simpson's Paradox. Explain the conflict between the conclusions made from the conditional and marginal distributions.

JUST CHECKING ANSWERS

1 50.0%

2 40.0%

3 25.0%

4 15.6% Nearsighted, 56.3% Farsighted, 28.1% Need Bifocals

5 18.8% Nearsighted, 62.5% Farsighted, 18.8% Need Bifocals

6 40% of the nearsighted customers are female, while 50% of customers are female.

7 Since nearsighted customers appear less likely to be female, it seems that they may not be independent. (But the numbers are small.)

3

Displaying and Describing Quantitative Data

AIG

The American International Group (AIG) was once the 18th largest corporation in the world. AIG was founded nearly 100 years ago by Cornelius Vander Starr, who opened an insurance agency in Shanghai, China. As the first Westerner to sell insurance to the Chinese, Starr grew his business rapidly until 1949 when Mao Zedong and the People's Liberation Army took over Shanghai. Starr moved the company to New York City, where it continued to grow, expanding its markets worldwide. In 2004, AIG stock hit an all-time high of $76.77, putting its market value at nearly $300 billion.

According to its own website, "By early 2007 AIG had assets of $1 trillion, $110 billion in revenues, 74 million customers and 116,000 employees in 130 countries and jurisdictions. Yet just 18 months later, AIG found itself on the brink of failure and in need of emergency government assistance." AIG was one of the largest beneficiaries of the U.S. government's Troubled Asset Relief Program (TARP), established in 2008 during the financial crisis to purchase assets and equity from financial institutions. TARP was an attempt to strengthen the financial sector and avoid a repeat of a depression as severe as the 1930s. Many banks quickly repaid the government part or all of the money given to them under the TARP program, but AIG, which received $170 billion, took until the end of 2012 to repay the government completely.

Even though AIG stock today is on solid financial footing, its stock price is only a fraction (adjusted for splits) of what it was before the 2008 crisis. Between 2007 and 2009 AIG stock lost more than 99% of its value, hitting $0.35 in early March. That same month AIG became embroiled in controversy when it disclosed that it had paid $218 million in bonuses to employees of its financial services division. AIG's drop in stock price represented a loss of nearly $300 billion for investors. Portfolio managers typically examine stock prices and volumes to determine stock volatility and to help them decide which stocks to buy and sell. Were there early warning signs in AIG's data?

Table 3.1 gives the monthly average stock price (in dollars) for the six years leading up to the company's crisis. Were there clues to warn of problems?

	Jan.	Feb.	Mar.	Apr.	May	June	July	Aug.	Sept.	Oct.	Nov.	Dec.
2002	77.26	72.95	73.72	71.57	68.42	65.99	61.22	64.10	58.04	60.26	65.03	59.96
2003	59.74	49.57	49.41	54.38	56.52	57.88	59.80	61.51	59.39	60.93	58.73	62.37
2004	69.02	73.25	72.06	74.21	70.93	72.61	69.85	69.58	70.67	62.31	62.17	65.33
2005	66.74	68.96	61.55	51.77	53.81	55.66	60.27	60.86	60.54	62.64	67.06	66.72
2006	68.33	67.02	67.15	64.29	63.14	59.74	59.40	62.00	65.25	67.02	69.86	71.35
2007	70.45	68.99	68.14	68.25	71.78	71.75	68.64	65.21	66.02	66.12	56.86	58.13

Table 3.1   Monthly stock price in dollars of AIG stock for the period 2002 through 2007.

It's hard to tell very much from tables of values like this. You might get a rough idea of how much the stock cost—usually somewhere around $60 or so, but that's about it.

3.1   Displaying Quantitative Variables

The first rule of data analysis is to make a picture. AIG's stock price is a *quantitative* variable, whose units are dollars, so a bar chart or pie chart won't work. For quantitative variables, there are no categories. Instead, we usually slice up all the possible values into **bins** and then count the number of cases that fall into each bin. The bins, together with these counts, give the **distribution** of the quantitative variable and provide the building blocks for the display of the distribution, called a **histogram**.

Histograms

Here are the monthly prices of AIG stock displayed in a histogram.

WHO	Months
WHAT	Monthly average price for AIG's stock (in dollars)
WHEN	2002 through 2007
WHERE	New York Stock Exchange
WHY	To examine AIG stock volatility

Figure 3.1   Monthly average prices of AIG stock. The histogram displays the distribution of prices by showing for each "bin" of prices, the number of months having prices in that bin. (Data in **AIG Monthly Prices**.)

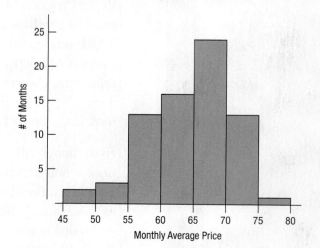

A histogram plots the bin counts as the heights of bars. It counts the number of cases that fall into each bin, and displays that count as the height of the corresponding bar. In this histogram of monthly average prices, each bin has a width of $5, so, for example, the height of the tallest bar says that there were 24 months whose average price of AIG stock was between $65 and $70. In this way, the histogram displays the entire distribution of prices. Unlike a bar chart, which puts gaps between bars to separate the categories, there are no gaps between the bars of a histogram unless there are actual gaps in the data. **Gaps** indicate a region where there are no values. Gaps can be important features of the distribution so watch out for them and point them out.

For categorical variables, each category got its own bar. The only choice was whether to combine categories for ease of display. For quantitative variables, we have to choose the width of the bins. It isn't hard to make a histogram by hand, but we almost always use technology. Many statistics programs allow you to adjust the bin width yourself.

From the histogram, we can see that in these months the AIG stock price was typically near $65 and usually between $55 and $75. Keep in mind that this histogram is a static picture. We have treated these prices simply as a collection of months, not as a time series, and shown their distribution. Later in the chapter we will discuss when this is appropriate and add time to the story.

Does the distribution look as you expected? It's often a good idea to imagine what the distribution might look like before making a display. That way you're less likely to be fooled by errors either in your display or in the data themselves.

The vertical axis of a histogram shows the number of cases falling in each bin. An alternative is to report the percentage of cases in each bin, creating a **relative frequency histogram**. The shape of the two histograms is the same; only the vertical axis and labels are different. A relative frequency histogram is faithful to the area principle by displaying the *percentage* of cases in each bin instead of the count.

Figure 3.2 A relative frequency histogram looks just like a frequency histogram except that the *y*-axis now shows the percentage of months in each bin.

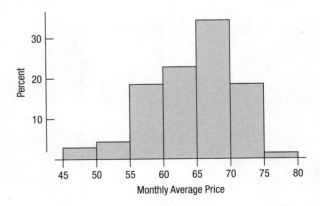

FOR EXAMPLE   Creating a histogram

As the chief financial officer of a music download site, you've just secured the rights to offer downloads of a new album. You'd like to see how well it's selling, so you collect the number of downloads per hour for the past 24 hours: (Data in **Downloads**.)

(continued)

Hour	Downloads	Hour	Downloads
12:00 a.m.	36	12:00 p.m.	25
1:00 a.m.	28	1:00 p.m.	22
2:00 a.m.	19	2:00 p.m.	17
3:00 a.m.	10	3:00 p.m.	18
4:00 a.m.	5	4:00 p.m.	20
5:00 a.m.	3	5:00 p.m.	23
6:00 a.m.	2	6:00 p.m.	21
7:00 a.m.	6	7:00 p.m.	18
8:00 a.m.	12	8:00 p.m.	24
9:00 a.m.	14	9:00 p.m.	30
10:00 a.m.	20	10:00 p.m.	27
11:00 a.m.	18	11:00 p.m.	30

QUESTION Make a histogram for this variable.

ANSWER Create a frequency table of bins of width five from 0 to 40 and put values at the ends of bins into the right bin:

Downloads	Number of Hours
0–5	2
5–10	2
10–15	3
15–20	5
20–25	6
25–30	3
30–35	2
35–40	1
Total	**24**

The histogram looks like this:

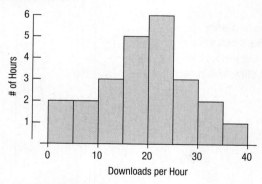

*Stem-and-Leaf Displays

Stem-and-leaf displays are like histograms, but they also show the individual values. They are easy to make by hand for data sets that aren't too large, so they're a great way to look at a small batch of values quickly.[1] Here's a stem-and-leaf display for the AIG stock data, alongside a histogram of the same data.

---

*Sections marked with an asterisk may be optional. Check with your Instructor.

[1] The authors like to make stem-and-leaf displays whenever data are presented (without a suitable display) at committee meetings or working groups. The insights from just that quick look at the distribution are often quite valuable.

Figure 3.3 The AIG monthly average stock prices displayed both by a histogram (left) and stem-and-leaf display (right). Stem-and-leaf displays are typically made by hand, so we are most likely to use them for small data sets. For much larger data sets, we use a histogram.

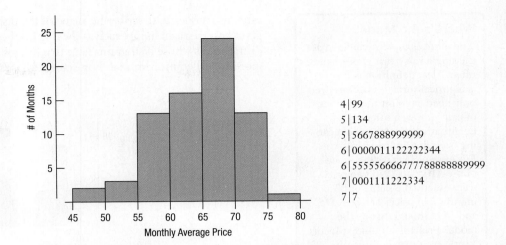

```
4|99
5|134
5|5667888999999
6|0000011122222344
6|5555566667777788888889999
7|0001111222334
7|7
```

How Stem-and-Leaf Displays Work

A stem-and-leaf display breaks each number into two parts: the stem shown to the left of the solid line and the leaf, to the right. For the AIG data, each price, for example, $67.02, is first truncated to two digits, $67. Then it is split into two components: 6|7. The line 5|134 displays the values $51, $53, and $54 and corresponds to the histogram bin from $50 to $55. The stem-and-leaf in Figure 3.3 uses a bin width of 5. Another choice would be to increase the bin size and put all the prices from $50 to $60 on one line:

$$5|1345667888999999$$

That would decrease the number of bins to 4, but makes the bin from $60 to $70 too crowded:

```
4|99
5|1345667888999999
6|000001112222234455555666677778888889999
7|00011112223347
```

Before making a stem-and-leaf display, or a histogram, you should check the **Quantitative Data Condition**: The data must be values of a quantitative variable whose units are known.

Although a bar chart and a histogram may look similar, they're not the same display. You can't display categorical data in a histogram or quantitative data in a bar chart. Always check the condition that confirms what type of data you have before making your display.

3.2 Shape

Once you've displayed the distribution in a histogram or stem-and-leaf display, what can you say about it? When you describe a distribution, you should pay attention to three things: its **shape**, its **center**, and its **spread**.

We describe the shape of a distribution in terms of its modes, its symmetry, and whether it has any gaps or outlying values.

Mode

Does the histogram have a single, central hump (or peak) or several, separated humps? These humps are called **modes**. Formally, the mode is the single, most frequent value, but we rarely use the term that way.[2] The AIG stock prices have a single mode around $65. We

---

[2] Technically, the mode is the value on the *x*-axis of the histogram below the highest peak, but when asked to identify the mode, most people would point to the peak itself.

often use modes to describe the shape of the distribution. A distribution whose histogram has one main hump, such as the one for the AIG stock prices, is called **unimodal**; distributions whose histograms have two humps are **bimodal**, and those with three or more are called **multimodal**. For example, here's a bimodal distribution.

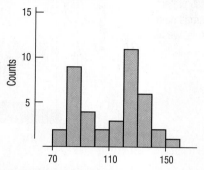

Figure 3.4   A bimodal distribution has two apparent modes.

A bimodal histogram is often an indication that there are two groups in the data. It's a good idea to investigate when you see bimodality. But don't get overly excited by minor fluctuations in the histogram, which may just be artifacts of where the bin boundaries fall. To be a true mode, the hump should still be there when you display the histogram with slightly different bin widths.

A distribution whose histogram doesn't appear to have any mode and in which all the bars are approximately the same height is called **uniform**. (Chapter 5 gives a more formal definition.)

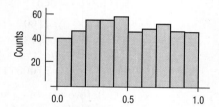

Figure 3.5   In a uniform distribution, bars are all about the same height. The histogram doesn't appear to have a mode.

Symmetry

Could you fold the histogram along a vertical line through the middle and have the edges match pretty closely, as in Figure 3.6, or are more of the values on one side, as in the histograms in Figure 3.7? A distribution is **symmetric** if the halves on either side of the center look, at least approximately, like mirror images.

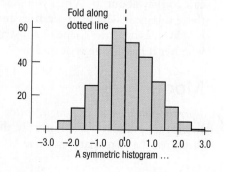
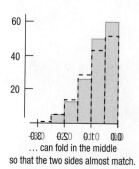

Figure 3.6   A symmetric histogram can fold in the middle so that the two sides almost match.

The (usually) thinner ends of a distribution are called the **tails**. If one tail stretches out farther than the other, the distribution is said to be **skewed** to the side of the longer tail.

Figure 3.7 Two skewed histograms showing the age (left) and hospital charges (right) for all female heart attack patients in New York State in one year. The histogram of Age (in blue) is skewed to the left, while the histogram of Charges (in purple) is skewed to the right.

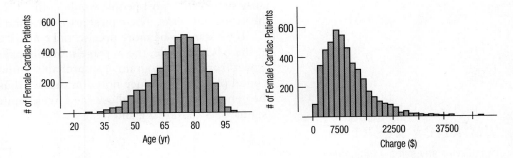

> **Skewed Right or Left?**
> Amounts of things (dollars, employees, waiting times) can't be negative so they run up against zero. But they have no natural upper limit. So, they often have distributions that are skewed to the right. Grades on a test where most students do well are often skewed to the left with many scoring near the top, but a few straggling off to the low end.

Outliers

Do any values appear to stick out? Often such values tell us something interesting or exciting about the data. You should always point out any stragglers or **outliers** that stand off away from the body of the distribution. For example, if you're studying the personal wealth of Americans and Bill Gates is in your sample, he would certainly be an outlier. Because his wealth would be so obviously atypical, you'd want to point it out as a special feature.

Outliers can affect almost every method we discuss in this book, so we'll always be on the lookout for them. An outlier can be the most informative part of your data, or it might just be an error. Either way, you shouldn't throw it away without comment. Treat it specially and discuss it when you report your conclusions about your data. (Or find the error and fix it if you can.)

Using Your Judgment

How you characterize a distribution is often a judgment call. Do the two humps in the histogram really reveal two subgroups, or will the shape look different if you change the bin width slightly? Are those observations at the high end of the histogram truly unusual, or are they just the largest ones at the end of a long tail? These are matters of judgment on which different people can legitimately disagree. There's no automatic calculation or rule of thumb that can make the decision for you. Understanding your data and how they arose can help. What should guide your decisions is an honest desire to understand what is happening in the data. That's what you'll need to make sound business decisions.

Viewing a histogram at several different bin widths can help you to see how persistent some of the features are. Some technologies offer ways to change the bin width interactively to get multiple views of the histogram. If the number of observations in each bin is small enough so that moving a couple of values to the next bin changes your assessment of how many modes there are, be careful. Be sure to think about the data, where they came from, and what kinds of questions you hope to answer from them.

> **FOR EXAMPLE** Describing the shape of a distribution
>
> **QUESTION** Describe the shape of the distribution of downloads from the example on page 51.
>
> **ANSWER** It is symmetric and unimodal with no outliers.

3.3   Center

Look again at the AIG prices in Figure 3.1. If you had to pick one number to describe a *typical* price, what would you pick? When a histogram is unimodal and fairly symmetric, most people would point to the center of the distribution, where the histogram peaks. The typical price is around $65.00.

If we want to be more precise and *calculate* a number, we can *average* the data. In the AIG example, the average monthly price is $64.48, about what we might expect from the histogram. You probably know how to average values, but this is a good place to introduce notation that we'll use throughout the book. We'll call the generic variable *x*, and use the Greek capital letter sigma, Σ, to mean "sum" (sigma is "S" in Greek), and write[3]:

$$\bar{x} = \frac{Total}{n} = \frac{\Sigma x}{n}.$$

NOTATION ALERT

A bar over any symbol indicates the mean of that quantity.

According to this formula, we add up all the values of the variable, *x*, and divide that sum (*Total*, or Σx) by the number of data values, *n*. We call the resulting value the **mean** of *x*.[4]

Although the mean is a natural summary for unimodal, symmetric distributions, it can be misleading for skewed data or for distributions with gaps or outliers. The histogram of AIG monthly prices in Figure 3.1 is unimodal, and nearly symmetric, with a slight left skew. A look at the total volume of AIG stock sold each month for the same 6 years tells a very different story. (Data in **AIG Monthly Prices**.) Figure 3.8 shows a unimodal but strongly right-skewed distribution with two gaps. The mean monthly volume was 170.1 million shares. Locate that value on the histogram. Does it seem a little high as a summary of a typical month's volume? In fact, more than two out of three months have volumes that are less than that value. It might be better to use the **median**—the value that splits the histogram into two equal *areas*. The median is commonly used for variables such as cost or

Figure 3.8   The median splits the area of the histogram in half at 135.9 million shares. The mean is the point on which the histogram would balance. Because the distribution is skewed to the right, the mean 170.1 million shares is *higher* than the median. The points at the right have pulled the mean toward them, away from the median.

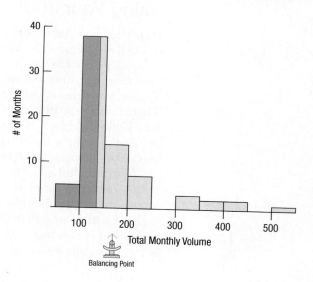

Total Monthly Volume

Balancing Point

---

[3] You may also see the variable called *y* and the equation written $\bar{y} = \frac{Total}{n} = \frac{\Sigma y}{n}$. We actually prefer to call a single variable *y* instead of *x*, because in the next chapter we'll need *x* to name a variable that predicts another (which we'll call *y*), but when you have only one variable either name is common. Most calculators call a single variable *x*.

[4] Once you've averaged the data, you might logically expect the result to be called the *average*. But average is used too colloquially as in the "average" home buyer, where we don't sum up anything. Even though average *is* sometimes used in the way we intend, as in the Dow Jones Industrial Average (which is actually a weighted average) or a batting average, we'll usually use the term *mean* throughout the book.

income, which are likely to be skewed. That's because the median is *resistant* to unusual observations and to the shape of the distribution. For the AIG monthly trading volumes, the median is 135.9 million shares, which seems like a more appropriate summary.

Does it really make a difference whether we choose a mean or a median? The mean monthly price for the AIG stock is $64.48. Because the distribution of the prices is roughly symmetric, we'd expect the mean and median to be close. In fact, we compute the median to be $65.23. But for variables with skewed distributions, the story is quite different. For a right-skewed distribution like the monthly volumes in Figure 3.8, the mean is larger than the median: 170.1 compared to 135.9. The two give quite different summaries. The difference is due to the overall shape of the distributions.

BY HAND Finding the Median

Finding the median of a batch of n numbers is easy as long as you remember to order the values first. If n is odd, the median is the middle value. Counting in from the ends, we find this value in the $\dfrac{n + 1}{2}$ position.

When n is even, there are two middle values. So, in this case, the median is the average of the two values in positions $\dfrac{n}{2}$ and $\dfrac{n}{2} + 1$.

Here are two examples:

Suppose the batch has the values 14.1, 3.2, 25.3, 2.8, -17.5, 13.9, and 45.8. First we order the values: -17.5, 2.8, 3.2, 13.9, 14.1, 25.3, and 45.8. There are 7 values, so the median is the $(7 + 1)/2 = $ 4th value counting from the top or bottom: 13.9.

Suppose we had the same batch with another value at 35.7. Then the ordered values are -17.5, 2.8, 3.2, 13.9, 14.1, 25.3, 35.7, and 45.8. The median is the average of the $8/2$, or 4th, and the $(8/2) + 1$, or 5th, values. So the median is $(13.9 + 14.1)/2 = 14.0$.

The mean is the point at which the histogram would balance. A value far from the center has more leverage, pulling the mean in its direction. It's hard to argue that a summary that's been pulled aside by only a few outlying values or by a long tail is what we mean by the center of the distribution. That's why the median is usually a better choice for skewed data.

However, when the distribution is unimodal and symmetric, the mean offers better opportunities to calculate useful quantities and draw interesting conclusions. It will be the summary-value we work with throughout the rest of the book.

FOR EXAMPLE Finding the mean and median

QUESTION From the data on page 52, what is a typical number of downloads per hour?

ANSWER The mean number is 18.7 downloads per hour. The median is 19.5 downloads per hour. Because the distribution is unimodal and roughly symmetric, we shouldn't be surprised that the two are close. There are a few more hours (in the middle of the night) with small numbers of downloads that pull the mean lower than the median, but either one seems like a reasonable summary to report.

3.4   Spread of the Distribution

We know that the typical price of the AIG stock is around $65, but knowing the mean or median alone doesn't tell us about the entire distribution. A stock whose price doesn't move away from its center isn't very interesting.[5] The more the data vary, the less a measure of center can tell us. We need to know how spread out the data are as well.

One simple measure of spread is the **range**, defined as the difference between the extremes:

$$\text{Range} = max - min.$$

For the AIG price data, the range is $77.26 - $49.41 = $27.85. Notice that the range is *a single number* that describes the spread of the data, not an interval of values—as you might think from its use in common speech. If there are any unusual observations in the data, the range is not resistant and will be influenced by them. Concentrating on the middle of the data avoids this problem.

The **lower quartile, Q1**, is defined as the value for which one quarter of the data lie below it and the **upper quartile, Q3**, is the value for which one quarter of the data lie above it. In this way, the quartiles frame the middle 50% of the data. The **interquartile range (IQR)** summarizes the spread by focusing on the middle half of the data. It's defined as the difference between the two quartiles:

$$\text{IQR} = Q3 - Q1.$$

BY HAND   Finding Quartiles

Quartiles are easy to find in theory, but more difficult in practice. The three quartiles, Q1 (lower quartile), Q2 (the median), and Q3 (the upper quartile) split the sorted data values into quarters. So, for example, 25% of the data values will lie at or below Q1. The problem lies in the fact that unless your sample size divides nicely by 4, there isn't just one way to split the data into quarters. The statistical software package SAS offers at least five different ways to compute quartiles. The differences are usually small, but can be annoying. Here are two of the most common methods for finding quartiles by hand or with a calculator:

1. The Tukey Method (as recommended by the prominent statistician John Tukey)

    Split the sorted data at the median. (If *n* is odd, include the median with each half.) Then find the median of each of these halves—use these as the quartiles.

    Example: The data set {14.1, 3.2, 25.3, 2.8, −17.5, 13.9, 45.8}

    First we order the values: {−17.5, 2.8, 3.2, 13.9, 14.1, 25.3, 45.8}. We found the median to be 13.9, so form two data sets: {−17.5, 2.8, 3.2, 13.9} and {13.9, 14.1, 25.3, 45.8}. The medians of these are 3.0 = (2.8 + 3.2)/2 and 19.7 = (14.1 + 25.3)/2. So we let Q1 = 3.0 and Q3 = 19.7.

2. The TI calculator method

    The same as the Tukey method, except we *don't* include the median with each half. So for {14.1, 3.2, 25.3, 2.8, −17.5, 13.9, and 45.8} we find the two data sets:

    {−17.5, 2.8, 3.2} and {14.1, 25.3, 45.8} by not including the median in either.

---

[5] And not much of an investment, either.

Now the medians of these are Q1 = 2.8 and Q3 = 25.3.

Notice the effect on the IQR. For Tukey:

IQR = Q3−Q1 = 19.7−3.0 = 16.7, but for TI,
IQR = 25.3−2.8 = 22.5.

For both of these methods, notice that the quartiles are either data values, or the average of two adjacent values. In Excel and other software, the quartiles are *interpolated*, so they may not be simple averages of two values. Be aware that there may be differences, but the idea is the same: the quartiles Q1, Q2, and Q3 split the data roughly into quarters.

Waiting in Line[7]

Why do banks favor a single line that feeds several teller windows rather than separate lines for each teller? It does make the average waiting time slightly shorter, but that improvement is very small. The real difference people notice is that the time you can expect to wait is less variable when there is a single line, and people prefer consistency.

For the AIG data,[6] Q1 = \$60.11, Q3 = \$69.01. So the IQR = Q3 − Q1 = \$69.01 − \$60.11 = \$8.90.

The IQR is a reasonable summary of spread, but because it uses only the two quartiles of the data, it ignores much of the information about how individual values vary. By contrast, the standard deviation, takes into account how far each value is from the mean. Like the mean, the standard deviation is appropriate only for symmetric data and can be influenced by outlying observations.[8]

As the name implies, the standard deviation uses the *deviations* of each data value from the mean. The average of the *squared* deviations is called the **variance** and is denoted by s^2:

$$s^2 = \frac{\sum (x - \bar{x})^2}{n - 1}.$$

The variance plays an important role in statistics, but as a measure of spread, it has a problem. Whatever the units of the original data, the variance is in *squared* units. We want measures of spread to have the same units as the data, so we usually take the square root of the variance. That gives the **standard deviation**.

$$s = \sqrt{\frac{\sum (x - \bar{x})^2}{n - 1}}.$$

For the AIG stock prices, $s = \$6.12$.

FOR EXAMPLE   Describing the spread

QUESTION   For the data on page 52, describe the spread of the number of downloads per hour.

ANSWER   The range of downloads is 36 − 2 = 34 downloads per hour. The quartiles are 13 and 24.5, so the IQR is 24.5 − 13 = 11.5 downloads per hour. The standard deviation is 8.94 downloads per hour.

---

[6] In general, we use the Tukey method in this book unless stated otherwise.

[7] If you are from New York or New Jersey, you may say wait "on line." Although most people now think that means you're on the Internet, this expression dates from immigrants who waited by standing literally on lines drawn on the floor of the Ellis Island processing center.

[8] For technical reasons, we divide by $n - 1$ instead of n to take this average. We'll discuss this more in Chapter 11.

JUST CHECKING

Thinking about Variation

1 The U.S. Census Bureau reports the median family income in its summary of census data. Why do you suppose they use the median instead of the mean? What might be the disadvantages of reporting the mean?

2 You've just bought a new car that claims to get a highway fuel efficiency of 31 miles per gallon (mpg). Of course, your mileage will "vary." If you had to guess, would you expect the IQR of gas mileage attained by all cars like yours to be 30 mpg, 3 mpg, or 0.3 mpg? Why?

3 A company selling a new MP3 player advertises that the player has a mean lifetime of 5 years. If you were in charge of quality control at the factory, would you prefer that the standard deviation of life spans of the players you produce be 2 years or 2 months? Why?

3.5   Shape, Center, and Spread—A Summary

What should you report about a quantitative variable? Report the shape of its distribution, and include a center and a spread. But which measure of center and which measure of spread? The guidelines are pretty easy.

- If the shape is skewed, point that out and report the median and IQR. You may want to include the mean and standard deviation as well, explaining why the mean and median differ. The fact that the mean and median do not agree is a sign that the distribution may be skewed. A histogram will help you make the point.

- If the shape is unimodal and symmetric, report the mean and standard deviation and possibly the median and IQR as well. For unimodal symmetric data, the IQR is usually a bit larger than the standard deviation. If that's not true for your data set, look again to make sure the distribution isn't skewed or multimodal and that there are no outliers.

- If there are multiple modes, try to understand why. If you can identify a reason for separate modes, it may be a good idea to split the data into separate groups.

- If there are any clearly unusual observations, point them out. If you are reporting the mean and standard deviation, report them computed with and without the unusual observations. The differences may be revealing.

- Always pair the median with the IQR and the mean with the standard deviation. It's not useful to report a measure of center without a corresponding measure of spread. Reporting a center without a spread can lead you to think you know more about the distribution than you do. Reporting only the spread omits important information.

FOR EXAMPLE   Summarizing data

QUESTION Report on the shape, center, and spread of the downloads data; see page 52.

ANSWER The distribution of downloads per hour over the past 24 hours is unimodal and roughly symmetric. The mean number of downloads per hour is 18.7 and the standard deviation is 8.94. There are several hours in the middle of the night with very few downloads, but none seem to be so unusual as to be considered outliers.

3.6   Standardizing Variables

A real estate agent in California covers two markets. One is near Stanford University, filled with old homes and tree lined streets in an area called Old Palo Alto. The other is a newer neighborhood, in Foster City, created when part of the San Francisco Bay was filled in to create space for more housing.

Here are summaries of the prices of a sample of houses in these two neighborhoods as found on Zillow.com:

Figure 3.9 Prices of houses from samples from Old Palo Alto (left) and Foster City (right). Note that the horizontal scale is quite different for the two neighborhoods. Prices are in $1000's. (Data in **Bay Area Prices**.)

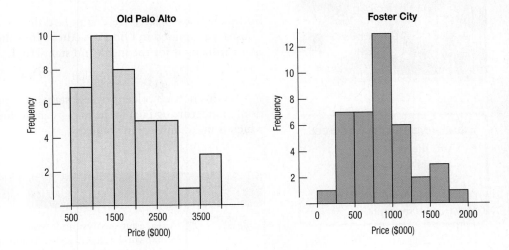

The average house in Old Palo Alto was worth $1,930,436 with a standard deviation of $914,523, while the average Foster City house cost $711,400 with a standard deviation of $318,177. So, a $1,000,000 home in Foster City is on the expensive side, but for Old Palo Alto, that's just over half the average cost.

Which would be more unusual, a $2M home in Foster City, or a $3.5M home in Old Palo Alto?

Using the standard deviation as a way to measure distance helps us answer this question. In Foster City, a $2M home is $1,288,600 over the average. In Old Palo Alto, a $3.5M home is $1,569,564 over its average. It might seem to be more unusual. But look at the standard deviations. That excess of $1,288,600 is over 4 standard deviations above the mean for Foster City. But in Old Palo Alto, the standard deviation is nearly $1M, so the $3.5M home is "only" 1.72 standard deviations above the mean. If you look at the histogram, you can see that $2M (2000) is off the right side for Foster City, but for Old Palo Alto, $3.5M (3500) is still in the histogram.

Using the mean and standard deviation this way gives us a way to standardize values in different distributions to compare them.

How Does Standardizing Work?

We first need to find the mean and standard deviation of each variable for the prices in each town.

	Mean ($000)	SD ($000)
Old Palo Alto	1930.4	914.5
Foster City	711.40	318.2

Next we measure how *far* each of our values is from the mean of its variable. We subtract the mean and then divide by the standard deviation:

$$z = (x - \bar{x})/s$$

We call the resulting value a **standardized value** and denote it with the letter z. Usually, we just call it a **z-score**. The z-score tells us how many standard deviations a value is from its mean.

Standardizing into z-Scores:
- Shifts the mean to 0.
- Changes the standard deviation to 1.
- Does not change the shape.
- Removes the units.

Let's look at Old Palo Alto first.

To compute the z-score for the \$3.5M house, take its value (3500 in \$000 units), subtract the mean (1930.4) and divide by the standard deviation, 914.5:

$$z = (3500 - 1930.4)/914.5 = 1.72$$

So this house's price is 1.72 standard deviations *above* the mean price of all the houses we sampled in Old Palo Alto. How about that \$2M home in Foster City? Standardizing it for the mean and standard deviation of Foster City prices, we find

$$z = (2000 - 711.4)/318.2 = 4.05$$

So this house's price is over 4 standard deviations above the mean for its location! Standardizing enables us to compare values from different distributions to see which is more unusual in context.

FOR EXAMPLE   Comparing values by standardizing

QUESTION   A real estate analyst finds from data on 350 recent sales, that the average price was \$175,000 with a standard deviation of \$55,000. The size of the houses (in square feet) averaged 2100 sq. ft. with a standard deviation of 650 sq. ft. Which is more unusual, a house in this town that costs \$340,000, or a 5000 sq. ft. house?

ANSWER   Compute the z-scores to compare. For the \$340,000 house:

$$z = \frac{x - \bar{x}}{s} = \frac{(340,000 - 175,000)}{55,000} = 3.0$$

The house price is 3 standard deviations above the mean.
For the 5000 sq. ft. house:

$$z = \frac{x - \bar{x}}{s} = \frac{(5000 - 2100)}{650} = 4.46$$

This house is 4.46 standard deviations above the mean in size. That's more unusual than the house that costs \$340,000.

3.7   Five-Number Summary and Boxplots

The **five-number summary** of a distribution reports its median, quartiles, and extremes (maximum and minimum). The five-number summary of the monthly trading volumes of AIG stock for the period 2002–2007 looks like this (in millions of shares).

Max	515.62
Q3	182.32
Median	135.87
Q1	121.04
Min	83.91

Table 3.2   The five-number summary of monthly trading volume of AIG shares (in millions of shares) for the period 2002–2007.

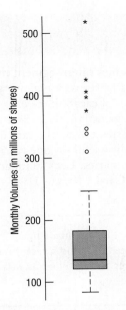

Figure 3.10 Boxplot of monthly volumes of AIG stock traded in the period 2002–2007 (in millions of shares).

The five-number summary provides a good overall look at the distribution. We can see that on half of the days the volume was between 121.04 and 182.32 million shares, and that it was never above 515.62 or below 83.91 million shares.

We can display the information from a five-number summary in a **boxplot** (see Figure 3.10).

A boxplot highlights several features of the distribution of a variable. The central box shows the middle half of the data, between the quartiles. Because the top of the box is at the upper quartile (Q3) and the bottom is at Q1, the height of the box is equal to Q3 − Q1 which is the IQR. (For the AIG data, it's 61.28.) The median is displayed as a horizontal line. If the median is roughly centered between the quartiles, then the middle half of the data is roughly symmetric. If it is not centered, the distribution is skewed. In extreme cases, the median can coincide with one of the quartiles.

The whiskers reach out from the box to the most extreme values that are not considered outliers. The boxplot nominates points as outliers if they fall farther than 1.5 IQRs beyond either quartile (for the AIG data, 1.5 IQR = 1.5 × 61.28 = 91.92). Outliers are displayed individually, both to keep them out of the way for judging skewness and to encourage you to give them special attention. They may be mistakes or they may be the most interesting cases in your data. This rule is not a definition of what makes a point an outlier. It just nominates cases for special attention. And it is not a substitute for careful analysis and thought about whether an extreme value deserves to be treated specially.

Boxplots are especially useful for comparing several distributions side by side. From the shape of the box in Figure 3.10, we can see that the central part of the distribution of volume is skewed to the right (upward here) and the dissimilar length of the two whiskers shows that the skewness continues into the tails of the distribution. We also see several high-volume and some extremely high-volume months. Those months may warrant some inquiries into why trading volume was so high.

Why Use 1.5 IQRs for Nominating Outliers?

Nominate a point as a potential outlier if it lies farther than 1.5 IQRs beyond either the lower (Q1) or upper (Q3) quartile. Some boxplots also designate points as "far" outliers if they lie more than 3 IQRs from the quartiles (as in Figure 3.10). The prominent statistician John W. Tukey, the originator of the boxplot, was asked (by one of the authors) why the outlier nomination rule cut at 1.5 IQRs beyond each quartile. He answered that the reason was simple—1 IQR would be too small and 2 IQRs would be too large.

FOR EXAMPLE   The boxplot rule for nominating outliers

QUESTION   From the histogram on page 52, we saw that no download times seemed to be so far from the center as to be considered outliers. Use the 1.5 IQR rule to see if it nominates any points as outliers.

ANSWER   The quartiles are 13 and 24.5 and the IQR is 11.5, and 1.5 × IQR = 17.25. A value would have to be larger than 24.5 + 17.25 = 41.75 downloads per hour or smaller than 13 − 17.25 = −4.25. The largest value was 36 downloads per hour and all values must be nonnegative, so there are no points nominated as outliers.

GUIDED EXAMPLE   Credit Card Bank Customers

To focus on the needs of particular customers, companies often segment their customers into groups with similar needs or spending patterns. A major credit card bank wanted to see how much a particular group of cardholders charged per month on their cards to understand the potential growth in their card use. The data for each customer was the amount he or she spent using the card during a recent three-month period. Boxplots are especially useful for one variable when combined with a histogram and numerical summaries. Let's summarize the spending of this market segment. (Data in **Credit Card Charges**.)

| **PLAN** | **Setup** Identify the *variable*, the time frame of the data, and the objective of the analysis. | We want to summarize the monthly charges (in dollars) made by 500 cardholders from a market segment of interest during a recent month. The data are quantitative, so we'll use histograms and boxplots, as well as numerical summaries. |

DO

Mechanics Select an appropriate display based on the nature of the data and what you want to know about it.

REALITY CHECK It is always a good idea to anticipate the shape of the distribution so you can check whether the histogram is close to what you expected. Are these data reasonable amounts for customers to charge on their cards in a month? A typical value is a few hundred dollars. That seems like the right ballpark.

Note that outliers are often easier to see with boxplots than with histograms, but the histogram provides more details about the shape of the distribution. The computer program that made this boxplot "jitters" the outliers in the boxplot so they don't lie on top of each other, making them easier to see.

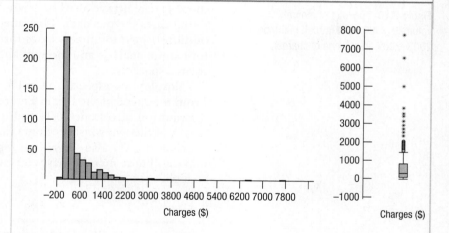

Both graphs show a distribution that is highly skewed to the right with several outliers and an extreme outlier near $7000.

Five-Number Summary of Monthly Charges	
Min	−18.33
Q1	73.84
Med	216.48
Q3	624.80
Max	7759.66

The mean (478.19) is much larger than the median. The data have a skewed distribution.

REPORT

Interpretation Describe the shape, center, and spread of the distribution. Be sure to report on the symmetry, number of modes, and any gaps or outliers.

Recommendation State a conclusion and any recommended actions or analysis.

MEMO

Re: Report on segment spending

The distribution of charges for this segment during this month is unimodal and skewed to the right. For that reason, we have summarized the data with the median and interquartile range (IQR).

The median amount charged was $216.48. The IQR is $550.96 with half the cardholders charging between $73.84 and $624.80 per month.

In addition, there are several high outliers, with one extreme value at $7759.66.

There are also a few negative values. We suspect that these are people who returned more than they charged in a month, but because the values might be data errors, we suggest that they be checked.

Future analyses should look at whether charges during this month were similar to charges in the rest of the year. We would also like to investigate if there is a seasonal pattern and, if so, whether it can be explained by our advertising campaigns or by other factors.

3.8   Comparing Groups

Stock prices can sometimes reflect turmoil within a company. Could an investor have seen signs of trouble in the AIG stock prices? Figure 3.11 shows the daily closing prices for the first two years of our data, 2002 and 2003: (Data in **AIG Daily Prices**.)

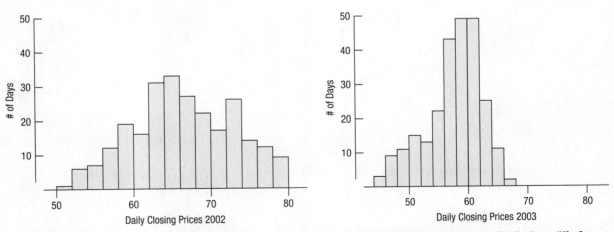

Figure 3.11   Daily closing prices of AIG on the NYSE for the two years 2002 and 2003. How do the two distributions differ?

Prices were generally lower in 2003 than 2002. The price distribution for 2002 appears to be symmetric with a center in the high $60s while the 2003 distribution is left skewed with a center below $60. For comparison, we displayed the two histograms on the same scale. Histograms with very different centers and spreads can appear similar unless you do that.

When we compare several groups, boxplots usually do a better job. Boxplots offer an ideal balance of information and simplicity, hiding the details while displaying the overall summary information. And we can plot them side by side, making it easy to compare multiple groups or categories.

When we place boxplots side by side, we can compare their centers and spreads. We can see past any outliers in making these comparisons because the outliers are displayed individually. We can also begin to look for trends in both the centers and the spreads.

GUIDED EXAMPLE AIG Stock Price

What really happened to the AIG stock price from the beginning of the period we've been studying through the financial crisis of 2008/2009? We will use the daily closing prices of AIG stock for these nine years.

PLAN	**Setup** Identify the variables, report the time frame of the data, and state the objective.	We want to compare the daily prices of AIG shares traded on the NYSE from 2002 through 2009.[9] The daily price is quantitative and measured in dollars. We can partition the values by year and use side-by-side boxplots to compare the daily prices across years.
DO	**Mechanics** Plot the side-by-side boxplots of the data.	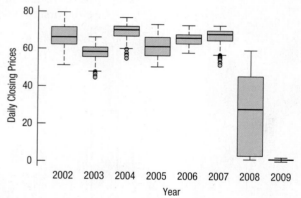
	Display any other plots suggested by the analysis so far.	What happened in 2008? We'd better look there with a finer partition. Here are boxplots by month for 2008. 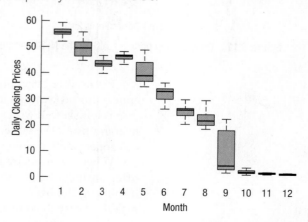

[9]On July 1, 2009, AIG stock had a 1/20 reverse split. To replicate this plot, use the *Adjusted Closing Price* in the data set.

REPORT

Conclusion

Report what you've learned about the data and any recommended action or analysis.

MEMO

Re: Research on price of AIG stock

We have examined the daily closing prices of AIG stock on the NYSE for the period 2002 through 2009. Prices were relatively stable for the period 2002 through 2007. Prices were a bit lower in 2003 but recovered and stayed generally above $60 for 2004 through 2007. Then throughout the first 9 months of 2008, prices dropped dramatically, and remained low throughout 2009. A boxplot by month during 2008 shows that the decline in price was sharpest in September. Most analysts point to that month as the beginning of the financial meltdown, but clearly there were signs in the price of AIG that trouble had been brewing for much longer. By October, and for the rest of the year, the price was very low with almost no variation.

FOR EXAMPLE Comparing boxplots

QUESTION   For the data on page 52, compare the a.m. downloads to the p.m. downloads by displaying the two distributions side-by-side with boxplots.

ANSWER   There are generally more downloads in the afternoon than in the morning. The median number of afternoon downloads is around 22 as compared with 14 for the morning hours. The p.m. downloads are also much more consistent. The entire range of the p.m. hours, 15, is about the size of the IQR for a.m. hours. Both distributions appear to be fairly symmetric, although the a.m. hour distribution has some high points which seem to give some asymmetry.

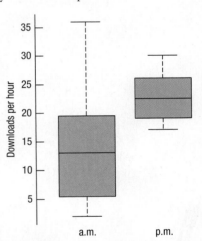

3.9   Identifying Outliers

We've just seen that the price of AIG shares dropped precipitously during the year 2008. Figure 3.12 (next page) shows the daily sales volume by month.

Cases that stand out from the rest of the data deserve our attention. Boxplots have a rule for nominating extreme cases to display as outliers, but that's just a rule of thumb—not a definition. The rule doesn't tell you what to do with them. It's never a substitute for careful thinking about the data and their context.

So, what *should* we do with outliers? The first thing to do is to try to understand them in the context of the data. Once you've identified likely outliers, you should always investigate them. Some outliers are not plausible and may simply be errors. A decimal point may have been misplaced, digits transposed, digits repeated or omitted, or the wrong value transcribed. Or, the units may be wrong. If you saw

Figure 3.12 In January, there was a high-volume day of 38 million shares that is nominated as an outlier for that month. In February there were three outliers with a maximum of over 100 million shares. In most months one or more high-volume days are identified as outliers for their month. But none of these high-volume days would have been considered unusual during September, when the median daily volume of AIG stock was 170 million shares. Days that may have seemed ordinary for September if placed in another month would have seemed extraordinary and *vice versa*. That high-volume day in January certainly wouldn't stand out in September or even October or November, but for January it was remarkable.

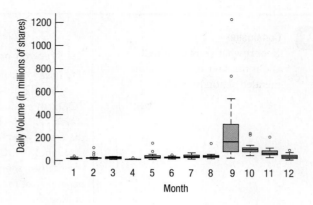

the number of AIG shares traded on the NYSE listed as 2 shares for a particular day, you'd know something was wrong. It could be that it was meant as 2 million shares, but you'd have to check to be sure. If you can identify the error, then you should certainly correct it.

Other outliers are not wrong; they're just different. These are the cases that often repay your efforts to understand them. You may learn more from the extraordinary cases than from summaries of the overall dataset.

What about those two days in September that stand out as extreme even during that volatile month? Those were September 15 and 16, 2008. On the 15th, 740 million shares of AIG stock were traded. That was followed by an incredible volume of over 1 billion shares of stock from a single company traded the following day. Here's how Barron's described the trading of September 16:

Record Volume for NYSE Stocks, Nasdaq Trades Surge
Beats Its July Record

Yesterday's record-setting volume of 8.14 billion shares traded of all stocks listed on the New York Stock Exchange was pushed aside today by 9.31 billion shares in NYSE Composite volume. The biggest among those trades was the buying and selling of American International Group, with 1.11 billion shares traded as of 4 p.m. today. The AIG trades were 12% of all NYSE Composite volume.

FOR EXAMPLE Identifying outliers and summarizing data

QUESTION A real estate report lists the following prices for sales of single-family homes in a small town in Virginia (rounded to the nearest thousand). Write a couple of sentences describing house prices in this town.

155,000	329,000	172,000	122,000	260,000
139,000	178,000	339,435,000	136,000	330,000
158,000	194,000	279,000	167,000	159,000
149,000	160,000	231,000	136,000	128,000

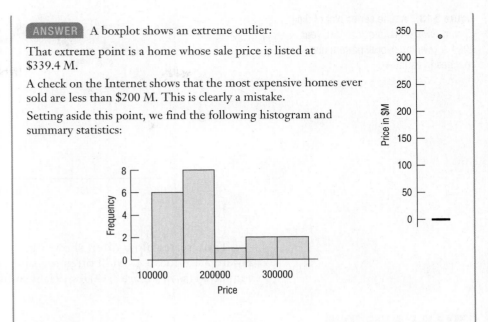

ANSWER   A boxplot shows an extreme outlier:

That extreme point is a home whose sale price is listed at $339.4 M.

A check on the Internet shows that the most expensive homes ever sold are less than $200 M. This is clearly a mistake.

Setting aside this point, we find the following histogram and summary statistics:

The distribution of prices is strongly skewed to the right. The median price is $160,000. The minimum is $122,000 and the maximum (without the outlier) is $330,000. The middle 50% of house prices lie between $144,000 and $212,500 with an IQR of $68,500. (Calculated using the Tukey method.)

3.10   Time Series Plots

A histogram can provide information about the distribution of a variable, but it can't show any pattern over time. Whenever we have time series data, it is a good idea to look for patterns by plotting the data in time order. The histogram we saw in the beginning of the chapter (Figure 3.1) was an appropriate display for the distribution of prices because during that period from 2002 to 2007 the monthly prices were fairly stable.

When a time series has no strong trend or change in variability we say that it is **stationary**.[10] A histogram can provide a useful summary of a stationary series. When your data are measured over time, you should look for patterns by plotting the data in time order. For example, when we examine the daily prices for the year 2007 (Figure 3.13 on the next page), we see that prices started to change during the last quarter.

A display of values against time is called a **time series plot**. This plot reveals a pattern that we were unable to see in either a histogram or a boxplot. Now we can see that although the price rallied in the spring of 2007, after July there were already signs that the price might not stay above $60. By October, that pattern was clear.

---

[10] Sometimes we separate the properties and say the series is stationary with respect to the mean (if there is no trend) or stationary with respect to the variance (if the spread doesn't change). But unless otherwise noted, we'll assume that all the statistical properties of a stationary series are constant over time.

Figure 3.13   A time series plot of daily closing *Price* of AIG stock for the year 2007 shows the overall pattern and changes in variation.

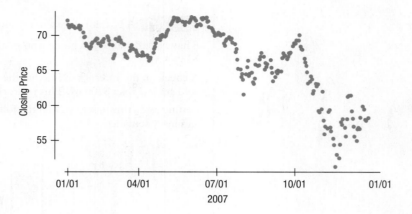

Time series plots often show a great deal of point-to-point variation, as Figure 3.13 does, so you'll often see time series plots drawn with all the points connected (as in Figure 3.14), especially in financial publications.

Figure 3.14   The *Daily Prices* of Figure 3.13, drawn with lines connecting all the points. Sometimes this can help us see an underlying pattern.

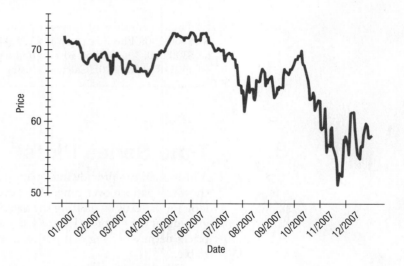

Often it is better to try to smooth out the local point-to-point variability. After all, we usually want to see past this variation to understand any underlying trend and think about how the values vary around that trend—the time series version of center and spread. There are many ways for computers to find a smooth trace through a time series plot.

A smooth trace can highlight long-term patterns and help us see them through the more local variation. Figure 3.15 shows the daily prices of Figures 3.13 and 3.14 with a typical smoothing function, available in many statistics programs. With the smooth trace, it's a bit easier to see a pattern. The trace helps our eye follow the main trend and alerts us to points that don't fit the overall pattern.

It is always tempting to try to extend what we see in a timeplot into the future. Sometimes that makes sense. Most likely, the NYSE volume follows some regular patterns throughout the year. It's probably safe to predict more volume on triple witching days (when contracts expire) and less activity in the week between Christmas and New Year's Day.

Figure 3.15 The *2007 Daily Prices* of Figure 3.13, with a smooth trace added to help your eye see the long-term pattern.

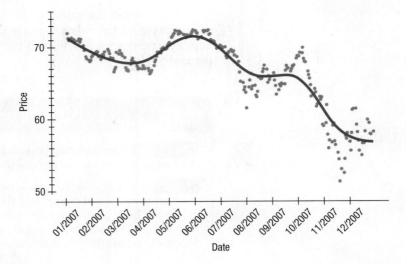

Other patterns are riskier to extend into the future. If a stock's price has been rising, how long will it continue to go up? No stock has ever increased in value indefinitely, and no stock analyst has consistently been able to forecast when a stock's value will turn around. Stock prices, unemployment rates, and other economic, social, or psychological measures are much harder to predict than physical quantities. The path a ball will follow when thrown from a certain height at a given speed and direction is well understood. The path that interest rates will take is much less clear.

Unless you have strong (nonstatistical) reasons for doing otherwise, you should resist the temptation to think that any trend you see will continue indefinitely. Statistical models often tempt those who use them to think beyond the data. We'll pay close attention later in this book to understanding when, how, and how much we can justify doing that.

Look at the prices in Figures 3.13 through 3.15 and try to guess what happened in the subsequent months. Was that drop from October to December a sign of trouble ahead, or was the increase in December back to around $60 where the stock had comfortably traded for several years a sign that stability had returned to AIG's stock price? Perhaps those who picked up the stock for $51 in early November really got a bargain. Let's look ahead to 2008:

Figure 3.16 A time series plot of daily AIG *Price* in 2008 shows a general decline followed by a sharp collapse in September.

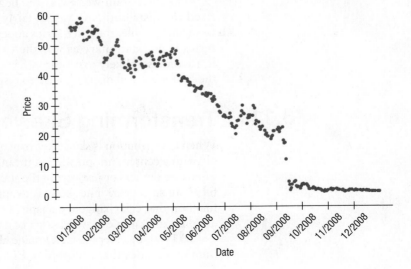

Even through the spring of 2008, although the price was gently falling, nothing prepared traders following only the time series plot for what was to follow. In September the stock lost nearly all of its value. But, by early 2015, it was trading in the mid $50's.

FOR EXAMPLE   Plotting time series data

QUESTION The download times from the example on page 51 are a time series. Plot the data by hour of the day and describe any patterns you see.

ANSWER For this day, downloads were highest at midnight with about 36 downloads per hour then dropped sharply until about 5–6 a.m. when they reached their minimum at 2–3 per hour. They gradually increased to about 20 per hour by noon, and then stayed in the twenties until midnight, with a slight increase during the evening hours. When we ignored the time order, as we did earlier, we missed this pattern entirely.

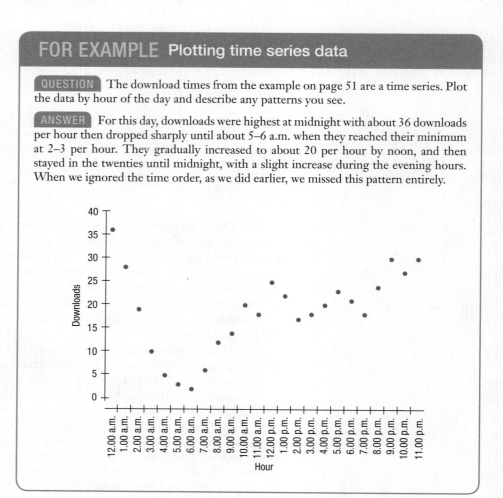

The histogram we saw in the beginning of the chapter (Figure 3.1) summarized the distribution of prices fairly well because during that period the prices were fairly stable; the price series appears to be stationary. However, when the time series is not stationary as was the case for AIG prices after 2007, be careful. A histogram is unlikely to capture what is really of interest. Then, a time series plot is the best graphical display to use to display the behavior of the data.

*3.11   Transforming Skewed Data

When a distribution is skewed, it may not be appropriate to summarize the data simply with a center and spread, and it can be hard to decide whether the most extreme values are outliers or just part of the stretched-out tail. How can we say anything useful about such data? The secret is to apply a simple function to each data value. One function that can change the shape of a distribution is the logarithm function. Let's examine an example in which a set of data is severely skewed.

In 1980, the average CEO made about 42 times the average worker's salary. In the two decades that followed, CEO compensation soared when compared with the average worker's pay; by 2013, the multiple had jumped to 331.[11] What does the distribution of the 500 largest companies' CEOs look like? Figure 3.17 shows a boxplot and a histogram of the CEO compensation from 2012.

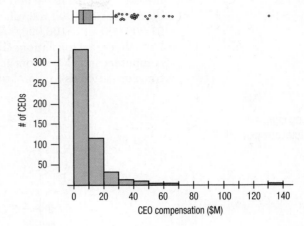

Figure 3.17 The total compensation for CEOs (in $M) of the 500 largest companies is skewed and includes some extraordinarily large values.

These values are reported in *millions* of dollars. The boxplot indicates that some of the 500 CEOs received extraordinarily high compensation. The first bin of the histogram, containing more than half the CEOs, covers the range $0 to $10,000,000. The reason that the histogram seems to leave so much of the area blank is that the largest observations are so far from the bulk of the data, as we can see from the boxplot. Both the histogram and boxplot make it clear that this distribution is very skewed to the right.

Total compensation for CEOs consists of their base salaries, bonuses, and extra compensation, usually in the form of stock or stock options. Data that add together several variables, such as the compensation data, can easily have skewed distributions. It's often a good idea to separate the component variables and examine them individually, but we don't have that information for the CEOs.

Skewed distributions are difficult to summarize. It's hard to know what we mean by the "center" of a skewed distribution, so it's not obvious what value to use to summarize the distribution. What would you say was a typical CEO total compensation? The mean value in 2012 was $10,475,830, while the median was "only" $6,967,500. Each tells something different about how the data are distributed.

One way to make a skewed distribution more symmetric is to **re-express**, or **transform**, the data by applying a simple function to all the data values. It is common to take the logarithm of variables like income, corporate earnings, and prices, which tend to be skewed to the right. Economists do this as a matter of course in many of their models.

Dealing with Logarithms

You may think of logarithms as something technical, but they are just a function that can make some values easier to work with. You have probably already seen logarithmic scales in decibels, Richter scale values, pH values, and others. You may not have realized that logs had been used. Base 10 logs are the easiest to understand, but natural logs are often used as well. (Either one is fine.) You can think of the base 10 log of a number as roughly one less than the number of digits you need to write that number. So 100, which is the smallest number to require 3 digits, has a $\log_{10}$ of 2. And 1000 has a $\log_{10}$ of 3. The $\log_{10}$ of 500 is between 2 and 3, but you'd need a calculator to find that it's approximately 2.7. All salaries of "six figures" have $\log_{10}$ between 5 and 6. Fortunately, with technology, it is easy to re-express data by logs.

---

[11] Source: www.aflcio.org/Corporate-Watch/Paywatch-2014/Terms-and-Data-Sources

The histogram of the logs of the total CEO compensations in Figure 3.18 is nearly symmetric, so we can say that a typical log compensation is between 6.0 and 7.0, which means that it lies between $1 million and $10 million. To be more precise, both the mean and median log_{10} values are 6.84 (that's about $6,920,000). Note that nearly all the values are between 6.0 and 8.0—in other words, between $1,000,000 and $100,000,000 per year. Logarithmic transformations are common, but other transformations like square root and reciprocal are also used. Because computers and calculators are available to do the calculating, you should consider transformation as a helpful tool whenever you have skewed data.

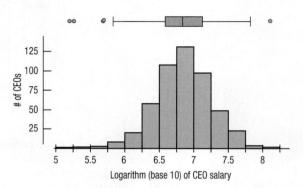

Figure 3.18   Taking logs makes the histogram of CEO total compensation nearly symmetric.

FOR EXAMPLE   Transforming skewed data

QUESTION   *Fortune* magazine publishes a list of the 100 best companies to work for (money.cnn.com/magazines/fortune/bestcompanies/2010/). One statistic often looked at is the average annual pay for the most common job title at the company. Can we characterize those pay values? Here is a histogram of the average annual pay values and a histogram of the logarithm of the pay values. Which would provide the better basis for summarizing pay?

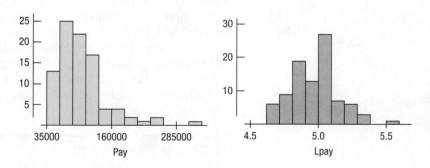

ANSWER   The pay values are skewed to the high end. The logarithmic transformation makes the distribution more nearly symmetric, making it more appropriate to summarize with a mean and standard deviation.

WHAT CAN GO WRONG?

A data display should tell a story about the data. To do that it must speak in a clear language, making plain what variable is displayed, what any axis shows, and what the values of the data are. And it must be consistent in those decisions.

The task of summarizing a quantitative variable requires that we follow a set of rules. We need to watch out for certain features of the data that make summarizing them with a number dangerous. Here's some advice:

- **Don't make a histogram of a categorical variable.** Just because the variable contains numbers doesn't mean it's quantitative. Here's a histogram of the insurance policy numbers of some workers. It's not very informative because the policy numbers are categorical. A histogram or stem-and-leaf display of a categorical variable makes no sense. A bar chart or pie chart may do better.

Figure 3.19 It's not appropriate to display categorical data like policy numbers with a histogram.

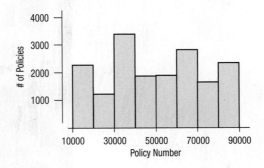

- **Choose a scale appropriate to the data.** Computer programs usually do a pretty good job of choosing histogram bin widths. Often, there's an easy way to adjust the width, sometimes interactively. Figure 3.20 shows the AIG price histogram with two other choices for the bin size. Neither seems to be the best choice.

Figure 3.20 Changing the bin width changes how the histogram looks. The AIG stock prices look very different with these two choices.

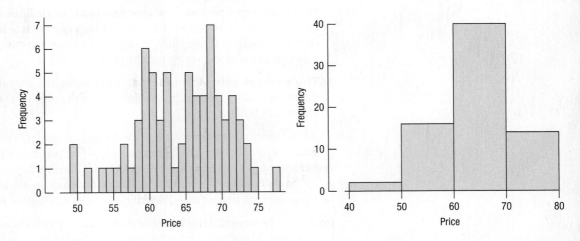

- **Avoid inconsistent scales.** Parts of displays should be mutually consistent—it's not fair to change scales in the middle or to plot two variables on different scales on the same display. When comparing two groups, be sure to draw them on the same scale.

- **Label clearly.** Variables should be identified clearly and axes labeled so a reader knows what the plot displays.

Here's a remarkable example of a plot gone wrong. It illustrated a news story about rising college costs. It uses time series plots, but it gives a misleading impression. First, think about the story you're being told by this display. Then try to figure out what has gone wrong.

What's wrong? Just about everything.

- **The horizontal scales are inconsistent.** Both lines show trends over time, but for what years? The tuition sequence starts in 1965, but rankings are graphed from 1989. Plotting them on the same (invisible) scale makes it seem that they're for the same years.

- **The vertical axis isn't labeled.** That hides the fact that it's using two different scales. Does it graph dollars (of tuition) or ranking (of Cornell University)?

This display violates every rule we can think of. And it's even worse than that. It violates a rule that we didn't even consider. The two inconsistent scales for the vertical axis don't point in the same direction! The line for Cornell's rank shows that it has "plummeted" from 15th place to 6th place in academic rank. Most of us think that's an *improvement*, but that's not the message of this graph.

- **Do a reality check.** Don't let the computer (or calculator) do your thinking for you. Make sure the calculated summaries make sense. For example, does the mean look like it is in the center of the histogram? Think about the spread. An IQR of 50 mpg would clearly be wrong for a family car. And no measure of spread can be negative. The standard deviation can take the value 0, but only in the very unusual case that all the data values equal the same number. If you see the IQR or standard deviation equal to 0, it's probably a sign that something's wrong with the data.

- **Don't compute numerical summaries of a categorical variable.** The mean ZIP code or the standard deviation of Social Security numbers is not meaningful. If the variable is categorical, you should instead report summaries such as percentages. It is easy to make this mistake when you let technology do the summaries for you. After all, the computer doesn't care what the numbers mean.

- **Watch out for multiple modes.** If the distribution—as seen in a histogram, for example—has multiple modes, consider separating the data into groups. If you cannot separate the data in a meaningful way, you should not summarize the center and spread of the variable.

- **Beware of outliers.** If the data have outliers but are otherwise unimodal, consider holding the outliers out of the further calculations and reporting them individually. If you can find a simple reason for the outlier (for instance, a data transcription error), you should remove or correct it. If you cannot do either of these, then choose the median and IQR to summarize the center and spread.

ETHICS IN ACTION

Beth Tully owns Zenna's Café, an independent coffee shop located in a small Midwestern city. Since opening Zenna's in 2002, she has been steadily growing her business and now distributes her custom coffee blends to a number of regional restaurants and markets. She operates a microroaster that offers specialty-grade Arabica coffees recognized by some as the best in the area.

In addition to providing the highest quality coffees, Beth also wants her business to be socially responsible. Toward that end, she pays fair prices to coffee farmers and donates funds to help charitable causes in Panama, Costa Rica, and Guatemala. In addition, she encourages her employees to get involved in the local community.

Recently, one of the well-known multinational coffeehouse chains announced plans to locate shops in her area. This chain is one of the few to offer Certified Free-Trade coffee products and work toward social justice in the global community. Consequently, Beth thought it might be a good idea for her to begin communicating Zenna's socially responsible efforts to the public, but with an emphasis on their commitment to the local community.

Three months ago she began collecting data on the number of volunteer hours donated by her employees per week. She has a total of 12 employees, of whom 10 are full time. Most employees volunteered less than 2 hours per week, but Beth noticed that one part-time employee volunteered more than 20 hours per week. She discovered that her employees collectively volunteered an average of 15 hours per month (with a median of 8 hours). She planned to report the average number and believed most people would be impressed with Zenna's level of commitment to the local community.

- Identify the ethical dilemma in this scenario.
- What are the undesirable consequences?
- Propose an ethical solution that considers the welfare of all stakeholders.

WHAT HAVE WE LEARNED?

Learning Objectives

Make and interpret histograms to display the distribution of a variable.
- We understand distributions in terms of their shape, center, and spread.

Describe the shape of a distribution.
- A **symmetric** distribution has roughly the same shape reflected around the center.
- A **skewed** distribution extends farther on one side than on the other.

- A **unimodal** distribution has a single major hump or mode; a bimodal distribution has two; multimodal distributions have more.
- **Outliers** are values that lie far from the rest of the data.

Compute the mean and median of a distribution, and know when it is best to use each to summarize the center.

- The **mean** is the sum of the values divided by the count. It is a suitable summary for unimodal, symmetric distributions.
- The median is the middle value; half the values are above and half are below the median. It is a better summary when the distribution is skewed or has outliers.

Compute the standard deviation and interquartile range (IQR), and know when it is best to use each to summarize the spread.

- The **standard deviation** is the square root of the average squared difference between each data value and the mean. It is the summary of choice for the spread of unimodal, symmetric variables.
- The **IQR** is the difference between the quartiles. It is often a better summary of spread for skewed distributions or data with outliers.

Standardize values and use them for comparisons of otherwise disparate variables.

- We standardize by finding **z-scores**. To convert a data value to its z-score, subtract the mean and divide by the standard deviation.
- z-scores have no units, so they can be compared to z-scores of other variables.
- The idea of measuring the distance of a value from the mean in terms of standard deviations is a basic concept in Statistics and will return many times later in the course.

Find a five-number summary and, using it, make a boxplot. Use the boxplot's outlier nomination rule to identify cases that may deserve special attention.

- A **five-number summary** consists of the median, the quartiles, and the extremes of the data.
- A **boxplot** shows the quartiles as the upper and lower ends of a central box, the median as a line across the box, and "whiskers" that extend to the most extreme values that are not nominated as outliers.
- Boxplots display separately any case that is more than 1.5 IQRs beyond each quartile. These cases should be considered as possible outliers.

Use boxplots to compare distributions.

- Boxplots facilitate comparisons of several groups. It is easy to compare centers (medians) and spreads (IQRs).
- Because boxplots show possible outliers separately, any outliers don't affect comparisons.

Make and interpret time plots for time series data.

- Look for the trend and any changes in the spread of the data over time.

Terms	
Bin	In a histogram, the range of possible values are split into intervals called bins, over which the frequencies are displayed.
Bimodal	Distributions with two modes.
Boxplot	A boxplot displays the five-number summary as a central box with whiskers that extend to the nonoutlying values. Boxplots are particularly effective for comparing groups.
Center	The middle of the distribution, usually summarized numerically by the mean or the median.
Distribution	The distribution of a variable gives:

- possible values of the variable
- frequency or relative frequency of each value or range of values

Five-number summary	A five-number summary for a variable consists of: • The minimum and maximum • The quartiles Q1 and Q3 • The median
Gap	A region of a distribution where there are no values
Histogram (relative frequency histogram)	A histogram uses adjacent bars to show the distribution of values in a quantitative variable. Each bar represents the frequency (relative frequency) of values falling in an interval of values.
Interquartile range (IQR)	The difference between the lower and upper quartiles. $IQR = Q3 - Q1$.
Mean	A measure of center found as $\bar{x} = \sum x / n$.
Median	The middle value with half of the data above it and half below it.
Mode	A peak or local high point in the shape of the distribution of a variable. The apparent location of modes can change as the scale of a histogram is changed.
Multimodal	Distributions with more than two modes.
Outliers	Extreme values that don't appear to belong with the rest of the data. They may be unusual values that deserve further investigation or just mistakes; there's no obvious way to tell.
Quartile	The lower quartile (Q1) is the value with a quarter of the data below it. The upper quartile (Q3) has a quarter of the data above it. The median and quartiles divide the data into four equal parts.
Range	The difference between the lowest and highest values in a data set: $Range = max - min$.
*Re-express or transform	To re-express or transform data, take the logarithm, square root, reciprocal, or some other mathematical operation on all values of the data set. Re-expression can make the distribution of a variable more nearly symmetric and the spread of groups more nearly alike.
Shape	The visual appearance of the distribution. To describe the shape, look for: • single vs. multiple modes • symmetry vs. skewness
Skewed	A distribution is skewed if one tail stretches out farther than the other.
Spread	The description of how tightly clustered the distribution is around its center. Measures of spread include the IQR and the standard deviation.
Standard deviation	A measure of spread found as $s = \sqrt{\dfrac{\sum (x - \bar{x})^2}{n - 1}}$.
Standardized value	We standardize a value by subtracting the mean and dividing by the standard deviation for the variable. These values, called z-scores, have no units.
Stationary	A time series is said to be stationary if its statistical properties don't change over time.
*Stem-and-leaf display	A stem-and-leaf display shows quantitative data values in a way that sketches the distribution of the data. It's best described in detail by example like the one on page 53.
Symmetric	A distribution is symmetric if the two halves on either side of the center look approximately like mirror images of each other.
Tail	The tails of a distribution are the parts that typically trail off on either side.
Time series plot	A time series plot displays the values of a time series plotted against time. Often, successive values are connected with lines to show trends more clearly.
Uniform	A distribution that's roughly flat is said to be uniform.
Unimodal	Having one mode. This is a useful term for describing the shape of a histogram when it's generally mound-shaped.
Variance	The standard deviation squared.
z-Score	A standardized value that tells how many standard deviations a value is from the mean; z-scores have a mean of 0 and a standard deviation of 1.

TECHNOLOGY HELP: Displaying and Summarizing Quantitative Variables

Almost any program that displays data can make a histogram, but some will do a better job of determining where the bars should start and how they should partition the span of the data (see the figure below).

Many statistics packages offer a prepackaged collection of summary measures. The result might look like this:

```
Variable: Weight
N = 234
Mean = 143.3    Median = 139
St. Dev = 11.1  IQR = 14
```

Alternatively, a package might make a table for several variables and summary measures:

```
Variable  N    mean   median  stdev  IQR
Weight    234  143.3  139     11.1   14
Height    234  68.3   68.1    4.3    5
Score     234  86     88      9      5
```

It is usually easy to read the results and identify each computed summary statistic. You should be able to read the summary statistics produced by any computer package.

Packages often provide many more summary statistics than you need. Of course, some of these may not be appropriate when the data are skewed or have outliers. It is your responsibility to check a histogram or stem-and-leaf display and decide which summary statistics to use.

It is common for packages to report summary statistics to many decimal places of "accuracy." Of course, it is rare to find data that have such accuracy in the original measurements. The ability to calculate to six or seven digits beyond the decimal point doesn't mean that those digits have any meaning. Generally, it's a good idea to round these values, allowing perhaps one more digit of precision than was given in the original data.

Displays and summaries of quantitative variables are among the simplest things you can do in most statistics packages.

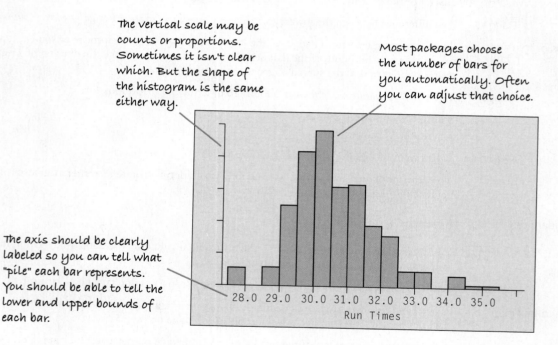

The vertical scale may be counts or proportions. Sometimes it isn't clear which. But the shape of the histogram is the same either way.

Most packages choose the number of bars for you automatically. Often you can adjust that choice.

The axis should be clearly labeled so you can tell what "pile" each bar represents. You should be able to tell the lower and upper bounds of each bar.

EXCEL

To make a histogram in Excel 2010 or 2013, use the Data Analysis add-in. If you have not installed that, you must do that first:

- On the File tab, click **Options,** and then click **Add-Ins.**
- Near the bottom of the Excel Options dialog box, select **Excel add-ins** in the Manage box, and then click **Go.**
- In the Add-Ins dialog box, select the check box for Analysis ToolPak, and then click **OK.**
- If Excel displays a message that states it can't run this add-in and prompts you to install it, click **Yes** to install the add-in.

To make a histogram,

- From Data, select the **Data Analysis add-in.**
- From its menu, select **Histograms.**
- Indicate the range of the data whose histogram you wish to draw.
- Indicate the bin ranges that are up to and including the right end points of each bin.
- Check **Labels** if your columns have names in the first cell.
- Check **Chart output** and click **OK.**
- Right-click on any bar of the resulting graph and, from the menu that drops down, select **Format Data Series ...**

- In the dialog box that opens, select **Series Options** from the sidebar.
- Slide the Gap Width slider to **No Gap**, and click **Close**.
- In the pivot table on the left, use your pointing tool to slide the bottom of the table up to get rid of the "more" bin.
- Edit the bin names in Column A to properly identify the contents of each bin.

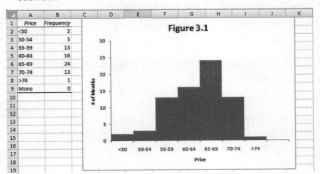

- You can right click on the legend or axis names to edit or remove them.
- Following these instructions, you can reproduce Figure 3.1 using the data set AIG stock series.

Alternatively, you can set up your own bin boundaries and count the observations falling within each bin using an Excel function such as FREQUENCY (Data array, Bins array). Consult your Excel manual or help files for details of how to do this.

To create a time series plot in Excel:

- Open a time-series data file sorted in ascending order by time.
- Highlight the column(s) holding the values of the quantitative variable(s) measured over a period of time.
- Choose **Insert** > **Charts** > **Line** > **2-D Line**.
- Click the chart and choose **Select Data** from the **Chart Tools** > **Design menu.**
- Choose **Select Data Source** > **Edit** under Horizontal (Category) Axis Labels and select the time data.

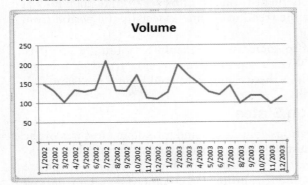

XLStat

To make a boxplot:

- Choose **Visualizing data**, and then select **Univariate plots**.
- Enter the cell range of your data in the **Quantitative** or **Qualitative data** field.
- Select the type of chart on the **Charts** tab.

NOTE: XLStat scales side-by-side boxplots individually, so they are not suitable for comparing groups.

JMP

To make a histogram and find summary statistics:

- Choose **Distribution** from the **Analyze** menu.
- In the **Distribution** dialog, drag the name of the variable that you wish to analyze into the empty window beside the label "**Y, Columns.**"
- Click **OK**. JMP computes standard summary statistics along with displays of the variables.

To make boxplots:

- Choose **Fit Y By X**. Assign a continuous response variable to **Y, Response** and a nominal group variable holding the group names to **X, Factor,** and click **OK**. JMP will offer (among other things) dotplots of the data. Click the red triangle and, under **Display Options**, select **Boxplots**. Note: If the variables are of the wrong type, the display options might not offer boxplots.

Alternatively

- Chose **Graph** > **Graph Builder.** Drag the quantitative variable to **Y** and the categorical variable to **X**. Right click on the points and change Points to **Box Plot**

To make a time series plot in JMP:

- From the **Analyze** menu, choose **Fit Y by X**.
- Move the y variable (measured over time) into the **Y, Response** box.
- Move the x variable (time) into the **X, Factor** box.
- Press **OK**.
  - To connect the points select **Fit Each Value** from the red triangle next to Bivariate Fit.
- To put a smooth through the points, select either **Kernel Smoother** or **Fit Spline** under the red triangle.

Comments

For either the **Kernel Smoother** or **Spline** a slider bar appears to adjust the amount of smoothing.

Alternatively,

- Select **Graph** > **Graph Builder**.
- Drag the y variable into the Y window and the x variable (time) into the X window.
- The default shows the points and a smoother.
  - Right click on the graph and select **Smoother** > **Change to** > **Line** to connect the points instead.

MINITAB

To make a histogram:

- Choose **Histogram** from the **Graph** menu.
- Select **Simple** for the type of graph and click **OK**.
- Enter the name of the quantitative variable you wish to display in the box labeled "Graph variables." Click **OK**.

To make a boxplot:

- Choose **Boxplot** from the **Graph** menu and specify your data format.

To calculate summary statistics:

- Choose **Basic Statistics** from the **Stat** menu. From the **Basic Statistics** submenu, choose **Display Descriptive Statistics**.
- Assign variables from the variable list box to the Variables box. MINITAB makes a Descriptive Statistics table.

For a quantitative variable X:

- **summary(X)** gives a five-number summary and the mean
- **mean(X)** gives the mean and **sd(X)** gives the standard deviation
- **hist(X)** produces a histogram and **boxplot(X)** makes a boxplot

Comments

Many other summaries are available, including **min( ), max( ), quantile(X,prob=p)** (where p is a probability between 0 and 1), and **median( ).** Both hist and boxplot have many options.

SPSS

To make a histogram or boxplot in SPSS open the Chart Builder from the Graphs menu.

- Click the **Gallery** tab.
- Choose **Histogram** or **Boxplot** from the list of chart types.
- Drag the icon of the plot you want onto the canvas.
- Drag a scale variable to the *y*-axis drop zone.
- Click **OK**.

To make side-by-side boxplots, drag a categorical variable to the *x*-axis drop zone and click **OK**.

To calculate summary statistics:

- Choose **Explore** from the **Descriptive Statistics** submenu of the **Analyze** menu. In the **Explore** dialog, assign one or more variables from the source list to the Dependent List and click the **OK** button.

Brief Case

Detecting the Housing Bubble

The S&P/Case-Shiller Home Price Indices provide measures of the U.S. residential housing market. They track changes in the value of residential real estate nationally and in 20 metropolitan regions. (Some of these indices are actually traded on the Chicago Mercantile Exchange.) The dataset **Case-Shiller** gives the monthly index values for each of the 20 cities tracked by the Case-Shiller index and two national composite series. Examine these values and write a report on them.

Some suggestions:
First consider the Composite.20 series, which combines (seasonally adjusted) data for the 20 cities. Describe the distribution of prices overall, then look at a time series plot and discuss the trend over time especially the period from 2007 to 2008.

Then select several cities to compare. For example, you might compare Miami, Boston, and Detroit. Write a report discussing how trends in housing prices changed over time and how these changes differed from city to city.

Socio-Economic Data on States

The dataset **States** contains various educational and economic measures of the 50 U.S. states, including the District of Columbia. Examine the variables, commenting on their shape, center, and spread and any unusual features. If you see any unusual cases, set them aside if appropriate, comment on why you took them out, and redo the analysis without them.

EXERCISES

SECTION 3.1

1. As part of the marketing team at an Internet music site, you want to understand who your customers are. You send out a survey to 25 customers (you use an incentive of $50 worth of downloads to guarantee a high response rate) asking for demographic information. One of the variables is the customer's age. For the 25 customers the ages are:

20	32	34	29	30
30	30	14	29	11
38	22	44	48	26
25	22	32	35	32
35	42	44	44	48

a) Make a histogram of the data using a bar width of 10 years.
b) Make a histogram of the data using a bar width of 5 years.
c) Make a relative frequency histogram of the data using a bar width of 5 years.
d) *Make a stem-and-leaf plot of the data using 10s as the stems and putting the youngest customers on the top of the plot.

2. As the new manager of a small convenience store, you want to understand the shopping patterns of your customers. You randomly sample 20 purchases from yesterday's records (all purchases in U.S. dollars):

39.05	2.73	32.92	47.51
37.91	34.35	64.48	51.96
56.95	81.58	47.80	11.72
21.57	40.83	38.24	32.98
75.16	74.30	47.54	65.62

a) Make a histogram of the data using a bar width of $20.
b) Make a histogram of the data using a bar width of $10.
c) Make a relative frequency histogram of the data using a bar width of $10.
d) *Make a stem-and-leaf plot of the data using $10 as the stems and putting the smallest amounts on top and round the data to the nearest $.

SECTION 3.2

3. For the histogram you made in Exercise 1a:

a) Is the distribution unimodal or multimodal?
b) Where is (are) the mode(s)?
c) Is the distribution symmetric?
d) Are there any outliers?

4. For the histogram you made in Exercise 2a:

a) Is the distribution unimodal or multimodal?
b) Where is (are) the mode(s)?
c) Is the distribution symmetric?
d) Are there any outliers?

SECTION 3.3

5. For the data in Exercise 1:

a) Would you expect the mean age to be smaller than, bigger than, or about the same size as the median? Explain.
b) Find the mean age.
c) Find the median age.

6. For the data in Exercise 2:

a) Would you expect the mean purchase to be smaller than, bigger than, or about the same size as the median? Explain.
b) Find the mean purchase.
c) Find the median purchase.

SECTION 3.4

7. For the data in Exercise 1:

a) Find the quartiles using your calculator.
b) Find the quartiles using Tukey's method (page 58).
c) Find the IQR using the quartiles from part b.
d) Find the standard deviation.

8. For the data in Exercise 2:

a) Find the quartiles using your calculator.
b) Find the quartiles using Tukey's method (page 58).
c) Find the IQR using the quartiles from part b.
d) Find the standard deviation.

SECTION 3.5

9. The histogram shows the December charges (in $) for 5000 customers from one marketing segment from a credit card company. (Negative values indicate customers who received more credits than charges during the month.)

a) Write a short description of this distribution (shape, center, spread, unusual features).
b) Would you expect the mean or the median to be larger? Explain.
c) Which would be a more appropriate summary of the center, the mean or the median? Explain.

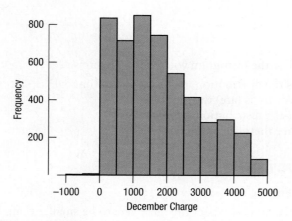

10. Adair Vineyard is a 10-acre vineyard in New Paltz, New York. The winery itself is housed in a 200-year-old historic Dutch barn, with the wine cellar on the first floor and the tasting room and gift shop on the second. Since they are relatively small and considering an expansion, they are curious about how their size compares to that of other vineyards. The histogram shows the sizes (in acres) of 36 wineries in upstate New York.

a) Write a short description of this distribution (shape, center, spread, unusual features).

b) Would you expect the mean or the median to be larger? Explain.

c) Which would be a more appropriate summary of the center, the mean or the median? Explain.

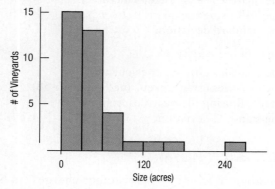

SECTION 3.6

11. Using the ages from Exercise 1:

a) Standardize the minimum and maximum ages using the mean from Exercise 5b and the standard deviation from Exercise 7d.

b) Which has the more extreme z-score, the min or the max?

c) How old would someone with a z-score of 3 be?

12. Using the purchases from Exercise 2:

a) Standardize the minimum and maximum purchase using the mean from Exercise 6b and the standard deviation from Exercise 8d.

b) Which has the more extreme z-score, the min or the max?

c) How large a purchase would a purchase with a z-score of 3.5 be?

SECTION 3.7

13. For the data in Exercise 1:

a) Draw a boxplot using the quartiles from Exercise 7b.

b) Does the boxplot nominate any outliers?

c) What age would be plotted as an outlier?

14. For the data in Exercise 2:

a) Draw a boxplot using the quartiles from Exercise 8b.

b) Does the boxplot nominate any outliers?

c) What purchase amount would be plotted as an outlier?

15. Here are summary statistics for the sizes (in acres) of upstate New York vineyards from Exercise 10.

Variable	N	Mean	StDev	Minimum	Q1	Median	Q3	Maximum
Acres	36	46.50	47.76	6	18.50	33.50	55	250

a) From the summary statistics, would you describe this distribution as symmetric or skewed? Explain.

b) From the summary statistics, are there any outliers? Explain.

c) Using these summary statistics, sketch a boxplot. What additional information would you need to complete the boxplot?

16. A survey of major universities asked what percentage of incoming freshmen usually graduate "on time" in 4 years. Use the summary statistics given to answer these questions.

	% on time
Count	48
Mean	68.35
Median	69.90
StdDev	10.20
Min	43.20
Max	87.40
Range	44.20
25th %tile	59.15
75th %tile	74.75

a) Would you describe this distribution as symmetric or skewed?

b) Are there any outliers? Explain.

c) Create a boxplot of these data.

SECTION 3.8

17. The survey from Exercise 1 had also asked the customers to say whether they were male or female. Here are the data:

Age	Sex	Age	Sex	Age	Sex	Age	Sex	Age	Sex
20	M	32	F	34	F	29	M	30	M
30	F	30	M	14	M	29	M	11	M
38	F	22	M	44	F	48	F	26	F
25	M	22	M	32	F	35	F	32	F
35	F	42	F	44	F	44	F	48	F

Construct boxplots to compare the ages of men and women and write a sentence summarizing what you find.

18. The store manager from Exercise 2 has collected data on purchases from weekdays and weekends. Here are some summary statistics (rounded to the nearest dollar):

Weekdays: n = 230

Min = 4, Q1 = 28, Median = 40, Q3 = 68, Max = 95

Weekends: n = 150

Min = 10, Q1 = 35, Median = 55, Q3 = 70, Max = 100

From these statistics, construct side-by-side boxplots and write a sentence comparing the two distributions.

19. Here are boxplots of the weekly sales (in USD) over a two-year period for a regional food store for two locations. Location #1 is a metropolitan area that is known to be residential where shoppers walk to the store. Location #2 is a suburban area where shoppers drive to the store. Assume that the two towns have similar populations and that the two stores are similar in square footage. Write a brief report discussing what these data show.

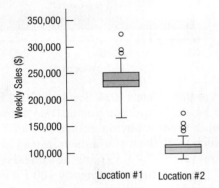

20. Recall the distributions of the weekly sales for the regional stores in Exercise 19. Following are boxplots of weekly sales for this same food store chain for three stores of similar size and location for two different states: Massachusetts (MA) and Connecticut (CT). Compare the distribution of sales for the two states and describe in a report.

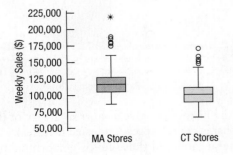

SECTION 3.9

21. The five-number summary for the total revenue (in $M) of the top 100 movies of 2012 looks like this:

Min	Q1	Med	Q3	Max
28.8	44.9	64.1	123.0	623.4

Are there any outliers in these data? How can you tell? What might your next steps in the analysis be?

22. The five-number summary for the ages of 100 respondents to a survey on cell phone use looks like this:

Min	Q1	Med	Q3	Max
13	24	38	49	256

Are there any outliers in these data? How can you tell? What might your next steps in the analysis be?

SECTION 3.10

23. Are the following data time series? If not, explain why.

a) Quarterly earnings of Microsoft Corp.
b) Unemployment in August 2010 by education level.
c) Time spent in training by workers in NewCo.
d) Numbers of e-mails sent by employees of SynCo each hour in a single day.

24. Are the following data time series? If not, explain why.

a) Reports from the Bureau of Labor Statistics on the number of U.S. adults who are employed full time in each major sector of the economy.
b) The quarterly Gross Domestic Product (GDP) of France from 1980 to the present.
c) The dates on which a particular employee was absent from work due to illness over the past two years.
d) The number of cases of flu reported by the CDC each week during a flu season.

SECTION 3.11

25. The histogram of the total revenues (in $M) of the movies in exercise 21 looks like this:

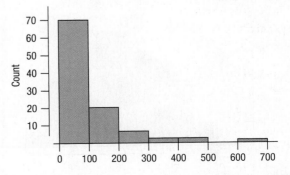

What might you suggest for the next step of the analysis?

26. The histogram of the ages of the respondents in exercise 22 looks like this:

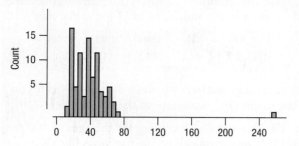

What might you suggest for the next step of the analysis?

CHAPTER EXERCISES

27. Statistics in business. Find a histogram that shows the distribution of a quantitative variable in a business publication (e.g., *The Wall Street Journal, Business Week*).

a) Does the article identify the Ws?
b) Discuss whether the display is appropriate for the data.
c) Discuss what the display reveals about the variable and its distribution.
d) Does the article accurately describe and interpret the data? Explain.

28. Statistics in business, part 2. Find a graph other than a histogram that shows the distribution of a quantitative variable in a business publication (e.g., *The Wall Street Journal, Business Week*).

a) Does the article identify the Ws?
b) Discuss whether the display is appropriate for the data.
c) Discuss what the display reveals about the variable and its distribution.
d) Does the article accurately describe and interpret the data? Explain.

T **29.** Shirt sizes. A clothing manufacturer wants to study men's neck sizes to plan how many shirts of different sizes to produce (shirt sizes are generally one-half inch larger than measured neck sizes and rounded to the nearest half inch). The following histogram shows the distribution of neck sizes of random volunteers for a health study conducted in Utah.

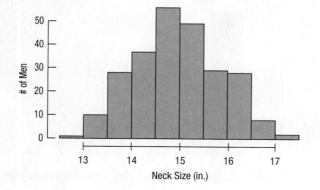

a) Write a short description of this distribution (shape, center, spread, unusual features).
b) What two bins hold the most values for neck size?
c) If we add half an inch to the neck size and round up to the nearest half inch to get the shirt size, what are the two most popular shirt sizes?

T **30.** Gas prices 2015. The website LosAngelesGasPrices .com has current gasoline prices all over the United States. In the week of January 19, 2015, the following histogram shows the gas prices at 37 stations in the San Francisco Bay Area. Describe the shape of this distribution (shape, center, spread, unusual features).

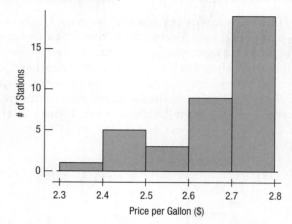

T **31.** Exchange-traded funds 2014. Stocks in 2014 had a good year. The S&P 500 average was up 11.5% for the year. Many people suggest buying exchange-traded funds (or ETFs), which track the major indices instead of individual stocks. However, not all ETFs are the same. Here is a histogram of the performance of nearly 100 ETFs for 2014 as reported by Morningstar (http://news.morningstar.com/ etf/Lists/ETFReturns.html).

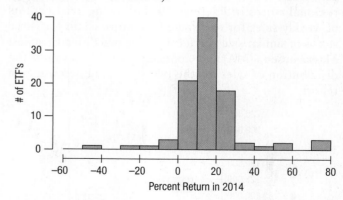

a) From the histogram, give a short summary of the distribution (shape, center, spread, unusual features).
b) In general, how did these funds perform compared to the S&P 500?

32. Car discounts. A researcher, interested in studying gender differences in negotiations, collects data on the prices that men and women pay for new cars. Here is a histogram of the discounts (the amount in $ below the list price) that men and women received at one car dealership for the last 100 transactions (54 men and 46 women). Give a short summary of this distribution (shape, center, spread, unusual features). What do you think might account for this particular shape?

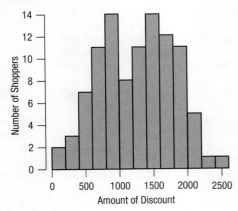

33. Exchange-traded funds 2014, part 2. Use the data for 12 month returns in of Exercise 31 to answer the following questions.

a) Find the five-number summary for these data.
b) Find appropriate measures of center and spread for these data.
c) Create a boxplot for these data.
d) What can you see, if anything, in the histogram that isn't clear in the boxplot?

34. Car discounts, part 2. Use the data set of Exercise 32 to answer the following questions.

a) Find the five-number summary for these data.
b) Create a boxplot for these data.
c) What can you see, if anything, in the histogram of Exercise 32 that isn't clear in the boxplot?

*35. Vineyards. The data set provided contains the data from Exercises 10 and 15. Create a stem-and-leaf display of the sizes of the vineyards in acres. Point out any unusual features of the data that you can see from the stem-and-leaf.

*36. Gas prices 2015, again. Create a stem-and-leaf display of the data on gas prices from Exercise 30. Point out any unusual features of the data that you can see from the stem-and-leaf.

37. Gretzky. During his 20 seasons in the National Hockey League, Wayne Gretzky scored 50% more points than anyone else who ever played professional hockey. He accomplished this amazing feat while playing in 280 fewer games than Gordie Howe, the previous record holder.

Here are the number of games Gretzky played during each season:

79, 80, 80, 80, 74, 80, 80, 79, 64, 78, 73, 78, 74, 45, 81, 48, 80, 82, 82, 70

*a) Create a stem-and-leaf display.
b) Sketch a boxplot.
c) Briefly describe this distribution.
d) What unusual features do you see in this distribution? What might explain this?

38. McGwire. In his 16-year career as a player in major league baseball, Mark McGwire hit 583 home runs, placing him eighth on the all-time home run list (as of 2008). Here are the number of home runs that McGwire hit for each year from 1986 through 2001:

3, 49, 32, 33, 39, 22, 42, 9, 9, 39, 52, 58, 70, 65, 32, 29

*a) Create a stem-and-leaf display.
b) Sketch a boxplot.
c) Briefly describe this distribution.
d) What unusual features do you see in this distribution? What might explain this?

39. Gretzky returns. Look once more at data of hockey games played each season by Wayne Gretzky, seen in Exercise 37.

a) Would you use the mean or the median to summarize the center of this distribution? Why?
b) Without actually finding the mean, would you expect it to be lower or higher than the median? Explain.
c) A student was asked to make a histogram of the data in Exercise 33 and produced the following. Comment.

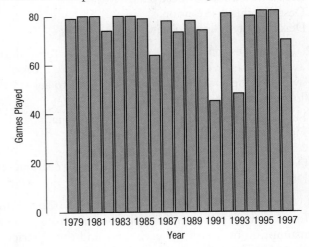

40. McGwire, again. Look once more at data of home runs hit by Mark McGwire during his 16-year career as seen in Exercise 38.

a) Would you use the mean or the median to summarize the center of this distribution? Why?
b) Find the median.

c) Without actually finding the mean, would you expect it to be lower or higher than the median? Explain.

d) A student was asked to make a histogram of the data in Exercise 38 and produced the following. Comment.

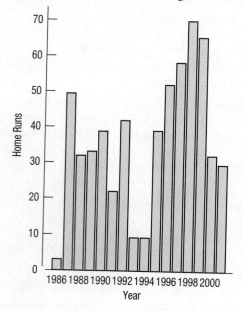

Alabama	239.34	Montana	222.39
Alaska	160.01	Nebraska	200.78
Arizona	177.87	Nevada	176.03
Arkansas	208.68	New Hampshire	237.93
California	169.12	New Jersey	199.89
Colorado	181.35	New Mexico	200.95
Connecticut	178.46	New York	121.30
Delaware	206.98	North Carolina	194.76
Dist. of Col.	66.92	North Dakota	273.81
Florida	188.48	Ohio	190.80
Georgia	216.57	Oklahoma	214.61
Hawaii	142.43	Oregon	165.65
Idaho	190.00	Pennsylvania	174.19
Illinois	159.92	Rhode Island	154.09
Indiana	202.23	South Carolina	251.14
Iowa	221.81	South Dakota	230.30
Kansas	197.61	Tennessee	216.68
Kentucky	216.21	Texas	212.65
Louisiana	218.55	Utah	166.95
Maine	247.26	Vermont	225.56
Maryland	212.07	Virginia	211.72
Massachusetts	183.23	Washington	172.81
Michigan	207.37	West Virginia	190.94
Minnesota	203.70	Wisconsin	189.88
Mississippi	240.01	Wyoming	264.17
Missouri	226.93		

T **41. Pizza prices.** The weekly prices of one brand of frozen pizza over a three-year period in Dallas are provided in the data file. Use the price data to answer the following questions.

a) Find the five-number summary for these data.
b) Find the range and IQR for these data.
c) Create a boxplot for these data.
d) Describe this distribution.
e) Describe any unusual observations.

T **42. Pizza prices, part 2.** The weekly prices of one brand of frozen pizza over a three-year period in Chicago are provided in the data file. Use the price data to answer the following questions.

a) Find the five-number summary for these data.
b) Find the range and IQR for these data.
c) Create a boxplot for these data.
d) Describe the shape (center and spread) of this distribution.
e) Describe any unusual observations.

T **43. Gasoline usage 2013.** The U.S. Energy Information Administration (EIA) collects data on the total energy consumption by sector for each state and the District of Columbia. The following data show the per capita consumption in the year 2013 used for transportation (converted to gallons of gasoline per person). Write a report on the gasoline usage by states in the year 2013, being sure to include appropriate graphical displays and summary statistics.

T **44. OECD 2013.** Established in Paris in 1961, the Organisation for Economic Co-operation and Development (OECD) (www.oecd.org) collects information on many economic and social aspects of countries around the world. Here are the 2013 GDP growth rates (in percentages) of 34 industrialized countries. Write a brief report on the 2013 GDP growth rates of these countries being sure to include appropriate graphical displays and summary statistics.

Country	2013 GDP Growth Rate	Country	2013 GDP Growth Rate
Australia	1.42	Luxembourg	1.003
Austria	0.732	Mexico	−0.757
Belgium	0.334	Netherlands	0.85
Canada	1.097	New Zealand	−0.892
Chile	3.484	Norway	0.043
Czech Republic	−0.351	Poland	1.923
Denmark	−0.071	Portugal	0.956
Estonia	1.446	Russia	1.344
Finland	0.736	Slovak Republic	3.225
France	0.474	Slovenia	−0.351
Germany	0.375	Spain	1.704
Greece	0.432	Sweden	0.985
Hungary	1.186	Switzerland	1.752
Iceland	0.271	Turkey	2.463
Israel	1.118	United Kingdom	−0.056
Italy	0.051	United States	0.405
Japan	1.444		
Korea	5.454		

T 45. Golf courses. A start-up company is planning to build a new golf course. For marketing purposes, the company would like to be able to advertise the new course as one of the more difficult courses in the state of Vermont. One measure of the difficulty of a golf course is its length: the total distance (in yards) from tee to hole for all 18 holes. Here are the histogram and summary statistics for the lengths of all the golf courses in Vermont.

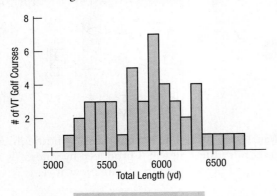

Count	45
Mean	5892.91 yd
StdDev	386.59
Min	5185
Q1	5585.75
Median	5928
Q3	6131
Max	6796

a) What is the range of these lengths?
b) Between what lengths do the central 50% of these courses lie?
c) What summary statistics would you use to describe these data?
d) Write a brief description of these data (shape, center, and spread).

46. Real estate. A real estate agent has surveyed houses in 20 nearby ZIP codes in an attempt to put together a comparison for a new property that she would like to put on the market. She knows that the size of the living area of a house is a strong factor in the price, and she'd like to market this house as being one of the biggest in the area. Here is a histogram and summary statistics for the sizes of all the houses in the area.

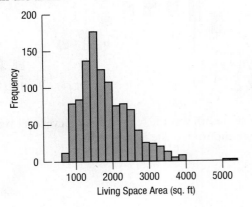

Count	1057
Mean	1819.498 sq. ft
StdDev	662.9414
Min	672
Q1	1342
Median	1675
Q3	2223
Max	5228
Missing	0

a) What is the range of these sizes?
b) Between what sizes do the central 50% of these houses lie?
c) What summary statistics would you use to describe these data?
d) Write a brief description of these data (shape, center, and spread).

T 47. Food sales. Sales (in $) for one week were collected for 18 stores in a food store chain in the northeastern United States. The stores and the towns they are located in vary in size.

a) Make a suitable display of the sales from the data provided.
b) Summarize the central value for sales for this week with a median and mean. Why do they differ?
c) Given what you know about the distribution, which of these measures does the better job of summarizing the stores' sales? Why?
d) Summarize the spread of the sales distribution with a standard deviation and with an IQR.
e) Given what you know about the distribution, which of these measures does the better job of summarizing the spread of stores' sales? Why?
f) If we were to remove the outliers from the data, how would you expect the mean, median, standard deviation, and IQR to change?

T 48. Insurance profits. Insurance companies don't know whether a policy they've written is profitable until the policy matures (expires). To see how they've performed recently, an analyst looked at mature policies and investigated the net profit to the company (in $).

a) Make a suitable display of the profits from the data provided.
b) Summarize the central value for the profits with a median and mean. Why do they differ?
c) Given what you know about the distribution, which of these measures might do a better job of summarizing the company's profits? Why?
d) Summarize the spread of the profit distribution with a standard deviation and with an IQR.
e) Given what you know about the distribution, which of these measures might do a better job of summarizing the spread in the company's profits? Why?

f) If we were to remove the outliers from the data, how would you expect the mean, median, standard deviation, and IQR to change?

49. iPod failures. In the early days of the iPod, MacInTouch (www.macintouch.com/reliability/ipodfailures.html) surveyed readers about reliability. Of the 8926 iPods owned at that time, 7510 were problem-free while the other 1416 failed. From the data, compute the failure rate for each of the 17 iPod models. Produce an appropriate graphical display of the failure rates and briefly describe the distribution. (To calculate the failure rate, divide the number failed by the sum of the number failed and the number OK for each model and then multiply by 100.)

50. Unemployment 2014. The data set provided contains 2014 (3rd quarter) unemployment rates for 33 developed countries (www.oecd.org). Produce an appropriate graphical display and briefly describe the distribution of unemployment rates. Report and comment on any outliers you may see.

51. Gas prices 2012. A driver has recorded and posted on the Internet (www.randomuseless.info/gasprice/gasprice .html) the price he paid for gasoline at every purchase from 1979 to 2012. Since 1984 all purchases were self-serve and all were for premium (92-93 octane) gas. He has also standardized the prices to April 1979 dollars. Here are boxplots for 2003, 2006, 2009, and 2012:

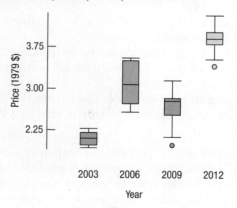

a) Compare the distribution of prices over the four years.
b) In which year were the prices least stable (most volatile)? Explain.

52. Fuel economy. American automobile companies are becoming more motivated to improve the fuel efficiency of the automobiles they produce. It is well known that fuel efficiency is impacted by many characteristics of the car. Describe what these boxplots tell you about the relationship between the number of cylinders a car's engine has and the car's fuel economy (mpg).

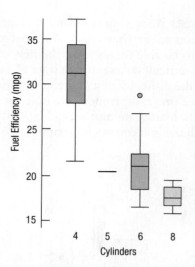

53. Wine prices 2013. The boxplots display bottle prices (in dollars) of dry Riesling wines produced by vineyards along three of the Finger Lakes in upstate New York.

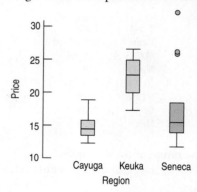

a) Which lake region produced the most expensive wine?
b) Which lake region produced the cheapest wine?
c) In which region were the wines generally more expensive?
d) Write a few sentences describing these prices.

54. Ozone. Historic ozone levels (in parts per billion, ppb) were recorded at sites in New Jersey monthly. Here are boxplots of the data for each month (over 46 years) lined up in order (January = 1).

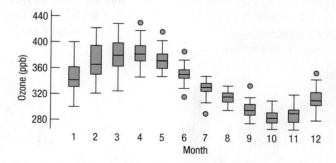

a) In what month was the highest ozone level ever recorded?
b) Which month has the largest IQR?

c) Which month has the smallest range?

d) Write a brief comparison of the ozone levels in January and June.

e) Write a report on the annual patterns you see in the ozone levels.

55. Derby speeds. How fast do horses run? Kentucky Derby winners top 30 mph, as shown in the graph. This graph shows the percentage of Kentucky Derby winners that have run *slower* than a given speed. Note that few have won running less than 33 mph, but about 85% of the winning horses have run less than 37 mph. (A cumulative frequency graph like this is called an **ogive**.)

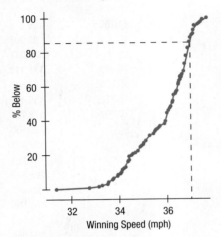

a) Estimate the median winning speed.

b) Estimate the quartiles.

c) Estimate the range and the IQR.

d) Create a boxplot of these speeds.

e) Write a few sentences about the speeds of the Kentucky Derby winners.

56. Mutual funds, historical. Here is an ogive of the distribution of monthly returns for a group of aggressive (or high growth) mutual funds over a period of 25 years. (Recall from Exercise 55 that an ogive, or cumulative relative frequency graph, shows the percent of cases at or below a certain value. Thus this graph always begins at 0% and ends at 100%.)

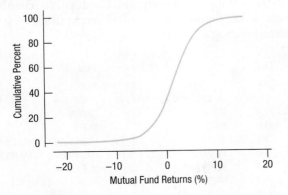

a) Estimate the median.

b) Estimate the quartiles.

c) Estimate the range and the IQR.

57. Test scores. Three Statistics classes all took the same test. Here are histograms of the scores for each class.

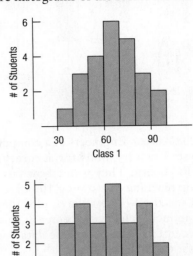

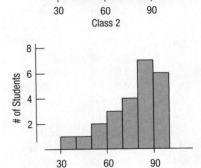

a) Which class had the highest mean score?

b) Which class had the highest median score?

c) For which class are the mean and median most different? Which is higher? Why?

d) Which class had the smallest standard deviation?

e) Which class had the smallest IQR?

58. Test scores, again. Look again at the histograms of test scores for the three Statistics classes in Exercise 57.

a) Overall, which class do you think performed better on the test? Why?

b) How would you describe the shape of each distribution?

c) Match each class with the corresponding boxplot.

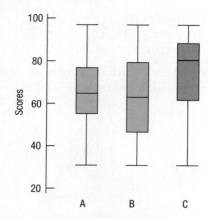

59. Quality control holes. Engineers at a computer production plant tested two methods for accuracy in drilling holes into a PC board. They tested how fast they could set the drilling machine by running 10 boards at each of two different speeds. To assess the results, they measured the distance (in inches) from the center of a target on the board to the center of the hole. The data and summary statistics are shown in the table.

	Fast	Slow
	0.000101	0.000098
	0.000102	0.000096
	0.000100	0.000097
	0.000102	0.000095
	0.000101	0.000094
	0.000103	0.000098
	0.000104	0.000096
	0.000102	0.975600
	0.000102	0.000097
	0.000100	0.000096
Mean	0.000102	0.097647
StdDev	0.000001	0.308481

Write a report summarizing the findings of the experiment. Include appropriate visual and verbal displays of the distributions, and make a recommendation to the engineers if they are most interested in the accuracy of the method.

60. Fire sale. A real estate agent notices that houses with fireplaces often fetch a premium in the market and wants to assess the difference in sales price of 60 homes that recently sold. The data and summary are shown in the table.

No Fireplace	Fireplace
142,212	134,865
206,512	118,007
50,709	138,297
108,794	129,470
68,353	309,808
123,266	157,946
80,248	173,723

(continued)

	No Fireplace	Fireplace
	135,708	140,510
	122,221	151,917
	128,440	235,105,000
	221,925	259,999
	65,325	211,517
	87,588	102,068
	88,207	115,659
	148,246	145,583
	205,073	116,289
	185,323	238,792
	71,904	310,696
	199,684	139,079
	81,762	109,578
	45,004	89,893
	62,105	132,311
	79,893	131,411
	88,770	158,863
	115,312	130,490
	118,952	178,767
		82,556
		122,221
		84,291
		206,512
		105,363
		103,508
		157,513
		103,861
Mean	116,597.54	7,061,657.74
Median	112,053	136,581

Write a report summarizing the findings of the investigation. Include appropriate visual and verbal displays of the distributions, and make a recommendation to the agent about the average premium that a fireplace is worth in this market.

61. Customer database. A philanthropic organization has a database of millions of donors that they contact by mail to raise money for charities. One of the variables in the database, *Title*, contains the title of the person or persons printed on the address label. The most common are Mr., Ms., Miss, and Mrs., but there are also Ambassador and Mrs., Your Imperial Majesty, and Cardinal, to name a few others. In all, there are over 100 different titles, each with a corresponding numeric code.

Code	Title	Code	Title
000	MR.	127	PRINCESS
001	MRS.	128	CHIEF
1002	MR. and MRS.	129	BARON
003	MISS	130	SHEIK
004	DR.	131	PRINCE AND PRINCESS
005	MADAME	132	YOUR IMPERIAL MAJESTY
006	SERGEANT	135	M. ET MME.
009	RABBI	210	PROF.
010	PROFESSOR	:	:
126	PRINCE		

An intern who was asked to analyze the organization's fundraising efforts presented these summary statistics for the variable *Title*.

Mean	54.41
StdDev	957.62
Median	1
IQR	2
n	94,649

a) What does the mean of 54.41 mean?
b) What are the typical reasons that cause measures of center and spread to be as different as those in this table?
c) Is that why these are so different?

62. CEOs. For each CEO, a code is listed that corresponds to the industry of the CEO's company. Here are a few of the codes and the industries to which they correspond:

Industry	Industry Code	Industry	Industry Code
Financial services	1	Energy	12
Food/drink/tobacco	2	Capital goods	14
Health	3	Computers/communications	16
Insurance	4	Entertainment/information	17
Retailing	6	Consumer non-durables	18
Forest products	9	Electric utilities	19
Aerospace/defense	11		

A recently hired investment analyst has been assigned to examine the industries and the compensations of the CEOs. To start the analysis, he produces the following histogram of industry codes.

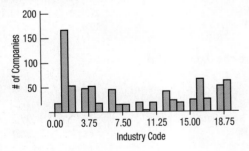

a) What might account for the gaps seen in the histogram?
b) What advice might you give the analyst about the appropriateness of this display?

T 63. Exchange-traded funds—by type. The 98 mutual funds of Exercise 31 are classified into five types: Health, Large Growth, Technology, Trading-Leveraged Equity, and Other Funds. Compare the 12-month returns of the five types of funds using an appropriate display and write a brief summary of the differences.

T 64. Car discounts, part 3. The discounts negotiated by the car buyers in Exercise 32 are classified by whether the buyer was Male (code = 0) or Female (code = 1). Compare the discounts of men vs. women using an appropriate display and write a brief summary of the differences.

65. Houses for sale. Each house listed on the multiple listing service (MLS) is assigned a sequential ID number. A recently hired real estate agent decided to examine the MLS numbers in a recent random sample of homes for sale by one real estate agency in nearby towns. To begin the analysis, the agent produces the following histogram of ID numbers.

a) What might account for the distribution seen in the histogram?
b) What advice might you give the analyst about the appropriateness of this display?

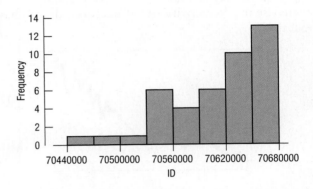

66. ZIP codes. Holes-R-Us, an Internet company that sells piercing jewelry, keeps transaction records on its sales. At a recent sales meeting, one of the staff presented the following histogram and summary statistics of the ZIP codes of the last 500 customers, so that the staff might understand where sales are coming from. Comment on the usefulness and appropriateness of this display.

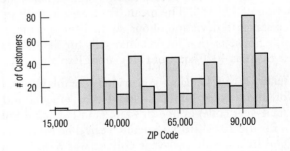

T 67. Hurricanes. Buying insurance for property loss from hurricanes has become increasingly difficult since Hurricane Katrina caused record property loss damage. Many companies have refused to renew policies or write new ones. The data set provided contains the total number of hurricanes by every full decade from 1851 to 2000 (from the National Hurricane Center). Some scientists claim that there has been an increase in the number of hurricanes in recent years.

a) Create a histogram of these data.
b) Describe the distribution.
c) Create a time series plot of these data.
d) Discuss the time series plot. Does this graph support the claim of these scientists, at least up to the year 2000?

68. Hurricanes, part 2. Using the hurricanes data set, examine the number of major hurricanes (category 3, 4, or 5) by every full decade from 1851 to 2000.

a) Create a histogram of these data.
b) Describe the distribution.
c) Create a timeplot of these data.
d) Discuss the timeplot. Does this graph support the claim of scientists that the number of major hurricanes has been increasing (at least up through the year 2000)?

69. Productivity study. The National Center for Productivity releases information on the efficiency of workers. In a recent report, they included the following graph showing a rapid rise in productivity. What questions do you have about this?

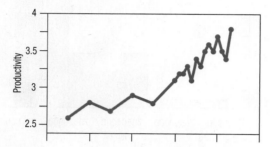

70. Productivity study revisited. A second report by the National Center for Productivity analyzed the relationship between productivity and wages. They used the graph from Exercise 69, with the x-axis labeled "wages". Comment on any problems you see with their analysis.

71. Real estate, part 2. The 1057 houses described in Exercise 46 have a mean price of $167,900, with a standard deviation of $77,158. The mean living area is 1819 sq. ft., with a standard deviation of 663 sq. ft. Which is more unusual, a house in that market that sells for $400,000 or a house that has 4000 sq. ft of living area? Explain.

72. Tuition 2015. The data set provided contains the average tuition and fees for public 2- and 4-year colleges and for private 4-year colleges in each state (and D.C. and Puerto Rico) for the 2014–2015 academic year. The mean tuition charged by a public two-year college was $3825, with a standard deviation of $1113. For public four-year colleges the mean was $8851, with a standard deviation of $2469. Which would be more unusual: a state whose average public two-year college is $1600 or a state whose average public four-year college tuition was $4200? Explain.

73. Food consumption. FAOSTAT, the Food and Agriculture Organization of the United Nations, collects information on the production and consumption of more than 200 food and agricultural products for 200 countries around the world. Here are two tables, one for meat consumption (per capita in kg per year) and one for alcohol consumption (per capita in gallons per year). The United States

leads in meat consumption with 267.30 pounds, while Ireland is the largest alcohol consumer at 55.80 gallons. Using z-scores, find which of these two countries is the larger consumer of both meat and alcohol together.

Country	Alcohol	Meat	Country	Alcohol	Meat
Australia	29.56	242.22	Luxembourg	34.32	197.34
Austria	40.46	242.22	Mexico	13.52	126.50
Belgium	34.32	197.34	Netherlands	23.87	201.08
Canada	26.62	219.56	New Zealand	25.22	228.58
Czech Republic	43.81	166.98	Norway	17.58	129.80
Denmark	40.59	256.96	Poland	20.70	155.10
Finland	25.01	146.08	Portugal	33.02	194.92
France	24.88	225.28	Slovakia	26.49	121.88
Germany	37.44	182.82	South Korea	17.60	93.06
Greece	17.68	201.30	Spain	28.05	259.82
Hungary	29.25	179.52	Sweden	20.07	155.32
Iceland	15.94	178.20	Switzerland	25.32	159.72
Ireland	55.80	194.26	Turkey	3.28	42.68
Italy	21.68	200.64	United Kingdom	30.32	171.16
Japan	14.59	93.28	United States	26.36	267.30

74. World Bank. The World Bank, through their Doing Business project (www.doingbusiness.org), ranks nearly 200 economies on the ease of doing business. One of their rankings measures the ease of starting a business and is made up (in part) of the following variables: number of required start-up procedures, average start-up time (in days), and average start-up cost (in % of per capita income). The following table gives the mean and standard deviations of these variables for 95 economies.

	Procedures (#)	Time (Days)	Cost (%)
Mean	7.9	27.9	14.2
SD	2.9	19.6	12.9

Here are the data for three countries.

	Procedures	Time	Cost
Spain	10	47	15.1
Guatemala	11	26	47.3
Fiji	8	46	25.3

a) Use z-scores to combine the three measures.
b) Which country has the best environment after combining the three measures? Be careful—a lower rank indicates a better environment to start up a business.

75. Regular gas 2015. The data set provided contains U.S. regular retail gasoline prices (cents/gallon) from August 20, 1990, to January 19, 2015, from a national sample of gasoline stations obtained from the U.S. Department of Energy.

a) Create a histogram of the data and describe the distribution.
b) Create a time series plot of the data and describe the trend.

c) Which graphical display seems the more appropriate for these data? Explain.

76. Home price index 2014. Standard and Poor's Case-Shiller® Home Price Index measures the residential housing market in metropolitan regions across the United States. The national index, Composite.10, is a composite of 10 regions, and can be found in the data set provided for the years 1987 to 2014.

a) Create a histogram of the data and describe the distribution.
b) Create a time series plot of the data and describe the trend.
c) Which graphical display seems the more appropriate for these data? Explain.

*77. Unemployment rate 2014. The histogram shows the monthly U.S. unemployment rate from January 2003 to December 2014 (data.bls.gov/timeseries/LNS14000000).

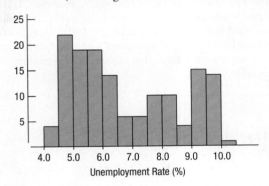

Here is the time series plot for the same data.

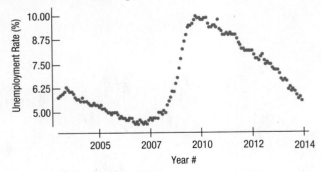

a) What features of the data can you see in the histogram that aren't clear in the time series plot?
b) What features of the data can you see in the time series plot that aren't clear in the histogram?
c) Which graphical display seems the more appropriate for these data? Explain.
d) Write a brief description of unemployment rates over this time period in the United States.

*78. Consumer Price Index (CPI) 2014. Here is a histogram of the monthly CPI as reported by the Bureau of Labor Statistics (ftp://ftp.bls.gov/pub/special.requests/cpi/cpiai.txt) from 2010 through 2014:

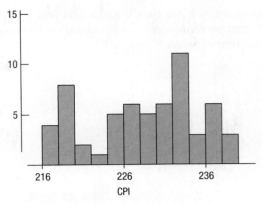

Here is the time series plot for the same data.

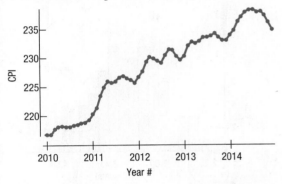

a) What features of the data can you see in the histogram that aren't clear from the time series plot?
b) What features of the data can you see in the time series plot that aren't clear in the histogram?
c) Which graphical display seems the more appropriate for these data? Explain.
d) Write a brief description of monthly CPI over this time period.

79. Assets. Here is a histogram of the assets (in millions of dollars) of 79 companies chosen from the *Forbes* list of the nation's top corporations.

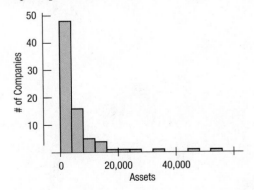

a) What aspect of this distribution makes it difficult to summarize, or to discuss, center and spread?
b) What would you suggest doing with these data if we want to understand them better?

80. Assets, again. Here are the same data you saw in Exercise 79 after re-expressions as the square root of assets and the logarithm of assets.

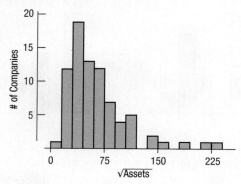

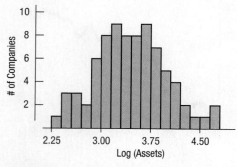

a) Which re-expression do you prefer? Why?
b) In the square root re-expression, what does the value 50 actually indicate about the company's assets?

JUST CHECKING ANSWERS

1 Incomes are probably skewed to the right and not symmetric, making the median the more appropriate measure of center. The mean will be influenced by the high end of family incomes and not reflect the "typical" family income as well as the median would. It will give the impression that the typical income is higher than it is.

2 An IQR of 30 mpg would mean that only 50% of the cars get gas mileages in an interval 30 mpg wide. Fuel economy doesn't vary that much. 3 mpg is reasonable. It seems plausible that 50% of the cars will be within about 3 mpg of each other. An IQR of 0.3 mpg would mean that the gas mileage of half the cars varies little from the estimate. It's unlikely that cars, drivers, and driving conditions are that consistent.

3 We'd prefer a standard deviation of 2 months. Making a consistent product is important for quality. Customers want to be able to count on the MP3 player lasting somewhere close to 5 years, and a standard deviation of 2 years would mean that life spans were highly variable.

4

Correlation and Linear Regression

Amazon.com

Amazon.com opened for business in July 1995, billing itself even then as "Earth's Biggest Bookstore," with an unusual business plan: They didn't plan to turn a profit for four to five years. Although some shareholders complained when the dotcom bubble burst, Amazon continued its slow, steady growth, becoming profitable for the first time in 2002. Since then, Amazon has remained profitable and has continued to grow. By 2011, sales had topped $48 billion, of which 44% were international sales. In 2012 Amazon was ranked the 20th most valuable brand by *Business Week*. Amazon's selection of merchandise has expanded to include almost anything you can imagine, from $400,000 necklaces, to yak cheese from Tibet, to the largest book in the world.

Amazon R&D is constantly monitoring and evolving their website to best serve their customers and maximize their sales performance. To make changes to the site, they experiment by collecting data and analyzing what works best. As Ronny Kohavi, former director of Data Mining and Personalization, said, "Data trumps intuition. Instead of using our intuition, we experiment on the live site and let our customers tell us what works for them."

WHO	Books sold by Amazon
WHAT	List Price and Weight
UNITS	$ and Ounces
WHEN	2012
WHERE	Online
WHY	Originally collected as a class project

E ven with the rapid growth of e-books (about 15% growth per year), sales of traditional ink on paper books have held their own. Of course, one difference between e-books and print books is that a print book has weight, and that can help predict its price. The weight can account for the materials in the book's manufacture as well as other related costs. Amazon makes a variety of facts available for most books it sells, including the shipping weight (in ounces), the number of pages, the dimensions, and of course, the price.

Figure 4.1 shows a plot of *List Price* ($) against (shipping) *Weight* (oz) for 307 books selected[1] from Amazon's offerings. Clearly price and weight are related. If you were asked to summarize the relationship, what would you say (Data in **Amazon**.)?

Figure 4.1 *List Price* ($) vs. *Weight* (oz) for 307 books sold by Amazon

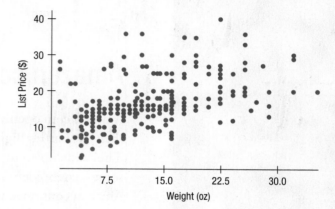

There is an overall trend; heavier books tend to cost more. But the relationship is far from perfect. This plot is an example of a **scatterplot**. It plots one quantitative variable against another. Just by looking at a scatterplot, you can see patterns, trends, relationships, and the occasional unusual cases standing apart from the general pattern. Scatterplots are the best way to start observing the relationship between two *quantitative* variables.

Relationships between variables are often at the heart of what we'd like to learn from data.

- Is consumer confidence related to oil prices?
- Are customers who consult online help sites as satisfied with a company's customer relations as those who speak with human customer support specialists?
- Is an increase in money spent on advertising related to sales?
- What is the relationship between a stock's sales volume and its price?

Questions such as these relate two quantitative variables and ask whether there is an **association** between them. Scatterplots are the ideal way to picture such associations.

4.1   Looking at Scatterplots

The Texas Transportation Institute, which studies the mobility provided by the nation's transportation system, issues an annual report on traffic congestion and its costs to society and business. Figure 4.2 shows a scatterplot of the annual *Congestion Cost per Person* of traffic delays (in dollars) in 65 cities in the United States against the *Peak Period Freeway Speed* (mph). (Data in **Congestion Cost**.)

---

[1] The books were selected by "walking" randomly through Amazon's offerings, starting at a variety of haphazardly selected books and then selecting a randomly chosen book from the "Customers who bought this item also bought" list, and continuing the randomly generated thread. Although it is not a random sample, for our purposes it can be regarded as a representative collection of books.

WHO	Cities in the United States
WHAT	*Congestion Cost per Person* and *Peak Period Freeway Speed*
UNITS	*Congestion Cost per Person* ($ per person per year); *Peak Period Freeway Speed* (mph)
WHEN	2000
WHERE	United States
WHY	To examine the relationship between congestion on the highways and its impact on society and business

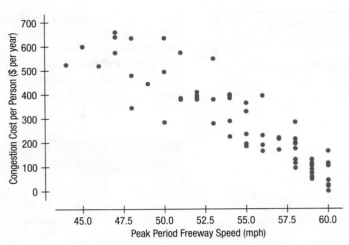

Figure 4.2 *Congestion Cost per Person* ($ per year) of traffic delays against *Peak Period Freeway Speed* (mph) for 65 U.S. cities.

Everyone looks at scatterplots. But many people would find it hard to say what to look for in a scatterplot. What do *you* see? Try to describe the scatterplot of *Congestion Cost* against *Freeway Speed*.

You might say that the **direction** of the association is important. As the peak freeway speed goes up, the cost of congestion goes down. A pattern that runs from

> Look for **Direction**: What's the sign—positive, negative, or neither?

the upper left to the lower right   is said to be **negative**. A pattern running

the other way   , as we saw for the price and weight of books, is called **positive**.

The second thing to look for in a scatterplot is its **form**. If there is a straight line relationship, it will appear as a cloud or swarm of points stretched out in a generally consistent, straight form. For example, the scatterplot of traffic congestion has an underlying **linear** form, although some points stray away from it.

Scatterplots can reveal many different kinds of patterns. Often they will not be straight, but straight line patterns are both the most common and the most useful for statistics.

If the relationship isn't straight, but curves gently, while still increasing or

decreasing steadily,   we can often find ways to straighten it out. But if it

> Look for **Form**: Straight, curved, something exotic, or no pattern?

curves sharply—up and then down, for example,   —then we'd need more advanced methods.

The third feature to look for in a scatterplot is the **strength** of the relationship.

At one extreme, do the points appear tightly clustered in a single stream (whether straight, curved, or bending all over the place)? Or, at the other extreme, do the points seem to be so variable and spread out that we can barely

> Look for **Strength**: How much scatter?

discern any trend or pattern?   The traffic congestion plot shows moderate scatter around a generally straight form. That indicates that there's a moderately strong linear relationship between cost and speed.

Finally, always look for the unexpected. Often the most interesting discovery in a scatterplot is something you never thought to look for. One example of such

> Look for **Unusual Features**:
> Are there unusual observations or subgroups?

a surprise is an unusual observation, or **outlier**, standing away from the overall pattern of the scatterplot. Such a point is almost always interesting and deserves special attention. You may see entire clusters or subgroups that stand away or show a trend in a different direction than the rest of the plot. That should raise questions about why they are different. They may be a clue that you should split the data into subgroups instead of looking at them all together.

FOR EXAMPLE   Creating a scatterplot

The first automobile crash in the United States occurred in New York City in 1896, when a motor vehicle collided with a "pedalcycle" rider. Cycle/car accidents are a serious concern for insurance companies. About 53,000 cyclists have died in traffic crashes in the United States since 1932. Demographic information such as this is often available from government agencies. It can be useful to insurers, who use it to set appropriate rates, and to retailers, who must plan what safety equipment to stock and how to present it to their customers. This becomes a more pressing concern when the demographic profiles change over time.

Here's data on the mean age of cyclists killed each year during the decade from 1998 to 2012. (Source: National Highway Transportation Safety Agency, found at www-nrd.nhtsa.dot.gov/Pubs/812151.pdf) (Data in **Cyclist 2012**.)

Year	Mean Age
1998	32
1999	33
2000	35
2001	36
2002	37
2003	36
2004	39
2005	39
2006	41
2007	40
2008	41
2009	41
2010	42
2011	43
2012	43

QUESTION   Make a scatterplot and summarize what it says.

ANSWER   The mean age of cyclist traffic deaths has been increasing almost linearly during this period. The trend is a strong one.

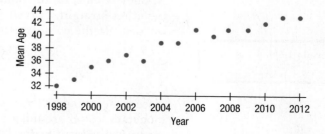

4.2 Assigning Roles to Variables in Scatterplots

Descartes was a philosopher and mathematician, famous for his statement *cogito, ergo sum*: I think, therefore I am.

Scatterplots were among the first modern mathematical displays. The idea of using two axes at right angles to define a field on which to display values can be traced back to René Descartes (1596–1650), and the playing field he defined in this way is formally called a *Cartesian plane*, in his honor.

The two axes Descartes specified characterize the scatterplot. The axis that runs up and down is, by convention, called the y-axis, and the one that runs from side to side is called the x-axis. These terms are standard.[2]

To make a scatterplot of two quantitative variables, assign one to the y-axis and the other to the x-axis. As with any graph, be sure to label the axes clearly, and indicate the scales of the axes with numbers. Scatterplots display *quantitative* variables. Each variable has units, and these should appear with the display—usually near each axis. Each point is placed on a scatterplot at a position that corresponds to values of these two variables. Its horizontal location is specified by its x-value, and its vertical location is specified by its y-value variable. Together, these are known as *coordinates* and written (x, y).

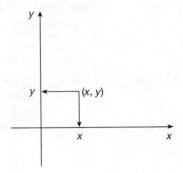

Scatterplots made by computer programs (such as the two we've seen in this chapter) often do not—and usually should not—show the *origin*, the point at $x = 0$, $y = 0$ where the axes meet. If both variables have values near or on both sides of zero, then the origin will be part of the display. If the values are far from zero, though, there's no reason to include the origin. In fact, it's far better to focus on the part of the Cartesian plane that contains the data. In our example about books, none of the books were free and all weighed something, so the computer drew the scatterplot in Figure 4.1 with axes that don't quite meet.

Which variable should go on the x-axis and which on the y-axis? What we want to know about the relationship can tell us how to make the plot. Amazon may have questions such as:

• How are prices related to sales volume?
• Are increased sales at Amazon reflected in its stock price?
• What offers will encourage shoppers to browse the Amazon site for a longer time?

In all of these examples, one variable plays the role of the **explanatory** or **predictor variable**, while the other takes on the role of the **response variable**. We place the explanatory variable on the x-axis and the response variable on the y-axis. When you make a scatterplot, you can assume that those who view it will think this way, so choose which variables to assign to which axes carefully.

---

[2] The axes are also called the "ordinate" and the "abscissa"—but we can never remember which is which because statisticians don't generally use these terms. In Statistics (and in all statistics computer programs) the axes are generally called "x" (abscissa) and "y" (ordinate) and are usually labeled with the names of the corresponding variables.

The roles that we choose for variables have more to do with how we *think* about them than with the variables themselves. Just placing a variable on the *x*-axis doesn't necessarily mean that it explains or predicts *anything*, and the variable on the *y*-axis may not respond to it in any way. The Amazon marketing department may want to predict prices leading to *Price* as the response variable in Figure 4.1. But the shipping department is likely to be more interested in predicting the weights of books, so for them, *Weight* would be a natural response variable.

The *x*- and *y*-variables are sometimes referred to as the **independent** and **dependent** variables, respectively. The idea is that the *y*-variable *depends* on the *x*-variable and the *x*-variable acts *independently* to make *y* respond. These names, however, conflict with other uses of the same terms in Statistics. Instead, we'll sometimes use the terms "explanatory" or "predictor variable" and "response variable" when we're discussing roles, but we'll often just say *x-variable* and *y-variable*.

> **FOR EXAMPLE**   Assigning roles to variables
>
> **QUESTION**   When examining the ages of victims in cycle/car accidents, why does it make the most sense to plot *year* on the *x*-axis and *mean age* on the *y*-axis? (See the example on page 100.)
>
> **ANSWER**   We are interested in how the age of accident victims might change over time, so we think of the year as the basis for prediction and the mean age of victims as the variable that is predicted.

4.3   Understanding Correlation

If you had to put a number (say, between 0 and 1) on the strength of the linear association between book prices and weights in Figure 4.1, what would it be? Your measure shouldn't depend on the choice of units for the variables. After all, Amazon sells books in euros as well as in dollars and book weights can be recorded in grams rather than ounces, but regardless of the units, the scatterplot would look the same. When we change units, the direction, form, and strength won't change, so neither should our measure of the association's (linear) strength.

We saw a way to remove the units in the previous chapter. We can standardize each of the variables, finding $z_x = \left(\dfrac{x - \bar{x}}{s_x}\right)$ and $z_y = \left(\dfrac{y - \bar{y}}{s_y}\right)$. With these, we can compute a measure of strength that you've probably heard of—the **correlation coefficient**:

$$r = \frac{\sum z_x z_y}{n - 1}.$$

Keep in mind that the *x*'s and *y*'s are paired. For each book we have a price and a weight. To find the correlation we multiply each standardized value by the standardized value it is paired with and add up those *cross products*. We divide the total by the number of pairs minus one, $n - 1$.[3]

---

[3] The same $n - 1$ we used for calculating the standard deviation.

There are alternative formulas for the correlation in terms of the variables x and y. Here are two of the more common:

$$r = \frac{\sum(x - \bar{x})(y - \bar{y})}{\sqrt{\sum(x - \bar{x})^2 \sum(y - \bar{y})^2}} = \frac{\sum(x - \bar{x})(y - \bar{y})}{(n - 1)s_x s_y}.$$

These formulas can be more convenient for calculating correlation by hand, but the form using z-scores is best for understanding what correlation means. No matter which formula you use, the correlation between *List Price* and *Weight* for the Amazon books is 0.498.

FOR EXAMPLE   Finding the correlation coefficient

QUESTION   What is the correlation of *mean age* and *year* for the cyclist accident data on page 100?

ANSWER   Working by hand:

$$\bar{x} = 2005, s_x = 4.47$$
$$\bar{y} = 38.47, s_y = 3.56$$

The sum of the cross product of the deviations is found as follows:

$$\sum(x - \bar{x})(y - \bar{y}) = 217$$

Putting the sum of the cross products in the numerator and $(n - 1) \times s_x \times s_y$ in the denominator, we get

$$\frac{217}{(15 - 1) \times 4.47 \times 3.56} = 0.97$$

For *mean age* and *year*, the correlation coefficient is 0.97. That indicates a strong linear association. Because this is a time series, we refer to it as a strong "trend."

Correlation Conditions

Correlation measures the strength of the *linear* association between two *quantitative* variables. Before you use correlation, you must check three *conditions*:

- **Quantitative Variables Condition:** Correlation applies only to quantitative variables. Don't apply correlation to categorical data masquerading as quantitative. Check that you know the variables' units and what they measure.
- **Linearity Condition:** Sure, you can *calculate* a correlation coefficient for any pair of variables. But correlation measures the strength only of the *linear* association and will be misleading if the relationship is not straight enough. What is "straight enough"? This question may sound too informal for a statistical condition, but that's really the point. We can't verify whether a relationship is linear or not. Very few relationships between variables are perfectly linear, even in theory, and scatterplots of real data are never perfectly straight. How nonlinear looking would the scatterplot have to be to fail the condition? This is a judgment call that you just have to think about. Do you think that the underlying relationship is curved? If so, then summarizing its strength with a correlation would be misleading.
- **Outlier Condition:** Unusual observations can distort the correlation and can make an otherwise small correlation look big or, on the other hand, hide a large correlation. It can even give an otherwise positive association a negative correlation coefficient (and vice versa). When you see an outlier, it's often a good idea to report the correlation both with and without the point.

Each of these conditions is easy to check with a scatterplot. Many correlations are reported without supporting data or plots. You should still think about the conditions. You should be cautious in interpreting (or accepting others' interpretations of) the correlation when you can't check the conditions for yourself.

Throughout this course, you'll see that doing Statistics right means selecting the proper methods. That means you have to think about the situation at hand. An important first step is to check that the type of analysis you plan is appropriate. These conditions are just the first of many such checks.

JUST CHECKING

For the years 1992 to 2002, the quarterly stock prices of the semiconductor companies Cypress and Intel have a correlation of 0.86.

1 Before drawing any conclusions from the correlation, what would you like to see? Why?

2 If your coworker tracks the same prices in euros, how will this change the correlation? Will you need to know the exchange rate between euros and U.S. dollars to draw conclusions?

3 If you standardize both prices, how will this affect the correlation?

4 In general, if on a given day the price of Intel is relatively low, is the price of Cypress likely to be relatively low as well?

5 If on a given day the price of Intel stock is high, is the price of Cypress stock definitely high as well?

GUIDED EXAMPLE   Customer Spending

A major credit card company sends an incentive to its best customers hoping that the customers will use the card more. Market analysts wonder how often they can offer the incentive. Will repeated offerings of the incentive result in repeated increased credit card use? To examine this question, an analyst took a random sample of 184 customers from their highest use segment and investigated the charges in each of the two months following the date that customers received the incentive.

 PLAN

Setup State the objective. Identify the quantitative variables to examine. Report the time frame over which the data have been collected and define each variable. (State the W's.)

Make the scatterplot and clearly label the axes to identify the scale and units.

Our objective is to investigate the association between the amount that a customer charges in each of the two months after they receive an incentive. The customers have been randomly selected from among the highest use segment of customers. The variables measured are the total credit card charges (in $) in the two months of interest.

✓ **Quantitative Variable Condition.** Both variables are quantitative. Both charges are measured in dollars.

Because we have two quantitative variables measured on the same cases, we can make a scatterplot.

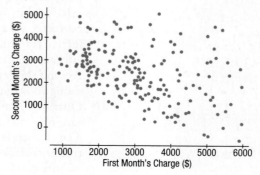

| | Check the conditions. | ✓ **Linearity Condition.** The scatterplot is straight enough. |
| | | ✓ **Outlier Condition.** There are no obvious outliers. |

| **DO** | **Mechanics** Once the conditions are satisfied, calculate the correlation with technology. | The correlation is −0.391. |
| | | The negative correlation coefficient confirms the impression from the scatterplot. |

REPORT	**Conclusion** Describe the direction, form, and the strength of the plot, along with any unusual points or features. Be sure to state your interpretation in the proper context.	**MEMO**
		Re: credit card spending
		We have examined some of the data from the incentive program. In particular, we looked at the charges made in each of the first two months of the program. We noted that there was a negative association between charges in the second month and charges in the first month. The correlation was −0.391, which is only moderately strong, and indicates substantial variation.
		We've concluded that although the observed pattern is negative, these data do not allow us to find the causes of this behavior. It is likely that some customers were encouraged by the offer to increase their spending in the first month, but then returned to former spending patterns. It is possible that others didn't change their behavior until the second month of the program, increasing their spending at that time. Without data on the customers' pre-incentive spending patterns it would be hard to say more.
		We suggest further research, and we suggest that the next trial extend for a longer period of time to help determine whether the patterns seen here persist.

Correlation Properties

Because correlation is so widely used as a measure of association it's a good idea to remember some of its basic properties. Here's a useful list of facts about the correlation coefficient:

- **The sign of a correlation coefficient gives the direction of the association.**
- **Correlation is always between −1 and +1.** Correlation *can* be exactly equal to −1.0 or +1.0, but watch out. These values are unusual in real data because they mean that all the data points fall *exactly* on a single straight line.
- **Correlation treats *x* and *y* symmetrically.** The correlation of *x* with *y* is the same as the correlation of *y* with *x*.
- **Correlation has no units.** This fact can be especially important when the data's units are somewhat vague to begin with (customer satisfaction, worker efficiency, productivity, and so on).
- **Correlation is not affected by changes in the center or scale of either variable.** Changing the units or baseline of either variable has no effect on the correlation coefficient because the correlation depends only on the *z*-scores.
- **Correlation measures the strength of the *linear* association between the two variables.** Variables can be strongly associated but still have a small correlation if the association is not linear.
- **Correlation is sensitive to unusual observations.** A single outlier can make a small correlation large or make a large one small.

How Strong Is Strong?

There's little agreement on what the terms "weak," "moderate," and "strong" mean. The same correlation might be strong in one context and weak in another. A correlation of 0.7 between an economic index and stock market prices would be exciting, but finding "only" a correlation of 0.7 between a drug dose and blood pressure might be seen as a failure by a pharmaceutical company. Use these terms cautiously and be sure to report the correlation and show a scatterplot so others can judge the strength for themselves.

Correlation Tables

Sometimes you'll see the correlations between each pair of variables in a dataset arranged in a table. The rows and columns of the table name the variables, and the cells hold the correlations.

	#Pages	Width	Thick	Pub year
#Pages	1.000			
Width	0.003	1.000		
Thick	0.813	0.074	1.000	
Pub year	0.253	0.012	0.309	1.000

Table 4.1   A correlation table for some variables collected on a sample of Amazon books.

Correlation tables are compact and give a lot of summary information at a glance. The diagonal cells of a correlation table always show correlations of exactly 1.000, and the upper half of the table is symmetrically the same as the lower half (can you see why?), so by convention, only the lower half is shown. A table like this can be an efficient way to start looking at a large dataset, but be sure to check for linearity and unusual observations or the correlations in the table may be misleading or meaningless. Can you be sure, looking at Table 4.1, that the variables are linearly associated? Correlation tables are often produced by statistical software packages. Fortunately, these same packages often offer simple ways to make all the scatterplots you need to look at.[4]

4.4   Lurking Variables and Causation

An educational researcher finds a strong association between height and reading ability among elementary school students in a nationwide survey. Taller children tend to have higher reading scores. Does that mean that students' height *causes* their reading scores to go up? No matter how strong the correlation is between two variables, there's no simple way to show from observational data that one variable causes the other. A high correlation just increases the temptation to think and to say that the *x*-variable *causes* the *y*-variable. Just to make sure, let's repeat the point again.

No matter how strong the association, no matter how large the *r* value, no matter how straight the form, there is no way to conclude from a high correlation *alone* that one variable causes the other. There's always the possibility that some third variable—a **lurking variable**—is affecting both of the variables you have observed. In the reading score example, you may have already guessed that the lurking variable is the age of the child. Older children tend to be taller and have stronger reading skills. But even when the lurking variable isn't as obvious, resist the temptation to think that a high correlation implies causation. Here's another example.

---

[4] A table of scatterplots arranged just like a correlation table is sometimes called a *scatterplot matrix*, or SPLOM, and is easily created using a statistics package.

Figure 4.3 *Life Expectancy* and number of *Doctors per Person* in 40 countries shows a fairly strong, positive, somewhat linear relationship with a correlation of 0.705.

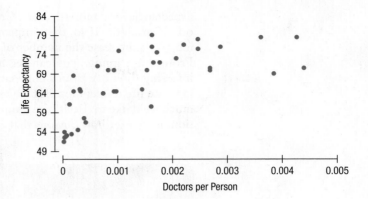

Figure 4.3 shows the *Life Expectancy* (average of men and women, in years) for each of 40 countries of the world, plotted against the number of *Doctors per Person* in each country. The strong positive association ($r = 0.705$) seems to confirm our expectation that more *Doctors per Person* improves health care, leading to longer lifetimes and a higher *Life Expectancy*. Perhaps we should send more doctors to developing countries to increase life expectancy. (Data in **Doctors and Life Expectancy**.)

If we increase the number of doctors, will the life expectancy increase? That is, would adding more doctors *cause* greater life expectancy? Could there be another explanation of the association? Figure 4.4 shows another scatterplot. *Life Expectancy* is still the response, but this time the predictor variable is not the number of doctors, but the number of *Televisions per Person* in each country. The positive association in this scatterplot looks even *stronger* than the association in the previous plot. If we wanted to calculate a correlation, we should straighten the plot first, but even from this plot, it's clear that higher life expectancies are associated with more televisions per person. Should we conclude that increasing the number of televisions extends lifetimes? If so, we should send televisions instead of doctors to developing countries. Not only is the association with life expectancy stronger, but televisions are cheaper than doctors.

Figure 4.4 *Life Expectancy* and number of *TVs per Person* shows a strong, positive (although clearly not linear) relationship.

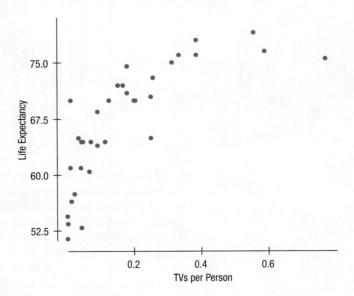

What's wrong with this reasoning? Maybe we were a bit hasty earlier when we concluded that doctors *cause* greater life expectancy. Maybe there's a lurking variable here. Countries with higher standards of living have both longer life expectancies *and* more doctors. Could higher living standards cause changes in the other variables? If so, then improving living standards might be expected to prolong lives, increase the number of doctors, and increase the number of televisions. From this example, you can see how easy it is to fall into the trap of mistakenly inferring causality from a correlation. For all we know, doctors (or televisions) *do* increase life expectancy. But we can't tell that from data like these no matter how much we'd like to. Resist the temptation to conclude that *x* causes *y* from a correlation, no matter how obvious that conclusion seems to you.

FOR EXAMPLE   Understanding causation

QUESTION   An insurance company analyst suggests that the data on ages of cyclist accident deaths are actually due to the entire population of cyclists getting older and not to a change in the safe riding habits of older cyclists (see page 100). What would we call the *mean cyclist age* if we had that variable available?

ANSWER   It would be a lurking variable. If the entire population of cyclists is aging then that would lead to the average age of cyclists in accidents increasing.

4.5   The Linear Model

> *"Statisticians, like artists, have the bad habit of falling in love with their models."*
>
> —George Box, Famous Statistician

Let's return to the relationship between Amazon book prices and weights. In Figure 4.1 (repeated here) we saw a moderate, positive, linear relationship, so we can summarize its strength with a correlation. For this relationship, the correlation is 0.498.

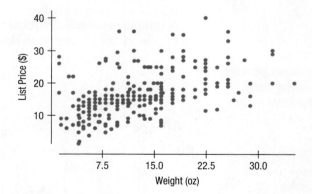

That's moderately strong, but the strength of the relationship is only part of the picture. Amazon's management might want a deeper understanding of prices for trade books to compare with prices in the rapidly growing e-book market. That's a reasonable business need but to meet it we'll need a model for the trend. The correlation says that there seems to be a linear association between the variables, but it doesn't tell us what that association is.

Of course, we can say more. We can model the relationship with a line and give the equation. Specifically, we can find a linear model to describe the relationship we saw in Figure 4.1 between *Price* and *Weight*. A **linear model** is just an equation of a straight line through the data. The points in the scatterplot don't all line up, but a straight line can summarize the general pattern with only a few parameters. This model can help us understand how the variables are associated.

NOTATION ALERT

"Putting a hat on it" is standard Statistics notation to indicate that something has been predicted by a model. Whenever you see a hat over a variable name or symbol, you can assume it is the predicted version of that variable or symbol.

Residuals

We know the model won't be perfect. No matter what line we draw, it won't go through many of the points. The best line might not even hit *any* of the points. Then how can it be the "best" line? We want to find the line that somehow comes *closer* to all the points than any other line. Some of the points will be above the line and some below. Any linear model can be written as $\hat{y} = b_0 + b_1 x$, where b_0 and b_1 are numbers estimated from the data and $\hat{y}$ (pronounced *y*-hat) is the **predicted value**. We use the *hat* to distinguish the predicted value from the observed value *y*. The difference between these two is called the **residual**:

$$e = y - \hat{y}.$$

The residual value tells us how far the model's prediction is from the observed value at that point. To find the residuals, we always subtract the predicted values from the observed ones.

Our question now is how to find the right line.

The Line of "Best Fit"

When we draw a line through a scatterplot, some residuals are positive, and some are negative. We can't assess how well the line fits by adding up all the residuals—the positive and negative ones would just cancel each other out. We need to find the line that's closest to all the points, and to do that, we need to make all the distances positive. We faced the same issue when we calculated a standard deviation to measure spread. And we deal with it the same way here: by squaring the residuals to make them positive. The sum of all the squared residuals tells us how well the line we drew fits the data—the smaller the sum, the better the fit. A different line will produce a different sum, maybe bigger, maybe smaller. The **line of best fit** is the line for which the sum of the squared residuals is smallest—often called the **least squares line**.

This line has the special property that the variation of the data around the model, as seen in the residuals, is the smallest it can be for any straight line model for these data. No other line has this property. Speaking mathematically, we say that this line minimizes the sum of the squared residuals. You might think that finding this "least squares line" would be difficult. Surprisingly, it's not, although it was an exciting mathematical discovery when Legendre published it in 1805.

4.6

Correlation and the Line

Any straight line can be written as:

$$y = b_0 + b_1 x.$$

If we were to plot all the (x, y) pairs that satisfy this equation, they'd fall exactly on a straight line. We'll use this form for our linear model. Of course, with real data, the points won't all fall on the line. So, we write our model as $\hat{y} = b_0 + b_1 x$, using $\hat{y}$ for the predicted values, because it's the predicted values (not the data values) that fall on the line. If the model is a good one, the data values will scatter closely around it.

For the Amazon book data, the line is:

$$\widehat{Price} = 10.35 + 0.477 \, Weight.$$

What does this mean? The **slope**, 0.477, says that we can expect a book that weighs an ounce more to cost about $0.48 more. Slopes are always expressed in *y*-units per *x*-units. They tell you how the response variable changes for a one unit step in the predictor variable. So we'd say that the slope is 0.477 dollars per ounce.

Interpreting the Intercept

Are e-books just "weightless" books? They don't have the paper, ink, binding, and other physical attributes that make up the weight of a book and are responsible for some of its cost. The typical price of e-books seems to be settling down near $10, so maybe our intercept can be interpreted in that way. But we should be cautious about extrapolating to a prediction that far from the data we have. (See "Let the Ebook Price Wars Begin: Three Ebook Pricing Predictions," *Forbes* 12.10.2012 found at www.forbes.com/sites/jeremygreenfield/2012/12/10/let-the-ebook-price-wars-begin-three-ebook-pricing-predictions/.)

The **intercept**, 10.35, is the value of the line when the *x*-variable is zero. Of course a weightless physical book isn't possible, so we wouldn't use these data to predict such a price, and we would choose to treat the intercept as just a "starting point" for our model.

JUST CHECKING

A scatterplot of sales per month (in thousands of dollars) vs. number of employees for all the outlets of a large computer chain shows a relationship that is straight, with only moderate scatter and no outliers. The correlation between *Sales* and *Employees* is 0.85, and the equation of the least squares model is:

$$\widehat{Sales} = 9.564 + 122.74\ Employees$$

6 What does the slope of 122.74 mean?

7 What are the units of the slope?

8 The outlet in Dallas, Texas, has 10 more employees than the outlet in Cincinnati. How much more *Sales* do you expect it to have?

How do we find the slope and intercept of the least squares line? The formulas are simple. The model is built from the summary statistics we've used before. We'll need the correlation (to tell us the strength of the linear association), the standard deviations (to give us the units), and the means (to tell us where to locate the line).

The slope of the line is computed as:

$$b_1 = r\frac{s_y}{s_x}.$$

We've already seen that the correlation tells us the sign and the strength of the relationship, so it should be no surprise to see that the slope inherits this sign as well. If the correlation is positive, the scatterplot runs from lower left to upper right, and the slope of the line is positive.

Correlations don't have units, but slopes do. How *x* and *y* are measured—what units they have—doesn't affect their correlation, but does change the slope. The slope gets its units from the ratio of the two standard deviations. Each standard deviation has the units of its respective variable. So, the units of the slope are a ratio, too, and are always expressed in units of *y* per unit of *x*.

How do we find the intercept? If you had to predict the *y*-value for a data point whose *x*-value was average, what would you say? The best fit line predicts $\bar{y}$ for points whose *x*-value is $\bar{x}$. Putting that into our equation and using the slope we just found gives:

$$\bar{y} = b_0 + b_1\bar{x}$$

and we can rearrange the terms to find:

$$b_0 = \bar{y} - b_1\bar{x}.$$

It's easy to use the estimated linear model to predict the price of a book from its weight. Consider, for example, Peter Drucker's book *Innovation and Entrepreneurship*. According to Amazon, the book weighs 6.4 ounces. Our model predicts its price as:

$$\widehat{Price} = 10.35 + 0.477 \times 6.4 = \$13.40$$

In fact, the book's list price is $16.99. The difference between the observed value and the value predicted by the regression equation is called the **residual**. For Drucker's book, the residual is $16.99 − 13.40 = $3.59.

Least squares lines are commonly called **regression lines**. Although this name is an accident of history (as we'll soon see), "regression" almost always means "the linear model fit by least squares." Clearly, regression and correlation are closely related. We'll need to check the same conditions before computing a regression as we did for correlation:

1. **Quantitative Variables Condition**
2. **Linearity Condition**
3. **Outlier Condition**

A little later in the chapter we'll add two more.

FOR EXAMPLE   Interpreting the equation of a linear model

QUESTION The data on cyclist accident deaths show a linear pattern. Find and interpret the equation of a linear model for that pattern. Refer to the values given in the answer to the example on page 103.

ANSWER

$$b_1 = 0.97 \times \frac{3.56}{4.47} = 0.77$$

$$b_a = 38.47 - 0.77 \times 2005 = -1505.38$$

$$\widehat{MeanAge} = -1505.38 + 0.77\,Year$$

The mean age of cyclists killed in vehicular accidents has increased by about 0.77 years of age (about 10 months) per year during the years observed by these data.[5]

Understanding Regression from Correlation

The correlation coefficient has no units. (It is found from z-scores, which have no units.) When we multiply r by the ratio of the standard deviations of the two variables to find the slope, we introduce their units. As a result, $b_1 = r\dfrac{s_y}{s_x}$ is in *y-units* per *x-unit*. For the Amazon books, the slope is in dollars per ounce.

What happens to the regression equation if we standardize both the predictor and response variables and regress z_y on z_x? Both standardized variables have standard deviation = 1 and mean = 0. So, the slope is just r, and the intercept is 0 (because both $\bar{y}$ and $\bar{x}$ are now 0), and we have the simple equation

$$\hat{z}_y = r\,z_x.$$

Although we don't usually standardize variables for regression, thinking in z-scores is a good way to understand what the regression equation is doing. The equation says that cases that deviate by one standard deviation from the mean in x are predicted to have a value of y that is r standard deviations away from the mean in y.

Let's be more specific. For the Amazon books the correlation is 0.498. So, we know immediately that:

$$\hat{z}_{Price} = 0.498\,z_{Weight}.$$

That means that a book that is one standard deviation heavier than the mean book is predicted by our model to cost about half a standard deviation more than the mean book.

---

[5] When calculated using a computer without rounding intermediate steps to fewer digits, the equation is $-1515.41 + 0.775\,Year$.

4.7   Regression to the Mean

Sir Francis Galton was the first to speak of "regression," although others had fit lines to data by the same method.

Suppose you were told of a book and, without any additional information, you were asked to guess its price. What would be your guess? A good guess would be the mean price of books. Now suppose you are also told that this book has an ISBN (International Standard Book Number) that is 2 standard deviations (SDs) above the mean ISBN. Would that change your guess? Probably not. The correlation between ISBN and *Price* is near 0, so knowing the ISBN doesn't tell you anything and doesn't move your guess. (And the standardized regression equation, $\hat{z}_y = r\,z_x$ tells us that as well, since it says that we should move 0×2 SDs from the mean.)

On the other hand, if you were told that, measured in euros, the book's price was 2 SDs above the mean, you'd know the price in dollars. There's a perfect correlation between *Price* in dollars and *Price* in euros ($r = 1$), so you know it's 2 SDs above mean *Price* in dollars as well.

What if you were told that the book was 2 SDs above the mean in number of pages? Would you still guess that its list price is average? You might guess that it costs more than average, since there's a positive correlation between Price and number of pages. But would you guess 2 SDs above the mean? When there was no correlation, we didn't move away from the mean at all. With a perfect correlation, we moved our guess the full 2 SDs. Any correlation between these extremes should lead us to move somewhere between 0 and 2 SDs above the mean. (To be exact, our best guess would be to move $r \times 2$ standard deviations away from the mean.)

Notice that if x is 2 SDs above its mean, we won't ever move more than 2 SDs away for y, since r can't be bigger than 1.0. So each predicted y tends to be closer to its mean (in standard deviations) than its corresponding x was. This property of the linear model is called **regression to the mean**. This is why the line is called the regression line.

The First Regression

Sir Francis Galton related the heights of sons to the heights of their fathers with a regression line. The slope of his line was less than 1. That is, sons of tall fathers were tall, but not as much above the average height as their fathers had been above their mean. Sons of short fathers were short, but generally not as far from their mean as their fathers. Galton interpreted the slope correctly as indicating a "regression" toward the mean height—and "regression" stuck as a description of the method he had used to find the line.

Harold Hotelling was a prominent statistician and economic theorist. He taught statistics to (among others) Nobel Prize winners Milton Friedman and Kenneth Arrow.

Business Tales about Regression to the Mean

During the Great Depression, Northwestern University professor Horace Secrist followed the fortunes of 49 department stores, measuring the ratio of net profit or loss to net sales. He divided the stores into four groups, took the average performance of each group, and followed those averages from 1920 to 1930. He found that stores that had been above average tended to perform worse, while those that were below average tended to improve.

A careful scientist, he then examined other types of business. All 73 of the different industries he examined showed the same pattern! He solicited comments and criticisms from economists and statisticians. He then published a book entitled *Triumph of Mediocrity in Business* (Secrist 1933) announcing his discovery. Initial reviews were favorable, but the prominent statistician Harold Hotelling pointed out in a scathing review that Secrist had simply rediscovered regression to the mean, which is a mathematical certainty and not a new principle of economics or business.

Although Secrist wrote 13 books on economics and statistics and was director of Northwestern University's bureau of business research, his name is still largely associated with the fallacious interpretation of regression to the mean.

More recently, the psychologist Daniel Kahneman, winner of the 2002 Nobel Prize in Economics, explained (see his book *Thinking, Fast and Slow*) that regression to the mean can make managers believe that praising employees who do well doesn't work, but punishing them when they do badly does. Employees who score a success are likely to do a bit less well the next time (just due to random fluctuation), whether they are praised or not. Those who make a big error are likely to do better next time, whether they were chastised or not. But managers like to believe that their management actions have consequences and will therefore conclude that praise causes workers to do worse, but criticism encourages workers to improve.

Checking the Model

The linear regression model may be the most widely used model in all of Statistics. It has everything we could want in a model: two easily estimated parameters, a meaningful measure of how well the model fits the data, and the ability to predict new values. It even provides a self-check in plots of the residuals to help us avoid all kinds of mistakes. For the linear model, we start by checking the same conditions we checked earlier in this chapter for using correlation.

Linear models only make sense for quantitative data. The **Quantitative Data Condition** is pretty easy to check, but don't be fooled by categorical data recorded as numbers. You probably don't want to predict zip codes from credit card account numbers.

The regression model *assumes* that the relationship between the variables is, in fact, linear—the **Linearity Assumption**. If you try to model a curved relationship with a straight line, you'll usually get what you deserve. We can't ever verify that the underlying relationship between two variables is truly linear, but an examination of the scatterplot will let you check the **Linearity Condition** as we did for correlation. If you don't judge the scatterplot to be straight enough, stop. You can't use a linear model for just *any* two variables, even if they are related. The two variables must have a *linear* association, or the model won't mean a thing and decisions you base on the model may be wrong. Some nonlinear relationships can be saved by re-expressing—or transforming—the data to make the scatterplot more linear. (See Section 4.11.)

Watch out for outliers. The linearity assumption also requires that no points lie far enough away to distort the line of best fit. Check the **Outlier Condition** to make sure no point needs special attention. Outlying values may have large residuals, and squaring makes their influence that much greater. Outlying points can dramatically change a regression model. Unusual observations can even change the sign of the slope, misleading us about the direction of the underlying relationship between the variables.

Another assumption that is usually made when fitting a linear regression is that the residuals are independent of each other. We don't strictly need this assumption to fit the line, but we will need it to draw conclusions beyond the data. We'll come back to it when we discuss inference. We can't be sure that the **Independence Assumption** is true, but we are more willing to believe that the cases are independent if the cases are a random sample from the population.

We can also check displays of the regression residuals for evidence of patterns, trends, or clumping, any of which would suggest a failure of independence. In the special case when we have a time series, a common violation of the Independence Assumption is for successive errors to be correlated with each other (autocorrelation). The error our model makes today may be similar to the one it made yesterday. We can check this violation by plotting the residuals against time and looking for patterns.

When our goal is just to explore and describe the relationship, independence isn't essential. However, when we want to go beyond the data at hand and make inferences for other situations (in Chapter 15) this will be a crucial assumption, so it's good practice to think about it even now, especially for time series.

We always check conditions with a scatterplot of the data, but we can learn even more after we've fit the regression model. There's extra information in the residuals that we can use to help us decide how reasonable our model is and how well the model fits. So, we plot the residuals and check the conditions again.

Who Was First?

One of history's most famous disputes of authorship was between Gauss and Legendre over the method of "least squares." Legendre was the first to publish the solution to finding the best fit line through data in 1805, at which time Gauss claimed to have known it for years. There is some evidence that, in fact, Gauss may have been right, but he hadn't bothered to publish it, and had been unable to communicate its importance to other scientists.[6] Gauss later referred to the solution as "*our* method" (principium *nostrum*), which certainly didn't help his relationship with Legendre.

Why *r* for *Correlation*?

In his original paper on correlation, Galton used *r* for the "index of correlation"—what we now call the correlation coefficient. He calculated it from the regression of *y* on *x* or of *x* on *y* after standardizing the variables, just as we have done. It's fairly clear from the text that he used *r* to stand for (standardized) regression.

Assumptions and Conditions

Most models are useful only when specific assumptions are true. Of course, assumptions are hard—often impossible—to check. That's why we *assume* them. But we should check to see whether the assumptions are *reasonable*. Fortunately, there are often *conditions* that we can check. Checking the conditions provides information about whether the assumptions are reasonable, and whether it's safe to proceed with the model.

Check the Scatterplot!

Check the scatterplot. The shape must be linear, or you can't use regression for the variables in their current form. And watch out for outliers.

---

[6] Stigler, Steven M., "Gauss and the Invention of Least Squares," **Annals of Statistics**, 9, (3), 1981, pp. 465–474.

The residuals are the part of the data that *hasn't* been modeled. We can write

$$Data = Predicted + Residual$$

or, equivalently,

$$Residual = Data - Predicted.$$

Or, as we showed earlier, in symbols:

$$e = y - \hat{y}.$$

A scatterplot of the residuals versus the *x*-values should be a plot without patterns. It shouldn't have any interesting features—no direction, no shape. It should stretch horizontally, showing no bends, and it should have no outliers. If you see nonlinearities, outliers, or clusters in the residuals, find out what the regression model missed.

Let's examine the residuals from our regression of Amazon book prices on weight.[7]

Figure 4.5 Residuals of the regression model predicting Amazon book prices from weights.

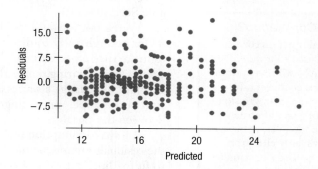

Not only can the residuals help check the conditions, but they can also tell us how well the model performs. The better the model fits the data, the less the residuals will vary around the line. The standard deviation of the residuals, s_e, gives us a measure of how much the points spread around the regression line. Of course, for this summary to make sense, the residuals should all share the same underlying spread. So we must *assume* that the standard deviation around the line is the same wherever we want the model to apply.

This new assumption about the standard deviation around the line gives us a new condition, called the **Equal Spread Condition**. The associated question to ask is does the plot have a consistent spread or does it fan out? We check to make sure that the spread of the residuals is about the same everywhere. We can check that either in the original scatterplot of *y* against *x* or in the scatterplot of residuals (or, preferably, in both plots). We estimate the **standard deviation of the residuals** in almost the way you'd expect:

$$s_e = \sqrt{\frac{\sum e^2}{n - 2}}$$

---

[7] Most computer statistics packages plot the residuals as we did in Figure 4.5, against the predicted values, rather than against *x*. When the slope is positive, the scatterplots are virtually identical except for the axes labels. When the slope is negative, the two versions are mirror images. Since all we care about is the patterns (or, better, lack of patterns) in the plot, either plot is useful.

We don't need to subtract the mean of the residuals because $\bar{e} = 0$. Why divide by $n - 2$ rather than $n - 1$? We used $n - 1$ for s when we estimated the mean. Now we're estimating both a slope and an intercept. Looks like a pattern—and it is. We subtract one more for each parameter we estimate.

When we predicted the price of Peter Drucker's book *Innovation and Entrepreneurship* our model made an error of $3.59. The standard deviation of the errors is $s_e = \$5.49$, so that's a fairly typical size for a residual because it's within one standard deviation.

FOR EXAMPLE **Examining the residuals**

Here is a scatterplot of the residuals for the linear model found in the example on page 111 plotted against the predicted values:

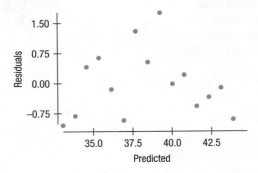

QUESTION Show how the plotted values were calculated. What does the plot suggest about the model?

ANSWER The predicted values are the values of $\widehat{MeanAge}$ found for each year by substituting the year value in the linear model. The residuals are the differences between the actual mean ages and the predicted values for each year.

The plot shows some remaining pattern in the form of four possible, nearly parallel, lower left to upper right trends. A further analysis may want to determine the reason for this pattern.

4.9 Variation in the Model and R^2

How is the thickness of a book related to the number of pages? We certainly expect books with more pages to be thicker, but can we find a model to relate these two variables? Figure 4.6 shows a scatterplot of the data.

Figure 4.6 Naturally enough, the thickness of a book grows with the number of pages.

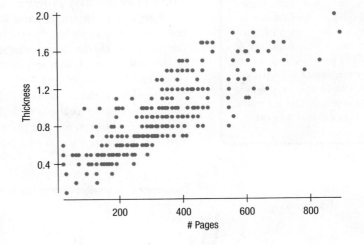

The regression model is

$$\widehat{Thickness} = 0.28 + 0.00189\ Pages$$

which says that books start out about 0.28 inches thick (due to covers and binding, perhaps) and then have about 0.00189 inches per page of thickness. We can see (in Figure 4.7) that the residuals vary less than the original thickness values did. That shows that we can get a better prediction of the thickness of a book by using a model rather than just using the mean to estimate the thickness of all the books.

Figure 4.7 Thickness and the regression residuals compared. The thickness values have their mean subtracted for the comparison. The smaller variation of the residuals shows the success of the regression model.

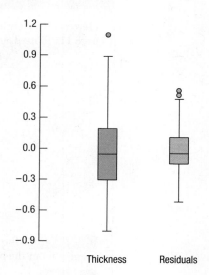

If the linear model were perfect, the residuals would all be zero and would have a standard deviation of 0. If knowing about the number of pages gave us no information about the thickness of a book, then we'd just use the mean thickness and not bother with the regression.

We can construct a measure that tells us where our model falls between being perfect and being useless. One measure we could use is the correlation between the data y and the predicted values $\hat{y}$. In a perfect regression model, the predictions would match the observed values and the correlation would be 1.0. In the worthless regression, we'd expect a correlation of 0.

All regression models fall somewhere between the two extremes of zero correlation and perfect correlation. We'd like to gauge where our model falls. But a regression model with correlation +0.5 is doing as well as one with correlation −0.5. They just have different directions. If we *square* the correlation coefficient, we'll get a value between 0 and 1, and the direction won't matter. But that's not the real reason for squaring the correlation. In fact, the squared correlation, r^2, gives the *fraction of the data's variation accounted for by the model*, and $1 - r^2$ is the fraction of the original variation left in the residuals. For the thickness and pages model, $r^2 = 0.822^2 = 0.676$, and $1 - r^2$ is 0.324, so 32.4% of the variability in *Thickness* has been left in the residuals.

All regression analyses include this statistic, although by tradition, it is written with a capital letter, R^2, and pronounced "**R-squared**." Because R^2 is a fraction of a whole, it is often given as a percentage.[8] An R^2 of 0% means that none of the

---

r and R^2

Is a correlation of 0.80 twice as strong as a correlation of 0.40? Not if you think in terms of R^2. A correlation of 0.80 means an R^2 of $0.80^2 = 64\%$. A correlation of 0.40 means an R^2 of $0.40^2 = 16\%$—only a quarter as much of the variability accounted for. A correlation of 0.80 gives an R^2 *four* times as strong as a correlation of 0.40 and accounts for four times as much of the variability.

---

[8] By contrast, we usually give correlation coefficients as decimal values between −1.0 and 1.0.

variance in the data is in the model; all of it is still in the residuals. An R^2 of 100% sounds great, but means you've probably made a mistake.[9]

We can see how R^2 relates to the variance. The variance of *Thickness* is 0.128. The variance of the residuals is 0.0415.[10] That's $0.0415/0.128 = 0.324$, or 32.4% of the variance of *Thickness* left behind in the residuals. So $100\% - 32.4\% = 67.6\%$ is the R^2 of the regression. The appropriate way to report this as part of a regression analysis is to say that 67.6% of the variance *is accounted for by the regression*.

JUST CHECKING

Let's go back to our regression of sales ($000) on number of employees again.

$$\widehat{Sales} = 9.564 + 122.74\,Employees$$

The R^2 value is reported as 72.25%.

9 What does the R^2 value mean about the relationship of *Sales* and *Employees*?

10 Is the correlation of *Sales* and *Employees* positive or negative? How do you know?

11 If we measured the *Sales* in thousands of euros instead of thousands of dollars, would the R^2 value change? How about the slope?

Some Extreme Tales

One major company developed a method to differentiate between proteins. To do so, they had to distinguish between regressions with R^2 of 99.99% and 99.98%. For this application, 99.98% was not high enough.

The president of a financial services company reports that although his regressions give R^2 below 2%, they are highly successful because those used by his competition are even lower.

How Big Should R^2 Be?

The value of R^2 is always between 0% and 100%. But what is a "good" R^2 value? The answer depends on the kind of data you are analyzing and on what you want to do with it. Just as with correlation, there is no value for R^2 that automatically determines that the regression is "good." Data from scientific experiments often have R^2 in the 80% to 90% range and even higher. Data from observational studies and surveys, though, often show relatively weak associations because it's so difficult to measure reliable responses. An R^2 of 30% to 50% or even lower might be taken as evidence of a useful regression. The standard deviation of the residuals can give us more information about the usefulness of the regression by telling us how much scatter there is around the line.

FOR EXAMPLE Understanding R^2

QUESTION Find and interpret the R^2 for the regression of cyclist death ages vs. time found in the example on page 100. (Hint: The calculation is a simple one.)

ANSWER We are given the correlation, $r = 0.97$. R^2 is the square of this, or 0.94. It tells us that 94% of the variation in the mean age of cyclist deaths can be accounted for by the trend of increasing age over time.

As we've seen, an R^2 of 100% is a perfect fit, with no scatter around the line. The s_e would be zero. All of the variance would be accounted for by the model with none left in the residuals. This sounds great, but it's too good to be true for real data.[11]

---

[9] Well, actually, it means that you have a perfect fit. But perfect models don't happen with real data unless you accidentally try to predict a variable from itself.

[10] This isn't quite the same as squaring s_e which we discussed previously, but it's very close.

[11] If you see an R^2 of 100%, it's a good idea to investigate what happened. You may have accidentally regressed two variables that measure the same thing.

4.10   Reality Check: Is the Regression Reasonable?

Statistics don't come out of nowhere. They are based on data. The results of a statistical analysis should make sense. If the results are surprising, then either you've learned something new about the world or your analysis is wrong.

Whenever you perform a regression, think about the coefficients and ask whether they make sense. Is the slope reasonable? Does the direction of the slope seem right? The small effort of asking whether the regression equation is plausible will be repaid whenever you catch errors or avoid saying something silly or absurd about the data. It's too easy to take something that comes out of a computer at face value and assume that it makes sense.

Always be skeptical and ask yourself if the answer is reasonable.

GUIDED EXAMPLE   Home Size and Price

Real estate agents know the three most important factors in determining the price of a house are *location*, *location*, and *location*. But what other factors help determine the price at which a house should be listed? Number of bathrooms? Size of the yard? A student amassed publicly available data on thousands of homes in upstate New York. We've drawn a random sample of 1000 homes from that larger dataset to examine house pricing. Among the variables she collected were the total living area (in square feet), number of bathrooms, number of bedrooms, size of lot (in acres), and age of house (in years). We will investigate how well the size of the house, as measured by living area, can predict the selling price.

PLAN	**Setup** State the objective of the study. Identify the variables and their context.	We want to find out how well the living area of a house in upstate NY can predict its selling price. We have two quantitative variables: the living area (in square feet) and the selling price ($). These data come from public records in upstate New York in 2006.

Model We need to check the same conditions for regression as we did for correlation. To do that, make a picture. Never fit a regression without looking at the scatterplot first.

✓ **Quantitative Variables Condition**

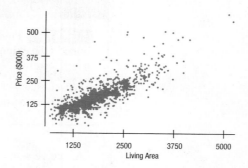

Check the **Linearity**, **Equal Spread**, and **Outlier Conditions**.

✓ **Linearity Condition** The scatterplot shows two variables that appear to have a fairly strong positive association. The plot appears to be fairly linear.

✓ **Equal Spread Condition** The scatterplot shows a consistent spread across the x-values.

✓ **Outlier Condition** There appear to be a few possible outliers, especially among large, relatively expensive houses. A few smaller houses are expensive for their size. We will check their influence on the model later.

We have two quantitative variables that appear to satisfy the conditions, so we will model this relationship with a regression line.

DO **Mechanics** Find the equation of the regression line using a statistics package. Remember to write the equation of the model using meaningful variable names.

Our software produces the following output.

```
Dependent variable is: Price
1000 total cases
R squared = 62.43%
s = 57930 with 1000 - 2 = 998 df
Variable        Coefficient
Intercept       6378.08
Living Area     115.13
```

Once you have the model, plot the residuals and check the Equal Spread Condition again.

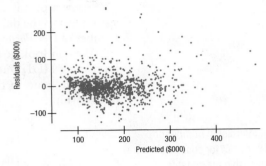

The residual plot appears generally patternless. The few relatively expensive small houses are evident, but setting them aside and refitting the model did not change either the slope or intercept very much so we left them in. There is a slight tendency for cheaper houses to have less variation, but the spread is roughly the same throughout.

REPORT **Conclusion** Interpret what you have found in the proper context.

MEMO

Re: Report on housing prices

We examined how well the size of a house could predict its selling price. Data were obtained from recent sales of 1000 homes in upstate New York. The model is:

$$\widehat{Price} = \$6378.08 + 115.13 \times Living\ Area$$

In other words, from a base of \$6378.08, houses cost about \$115.13 per square foot in upstate NY.

This model appears reasonable from both a statistical and real estate perspective. Although we know that size is not the only factor in pricing a house, the model accounts for 62.4% of the variation in selling price.

As a reality check, we checked with several real estate pricing sites (www .realestateabc.com, www.zillow.com) and found that houses in this region were averaging \$100 to \$150 per square foot, so our model is plausible.

Of course, not all house prices are predicted well by the model. We computed the model without several of these houses, but their impact on the regression model was small. We believe that this is a reasonable place to start to assess whether a house is priced correctly for this market. Future analysis might benefit by considering other factors.

4.11   Nonlinear Relationships

Everything we've discussed in this chapter requires that the underlying relationship between two variables be linear. But what should we do when the relationship is nonlinear and we can't use the correlation coefficient or a linear model? There are three basic approaches, each with its advantages and disadvantages.

Let's consider an example. The Human Development Index (HDI) was developed by the United Nations as a general measure of quality of life in countries around the world. It combines economic information (GDP), life expectancy, and education. The growth of cell phone usage has been phenomenal worldwide. Is cell phone usage related to the developmental state of a country? Figure 4.8 shows a scatterplot of number of *Cell Phones* vs. *HDI* for 154 countries of the world. (Data in **HDI**.)

We can look at the scatterplot and see that cell phone usage increases with increasing HDI. But the relationship is not straight. In Figure 4.8, we can easily see the bend in the form. But that doesn't help us summarize or model the relationship.

Figure 4.8 The scatterplot of number of *Cell Phones* (000s) vs. *HDI* for countries shows a bent relationship not suitable for correlation or regression.

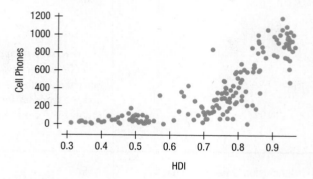

You might think that we should just fit some curved function such as an exponential or quadratic to a shape like this. But using curved functions is complicated, and the resulting model can be difficult to interpret. And many of the convenient associated statistics are not appropriate for such models. So this approach isn't often used.

A better approach to a nonlinear relationship is to transform or re-express one or both of the variables by a function such as the square root, logarithm, or reciprocal. We saw in Chapter 3 that a transformation can improve the symmetry of the distribution of a single variable. In the same way—and often with the same transforming function—transformations can make a relationship more nearly linear. Figure 4.9, for example, shows the relationship between the log of the number of *Cell Phones* and the *HDI* for the same countries.

Figure 4.9 Taking the logarithm of *Cell Phones* results in a relationship that is more linear than the original one in Figure 4.8. Notice also that two countries with very low numbers of Cell Phones for their HDI are now apparent..

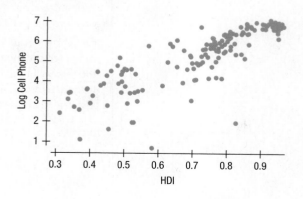

The advantage of re-expressing variables is that we *can* use regression models, along with all the supporting statistics still to come. The disadvantage is that we must interpret our results in terms of the re-expressed data, and it can be difficult to explain what we mean by the logarithm of the number of cell phones in a country. We can, of course, reverse the transformation to transform a predicted value or residual back to the original units. (In the case of a logarithmic transformation, calculate 10^y to get back to the original units.)

Which approach you choose is likely to depend on the situation and your needs. Statisticians, economists, and scientists generally prefer to transform their data, and many of their laws and theories include transforming functions.[12] But for just understanding the shape of a relationship, a scatterplot does a fine job.

FOR EXAMPLE Re-expressing for linearity

Consider the relationship between a company's *Assets* and its *Sales* as reported in annual financial statements. Here's a scatterplot of those variables for 79 of the largest companies:

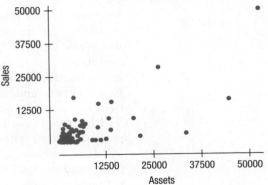

The correlation is 0.746. Taking the logarithm of both variables produces the following scatterplot:

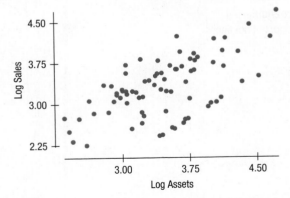

QUESTION What should we say about the relationship between *Assets* and *Sales*?

ANSWER The correlation is not appropriate because the scatterplot of the data is not linear. The scatterplot of the log transformed variables is linear and shows a strong pattern. We could find a linear model for this relationship, but we'd have to interpret it in terms of *log Sales* and *log Assets*.

---

[12] In fact, the HDI itself includes such transformed variables in its construction.

WHAT CAN GO WRONG?

- **Don't say "correlation" when you mean "association."** How often have you heard the word "correlation"? Chances are pretty good that when you've heard the term, it's been misused. It's one of the most widely misused Statistics terms, and given how often Statistics are misused, that's saying a lot. One of the problems is that many people use the specific term *correlation* when they really mean the more general term *association*. Association is a deliberately vague term used to describe the relationship between two variables.

  Correlation is a precise term used to describe the strength and direction of a linear relationship between quantitative variables.

- **Don't correlate categorical variables.** Be sure to check the Quantitative Variables Condition. It makes no sense to compute a correlation of categorical variables.

- **Make sure the association is linear.** Not all associations between quantitative variables are linear. Correlation can miss even a strong nonlinear association. And linear regression models are never appropriate for relationships that are not linear. A company, concerned that customers might use ovens with imperfect temperature controls, performed a series of experiments[13] to assess the effect of baking temperature on the quality of brownies made from their freeze-dried reconstituted brownies. The company wants to understand the sensitivity of brownie quality to variation in oven temperatures around the recommended baking temperature of 325°F. The lab reported a correlation of −0.05 between the scores awarded by a panel of trained taste-testers and baking temperature and a regression slope of −0.02, so they told management that there is no relationship. Before printing directions on the box telling customers not to worry about the temperature, a savvy intern asks to see the scatterplot (see Figure 4.10).

Figure 4.10 The relationship between brownie taste *Score* and *Baking Temperature* is strong, but not linear.

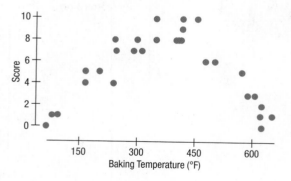

The plot actually shows a strong association—but not a linear one. Don't forget to check the Linearity Condition.

- **Beware of outliers.** You can't interpret a correlation coefficient or a regression model safely without a background check for unusual observations. Here's

---

[13] Experiments designed to assess the impact of environmental variables outside the control of the company on the quality of the company's products were advocated by the Japanese quality expert Dr. Genichi Taguchi starting in the 1980s in the United States.

an example. The relationship between *IQ* and *Shoe Size* among comedians shows a surprisingly strong positive correlation of 0.50. To check assumptions, we look at the scatterplot.

Figure 4.11 *IQ* vs. *Shoe Size*

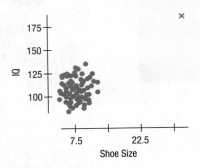

From this "study," what can we say about the relationship between the two? The correlation is 0.50. But who *does* that point in the upper right-hand corner belong to? The outlier is Bozo the Clown, known for his large shoes and widely acknowledged to be a comic "genius." Without Bozo the correlation is near zero.

Even a single unusual observation can dominate the correlation value. That's why you need to check the Unusual Observations Condition.

- **Don't confuse correlation with causation.** Once we have a strong correlation, it's tempting to try to explain it by imagining that the predictor variable has *caused* the response to change. Putting a regression line on a scatterplot tempts us even further. Humans are like that; we tend to see causes and effects in everything. Just because two variables are related does not mean that one *causes* the other.

Does Cancer Cause Smoking?

Even if the correlation of two variables is due to a causal relationship, the correlation itself cannot tell us what causes what.

Sir Ronald Aylmer Fisher (1890–1962) was one of the greatest statisticians of the 20th century. Fisher testified in court (paid by the tobacco companies) that a causal relationship might underlie the correlation of smoking and cancer:

> Is it possible, then, that lung cancer . . . is one of the causes of smoking cigarettes? I don't think it can be excluded . . . the pre-cancerous condition is one involving a certain amount of slight chronic inflammation. . . .
> A slight cause of irritation . . . [is] commonly accompanied by pulling out a cigarette, and getting a little compensation for life's minor ills in that way. And . . . is not unlikely to be associated with smoking more frequently . . . And to take the poor chap's cigarettes away from him would . . . make an already unhappy person a little more unhappy than he need be.

Ironically, the proof that smoking indeed is the cause of many cancers came from experiments conducted following the principles of experiment design and analysis that Fisher himself developed.

Scatterplots, correlation coefficients, and regression models never prove causation. This is, for example, partly why it took so long for the U.S. Surgeon General to get warning labels on cigarettes. Although there was plenty of evidence that increased smoking was *associated* with increased levels of lung cancer, it took years to provide evidence that smoking actually *causes* lung cancer. (The tobacco companies used this to great advantage.)

(continued)

- **Watch out for lurking variables.** A scatterplot of the damage (in dollars) caused to a house by fire would show a strong correlation with the number of firefighters at the scene. Surely the damage doesn't cause firefighters. And firefighters actually do cause damage, spraying water all around and chopping holes, but does that mean we shouldn't call the fire department? Of course not. There is an underlying variable that leads to both more damage and more firefighters—the size of the blaze. You can often debunk claims made about data by finding a lurking variable behind the scenes.

- **Don't fit a straight line to a nonlinear relationship.** Linear regression is suited only to relationships that are, in fact, linear.

- **Beware of extraordinary points.** Data values can be extraordinary or unusual in a regression in two ways. They can have y-values that stand off from the linear pattern suggested by the bulk of the data. These are what we have been calling outliers; although with regression, a point can be an outlier by being far from the linear pattern even if it is not the largest or smallest y-value. Points can also be extraordinary in their x-values. Such points can exert a strong influence on the line. Both kinds of extraordinary points require attention.

- **Don't predict far beyond the data.** A linear model will often do a reasonable job of summarizing a relationship in the range of observed x-values. Once we have a working model for the relationship, it's tempting to use it. But beware of predicting y-values for x-values that lie too far outside the range of the original data. The model may no longer hold there, so such extrapolations too far from the data are dangerous.

- **Don't choose a model based on R^2 alone.** Although R^2 measures the *strength* of the linear association, a high R^2 does not demonstrate the *appropriateness* of the regression. A single unusual observation, or data that separate into two groups, can make the R^2 seem quite large when, in fact, the linear regression model is simply inappropriate. Conversely, a low R^2 value may be due to a single outlier. It may be that most of the data fall roughly along a straight line, with the exception of a single point. Always look at the scatterplot.

ETHICS IN ACTION

Rebekkah Greene, owner of Up with Life Café and Marketplace, is a true believer in the health and healing benefits of food. She offers her customers pure, wholesome, and locally sourced products. Recently, she decided to add a line of hearty "made to order" cereals that she prepares according to each customer's expressed preferences. To do this, she keeps a wide variety of ingredients on hand from which her customers can choose to "design" their own unique cereal mix. These not only include organic grains, nuts, and dried fruits typically found in cereals, like oat bran and wheat germ, but also a wide assortment of sprouted grains.

Rebekkah is following the lead of food innovators who understand that sprouting grains activates food enzymes, increases vitamin content, and decreases starch—all of which improve digestion and absorption. She finds that sprouted grains are particularly delicious choices for warm breakfast cereals. Being a purist, Rebekkah decided that she would take care of the sprouting process herself. She therefore invested in the equipment and designed a "sprouting" space with appropriate temperature and humidity controls.

At the onset, her "made to order" cereal mixes were a big hit. But after the novelty wore off, she noticed that sales began to decline. Since most of her regular customers are young to middle-aged women, Rebekkah now considered the possibility that most are weight conscious as well as health conscious. She suspects that many ultimately decided to give up eating cereal for breakfast, even if it can provide superior health benefits, to opt for lower carb alternatives to maintain or lose weight.

Given her sizeable investment in sprouting grains, Rebekkah realizes she needs to do something to get more of her cereals back on the breakfast table! She began to do some research on the topic to better educate her customers. While she found a number of studies suggesting that sprouted grains are particularly healthful, she focused her attention on findings that emphasize the relationship between eating breakfast (specifically cereal) and weight loss. In her weekly flyer on "Up with

Life Food Facts" she cited one study she found in a dietetic association journal that showed regular cereal eaters had fewer weight problems than infrequent cereal eaters. She stressed this positive correlation in her advice to customers. *The more often you eat cereal for breakfast the more weight you can lose . . . more cereal = more weight lost!* She did fail to mention, however, that this particular study was funded by the big cereal companies.

- Identify the ethical dilemma in this scenario.

- What are the undesirable consequences?

- Propose an ethical solution that considers the welfare of all stakeholders.

WHAT HAVE WE LEARNED?

Learning Objectives

Make a scatterplot to display the relationship between two quantitative variables.
- Look at the direction, form, and strength of the relationship, and any outliers that stand away from the overall pattern.

Provided the form of the relationship is linear, summarize its strength with a correlation, *r*.
- The sign of the correlation gives the direction of the relationship.
- $-1 \leq r \leq 1$; A correlation of 1 or -1 is a perfect linear relationship. A correlation of 0 is a lack of linear relationship.
- Correlation has no units, so shifting or scaling the data, standardizing, or even swapping the variables has no effect on the numerical value.
- A large correlation is not a sign of a causal relationship

Model a linear relationship with a least squares regression model.
- The regression (best fit) line doesn't pass through all the points, but it is the best compromise in the sense that the sum of squares of the residuals is the smallest possible.
- The slope tells us the change in y per unit change in x.
- The R^2 gives the fraction of the variation in y accounted for by the linear regression model.

Recognize regression to the mean when it occurs in data.
- A deviation of one standard deviation from the mean in one variable is predicted to correspond to a deviation of r standard deviations from the mean in the other. Because r is never more than 1, we predict a change toward the mean.

Examine the residuals from a linear model to assess the quality of the model.
- When plotted against the predicted values, the residuals should show no pattern and no change in spread.

Terms

Association
- **Direction:** A positive direction or association means that, in general, as one variable increases, so does the other. When increases in one variable generally correspond to decreases in the other, the association is negative.

- **Form:** The form we care about most is straight, but you should certainly describe other patterns you see in scatterplots.
- **Strength:** A scatterplot is said to show a strong association if there is little scatter around the underlying relationship.

Correlation coefficient
A numerical measure of the direction and strength of a linear association.

$$r = \frac{\sum z_x z_y}{n - 1}$$

Explanatory or independent variable (x-variable)
The variable that accounts for, explains, predicts, or is otherwise responsible for the y-variable.

Intercept
The intercept, b_0, gives a starting value in y-units. It's the $\hat{y}$ value when x is 0.

$$b_0 = \bar{y} - b_1\bar{x}$$

Least squares
A criterion that specifies the unique line that minimizes the variance of the residuals or, equivalently, the sum of the squared residuals.

Linear model (Line of best fit)
The linear model of the form $\hat{y} = b_0 + b_1x$ fit by least squares. Also called the regression line. To interpret a linear model, we need to know the variables and their units.

Lurking variable
A variable other than x and y that simultaneously affects both variables, accounting for the correlation between the two.

Outlier
A point that does not fit the overall pattern seen in the scatterplot.

Predicted value
The prediction for y found for each x-value in the data. A predicted value, $\hat{y}$, is found by substituting the x-value in the regression equation. The predicted values are the values on the fitted line; the points $(x, \hat{y})$ lie exactly on the fitted line.

Predictor
A variable used on the right hand side of a regression equation. An x-variable in a regression. Also sometimes called an *independent variable*.

Re-expression or transformation
Re-expressing one or both variables using functions such as log, square root, or reciprocal can improve the straightness of the relationship between them.

Residual
The difference between the actual data value and the corresponding value predicted by the regression model—or, more generally, predicted by any model:

$$e = y - \hat{y}.$$

Regression line
The particular linear equation that satisfies the least squares criterion, often called the line of best fit.

Regression to the mean
Because the correlation is always less than 1.0 in magnitude, each predicted y tends to be fewer standard deviations from its mean than its corresponding x is from its mean.

Response or dependent variable (y-variable)
The variable that the model is intended to explain or predict.

R^2
- The square of the correlation between y and x
- The fraction of the variability of y accounted for by the least squares linear regression on x
- An overall measure of how successful the regression is in linearly relating y to x

Scatterplot
A graph that shows the relationship between two quantitative variables measured on the same cases.

Slope
The slope, b_1, is given in y-units per x-unit. Differences of one unit in x are associated with differences of b_1 units in predicted values of y:

$$b_1 = r\frac{s_y}{s_x}.$$

Standard deviation of the residuals
s_e is found by:

$$s_e = \sqrt{\frac{\sum e^2}{n - 2}}.$$

TECHNOLOGY HELP: Correlation and Regression

All statistics packages make a table of results for a regression. These tables may differ slightly from one package to another, but all are essentially the same—and all include much more than we need to know for now. Every computer regression table includes a section that looks something like this:

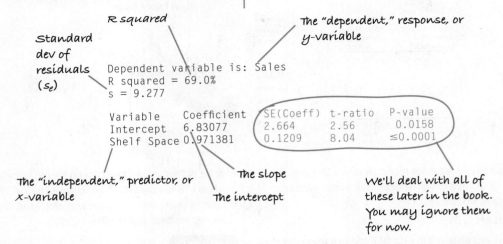

Standard dev of residuals (s_e)

R squared

The "dependent," response, or *y*-variable

The "independent," predictor, or *x*-variable

The slope

The intercept

We'll deal with all of these later in the book. You may ignore them for now.

The slope and intercept coefficient are given in a table such as this one. Usually the slope is labeled with the name of the x-variable, and the intercept is labeled "Intercept" or "Constant." So the regression equation shown here is

$$\widehat{Sales} = 6.83077 + 0.971381\,Shelf\,Space.$$

EXCEL

To make a scatterplot in Excel:

- Arrange data in worksheet so that *x*-variable and *y*-variable are in columns next to each other in that order.
- Highlight data in *x*-variable and *y*-variable columns.
- Navigate to **Insert > Scatter.** (Mac users choose **Marked Scatter.**)
- Choose **Scatter With Only Markers.**
- The scatterplot appears. Shown here is a scatterplot of the Real Estate data.

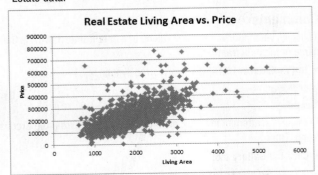

- To move the scatterplot to a new worksheet choose **Chart Tools > Design > Move Chart.**
- To add a linear regression Line to a scatterplot, choose **Chart Tools > Layout > Trendline > Linear Trendline.** (Mac users choose **Analysis > Trendline > Linear Trendline.**)

- Design, Layout, and Format options now show at the top of the screen as Chart Tools. Use these to change chart layouts, labels, and colors.
- By default, Excel includes in the intercept in the plot. If your data values are all far from zero, you may need to re-format your scatterplot. (See the discussion of Format axis below.)

To carry out a Linear Regression in Excel:

- First, make sure that you've installed the Data Analysis add-in, as follows:
  - On the **File** tab, click **Options**, and then click **Add-Ins.**
  - Near the bottom of the **Excel Options** dialog box, select **Excel Add-ins** in the **Manage** box, and then click **Go.**
  - In the **Add-Ins** dialog box, select the check box for **Analysis ToolPak**, and then click **OK.**
  - If Excel displays a message that states it can't run this add-in and prompts you to install it, click **Yes** to install the add-in.
- Navigate to **Data > Data Analysis.**
- Select **Regression.**

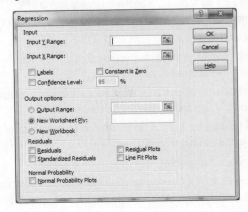

- Choose the cells of the spreadsheet holding the *x* and *y* variables.
- Check the **Labels** box if your data columns have variable names in the first row.

- Select where the output will appear.
- Check the **Line Fit Plots** box to display a scatterplot of the data with the Least Squares Regression Line.

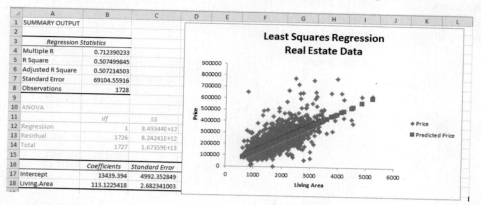

- The series displayed in red is the predicted values for price based on the Linear Regression Model. These points make up the Least Squares Regression Line.
- The R^2 value is in cell **B5.**
- The *y*-intercept and slope are in cells **B17** and **B18** respectively.
- The Design, Layout, and Format options now show at the top of the screen as Chart Tools. Use these to change chart layouts, labels, and colors.

But we aren't quite done yet. Excel always scales the axes of a scatterplot to show the origin (0, 0). But most data are not near the origin, so you may get a plot that, like this one, is bunched up in one corner.

- Right-click on the *x*-axis labels. From the menu that drops down, choose **Format axis**…
- Choose **Scale**.
- Set the *x*-axis minimum value. One useful trick is to use the dialog box itself as a straightedge to read over to the *x*-axis so you can estimate a good minimum value. Here 500 seems appropriate.
- Repeat the process with the *y*-axis if necessary.

JMP

To make a scatterplot and compute correlation:
- Choose **Fit Y by X** from the **Analyze** menu.
- In the **Fit Y by X** dialog, drag the Y variable into the **"Y, Response"** box, and drag the X variable into the **"X, Factor"** box.
- Click the **OK** button.

Once JMP has made the scatterplot, click on the red triangle next to the plot title to reveal a menu of options.
- Select **Density Ellipse** and select **.95**. JMP draws an ellipse around the data and reveals the Correlation tab.
- Click the blue triangle next to Correlation to reveal a table containing the correlation coefficient.

To compute a regression,
- Choose **Fit Y by X** from the **Analyze** menu. Specify the Y variable in the — Select Columns box and click the **Y, Response** button.
- Specify the *x*-variable and click the **X, Factor** button.
- Click **OK** to make a scatterplot.

- In the scatterplot window, click on the red triangle beside the heading labeled **"Bivariate Fit…"** and choose **Fit Line**. JMP draws the least squares regression line on the scatterplot and displays the results of the regression in tables below the plot.

MINITAB

To make a scatterplot,
- Choose **Scatterplot** from the Graph menu.
- Choose **Simple** for the type of graph. Click **OK**.
- Enter variable names for the *Y*-variable and *X*-variable into the table. Click **OK**.

To compute a correlation coefficient,
- Choose **Basic Statistics** from the Stat menu.
- From the Basic Statistics submenu, choose **Correlation**. Specify the names of at least two quantitative variables in the "Variables" box.
- Click **OK** to compute the correlation table.

R

- **lm(Y~X)** produces the linear model.
- **summary (lm(Y~X))** produces more information about the model.

Comments
Typically, your variables X and Y will be in a data frame. If DATA is the name of the data frame, then
- **lm(Y~X, data=DATA)** is the preferred syntax.

SPSS

To make a scatterplot in SPSS, open the Chart Builder from the Graphs menu. Then
- Click the **Gallery** tab.
- Choose **Scatterplot** from the list of chart types.
- Drag the scatterplot onto the canvas.
- Drag a scale variable you want as the response variable to the *y*-axis drop zone.
- Drag a scale variable you want as the factor or predictor to the *x*-axis drop zone.
- Click **OK**.

To compute a correlation coefficient,

- Choose **Correlate** from the Analyze menu.
- From the Correlate submenu, choose **Bivariate**.
- In the Bivariate Correlations dialog, use the arrow button to move variables between the source and target lists. Make sure the **Pearson** option is selected in the Correlation Coefficients field.

To compute a regression, from the Analyze menu, choose

- **Regression > Linear** … In the Linear Regression dialog, specify the Dependent (*y*), and Independent (*x*) variables.
- Click the **Plots** button to specify plots and Normal Probability Plots of the residuals. Click **OK**.

Brief Case

Fuel Efficiency

Both drivers and auto companies are motivated to raise the fuel efficiency of cars. Recent information posted by the U.S. government proposes some simple ways to increase fuel efficiency (see www.fueleconomy.gov): avoid rapid acceleration, avoid driving over 60 mph, reduce idling, and reduce the vehicle's weight. An extra 100 pounds can reduce fuel efficiency (mpg) by up to 2%. A marketing executive is studying the relationship between the fuel efficiency of cars (as measured in miles per gallon) and their weight to design a new compact car campaign. In the file **Fuel Efficiency** you'll find data on the variables below.[14]

- Model of car
- Engine size
- Cylinders
- MSRP (Manufacturer's Suggested Retail Price in $)
- City (mpg)

- Highway (mpg)
- Carbon footprint
- Transmission type (Automatic or Manual)
- Fuel type (Regular or premium)

Describe the relationship of *MSRP* and *Engine Size* with *Fuel Efficiency* (both City and Highway) in a written report. Only in the U.S. is fuel efficiency measured in miles per gallon. The rest of the world uses liters per 100 kilometers. To convert mpg to l/100 km, divide 235.215 by the mpg value. Try that form of the variable and compare the resulting models. Be sure to plot the residuals.

Cost of Living

The Numbeo website (www.numbeo.com) provides access to a variety of data. One table lists prices of certain items in selected cities around the world. They also report an overall cost-of-living index for each city compared to the costs of hundreds of items in New York City. For example, London at 110.69 is 10.69% more expensive than New York. You'll find the data for 322 cities as of March 23, 2013, in the file **Cost of living 2013**. Included are the *Cost of Living Index*, a *Rent Index*, a *Groceries Index*, a *Restaurant price Index*, and a *Local Purchasing Power Index* that measures the ability of the average wage earner in a city to buy goods and services. All indices are measured relative to New York City, which is scored 100.

Examine the relationship between the *Cost of Living Index* and the *Cost Index* for each of these individual items. Verify the necessary conditions and describe the relationship in as much detail as possible. (Remember to look at direction, form, and strength.) Identify any unusual observations.

Based on the correlations and linear regressions, which item would be the best predictor of overall cost in these cities? Which would be the worst? Are there any surprising relationships? Write a short report detailing your conclusions.

(continued)

---

[14] Data are from the 2004 model year and were compiled from www.Edmonds.com.

Mutual Funds

According to the U.S. Securities and Exchange Commission (SEC), a mutual fund is a professionally managed collection of investments for a group of investors in stocks, bonds, and other securities. The fund manager manages the investment portfolio and tracks the wins and losses. Eventually the dividends are passed along to the individual investors in the mutual fund. The first group fund was founded in 1924, but the spread of these types of funds was slowed by the stock market crash in 1929. Congress passed the Securities Act in 1933 and the Securities Exchange Act in 1934 to require that investors be provided disclosures about the fund, the securities, and the fund manager. The SEC drafted the Investment Company Act, which provided guidelines for registering all funds with the SEC. By the end of the 1960s, funds reported $48 billion in assets and, by October 2007 there were over 8,000 mutual funds with combined assets under management of over $12 trillion.

Investors often choose mutual funds on the basis of past performance, and many brokers, mutual fund companies, and other websites offer such data. In the file **Mutual fund returns 2013**, you'll find the 3-month return, the 6-month return, the annualized 1-year, 3-year, 5-year, and 10-year returns, and the return since inception of 64 funds of various types. Which data from the past provides the best predictions of the recent 3 months? Examine the scatterplots and regression models for predicting 3-month returns and write a short report containing your conclusions.

EXERCISES

The calculations for correlation and regression models can be very sensitive to how intermediate results are rounded. If you find your answers using a calculator and writing down intermediate results, you may obtain slightly different answers than you would have had you used statistics software. Different programs can also yield different results. So your answers may differ in the trailing digits from those in the Appendix. That should not concern you. The meaningful digits are the first few; the trailing digits may be essentially random results of the rounding of intermediate results.

SECTION 4.1

1. Consider the following data from a small bookstore.

Number of Sales People Working	Sales (in $1000)
2	10
3	11
7	13
9	14
10	18
10	20
12	20
15	22
16	22
20	26
$\bar{x} = 10.4$	$\bar{y} = 17.6$
$SD(x) = 5.64$	$SD(y) = 5.34$

a) Prepare a scatterplot of *Sales* against *Number of Sales People Working*.
b) What can you say about the direction of the association?
c) What can you say about the form of the relationship?
d) What can you say about the strength of the relationship?
e) Does the scatterplot show any outliers?

2. Disk drives have been getting larger. Their capacity is now often given in *terabytes* (TB) where 1 TB = 1000 gigabytes, or about a trillion bytes. A survey of prices for external disk drives found the following data:

Capacity (in TB)	Price (in $)
0.150	35.00
0.200	299.00
0.250	39.95
0.320	49.95
1.0	75.00
2.0	110.00
3.0	140.00
4.0	325.00

a) Prepare a scatterplot of *Price* against *Capacity*.
b) What can you say about the direction of the association?
c) What can you say about the form of the relationship?
d) What can you say about the strength of the relationship?
e) Does the scatterplot show any outliers?

SECTION 4.2

3. The human resources department at a large multinational corporation wants to be able to predict average salary for a given number of years' experience. Data on salary (in $1000s) and years of experience were collected for a sample of employees.

a) Which variable is the explanatory or predictor variable?
b) Which variable is the response variable?
c) Which variable would you plot on the *y*-axis?

4. A company that relies on Internet-based advertising linked to key search terms wants to understand the relationship between the amount it spends on this advertising and revenue (in $).

a) Which variable is the explanatory or predictor variable?
b) Which variable is the response variable?
c) Which variable would you plot on the *x*-axis?

SECTION 4.3

5. If we assume that the conditions for correlation are met, which of the following are true? If false, explain briefly.

a) A correlation of -0.98 indicates a strong, negative association.
b) Multiplying every value of *x* by 2 will double the correlation.
c) The units of the correlation are the same as the units of *y*.

6. If we assume that the conditions for correlation are met, which of the following are true? If false, explain briefly.

a) A correlation of 0.02 indicates a strong positive association.
b) Standardizing the variables will make the correlation 0.
c) Adding an outlier can dramatically change the correlation.

SECTION 4.4

7. A larger firm is considering acquiring the bookstore of Exercise 1. An analyst for the firm, noting the relationship seen in Exercise 1, suggests that when they acquire the store they should hire more people because that will drive higher sales. Is his conclusion justified? What alternative explanations can you offer? Use appropriate statistics terminology.

8. A study finds that during blizzards, online sales are highly associated with the number of snow plows on the road; the more plows, the more online purchases. The director of an association of online merchants suggests that the organization should encourage municipalities to send out more plows whenever it snows because, he says, that will increase business. Comment.

SECTION 4.5

9. True or False. If False, explain briefly.

a) We choose the linear model that passes through the most data points on the scatterplot.
b) The residuals are the observed *y*-values minus the *y*-values predicted by the linear model.
c) Least squares means that the square of the largest residual is as small as it could possibly be.

10. True or False. If False, explain briefly.

a) Some of the residuals from a least squares linear model will be positive and some will be negative.
b) Least Squares means that some of the squares of the residuals are minimized.
c) We write $\hat{y}$ to denote the predicted values and *y* to denote the observed values.

SECTION 4.6

11. For the bookstore sales data in Exercise 1, the correlation of number of sales people and sales is 0.965.

a) If the number of people working is 2 standard deviations above the mean, how many standard deviations above or below the mean do you expect sales to be?
b) What value of sales does that correspond to?
c) If the number of people working is 1 standard deviation below the mean, how many standard deviations above or below the mean do you expect sales to be?
d) What value of sales does that correspond to?

12. For the hard drive data in Exercise 2, some research on the prices discovered that the 200 GB hard drive was a special "hardened" drive designed to resist physical shocks and work under water. Because it is completely different from the other drives, it was removed from the data. For the remaining 7 drives, the correlation is now 0.927 and other summary statistics are:

Capacity (in TB)	Price (in $)
$\bar{x} = 1.531$	$\bar{y} = 110.70$
$SD(x) = 1.515$	$SD(y) = 102.05$

a) If a drive has a capacity of 3.046 TB (or 1 SD above the mean of 1.531 TB), how many standard deviations above or below the mean price of $110.70 do you expect the drive to cost?
b) What price does that correspond to?

13. For the bookstore of Exercise 1, the manager wants to predict *Sales* from *Number of Sales People Working*.

a) Find the slope estimate, b_1.
b) What does it mean, in this context?

c) Find the intercept, b_0.
d) What does it mean, in this context? Is it meaningful?
e) Write down the equation that predicts *Sales* from *Number of Sales People Working*.
f) If 18 people are working, what *Sales* do you predict?
g) If sales are actually $25,000 when 18 people are working, what is the value of the residual?
h) Have we overestimated or underestimated the sales?

14. For the disk drives in Exercise 2 (as corrected in Exercise 12), we want to predict *Price* from *Capacity*.

a) Find the slope estimate, b_1.
b) What does it mean, in this context?
c) Find the intercept, b_0.
d) What does it mean, in this context? Is it meaningful?
e) Write down the equation that predicts *Price* from *Capacity*.
f) What would you predict for the price of a 3.0 TB disk?
g) You have found a 3.0 TB drive for $175. Is this a good buy? How *much would you* save compared to what you expected to pay?
h) Did the model overestimate or underestimate the pricing?

SECTION 4.7

15. A CEO complains that the winners of his "rookie junior executive of the year" award often turn out to have less impressive performance the following year. He wonders whether the award actually encourages them to slack off. Can you offer a better explanation?

16. An online investment blogger advises investing in mutual funds that have performed badly the past year because "regression to the mean tells us that they will do well next year." Is he correct?

SECTION 4.8

17. Here are the residuals for a regression of *Sales* on *Number of Sales People Working* for the bookstore of Exercise 1:

Sales People Working	Residual
2	0.07
3	0.16
7	−1.49
9	−2.32
10	0.77
10	2.77
12	0.94
15	0.20
16	−0.72
20	−0.37

a) What are the units of the residuals?
b) Which residual contributes the most to the sum that was minimized according to the Least Squares Criterion to find this regression?
c) Which residual contributes least to that sum?

18. Here are residual plots (residuals plotted against predicted values) for three linear regression models. Indicate

which condition appears to be violated (linearity, outlier, or equal spread) in each case.

a)

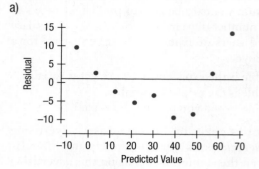

b)

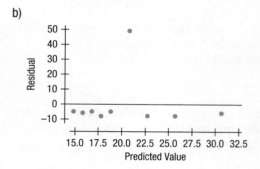

c)

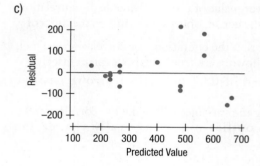

SECTION 4.9

19. For the regression model for the bookstore of Exercise 1, what is the value of R^2 and what does it mean?

20. For the disk drive data of Exercise 2 (as corrected in Exercise 12), find and interpret the value of R^2.

SECTION 4.11

21. When analyzing data on the number of employees in small companies in one town, a researcher took square roots of the counts. Some of the resulting values, which are reasonably symmetric were:

$$4, 4, 6, 7, 7, 8, 10$$

What were the original values, and how are they distributed?

22. You wish to explain to your boss what effect taking the base-10 logarithm of the salary values in the company's database will have on the data. As simple, example values you compare a salary of $10,000 earned by a part-time shipping clerk, a salary of $100,000 earned by a manager, and the CEO's $1,000,000 compensation package. Why might the average of these values be a misleading summary? What would the logarithms of these three values be?

CHAPTER EXERCISES

23. Association. Suppose you were to collect data for each pair of variables. You want to make a scatterplot. Which variable would you use as the explanatory variable and which as the response variable? Why? What would you expect to see in the scatterplot? Discuss the likely direction and form.

a) Cell phone bills: number of text messages, cost.
b) Automobiles: fuel efficiency (mpg), sales volume (number of autos).
c) For each week: ice cream cone sales, air conditioner sales.
d) Product: price ($), demand (number sold per day).

24. Association, part 2. Suppose you were to collect data for each pair of variables. You want to make a scatterplot. Which variable would you use as the explanatory variable and which as the response variable? Why? What would you expect to see in the scatterplot? Discuss the likely direction and form.

a) T- shirts at a store: price each, number sold.
b) Real estate: house price, house size (square footage).
c) Economics: Interest rates, number of mortgage applications.
d) Employees: salary, years of experience.

25. Scatterplots. Which of the scatterplots show:

a) Little or no association?
b) A negative association?
c) A linear association?
d) A moderately strong association?
e) A very strong association?

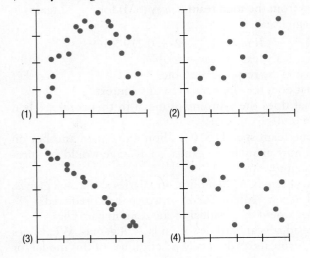

26. Scatterplots, part 2. Which of the scatterplots show:

a) Little or no association?
b) A negative association?
c) A linear association?
d) A moderately strong association?
e) A very strong association?

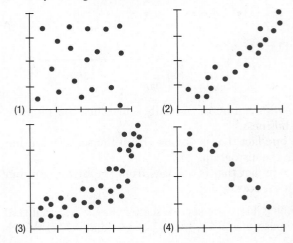

27. Manufacturing. A ceramics factory can fire eight large batches of pottery a day. Sometimes a few of the pieces break in the process. In order to understand the problem better, the factory records the number of broken pieces in each batch for three days and then creates the scatterplot shown.

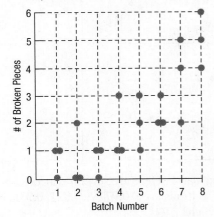

a) Make a histogram showing the distribution of the number of broken pieces in the 24 batches of pottery examined.
b) Describe the distribution as shown in the histogram. What feature of the problem is more apparent in the histogram than in the scatterplot?
c) What aspect of the company's problem is more apparent in the scatterplot?

28. Coffee sales. Owners of a new coffee shop tracked sales for the first 20 days and displayed the data in a scatterplot (by day).

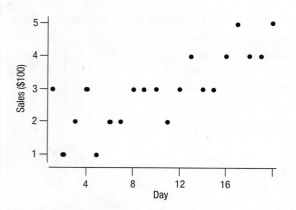

a) Make a histogram of the daily sales since the shop has been in business.

b) State one fact that is obvious from the scatterplot, but not from the histogram.

c) State one fact that is obvious from the histogram, but not from the scatterplot.

29. Matching. Here are several scatterplots. The calculated correlations are −0.923, −0.487, 0.006, and 0.777. Which is which?

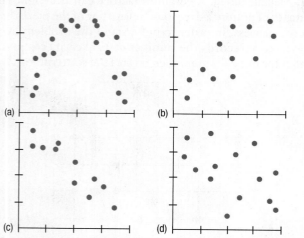

30. Matching, part 2. Here are several scatterplots. The calculated correlations are −0.977, −0.021, 0.736, and 0.951. Which is which?

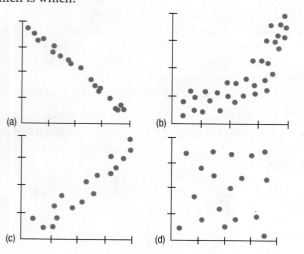

T 31. Pizza sales and price. A linear model fit to predict weekly *Sales* of frozen pizza (in pounds) from the average *Price* ($/unit) charged by a sample of stores in the city of Dallas in 39 recent weeks is:

$$\widehat{Sales} = 141{,}865.53 - 24{,}369.49 \, Price.$$

a) What is the explanatory variable?
b) What is the response variable?
c) What does the slope mean in this context?
d) What does the *y*-intercept mean in this context? Is it meaningful?
e) What do you predict the sales to be if the average price charged was $3.50 for a pizza?
f) If the sales for a price of $3.50 turned out to be 60,000 pounds, what would the residual be?

T 32. Used Honda prices 2013. A linear model to predict the *Price* of a 2012 Honda Civic EX (in $) from its *Mileage* (in miles) was fit to 18 cars that were available during the week of March 1, 2013 (Kelly's Blue Book, www.kbb.com) within 200 miles of San Francisco, CA. The model was:

$$\widehat{Price} = 21{,}253.58 - 0.11097 \, Mileage.$$

a) What is the explanatory variable?
b) What is the response variable?
c) What does the slope mean in this context?
d) What does the *y*-intercept mean in this context? Is it meaningful?
e) What do you predict the price to be for a car with 50,000 miles on it?
f) If the price for a car with 50,000 miles on it was selling for $14,000, what would the residual be?
g) Would that car for $14,000 and 50,000 miles seem like a good deal or a bad deal? Explain.

T 33. Football salaries 2013. Is there a relationship between total team salary and the performance of teams in the National Football League (NFL)? For the 2012–2013 season, a linear model predicting *Wins* (out of 16 regular season games) from the total team *Salary* ($M) for the 32 teams in the league is:

$$\widehat{Wins} = -16.32 + 0.219 \, Salary$$

a) What is the explanatory variable?
b) What is the response variable?
c) What does the slope mean in this context?
d) What does the *y*-intercept mean in this context? Is it meaningful?
e) If one team spends $10 million more than another on salary, how many more games on average would you predict them to win?
f) If a team spent $120 million on salaries and won 8 games, would they have done better or worse than predicted?
g) What would the residual of the team in part f be?
The residual standard deviation is 2.78 games. What does that tell you about the likely practical use of this model for predicting wins?

T **34.** **Baseball salaries 2012.** In 2012, the New York Yankees won 95 games and spent $198 million on salaries for their players (*USA Today*). Is there a relationship between salary and team performance in Major League Baseball? For the 2012 season, a linear model fit to the number of *Wins* (out of 162 regular season games) from the team *Salary* ($M) for the 30 teams in the league is:

$$\widehat{Wins} = 76.46 + 0.046\,Salary.$$

a) What is the explanatory variable?
b) What is the response variable?
c) What does the slope mean in this context?
d) What does the *y*-intercept mean in this context? Is it meaningful?
e) If one team spends $10 million more than another on salaries, how many more games on average would you predict them to win?
f) If a team spent $110 million on salaries and won half (81) of their games, would they have done better or worse than predicted?
g) What would the residual of the team in part f be?
h) The R^2 for this model is 2.05% and the residual standard deviation is 12.0 games. How useful is this model likely to be for predicting the number of wins?

T **35.** **Pizza sales and price, part 2.** For the data in Exercise 31, the average *Sales* was 52,697 pounds (SD = 10,261 pounds), and the correlation between *Price* and *Sales* was = −0.547. If the *Price* in a particular week was one SD higher than the mean *Price*, how much pizza would you predict was sold that week?

T **36.** **Used Honda prices 2013, part 2.** The 18 cars in Exercise 32 had an average *Price* of $19,843.50 (SD = $1853.59), and the correlation between *Price* and *Mileage* was = −0.889. If the *Mileage* of a 2012 Honda Civic EX was 1 SD below the average number of miles, what *Price* would you predict for it?

37. **Packaging.** A CEO announces at the annual shareholders meeting that the new see-through packaging for the company's flagship product has been a success. In fact, he says, "There is a strong correlation between packaging and sales." Criticize this statement on statistical grounds.

38. **Insurance.** Insurance companies carefully track claims histories so that they can assess risk and set rates appropriately. The National Insurance Crime Bureau reports that Honda Accords, Honda Civics, and Toyota Camrys are the cars most frequently reported stolen, while Ford Tauruses, Pontiac Vibes, and Buick LeSabres are stolen least often. Is it correct to say that there's a correlation between the type of car you own and the risk that it will be stolen?

39. **Sales by region.** A sales manager for a major pharmaceutical company analyzes last year's sales data for her 96 sales representatives, grouping them by region (1 = East Coast United States; 2 = Midwest United States; 3 = West United States; 4 = South United States; 5 = Canada; 6 = Rest of World). She plots *Sales* (in $1000) against *Region* (1–6) and sees a strong negative correlation.

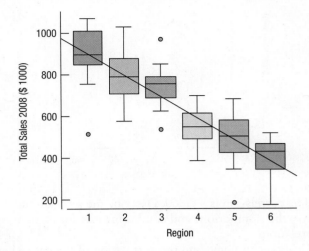

She fits a regression to the data and finds:

$$\widehat{Sales} = 1002.5 - 102.7\,Region$$

The R^2 is 70.5%.

Write a few sentences interpreting this model and describing what she can conclude from this analysis.

40. **Salary by job type.** At a small company, the head of human resources wants to examine salary to prepare annual reviews. He selects 28 employees at random with job types ranging from 01 = Stocking clerk to 99 = President. He plots *Salary* ($) against *Job Type* and finds a strong linear relationship with a correlation of 0.96.

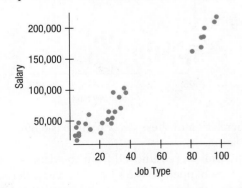

The regression output gives:

$$\widehat{Salary} = 15827.9 + 1939.1\,Job\ Type$$

Write a few sentences interpreting this model and describing what he can conclude from this analysis.

T **41.** **Carbon footprint 2015.** The scatterplot shows, for 2015 cars, the carbon footprint (tons of CO_2 per mile) vs. the new Environmental Protection Agency (EPA) highway mileage for 69 family sedans as reported by the U.S. government (www.fueleconomy.gov/feg/byclass.htm) the cars in the lower right of the scatterplot plotted with red x's are all hybrids.

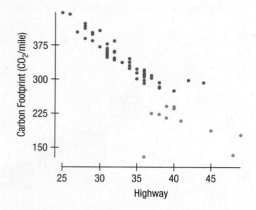

a) The correlation is −0.860. Describe the association.
b) Are the assumptions and conditions met for computing correlation?
c) Using technology, find the correlation of the data when the hybrid cards are not included with the others. Can you explain why it changes in that way?

T 42. EPA mpg 2013. In 2008, the EPA revised their methods for estimating the fuel efficiency (mpg) of cars—a factor that plays an increasingly important role in car sales. How do the new highway and city estimated mpg values relate to each other? Here's a scatterplot for 76 family sedans as reported by the U.S. government.

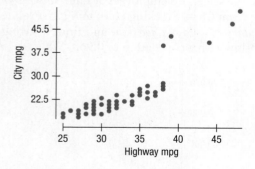

a) The correlation of these two variables is 0.896. Describe the association.
b) If the hybrids were removed from the data, what would you expect to happen to the slope (increase, decrease, or stay pretty much the same) and to the correlation (increase, decrease, the same)? Try it using technology. Report and discuss what you find.

43. Real estate. Is the number of total rooms in the house associated with the price of a house? Here is the scatterplot of a random sample of homes for sale:

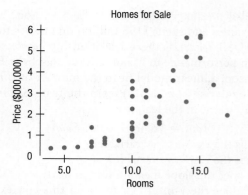

a) Is there an association?
b) Check the assumptions and conditions for correlation.

T 44. Economic analysis 2012. An economics student is studying the American economy and finds that the correlation between the inflation-adjusted Dow Jones Industrial Average and the Gross Domestic Product (GDP) (also inflation adjusted) is 0.81 for the years 1946 to 2011 (www.measuringworth.com). From that he concludes that there is a strong linear relationship between the two series and predicts that a drop in the GDP will make the stock market go down. Here is a scatterplot of the adjusted DJIA against the GDP (in the years 1946 to 2011). Describe the relationship and comment on the student's conclusions.

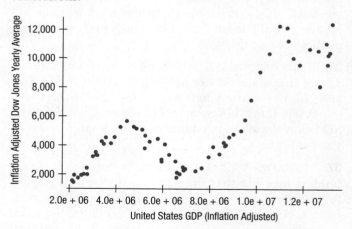

T 45. GDP growth 2012. Is economic growth in the developing world related to growth in the industrialized countries? Here's a scatterplot of the growth (in % of Gross Domestic Product) of 180 developing countries vs. the growth of 33 developed countries as grouped by the World Bank (www.ers.usda.gov/data/macroeconomics). Each point represents one of the years from 1970 to 2011. The output of a regression analysis follows.

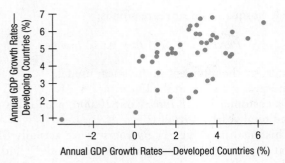

Dependent variable: GDP Growth Developing
Countries
$R^2 = 31.64\%$
s = 1.201

Variable	Coefficient
Intercept	3.38
GDP Growth Developed Countries	0.468

a) Check the assumptions and conditions for the linear model.
b) Explain the meaning of R^2 in this context.
c) What are the cases in this model?

46. European GDP growth 2012. Is economic growth in Europe related to growth in the United States? Here's a scatterplot of the average growth in 25 European countries (in % of Gross Domestic Product) vs. the growth in the United States. Each point represents one of the years from 1970 to 2011.

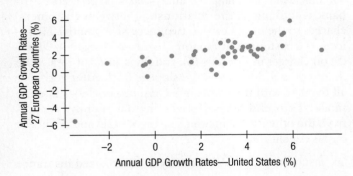

Dependent variable: European Countries GDP
Growth
$R^2 = 44.92\%$
s = 1.352

Variable	Coefficient
Intercept	0.693
U.S. GDP Growth	0.534

a) Check the assumptions and conditions for the linear model.
b) Explain the meaning of R^2 in this context.

47. GDP growth 2012, part 2. From the linear model fit to the data on GDP growth in Exercise 45:

a) Write the equation of the regression line.

b) What is the meaning of the intercept? Does it make sense in this context?
c) Interpret the meaning of the slope.
d) In a year in which the developed countries grow 4%, what do you predict for the developing world?
e) In 2007, the developed countries experienced a 2.65% growth, while the developing countries grew at a rate of 6.09%. Is this more or less than you would have predicted?
f) What is the residual for this year?

48. European GDP growth 2012, part 2. From the linear model fit to the data on GDP growth of Exercise 46:

a) Write the equation of the regression line.
b) What is the meaning of the intercept? Does it make sense in this context?
c) Interpret the meaning of the slope.
d) In a year in which the United States grows at 0%, what do you predict for European growth?
e) In 2010, the United States experienced a 3.00% growth, while Europe grew at a rate of 1.78%. Is this more or less than you would have predicted?
f) What is the residual for this year?

49. Attendance 2014. American League baseball games are played under the designated hitter rule, meaning that weak-hitting pitchers do not come to bat. Baseball owners believe that the designated hitter rule means more runs scored, which in turn means higher attendance. Is there evidence that more fans attend games if the teams score more runs? Data collected from Major League games from both major leagues during the season have a correlation of 0.233 between *Runs Scored* and the *Attendance* (www.baseball-reference.com).

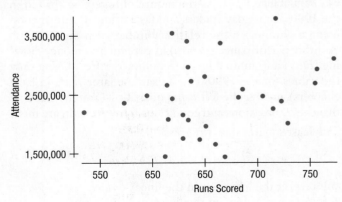

a) Does the scatterplot indicate that it's appropriate to calculate a correlation? Explain.
b) Describe the association between attendance and runs scored.
c) Does this association prove that the owners are right that more fans will come to games if the teams score more runs?

T 50. Attendance 2014, part 2. Perhaps fans are just more interested in teams that win. Here are displays of other variables in the dataset of exercise 49 (www.baseball-reference.com). Are the teams that win necessarily those that score the most runs?

	Correlation		
	Wins	**Runs**	**Attend**
Wins	1.000		
Runs	0.410	1.000	
Attend	0.346	0.233	1.000

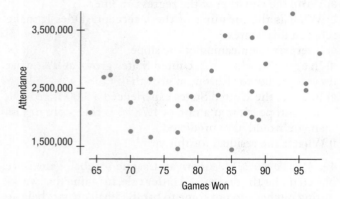

a) Do winning teams generally enjoy greater attendance at their home games? Describe the association.
b) Is attendance more strongly associated with winning or scoring runs? Explain.
c) How strongly is scoring more runs associated with winning more games?

51. Mutual fund flows. As the nature of investing shifted in the 1990s (more day traders and faster flow of information using technology), the relationship between mutual fund monthly performance (*Return*) in percent and money flowing (*Flow*) into mutual funds ($ million) shifted. Using only the values for the 1990s (we'll examine later years in later chapters), answer the following questions. (You may assume that the assumptions and conditions for regression are met.)

The least squares linear regression is:

$$\widehat{Flow} = 9747 + 771\ Return.$$

a) Interpret the intercept in the linear model.
b) Interpret the slope in the linear model.
c) What is the predicted fund *Flow* for a month that had a market *Return* of 0%?
d) If during this month, the recorded fund *Flow* was $5 billion, what is the residual using this linear model? Did the model provide an underestimate or overestimate for this month?

52. Online clothing purchases. An online clothing retailer examined their transactional database to see if total yearly *Purchases* ($) were related to customers' *Incomes* ($). (You may assume that the assumptions and conditions for regression are met.)

The least squares linear regression is:

$$\widehat{Purchases} = -31.6 + 0.012\ Income.$$

a) Interpret the intercept in the linear model.
b) Interpret the slope in the linear model.
c) If a customer has an *Income* of $20,000, what is his predicted total yearly *Purchases*?
d) This customer's yearly *Purchases* were actually $100. What is the residual using this linear model? Did the model provide an underestimate or overestimate for this customer?

53. Residual plots. Tell what each of the following residual plots indicates about the appropriateness of the linear model that was fit to the data.

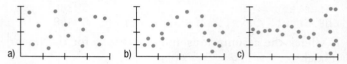

54. Residual plots, again. Tell what each of the following residual plots indicates about the appropriateness of the linear model that was fit to the data.

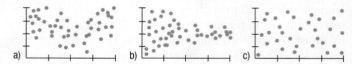

T 55. Consumer spending. An analyst at a large credit card bank is looking at the relationship between customers' charges to the bank's card in two successive months. He selects 150 customers at random, regresses charges in *March* ($) on charges in *February* ($), and finds an R^2 of 79%. The intercept is $730.20, and the slope is 0.79. After verifying all the data with the company's CPA, he concludes that the model is a useful one for predicting one month's charges from the other. Examine the data on the CD and comment on his conclusions.

T 56. Insurance policies. An actuary at a mid-sized insurance company is examining the sales performance of the company's sales force. She has data on the average size of the policy ($) written in two consecutive years by 200 salespeople. She fits a linear model and finds the slope to be 3.00 and the R^2 is 99.92%. She concludes that the predictions for next year's policy size will be very accurate. Examine the data on the CD and comment on her conclusions.

57. What slope? If you create a regression model for predicting the sales ($ million) from money spent on advertising the prior month ($ thousand), is the slope most likely to be closer to 0.03, 300, or 3000? Explain.

58. What slope, part 2? If you create a regression model for estimating a student's business school GPA (on a scale of 1–5) based on his math SAT (on a scale of 200–800), is the slope most likely to be closer to 0.01, 1, or 10? Explain.

59. Misinterpretations. An advertising agent who created a regression model using amount spent on *Advertising* to predict annual *Sales* for a company made these two statements. Assuming the calculations were done correctly, explain what is wrong with each interpretation.

a) My R^2 of 93% shows that this linear model is appropriate.
b) If this company spends $1.5 million on advertising, then annual sales will be $10 million.

60. More misinterpretations. An economist investigated the association between a country's *Literacy Rate* and *Gross Domestic Product (GDP)* and used the association to draw the following conclusions. Explain why each statement is incorrect. (Assume that all the calculations were done properly.)

a) The *Literacy Rate* determines 64% of the *GDP* for a country.
b) The slope of the line shows that an increase of 5% in *Literacy Rate* will produce a $1 billion improvement in *GDP*.

61. Business admissions. An analyst at a business school's admissions office claims to have developed a valid linear model predicting success (measured by starting salary ($) at time of graduation) from a student's undergraduate performance (measured by GPA). Describe how you would check each of the four regression conditions in this context.

62. School rankings. A popular magazine annually publishes rankings of both U.S. business programs and international business programs. The latest issue claims to have developed a linear model predicting the school's ranking (with "1" being the highest ranked school) from its financial resources (as measured by size of the school's endowment). Describe how you would apply each of the four regression conditions in this context.

63. Used BMW prices 2013. A business student needs cash, so he decides to sell his car. The car is a valuable BMW 850CSi that was only made over the course of a few years in the 1990s. He would like to sell it on his own, rather than through a dealer so he'd like to predict the price he'll get for his car's model year.

a) Make a scatterplot for the data on used BMW 850SCi's provided.
b) Describe the association between year and price.
c) Do you think a linear model is appropriate?
d) Computer software says that $R^2 = 57.3\%$. What is the correlation between year and price?
e) Explain the meaning of R^2 in this context.
f) Why doesn't this model explain 100% of the variability in the price of a used BMW 850CSi?

64. Used BMW prices 2013, part 2. Use the advertised prices for BMW 850s given in Exercise 63 to create a linear model for the relationship between a car's *Model Year* and its *Price*.

a) Find the equation of the regression line.
b) Explain the meaning of the slope of the line.

c) Explain the meaning of the intercept of the line.
d) If you wanted to sell a 1997 BMW 850, what price seems appropriate?
e) You have a chance to buy one of two cars. They are about the same age and appear to be in equally good condition. Would you rather buy the one with a positive residual or the one with a negative residual? Explain.

65. Expensive cities. The *Worldwide Cost of Living Survey City Rankings* determine the cost of living in the most expensive cities in the world as an index. This index scales New York City as 100 and expresses the cost of living in other cities as a percentage of the New York cost. For example, in 2007, the cost of living index in Tokyo was 122.1, which means that it was 22% higher than New York. The scatterplot shows the index for 2013 plotted against the 2007 index for the 15 most expensive cities of 2007.

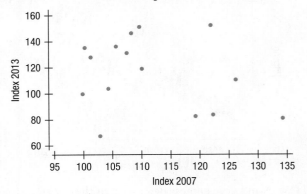

a) Describe the association between cost of living indices in 2007 and 2013.
b) The R^2 for the regression equation is 0.070. Interpret the value of R^2.
c) Find the correlation.
d) Using the data provided, find the least squares fit of the 2013 index to the 2007 index.
e) Predict the 2013 cost of living index of Moscow and find its residual.

66. Lobster prices 2012. The demand for lobster has grown steadily for several decades. The Maine lobster fishery is carefully controlled to protect the lobster population from over-fishing. The number of fishing licenses and the number of traps are both limited. During the years from 1974 to 2006, the price of lobster also grew, as shown in this plot.

a) Describe the increase in the *Price* of lobster during this period.
b) The R^2 for the regression equation is 87.42%. Interpret the value of R^2.
c) Find the correlation.
d) Find the linear model.

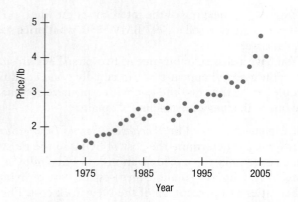

The years from 2007 to 2012 have seen a change in this pattern. Here is the scatterplot.

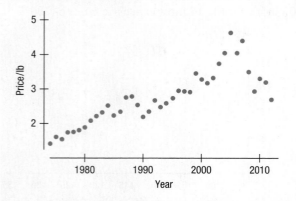

e) How would you suggest dealing with these new cases to model the change in prices?

67. El Niño. Concern over the weather associated with El Niño has increased interest in the possibility that the climate on Earth is getting warmer. The most common theory relates an increase in atmospheric levels of carbon dioxide (CO_2), a greenhouse gas, to increases in temperature. Here is a scatterplot showing the mean annual CO_2 concentration in the atmosphere, measured in parts per million (ppm) at the top of Mauna Loa in Hawaii, and the mean annual air temperature over both land and sea across the globe, in degrees Celsius (°C).

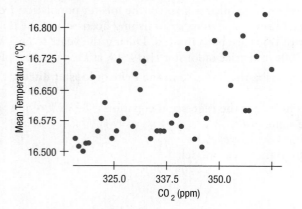

A regression predicting *Mean Temperature* from CO_2 produces the following output table (in part).

```
Dependent variable: Temperature
R-squared = 33.4%
Variable            Coefficient
Intercept           15.3066
CO₂                 0.004
```

a) What is the correlation between CO_2 and *Mean Temperature*?
b) Explain the meaning of R-squared in this context.
c) Give the regression equation.
d) What is the meaning of the slope in this equation?
e) What is the meaning of the intercept of this equation?
f) Here is a scatterplot of the residuals vs. CO_2. Does this plot show evidence of the violations of any of the assumptions of the regression model? If so, which ones?

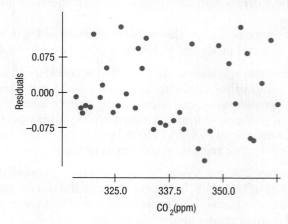

g) CO_2 levels may reach 364 *ppm* in the near future. What *Mean Temperature* does the model predict for that value?

T **68. U.S. birthrates.** The table shows the number of live births per 1000 women aged 15–44 years in the United States, starting in 1965. (National Center for Health Statistics, www.cdc.gov/nchs/)

Year	1965	1970	1975	1980	1985
Rate	19.4	18.4	14.8	15.9	15.6

Year	1990	1995	2000	2005	2010
Rate	16.4	14.8	14.4	14.0	13.0

a) Make a scatterplot and describe the general trend in *Birthrates*. (Enter *Year* as years since 1900: 65, 70, 75, etc.)
b) Find the equation of the regression line.
c) Check to see if the line is an appropriate model. Explain.
d) Interpret the slope of the line.
e) The table gives rates only at 5-year intervals. Estimate what the rate was in 1978.
f) In 1978, the birthrate was actually 15.0. What was the residual?
g) Predict what the *Birthrate* will be in 2012. Comment on your faith in this prediction.
h) Predict the *Birthrate* for 2050. Comment on your faith in this prediction.

JUST CHECKING ANSWERS

1. We know the scores are quantitative. We should check to see if the *Linearity Condition* and the *Outlier Condition* are satisfied by looking at a scatterplot of the two scores.

2. It won't change.

3. It won't change.

4. They are more likely to do poorly. The positive correlation means that low closing prices for Intel are associated with low closing prices for Cypress.

5. No, the general association is positive, but daily closing prices may vary.

6. For each additional employee, monthly sales increase, on average, $122,740.

7. Thousands of $ per employee.

8. $1,227,400 per month.

9. Differences in the number of employees account for about 72.25% of the variation in the monthly sales.

10. It's positive. The correlation and the slope have the same sign.

11. R^2, No. Slope, Yes.

Case **Study**

Paralyzed Veterans of America

Philanthropic organizations often rely on contributions from individuals to finance the work that they do, and a national veterans' organization is no exception. The Paralyzed Veterans of America (PVA) was founded as a congressionally chartered veterans' service organization more than 60 years ago. It provides a range of services to veterans who have experienced spinal cord injury or dysfunction. Some of the services offered include medical care, research, education, and accessibility and legal consulting. In 2008, this organization had total revenue of more than $135 million, with more than 99% of this revenue coming from contributions.

An organization that depends so heavily on contributions needs a multifaceted fundraising program, and PVA solicits donations in a number of ways. From its website (www.pva.org), people can make a onetime donation, donate monthly, donate in honor or in memory of someone, and shop in the PVA online store. People can also support one of the charity events, such as its golf tournament, National Veterans Wheelchair Games, and Charity Ride.

Traditionally, one of PVA's main methods of soliciting funds was the use of return address labels and greeting cards (although still used, this method has declined in recent years). Typically, these gifts were sent to potential donors about every six weeks with a request for a contribution. From its established donors, PVA could expect a response rate of about 5%, which, given the relatively small cost to produce and send the gifts, kept the organization well funded.

But fundraising accounts for 28% of expenses, so PVA wanted to know who its donors are, what variables might be useful in predicting whether a donor is likely to give to an upcoming campaign, and what the size of that gift might be. The dataset **PVA** includes data designed to be very similar to part of the data that this organization works with. Here is a description of some of the variables. Keep in mind, however, that in the real dataset, there would be hundreds more variables given for each donor.

Variable Name	Units (if applicable)	Description	Remarks
Age	Years		
Own Home?	H = Yes; U = No or unknown		
Children	Counts		
Income		1 = Lowest ; 7 = Highest	Based on national medians and percentiles
Sex	M = Male; F = Female		
Total Wealth		0 = Lowest; 9 = Highest	Based on national medians and percentiles
Gifts to Other Orgs	Counts	Number of Gifts (if known) to other philanthropic organizations in the same time period	
Number of Gifts	Counts	Number of Gifts to this organization in this time period	
Time Between Gifts	Months	Time between first and second gifts	
Smallest Gift	$	Smallest Gift (in $) in the time period	See also Sqrt(Smallest Gift)
Largest Gift	$	Largest Gift (in $) in the time period	See also Sqrt(Largest Gift)
Previous Gift	$	Gift (in $) for previous campaign	See also Sqrt(Previous Gift)

Variable Name	Units (if applicable)	Description	Remarks
Average Gift	$	Total amount donated divided by total number of gifts	See also Sqrt(Average Gift)
Current Gift	$	Gift (in $) to organization this campaign	See also Sqrt(Current Gift)
Sqrt(Smallest Gift)	Sqrt($)	Square Root of Smallest Gift in $	
Sqrt(Largest Gift)	Sqrt($)	Square Root of Largest Gift in $	
Sqrt(Previous Gift)	Sqrt($)	Square Root of Previous Gift in $	
Sqrt(Average Gift)	Sqrt($)	Square Root of Average Gift in $	
Sqrt(Current Gift)	Sqrt($)	Square Root of Current Gift in $	

Let's see what the data can tell us. Are there any interesting relationships between the current gift and other variables? Is it possible to use the data to predict who is going to respond to the next direct-mail campaign?

> Recall that when variables are highly skewed or the relationship between variables is not linear, reporting a correlation coefficient is not appropriate. You may want to consider a transformed version of those variables (square roots are provided for all the variables concerning gifts) or a correlation based on the ranks of the values rather than the values themselves.

Suggested Study Plan and Questions

Write a report of what you discover about the donors to this organization. Be sure to follow the Plan, Do, Report outline for your report. Include a basic description of each variable (shape, center, and spread), point out any interesting features, and explore the relationships between the variables. In particular you should describe any interesting relationships between the current gift and other variables. Use these questions as a guide:

- Is the age distribution of the clients a typical one found in most businesses?

- Do people who give more often make smaller gifts on average?

- Do people who give to other organizations tend to give to this organization?

Describe the relationship between the Income and Wealth rankings. How do you explain this relationship (or lack of one)? (*Hint*: Look at the age distribution.)

What variables (if any) seem to have an association with the Current Gift? Do you think the organization can use any of these variables to predict the gift for the next campaign?

Optional: This file includes people who did not give to the current campaign. Do your answers to any of the questions above change if you consider only those who gave to this campaign?

5

Randomness and Probability

Credit Reports and the Fair Isaacs Corporation

You've probably never heard of the Fair Isaacs Corporation, but they probably know you. Whenever you apply for a loan, a credit card, or even a job, your credit "score" is used to determine whether you are a good risk. And because the most widely used credit scores are Fair Isaacs' FICO® scores, the company may well be involved in the decision. The Fair Isaacs Corporation (FICO) was founded in 1956, with the idea that data, used intelligently, could improve business decision making. Today, Fair Isaacs claims that their services provide companies around the world with information for more than 180 billion business decisions a year.

Your credit score is a number between 350 and 850 that summarizes your credit "worthiness." It's a snapshot of credit risk today based on your credit history and past behavior. Lenders of all kinds use credit scores to predict behavior, such as how likely you are to make your loan payments on time or to default on a loan. Lenders use the score to determine not only whether to give credit, but also the cost of the credit that they'll offer. There are no established boundaries, but generally scores over 750 are considered excellent, and applicants with those scores get the best rates. An applicant with a score below 620 is generally considered to be a poor risk. Those with very low scores may be denied credit outright or only offered "subprime" loans at substantially higher rates.

It's important that you be able to verify the information that your score is based on, but until recently, you could only hope that your score was based on correct information. That changed in 2000, when a California law gave mortgage applicants the right to see their credit scores. Today, the credit industry is more open about giving consumers access to their scores and the U.S. government, through the Fair and Accurate Credit Transaction Act (FACTA), now guarantees that you can access your credit report at no cost, at least once a year.[1]

Companies have to manage risk to survive, but by its nature, risk carries uncertainty. A bank can't know for certain that you'll pay your mortgage on time—or at all. What can they do with events they can't predict? They start with the fact that, although individual outcomes cannot be anticipated with certainty, random phenomena do, in the long run, settle into patterns that are consistent and predictable. It's this property of random events that makes Statistics practical.

5.1   Random Phenomena and Probability

When a customer calls the 800 number of a credit card company, he or she is asked for a card number before being connected with an operator. As the connection is made, the purchase records of that card and the demographic information of the customer are retrieved and displayed on the operator's screen. If the customer's FICO score is high enough, the operator may be prompted to "cross-sell" another service—perhaps a new "platinum" card for customers with a credit score of at least 750.

Of course, the company doesn't know which customers are going to call. Call arrivals are an example of a random phenomenon. With **random phenomena**, we can't predict the individual outcomes, but we can hope to understand characteristics of their long-run behavior. We don't know whether the *next* caller will qualify for the platinum card, but as calls come into the call center, the company will find that the percentage of platinum callers who qualify for cross-selling will settle into a pattern, like that shown in the graph in Figure 5.1.

Part of a call center operator's earnings might be based on the number of platinum cards she sells. To figure out what her potential bonus might be, an operator might first want to know what percentage of all callers qualifies. She decides to write down whether the caller from each call she gets today qualifies or not. The first caller today qualified. Then the next five callers' qualifications were no, yes, yes, no, and no. If we plot the percentage who qualify against the number of calls she's made so far the graph would start at 100% because the first caller qualified (1 out of 1, for 100%). The next caller didn't qualify, so the accumulated percentage dropped to 50% (1 out of 2). The third caller qualified (2 out of 3, or 67%), then yes again (3 out of 4, or 75%), then no twice in a row (3 out of 5, for 60%, and then 3 out of 6, for 50%), and so on (Table 5.1). Each new call is a smaller fraction of the total number, so the percentages change less after each call. After a while, the graph starts to settle down and we can see that the fraction of customers who qualify is about 35% (Figure 5.1).

When talking about long-run behavior, it helps to define our terms. For any random phenomenon, each attempt, or **trial**, generates an **outcome**. For the call center, each call is a trial. Something happens on each trial, and we call whatever happens the outcome. Here the outcome is whether the caller qualifies or not. We use the more general term **event** to refer to outcomes or combinations of outcomes. For example, suppose we categorize callers into 6 risk categories and number these outcomes from 1 to 6 (of increasing credit worthiness). The three outcomes 4, 5, or 6 could make up the event "caller is at least a category 4."

We sometimes talk about the collection of *all possible outcomes*, a special event that we'll refer to as the **sample space**. We denote the sample space **S**; you may

> A **random phenomenon** consists of **trials**. Each trial has an **outcome**. Outcomes combine to make **events**.

---

[1] However, the score you see in your report will be an "educational" score intended to show consumers how scoring works. You still have to pay a "reasonable fee" to see your FICO score.

Figure 5.1 The percentage of credit card customers who qualify for the premium card.

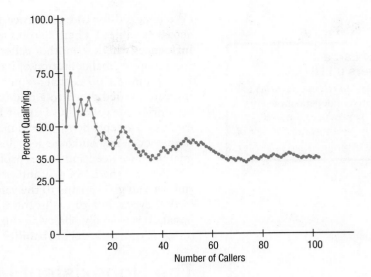

Table 5.1 Data on the first six callers showing their FICO score, whether they qualified for the platinum card offer, and a running percentage of number of callers who qualified.

Call	FICO Score	Qualify?	Running % Qualify
1	750	Yes	100
2	640	No	50
3	765	Yes	66.7
4	780	Yes	75
5	680	No	60
6	630	No	50
⋮	⋮		⋮

also see the Greek letter Ω used. But whatever symbol we use, the sample space is the set that contains all the possible outcomes. For the calls, if we let Q = qualified and N = not qualified, the sample space is simple: $S = \{Q, N\}$. If we look at two calls together, the sample space has four outcomes: $S = \{QQ, QN, NQ, NN\}$. If we were interested in at least one qualified caller from the two calls, we would be interested in the event (call it A) consisting of the three outcomes QQ, QN, and NQ, and we'd write $A = \{QQ, QN, NQ\}$.

Although we may not be able to predict a *particular* individual outcome, such as which incoming call represents a potential upgrade sale, we can say a lot about the long-run behavior. Look back at Figure 5.1. If you were asked for the probability that a random caller will qualify, you might say that it was 35% because, in the *long run*, the percentage of the callers who qualify is about 35%. That's exactly what we mean by **probability**.

When we think about what happens with a series of trials, it really simplifies things if the individual trials are independent. Roughly speaking, **independence** means that the outcome of one trial doesn't influence or change the outcome of another. Recall, that in Chapter 2, we called two variables *independent* if the value of one categorical variable did not influence the value of another categorical variable.

> **Probability as Long-Run Frequency**
>
> The **probability** of an event is its long-run relative frequency. A relative frequency is a fraction, so we can write it as $\frac{35}{100}$, as a decimal, 0.35, or as a percentage, 35%.

(We checked for independence by comparing relative frequency distributions across variables.) There's no reason to think that whether the one caller qualifies influences whether another caller qualifies, so these are independent trials. We'll see a more formal definition of independence later in the chapter.

You might think that we just got lucky when the percentage of the qualifying calls settled down to a number. But for independent events, we can depend on a principle called the **Law of Large Numbers (LLN)**, which states that if the events are independent, then as the number of trials increases, the long-run relative frequency of any outcome gets closer and closer to a single value. This gives us the guarantee we need and makes probability a useful concept.

Because the LLN guarantees that relative frequencies settle down in the long run, we can give a name to the value that they approach. We call it the probability of that event. For the call center, we can write $P(\text{qualified}) = 0.35$. Because it is based on repeatedly observing the event's outcome, this definition of probability is often called **empirical probability**.

5.2   The Nonexistent Law of Averages

The Law of Large Numbers is often misunderstood to be a "law of averages." Many people believe, for example, that an outcome of a random event that hasn't occurred in many trials is "due" to occur. The original "dogs of the Dow" strategy for buying stocks recommended buying the 10 worst performing stocks of the 30 that make up the Dow Jones Industrial Average, figuring that these "dogs" were bound to do better next year. The thinking was that, since the relative frequency will settle down to the probability of that outcome in the long run, we'll have some "catching up" to do. That may seem logical, but random events don't work that way. In fact, Louis Rukeyser (the former host of *Wall Street Week*) said of the "dogs of the Dow" strategy, "that theory didn't work as promised."

Here's why. We actually know very little about the behavior of random events in the short run. The fact that we are seeing independent random events makes each individual result impossible to predict. Relative frequencies even out *only* in the long run. And the long run referred to in the LLN is really long. The "Large" in the law's name means *infinitely* large. Sequences of random events don't compensate in the short run and don't need to do so to get back to the right long-run probability. Any short-run deviations will be overwhelmed in the long run. If the probability of an outcome doesn't change and the events are independent, the probability of any outcome in another trial never changes, no matter what has happened in other trials.

Many people confuse the Law of Large numbers with the so-called Law of Averages, which says that things have to even out in the short run. But even though the Law of Averages doesn't exist at all, you'll hear people talk about it as if it does. Is a good hitter in baseball who has struck out the last six times *due* for a hit his next time up? If the stock market has been down for the last three sessions, is it *due* to increase today? No. This isn't the way random phenomena work. There is no Law of Averages for short runs—no "Law of Small Numbers." A belief in such a "law" can lead to poor business decisions.

"Slump? I ain't in no slump. I just ain't hittin'."

—Yogi Berra

You may think it's obvious that the frequency of repeated events settles down in the long run to a single number. The discoverer of the Law of Large Numbers thought so, too. The way Jacob Bernoulli put it was: *"For even the most stupid of men is convinced that the more observations have been made, the less danger there is of wandering from one's goal."*

Keno and the Law of Averages

Of course, sometimes an apparent drift from what we expect means that the probabilities are, in fact, not what we thought. If you get 10 heads in a row, maybe the coin has heads on both sides! Here's a true story that illustrates this.

Keno is a simple casino game in which numbers from 1 to 80 are chosen. The numbers, as in most lottery games, are supposed to be equally likely. Payoffs are

made depending on how many of those numbers you match on your card. A group of graduate students from a Statistics department decided to take a field trip to Reno. They (*very* discreetly) wrote down the outcomes of the games for a couple of days, then drove back to test whether the numbers were, in fact, equally likely. It turned out that some numbers were *more likely* to come up than others. Rather than bet on the Law of Averages and put their money on the numbers that were "due," the students put their faith in the LLN—and all their (and their friends') money on the numbers that had come up before. After they pocketed more than $50,000, they were escorted off the premises and invited never to show their faces in that casino again. Not coincidentally, the ringleader of that group currently makes his living on Wall Street.

> *"In addition, in time, if the roulette-betting fool keeps playing the game, the bad histories [outcomes] will tend to catch up with him."*
> —Nassim Nicholas Taleb in
> *Fooled by Randomness*

The Law of Averages Debunked

You've just flipped a fair coin and seen six heads in a row. Does the coin "owe" you some tails? Suppose you spend that coin and your friend gets it in change. When she starts flipping the coin, should she expect a run of tails? Of course not. Each flip is a new event. The coin can't "remember" what it did in the past, so it can't "owe" any particular outcomes in the future. Just to see how this works in practice, we simulated 100,000 flips of a fair coin on a computer. In our 100,000 "flips," there were 2981 streaks of at least 5 heads. The "Law of Averages" suggests that the next flip after a run of 5 heads should be tails more often to even things out. Actually, in this particular simulation the next flip was heads more often than tails: 1550 times to 1431 times. That's 51.9% heads. You can perform a similar simulation easily.

JUST CHECKING

1 It has been shown that the stock market fluctuates randomly. Nevertheless, some investors believe that they should buy right after a day when the market goes down because it is bound to go up soon. Explain why this is faulty reasoning.

5.3   Different Types of Probability

Model-Based (Theoretical) Probability

> **Model-Based Probability**
> We can write:
> $$P(\mathbf{A}) = \frac{\# \text{ of outcomes in } \mathbf{A}}{\text{total } \# \text{ of outcomes}}$$
> and call this the **(theoretical) probability** of the event.

We've discussed *empirical probability*—the relative frequency of an event's occurrence as the probability of an event. There are other ways to define probability as well. Probability was first studied extensively by a group of French mathematicians who were interested in games of chance. Rather than experiment with the games and risk losing their money, they developed mathematical models of probability. To make things simple (as we usually do when we build models), they started by looking at games in which the different outcomes were equally likely. Fortunately, many games of chance are like that. Any of 52 cards is equally likely to be the next one dealt from a well-shuffled deck. Each face of a die is equally likely to land up (or at least it should be).

When we have equally likely outcomes, we write the **(theoretical) probability** of an event **A**, as $P(\mathbf{A}) = \#$ of outcomes in **A**/total # of outcomes possible.

When outcomes are equally likely, the probability that one of them occurs is easy to compute—it's just 1 divided by the number of possible outcomes. So the probability of rolling a 3 with a fair die is one in six, which we write as 1/6. The probability of picking the ace of spades from a well-shuffled deck is 1/52.

It's almost as simple to find probabilities for events that are made up of several equally likely outcomes. We just count all the outcomes that the event contains. The probability of the event is the number of outcomes in the event divided by the total number of possible outcomes.

For example, Pew Research[2] reports that of 10,190 randomly generated working phone numbers called for a survey, the initial results of the calls were as follows:

Result	Number of Calls
No Answer	311
Busy	61
Answering Machine	1336
Callbacks	189
Other Non-Contacts	893
Contacted Numbers	7400

> **Equally Likely?**
>
> In an attempt to understand why someone would buy a lottery ticket, an interviewer asked someone who had just purchased one "What do you think your chances are of winning the lottery?" The reply was, "Oh, about 50–50." The shocked interviewer asked, "How do you get that?" to which the response was, "Well, the way I figure it, either I win or I don't!" The moral of this story is that events are not always equally likely.

The phone numbers were generated randomly, so each was equally likely. To find the probability of a contact, we just divide the number of contacts by the number of calls: $7400/10,190 = 0.7262$.

But don't get trapped into thinking that random events are always equally likely. The chance of winning a lottery—especially lotteries with very large payoffs—is small. Regardless, people continue to buy tickets.

Personal Probability

What's the probability that gold will sell for more than $2000 an ounce at the end of next year? You may be able to come up with a number that seems reasonable. Of course, no matter what your guess is, your probability should be between 0 and 1. In our discussion of probability, we've defined probability in two ways: (1) in terms of the relative frequency—or the fraction of times—that an event occurs in the long run or (2) as the number of outcomes in the event divided by the total number of outcomes. Neither situation applies to your assessment of gold's chances of selling for more than $2000.

We use the *language* of probability in everyday speech to express a degree of uncertainty without necessarily basing it on long-run relative frequencies. Your personal assessment of an event expresses your uncertainty about the outcome. We call this kind of probability a subjective, or **personal probability**.

Although personal probabilities may be based on experience, they are typically not based on long-run relative frequencies or on equally likely events. But, like the two other probabilities we defined, they need to satisfy the same rules as both empirical and theoretical probabilities that we'll discuss in the next section.

---

[2] www.pewinternet.org/pdfs/PIP_Digital_Footprints.pdf.

5.4   Probability Rules

For some people, the phrase "50/50" means something vague like "I don't know" or "whatever." But when we discuss probabilities, 50/50 has the precise meaning that two outcomes are *equally likely*. Speaking vaguely about probabilities can get you into trouble, so it's wise to develop some formal rules about how probability works. These rules apply to probability whether we're dealing with empirical, theoretical, or personal probability.

Rule 1. If the probability of an event occurring is 0, the event won't occur; likewise if the probability is 1, the event will *always* occur. Even if you think an event is very unlikely, its probability can't be negative, and even if you're sure it will happen, its probability can't be greater than 1. So we require that:

> **A probability is a number between 0 and 1.**
> **For any event A, $0 \leq P(A) \leq 1$.**

Rule 2. If a random phenomenon has only one possible outcome, it's not very interesting (or very random). So we need to distribute the probabilities among all the outcomes a trial can have. How can we do that so that it makes sense? For example, consider the behavior of a certain stock. The possible daily outcomes might be:

A: The stock price goes up.
B: The stock price goes down.
C: The stock price remains the same.

When we assign probabilities to these outcomes, we should be sure to distribute all of the available probability. Something always occurs, so the probability of *something* happening is 1. This is called the **Probability Assignment Rule**:

> **The probability of the set of all possible outcomes must be 1.**
> $$P(S) = 1$$

where **S** is the sample space.

Rule 3. Suppose the probability that you get to class on time is 0.8. What's the probability that you don't get to class on time? Yes, it's 0.2. The set of outcomes that are *not* in the event **A** is called the "complement" of **A** and is denoted A^C. This leads to the **Complement Rule**:

> **The probability of an event occurring is 1 minus the probability**
> **that it doesn't occur.**
> $$P(A) = 1 - P(A^C)$$

"Baseball is 90% mental.
The other half is physical."
—Yogi Berra

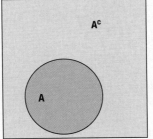

The set **A** and its complement A^C. Together, they make up the entire sample space **S**.

FOR EXAMPLE   Applying the Complement Rule

Lee's Lights sells lighting fixtures. Some customers are there only to browse, so Lee records the behavior of all customers for a week to assess how likely it is that a customer will make a purchase. Lee finds that of 1000 customers entering the store during the week, 300 make purchases. Lee concludes that the probability of a customer making a purchase is 0.30.

QUESTION   If $P(\text{purchase}) = 0.30$, what is the probability that a customer *doesn't* make a purchase?

ANSWER   Because "no purchase" is the complement of "purchase,"

$$P(\text{no purchase}) = 1 - P(\text{purchase})$$
$$= 1 - 0.30 = 0.70$$

There is a 70% chance a customer won't make a purchase.

Rule 4. Whether or not a caller qualifies for a platinum card is a random outcome. Suppose the probability of qualifying is 0.35. What's the chance that the next two callers qualify? The **Multiplication Rule** says that to find the probability that two independent events occur, we multiply the probabilities. For two independent events **A** and **B**, the probability that both **A** *and* **B** occur is the product of the probabilities of the two events:

$$P(A \text{ and } B) = P(A) \times P(B), \text{ provided that A and B are independent.}$$

Thus if **A** = {customer 1 qualifies} and **B** = {customer 2 qualifies}, the chance that both qualify is:

$$0.35 \times 0.35 = 0.1225$$

Of course, to calculate this probability, we have used the assumption that the two events are independent. We'll expand the multiplication rule to be more general later in this chapter.

FOR EXAMPLE   Using the Multiplication Rule

Lee knows that the probability that a customer will make a purchase is 30%.

QUESTION If we can assume that customers behave independently, what is the probability that the next two customers entering Lee's Lights both make purchases?

ANSWER Because the events are independent, we can use the multiplication rule.

$P(\text{first customer makes a purchase } and \text{ second customer makes a purchase})$

$\quad = P(\text{purchase}) \times P(\text{purchase})$

$\quad = 0.30 \times 0.30 = 0.09$

There's about a 9% chance that the next two customers will both make purchases.

Rule 5. Suppose the card center operator has more options. She can **A**: offer a special travel deal, **B**: offer a platinum card, or **C**: decide to send information about a new affinity card. If she can do one, but only one, of these, then these outcomes are **disjoint** (or **mutually exclusive**). To see whether two events are disjoint, we separate them into their component outcomes and check whether they have any outcomes in common. For example, if the operator can choose to both offer the travel deal and send the affinity card information, those events would not be disjoint. The **Addition Rule** allows us to add the probabilities of disjoint events to get the probability that *either* event occurs:

$$P(A \text{ or } B) = P(A) + P(B), \text{ provided that A and B are disjoint.}$$

Thus the probability that the caller *either* is offered a platinum card *or* is sent the affinity card information is the sum of the two probabilities, since the events are disjoint.

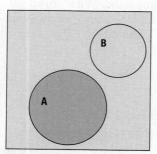

Two disjoint sets, **A** and **B**.

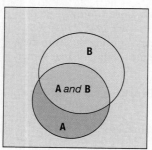

Two sets **A** and **B** that are not disjoint. The event (**A** and **B**) is their intersection.

FOR EXAMPLE   Using the Addition Rule

Some customers prefer to see the merchandise but then make their purchase later using Lee's Lights' website. Lee offers a promotion to attempt to track customer behavior. Customers leaving the store without making a purchase are offered a "bonus code" to use at the Internet site. Using these codes, Lee determines that there's a 9% chance of a customer making a purchase using the code later. We know that about 30% of customers make purchases when they enter the store.

> **QUESTION** What is the probability that a customer who enters the store will not make a purchase at all?
>
> **ANSWER** We can use the Addition Rule because the alternatives "no purchase at all," "purchase in the store," and "purchase online" are disjoint events.
>
> $$P(\text{purchase in the store } or \text{ online}) = P(\text{purchase in store}) + P(\text{purchase online})$$
> $$= 0.30 + 0.09 = 0.39$$
>
> $$P(\text{no purchase at all}) = P(\text{not (purchase in the store } or \text{ purchase online}))$$
> $$= 1 - P(\text{in store } or \text{ online})$$
> $$= 1 - 0.39 = 0.61$$

NOTATION ALERT

You may see the event (**A** or **B**) written as (**A** $\cup$ **B**). The symbol $\cup$ means "union" and represents the outcomes in event **A** or event **B**. Similarly the symbol $\cap$ means intersection and represents outcomes that are in *both* event **A** and event **B**. You may see the event (**A** *and* **B**) written as (**A** $\cap$ **B**).

Rule 6. Suppose we would like to know the probability that either of the next two callers qualifies for a platinum card? We know $P(A) = P(B) = 0.35$, but $P(A \text{ or } B)$ is not simply the sum $P(A) + P(B)$ because the events **A** and **B** are not disjoint in this case. Both customers could qualify. So we need a new probability rule.

We can't simply add the probabilities of **A** and **B** because that would count the outcome of *both* customers qualifying twice. So, if we started by adding the two probabilities, we could compensate by subtracting out the probability that both qualify. In other words,

$P(\text{customer A } or \text{ customer B qualifies}) =$

$P(\text{customer A qualifies}) + P(\text{customer B qualifies}) - P(\text{both customers qualify})$

$= (0.35) + (0.35) - (0.35 \times 0.35) \text{ (since events are independent)}$

$= (0.35) + (0.35) - (0.1225)$

$= 0.5775$

It turns out that this method works in general. We add the probabilities of two events and then subtract out the probability of their intersection. This gives us the **General Addition Rule**, which does not require disjoint events:

$$P(A \text{ } or \text{ } B) = P(A) + P(B) - P(A \text{ } and \text{ } B) \text{ for any two events A and B.}$$

> ## FOR EXAMPLE   Using the General Addition Rule
>
> Lee notices that when two customers enter the store together, their purchases are not disjoint. In fact, there's a 20% chance they'll *both* make a purchase.
>
> **QUESTION** When two customers enter the store together, what is the probability that *at least one* of them makes a purchase?
>
> **ANSWER** Now we know that the events are not disjoint, so we must use the General Addition Rule.
>
> $$P(\text{at least one purchases}) = P(A \text{ purchases } or \text{ B purchases})$$
> $$= P(A \text{ purchases}) + P(B \text{ purchases})$$
> $$- P(A \text{ and B both purchase})$$
> $$= 0.30 + 0.30 - 0.20 = 0.40$$

JUST CHECKING

2 Even successful companies sometimes make products with high failure rates. One (in)famous example is the Apple 40GB click wheel iPod, which used a tiny disk drive for storage. According to Macintouch.com, 30% of those devices eventually failed. It is reasonable to assume that the failures were independent. What would a store that sold these devices have seen?

a) What is the probability that a particular 40GB click wheel iPod failed?

b) What is the probability that two 40GB click wheel iPods sold together *both* failed?

c) What is the probability that the store's first failure problem was the third one they sold?

d) What is the probability the store had a failure problem with at least one of the five that they sold on a particular day?

GUIDED EXAMPLE   M&M's Modern Market Research

In 1941, when M&M's® milk chocolate candies were introduced to American GIs in World War II, there were six colors: brown, yellow, orange, red, green, and violet. Mars®, the company that manufactures M&M's, has used the introduction of a new color as a marketing and advertising event several times in the years since then. In 1980, the candy went international adding 16 countries to their markets. In 1995, the company conducted a "worldwide survey" to vote on a new color. Over 10 million people voted to add blue. They even got the lights of the Empire State Building in New York City to glow blue to help announce the addition. In 2002, they used the Internet to help pick a new color. Children from over 200 countries were invited to respond via the Internet, telephone, or mail. Millions of voters chose among purple, pink, and teal. The global winner was purple, and for a brief time, purple M&M's could be found in packages worldwide (although in 2013, the colors were brown, yellow, red, blue, orange, and green). In the United States, 42% of those who voted said purple, 37% said teal, and only 19% said pink. But in Japan the percentages were 38% pink, 36% teal, and only 16% purple. Let's use Japan's percentages to ask some questions.

1. What's the probability that a Japanese M&M's survey respondent selected at random preferred either pink or teal?
2. If we pick two respondents at random, what's the probability that they *both* selected purple?
3. If we pick three respondents at random, what's the probability that *at least one* preferred purple?

PLAN   **Setup** The probability of an event is its long-term relative frequency. This can be determined in several ways: by looking at many replications of an event, by deducing it from equally likely events, or by using some other information. Here, we are told the relative frequencies of the three responses.

The M&M's website reports the proportions of Japanese votes by color. These give the probability of selecting a voter who preferred each of the colors:

$$P(pink) = 0.38$$
$$P(teal) = 0.36$$
$$P(purple) = 0.16$$

Make sure the probabilities are legitimate. Here, they're not. Either there was a mistake or the other voters must have chosen a color other than the three given. A check of other countries shows a similar deficit, so probably we're seeing those who had no preference or who wrote in another color.

Each is between 0 and 1, but these don't add up to 1. The remaining 10% of the voters must have not expressed a preference or written in another color. We'll put them together into "other" and add
$P(other) = 0.10$
With this addition, we have a legitimate assignment of probabilities.

Question 1: What's the probability that a Japanese M&M's survey respondent selected at random preferred either pink or teal?

PLAN	**Setup** Decide which rules to use and check the conditions they require.	The events "pink" and "teal" are individual outcomes (a respondent can't choose both colors), so they are disjoint. We can apply the General Addition Rule anyway.
DO	**Mechanics** Show your work.	$P(pink\ or\ teal) = P(pink) + P(teal)$ $- P(pink\ and\ teal)$ $= 0.38 + 0.36 - 0 = 0.74$ The probability that both pink and teal were chosen is zero, since respondents were limited to one choice.
REPORT	**Conclusion** Interpret your results in the proper context.	The probability that the respondent said pink or teal is 0.74.

Question 2: If we pick two respondents at random, what's the probability that they both said purple?

PLAN	**Setup** The word "both" suggests we want $P(\mathbf{A}\ and\ \mathbf{B})$, which calls for the Multiplication Rule. Check the required condition.	**Independence** It's unlikely that the choice made by one respondent affected the choice of the other, so the events seem to be independent. We can use the Multiplication Rule.
DO	**Mechanics** Show your work. For both respondents to pick purple, each one has to pick purple.	$P(both\ purple)$ $= P(first\ respondent\ picks\ purple\ and$ $second\ respondent\ picks\ purple)$ $= P(first\ respondent\ picks\ purple)$ $\times P(second\ respondent\ picks\ purple)$ $= 0.16 \times 0.16 = 0.0256$
REPORT	**Conclusion** Interpret your results in the proper context.	The probability that both respondents pick purple is 0.0256.

(continued)

Question 3: If we pick three respondents at random, what's the probability that at least one preferred purple?

PLAN	**Setup** The phrase "at least one" often flags a question best answered by looking at the complement, and that's the best approach here. The complement of "at least one preferred purple" is "none of them preferred purple."	$P(\text{at least one picked purple})$ $= P(\{\text{none picked purple}\}^c)$ $= 1 - P(\text{none picked purple}).$
	Check the conditions.	**Independence.** These are independent events because they are choices by three random respondents. We can use the Multiplication Rule.
DO	**Mechanics** We calculate $P(\text{none purple})$ by using the Multiplication Rule.	$P(\text{none picked purple}) = P(\text{first not purple})$ $\times P(\text{second not purple})$ $\times P(\text{third not purple})$ $= [P(\text{not purple})]^3.$ $P(\text{not purple}) = 1 - P(\text{purple})$ $= 1 - 0.16 = 0.84.$ So $P(\text{none picked purple}) = (0.84)^3 = 0.5927.$ $P(\text{at least 1 picked purple})$ $= 1 - P(\text{none picked purple})$ $= 1 - 0.5927 = 0.4073.$
	Then we can use the Complement Rule to get the probability we want.	
REPORT	**Conclusion** Interpret your results in the proper context.	There's about a 40.7% chance that at least one of the respondents picked purple.

5.5 Joint Probability and Contingency Tables

As part of a Pick Your Prize Promotion, a chain store invited customers to choose which of three prizes they'd like to win (while providing name, address, phone number, and e-mail address). At one store, the responses could be placed in the contingency table in Table 5.2.

		Prize Preference			
		MP3	**Camera**	**Bike**	**Total**
Sex	**Man**	117	50	60	**227**
	Woman	130	91	30	**251**
	Total	**247**	**141**	**90**	**478**

Table 5.2 Prize preference for 478 customers.

If the winner is chosen at random from these customers, the probability we select a woman is just the corresponding relative frequency (since we're equally likely to select any of the 478 customers). There are 251 women in the data out of a total of 478, giving a probability of:

$$P(\text{woman}) = 251/478 = 0.525$$

This is called a **marginal probability** because it depends only on totals found in the margins of the table. The same method works for more complicated events. For example, what's the probability of selecting a woman whose preferred prize is the camera? Well, 91 women named the camera as their preference, so the probability is:

$$P(\text{woman } and \text{ camera}) = 91/478 = 0.190$$

> A **marginal probability** uses a marginal frequency (from either the Total row or Total column) to compute the probability.

Probabilities such as these are called **joint probabilities** because they give the probability of two events occurring together.

The probability of selecting a customer whose preferred prize is a bike is:

$$P(\text{bike}) = 90/478 = 0.188$$

FOR EXAMPLE **Marginal probabilities**

Lee suspects that men and women make different kinds of purchases at Lee's Lights (see the example on pages 152–153). The table shows the purchases made by the last 100 customers.

	Utility Lighting	Fashion Lighting	Total
Men	40	20	60
Women	10	30	40
Total	50	50	100

QUESTION   What's the probability that one of Lee's customers is a woman? What is the probability that a random customer is a man who purchases fashion lighting?

ANSWER   From the marginal totals we can see that 40% of Lee's customers are women, so the probability that a customer is a woman is 0.40. The cell of the table for Men who purchase Fashion lighting has 20 of the 100 customers, so the probability of that event is 0.20.

5.6 Conditional Probability

Since our sample space is these 478 customers, we can recognize the relative frequencies as probabilities. What if we are given the information that the selected customer is a woman? Would that change the probability that the selected customer's preferred prize is a bike? You bet it would! The pie charts in Figure 5.2 on the next page show that women are much less likely to say their preferred prize is a bike than are men. When we restrict our focus to women, we look only at the women's row of the table, which gives the conditional distribution of preferred prizes given "woman." Of the 251 women, only 30 of them said their preferred prize was a bike. We write the probability that a selected customer wants a bike *given* that we have selected a woman as:

$$P(\text{bike}|\text{woman}) = 30/251 = 0.120$$

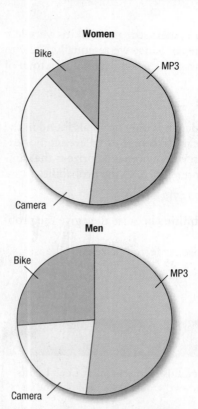

Figure 5.2   Conditional distributions of *Prize Preference* for *Women* and for *Men*.

For men, we look at the conditional distribution of preferred prizes given "man" shown in the top row of the table. There, of the 227 men, 60 said their preferred prize was a bike. So, $P(\text{bike}|\text{man}) = 60/227 = 0.264$, more than twice the women's probability (see Figure 5.2).

In general, when we want the probability of an event from a *conditional* distribution, we write $P(\mathbf{B}|\mathbf{A})$ and pronounce it "the probability of **B** *given* **A**." A probability that takes into account a given *condition* such as this is called a **conditional probability**.

Let's look at what we did. We worked with the counts, but we could work with the probabilities just as well. There were 30 women who selected a bike as a prize, and there were 251 women customers. So we found the probability to be 30/251. To find the probability of the event **B** *given* the event **A**, we restrict our attention to the outcomes in **A**. We then find in what fraction of *those* outcomes **B** also occurred. Formally, we write:

$$P(\mathbf{B}|\mathbf{A}) = \frac{P(\mathbf{A} \text{ and } \mathbf{B})}{P(\mathbf{A})}$$

We can use the formula directly with the probabilities derived from the contingency table (Table 5.2) to find:

$$P(\text{bike}|\text{woman}) = \frac{P(\text{bike and woman})}{P(\text{woman})} = \frac{30/478}{251/478} = \frac{30}{251} = 0.120 \text{ as before.}$$

The formula for conditional probability requires one restriction. The formula works only when the event that's given has probability greater than 0. The formula doesn't work if $P(\mathbf{A})$ is 0 because that would mean we had been "given" the fact that **A** was true even though the probability of **A** is 0, which would be a contradiction.

Rule 7. Remember the Multiplication Rule for the probability of **A** *and* **B**? It said

$$P(\mathbf{A} \text{ and } \mathbf{B}) = P(\mathbf{A}) \times P(\mathbf{B})$$

when **A** and **B** are independent. Now we can write a more general rule that doesn't require independence. In fact, we've already written it. We just need to rearrange the equation a bit.

The equation in the definition for conditional probability contains the probability of **A** *and* **B**. Rearranging the equation gives the **General Multiplication Rule** for compound events that does not require the events to be independent:

$$P(\mathbf{A} \text{ and } \mathbf{B}) = P(\mathbf{A}) \times P(\mathbf{B}|\mathbf{A}) \text{ for any two events } \mathbf{A} \text{ and } \mathbf{B}.$$

The probability that two events, **A** and **B**, both occur is the probability that event **A** occurs multiplied by the probability that event **B** also occurs *given* that event **A** occurs.

Of course, there's nothing special about which event we call **A** and which one we call **B**. We should be able to state this the other way around. Indeed we can. It is equally true that:

$$P(\mathbf{A} \text{ and } \mathbf{B}) = P(\mathbf{B}) \times P(\mathbf{A}|\mathbf{B}).$$

Let's return to the question of just what it means for events to be independent. We said informally in Chapter 2 that what we mean by independence is that the outcome of one event does not influence the probability of the other. With our new notation for conditional probabilities, we can write a formal definition. Events **A** and **B** are **independent** whenever:

$$P(\mathbf{B}|\mathbf{A}) = P(\mathbf{B}).$$

Now we can see that the Multiplication Rule for independent events is just a special case of the General Multiplication Rule. The general rule says

$$P(\mathbf{A} \textit{ and } \mathbf{B}) = P(\mathbf{A}) \times P(\mathbf{B}|\mathbf{A})$$

whether the events are independent or not. But when events **A** and **B** are independent, we can write $P(\mathbf{B})$ for $P(\mathbf{B}|\mathbf{A})$ and we get back our simple rule:

$$P(\mathbf{A} \textit{ and } \mathbf{B}) = P(\mathbf{A}) \times P(\mathbf{B}).$$

Sometimes people use this statement as the definition of independent events, but we find the other definition more intuitive. When events are independent, the fact that one has occurred does not affect the probability of the other.

Using our earlier example, is the probability of the event *choosing a bike* independent of the sex of the customer? We need to check whether

$$P(\text{bike}|\text{man}) = \frac{P(\text{bike } \textit{and } \text{man})}{P(\text{man})} = \frac{0.126}{0.475} = 0.264$$

is the same as $P(\text{bike}) = 0.188$.

Because these probabilities aren't equal, we can say that prize preference is *not* independent of the sex of the customer. Whenever at least one of the joint probabilities in the table is *not* equal to the product of the marginal probabilities, we say that the variables are not independent.

Independent *vs.* Disjoint

Are disjoint events independent? Both concepts seem to have similar ideas of separation and distinctness about them, but in fact disjoint events *cannot* be independent.[3] Let's see why. Consider the two disjoint events {you get an A in this course} and {you get a B in this course}. They're disjoint because they have no outcomes in common. Suppose you learn that you *did* get an A in the course. Now what is the probability that you got a B? You can't get both grades, so it must be 0.

Think about what that means. The fact that the first event (getting an A) occurred changed the probability for the second event (down to 0). So these events aren't independent.

Mutually exclusive events can never be independent. They have no outcomes in common, so knowing that one occurred means the other didn't. A common error is to treat disjoint events as if they were independent and apply the Multiplication Rule for independent events. Don't make that mistake.

Are events A and B independent or disjoint?	
Independent	Check whether $P(\mathbf{B}|\mathbf{A}) = P(\mathbf{B})$ or Check whether $P(\mathbf{A}|\mathbf{B}) = P(\mathbf{A})$ or Check whether $P(\mathbf{A} \textit{ and } \mathbf{B}) = P(\mathbf{A}) \times P(\mathbf{B})$
Disjoint	Check whether $P(\mathbf{A} \textit{ and } \mathbf{B}) = 0$ or Check whether events A and B overlap in a sample space diagram or Check whether the two events can occur together

---

[3] Technically two disjoint events *can* be independent, but only if the probability of one of the events is 0. For practical purposes, we can ignore this case, since we don't anticipate collecting data about things that don't happen.

FOR EXAMPLE   Conditional probability

QUESTION   Using the table from the example on page 157, if a customer purchases a Fashion light, what is the probability that the customer is a woman?

ANSWER $P(\text{Woman} \mid \text{Fashion}) = P(\text{Woman } and \text{ Fashion})/P(\text{Fashion})$

$$= 0.30/0.50 = 0.60$$

5.7   Constructing Contingency Tables

Sometimes we're given probabilities without a contingency table. You can often construct a simple table to correspond to the probabilities.

A survey of real estate in upstate New York classified homes into two price categories (Low—less than \$175,000 and High—over \$175,000). It also noted whether the houses had at least 2 bathrooms or not (True or False). We are told that 56% of the houses had at least 2 bathrooms, 62% of the houses were Low priced, and 22% of the houses were both. That's enough information to fill out the table. Translating the percentages to probabilities, we have:

		At Least 2 Bathrooms		
		True	**False**	**Total**
Price	**Low**	0.22		**0.62**
	High			
	Total	**0.56**		**1.00**

The 0.56 and 0.62 are marginal probabilities, so they go in the margins. What about the 22% of houses that were both low priced and had at least 2 bathrooms? That's a *joint* probability, so it belongs in the interior of the table.

Because the cells of the table show disjoint events, the probabilities always add to the marginal totals going across rows or down columns.

		At Least 2 Bathrooms		
		True	**False**	**Total**
Price	**Low**	0.22	0.40	**0.62**
	High	0.34	0.04	**0.38**
	Total	**0.56**	**0.44**	**1.00**

Now, finding any other probability is straightforward. For example, what's the probability that a high-priced house has at least 2 bathrooms?

$P(\text{at least 2 bathrooms} \mid \text{high-priced})$

$\quad = P(\text{at least 2 bathrooms } and \text{ high-priced})/P(\text{high-priced})$

$\quad = 0.34/0.38 = 0.895 \text{ or } 89.5\%.$

JUST CHECKING

3 Suppose a supermarket is conducting a survey to find out the busiest time and day for shoppers. Survey respondents are asked (1) whether they shopped at the store on a weekday or on the weekend and (2) whether they shopped at the store before or after 5 p.m. The survey revealed that:

- 48% of shoppers visited the store before 5 p.m. 27% of shoppers visited the store on a weekday (Mon.–Fri.)

- 7% of shoppers visited the store before 5 p.m. on a weekday.

a) Make a contingency table for the variables *time of day* and *day of week*.

b) What is the probability that a randomly selected shopper who shops on a weekday also shops before 5 p.m.?

c) Are time and day of the week disjoint events?

d) Are time and day of the week independent events?

5.8   Probability Trees

Some business decisions involve more subtle evaluation of probabilities. Given the probabilities of various outcomes, we can use a picture called a probability tree or **tree diagram** to help think through the decision-making process. A tree shows sequences of events as paths that look like branches of a tree. This can enable us to compare several possible scenarios. Here's a manufacturing example.

Personal electronic devices, such as smart phones and tablets, are getting more capable all the time. Manufacturing components for these devices is a challenge, and at the same time, consumers are demanding more and more functionality and increasing sturdiness. Microscopic and even submicroscopic flaws that can cause intermittent performance failures can develop during their fabrication. Defects will always occur, so the quality engineer in charge of the production process must monitor the number of defects and take action if the process seems out of control.

Let's suppose that the engineer is called down to the production line because the number of defects has crossed a threshold and the process has been declared to be out of control. She must decide between two possible actions. She knows that a small adjustment to the robots that assemble the components can fix a variety of problems, but for more complex problems, the entire production line needs to be shut down in order to pinpoint the problem. The adjustment requires that production be stopped for about an hour. But shutting down the line takes at least an entire shift (8 hours). Naturally, her boss would prefer that she make the simple adjustment. But without knowing the source or severity of the problem, she can't be sure whether that will be successful.

If the engineer wants to predict whether the smaller adjustment will work, she can use a probability tree to help make the decision. Based on her experience, the engineer thinks that there are three possible problems: (1) the motherboards could have faulty connections, (2) the memory could be the source of the faulty connections, or (3) some of the cases may simply be seating incorrectly in the assembly line. She knows from past experience how often these types of problem crop up and how likely it is that just making an adjustment will fix each type of problem. *Motherboard* problems are rare (10%), *memory* problems have been showing up about 30% of the time, and *case* alignment issues occur most often (60%). We can put those probabilities on the first set of branches in Figure 5.3 on the next page.

Figure 5.3   Possible problems and their probabilities.

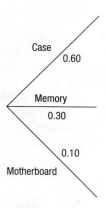

Notice that we've covered all the possibilities, and so the probabilities sum to one. To this diagram we can now add the *conditional* probabilities that a minor adjustment will fix each type of problem. Most likely the engineer will rely on her experience or assemble a team to help determine these probabilities. For example, the engineer knows that motherboard connection problems are not likely to be fixed with a simple adjustment: $P(\text{Fix}|\text{Motherboard}) = 0.10$. After some discussion, she and her team determine that $P(\text{Fix}|\text{Memory}) = 0.50$ and $P(\text{Fix}|\text{Case alignment}) = 0.80$. At the end of each branch representing the problem type, we draw two possible outcomes (*Fixed* or *Not Fixed*) and write the conditional probabilities on the branches.

Figure 5.4   Extending the tree diagram, we can show both the problem class and the outcome probabilities. The outcome (Fixed or Not fixed) probabilities are conditional on the problem type, and they change depending on which branch we follow.

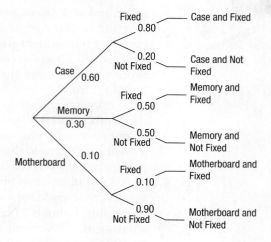

At the end of each second branch, we write the *joint event* corresponding to the combination of the two branches. For example, the top branch is the combination of the problem being Case alignment, and the outcome of the small adjustment is that the problem is now Fixed. For each of the joint events, we can use the general multiplication rule to calculate their joint probability. For example:

$$P(\textit{Case and Fixed}) = P(\textit{Case}) \times P(\textit{Fixed}\,|\,\textit{Case})$$
$$= 0.60 \times 0.80 = 0.48$$

We write this probability next to the corresponding event. Doing this for all branch combinations gives us Figure 5.5.

Figure 5.5 We can find the probabilities of compound events by multiplying the probabilities along the branch of the tree that leads to the event, just the way the General Multiplication Rule specifies.

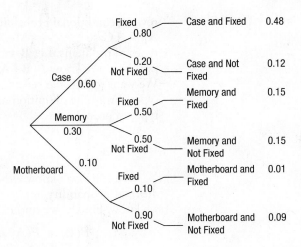

All the outcomes at the far right are disjoint because at every node, all the choices are disjoint alternatives. And those alternatives are *all* the possibilities, so the probabilities on the far right must add up to one.

Because the final outcomes are disjoint, we can add up any combination of probabilities to find probabilities for compound events. In particular, the engineer can answer her question: What's the probability that the problem will be fixed by a simple adjustment? She finds all the outcomes on the far right in which the problem was fixed. There are three (one corresponding to each type of problem), and she adds their probabilities: 0.48 + 0.15 + 0.01 = 0.64. So 64% of all problems are fixed by the simple adjustment. The other 36% require a major investigation.

*5.9   Reversing the Conditioning: Bayes' Rule

The engineer in our story decided to try the simple adjustment and, fortunately, it worked. Now she needs to report to the quality engineer on the next shift what she thinks the problem was. Was it more likely to be a case alignment problem or a motherboard problem? We know the probabilities of those problems beforehand, but they change now that we have more information. What are the likelihoods that each of the possible problems was, in fact, the one that occurred?

Unfortunately, we can't read those probabilities from the tree in Figure 5.5. For example, the tree gives us $P(\textit{Fixed and Case}) = 0.48$, but we want $P(\textit{Case}|\textit{Fixed})$. We know $P(\textit{Fixed}|\textit{Case}) = 0.80$, but that's not the same thing. It isn't valid to reverse the order of conditioning in a conditional probability statement. To "turn the probability around," we need to go back to the definition of conditional probability.

$$P(\textit{Case}\,|\textit{Fixed}) = \frac{P(\textit{Case and Fixed})}{P(\textit{Fixed})}$$

We can read the probability in the numerator from the tree, and we've already calculated the probability in the denominator by adding all the probabilities on the final branches that correspond to the event *Fixed*. Putting those values in the formula, the engineer finds:

$$P(\textit{Case}\,|\textit{Fixed}) = \frac{0.48}{0.48 + 0.15 + 0.01} = 0.75$$

She knew that 60% of all problems were due to case alignment, but now that she knows the problem has been fixed, she knows more. Given the additional information that a simple adjustment was able to fix the problem, she now can increase the probability that the problem was case alignment to 0.75.

It's usually easiest to solve problems like this by reading the appropriate probabilities from the tree. However, we can write a general formula for finding the reverse conditional probability. To understand it, let's review our example again. Let $A_1 = \{Case\}$, $A_2 = \{Memory\}$, and $A_3 = \{Motherboard\}$ represent the three types of problems. Let $B = \{Fixed\}$, meaning that the simple adjustment fixed the problem. We know $P(B \mid A_1) = 0.80$, $P(B \mid A_2) = 0.50$, and $P(B \mid A_3) = 0.10$. We want to find the reverse probabilities, $P(A_i \mid B)$, for the three possible problem types. From the definition of conditional probability, we know (for any of the three types of problems):

$$P(A_i \mid B) = \frac{P(A_i \, and \, B)}{P(B)}$$

We still don't know either of these quantities, but we use the definition of conditional probability again to find $P(A_i \, and \, B) = P(B \mid A_i)P(A_i)$, both of which we know. Finally, we find $P(B)$ by adding up the probabilities of the three events.

$$P(B) = P(A_1 \, and \, B) + P(A_2 \, and \, B) + P(A_3 \, and \, B) = $$
$$P(B \mid A_1)P(A_1) + P(B \mid A_2)P(A_2) + P(B \mid A_3)P(A_3)$$

In general, we can write this for n events A_i that are mutually exclusive (each pair is disjoint) and exhaustive (their union is the whole space). Then:

$$P(A_i \mid B) = \frac{P(B \mid A_i)P(A_i)}{\sum_j P(B \mid A_j)P(A_j)}$$

This formula is known as Bayes' rule, after the Reverend Thomas Bayes (1702–1761), even though historians don't really know if Bayes first came up with the reverse conditioning probability. When you need to find reverse conditional probabilities, we recommend drawing a tree and finding the appropriate probabilities as we did at the beginning of the section, but the formula gives the general rule.

WHAT CAN GO WRONG?

- **Beware of probabilities that don't add up to 1.** To be a legitimate assignment of probability, the sum of the probabilities for all possible outcomes must total 1. If the sum is less than 1, you may need to add another category ("other") and assign the remaining probability to that outcome. If the sum is more than 1, check that the outcomes are disjoint. If they're not, then you can't assign probabilities by counting relative frequencies. (And if they are, you must locate the error.)

- **Don't add probabilities of events if they're not disjoint.** Events must be disjoint to use the Addition Rule. The probability of being under 80 *or* a female is not the probability of being under 80 *plus* the probability of being female. That sum may be more than 1.

- **Don't multiply probabilities of events if they're not independent.** The probability of selecting a customer at random who is over 70 years old *and* retired is not the probability the customer is over 70 years old *times* the probability the customer is retired. Knowing that the customer is over 70 changes the probability of his or her being retired. You can't multiply these probabilities. The multiplication of probabilities of events that are not independent is one of the most common errors people make in dealing with probabilities.

- **Don't confuse disjoint and independent.** Disjoint events *can't* be independent. If $A = \{you \, get \, a \, promotion\}$ and $B = \{you \, don't \, get \, a \, promotion\}$, A and B are disjoint. Are they independent? If you find out that A is true, does that change the probability of B? Yes, if A is true, then B cannot be true, so they are not independent.

ETHICS IN ACTION

Fabrizio Rivetti is an entrepreneur who has recently started a wine importing business. While he currently has an exclusive relationship with only one premier winery in Tuscany, he is hoping to expand his importing business to include other wineries as well as artisan Italian food products, such as cheeses and specialty meats.

With plans to expand, Fabrizio is in need of extra funds. As a first step, he approaches a friend and fellow entrepreneur who has considerable experience dealing with angel investors, Chas Mulligan. Chas has successfully obtained funds from angel investors for his social networking start-up company, so Fabrizio is hopeful that Chas can provide some sound advice. Chas explains to Fabrizio that most angel investors bear considerable risk and consequently favor ventures that are in high-growth areas, such as software, healthcare, and biotech. He also mentions that many angel investors, like venture capitalists, want to exercise some control over the start-up companies in which they invest, either by securing a seat on the company's board of directors or having veto power.

Fabrizio is now a bit unsure about seeking angel investments, so Chas puts him in contact with a consultant, Paula Foxx, who can help him make the right decision. Paula is well connected with a network of angel investors, understands the types of start-ups they prefer to invest in, and, most importantly, knows how to prepare the perfect pitch. At their first meeting, Paula is quick to inform Fabrizio of her consultancy fee schedule. Next, she assures Fabrizio that she is acquainted with a number of angels whom she believes might be interested in his wine importing business. Fabrizio expresses to Paula his reservations about sharing too much control of his start-up with investors, and is particularly wary of granting investors veto power.

He is reluctant to hire Paula on the spot, so Paula suggests they meet again after she has had the opportunity to pull together data on some of her most successful clients. Paula's objective is to direct clients to angels who tend to make large initial investments. In this way, her clients reach their goals more quickly and she can spend less time with each client. She decides to compile some data only for angel investors who have made significant initial investments in her clients' start-ups (in excess of $250,000). She came up with the following contingency table for this group of investors.

		Veto Power?		
		Yes	No	Total
Board Seat?	Yes	.05	.45	.50
	No	.45	.05	.50
	Total	.50	.50	1.00

She was happy to find that 50% did not get veto power and 50% did not sit on the board. By multiplying these two probabilities, she arrived at a figure she thought would help persuade Fabrizio to pursue angels and hire her to do so. She planned to tell him that 25% of angels who make large investments are not interested in either veto power or a seat on the board in the start-ups they fund. She called her administrative assistant to arrange another meeting with Fabrizio as soon as possible.

- Identify the ethical dilemma in this scenario.

- What are the undesirable consequences?

- Propose an ethical solution that considers the welfare of all stakeholders

WHAT HAVE WE LEARNED?

Learning Objectives

Apply the facts about probability to determine whether an assignment of probabilities is legitimate.
- Probability is long-run relative frequency.
- Individual probabilities must be between 0 and 1.
- The sum of probabilities assigned to all outcomes must be 1.

Understand the Law of Large Numbers and that the common understanding of the "Law of Averages" is false.

Know the rules of probability and how to apply them.
- The **Complement Rule** says that $P(not\ \mathbf{A}) = P(\mathbf{A^C}) = 1 - P(\mathbf{A})$.
- The **Multiplication Rule** for independent events says that $P(\mathbf{A}\ and\ \mathbf{B}) = P(\mathbf{A}) \times P(\mathbf{B})$ provided events $\mathbf{A}$ and $\mathbf{B}$ are independent.

- The **General Multiplication Rule** says that $P(A \text{ and } B) = P(A) \times P(B|A)$ for any events **A** and **B**.
- The **Addition Rule** for disjoint events says that $P(A \text{ or } B) = P(A) + P(B)$ provided events A and B are disjoint.
- The **General Addition Rule** says that $P(A \text{ or } B) = P(A) + P(B) - P(A \text{ and } B)$ for any events **A** and **B**.

Know how to construct and read a contingency table.

Know how to define and use independence.

- Events A and B are independent if $P(A|B) = P(A)$.

Know how to construct tree diagrams and use them to calculate and understand conditional probabilities.

Know how to use Bayes' Rule to compute conditional probabilities.

Terms

Addition Rule	If **A** and **B** are disjoint events, then the probability of **A** *or* **B** is $$P(A \text{ or } B) = P(A) + P(B).$$
Complement Rule	The probability of an event occurring is 1 minus the probability that it doesn't occur: $$P(A) = 1 - P(A^C).$$
Conditional probability	$$P(B|A) = \frac{P(A \text{ and } B)}{P(A)}.$$ $P(B|A)$ is read "the probability of **B** *given* **A**."
Disjoint (or Mutually Exclusive) Events	Two events are disjoint if they have no outcomes in common. If **A** and **B** are disjoint, then the fact that **A** occurs tells us that **B** cannot occur. Disjoint events are also called "mutually exclusive."
Empirical probability	When the probability comes from the long-run relative frequency of the event's occurrence, it is an empirical probability.
Event	A collection of outcomes. Usually, we identify events so that we can attach probabilities to them. We denote events with bold capital letters such as **A**, **B**, or **C**.
General Addition Rule	For any two events, **A** and **B**, the probability of **A** *or* **B** is: $$P(A \text{ or } B) = P(A) + P(B) - P(A \text{ and } B).$$
General Multiplication Rule	For any two events, **A** and **B**, the probability of **A** *and* **B** is: $$P(A \text{ and } B) = P(A) \times P(B|A).$$
Independence (informally)	Two events are *independent* if the fact that one event occurs does not change the probability of the other.
Independence (used formally)	Events **A** and **B** are independent when $P(B|A) = P(B)$.
Joint probabilities	The probability that two events both occur.
Law of Large Numbers (LLN)	The Law of Large Numbers states that the *long-run relative frequency* of repeated, independent events settles down to the *true relative frequency* as the number of trials increases.
Marginal probability	In a joint probability table a marginal probability is the probability distribution of either variable separately, usually found in the rightmost column or bottom row of the table.
Multiplication Rule	If **A** and **B** are independent events, then the probability of **A** *and* **B** is: $$P(A \text{ and } B) = P(A) \times P(B).$$

Outcome	The outcome of a trial is the value measured, observed, or reported for an individual instance of that trial.
Personal probability	When the probability is subjective and represents one's personal degree of belief, it is called a personal probability.
Probability	The probability of an event is a number between 0 and 1 that reports the likelihood of the event's occurrence. A probability can be derived from a model (such as equally likely outcomes), from the long-run relative frequency of the event's occurrence, or from subjective degrees of belief. We write $P(\mathbf{A})$ for the probability of the event $\mathbf{A}$.
Probability Assignment Rule	The probability of the entire sample space must be 1: $$P(S) = 1.$$
Random phenomenon	A phenomenon is random if we know what outcomes *could* happen, but not which particular values *will* happen in any given trial.
Sample space	The collection of all possible outcome values. The sample space has a probability of 1.
Theoretical probability	When the probability comes from a mathematical model (such as, but not limited to, equally likely outcomes), it is called a theoretical probability.
Trial	A single attempt or realization of a random phenomenon.
Tree diagram (or probability tree)	A display of conditional events or probabilities that is helpful in thinking through conditioning.

TECHNOLOGY HELP: Generating Random Numbers

Most statistics packages generate single or lists of random numbers. You may find them useful for introducing randomness in a study or drawing a random sample. Excel can generate random numbers with the **RAND( )** function.

EXCEL

To generate a random number in Excel:

- In a cell, type **= RAND( )**. A random number between 0 and 1 (a real number to 9 decimal places) appears in the cell.

- To generate more random numbers, copy and paste this cell or select it and **Fill Down** to obtain more random values.

- To generate a random number within a range, type **= RAND( )*(b − a) + a** into the formula bar where **a** is the number at the low end of the range and **b** is the number at the high end of the range.

- You can also use the function **= RANDBETWEEN (a, b)** to generate an integer between a and b.

Random numbers are re-generated each time a change is made to the spreadsheet. To avoid this:

- Highlight the cell containing the random number.

- Copy the value and paste into same cell using the **Paste Values: Values** command.

Brief Case

Global Markets

A global survey firm reports data from surveys taken in several countries. The data file **Global** holds data for 800 respondents in each of five countries. The variables provide demographic information (sex, age, education, marital status) and responses to questions of interest to marketers on personal finance and purchasing.

Write a report that discusses how decisions about personal finance and shopping vary by country and by sex. You'll want to make contingency tables of some variables and consider the contingent probabilities that they show. You may also want to restrict your attention to one country and then consider relationships between variables within that country.

EXERCISES -

SECTION 5.1

1. Indicate which of the following represent independent events. Explain briefly.

a) The gender of customers using an ATM machine.
b) The last digit of the social security numbers of students in a class.
c) The scores you receive on the first midterm, second midterm, and the final exam of a course.

2. Indicate which of the following represent independent events. Explain briefly.

a) Prices of houses on the same block.
b) Successive measurements of your heart rate as you exercise on a treadmill.
c) Measurements of the heart rates of all students in the gym.

SECTION 5.2

3. In many state lotteries, you can choose which numbers to play. Consider a common form in which you choose 5 numbers. Which of the following strategies can improve your chance of winning? If the method works, explain why. If not, explain why using appropriate statistics terms.

a) Always play 1, 2, 3, 4, 5.
b) Choose the numbers that did come up in the most recent lottery drawing because they are "hot."

4. For the same kind of lottery as in Exercise 3, which of the following strategies can improve your chance of winning? If the method works, explain why. If not, explain why using appropriate statistics terms.

a) Choose randomly from among the numbers that have *not* come up in the last 3 lottery drawings because they are "due."
b) Generate random numbers using a computer or calculator and play those.

SECTION 5.4

5. A recent survey found that, despite airline requests, about 40% of passengers don't fully turn off their cell phones during takeoff and landing (although they may put them in "airplane mode"). The two passengers across the aisle (in seats A and B) clearly do not know each other.

a) What is the probability that the passenger in seat A does not turn off his phone?
b) What is the probability that he does turn off his phone?
c) What is the probability that both of them turn off their phones?
d) What is the probability that at least one of them turns off his or her phone?

6. At your school, 10% of the class are marketing majors. If you are randomly assigned to two partners in your statistics class,

a) What is the probability that the first partner will be a marketing major?
b) What is the probability that the first partner won't be a marketing major?
c) What is the probability that both will be marketing majors?
d) What is the probability that at least one will be a marketing major?

SECTION 5.5

7. The following contingency table shows opinion about global warming among U.S. adults, broken down by political party affiliation (based on a poll in October 2012 by Pew Research found at http://www.people-press.org/2012/10/15 /more-say-there-is-solid-evidence-of-global-warming/).

		Opinion on Global Warming		
		Nonissue	Serious Concern	Total
Political Party	Democratic	85	415	**500**
	Republican	290	210	**500**
	Independent	70	130	**200**
	Total	**445**	**755**	**1200**

a) What is the probability that a U.S. adult selected at random from these 1200 respondents believes that global warming is a serious issue?
b) What type of probability did you find in part a?
c) What is the probability that a U.S. adult selected at random is a Republican and believes that global warming is a serious issue?
d) What type of probability did you find in part c?

8. Multigenerational families can be categorized as having two adult generations such as parents living with adult children, "skip" generation families, such as grandparents living with grandchildren, and three or more generations living in the household. Pew Research surveyed multigenerational households. This table is based on their reported results.

	2 Adult Gens	2 Skip Gens	3 or More Gens	
White	509	55	222	**786**
Hispanic	139	11	142	**292**
Black	119	32	99	**250**
Asian	61	1	48	**110**
	828	**99**	**511**	**1438**

a) What is the probability that a multigenerational family is Hispanic?
b) What is the probability that a multigenerational family selected at random is a Black, two-adult-generation family?
c) What type of probability did you find in parts a and b?

SECTION 5.6

9. Using the table from Exercise 7,

a) What is the probability that a randomly selected U.S. adult who is a Republican believes that global warming is a serious issue?
b) What is the probability that a randomly selected U.S. adult is a Republican given that he or she believes global warming is a serious issue?
c) What is $P(\text{Serious Concern} \mid \text{Democratic})$?

10. Using the table from Exercise 8,

a) What is the probability that a randomly selected Black multigenerational family is a two-adult-generation family?
b) What is the probability that a randomly selected multigenerational family is White, given that it is a "skip" generation family?
c) What is $P(3 \text{ or more Generations} \mid \text{Asian})$?

SECTION 5.7

11. A national survey indicated that 30% of adults conduct their banking online. It also found that 40% are under the age of 50, and that 25% are under the age of 50 and conduct their banking online.

a) What percentage of adults do not conduct their banking online?
b) What type of probability is the 25% mentioned above?
c) Construct a contingency table showing all joint and marginal probabilities.
d) What is the probability that an individual conducts banking online given that the individual is under the age of 50?
e) Are *Banking online* and *Age* independent? Explain.

12. Facebook reports that 70% of their users are from outside the United States and that 50% of their users log on to Facebook every day. Suppose that 20% of their users are United States users who log on every day.

a) What percentage of Facebook's users are from the United States?
b) What type of probability is the 20% mentioned above?
c) Construct a contingency table showing all the joint and marginal probabilities.
d) What is the probability that a user is from the United States given that he or she logs on every day?
e) Are *From United States* and *Log on Every Day* independent? Explain.

SECTION 5.8

13. Summit Projects provides marketing services and website management for many companies that specialize in outdoor products and services (www.summitprojects.com). To understand customer Web behavior, the company experiments with different offers and website design. The results of such experiments can help to maximize the probability that customers purchase products during a visit to a website. Possible actions by the website include offering the customer an instant discount, offering the customer free shipping, or doing nothing. A recent experiment found that customers make purchases 6% of the time when offered the instant discount, 5% when offered free shipping, and 2% when no special offer was given. Suppose 20% of the customers are offered the discount and an additional 30% are offered free shipping.

a) Construct a probability tree for this experiment.
b) What percent of customers who visit the site made a purchase?
c) Given that a customer made a purchase, what is the probability that they were offered free shipping?

14. The company in Exercise 13 performed another experiment in which they tested three website designs to see which one would lead to the highest probability of purchase. The first (design A) used enhanced product information, the second (design B) used extensive iconography, and the third (design C) allowed the customer to submit their own product ratings. After 6 weeks of testing, the designs delivered probabilities of purchase of 4.5%, 5.2%, and 3.8%, respectively. Equal numbers of customers were sent randomly to each website design.

a) Construct a probability tree for this experiment.
b) What percent of customers who visited the site made a purchase?
c) What is the probability that a randomly selected customer was sent to design C?
d) Given that a customer made a purchase, what is the probability that the customer had been sent to design C?

SECTION 5.9

15. According to U.S. Census data, 68% of the civilian U.S. labor force self-identifies as White, 11% as Black, and the remaining 21% as Hispanic/Latino or Other. Among Whites in the labor force, 54% are Male, and 46% Female. Among Blacks, 52% are Male and 48% Female, and among Hispanic/Latino/Other, 58% are Male and 42% are Female.

a) Polling companies need to sample an appropriate number of respondents of each gender from each ethnic group. For a randomly selected U.S. worker, fill in the probabilities in this tree:

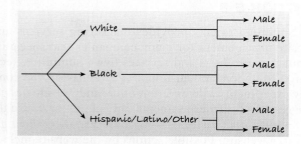

b) What is the probability that a randomly selected worker is a Black Female?
c) For a randomly selected worker, what is $P(Female \mid White)$?
d) For a randomly selected worked what is $P(White \mid Female)$?

16. U.S. Customs and Border Protection has been testing automated kiosks that may be able to detect lies (www.wired.com/threatlevel/2013/01/ff-lie-detector/all/). One measurement used (among several) is involuntary eye movements. Using this method alone, tests show that it can detect 60% of lies, but incorrectly identifies 15% of true statements as lies. Suppose that 95% of those entering the country tell the truth. The immigration kiosk asks questions such as "Have you ever been arrested for a crime?" Naturally, all the applicants answer "No," but the kiosk identifies some of those answers as lies, and refers the entrant to a human interviewer.

a) Here is the outline of a probability tree for this situation. Fill in the probabilities:

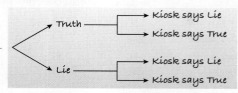

b) What is the probability that a random person will be telling the truth and will be cleared by the Kiosk?
c) What is the probability that a person who is rejected by the kiosk was actually telling the truth?

CHAPTER EXERCISES

17. What does it mean? Part 1. Respond to the following questions:

a) A casino claims that its roulette wheel is truly random. What should that claim mean?
b) A reporter on *Market Place* says that there is a 50% chance that the NASDAQ will hit a new high in the next month. What is the meaning of such a phrase?

18. What does it mean? Part 2. Respond to the following questions:

a) After an unusually dry autumn, a radio announcer is heard to say, "Watch out! We'll pay for these sunny days later on this winter." Explain what he's trying to say, and comment on the validity of his reasoning.
b) A batter who had failed to get a hit in seven consecutive times at bat then hits a game-winning home run. When talking to reporters afterward, he says he was very confident that last time at bat because he knew he was "due for a hit." Comment on his reasoning.

19. Airline safety. Even though commercial airlines have excellent safety records, in the weeks following a crash, airlines often report a drop in the number of passengers, probably because people are afraid to risk flying.

a) A travel agent suggests that since the law of averages makes it highly unlikely to have two plane crashes within a few weeks of each other, flying soon after a crash is the safest time. What do you think?
b) If the airline industry proudly announces that it has set a new record for the longest period of safe flights, would you be reluctant to fly? Are the airlines due to have a crash?

20. Economic predictions. An investment newsletter makes general predictions about the economy to help their clients make sound investment decisions.

a) Recently they said that because the stock market had been up for the past three months in a row that it was "due for a correction" and advised their client to reduce their holdings. What "law" are they applying? Comment.
b) They advised buying a stock that had gone down in the past four sessions because they said that it was clearly "due to bounce back." What "law" are they applying? Comment.

21. Fire insurance. Insurance companies collect annual payments from homeowners in exchange for paying to rebuild houses that burn down.

a) Why should you be reluctant to accept a $3000 payment from your neighbor to replace his house should it burn down during the coming year?

b) Why can the insurance company make that offer?

22. Casino gambling. Recently, the International Gaming Technology company issued the following press release:

(LAS VEGAS, Nev.)—Cynthia Jay was smiling ear to ear as she walked into the news conference at the Desert Inn Resort in Las Vegas today, and well she should. Last night, the 37-year-old cocktail waitress won the world's largest slot jackpot—$34,959,458—on a Megabucks machine. She said she had played $27 in the machine when the jackpot hit. Nevada Megabucks has produced 49 major winners in its 14-year history. The top jackpot builds from a base amount of $7 million and can be won with a 3-coin ($3) bet.

a) How can the Desert Inn afford to give away millions of dollars on a $3 bet?

b) Why did the company issue a press release? Wouldn't most businesses want to keep such a huge loss quiet?

23. Toy company. A toy company is preparing to market an electronic game for young children that "randomly" generates a color. They suspect, however, that the way the random color is determined may not be reliable, so they ask the programmers to perform tests and report the frequencies of each outcome. Are each of the following probability assignments possible? Why or why not?

	Probabilities of …			
	Red	Yellow	Green	Blue
a)	0.25	0.25	0.25	0.25
b)	0.10	0.20	0.30	0.40
c)	0.20	0.30	0.40	0.50
d)	0	0	1.00	0
e)	0.10	0.20	1.20	−1.50

24. Store discounts. Many stores run "secret sales": Shoppers receive cards that determine how large a discount they get, but the percentage is revealed by scratching off that black stuff (what *is* that?) only after the purchase has been totaled at the cash register. The store is required to reveal (in the fine print) the distribution of discounts available. Are each of these probability assignments plausible? Why or why not?

	Probabilities of …			
	10% Off	20% Off	30% Off	50% Off
a)	0.20	0.20	0.20	0.20
b)	0.50	0.30	0.20	0.10
c)	0.80	0.10	0.05	0.05
d)	0.75	0.25	0.25	−0.25
e)	1.00	0	0	0

25. Quality control. A tire manufacturer recently announced a recall because 2% of its tires are defective. If you just bought a new set of four tires from this manufacturer, what is the probability that at least one of your new tires is defective?

26. Pepsi promotion. For a sales promotion, the manufacturer places winning symbols under the caps of 10% of all Pepsi bottles. If you buy a six-pack of Pepsi, what is the probability that you win something?

27. Auto warranty. In developing their warranty policy, an automobile company estimates that over a 1-year period 17% of their new cars will need to be repaired once, 7% will need repairs twice, and 4% will require three or more repairs. If you buy a new car from them, what is the probability that your car will need:

a) No repairs?

b) No more than one repair?

c) Some repairs?

28. Consulting team. You work for a large global management consulting company. Of the entire work force of analysts, 55% have had no experience in the telecommunications industry, 32% have had limited experience (less than 5 years), and the rest have had extensive experience (5 years or more). On a recent project, you and two other analysts were chosen at random to constitute a team. It turns out that part of the project involves telecommunications. What is the probability that the first teammate you meet has:

a) Extensive telecommunications experience?

b) Some telecommunications experience?

c) No more than limited telecommunications experience?

29. Auto warranty, part 2. Consider again the auto repair rates described in Exercise 27. If you bought two new cars, what is the probability that:

a) Neither will need repair?

b) Both will need repair?

c) At least one car will need repair?

30. Consulting team, part 2. You are assigned to be part of a team of three analysts of a global management consulting company as described in Exercise 28. What is the probability that of your other two teammates:

a) Neither has any telecommunications experience?

b) Both have some telecommunications experience?

c) At least one has had extensive telecommunications experience?

31. Auto warranty, again. You used the Multiplication Rule to calculate repair probabilities for your cars in Exercise 29.

a) What must be true about your cars in order to make that approach valid?

b) Do you think this assumption is reasonable? Explain.

32. Final consulting team project. You used the Multiplication Rule to calculate probabilities about the telecommunications experience of your consulting teammates in Exercise 30.

a) What must be true about the groups in order to make that approach valid?

b) Do you think this assumption is reasonable? Explain.

33. Real estate. In a sample of real estate ads, 64% of homes for sale had garages, 21% have swimming pools, and 17% have both features. What is the probability that a home for sale has:

a) A pool, a garage, or both?

b) Neither a pool nor a garage?

c) A pool but no garage?

34. Human resource data. Employment data at a large company reveal that 72% of the workers are married, 44% are college graduates, and half of the college grads are married. What's the probability that a randomly chosen worker is:

a) Neither married nor a college graduate?

b) Married but not a college graduate?

c) Married or a college graduate?

35. Mars product information. The Mars company says that before the introduction of purple, yellow made up 20% of their plain M&M candies, red made up another 20%, and orange, blue, and green each made up 10%. The rest were brown.

a) If you picked an M&M at random from a pre-purple bag of candies, what is the probability that it was:

　i) Brown?

　ii) Yellow or orange?

　iii) Not green?

　iv) Striped?

b) Assuming you had an infinite supply of M&M's with the older color distribution, if you picked three M&M's in a row, what is the probability that:

　i) They are all brown?

　ii) The third one is the first one that's red?

　iii) None are yellow?

　iv) At least one is green?

36. American Red Cross. The American Red Cross must track their supply and demand for various blood types. They estimate that about 45% of the U.S. population has Type O blood, 40% Type A, 11% Type B, and the rest Type AB.

a) If someone volunteers to give blood, what is the probability that this donor:

　i) Has Type AB blood?

　ii) Has Type A or Type B blood?

　iii) Is not Type O?

b) Among four potential donors, what is the probability that:

　i) All are Type O?

　ii) None have Type AB blood?

　iii) Not all are Type A?

　iv) At least one person is Type B?

37. More Mars product information. In Exercise 35, you calculated probabilities of getting various colors of M&M's.

a) If you draw one M&M, are the events of getting a red one and getting an orange one disjoint or independent or neither?

b) If you draw two M&M's one after the other, are the events of getting a red on the first and a red on the second disjoint or independent or neither?

c) Can disjoint events ever be independent? Explain.

38. American Red Cross, part 2. In Exercise 36, you calculated probabilities involving various blood types.

a) If you examine one donor, are the events of the donor being Type A and the donor being Type B disjoint or independent or neither? Explain your answer.

b) If you examine two donors, are the events that the first donor is Type A and the second donor is Type B disjoint or independent or neither?

c) Can disjoint events ever be independent? Explain.

39. Tax accountant. A recent study of IRS audits showed that, for estates worth less than $5 million, about 1 out of 7 of all estate tax returns are audited, but that probability increases to 50% for estates worth over $5 million. Suppose a tax accountant has three clients who have recently filed returns for estates worth more than $5 million. What are the probabilities that:

a) All three will be audited?

b) None will be audited?

c) At least one will be audited?

d) What did you assume in calculating these probabilities?

40. Casinos. Because gambling is big business, calculating the odds of a gambler winning or losing in every game is crucial to the financial forecasting for a casino. A standard slot machine has three wheels that spin independently. Each has 10 equally likely symbols: 4 bars, 3 lemons, 2 cherries, and a bell. If you play once, what is the probability that you will get:

a) 3 lemons?

b) No fruit symbols?

c) 3 bells (the jackpot)?

d) No bells?

e) At least one bar (an automatic loser)?

41. Spam filter. A company has recently replaced their e-mail spam filter because investigations had found that the volume of spam e-mail was interrupting productive work on about 15% of workdays. To see how bad the situation was, calculate the probability that during a 5-day work week, e-mail spam would interrupt work:

a) On Monday and again on Tuesday?

b) For the first time on Thursday?

c) Every day?

d) At least once during the week?

42. Tablet tech support. The technical support desk at a college has set up a special service for tablets. A survey shows that 54% of tablets on campus run Apple's iOS, 43% run Google's Android OS, and 3% run Microsoft's Windows. Assuming that users of each of the operating systems are equally likely to call in for technical support what is the probability that of the next three calls:

a) All are iOS?
b) None are Android?
c) At least one is a Windows machine?
d) All are Windows machines?

43. Casinos, part 2. In addition to slot machines, casinos must understand the probabilities involved in card games. Suppose you are playing at the blackjack table, and the dealer shuffles a deck of cards. The first card shown is red. So is the second and the third. In fact, you are surprised to see 5 red cards in a row. You start thinking, "The next one is due to be black!"

a) Are you correct in thinking that there's a higher probability that the next card will be black than red? Explain.
b) Is this an example of the Law of Large Numbers? Explain.

44. Inventory. A shipment of road bikes has just arrived at The Spoke, a small bicycle shop, and all the boxes have been placed in the back room. The owner asks her assistant to start bringing in the boxes. The assistant sees 20 identical-looking boxes and starts bringing them into the shop at random. The owner knows that she ordered 10 women's and 10 men's bicycles, and so she's surprised to find that the first six are all women's bikes. As the seventh box is brought in, she starts thinking, "This one is bound to be a men's bike."

a) Is she correct in thinking that there's a higher probability that the next box will contain a men's bike? Explain.
b) Is this an example of the Law of Large Numbers? Explain.

45. U.S. economic conditions 2013. A Gallup Poll in March 2013 (www.gallup.com/poll/110821/gallup-daily-us-economic-conditions.aspx) asked 1500 U.S. adults to rate economic conditions in the country today as "excellent," "good," "only fair," or "poor." The results are below:

Current Economic Conditions	
Response	**Number of Respondents**
Excellent/Good	270
Only Fair	645
Poor	585
Total	**1500**

If we select a person at random from this sample of 1500 adults:

a) What is the probability that the person responded "Poor"?
b) What is the probability that the person responded "Fair" or "Poor"?

46. More economic conditions 2013. Exercise 45 shows the results of a Gallup Poll about U.S. economic conditions. Suppose we select three adults at random from this sample.

a) What is the probability that all three responded "Poor"?
b) What is the probability that none responded "Poor"?
c) What assumption did you make in computing these probabilities?
d) Explain why you think that assumption is reasonable.

47. Owning guns. The General Social Survey, run annually, asked respondents "Do you have in your home (or garage) any guns or revolvers?" The responses are given in the table (sda.berkeley.edu/cgi-bin/hsda?harcsda+gss10).

Gun Ownership	
Response	**Number of Respondents**
Yes	410
No	815
Don't Know/Refused	45
Total	**1270**

a) If we select a random person from this sample of 1270 adults, what is the probability that their response will be "No"?
b) What is the probability that their response will be "Don't know/refused"?
c) Show another way to calculate the probability in part b. (Hint: Use the complement.)

48. Gun ownership, part 2. Exercise 47 shows the results of a poll that asked about gun ownership. Suppose we select three adults at random from this sample.

a) What is the probability that all three respond "Yes"?
b) What is the probability that none responded "Yes"?
c) What assumption did you make in computing these probabilities?
d) Explain why you think that assumption is reasonable.

49. Contract bidding. As manager for a construction firm, you are in charge of bidding on two large contracts. You believe the probability you get contract #1 is 0.8. If you get contract #1, the probability you also get contract #2 will be 0.2, and if you do not get #1, the probability you get #2 will be 0.4.

a) Sketch the probability tree.
b) What is the probability you will get both contracts?
c) Your competitor hears that you got the second contract but hears nothing about the first contract. Given that you got the second contract, what is the probability that you also got the first contract?

50. Extended warranties. A company that manufactures and sells consumer video cameras sells two versions of their popular hard disk camera, a basic camera for $750, and a deluxe version for $1250. About 75% of customers select the basic camera. Of those, 60% purchase the extended warranty for an additional $200. Of the people who buy the deluxe version, 90% purchase the extended warranty.

a) Sketch the probability tree for total purchases.
b) What is the percentage of customers who buy an extended warranty?
c) What is the expected revenue of the company from a camera purchase (including warranty if applicable)?
d) Given that a customer purchases an extended warranty, what is the probability that he or she bought the deluxe version?

51. Tweeting. According to the Pew 2012 News Consumption survey, 50% of adults who post news on Twitter ("tweet") are younger than 30. But according to the U.S. Census, only 23% of adults are less than 30 years old. A separate survey by Pew in 2012 found that 15% of adults tweet.

a) Find the probability that a random adult is both less than 30 years old and a Twitter poster. That is, find $P(\text{Tweet and} <30)$.
b) For a random young (<30) adult, what is the probability he or she is a tweeter? That is, find $P(\text{Tweet}|<30)$.

52. Titanic survival. Of the 2201 people on the RMS Titanic, only 711 survived. The practice of "women and children first" was first used to describe the chivalrous actions of the sailors during the sinking of the HMS Birkenhead in 1852, but became popular after the sinking of the Titanic, during which 53% of the children and 73% of the women survived, but only 21% of the men survived. Part of the protocol stated that passengers enter lifeboats by ticket class as well. Here is a table showing survival by ticket class.

	First	Second	Third	Crew	Total
Alive	203	118	178	212	711
	28.6%	16.6%	25.0%	29.8%	100%
Dead	122	167	528	673	1490
	8.2%	11.2%	35.4%	45.2%	100%

a) Find the conditional probability of survival for each type of ticket.
b) Draw a probability tree for this situation.
c) Given that a passenger survived, what is the probability they had a first-class ticket?

53. Coffeehouse survey. A 2011 Mintel report on coffeehouses asked consumers if they were spending more time in coffeehouses. The table below gives the responses classified by age:

a) What is the probability that a randomly selected respondent is spending more time at coffeehouses and donut shops this year than last year?
b) What is the probability that the person is younger than 25 years old?
c) What is the probability that the person is younger than 25 years old and is spending more time at coffeehouses and donut shops compared to last year?
d) What is the probability that the person is younger than 25 years old *or* is spending more time at coffeehouses and donut shops compared to last year?

	Age						
	18–24	25–34	35–44	45–54	55–64	65 +	Total
I am spending less time at coffeehouses and donut shops this year than last year.	78	93	102	104	68	48	493
I am spending about the same time at coffeehouses and donut shops this year as last year.	82	109	106	89	75	67	528
I am spending more time at coffeehouses and donut shops this year than last year.	30	30	18	19	11	6	114
Total	190	232	226	212	154	121	1135

Source: 2011 Mintel Report. Reprinted by permission of Mintel, a leading market research company (www.mintel.com).

54. Electronic communications. A Mintel study asked consumers if electronic communications devices influenced whether or not they bought a certain car. The table on the next page gives the results classified by household income: If we select a person at random from this sample:

a) What is the probability that electronic communication devices somewhat influenced their decisions?

b) What is the probability that the person is earning at least $100K?
c) What is the probability that the person was somewhat influenced by electronic communications *and* earns at least $100K?
d) What is the probability that electronic communications somewhat influenced the purchase *or* that the person earns at least $100K?

Communications influence on car purchase, by household income, July 2011

Communication (e.g., hands-free calling):	Income			
	<$50K	$50K–99.9K	$100K +	Total
Very Much	30	57	41	128
Somewhat	26	39	62	127
Not At All	23	39	35	97
Total	79	135	138	352

Source: Mintel.

55. Red Cross Rh. Exercises 36 and 38 discussed the challenges faced by the Red Cross in finding enough blood of various types. But blood typing also depends on the Rh factor, which can be negative or positive. Here is a table of the estimated proportions worldwide for blood types categorized on both type and Rh factor:

		Blood Type			
		O	A	B	AB
Rh	+	36.44%	28.27%	20.59%	5.06%
	−	4.33%	3.52%	1.39%	0.45%

For a randomly selected human, what is the probability that he …

a) Is Rh negative given that he is type O?
b) Is type O given that he is Rh negative?
c) A person with Type A⁻ blood can accept donated blood only of types A⁻ and O⁻. What is the probability that a randomly selected donor can donate to a recipient given that the recipient's blood type is A⁻?

56. Automobile inspection. Twenty percent of cars that are inspected have faulty pollution control systems. The cost of repairing a pollution control system exceeds $100 about 40% of the time. When a driver takes her car in for inspection, what's the probability that she will end up paying more than $100 to repair the pollution control system?

57. Pharmaceutical company. A U.S. pharmaceutical company is considering manufacturing and marketing a pill that will help to lower both an individual's blood pressure and cholesterol. The company is interested in understanding the demand for such a product. The joint probabilities that an adult American man has high blood pressure, high cholesterol, or both are shown in the table.

		Blood Pressure	
		High	OK
Cholesterol	High	0.11	0.21
	OK	0.16	0.52

a) What's the probability that an adult American male has both conditions?
b) What's the probability that an adult American male has high blood pressure?

c) What's the probability that an adult American male with high blood pressure also has high cholesterol?
d) What's the probability that an adult American male has high blood pressure if it's known that he has high cholesterol?

58. International relocation. A European department store is developing a new advertising campaign for their new U.S. location, and their marketing managers need to understand their target market better. A survey of adult shoppers found the probabilities that an adult would shop at their new U.S. store classified by age is shown below.

		Shop		
		Yes	No	Total
Age	<20	0.26	0.04	0.30
	20–40	0.24	0.10	0.34
	>40	0.12	0.24	0.36
	Total	0.62	0.38	1.00

a) What's the probability that a survey respondent will shop at the U.S. store?
b) What is the probability that a survey respondent will shop at the store given that they are younger than 20 years old?
c) What is the probability that a survey respondent who is older than 40 shops at the store?
d) What is the probability that a survey respondent is younger than 20 or will shop at the store?

59. Pharmaceutical company, again. Given the table of probabilities compiled for marketing managers in Exercise 57, are high blood pressure and high cholesterol independent? Explain.

60. International relocation, again. Given the table of probabilities compiled for a department store chain in Exercise 58, are age and shopping at the department store independent? Explain.

61. Coffeehouse survey, part 2. Look again at the data from the coffeehouse survey in Exercise 53.

a) If we select a person at random, what's the probability we choose a person between 18 and 24 years old who is spending more time at coffeehouses?

b) Among the 18- to 24-year olds, what is the probability that the person responded that they are not spending more time at coffeehouses?

c) What's the probability that a person who spends the same amount of time at coffeehouses is between 35 and 44 years old?

d) If the person responded that they spend more time, what's the probability that they are at least 65 years old?

e) What's the probability that a person at least 65 years old spends the same amount of time?

f) Are the responses to the question and age independent?

62. Electronic communications, part 2. Look again at the data in the electronic communications in Exercise 54.

a) If we select a respondent at random, what's the probability that we choose a person earning less than $50 K and responded "somewhat"?

b) Among those earning $50–99.9K, what is the probability that the person responded "not at all"?

c) What's the probability that a person who responded "very much" was earning at least $100K?

d) If the person responded "very much," what is the probability that they earn between $50K and 99.9K?

e) Are the responses to the question and income level independent?

63. Real estate, part 2. In the real estate research described in Exercise 33, 64% of homes for sale have garages, 21% have swimming pools, and 17% have both features.

a) What is the probability that a home for sale has a garage, but not a pool?

b) If a home for sale has a garage, what's the probability that it has a pool, too?

c) Are having a garage and a pool independent events? Explain.

d) Are having a garage and a pool mutually exclusive? Explain.

64. Polling. Professional polling organizations face the challenge of selecting a representative sample of U.S. adults by telephone. This has been complicated by people who only use cell phones and by others whose landline phones are unlisted. A careful survey by Democracy Corps determined the following proportions:

Cell phone only	39%
Both cell and landline	29%
Landline only listed	22%
Landline only unlisted	7%

a) What's the probability that a randomly selected U.S. adult has a landline?

b) What's the probability that a U.S. adult has a landline given that he or she has a cell phone?

c) Are having a cell phone and a landline independent? Explain.

d) Are having a cell phone and a landline disjoint? Explain.

65. Property values. The following table shows a sample of property listings and values from one neighborhood (one ZIP code) in the Washington, DC, area in December 2011:

Property Values in this Neighborhood of Washington, DC

Price	Number of Bedrooms	Number of Stories	Garage	Number of Baths
$1.8MM	3	3	1	3
$1.1MM	2	2	1	3
$430K	1	1	0	1
$339K	1	1	0	1
$255K	0	1	0	1
$200K	0	1	0	1
$169K	0	1	0	1

a) In this sample, what proportion of homes is valued at $500K or less?

b) Are the number of bedrooms and property values independent? Explain.

66. Property values, part 2. A sample of 1800 homes in a different neighborhood of Washington, DC, in 2011 produced the data in the table for the number of bedrooms and house price. Is the price of the house independent of whether it has 3 or more bedrooms?

Property Values in the Washington, DC, Area

	3 or More Bedrooms	
House Price	Yes	No
Less than $300K	337	491
$300–450K	202	334
$450–600K	139	297

67. Used cars. A business student is searching for a used car to purchase, so she posts an ad to a website saying she wants to buy a used Jeep between $18,000 and $20,000. From Kelly's BlueBook.com, she learns that there are 149 cars matching that description within a 30-mile radius of her home. If we assume that those are the people who will call her and that they are equally likely to call her:

	Price		
Car Make	$18,000–$18,999	$19,000–$19,999	Total
Commander	3	6	9
Compass	6	1	7
Grand Cherokee	33	33	66
Liberty	17	6	23
Wrangler	33	11	44
Total	92	57	149

a) What is the probability that the first caller will be a Jeep Liberty owner?
b) What is the probability that the first caller will own a Jeep Liberty that costs between $18,000 and $18,999?
c) If the first call offers her a Jeep Liberty, what is the probability that it costs less than $19,000?
d) Suppose she decides to ignore calls with cars whose cost is ≥$19,000. What is the probability that the first call she takes will offer to sell her a Jeep Liberty?

68. CEO relocation. The CEO of a mid-sized company has to relocate to another part of the country. To make it easier, the company has hired a relocation agency to help purchase a house. The CEO has 5 children and so has specified that the house have at least 5 bedrooms, but hasn't put any other constraints on the search. The relocation agency has narrowed the search down to the houses in the table and has selected one house to showcase to the CEO and family on their trip out to the new site. The agency doesn't know it, but the family has its heart set on a Cape Cod house with a fireplace. If the agency selected the house at random, without regard to this:

		Fireplace?		
		No	Yes	Total
House Type	Cape Cod	7	2	9
	Colonial	8	14	22
	Other	6	5	11
	Total	21	21	42

a) What is the probability that the selected house is a Cape Cod?
b) What is the probability that the house is a Colonial with a fireplace?
c) If the house is a Cape Cod, what is the probability that it has a fireplace?
d) What is the probability that the selected house is what the family wants?

*69. Computer reliability. Laptop computers have been growing in popularity according to a study by Current Analysis Inc. Laptops now represent more than half the computer sales in the United States. A campus bookstore sells both types of computers and in the last semester sold 56% laptops and 44% desktops. Reliability rates for the two types of machines are quite different, however. In the first year, 5% of desktops require service, while 15% of laptops have problems requiring service.

a) Sketch a probability tree for this situation.
b) What percentage of computers sold by the bookstore last semester required service?
c) Given that a computer required service, what is the probability that it was a laptop?

JUST CHECKING ANSWERS

1 The probability of going up on the next day is not affected by the previous day's outcome.

2 a) 0.30
   b) $0.30(0.30) = 0.09$
   c) $(1 - 0.30)^2(0.30) = 0.147$
   d) $1 - (1 - 0.30)^5 = 0.832$

3 a)

		Weekday		
		Yes	No	Total
Before Five	Yes	0.07	0.41	0.48
	No	0.20	0.32	0.52
	Total	0.27	0.73	1.00

   b) $P(\mathbf{BF}\,|\,\mathbf{WD}) = P(\mathbf{BF}\ and\ \mathbf{WD})/P(\mathbf{WD}) = 0.07/0.27 = .259$
   c) No, shoppers can do both (and 7% do).
   d) To be independent, we'd need $P(\mathbf{BF}\,|\,\mathbf{WD}) = P(\mathbf{BF})$. $P(\mathbf{BF}\,|\,\mathbf{WD}) = 0.259$, but $P(\mathbf{BF}) = 0.48$. They do not appear to be independent.

6

Random Variables and Probability Models

Metropolitan Life Insurance Company

In 1863, at the height of the U.S. Civil War, a group of business-men in New York City decided to form a new company to insure Civil War soldiers against disabilities and injuries suffered from the war. After the war ended, they changed direction and decided to focus on selling life insurance. The new company was named Metropolitan Life (MetLife) because the bulk of the company's clients were in the "metropolitan" area of New York City.

Although an economic depression in the 1870s put many life insurance companies out of business, MetLife survived, modeling their business on similar successful programs in England. Taking advantage of spreading industrialism and the selling methods of British insurance agents, the company soon was enrolling as many as 700 new policies per day. By 1909, MetLife was the nation's largest life insurer in the United States.

During the Great Depression of the 1930s, MetLife expanded their public service by promoting public health campaigns, focusing on educating the urban poor in major U.S. cities about the risk of tuberculosis. Because the company invested primarily in urban and farm mortgages, as opposed to the stock market, they survived the crash of 1929 and ended up investing heavily in the postwar U.S. housing boom. They were the principal investors in both the Empire State Building (1929) and Rockefeller Center (1931). During World War II, the company was the single largest contributor to the Allied cause, investing more than half of their total assets in war bonds.

Today, in addition to life insurance, MetLife manages pensions and investments. In 2000, the company held an initial public offering and entered the retail banking business in 2001 with the launch of MetLife Bank. In 2012, MetLife Bank failed a Federal

Reserve stress test to determine how well it could handle a worst-case economic scenario. As a result, MetLife sold its banking unit to GE Capital. The company's public face is well known because of their use of Snoopy, the dog from the cartoon strip "Peanuts." In 2006, MetLife signed an international agreement with Peanuts Worldwide to gain exclusive rights in the financial services category to use Peanuts characters. The contract was set to expire in 2014. As of 2015, the contract appears to be in force.

Insurance companies earn their living by making bets. For example, they bet that you're going to live a long life. Ironically, you bet that you're going to die sooner. Both you and the insurance company want the company to stay in business, so it's important to find a "fair price" for your bet.

Of course, the right price for *you* depends on many factors, but the company can average its bets over many customers. By modeling lifetimes with probability models, it can make reasonably accurate estimates of the amount it can expect to pay out on a collection of policies. Using probability models, the company can find the fair price of almost any situation involving risk and uncertainty.

Here's a simple example. An insurance company offers a policy that pays $100,000 when a client dies or $50,000 if the client is permanently disabled. It charges a premium of $500 per year for this benefit. Is the company likely to make a profit selling such a plan? From a model based on actuarial information the company can calculate the expected value of this policy and the price it needs to charge to make a profit.

6.1   Expected Value of a Random Variable

To model the insurance company's risk, we need to define a few terms. The amount the company pays out on an individual policy is an example of a **random variable**, called that because its value is based on the outcome of a random event. We use a capital letter, in this case, X, to denote a random variable. We'll denote a particular *value* that it can have by the corresponding lowercase letter, in this case, x. For the insurance company, x can be $100,000 (if you die that year), $50,000 (if you are disabled), or $0 (if neither occurs). Because we can list all the outcomes, we call this random variable a **discrete random variable**. A random variable that can take on any value (possibly bounded on one or both sides) is called a **continuous random variable**. Continuous random variables are common in business applications for modeling physical quantities like heights and weights, and monetary quantities such as profits, revenues, and spending.

Sometimes it is obvious whether to treat a random variable as discrete or continuous, but at other times the choice is more subtle. Age, for example, might be viewed as discrete if it is measured only to the nearest decade with possible values 10, 20, 30, ... In a scientific context, however, it might be measured more precisely and treated as continuous.

For both discrete and continuous variables, the collection of all the possible values and the probabilities associated with them is called the **probability model** for the random variable.[1] For a discrete random variable, we can list the probability of all possible values in a table, or describe it by a formula. For example, to model the

NOTATION ALERT

The most common letters for random variables are X, Y, and Z, but any capital letter might be used.

---

[1] The probability model for a discrete random variable is also called the **probability mass function (pmf)** because it places some "mass" of probability at each possible value.

possible outcomes of a fair die, we can let X be the number showing on the face. The probability model for X is simply:

$$P(X = x) = \begin{cases} 1/6 & \textit{if } x = 1, 2, 3, 4, 5, \textit{ or } 6 \\ 0 & \textit{otherwise} \end{cases}$$

Sometimes it's useful to add up (or accumulate) the probabilities for all values up to x. The function $F(x) = P(X \leq x)$ is called the **cumulative distribution function (cdf)** for both discrete and continuous random variables.

Suppose in our insurance risk example that the death rate in any year is 1 out of every 1000 people and that another 2 out of 1000 suffer some kind of disability. The loss, which we'll denote as X, is a discrete random variable because it takes on only 3 possible values. We can display the probability model for X in a table, as in Table 6.1.

Policyholder Outcome	Payout x (cost)	Probability $P(X = x)$
Death	100,000	$\dfrac{1}{1000}$
Disability	50,000	$\dfrac{2}{1000}$
Neither	0	$\dfrac{997}{1000}$

Table 6.1   Probability model for an insurance policy.

Of course, we can't predict exactly what *will* happen during any given year, but we can say what we *expect* to happen—in this case, what we expect the profit of a policy will be. The expected value of a policy is a **parameter** (a numerically valued attribute) of the probability model. In fact, it's the mean. We'll signify this with the notation $E(X)$, for expected value (or sometimes μ to indicate that it is a mean). This isn't an average of data values, so we won't estimate it. Instead, we calculate it directly from the probability model for the random variable. Because it comes from a model and not data, we use the parameter μ to denote it (and *not* $\bar{y}$ or $\bar{x}$).

To see what the insurance company can expect, think about some convenient number of outcomes. For example, imagine that they have exactly 1000 clients and that the outcomes in one year followed the probability model exactly: 1 died, 2 were disabled, and 997 survived unscathed. Then our expected payout would be:

$$\mu = E(X) = \frac{100,000(1) + 50,000(2) + 0(997)}{1000} = 200$$

So our expected payout comes to $200 per policy.

Instead of writing the expected value as one big fraction, we can rewrite it as separate terms, each divided by 1000.

$$\mu = E(X) = \$100,000\left(\frac{1}{1000}\right) + \$50,000\left(\frac{2}{1000}\right) + \$0\left(\frac{997}{1000}\right)$$
$$= \$200$$

Writing it this way, we can see that for each policy, there's a $1/1000$ chance that we'll have to pay $100,000 for a death and a $2/1000$ chance that we'll have to pay $50,000 for a disability. Of course, there's a $997/1000$ chance that we won't have to pay anything.

So the **expected value** of a (discrete) random variable is found by multiplying each possible value of the random variable by the probability that it occurs and then summing all those products. This gives the general formula for the expected value of a discrete random variable:[2]

$$E(X) = \sum x\, P(x).$$

Be sure that *every* possible outcome is included in the sum. Verify that you have a valid probability model to start with—the probabilities should each be between 0 and 1 and should sum to one. (Recall the rules of probability in Chapter 5.)

FOR EXAMPLE　Calculating the expected value of a random variable

QUESTIONS　A fund-raising lottery offers 500 tickets for $3 each. If the grand prize is $250 and 4 second prizes are $50 each, what is the expected value of a single ticket? (Don't count the cost of the ticket in this yet.) Now, including its cost, what is the expected value of the ticket? (Knowing this value, does it make any "sense" to buy a lottery ticket?) The fund-raising group has a target of $1000 to be raised by the lottery. Can they expect to make this much?

ANSWERS　Each ticket has a $1/500$ chance of winning the grand prize of $250, a $4/500$ chance of winning $50, and a $495/500$ chance of winning nothing. So $E(X) = (1/500) \times \$250 + (4/500) \times \$50 + (495/500) \times \$0 = \$0.50 + \$0.40 + \$0.00 = \$0.90$. Including the cost, the expected value for a ticket holder is $\$0.90 - \$3 = -\$2.10$. Although no single person will lose $2.10 (they either lose $3 or win $50 or $250), $2.10 is the amount, on average, that the lottery gains per ticket. Therefore, they can expect to make $500 \times \$2.10 = \$1050$.

6.2　Standard Deviation of a Random Variable

Of course, this expected value (or mean) is not what actually happens to any *particular* policyholder. No individual policy actually costs the company $200. We are dealing with random events, so some policyholders receive big payouts and others nothing. Because the insurance company must anticipate this variability, it needs to know the standard deviation of the random variable.

For data, we calculate the standard deviation by first computing the deviation of each data value from the mean and squaring it. We perform a similar calculation when we compute the **standard deviation** of a (discrete) random variable as well. First, we find the deviation of each payout from the mean (expected value). (See Table 6.2.)

Policyholder Outcome	Payout x (cost)	Probability $P(X = x)$	Deviation $(x - EV)$
Death	100,000	$\dfrac{1}{1000}$	$(100{,}000 - 200) = 99{,}800$
Disability	50,000	$\dfrac{2}{1000}$	$(50{,}000 - 200) = 49{,}800$
Neither	0	$\dfrac{997}{1000}$	$(0 - 200) = -200$

Table 6.2　Deviations between the expected value and each payout (cost).

---

[2] The concept of expected values for continuous random variables is similar, but the calculation requires calculus and is beyond the scope of this text.

Next, we square each deviation. The **variance** is the expected value of those squared deviations. To find it, we multiply each by the appropriate probability and sum those products:

$$Var(X) = 99,800^2\left(\frac{1}{1000}\right) + 49,800^2\left(\frac{2}{1000}\right) + (-200)^2\left(\frac{997}{1000}\right)$$
$$= 14,960,000.$$

Finally, we take the square root to get the standard deviation:

$$SD(X) = \sqrt{14,960,000} \approx \$3867.82$$

The insurance company can expect an average payout of $200 per policy, with a standard deviation of $3867.82.

Think about that. The company charges $500 for each policy and expects to pay out $200 per policy. Sounds like an easy way to make $300. (In fact, most of the time—probability 997/1000—the company pockets the entire $500.) But would you be willing to take on this risk yourself and sell all your friends policies like this? The problem is that occasionally the company loses big. With a probability of 1/1000, it will pay out $100,000, and with a probability of 2/1000, it will pay out $50,000. That may be more risk than you're willing to take on. The standard deviation of $3867.82 gives an indication of the uncertainty of the profit, and that seems like a fairly large spread (and risk) for an average profit of $300.

Here are the formulas for these arguments. Because these are parameters of our probability model, the variance and standard deviation can also be written as σ^2 and σ, respectively (sometimes with the name of the random variable as a subscript). You should recognize both kinds of notation:

$$\sigma^2 = Var(X) = \sum (x - \mu)^2 P(x) = \sum (x - E(X))^2 P(x), \text{ and}$$

$$\sigma = SD(X) = \sqrt{Var(X)}.$$

FOR EXAMPLE Calculating the standard deviation of a random variable

QUESTION In the lottery example on page 182, we found the expected gain per ticket for the lottery to be $2.10. What is the standard deviation? What does it say about your chances in the lottery? Comment.

ANSWER

$$\sigma^2 = Var(X) = \sum (x - E(X))^2 P(x) = \sum (x - 2.10)^2 P(x)$$

$$= (250 - 2.10)^2 \frac{1}{500} + (50 - 2.10)^2 \frac{4}{500} + (0 - 2.10)^2 \frac{495}{500}$$

$$= 61,454.41 \times \frac{1}{500} + 2,294.41 \times \frac{4}{500} + 4.41 \times \frac{495}{500}$$

$$= 145.63$$

so $\sigma = \sqrt{145.63} = \$12.07$

That's a lot of variation for a mean of $2.10, which reflects the fact that there is a small chance that you'll win a lot but a large chance you'll win nothing.

GUIDED EXAMPLE   Computer Inventory

As the head of inventory for a computer company, you've had a challenging couple of weeks. One of your warehouses recently had a fire, and you had to flag all the computers stored there to be recycled. On the positive side, you were thrilled that you had managed to ship two computers to your biggest client last week. But then you discovered that your assistant hadn't heard about the fire and had mistakenly transported a whole truckload of computers from the damaged warehouse into the shipping center. It turns out that 30% of all the computers shipped last week were damaged. You don't know whether your biggest client received two damaged computers, two undamaged ones, or one of each. Computers were selected at random from the shipping center for delivery.

If your client received two undamaged computers, everything is fine. If the client gets one damaged computer, it will be returned at your expense—$100—and you can replace it. However, if both computers are damaged, the client will cancel all other orders this month, and you'll lose $10,000.

Question: What is the expected value and the standard deviation of your loss under this scenario?

PLAN	**Setup** State the problem.	We want to analyze the potential consequences of shipping damaged computers to a large client. We'll look at the expected value and standard deviation of the amount we'll lose.
		Let X = amount of loss. We'll denote the receipt of an undamaged computer by **U** and the receipt of a damaged computer by **D**. The three possibilities are: two undamaged computers (**U** and **U**), two damaged computers (**D** and **D**), and one of each (**UD** or **DU**). Because the computers were selected randomly and the number in the warehouse is large, we can assume independence.

| **DO** | **Model** List the possible values of the random variable, and compute all the values you'll need to determine the probability model. | Because the events are independent, we can use the multiplication rule (Chapter 5) and find: |

$$P(UU) = P(U) \times P(U)$$
$$= 0.7 \times 0.7 = 0.49$$
$$P(DD) = P(D) \times P(D)$$
$$= 0.3 \times 0.3 = 0.09$$

So, $P(UD \text{ or } DU) = 1 - (0.49 + 0.09) = 0.42$

We have the following model for all possible values of X.

Outcome	x	P(X = x)
Two damaged	10,000	$P(DD) = 0.09$
One damaged	100	$P(UD \text{ or } DU) = 0.42$
Neither damaged	0	$P(UU) = 0.49$

Mechanics Find the expected value.

$$E(X) = 0(0.49) + 100(0.42) + 10,000(0.09)$$
$$= \$942.00$$

Find the variance.

$$Var(X) = (0 - 942)^2 \times (0.49)$$
$$+ (100 - 942)^2 \times (0.42)$$
$$+ (10,000 - 942)^2 \times (0.09)$$
$$= 8,116,836$$

Find the standard deviation.

$$SD(X) = \sqrt{8,116,836} = \$2849.01$$

REPORT   **Conclusion** Interpret your results in context.

Reality check

> MEMO
>
> **Re: Damaged computers**
>
> The recent shipment of two computers to our large client may have some serious problems. Even though there is about a 50% chance that they will receive two perfectly good computers, there is a 9% chance that they will receive two damaged computers and will cancel the rest of their monthly order. We have analyzed the expected loss to the firm as $942 with a standard deviation of $2849.01. The large standard deviation reflects the fact that there is a real possibility of losing $10,000 from the mistake.
>
> Both numbers seem reasonable. The expected value of $942 is between the extremes of $0 and $10,000, and there's great variability in the outcome values.

6.3   Properties of Expected Values and Variances

Our example insurance company expected to pay out an average of $200 per policy, with a standard deviation of about $3868. The expected profit then was $500 − $200 = $300 per policy. Suppose that the company decides to lower the price of the premium by $50 to $450. It's pretty clear that the expected profit would drop an average of $50 per policy, to $450 − $200 = $250.

What about the standard deviation? We know that adding or subtracting a constant from data shifts the mean but doesn't change the variance or standard deviation. The same is true of random variables:[3]

$$E(X \pm c) = E(X) \pm c,$$
$$Var(X \pm c) = Var(X), \text{and}$$
$$SD(X \pm c) = SD(X).$$

What if the company decides to *double* all the payouts—that is, pay $200,000 for death and $100,000 for disability? This would double the average payout per policy and also increase the variability in payouts. In general, multiplying each value of a random variable by a constant multiplies the mean by that constant and multiplies the variance by the *square* of the constant:

$$E(aX) = aE(X), \text{and}$$
$$Var(aX) = a^2Var(X).$$

Taking square roots of the last equation shows that the standard deviation is multiplied by the absolute value of the constant:

$$SD(aX) = |a|SD(X).$$

What happens to the mean and variance when we have a collection of customers? The profit on a group of customers is the *sum* of the individual profits, so we'll need to know how to find expected values and variances for sums. To start, consider a simple case with just two customers who we'll call Mr. Ecks and Ms. Wye. With an expected payout of $200 on each policy, we might expect a total of $200 + $200 = $400 to be paid out on the two policies—nothing

---

[3] The rules in this section are true for both discrete *and* continuous random variables.

surprising there. In other words, we have the **Addition Rule for Expected Values of Random Variables**: *The expected value of the sum (or difference) of random variables is the sum (or difference) of their expected values*:

$$E(X \pm Y) = E(X) \pm E(Y).$$

The variability is another matter. Is the risk of insuring two people the same as the risk of insuring one person for twice as much? We wouldn't expect both clients to die or become disabled in the same year. In fact, because we've spread the risk, the standard deviation should be smaller. Indeed, this is the fundamental principle behind insurance. By spreading the risk among many policies, a company can keep the standard deviation quite small and predict costs more accurately. It's much less risky to insure thousands of customers than one customer when the total expected payout is the same, assuming that the events are independent. Catastrophic events such as hurricanes or earthquakes that affect large numbers of customers at the same time destroy the independence assumption, and often the insurance company along with it.

But how much smaller is the standard deviation of the sum? It turns out that, if the random variables are independent, we have the **Addition Rule for Variances of (Independent) Random Variables**: *The variance of the sum or difference of two independent random variables is the sum of their individual variances*:

$$Var(X \pm Y) = Var(X) + Var(Y)$$

if X and Y are independent.

MATH BOX Pythagorean Theorem of Statistics

We often use the standard deviation to measure variability, but when we add independent random variables, we use their variances. Think of the Pythagorean Theorem. In a right triangle (only), the *square* of the length of the hypotenuse is the sum of the *squares* of the lengths of the other two sides:

$$c^2 = a^2 + b^2.$$

For independent random variables (only), the *square* of the standard deviation of their sum is the sum of the *squares* of their standard deviations:

$$SD^2(X + Y) = SD^2(X) + SD^2(Y).$$

It's simpler to write this with *variances*:

$$Var(X + Y) = Var(X) + Var(Y),$$

but we'll use the standard deviation formula often as well:

$$SD(X + Y) = \sqrt{Var(X) + Var(Y)}.$$

For Mr. Ecks and Ms. Wye, the insurance company can expect their outcomes to be independent, so (using X for Mr. Ecks's payout and Y for Ms. Wye's):

$$Var(X + Y) = Var(X) + Var(Y)$$
$$= 14,960,000 + 14,960,000$$
$$= 29,920,000.$$

Let's compare the variance of writing two independent policies to the variance of writing only one for twice the size. If the company had insured only Mr. Ecks for twice as much, the variance would have been

$$Var(2X) = 2^2 Var(X) = 4 \times 14,960,000 = 59,840,000, \text{ or}$$

twice as big as with two independent policies, even though the expected payout is the same.

Of course, variances are in squared units. The company would prefer to know standard deviations, which are in dollars. The standard deviation of the payout for two independent policies is $SD(X + Y) = \sqrt{Var(X + Y)} = \sqrt{29,920,000} = \$5469.92$. But the standard deviation of the payout for a single policy of twice the size is twice the standard deviation of a single policy: $SD(2X) = 2SD(X) = 2(\$3867.82) = \$7735.64$, or about 40% more than the standard deviation of the sum of the two independent policies.

If the company has two customers, then it will have an expected annual total payout (cost) of \$400 with a standard deviation of about \$5470. If they write one policy with an expected annual payout of \$400, they increase the standard deviation by about 40%. Spreading risk by insuring many independent customers is one of the fundamental principles in insurance and finance.

Let's review the rules of expected values and variances for sums and differences.

- The expected value of the sum of two random variables is the sum of the expected values.
- The expected value of the difference of two random variables is the difference of the expected values:

$$E(X \pm Y) = E(X) \pm E(Y).$$

- If the random variables are independent, the variance of their sum or difference is always the sum of the variances:

$$Var(X \pm Y) = Var(X) + Var(Y).$$

Do we always *add* variances? Even when we take the *difference* of two random quantities? Yes! Think about the two insurance policies. Suppose we want to know the mean and standard deviation of the *difference* in payouts to the two clients. Since each policy has an expected payout of \$200, the expected difference is \$200 − \$200 = \$0. If we computed the variance of the difference by subtracting variances, we would get \$0 for the variance. But that doesn't make sense. Their difference won't always be exactly \$0. In fact, the difference in payouts could range from \$100,000 to −\$100,000, a spread of \$200,000. The variability in differences *increases* as much as the variability in sums. If the company has two customers, the difference in payouts has a mean of \$0 and a standard deviation of about \$5470.

For Random Variables, Does X + X + X = 3X?

Maybe, but be careful. As we've just seen, insuring one person for \$300,000 is not the same risk as insuring three people for \$100,000 each. When each instance represents a different outcome for the same random variable, though, it's easy to fall into the trap of writing all of them with the same symbol. Don't make this common mistake. Make sure you write each instance as a *different* random variable. Just because each random variable describes a similar situation doesn't mean that each random outcome will be the same. What you really mean is $X_1 + X_2 + X_3$. Written this way, it's clear that the sum shouldn't necessarily equal 3 times *anything*.

FOR EXAMPLE Sums of random variables

You are considering investing \$1000 into one or possibly two different investment funds. Historically, each has delivered 5% a year in profit with a standard deviation of 3%. So, a \$1000 investment would produce \$50 with a standard deviation of \$30.

(continued)

> **QUESTION** Assuming the two funds are independent, what are the relative advantages and disadvantages of putting $1000 into one, or splitting the $1000 and putting $500 into each? Compare the means and SDs of the profit from the two strategies.
>
> **ANSWER** Let X = amount gained by putting $1000 into one
>
> $$E(X) = 0.05 \times 1000 = \$50 \text{ and } SD(X) = 0.03 \times 1000 = \$30.$$
>
> Let W = amount gained by putting $500 into each. W_1 and W_2 are the amounts from each fund respectively. $E(W_1) = E(W_2) = 0.05 \times 500 = \$25$. So $E(W) = E(W_1) + E(W_2) = \$25 + \$25 = \$50$. The expected values of the two strategies are the same. You expect on average to earn $50 on $1000.
>
> $$\begin{aligned} SD(W) &= \sqrt{SD^2(W_1) + SD^2(W_2)} \\ &= \sqrt{(0.03 \times 500)^2 + (0.03 \times 500)^2} \\ &= \sqrt{15^2 + 15^2} \\ &= \$21.213 \end{aligned}$$
>
> The standard deviation of the amount earned is $21.213 by splitting the investment amount compared to $30 for investing in one. The expected values are the same. Spreading the investment into more than one vehicle *reduces* the variation. On the other hand, keeping it all in one vehicle increases the chances of both extremely good and extremely bad returns. Which one is better depends on an individual's appetite for risk.[4]

JUST CHECKING

1 Suppose that the time it takes a customer to get and pay for seats at the ticket window of a baseball park is a random variable with a mean of 100 seconds and a standard deviation of 50 seconds. When you get there, you find only two people in line in front of you.

a) How long do you expect to wait for your turn to get tickets?

b) What's the standard deviation of your wait time?

c) What assumption did you make about the two customers in finding the standard deviation?

6.4   Bernoulli Trials

When Google Inc. designs a new version of their web browser *Chrome*, they work hard to minimize the probability that the browser has trouble displaying a website. Before releasing the product, they test many websites to discover those that might fail. Although web browsers are relatively new, *quality control inspection* such as this is common throughout manufacturing worldwide and has been in use in industry for nearly 100 years.

The developers of *Chrome* sample websites, recording whether the browser displays the website correctly or has a problem. We call the act of testing a website a trial. There are two possible outcomes—either the website is displayed correctly or it isn't. The developers assume that the outcome of the test on any particular website is independent from what happens on other sites. Situations like this occur often and are called **Bernoulli trials**. To summarize, trials are Bernoulli if:

* There are only two possible outcomes (called *success* and *failure*) for each trial.
* The probability of success, denoted p, is the same on every trial. (The probability of failure, $1 - p$ is often denoted q.)
* The trials are independent. Finding that one website does not display correctly does not change what might happen with the next website.

---

[4] The assumption of independence is crucial, but not always (or ever) reasonable. As a March 3, 2010, article on *CNN Money* stated:

"It's only when economic conditions start to return to normal . . . that investors, and investments, move independently again. That's when diversification reasserts its case. . . ."

Common examples of Bernoulli trials include tossing a coin, collecting responses on Yes/No questions from surveys, or even shooting free throws in a basketball game. Bernoulli trials are remarkably versatile and can be used to model a wide variety of real-life situations. The specific question you might ask in different situations will give rise to different random variables that, in turn, have different probability models.

Of course, the *Chrome* developers want to find websites that don't display correctly so they can fix any problems in the browser. So for them a "success" is finding a failed website. The labels "success" and "failure" are often applied arbitrarily, so be sure you know what they mean in any particular situation.

One of the important requirements for Bernoulli trials is that the trials be independent. Sometimes that's a reasonable assumption. Is it true for our example? It's easy to imagine that related sites might have similar problems, but if the sites are selected at random, whether one has a problem should be independent of others.

The 10% Condition: Bernoulli trials must be independent. In theory, we need to sample from a population that's infinitely big. If the population is finite, our simple model will still work as long as the sample is smaller than about 10% of the population. If not, the variance needs to be adjusted to account for the finite population. In Google's case, they have a directory of millions of websites, so most samples will easily satisfy the 10% condition.

6.5 Discrete Probability Models

Sam Savage, Professor at Stanford University, says in his book *The Flaw of Averages* that plans based only on averages are, on average, wrong. Unfortunately, many business owners make decisions based on averages—the average amount sold last year, the average number of customers seen last month, etc. But averages are just too simple to represent real-world business practice. Fortunately, we can do better by modeling business situations with a probability model. Probability models can play an important role in helping decision makers predict both the outcomes and the consequences of their decision alternatives. In this section we'll see that some fairly simple models let us model a wide variety of business phenomena.

The Uniform Model

We'll start with the simplest probability model of all, the Uniform model. When we first studied probability in Chapter 5, we saw that equally likely events were the simplest case. For example, a single die can turn up 1, 2, . . . , 6 on one toss. A probability model for the toss is Uniform because each of the outcomes has the same probability (1/6) of occurring. Similarly if X is a random variable with possible outcomes 1, 2, . . . , n and $P(X = i) = 1/n$ for each value of i, then we say X has a **discrete Uniform distribution, $U[1, . . . , n]$**.

Unfortunately, some business decision makers take only one step away from averages and assume that all their unknown outcomes are equally likely. That can put them with the lottery ticket purchaser who thought her chances were 50/50: "either I win or I don't." Let's look at some more realistic (and more useful) probability models.

The Geometric Model

What's the probability that when Google tests *Chrome* on new websites, the first website that fails to display is the second one that they test? They can use Bernoulli trials to build a probability model. Let X denote the number of trials (websites) until the first such "success."[5] For X to be 2, the first website must have displayed correctly

Daniel Bernoulli (1700–1782) was the nephew of Jacob, whom you saw in Chapter 5. He was the first to work out the mathematics for what we now call Bernoulli trials.

---

[5] This is an example of applying the term "success" to something we care about—a *failure* of the browser. Don't be confused.

(which has probability $1 - p$), and then the second one must have not displayed correctly—a success, with probability p. Since the trials are independent, these probabilities can be multiplied, and so $P(X = 2) = (1 - p)(p)$ or qp. Maybe Google won't find a success until the fifth trial. What are the chances of that? *Chrome* would have to display the first four websites correctly and then choke on the fifth one, so $P(X = 5) = (1 - p)^4(p) = q^4p$.

Whenever the question is how long (how many trials) it will take to achieve the first success, the model that gives this probability is the **geometric probability model**. Geometric models are completely specified by one parameter, p, the probability of success. We denote them Geom(p).

Geometric Probability Model for Bernoulli Trials: Geom(p)

p = probability of success (and $q = 1 - p$ = probability of failure)
X = number of trials until the first success occurs

$$P(X = x) = q^{x-1}p$$

Expected value: $\mu = \dfrac{1}{p}$

Standard deviation: $\sigma = \sqrt{\dfrac{q}{p^2}}$

The geometric distribution can tell Google something important about its software. No large complex program is entirely free of bugs. So before releasing a program or upgrade, developers typically ask not whether it is free of bugs, but how long it is likely to be until the next bug is discovered. If the expected number of trials until the next bug discovery is high enough, then it makes business sense to ship the product rather than wait for that next bug report.

The Binomial Model

Suppose Google tests 5 websites. What's the probability that *exactly* 2 of them have problems (2 "successes")? The geometric model tells how long it should take until the first success. Now we want to find the probability of getting exactly 2 successes among the 5 trials. We are still talking about Bernoulli trials, but we're asking a different question.

This time we're interested in the *number of successes* in the 5 trials, which we'll denote by X. We want to find $P(X = 2)$. Whenever the random variable of interest is the number of successes in a series of Bernoulli trials, it's called a **Binomial random variable**. It takes two parameters to define this Binomial probability model: the number of trials, n, and the probability of success, p. We denote this model Binom(n, p).

Suppose that in an early phase of development, 10% of the sites exhibited some sort of problem so that $p = 0.10$. Exactly 2 successes in 5 trials means 2 successes and 3 failures. It seems logical that the probability should be $(p)^2(1 - p)^3$. Unfortunately, it's not *quite* that easy. That calculation would give you the probability of finding two successes and then three failures—*in that order*. But you could find the two successes in a lot of other ways, for example in the 2nd and 4th website you test. The probability of that sequence is $(1 - p)p(1 - p)p(1 - p)$ which is also $p^2(1 - p)^3$. In fact, as long as there are two successes and three failures, the probability will always be the same, regardless of the order of the sequence of successes and failures. The probability will be $(p)^2(1 - p)^3$. To find the probability of getting 2 successes in 5 trials in any order, we just need to know how many ways that outcome can occur.

Fortunately, all the sequences that lead to the same number of successes are *disjoint*. (For example, if your successes came on the first two trials, they couldn't come on the last two.) So once we find all the different sequences, we can add up their probabilities. And since the probabilities are all the same, we just need to find how many sequences there are and multiply $(p)^2(1-p)^3$ by that number.

Each different order in which we can have k successes in n trials is called a "combination." The total number of ways this can happen is written $\binom{n}{k}$ or $_nC_k$ and pronounced "n choose k":

$$\binom{n}{k} = {_nC_k} = \frac{n!}{k!(n-k)!} \text{ where } n! = n \times (n-1) \times \cdots \times 1.$$

For 2 successes in 5 trials,

$$\binom{5}{2} = \frac{5!}{2!(5-2)!} = \frac{(5 \times 4 \times 3 \times 2 \times 1)}{(2 \times 1 \times 3 \times 2 \times 1)} = \frac{(5 \times 4)}{(2 \times 1)} = 10.$$

So there are 10 ways to get 2 successes in 5 websites, and the probability of each is $(p)^2(1-p)^3$. To find the probability of exactly 2 successes in 5 trials, we multiply the probability of any particular order by the number of possible different orders:

$$P(\textit{exactly 2 successes in 5 trials}) = 10p^2(1-p)^3 = 10(0.10)^2(0.90)^3 = 0.0729$$

In general, we can write the probability of exactly k successes in n trials as

$$P(X = k) = \binom{n}{k}p^k q^{n-k}.$$

If the probability that any single website has a display problem is 0.10, what's the expected number of websites with problems if we test 100 sites? You probably said 10. We suspect you didn't use the formula for expected value that involves multiplying each value times its probability and adding them up. In fact, there is an easier way to find the expected value for a Binomial random variable. You just multiply the probability of success by n. In other words, $E(X) = np$. We prove this in the next Math Box.

The standard deviation is less obvious and you can't just rely on your intuition. Fortunately, the formula for the standard deviation also boils down to something simple: $SD(X) = \sqrt{npq}$. If you're curious to know where that comes from, it's in the Math Box, too.

In our website example, with $n = 100$, $E(X) = np = 100(0.10) = 10$ so we expect to find 10 successes out of the 100 trials. The standard deviation is

$$\sqrt{100 \times 0.10 \times 0.90} = 3 \text{ websites.}$$

Binomial Model for Bernoulli Trials: Binom(n, p)

n = number of trials
p = probability of success (and $q = 1 - p$ = probability of failure)
X = number of successes in n trials

$$P(X = x) = \binom{n}{x} p^x q^{n-x}, \text{ where } \binom{n}{x} = \frac{n!}{x!(n-x)!}$$

Mean: $\mu = np$
Standard deviation: $\sigma = \sqrt{npq}$

MATH BOX Mean and Standard Deviation of the Binomial Model

To derive the formulas for the mean and standard deviation of the Binomial model we start with the most basic situation.

Consider a single Bernoulli trial with probability of success p. Let's find the mean and variance of the number of successes.

Here's the probability model for the number of successes:

x	0	1
$P(X = x)$	q	p

Find the expected value:

$$E(X) = 0q + 1p$$
$$E(X) = p$$

Now the variance:

$$Var(X) = (0 - p)^2 q + (1 - p)^2 p$$
$$= p^2 q + q^2 p$$
$$= pq(p + q)$$
$$= pq(1)$$
$$Var(X) = pq$$

What happens when there is more than one trial? A Binomial model simply counts the number of successes in a series of n independent Bernoulli trials. That makes it easy to find the mean and standard deviation of a binomial random variable, Y.

$$Let\ Y = X_1 + X_2 + X_3 + \cdots + X_n$$
$$E(Y) = E(X_1 + X_2 + X_3 + \cdots + X_n)$$
$$= E(X_1) + E(X_2) + E(X_3) + \cdots + E(X_n)$$
$$= p + p + p + \cdots + p\ (\text{There are } n \text{ terms.})$$

So, as we thought, the mean is $E(Y) = np$.

And since the trials are independent, the variances add:

$$Var(Y) = Var(X_1 + X_2 + X_3 + \cdots + X_n)$$
$$= Var(X_1) + Var(X_2) + Var(X_3) + \cdots + Var(X_n)$$
$$= pq + pq + pq + \cdots + pq\ (\text{Again, } n \text{ terms.})$$
$$Var(Y) = npq$$

Voila! The standard deviation is $SD(Y) = \sqrt{npq}$.

GUIDED EXAMPLE   The American Red Cross

Every two seconds someone in America needs blood.

The American Red Cross is a nonprofit organization that runs like a large business. It serves over 3000 hospitals around the United States, providing a wide range of high-quality blood products and blood donor and patient testing services. It collects blood from over 4 million donors, provides blood to millions of patients, and is dedicated to meeting customer needs.

Balancing supply and demand is complicated not only by the logistics of finding donors that meet health criteria, but also by the fact that the blood type of donor and patient must be matched. People with O-negative blood are called "universal donors" because O-negative blood can be given to patients with any blood type, but only about 6% of people have O-negative blood. Unlike a manufacturer who can balance supply through production or purchasing, the Red Cross gets its supply from volunteer donors who show up more-or-less at random (at least in terms of blood type). Modeling the arrival of samples with various blood types helps the Red Cross managers to plan their blood allocations.

Here's a small example of the kind of planning required. Of the next 20 donors to arrive at a blood donation center, how many universal donors can be expected? Specifically, what are the mean and standard deviation of the number of universal donors? What is the probability that there are 2 or 3 universal donors?

Question 1: What are the mean and standard deviation of the number of universal donors?

Question 2: What is the probability that there are exactly 2 or 3 universal donors out of the 20 donors?

PLAN	**Setup** State the question.	We want to know the mean and standard deviation of the number of universal donors among 20 people and the probability that there are 2 or 3 of them.
	Check to see that these are Bernoulli trials.	✓ There are two outcomes: success = O-negative failure = other blood types ✓ $p = 0.06$ ✓ **10% Condition:** Fewer than 10% of all possible donors have shown up.
	Variable Define the random variable. **Model** Specify the model.	Let X = number of O-negative donors among $n = 20$ people. We can model X with a Binom(20, 0.06).

DO	**Mechanics** Find the expected value and standard deviation. Calculate the probability of 2 or 3 successes.	$E(X) = np = 20(0.06) = 1.2$ $SD(X) = \sqrt{npq} = \sqrt{20(0.06)(0.94)} \approx 1.06$ $P(X = 2 \text{ or } 3) = P(X = 2) + P(X = 3)$ $\qquad = \binom{20}{2}(0.06)^2(0.94)^{18}$ $\qquad\quad + \binom{20}{3}(0.06)^3(0.94)^{17}$ $\qquad \approx 0.2246 + 0.0860$ $\qquad = 0.3106$

REPORT	**Conclusion** Interpret your results in context.	MEMO **Re: Blood Drive** In groups of 20 randomly selected blood donors, we'd expect to find an average of 1.2 universal donors, with a standard deviation of 1.06. About 31% of the time, we'd expect to find exactly 2 or 3 universal donors among the 20 people.

*The Poisson Model

Not all discrete events can be modeled as Bernoulli trials. Sometimes we're interested simply in the number of events that occur over a given interval of time or space. For example, we might want to model the number of customers arriving in our store in the next ten minutes, the number of visitors to our website in the next minute, or the number of defects that occur in a computer monitor of a certain size. In cases like these, the number of occurrences can be modeled by a **Poisson model**. The Poisson's parameter, the mean of the distribution, is usually denoted by λ.

Poisson Probability Model for Occurrences: Poisson(λ)

λ = mean number of occurrences
X = number of occurrences

$$P(X = x) = \frac{e^{-\lambda}\lambda^x}{x!}$$

Expected value: $E(X) = \lambda$
Standard deviation: $SD(X) = \sqrt{\lambda}$

For example, data show an average of about 4 hits per minute to a small business website during the afternoon hours from 1:00 to 5:00 p.m. We can use the Poisson model to find the probability of any number of hits arriving. For example, if we let X be the number of hits arriving in the next minute, then $P(X = x) = \frac{e^{-\lambda}\lambda^x}{x!} = \frac{e^{-4}4^x}{x!}$, using the given average rate of 4 per minute. So, the probability of no hits during the next minute would be $P(X = 0) = \frac{e^{-4}4^0}{0!} = e^{-4} = 0.0183$ (The constant e is the base of the natural logarithms and is approximately 2.71828 and $0! = 1$).

One interesting and useful feature of the Poisson model is that it scales according to the interval size. For example, suppose we want to know the probability of no hits to our website in the next 30 seconds. Since the mean rate is 4 hits per minute, it's 2 hits per 30 seconds, so we can use the model with $\lambda = 2$ instead. If we let Y be the number of hits arriving in the next 30 seconds, then:

$$P(Y = 0) = \frac{e^{-2}2^0}{0!} = e^{-2} = 0.1353.$$

(Recall that $0! = 1$.) The Poisson model has been used to model phenomena such as customer arrivals, hot streaks in sports, and disease clusters.

Whenever or wherever rare events happen closely together, people want to know whether the occurrence happened by chance or whether an underlying change caused the unusual occurrence. The Poisson model can be used to find the probability of the occurrence and can be the basis for making the judgment.

---

e and Compound Interest

The constant e equals 2.7182818... (to 7 decimal places). One of the places e originally turned up was in calculating how much money you'd earn if you could get interest compounded more often. If you earn 100% per year simple interest, at the end of the year, you'd have twice as much money as when you started. But if the interest were compounded and paid at the end of every month, each month you'd earn 1/12 of 100% interest. At the year's end you'd have $(1 + 1/12)^{12} = 2.613$ times as much instead of 2. If the interest were paid every day, you'd get $(1 + 1/365)^{365} = 2.715$ times as much. If the interest were paid every second, you'd get $(1 + 1/3153600)^{3153600} = 2.7182812$ times as much. This is where e shows up. If you could get the interest compounded continually, you'd get e times as much. In other words, as n gets large, the limit of $(1 + 1/n)^n = e$. This unexpected result was discovered by Jacob Bernoulli in 1683.

JUST CHECKING

Roper Worldwide reports that they are able to contact 76% of the randomly selected households drawn for a telephone survey.

2 Explain why these phone calls can be considered Bernoulli trials.

3 Which of the models of this chapter (Geometric, Binomial, or Poisson) would you use to model the number of successful contracts from a list of 1000 sampled households?

4 Roper also reports that even after they contacted a household, only 38% of the contacts agreed to be interviewed. So the probability of getting a completed interview from a randomly selected household is only 0.29 (38% of 76%). Which of the models of this chapter would you use to model the number of households Roper has to call before they get the first completed interview?

FOR EXAMPLE Probability models

A venture capital firm has a list of potential investors who have previously invested in new technologies. On average, these investors invest about 5% of the time. A new client of the firm is interested in finding investors for a mobile phone application that enables financial transactions, an application that is finding increasing acceptance in much of the developing world. An analyst at the firm starts calling potential investors.

QUESTIONS

1. What is the probability that the first person she calls will want to invest?
2. What is the probability that none of the first five people she calls will be interested?
3. How many people will she have to call until the probability of finding someone interested is at least 0.50?
4. How many investors will she have to call, on average, to find someone interested?
5. If she calls 10 investors, what is the probability that exactly 2 of them will be interested?
6. What assumptions are you making to answer these questions?

ANSWERS

1. Each investor has a 5% or 1/20 chance of wanting to invest, so the chance that the first person she calls is interested is 1/20.
2. P(first one not interested) $= 1 - 1/20 = 19/20$. Assuming the trials are independent, P(none are interested) $= P$ (1st not interested) $\times P$ (2nd not interested) $\times \cdots \times P$ (5th not interested) $= (19/20)^5 = 0.774$.
3. By trial and error, $(19/20)^{13} = 0.513$ and $(19/20)^{14} = 0.488$, so she would need to call 14 people to have the probability of *no one* interested drop below 0.50, therefore making the probability that someone is interested greater than 0.50.
4. This uses a geometric model. Let $X =$ number of people she calls until the first interested person. $E(X) = 1/p = 1/(1/20) = 20$ people.
5. Using the Binomial model, let $Y =$ number of people interested in 10 calls, then

$$P(Y = 2) = \binom{10}{2}p^2(1 - p)^8 = \frac{10 \times 9}{2}(1/20)^2(19/20)^8 = 0.0746.$$

6. We are assuming that the trials are independent and that the probability of being interested in investing is the same for all potential investors.

WHAT CAN GO WRONG?

- **Probability models are still just models.** Models can be useful, but they are not reality. Think about the assumptions behind your models. Question probabilities as you would data.

- **If the model is wrong, so is everything else.** Before you try to find the mean or standard deviation of a random variable, check to make sure the probability model is reasonable. As a start, the probabilities should all be between 0 and 1 and they should add up to 1. If not, you may have calculated a probability incorrectly or left out a value of the random variable.

- **Watch out for variables that aren't independent.** You can add expected values of any two random variables, but you can only add variances of independent random variables. Suppose a survey includes questions about the number of hours of sleep people get each night and also the number of hours they are awake each day. From their answers, we find the mean and standard deviation of hours asleep and hours awake. The expected total must be 24 hours; after all, people are either asleep or awake. The means still add just fine. Since all the totals are exactly 24 hours, however, the standard deviation of the total will be 0. We can't add variances here because the number of hours you're awake depends on the number of hours you're asleep. Be sure to check for independence before adding variances.

- **Don't write independent instances of a random variable with notation that looks like they are the same variables.** Make sure you write each instance as a different random variable. Just because each random variable describes a similar situation doesn't mean that each random outcome will be the same. These are *random* variables, not the variables you saw in Algebra. Write $X_1 + X_2 + X_3$ rather than $X + X + X$.

- **Don't forget:** Variances of independent random variables add. Standard deviations don't.

- **Don't forget:** Variances of independent random variables add, even when you're looking at the difference between them.

- **Be sure you have Bernoulli trials.** Be sure to check the requirements first: two possible outcomes per trial ("success" and "failure"), a constant probability of success, and independence. Remember that the 10% Condition provides a reasonable substitute for independence.

ETHICS IN ACTION

Kurt Williams was about to open a new SEP IRA account and was interested in exploring various investment options. Although he had some ideas about how to invest his money, Kurt thought it best to seek the advice of a professional, so he made an appointment with Keith Klingman, a financial advisor at James, Morgan, and Edwards, LLC.

Prior to their first meeting, Kurt told Keith that he preferred to keep his investments simple and wished to allocate his money to only two funds. Also, he mentioned that while he was willing to take on some risk to yield higher returns, he was concerned about taking on too much risk given the recent volatility in the markets. After their conversation, Keith began to prepare for their first meeting.

Because Kurt was interested in investing his SEP IRA money in only two funds, Keith decided to compile figures on the expected annual return and standard deviation (a measure of risk) for a potential SEP IRA account consisting of different combinations of two funds. If *X* and *Y* represent the annual returns for two different funds, Keith knew he could represent the expected annual return for any combination of funds as $aX + (1 - a)Y$, where *a* is the fraction of funds Kurt will allocate to *X*.

Keith calculated the expected annual return using the formula $E(aX + (1 - a)Y) = aE(X) + (1 - a)E(Y)$. Keith knew that this formula would be true for all funds *X* and *Y* even if their performances were correlated. To find the variance if the combined investment he calculated $Var(aX + (1 - a)Y) = a^2 Var(X) + (1 - a)^2 Var(Y)$.

Keith knew that the variance calculation assumed that the two funds were independent, but he figured that the formula was close enough even if the funds performances were correlated, and he wanted to keep the presentation to Kurt simple.

Keith presented a variety of combinations of funds and allocations to Kurt. Because some equity funds delivered the best expected return, Keith advised Kurt to put all his money in two equity funds (funds that also generated higher brokerage fees) rather than allocating any money to a simple fixed income fund. Kurt was surprised to see that even under various market conditions, all the equity fund combinations seemed fairly safe in terms of volatility as evidenced by the fairly low standard deviations of the combined funds, and Keith assured him that these scenarios were realistic.

- Identify the ethical dilemma in this scenario.
- What are the undesirable consequences?
- Propose an ethical solution that considers the welfare of all stakeholders.

WHAT HAVE WE LEARNED?

Learning Objectives

Understand how probability models relate values to probabilities.
- For discrete random variables, probability models assign a probability to each possible outcome.

Know how to find the mean, or expected value, of a discrete probability model from $\mu = \Sigma x P(X = x)$ and the standard deviation from $\sigma = \sqrt{\Sigma(x - \mu)^2 P(x)}$.

Foresee the consequences of shifting and scaling random variables, specifically

$$E(X \pm c) = E(X) \pm c \qquad E(aX) = aE(X)$$
$$Var(X \pm c) = Var(X) \qquad Var(aX) = a^2 Var(X)$$
$$SD(X \pm c) = SD(X) \qquad SD(aX) = |a| SD(X)$$

Understand that when adding or subtracting random variables the expected values add or subtract as well: $E(X \pm Y) = E(X) \pm E(Y)$. However, when adding or subtracting independent random variables, the variances *add*:

$$Var(X \pm Y) = Var(X) + Var(Y)$$

Be able to explain the properties and parameters of the Uniform, the Binomial, the Geometric, and the Poisson distributions.

Terms

Addition Rule for Expected Values of Random Variables

$$E(X \pm Y) = E(X) \pm E(Y)$$

Addition Rule for Variances of (Independent) Random Variables

(Pythagorean Theorem of Statistics)

If *X* and *Y* are *independent*: $Var(X \pm Y) = Var(X) + Var(Y)$, and $SD(X \pm Y) = \sqrt{Var(X) + Var(Y)}$.

Bernoulli trials	A sequence of n trials are called Bernoulli trials if: 1. There are exactly two possible outcomes (usually denoted *success* and *failure*). 2. The probability of success is constant. 3. The trials are independent.
Binomial probability model	A Binomial model is appropriate for a random variable that counts the number of successes in a series of Bernoulli trials.
Changing a random variable by a constant	$$E(X \pm c) = E(X) \pm c \qquad Var(X \pm c) = Var(X) \qquad SD(X \pm c) = SD(X)$$ $$E(aX) = aE(X) \qquad Var(aX) = a^2 Var(X) \qquad SD(aX) = |a|SD(X)$$
Continuous random variable	A random variable that can take on any value (possibly bounded on one or both sides).
Cumulative distribution function (cdf)	The cumulative distribution function, $F(x)$, of a random variable X gives the probability that the random variable X is less than or equal to x. $F(X) = P(X \le x)$.
Discrete random variable	A random variable that can take one of a finite number[6] of distinct outcomes.
Expected value	The expected value of a random variable is its theoretical long-run average value, the center of its model. Denoted μ or $E(X)$, it is found (if the random variable is discrete) by summing the products of variable values and probabilities: $$\mu = E(X) = \sum x P(x)$$
Geometric probability model	A model appropriate for a random variable that counts the number of Bernoulli trials until the first success.
Parameter	A numerically valued attribute of a model, such as the values of μ and σ representing the mean and standard deviation.
Poisson model	A discrete model often used to model the number of arrivals of events such as customers arriving in a queue or calls arriving into a call center.
Probability model	A function that associates a probability P with each value of a discrete random variable X, denoted $P(X = x)$, or with any interval of values of a continuous random variable. For a discrete random variable, the model is also called a probability mass function (pmf).
Random variable	Assumes any of several different values as a result of some random event. Random variables are denoted by a capital letter, such as X.
Standard deviation of a random variable	Describes the spread in the model and is the square root of the variance.
Uniform model, Uniform distribution	For a discrete uniform distribution over a set of n values, each value has probability $1/n$.
Variance of a random variable	The variance of a random variable is the expected value of the squared deviations from the mean. For discrete random variables, it can be calculated as: $$\sigma^2 = Var(X) = \sum (x - \mu)^2 P(x).$$

---

[6]Technically, there could be an infinite number of outcomes as long as they're *countable*. Essentially, that means we can imagine listing them all in order, like the counting numbers 1, 2, 3, 4, 5, . . .

TECHNOLOGY HELP: Random Variables and Probability Models

Most statistics packages (and graphics calculators) offer functions that compute probabilities for various probability models. The important differences among these functions are in what they are named and the order of their arguments. In these functions, "pmf" stands for "probability mass function"—what we've been calling a probability model. The letters "cdf" stand for "cumulative distribution function." These technical terms show up in many of the function names. SPSS uses the term pdf (for probability density function), a term that should apply only to continuous random variables (see Chapter 7). Many packages allow the computation of a probability given a value based on a given distribution and also the calculation of a value based on the probability.

For example, in Excel, Binomdist (*x*, *n*, prob, cumulative) computes Binomial probabilities. If cumulative is set to false, the calculation is only for one value of *x*.

EXCEL

The following commands can be used to calculate Binomial and Poisson distribution probabilities. In Excel, the value for "cumulative" will give either a cdf (cumulative = TRUE) or a pmf (cumulative = FALSE). The commands can either be typed directly into a cell or into the function bar at the top of the screen.

Excel SYNTAX	EXAMPLE SYNTAX	RESULT	Probability Statement
BINOM.DIST(number of successes, trials, probability of success, cumulative)	=BINOM.DIST(5,20,0.06,TRUE)	0.9991	P (X≤5)
	=BINOM.DIST(5,20,0.06,FALSE)	0.0048	P (X=5)
BINOM.INV(trials, probability of success, probability)	=BINOM.INV(20,0.06,0.9991)	5	P(X=?)=0.9991
POISSON.DIST(number of occurrences, mean number of occurrences, cumulative)	=POISSON.DIST(2,4,TRUE)	0.2381	P (X≤2)
	=POISSON.DIST(2,4,FALSE)	0.1465	P (X=2)

R

The standard library stats contains pmf's, pdf's and cdf's of many common models:

- **dgeom(x,p)** # Gives $P(X = x)$ for the Geometric with probability p
- **pgeom(x,p)** # Gives the cdf $P(X \le x)$ for the Geometric with probability p
- **dbinom(x,n,p)** # Gives $P(X = x)$ for the Binomial with n and p
- **pbinom(x,n,p)** # Gives the cdf $P(X \le x)$ for the Binomial with n and p
- **dpois(x,lambda)** # Gives $P(X = x)$ for the Poisson with mean lambda
- **ppois(x,lambda)** # Gives the cdf $P(X \le x)$ for the Poisson with mean lambda
- **dnorm(x,mu,sig)** # Gives the pmf for the Normal model
- **pnorm(x,mu,sig)** # Gives the cdf for the Normal model

JMP

- Create a new data table: **File > New > New Data Table.**
- Right click on the header **Column 1** and select **Formula.**
- Select **Discrete Probability.**
- Select any of:
- Binomial Probability (prob, n, k) for the pmf
- Binomial Distribution (prob, n, k) for the cdf
- Poisson Probability (λ) for the pmf
- Poisson Distribution (λ) for the cdf
- Click OK twice. The probability will be displayed in the first row of Column 1.

MINITAB

- Choose **Probability Distributions** from the Calc menu.
- Choose **Binomial** from the Probability Distributions submenu.
- To calculate the probability of getting *x* successes in *n* trials, choose **Probability**.
- To calculate the probability of getting *x* or fewer successes among *n* trials, choose **Cumulative Probability**.
- For Poisson, choose **Poisson** from the Probability Distribution submenu.

SPSS

- In Data View, type values of the parameters for the desired distribution in the first row. For example, for a binomial pmf, type the number of successes in the first column, first row; trials in second column, first row; and probability of success in the third column, first row.
- Choose **Transform: Compute Variable**.
- Type a name for the variable that will contain the result.
- In the box under "Numeric Expression," type the desired probability to be calculated, using the labels for the variables where the parameter values are stored:
- PDF.BINOM(x, n, prob)
- CDF.BINOM(x, n, prob)
- PDF.Poisson(x, mean)
- CDF.Poisson(x, mean)
- Click **OK** and the probability will be calculated and stored in the next column. Adjust the column width to show more decimals—the value will be rounded to 2 decimal places by default.

Brief Case

Investment Options

A young entrepreneur has just raised $30,000 from investors, and she would like to invest it while she continues her fund-raising in hopes of starting her company one year from now. She wants to do due diligence and understand the risk of each of her investment options. After speaking with her colleagues in finance, she believes that she has three choices: (1) she can purchase a $30,000 certificate of deposit (CD); (2) she can invest in a mutual fund with a balanced portfolio; or (3) she can invest in a growth stock that has a greater potential payback but also has greater volatility. Each of her options will yield a different payback on her $30,000, depending on the state of the economy.

During the next year, she knows that the CD yields a constant annual percentage rate, regardless of the state of the economy. If she invests in a balanced mutual fund, she estimates that she will earn as much as 12% if the economy remains strong, but could possibly lose as much as 4% if the economy takes a downturn. Finally, if she invests all $30,000 in a growth stock, experienced investors tell her that she can earn as much as 40% in a strong economy, but may lose as much as 40% in a poor economy.

Estimating these returns, along with the likelihood of a strong economy, is challenging. Therefore, a "sensitivity analysis" is often conducted, where figures are computed using a range of values for each of the uncertain parameters in the problem. Following this advice, this investor decides to compute measures for a range of interest rates for CDs, a range of returns for the mutual fund, and a range of returns for the growth stock. In addition, the likelihood of a strong economy is unknown, so she will vary these probabilities as well.

Assume that the probability of a strong economy over the next year is 0.3, 0.5, or 0.7. To help this investor make an informed decision, evaluate the expected value and volatility of each of her investments using the following ranges of rates of growth:

CD: Look up the current annual rate for the return on a 3-year CD and use this value ±0.5%.

Mutual Fund: Use values of 8%, 10%, and 12% for a strong economy and values of 0%, −2%, and −4% for a weak economy.

Growth Stock: Use values of 10%, 25%, and 40% in a strong economy and values of −10%, −25%, and −40% in a weak economy.

Discuss the expected returns and uncertainty of each of the alternative investment options for this investor in each of the scenarios you analyzed. Be sure to compare the volatility of each of her options.

EXERCISES

SECTION 6.1

1. A company's employee database includes data on whether or not the employee includes a dependent child in his or her health insurance.

a) Is this variable discrete or continuous?
b) What are the possible values it can take on?

2. The database also, of course, includes each employee's compensation.

a) Is this variable discrete or continuous?
b) What are the possible values it can take on?

3. Suppose that the probabilities of a customer purchasing 0, 1, or 2 books at a book store are 0.5, 0.3, and 0.2, respectively. What is the expected number of books a customer will purchase?

4. A day trader buys an option on a stock that will return $100 profit if the stock goes up today and lose $400 if it goes down. If the trader thinks there is a 75% chance that the stock will go up,

a) What is her expected value of the option's profit?

b) What do you think of this option?

SECTION 6.2

5. Find the standard deviation of the book purchases in Exercise 3.

6. Find the standard deviation of the day trader's option value in Exercise 4.

7. An orthodontist has three financing packages, and each has a different service charge. He estimates that 30% of patients use the first plan, which has a $10 finance charge; 50% use the second plan, which has a $20 finance charge; and 20% use the third plan, which has a $30 finance charge.

a) Find the expected value of the service charge.

b) Find the standard deviation of the service charge.

8. A marketing agency has developed three vacation packages to promote a timeshare plan at a new resort. They estimate that 20% of potential customers will choose the Day Plan, which does not include overnight accommodations; 40% will choose the Overnight Plan, which includes one night at the resort; and 40% will choose the Weekend Plan, which includes two nights.

a) Find the expected value of the number of nights potential customers will need.

b) Find the standard deviation of the number of nights potential customers will need.

SECTION 6.3

9. Given independent random variables, X and Y, with means and standard deviations as shown, find the mean and standard deviation of each of the variables in parts a to d.

a) $3X$

b) $Y + 6$

c) $X + Y$

d) $X - Y$

	Mean	SD
X	10	2
y	20	5

10. Given independent random variables, X and Y, with means and standard deviations as shown, find the mean and standard deviation of each of the variables in parts a to d.

a) $X - 20$

b) $0.5Y$

c) $X + Y$

d) $X - Y$

	Mean	SD
X	80	12
y	12	3

11. A broker has calculated the expected values of two different financial instruments X and Y. Suppose that $E(X) = \$100, E(Y) = \$90, SD(X) = \$12,$ and $SD(Y) = \$8$. Find each of the following.

a) $E(X + 10)$ and $SD(X + 10)$

b) $E(5Y)$ and $SD(5Y)$

c) $E(X + Y)$ and $SD(X + Y)$

d) What assumption must you make in part c?

12. A company selling glass ornaments by mail-order expects, from previous history, that 6% of the ornaments it ships will break in shipping. You purchase two ornaments as gifts and have them shipped separately to two different addresses. What is the probability that both arrive safely? What did you assume?

SECTION 6.4

13. Which of these situations fit the conditions for using Bernoulli trials? Explain.

a) You are rolling 5 dice and need to get at least two 6s to win the game.

b) We record the distribution of home states of customers visiting our website.

c) A committee consisting of 11 men and 8 women selects a delegation of 4 to attend a professional meeting at random. What is the probability they choose all women?

d) A study (softwaresecure.typepad.com/multiple_choice/2007/05/cheat_cheat_nev.html) found that 56% of M.B.A. students admit to cheating. A business school dean surveys all the students in the graduating class and gets responses that admit to cheating from 250 of 481 students.

14. At the airport entry sites, a computer is used to randomly decide whether a traveler's baggage should be opened for inspection. If the chance of being selected is 12%, can you model your chance of having your baggage opened with a Bernoulli model? Check each of the conditions specifically.

SECTION 6.5

15. At many airports, a traveler entering the United States is sent randomly to one of several stations where his passport and visa are checked. If each of the 6 stations is equally likely, can the probabilities of which station a traveler will be sent be modeled with a Uniform model?

16. Through the career services office, you have arranged preliminary interviews at four companies for summer jobs. Each company will either ask you to come to their site for a follow-up interview or not. Let X be the random variable equal to the total number of follow-up interviews that you might have.

a) List all the possible values of X.

b) Is the random variable discrete or continuous?

c) Do you think a uniform distribution might be appropriate as a model for this random variable? Explain briefly.

17. The U.S. Census Bureau's 2007 Survey of Business Owners showed that 28.7% of all non-farm businesses are owned by women. You are phoning local businesses and assume that the national percentage is true in your area. You wonder how many calls you will have to make before you find one owned by a woman. What probability model should you use? (Specify the parameters as well.)

18. As in Exercise 17, you are phoning local businesses. You call three firms. What is the probability that all three are owned by women?

*19. A manufacturer of clothing knows that the probability of a button flaw (broken, sewed on incorrectly, or missing) is 0.002. An inspector examines 50 shirts in an hour, each with 6 buttons. Using a Poisson probability model:

a) What is the probability that she finds no button flaws?
b) What is the probability that she finds at least one?

*20. Replacing the buttons with snaps increases the probability of a flaw to 0.003, but the inspector can check 70 shirts an hour (still with 6 snaps each). Now what is the probability she finds no snap flaws?

CHAPTER EXERCISES

21. New website. You have just launched the website for your company that sells nutritional products online. Suppose X = the number of different pages that a customer hits during a visit to the website.

a) Assuming that there are n different pages in total on your website, what are the possible values that this random variable may take on?
b) Is the random variable discrete or continuous?

22. New website, part 2. For the website described in Exercise 21, let Y = the total time (in minutes) that a customer spends during a visit to the website.

a) What are the possible values of this random variable?
b) Is the random variable discrete or continuous?

23. Repairs. The probability model below describes the number of repair calls that an appliance repair shop may receive during an hour.

Repair Calls	0	1	2	3
Probability	0.1	0.3	0.4	0.2

a) How many calls should the shop expect per hour?
b) What is the standard deviation?

24. Software company. A small software company will bid on a major contract. It anticipates a profit of $50,000 if it gets it, but thinks there is only a 30% chance of that happening.

a) What's the expected profit?
b) Find the standard deviation for the profit.

25. Commuting to work. A commuter must pass through five traffic lights on her way to work and will have to stop at each one that is red. After keeping a record for several months, she developed the following probability model for the number of red lights she hits:

X = # of Red	0	1	2	3	4	5
p(X = x)	0.05	0.25	0.35	0.15	0.15	0.05

a) How many red lights should she expect to hit each day?
b) What's the standard deviation?

26. Defects. A consumer organization inspecting new cars found that many had appearance defects (dents, scratches, paint chips, etc.). While none had more than three of these defects, 7% had three, 11% had two, and 21% had one defect.

a) Find the expected number of appearance defects in a new car.
b) What is the standard deviation?

27. Fishing tournament. A sporting goods manufacturer was asked to sponsor a local boy in two fishing tournaments. They claim the probability that he will win the first tournament is 0.4. If he wins the first tournament, they estimate the probability that he will also win the second is 0.2. They guess that if he loses the first tournament, the probability that he will win the second is 0.3.

a) According to their estimates, are the two tournaments independent? Explain your answer.
b) What's the probability that he loses both tournaments?
c) What's the probability he wins both tournaments?
d) Let random variable X be the number of tournaments he wins. Find the probability model for X.
e) What are the expected value and standard deviation of X?

28. Contracts. Your company bids for two contracts. You believe the probability that you get contract #1 is 0.8. If you get contract #1, the probability that you also get contract #2 will be 0.2, and if you do not get contract #1, the probability that you get contract #2 will be 0.3.

a) Are the outcomes of the two contract bids independent? Explain.
b) Find the probability you get both contracts.
c) Find the probability you get neither contract.
d) Let X be the number of contracts you get. Find the probability model for X.
e) Find the expected value and standard deviation of X.

29. Battery recall. A company has discovered that a recent batch of batteries had manufacturing flaws, and has issued a recall. You have 10 batteries covered by the recall, and 3 are dead. You choose 2 batteries at random from your package of 10.

a) Has the assumption of independence been met? Explain.
b) Create a probability model for the number of good batteries chosen.

c) What's the expected number of good batteries?
d) What's the standard deviation?

30. Grocery supplier. A grocery supplier believes that the mean number of broken eggs per dozen is 0.6, with a standard deviation of 0.5. You buy 3 dozen eggs without checking them.

a) How many broken eggs do you expect to get?
b) What's the standard deviation?
c) Is it necessary to assume the cartons of eggs are independent? Why?

31. Commuting, part 2. A commuter finds that she waits an average of 14.8 seconds at each of five stoplights, with a standard deviation of 9.2 seconds. Find the mean and the standard deviation of the total amount of time she waits at all five lights. What, if anything, did you assume?

32. Defective pixels. For warranty purposes, analysts want to model the number of defects on a screen of the new tablet they are manufacturing. Let X = the number of defective pixels per screen. If X can be modeled by:

X = # of Defective Pixels	0	1	2	3	4 or more
P(X = x)	0.95	0.04	0.008	0.002	0

a) What is the expected number of defective pixels per screen?
b) What is the standard deviation of the number of defective pixels per screen?
c) What is the expected number of defective pixels in the next 100 screens?
d) What is the standard deviation of the number of defective pixels in the next 100 screens?

33. Repair calls. Suppose that the appliance shop in Exercise 23 plans an 8-hour day.

a) Find the mean and standard deviation of the number of repair calls they should expect in a day.
b) What assumption did you make about the repair calls?
c) Use the mean and standard deviation to describe what a typical 8-hour day will be like.
d) At the end of a day, a worker comments "Boy, I'm tired. Today was sure unusually busy!" How many repair calls would justify such an observation.

34. Casino. At a casino, people play the slot machines in hopes of hitting the jackpot, but most of the time, they lose their money. A certain machine pays out an average of $0.92 (for every dollar played), with a standard deviation of $120.

a) Why is the standard deviation so large?
b) If a gambler plays 5 times, what are the mean and standard deviation of the casino's profit?
c) If gamblers play this machine 1000 times in a day, what are the mean and standard deviation of the casino's profit?

35. Bike sale. A bicycle shop plans to offer 2 specially priced children's models at a sidewalk sale. The basic model will return a profit of $120 and the deluxe model $150. Past experience indicates that sales of the basic model will have a mean of 5.4 bikes with a standard deviation of 1.2, and sales of the deluxe model will have a mean of 3.2 bikes with a standard deviation of 0.8 bikes. The cost of setting up for the sidewalk sale is $200.

a) Define random variables and use them to express the bicycle shop's net profit.
b) What's the mean of the net profit?
c) What's the standard deviation of the net profit?
d) Do you need to make any assumptions in calculating the mean? How about the standard deviation?

36. Farmers' market. A farmer has 100 lb of apples and 50 lb of potatoes for sale. The market price for apples (per pound) each day is a random variable with a mean of 0.5 dollars and a standard deviation of 0.2 dollars. Similarly, for a pound of potatoes, the mean price is 0.3 dollars and the standard deviation is 0.1 dollars. It also costs him 2 dollars to bring all the apples and potatoes to the market. The market is busy with eager shoppers, so we can assume that he'll be able to sell all of each type of produce at that day's price.

a) Define your random variables, and use them to express the farmer's net income.
b) Find the mean of the net income.
c) Find the standard deviation of the net income.
d) Do you need to make any assumptions in calculating the mean? How about the standard deviation?

37. Cancelled flights. Mary is deciding whether to book the cheaper flight home college after her final exams, but she's unsure when her last exam will be. She thinks there is only a 20% chance that the exam will be scheduled after the last day she can get a seat on the cheaper flight. If it is and she has to cancel the flight, she will lose $150. If she can take the cheaper flight, she will save $100.

a) If she books the cheaper flight, what can she expect to gain, on average?
b) What is the standard deviation?

38. Day trading. An option to buy a stock is priced at $200. If the stock closes above 30 on May 15, the option will be worth $1000. If it closes below 20, the option will be worth nothing, and if it closes between 20 and 30 (inclusively), the option will be worth $200. A trader thinks there is a 50% chance that the stock will close in the 20–30 range, a 20% chance that it will close above 30, and a 30% chance that it will fall below 20 on May 15.

a) How much does she expect to gain?
b) What is the standard deviation of her gain?
c) Should she buy the stock option? Discuss the pros and cons in terms of your answers to (a) and (b).

39. eBay. A collector purchased a quantity of action figures and is going to sell them on eBay. He has 19 Hulk figures. In recent auctions, the mean selling price of similar figures has been $12.11, with a standard deviation of $1.38. He also has 13 Iron Man figures which have had a mean selling price of $10.19, with a standard deviation of $0.77. His insertion fee will be $0.55 on each item, and the closing fee will be 8.75% of the selling price. He assumes all will sell without having to be relisted.

a) Define your random variables, and use them to create a random variable for the collector's net revenue.
b) Find the mean (expected value) of the net revenue.
c) Find the standard deviation of the net revenue.
d) Do you have to assume independence for the sales on eBay? Explain.

40. Real estate. A real-estate broker in Washington, DC, purchased 3 two-bedroom houses in a depressed market for a combined cost of $1,000,000. He expects the cleaning and repair costs on each house to average $100,000 with a standard deviation of $15,000. When he sells them, after subtracting taxes and other closing costs, he expects to realize an average of $475,000 per house, with a standard deviation of $12,500.

a) Define your random variables, and use them to create a random variable for the broker's net profit.
b) Find the mean (expected value) of the net profit.
c) Find the standard deviation of the net profit.
d) Do you have to assume independence for the repairs and sale prices of the houses? Explain.

41. Bernoulli. Can we use probability models based on Bernoulli trials to investigate the following situations? Explain.

a) Each week a doctor rolls a single die to determine which of his six office staff members gets the preferred parking space.
b) A medical research lab has samples of blood collected from 120 different individuals. How likely is it that the majority of them are Type A blood, given that Type A is found in 43% of the population?
c) From a workforce of 13 men and 23 women, all five promotions go to men. How likely is that, if promotions are based on qualifications rather than gender?
d) We poll 500 of the 3000 stockholders to see how likely it is that the proposed budget will pass.
e) A company realizes that about 10% of its packages are not being sealed properly. In a case of 24 packages, how likely is it that more than 3 are unsealed?

42. Bernoulli, part 2. Can we use probability models based on Bernoulli trials to investigate the following situations? Explain.

a) You survey 500 potential customers to determine their color preference.
b) A manufacturer recalls a doll because about 3% have buttons that are not properly attached. Customers return 37 of these dolls to the local toy store. How likely are they to find any buttons not properly attached?
c) A city council of 11 Republicans and 8 Democrats picks a committee of 4 at random. How likely are they to choose all Democrats?
d) An executive reads that 74% of employees in his industry are dissatisfied with their jobs. How many dissatisfied employees can he expect to find among the 481 employees in his company?

43. Closing sales. A salesman normally makes a sale (closes) on 80% of his presentations. Assuming the presentations are independent, find the probability of each of the following.

a) He fails to close for the first time on his fifth attempt.
b) He closes his first presentation on his fourth attempt.
c) The first presentation he closes will be on his second attempt.
d) The first presentation he closes will be on one of his first three attempts.

44. Computer chip manufacturer. Suppose a computer chip manufacturer rejects 2% of the chips produced because they fail presale testing. Assuming the bad chips are independent, find the probability of each of the following.

a) The fifth chip they test is the first bad one they find.
b) They find a bad one within the first 10 they examine.
c) The first bad chip they find will be the fourth one they test.
d) The first bad chip they find will be one of the first three they test.

45. Side effects. Researchers testing a new medication find that 7% of users have side effects. To how many patients would a doctor expect to prescribe the medication before finding the first one who has side effects?

46. Credit cards. College students are a major target for advertisements for credit cards. At a university, 65% of students surveyed said they had opened a new credit card account within the past year. If that percentage is accurate, how many students would you expect to survey before finding one who had not opened a new account in the past year?

*47. Missing pixels. A company that manufactures large LCD screens knows that not all pixels on their screen light, even if they spend great care when making them. In a sheet 6 ft by 10 ft (72 in. by 120 in.) that will be cut into smaller screens, they find an average of 4.7 blank pixels. They believe that the occurrences of blank pixels are independent. Their warranty policy states that they will replace any screen sold that shows more than 2 blank pixels.

a) What is the mean number of blank pixels per square foot?
b) What is the standard deviation of blank pixels per square foot?
c) What is the probability that a 2 ft by 3 ft screen will have at least one defect?

d) What is the probability that a 2 ft by 3 ft screen will be replaced because it has too many defects?

*48. Bean bags. Cellophane that is going to be formed into bags for items such as dried beans or bird seed is passed over a light sensor to test if the alignment is correct before it passes through the heating units that seal the edges. Small adjustments can be made by the machine automatically. But if the alignment is too bad, the process is stopped and an operator has to manually adjust it. These misalignment stops occur randomly and independently. On one line, the average number of stops is 52 per 8-hour shift.

a) What is the mean number of stops per hour?
b) What is the standard deviation of stops per hour?

49. Hurricane insurance. An insurance company needs to assess the risks associated with providing hurricane insurance. During the 22 years from 1990 through 2011, Florida was hit by 27 major hurricanes (level 3 and above). If hurricanes are independent and the mean has not changed, what is the probability of having a year in Florida with each of the following?

a) No hits?
b) Exactly 1 hit?
c) More than 1 hit?

*50. Hurricane insurance, part 2. During the 18 years from 1995 through 2012, there were 144 hurricanes in the Atlantic basin. Assume that hurricanes are independent and the mean has not changed.

a) What is the mean number of major hurricanes per year?
b) What is the standard deviation of the annual frequency of major hurricanes?
c) What is the probability of having a year with no major hurricanes?
d) What is the probability of going three years in a row without a major hurricane?

51. Lefties. A manufacturer of game controllers is concerned that their controller may be difficult for left-handed users. They set out to find lefties to test. About 13% of the population is left-handed. If they select a sample of five customers at random in their stores, what is the probability of each of these outcomes?

a) The first lefty is the fifth person chosen.
b) There are some lefties among the 5 people.
c) The first lefty is the second or third person.
d) There are exactly 3 lefties in the group.
e) There are at least 3 lefties in the group.
f) There are no more than 3 lefties in the group.

52. Arrows. An Olympic archer is able to hit the bull's-eye 80% of the time. Assume each shot is independent of the others. If she shoots 6 arrows, what's the probability of each of the following results?

a) Her first bull's-eye comes on the third arrow.
b) She misses the bull's-eye at least once.

c) Her first bull's-eye comes on the fourth or fifth arrow.
d) She gets exactly 4 bull's-eyes.
e) She gets at least 4 bull's-eyes.
f) She gets at most 4 bull's-eyes.

53. Satisfaction survey. A cable provider wants to contact customers in a particular telephone exchange to see how satisfied they are with the new digital TV service the company has provided. All numbers are in the 452 exchange, so there are 10,000 possible numbers from 452-0000 to 452-9999. If they select the numbers with equal probability:

a) What distribution would they use to model the selection?
b) What is the probability the number selected will be an even number?
c) What is the probability the number selected will end in 000?

54. Manufacturing quality. In an effort to check the quality of their cell phones, a manufacturing manager decides to take a random sample of 10 cell phones from yesterday's production run, which produced cell phones with serial numbers ranging (according to when they were produced) from 43005000 to 43005999. If each of the 1000 phones is equally likely to be selected:

a) What distribution would they use to model the selection?
b) What is the probability that a randomly selected cell phone will be one of the last 100 to be produced?
c) What is the probability that the first cell phone selected is either from the last 200 to be produced or from the first 50 to be produced?
d) What is the probability that the first two cell phones are both from the last 100 to be produced?

*55. Web visitors. A website manager has noticed that during the evening hours, about 3 people per minute check out from their shopping cart and make an online purchase. She believes that each purchase is independent of the others and wants to model the number of purchases per minute.

a) What model might you suggest to model the number of purchases per minute?
b) What is the probability that in any 1 minute at least one purchase is made?
c) What is the probability that no one makes a purchase in the next 2 minutes?

*56. Quality control. The manufacturer in Exercise 54 has noticed that the number of faulty cell phones in a production run of cell phones is usually small and that the quality of one day's run seems to have no bearing on the next day.

a) What model might you use to model the number of faulty cell phones produced in one day?
b) If the mean number of faulty cell phones is 2 per day, what is the probability that no faulty cell phones will be produced tomorrow?
c) If the mean number of faulty cell phones is 2 per day, what is the probability that 3 or more faulty cell phones were produced in today's run?

57. Lefties, redux. Consider our group of 5 people from Exercise 51.

a) How many lefties do you expect?
b) With what standard deviation?
c) If we keep picking people until we find a lefty, how long do you expect it will take?

58. More arrows. Consider our archer from Exercise 52.

a) How many bull's-eyes do you expect her to get?
b) With what standard deviation?
c) If she keeps shooting arrows until she hits the bull's-eye, how long do you expect it will take?

59. Still more lefties. Suppose we choose 12 people instead of the 5 chosen in Exercise 57

a) Find the mean and standard deviation of the number of right-handers in the group.
b) What's the probability that they're not all right-handed?
c) What's the probability that there are no more than 10 righties?
d) What's the probability that there are exactly 6 of each?
e) What's the probability that the majority is right-handed?

60. Still more arrows. Suppose the archer from Exercise 58 shoots 10 arrows.

a) Find the mean and standard deviation of the number of bull's-eyes she may get.
b) What's the probability that she never misses?
c) What's the probability that there are no more than 8 bull's-eyes?
d) What's the probability that there are exactly 8 bull's-eyes?
e) What's the probability that she hits the bull's-eye more often than she misses?

JUST CHECKING ANSWERS

1. a) $100 + 100 = 200$ seconds
   b) $\sqrt{50^2 + 50^2} = 70.7$ seconds
   c) The times for the two customers are independent.

2. There are two outcomes (contact, no contact), the probability of contact stays constant at 0.76, and random calls should be independent.

3. Binomial

4. Geometric

7

The Normal and Other Continuous Distributions

The NYSE

The New York Stock Exchange (NYSE) was founded in 1792 by 24 stockbrokers who signed an agreement under a buttonwood tree on Wall Street in New York. The first offices were in a rented room at 40 Wall Street. In the 1830s traders who were not part of the Exchange did business in the street. They were called "curbstone brokers." It was the curbstone brokers who first made markets in gold and oil stocks and, after the Civil War, in small industrial companies such as the emerging steel, textile, and chemical industries.

By 1903 the NYSE was established at its current home at 18 Broad Street. The curbstone brokers finally moved indoors in 1921 to a building on Greenwich Street in lower Manhattan. In 1953 the curb market changed its name to the American Stock Exchange. In 1993 the American Stock Exchange pioneered the market for derivatives by introducing the first exchange-traded fund, Standard & Poor's Depositary Receipts (SPDRs).

The NYSE Euronext holding company was created in 2007 as a combination of the NYSE Group, Inc., and Euronext N.V. And in 2008, NYSE Euronext merged with the American Stock Exchange. The combined exchange is the world's largest and most liquid exchange group.

7.1   The Standard Deviation as a Ruler

WHO	Months
WHAT	CAPE10 values for the NYSE
WHEN	1880 through early 2015
WHY	Investment guidance

Investors have always sought ways to help them decide when to buy and when to sell. Such measures rely on identifying when the stock market is in an unusual state—either unusually undervalued (buy!) or unusually overvalued (sell!). One measure is the Cyclically Adjusted Price/Earnings Ratio (CAPE10) developed by Yale professor Robert Shiller. The CAPE10 is based on the standard Price/Earnings (P/E) ratio of stocks, but designed to smooth out short-term fluctuations by "cyclically adjusting" them. The CAPE10 has been as low as 4.78, in 1920, and as high as 44.20, in late 1999. The long-term average CAPE10 (since year 1881) is 16.58.

Investors who follow the CAPE10 use the metric to signal times to buy and sell. One mutual fund strategy buys only when the CAPE10 is 33% lower than the long-term average and sells (or "goes into cash") when the CAPE10 is 50% higher than the long-term average. Between January 1, 1971, and October 23, 2009, this strategy would have outperformed such standard measures as the Wiltshire 5000 in both average return and volatility, but it is important to note that the strategy would have been completely in cash from just before the stock market crash of 1987 all the way to March of 2009! Shiller popularized the strategy in his book *Irrational Exuberance.* Figure 7.1 shows a time series plot of the CAPE10 values for the New York Stock Exchange from 1880 until the beginning of 2015. Generally, the CAPE10 hovers around 15. But occasionally, it can take a large excursion. One such time was in 1999 and 2000, when the CAPE10 exceeded 40. But was this just a random peak or were these values really extraordinary?

Figure 7.1   CAPE10 values for the NYSE from 1880 to 2015.

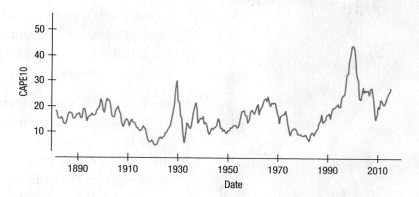

To answer this question, we can look at the overall distribution of CAPE10 values. Figure 7.2 shows a histogram of the same values. Now we don't see patterns over time, but we may be able to make a better judgment of whether values are extraordinary.

Overall, the main body of the distribution looks unimodal and reasonably symmetric. But then there's a tail of values that trails off to the high end. How can we assess how extraordinary they are?

Investors follow a wide variety of measures that record various aspects of stocks, bonds, and other investments, looking for times when these measures of stock performance are extraordinary because those often represent times of increased risk or opportunity. But these measures are quantitative values, not categories. Random variables that can take on any value in a range of values are continuous. The distributions of Chapter 6 won't be suitable to model them, but many of the basic concepts still apply.

We saw in Chapter 3 that z-scores provide a standard way to compare values. We use the standard deviation as a ruler, asking how many standard deviations a

Figure 7.2 The distribution of the CAPE10 values shown in Figure 7.1.

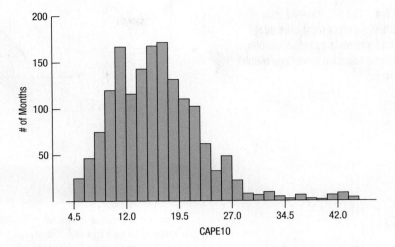

value is from the mean. A z-score reports the number of standard deviations away from the mean. We can convert the CAPE10 values to z-scores by subtracting their mean (16.58) and dividing by their standard deviation (6.58). Figure 7.3 shows the resulting distribution.

Figure 7.3 The CAPE10 values as z-scores.

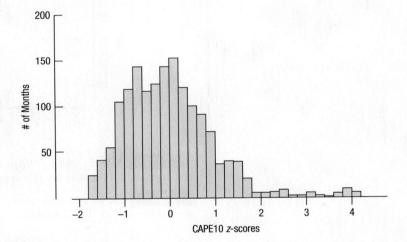

It's easy to see that the z-scores have the same distribution as the original values, but now we can also see that the largest of them is above 4. How extraordinary is it for a value to be four standard deviations away from the mean? Fortunately, there's a fact about unimodal, symmetric distributions that can guide us.[1]

The 68–95–99.7 Rule

In a unimodal, symmetric distribution, about 68% of the values fall within 1 standard deviation of the mean, about 95% fall within 2 standard deviations of the mean, and about 99.7%—almost all—fall within 3 standard deviations of the mean. Calling this rule the **68–95–99.7 Rule** provides a mnemonic for these three values.[2]

---

[1] All of the CAPE10 values in the right tail occurred after 1993. Until that time the distribution of CAPE10 values was quite symmetric and clearly unimodal.

[2] This rule is also called the "Empirical Rule" because it originally was observed without any proof. It was first published by Abraham de Moivre in 1733, 75 years before the underlying reason for it—which we're about to see—was known.

Figure 7.4   The 68–95–99.7 Rule tells us how much of most unimodal, symmetric models is found within one, two, or three standard deviations of the mean.

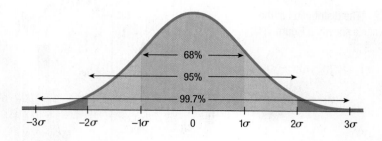

FOR EXAMPLE   An extraordinary day for the Dow?

After the financial crisis of 2007/2008, the Dow Jones Industrial Average (DJIA) improved from a low of 7278 on March 20, 2009, to new records over 18,000 just 6 years later. But on August 8, 2011, the Dow dropped 634.8 points. Although that wasn't the most ever lost in a day, it sent shock waves through the financial community. During the year from mid-2011 to mid-2012, the mean daily change in the DJIA was 1.87, with a standard deviation of 155.28 points. A histogram of day-to-day changes in the DJIA looked like this:

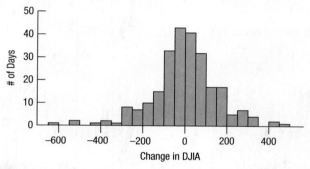

QUESTION   Use the 68–95–99.7 Rule to characterize how extraordinary the change on August 8, 2011 was. Is the rule appropriate?

ANSWER   The histogram is unimodal and symmetric, so the 68–95–99.7 Rule is an appropriate model. The z-score corresponding to the August 8 change is

$$\frac{-634.8 - 1.87}{155.28} = -4.10$$

A z-score bigger than 3 in magnitude will occur with a probability of less than 0.0015, or about once every 3 years for daily values. A z-score of 4 is even less likely. This was a truly extraordinary event.

7.2   The Normal Distribution

The 68–95–99.7 Rule is useful in describing how unusual a z-score is. But often in business we want a more precise answer than one of these three values. To say more about how big we expect a z-score to be, we need to *model* the data's distribution.

There is no universal standard for z-scores, but there is a model that shows up over and over in Statistics. You've probably heard of "bell-shaped curves." Statisticians call them Normal (or Gaussian) distributions. **Normal distributions** are appropriate models for distributions whose shapes are unimodal and roughly

"All models are wrong—but some are useful."
—GEORGE BOX, FAMOUS
STATISTICIAN

NOTATION ALERT

$N(\mu, \sigma)$ always denotes a Normal. The μ, pronounced "mew," is the Greek letter for "m," and always represents the mean in a model. The σ, sigma, is the lowercase Greek letter for "s," and always represents the standard deviation in a model.

symmetric. There is a Normal distribution for every possible combination of mean and standard deviation. We write $N(\mu, \sigma)$ to represent a Normal distribution with a mean of μ and a standard deviation of σ. We use Greek symbols here because this mean and standard deviation are parameters of the model, not summaries based on data. We can compute z-scores based on this model by using the **parameters** μ and σ. We still call these standardized values z-**scores**. We write

$$z = \frac{y - \mu}{\sigma}.$$

Standardized values have mean 0 and standard deviation 1, so by doing this to our values, we'll need only one model—the model $N(0,1)$. The Normal distribution with mean 0 and standard deviation 1 is called the **standard Normal distribution** (or the **standard Normal model**).

You shouldn't use a Normal model for just any data set. Remember that standardizing won't change the shape of the distribution. If the distribution is not unimodal and symmetric to begin with, standardizing won't make it Normal.

Is Normal Normal?

Don't be misled. The name "Normal" doesn't mean that these are the *usual* shapes for histograms. The name follows a tradition of positive thinking in Mathematics and Statistics in which functions, equations, and relationships that are easy to work with or have other nice properties are called "normal," "common," "regular," "natural," or similar terms. It's as if by calling them ordinary, we could make them actually occur more often and make our lives simpler.

JUST CHECKING

1 Your Accounting teacher has announced that the lower of your two tests will be dropped. You got a 90 on test 1 and an 80 on test 2. You're all set to drop the 80 until she announces that she grades "on a curve." She standardized the scores in order to decide which is the lower one. If the mean on the first test was 88 with a standard deviation of 4 and the mean on the second was 75 with a standard deviation of 5,

   a) Which one will be dropped?

   b) Does this seem "fair"?

Is the Standard Normal a Standard?

Yes. We call it the "Standard Normal" because it models standardized values. It is also a "standard" because this is the particular Normal model that we almost always use.

Unlike the discrete probability distributions we saw in Chapter 6, the Normal distribution can take on any value. So, it is a **continuous random variable**. The distribution of any continuous random variable can be shown with a curve called its **probability density function (pdf),** usually denoted as $f(x)$. The curve we use to work with the Normal distribution is called the Normal probability density function.

The probability density function (pdf) doesn't give the probability for specific values, as the probability models for discrete random variables did. Instead the pdf gives the probability as the area below its curve in some interval. For the standard Normal, shown in Figure 7.5, the area below the curve between -1 and 1 is about 68%, which is where the 68–95–99.7 Rule comes from.

Figure 7.5 The standard Normal density function (with mean 0 and standard deviation 1). The probability of finding a z-score in any interval is the area over that interval under the curve. For example, the probability that the z-score falls between -1 and 1 is about 68%, which can be seen approximately from the density function or found more precisely from a table or technology.

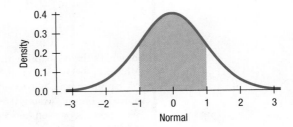

Keep in mind that the probability density function $f(x)$ isn't equal to $P(X = x)$. In fact, for a continuous random variable, X, $P(X = x)$ is 0 for every value of x! That may seem strange at first, but the probability is the area under the curve over an interval, so smaller intervals have smaller probability. If the interval is just a point, there is no area—and no probability. (See the box below.)

How Can *Every* Value Have Probability 0?

We can find a probability for any interval of z-scores. But the probability for a single z-score is zero. How can that be? Let's look at the standard Normal random variable, Z. We could find (from a table, website, or computer program) that the probability that Z lies between 0 and 1 is 0.3413.

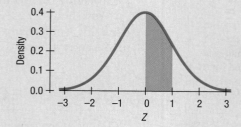

That's the area under the Normal pdf (in red) between the values 0 and 1.

So, what's the probability that Z is between 0 and 1/10?

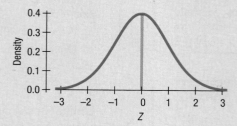

That area is only 0.0398. What is the chance then that Z will fall between 0 and 1/100? There's not much area—the probability is only 0.0040. If we kept going, the probability would keep getting smaller. The probability that Z is between 0 and 1/100,000 is less than 0.0001.

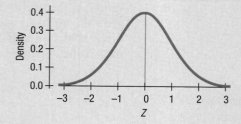

So, what's the probability that Z is *exactly* 0? Well, there's *no* area under the curve right at $x = 0$, so the probability is 0. It's only intervals that have positive probability, but that's OK. In real life we never mean exactly 0.0000000000 or any other value. If you say "exactly 164 pounds," you might really mean between 163.5 and 164.5 pounds or even between 163.99 and 164.01 pounds, but realistically not 164.000000000 . . . pounds.

Finding Normal Percentiles

We can use the standard Normal model to find the probability of finding a value in any interval. Table Z at the back of this text provides one simple way, but these days, it is more common to use an app on your smart phone, a statistics program, or even a spreadsheet. The technology solutions usually require a mean and standard deviation. Tables work with the standard Normal, so you must convert to z-scores first. Our value of 1.8 SD above the mean is a z-score of 1.80. In the table, look down the left column for the first two digits (1.8) and then read across for the third digit (0) (Figure 7.6). The table gives the area under the curve as 0.9641.

Figure 7.6 A table of Normal percentiles (Table Z in Appendix B) lets us find the percentage of individuals in a standard Normal distribution falling below any specified z-score value.

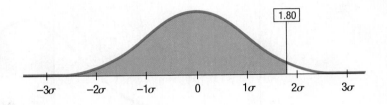

z	.00	.01
1.7	0.9554	0.9564
1.8	0.9641	0.9649
1.9	0.9713	0.9719

That means that 96.4% of the z-scores are less than 1.80. Since the total area is always 1, and $1 - 0.9641 = 0.0359$ we know that only 3.6% of all observations from a Normal distribution have z-scores higher than 1.80.

Finding Normal Percentiles in Excel

In Excel, two functions make it easy to do these calculations:

Normdist(x, mu, sigma, TRUE) returns the area under the Normal model with the parameters given, to the left of the value of x. (The function is Norm.dist() in some versions of Excel.)

NormInv(p, mu, sigma) returns the value of x such that the area to the left of x in a Normal model with the given parameters will be p. (The function is Norm.Inv() in some versions of Excel.)

But if you decide to use these functions, you should always first **make a picture** and color in the region of interest.

In Excel, type
=NORM.DIST(600,500,100,1)

FOR EXAMPLE   GMAT scores and the Normal model

The Graduate Management Admission Test (GMAT) has scores from 200 to 800. Scores are supposed to follow a distribution that is roughly unimodal and symmetric and is designed to have an overall mean of 500 and a standard deviation of 100. In any one year, the mean and standard deviation may differ from these target values by a small amount, but we can use these values as good overall approximations.

QUESTION   Suppose you earned a 600 on your GMAT test. From that information and the 68–95–99.7 Rule, where do you stand among all students who took the GMAT?

ANSWER   Because we're told that the distribution is unimodal and symmetric, we can approximate the distribution with a Normal model. We are also told the scores have a mean of 500 and an SD of 100. So, we'll use a $N(500,100)$. It's good practice at this point to draw the distribution. Find the score whose percentile you want to know and locate it on the picture. When you finish the calculation, you should check to make sure that it's a reasonable percentile from the picture.

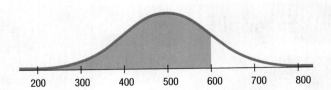

A score of 600 is 1 SD above the mean. That corresponds to one of the points in the 68–95–99.7 Rule. About 32% ($100\% - 68\%$) of those who took the test were more than one standard deviation from the mean, but only half of those were on the high side. So about 16% (half of 32%) of the test scores were better than 600 and about 84% were below that value, so you were in the 84th percentile.

in Excel, type
=NORM.DIST(600,500,100,1)-
NORM.DIST(450,500,100,1)

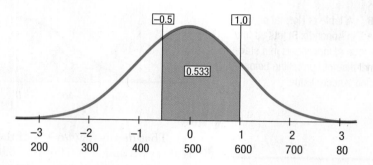

FOR EXAMPLE   More GMAT scores

QUESTION   Assuming the GMAT scores are nearly Normal with $N(500,100)$, what proportion of GMAT scores falls between 450 and 600?

ANSWER   The first step is to find the z-scores associated with each value. Standardizing the scores we are given, we find that for 600, $z = (600 - 500)/100 = 1.0$ and for 450, $z = (450 - 500)/100 = -0.50$. Then, we can label the axis below the picture either in the original values or the z-scores or even use both scales as the following picture shows.

From Table Z, we find the area $z \leq 1.0 = 0.8413$, which means that 84.13% of scores fall below 1.0, and the area $z \leq -0.50 = 0.3085$, which means that 30.85% of the values fall below -0.5, so the proportion of z-scores *between* them is $84.13\% - 30.85\% = 53.28\%$. So, the Normal model estimates that about 53.3% of GMAT scores fall between 450 and 600.

Finding areas from z-scores is the simplest way to work with the Normal distribution. Sometimes we start with areas and need to work backward to find the corresponding z-score or even the original data value. For instance, what z-score represents the first quartile, Q1, in a Normal distribution? In our first set of examples, we knew the z-score and used the table or technology to find the percentile. Now we want to find the cut point for the 25th percentile. Make a picture, shading the leftmost 25% of the area. Look in Table Z for an area of 0.2500. The exact area is not there, but 0.2514 is the closest number. That shows up in the table with -0.6 in the left margin and 0.07 in the top margin. The z-score for Q1, then, is approximately $z = -0.67$. Computers and calculators can determine the cut point more precisely (and more easily).[3]

---

[3] We'll often use those more precise values in our examples. If you're finding the values from the table you may not get *exactly* the same number to all decimal places as your classmate who's using a computer package.

GUIDED EXAMPLE   Cereal Company

A cereal manufacturer has a machine that fills the boxes. Boxes are labeled "16 oz," so the company wants to have that much cereal in each box. But no packaging process is perfect. If the machine is set at exactly 16 oz and the distribution is roughly symmetric, then about half of the boxes will be underweight, making consumers unhappy and exposing the company to bad publicity and possible lawsuits. To prevent underweight boxes, the manufacturer has to set the mean a little higher than 16.0 oz. Based on their experience with the packaging machine, the company believes that the amount of cereal in the boxes fits a Normal distribution with a standard deviation of 0.2 oz. The manufacturer decides to set the machine to put an average of 16.3 oz in each box. Let's use that model to answer a series of questions about these cereal boxes.

Question 1: What fraction of the boxes will be underweight?

PLAN	**Setup** State the variable and the objective.	The variable is weight of cereal in a box. We want to determine what fraction of the boxes risk being underweight.
	Model Check to see if a Normal distribution is appropriate.	We have no data, so we cannot make a histogram. But we are told that the company believes the distribution of weights from the machine is Normal.
	Specify which Normal distribution to use.	We use an $N(16.3, 0.2)$ model.
DO	**Mechanics** Make a graph of this Normal distribution. Locate the value you're interested in on the picture, label it, and shade the appropriate region.	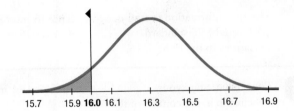
► REALITY CHECK	Estimate from the picture the percentage of boxes that are underweight. (This will be useful later to check that your answer makes sense.)	(It looks like a low percentage—maybe less than 10%.) We want to know what fraction of the boxes will weigh less than 16 oz.
	Convert your cutoff value into a z-score.	$$z = \frac{y - \mu}{\sigma} = \frac{16 - 16.3}{0.2} = -1.50.$$
	Look up the area in the Normal table, or use technology.	$$P(y < 16) = P(z < -1.50) = 0.0668$$
REPORT	**Conclusion** State your conclusion in the context of the problem.	We estimate that approximately 6.7% of the boxes will contain less than 16 oz of cereal.

Question 2: The company's lawyers say that 6.7% is too high. They insist that no more than 4% of the boxes can be underweight. So the company needs to set the machine to put a little more cereal in each box. What mean setting do they need?

PLAN	**Setup** State the variable and the objective.	The variable is weight of cereal in a box. We want to determine a setting for the machine.
	Model Check to see if a Normal model is appropriate.	We have no data, so we cannot make a histogram. But we are told that a Normal model applies.
	Specify which Normal distribution to use. This time you are not given a value for the mean!	We don't know μ, the mean amount of cereal. The standard deviation for this machine is 0.2 oz. The model, then, is $N(\mu, 0.2)$.

(continued)

REALITY CHECK	We found out earlier that setting the machine to $\mu = 16.3$ oz made 6.7% of the boxes too light. We'll need to raise the mean a bit to reduce this fraction.	We are told that no more than 4% of the boxes can be below 16 oz.

DO	**Mechanics** Make a graph of this Normal distribution. Center it at μ (since you don't know the mean) and shade the region below 16 oz.	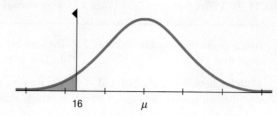
	Using the Normal table, a calculator, or software, find the z-score that cuts off the lowest 4%.	The z-score that has 0.04 area to the left of it is $z = -1.75$.
	Use this information to find μ. It's located 1.75 standard deviations to the right of 16.	Since 16 must be 1.75 standard deviations below the mean, we need to set the mean at $16 + 1.75 \cdot 0.2 = 16.35$.

REPORT	**Conclusion** State your conclusion in the context of the problem.	The company must set the machine to average 16.35 oz of cereal per box.

Question 3: The company president vetoes that plan, saying the company should give away less free cereal, not more. Her goal is to set the machine no higher than 16.2 oz and still have only 4% underweight boxes. The only way to accomplish this is to reduce the standard deviation. What standard deviation must the company achieve, and what does that mean about the machine?

PLAN	**Setup** State the variable and the objective.	The variable is weight of cereal in a box. We want to determine the necessary standard deviation to have only 4% of boxes underweight.
	Model Check that a Normal model is appropriate.	The company believes that the weights are described by a Normal distribution.
	Specify which Normal distribution to use. This time you don't know σ.	Now we know the mean, but we don't know the standard deviation. The model is therefore $N(16.2, \sigma)$.
REALITY CHECK	We know the new standard deviation must be less than 0.2 oz.	

DO	**Mechanics** Make a graph of this Normal distribution. Center it at 16.2, and shade the area you're interested in. We want 4% of the area to the left of 16 oz.	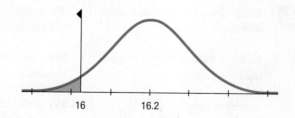

Find the *z*-score that cuts off the lowest 4%.	We already know that the *z*-score with 4% below it is $z = -1.75$.
Solve for σ. (Note that we need 16 to be 1.75 σ's below 16.2, so 1.75 σ must be 0.2 oz. You could just start with that equation.)	$$z = \frac{y - \mu}{\sigma}$$ $$-1.75 = \frac{16 - 16.2}{\sigma}$$ $$1.75\sigma = 0.2$$ $$\sigma = 0.114.$$

REPORT	**Conclusion** State your conclusion in the context of the problem.	The company must get the machine to box cereal with a standard deviation of only 0.114 oz. This means the machine must be more consistent (by nearly a factor of 2) in filling the boxes.
	As we expected, the standard deviation is lower than before— actually, quite a bit lower.	

JUST CHECKING

2 As a group, the Dutch are among the tallest people in the world. The average Dutch man is 184 cm tall—just over 6 feet (and the average Dutch woman is 170.8 cm tall—just over 5′7″). If a Normal model is appropriate and the standard deviation for men is about 8 cm, what percentage of all Dutch men will be over 2 meters (almost 6′7″) tall?

3 Suppose it takes you 20 minutes, on average, to drive to work, with a standard deviation of 2 minutes. Suppose a

Normal model is appropriate for the distributions of driving times.
a) How often will you arrive at work in less than 22 minutes?
b) How often will it take you more than 24 minutes?
c) Do you think the distribution of your driving times is unimodal and symmetric?
d) What does this say about the accuracy of your prediction? Explain.

7.3 Normal Probability Plots

Before using a Normal model you should check that the data follow a distribution that is at least close to Normal. You can check that the histogram is unimodal and symmetric, but there is also a specialized graphical display that can help you to decide whether the Normal model is appropriate: the **Normal probability plot**. If the distribution of the data is roughly Normal, the plot is roughly a diagonal straight line. Deviations from a straight line indicate that the distribution is not Normal. This plot is usually able to show deviations from Normality more clearly than the corresponding histogram, but it's usually easier to understand how a distribution fails to be Normal by looking at its histogram. Normal probability plots are difficult to make by hand, but are provided by most statistics software.

Some data on a car's fuel efficiency provide an example of data that are nearly Normal. The overall pattern of the Normal probability plot is straight. The two trailing low values correspond to the values in the histogram that trail off the low end. They're not quite in line with the rest of the data set. The Normal probability plot shows us that they're a bit lower than we'd expect of the lowest two values in a Normal distribution.

Figure 7.7 Histogram and Normal probability plot for gas mileage (mpg) recorded for a Nissan Maxima. The vertical axes are the same, so each dot on the probability plot would fall into the bar on the histogram immediately to its left.

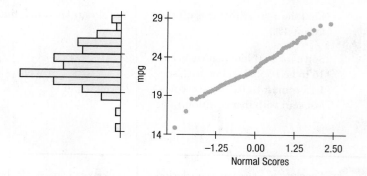

By contrast, the Normal probability plot of a sample of men's *Weights* in Figure 7.8 from a study of lifestyle and health is far from straight. The weights are skewed to the high end, and the plot is curved. We'd conclude from these pictures that approximations using the Normal model for these data would not be very accurate.

Figure 7.8 Histogram and Normal probability plot for men's weights. Note how a skewed distribution corresponds to a bent probability plot.

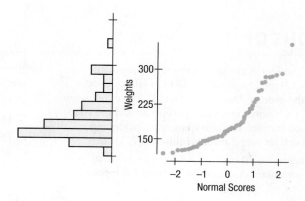

FOR EXAMPLE   Using a Normal probability plot

A Normal probability plot of the CAPE10 prices from page 208 looks like this:

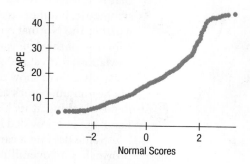

QUESTION   What does this plot say about the distribution of the CAPE10 scores?

ANSWER   The bent shape of the probability plot indicates a deviation from Normality. The upward bend is because the distribution is skewed to the high end. The "kink" in that bend suggest a collection of values that don't continue that skewness consistently. We should probably not use a Normal model for these data.

7.4 The Distribution of Sums of Normals

Another reason Normal models show up so often is that they have some special properties. An important one is that the sum or difference of two independent Normal random variables is also Normal. The mean of the sum or difference will be the sum or difference of the means of the separate distributions. And we know that variances add: the variance of the sum or difference will be the sum of the two variances. All we need is that the variables be independent.

GUIDED EXAMPLE   Packaging Stereos

Consider the company that manufactures and ships small stereo systems.

If the time required to pack the stereos can be described by a Normal distribution, with a mean of 9 minutes and standard deviation of 1.5 minutes, and the times for boxing and shipping can be modeled as Normal, with a mean of 6 minutes and standard deviation of 1 minute, what is the probability that packing an order of two systems takes over 20 minutes? What percentage of the stereo systems takes longer to pack than to box?

Question 1: What is the probability that packing an order of two systems takes more than 20 minutes?

PLAN	**Setup** State the problem.	We want to estimate the probability that packing an order of two systems takes more than 20 minutes.
	Variables Define your random variables.	Let P_1 = time for packing the first system $\quad P_2$ = time for packing the second system $\quad T$ = total time to pack two systems $\quad T = P_1 + P_2$
	Write an appropriate equation for the variables you need. Think about the model assumptions.	✓ **Normal Model Assumption.** We are told that packing times are well modeled by a Normal model, and we know that the sum of two Normal random variables is also Normal. ✓ **Independence Assumption.** There is no reason to think that the packing time for one system would affect the packing time for the next, so we can reasonably assume the two are independent.
DO	**Mechanics** Find the expected value. (Expected values always add.)	$$\begin{aligned} E(T) &= E(P_1 + P_2)\\ &= E(P_1) + E(P_2)\\ &= 9 + 9 = 18 \; minutes \end{aligned}$$
	Find the variance. For sums of independent random variables, variances add. (In general, we don't need the variables to be Normal for this to be true—just independent.) Find the standard deviation.	Since the times are independent, $$\begin{aligned} Var(T) &= Var(P_1 + P_2)\\ &= Var(P_1) + Var(P_2)\\ &= 1.5^2 + 1.5^2\\ Var(T) &= 4.50\\ SD(T) &= \sqrt{4.50} \approx 2.12 \; minutes \end{aligned}$$

(continued)

Now we use the fact that both random variables follow Normal distributions to say that their sum is also Normal.

We can model the time, T, with a $N(18, 2.12)$ model.

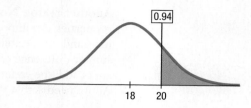

Sketch a picture of the Normal distribution for the total time, shading the region representing over 20 minutes.

Find the z-score for 20 minutes.

Use technology or a table to find the probability.

$$z = \frac{20 - 18}{2.12} = 0.94$$

$$P(T > 20) = P(z > 0.94) = 0.1736$$

REPORT   **Conclusion** Interpret your result in context.

MEMO

Re: Computer systems packing

Using past history to build a model, we find slightly more than a 17% chance that it will take more than 20 minutes to pack an order of two stereo systems.

Question 2: What percentage of stereo systems take longer to pack than to box?

PLAN   **Setup** State the question.

We want to estimate the percentage of the stereo systems that takes longer to pack than to box.

Variables Define your random variables.

Let P = time for packing a system

B = time for boxing a system

D = difference in times to pack and box a system

Write an appropriate equation.

$D = P - B$

What are we trying to find? Notice that we can tell which of two quantities is greater by subtracting and asking whether the difference is positive or negative.

A system that takes longer to pack than to box will have $P > B$, and so D will be positive. We want to find $P(D > 0)$.

Remember to think about the assumptions.

✓ **Normal Model Assumption.** We are told that both random variables are well modeled by Normal distributions, and we know that the difference of two Normal random variables is also Normal.

✓ **Independence Assumption.** There is no reason to think that the packing time for a system will affect its boxing time, so we can reasonably assume the two are independent.

DO   **Mechanics** Find the expected value.

$$E(D) = E(P - B)$$
$$= E(P) - E(B)$$
$$= 9 - 6 = 3 \text{ minutes}$$

For the difference of independent random variables, the variance is the sum of the individual variances.

Since the times are independent,

$$Var(D) = Var(P - B)$$
$$= Var(P) + Var(B)$$
$$= 1.5^2 + 1^2$$

Find the standard deviation. State what model you will use.	$Var(D) = 3.25$ $SD(D) = \sqrt{3.25} \approx 1.80 \; minutes$ We can model D with $N(3, 1.80)$.
Sketch a picture of the Normal distribution for the difference in times and shade the region representing a difference greater than zero.	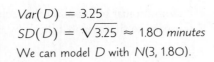
Find the z-score. Then use a table or technology to find the probability.	$$z = \dfrac{0 - 3}{1.80} = -1.67$$ $$P(D > 0) = P(z > -1.67) = 0.9525$$

REPORT   **Conclusion** Interpret your result in context.

MEMO

Re: Computer systems packing

In our second analysis, we found that just over 95% of all the stereo systems will require more time for packing than for boxing.

7.5   The Normal Approximation for the Binomial

In the previous chapter we modeled the number of successes of a series of trials with a Binomial. Suppose we send out 1000 flyers advertising a free cup of coffee at our new cafe and we think that the probability that someone will come is about 0.10. We might want to know the chance that at least 120 people will come to claim their coffee. We could use the binomial to calculate that with $n = 1000$ and $p = 0.10$. We know that the probability that exactly 120 people will come is $\dbinom{1000}{120} \times (0.10)^{120} \times (0.90)^{880}$ (about 0.005). But that's not the answer. We want to know the probability that *at least* 120 will show up, so we have to calculate a probability for 121, 122, 123, . . . and all the way up to 1000. There must be a better way. And there is. The Normal distribution can approximate the Binomial.

<aside>
Recall That This Notation:
$$\binom{1000}{120}$$
means "1000 choose 120". We first saw this notation in Chapter 6 on page 191. Look back there if you need a reminder.
</aside>

The Binomial model for our cafe has mean $np = 100$ and standard deviation $\sqrt{npq} \approx 9.5$. We might just try to approximate its distribution with a Normal distribution using the same mean and standard deviation. Remarkably enough, that turns out to be a very good approximation. Using that mean and standard deviation, we can find the *probability*:

$$P(X \geq 120) = P\left(z \geq \frac{120 - 100}{9.5}\right) \approx P(z \geq 2.11) \approx 0.0174$$

There seems to be only about a 1.7% chance that at least 120 people will show up. (Adding up all 881 probabilities using the Binomial agrees with this to 3 decimal places!)

We can't always use a Normal distribution to make estimates of Binomial probabilities. The success of the approximation depends on the sample size. Suppose we are searching for a prize in cereal boxes, where the probability of finding a prize is 20%. If we buy five boxes, the actual Binomial probabilities that we get 0, 1, 2, 3, 4, or 5 prizes are 33%, 41%, 20%, 5%, 1%, and 0.03%, respectively. The histogram just below shows that this probability model is skewed. We shouldn't try to estimate these probabilities by using a Normal model.

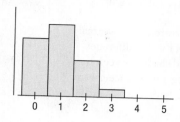

But if we open 50 boxes of this cereal and count the number of prizes we find, we'll get the histogram below. It is centered at $np = 50(0.2) = 10$ prizes, as expected, and it appears to be fairly symmetric around that center.

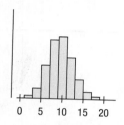

A Normal distribution is a close enough approximation to the Binomial only for a large enough number of trials. And what we mean by "large enough" depends on the probability of success. We'd need a larger sample if the probability of success were very low (or very high). It turns out that a Normal distribution works pretty well if we expect to see at least 10 successes and 10 failures. We can check the Success/Failure Condition.

Success/Failure Condition: A Binomial model is approximately Normal if we expect at least 10 successes and 10 failures:

$$np \geq 10 \text{ and } nq \geq 10.$$

Why 10? Well, actually it's 9, as revealed in the following Math Box.

MATH BOX Why Check $np \geq 10$?

It's easy to see where the magic number 10 comes from. You just need to remember how Normal models work. The problem is that a Normal model extends infinitely in both directions. But a Binomial model must have between 0 and n successes, so if we use a Normal to approximate a Binomial, we have to cut off its tails. That's not very important if the center of the Normal model is so far from 0 and n that the lost tails have only a negligible area. More than three standard deviations should do it because a Normal model has little probability past that.

So the mean needs to be at least 3 standard deviations from 0 and at least 3 standard deviations from n. Let's look at the 0 end.

We require:	$\mu - 3\sigma > 0$
Or, in other words:	$\mu > 3\sigma$
For a Binomial that's:	$np > 3\sqrt{npq}$
Squaring yields:	$n^2p^2 > 9npq$

Now simplify: $np > 9q$

Since $q \leq 1$, we require: $np > 9$

For simplicity we usually demand that np (and nq for the other tail) be at least 10 to use the Normal approximation which gives the Success/Failure Condition.[4]

---

*The Continuity Correction

When we use a continuous model to model a set of discrete events, we may need to make an adjustment called the **continuity correction**. We approximated the Binomial distribution $(50, 0.2)$ with a Normal distribution. But what does the Normal distribution say about the probability that $X = 10$? Every specific value in the Normal probability model has probability 0. That's not the answer we want.

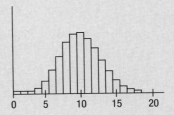

Because X is really discrete, it takes on the exact values $0, 1, 2, \ldots, 50$, each with positive probability. The histogram holds the secret to the correction. Look at the bin corresponding to $X = 10$ in the histogram. It goes from 9.5 to 10.5. What we really want is to find the area under the Normal curve *between* 9.5 and 10.5. So when we use the Normal distribution to approximate discrete events, we go halfway to the next value on the left and/or the right. We approximate $P(X = 10)$ by finding $P(9.5 \leq X \leq 10.5)$. For a Binom$(50, 0.2)$, $\mu = 10$ and $\sigma = 2.83$.

$$\text{So } P(9.5 \leq X \leq 10.5) \approx P\left(\frac{9.5 - 10}{2.83} \leq z \leq \frac{10.5 - 10}{2.83}\right)$$

$$= P(-0.177 \leq z \leq 0.177)$$

$$= 0.1405$$

By comparison, the Binomial probability is 0.1398 (to four decimal places).

7.6   Other Continuous Random Variables

Many phenomena in business can be modeled by continuous random variables. The Normal model is important, but it is only one of many different models. Entire courses are devoted to studying which models work well in different situations, but we'll introduce just one more that is commonly used: the uniform.

The Uniform Distribution

We've already seen the discrete version of the uniform probability model. A continuous uniform shares the principle that all events should be equally likely, but with a

---

[4] Looking at the final step, we see that we need $np > 9$ in the worst case, when q (or p) is near 1, making the Binomial model quite skewed. When q and p are near 0.5—for example, between 0.4 and 0.6—the Binomial model is nearly symmetric, and $np > 5$ ought to be safe enough. Although we'll always check for 10 expected successes and failures, keep in mind that for values of p near 0.5, we can be somewhat more forgiving.

continuous distribution we can't talk about the probability of a particular value because each value has probability zero. Instead, for a continuous random variable X, we say that the probability that X lies in any interval depends only on the length of that interval. Not surprisingly the density function of a continuous uniform random variable looks flat. It can be defined by the formula

$$f(x) = \begin{cases} \dfrac{1}{b-a} & if\ a \le x \le b \\ 0 & otherwise \end{cases}$$

Figure 7.9   The density function of a continuous uniform random variable on the interval from *a* to *b*.

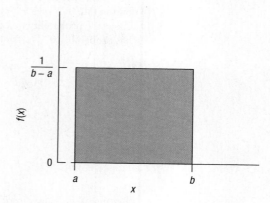

From Figure 7.9, it's easy to see that the probability that X lies in any interval between a and b is the same as any other interval of the same length. In fact, the probability is just the ratio of the length of the interval to the total length: $b - a$. In other words:

For values c and d $(c \le d)$ both within the interval $[a, b]$:

$$P(c \le X \le d) = \frac{(d-c)}{(b-a)}$$

As an example, suppose you arrive at a bus stop and want to model how long you'll wait for the next bus. The sign says that busses arrive about every 20 minutes, but no other information is given. You might assume that the arrival is equally likely to be anywhere in the next 20 minutes, and so the density function would be

$$f(x) = \begin{cases} \dfrac{1}{20} & if\ \ 0 \le x \le 20 \\ 0 & otherwise \end{cases}$$

It is not too surprising that the expected value is:

$$E(X) = \frac{a+b}{2}$$

The variance and standard deviation are less intuitive:

$$Var(X) = \frac{(b-a)^2}{12};\ SD(X) = \sqrt{\frac{(b-a)^2}{12}}.$$

WHAT CAN GO WRONG?

- **Probability models are still just models.** Models can be useful, but they are not reality. Think about the assumptions behind your models. Question probabilities as you would data.

- **Don't assume everything's Normal.** Just because a random variable is continuous or you happen to know a mean and standard deviation doesn't mean that a Normal model will be useful. You must think about whether the **Normality Assumption** is justified. Using a Normal model when it really does not apply will lead to wrong answers and misleading conclusions.

   A sample of CEOs has a mean total compensation of \$10,307,311.87 with a standard deviation of \$17,964,615.16. Using the Normal model rule, we should expect about 68% of the CEOs to have compensations between −\$7,657,303.29 and \$28,271,927.03. In fact, more than 90% of the CEOs have annual compensations in this range. What went wrong? The distribution is skewed, not symmetric. Using the 68–95–99.7 Rule for data like these will lead to silly results.

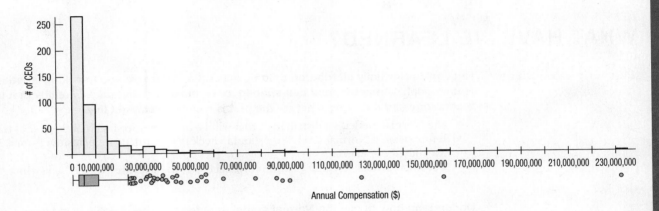

- **Don't use the Normal approximation with small *n*.** To use a Normal approximation in place of a Binomial model, there must be at least 10 expected successes and 10 expected failures.

ETHICS IN ACTION

Green River Army Depot's main business is the repair and refurbishment of electronics, mainly satellite and communication systems, in partnership with the Department of Defense (DOD). Recently, DOD has put a great deal of effort into continuous quality improvement, focusing on the length of time it takes to complete a project.

Dave Smith, head of the Productivity and Quality Improvement (PQI) directorate, is responsible for (among other things) facilitating lean improvement events throughout the Depot. These events bring together cross-functional teams with the goal of streamlining processes. PQI staff guide these teams in value stream mapping, identifying non–value added activities, and redesigning processes to eliminate waste. Dave is concerned that his group may not meet the new standards for quality which have specified that 97% of all projects must be completed within 60 days of their start dates.

(continued)

In preparation for a meeting with the Depot commander, Dave decides to review data on lean improvement events. He finds that of the past 137 projects, 30% went beyond 60 days. The commander suggests that Dave be "creative" in his presentation of the statistics to make it look like the group is actually in compliance with the DOD.

The completion times are very skewed to the high end, which doesn't surprise Dave in the least. After all, no project can take less than 0 time, but a few always seem to go on for a much longer time than planned. He is pleased to find that the average completion time is only 40 days and that the standard deviation is 10 days. Dave knows that a Normal model is a poor representation of the completion times because of the skewness, but decides to use a $N(40,10)$ model. With this model, only about 2.5% of the projects would be expected to take more than 60 days!

He explains his model to this supervisor who is pleased with the results. Even though they know the data don't fit the model well, they also know that the DOD analysts are very familiar with Normal models and will be pleased to know that they are in compliance using it.

- Identify the ethical dilemma in this scenario.

- What are the undesirable consequences?

- Propose an ethical solution that considers the welfare of all stakeholders.

WHAT HAVE WE LEARNED?

Learning Objectives

Recognize Normally distributed data by making a histogram and checking whether it is unimodal, symmetric, and bell-shaped, or by making a Normal probability plot using technology and checking whether the plot is roughly a straight line.

- The Normal model is a distribution that will be important for much of the rest of this course.
- Before using a Normal model, we should check that our data are plausibly from a Normally distributed population.
- A Normal probability plot provides evidence that the data are Normally distributed if it is linear.

Understand how to use the Normal model to judge whether a value is extreme.

- Standardize values to make z-scores and obtain a standard scale. Then refer to a standard Normal distribution.
- Use the 68–95–99.7 Rule as a rule-of-thumb to judge whether a value is extreme.

Know how to refer to tables or technology to find the probability of a value randomly selected from a Normal model falling in any interval.

- Know how to perform calculations about Normally distributed values and probabilities.

Recognize when independent random Normal quantities are being added or subtracted.

- The sum or difference will also follow a Normal model.
- The *variance* of the sum or difference will be the sum of the individual variances.
- The mean of the sum or difference will be the sum or difference, respectively, of the means.

Recognize when other continuous probability distributions are appropriate models.

Terms

68–95–99.7 Rule (or Empirical Rule)   In a Normal model, 68% of values fall within one standard deviation of the mean, 95% fall within two standard deviations of the mean, and 99.7% fall within three standard deviations of the mean. This is also approximately true for most unimodal, symmetric distributions.

Continuous random variable	A random variable that can take any numeric value within a range of values. The range may be infinite or bounded at either or both ends.
Normal Distribution	A unimodal, symmetric, "bell-shaped" distribution that appears throughout Statistics.
Normal percentile	The Normal percentile corresponding to a z-score gives the percentage of values in a standard Normal distribution found at that z-score or below.
Normal probability plot	A display to help assess whether a distribution of data is approximately Normal. If the plot is nearly straight, the data satisfy the Nearly Normal Condition.
Probability Density Function (pdf)	A function for any continuous probability model that gives the probability of a random value falling between any two values as the area under the pdf between those two values.
Standard Normal model or Standard Normal distribution	A Normal model, $N(\mu, \sigma)$ with mean $\mu = 0$ and standard deviation $\sigma = 1$.
Uniform Distribution	A continuous distribution that assigns a probability to any range of values (between 0 and 1) proportional to the difference between the values.

TECHNOLOGY HELP: Probability Calculations and Plots

The best way to tell whether your data can be modeled well by a Normal model is to make a picture or two. We've already talked about making histograms. Normal probability plots are almost never made by hand because the values of the Normal scores are tricky to find. But most statistics software can make Normal plots, though various packages call the same plot by different names and array the information differently.

EXCEL

Excel offers a "Normal probability plot" as part of the Regression command in the Data Analysis extension, but (as of this writing) it is not a correct Normal probability plot and should not be used.

To calculate Continuous Distribution Probabilities in Excel:

As discussed in Chapter 6, functions that calculate probabilities for continuous probability distributions will calculate either pdf ("probability density function"—what we've been calling a probability model) or cdf ("cumulative distribution function"—accumulate probabilities over a range of values). Excel uses the "cumulative" part of the command to determine whether you want a probability as your result (cumulative = true; this is the cdf) or a number as your result given a probability (cumulative = false; this is the pdf).

Distribution	Excel Syntax	Example Syntax	Result	Probability Statement
Normal (μ, σ)	NORM.DIST(x, mean, standard dev., cumulative)	=NORM.DIST(16,16.3,0.2,TRUE)	0.0668	P(X<16), N(16.3, 0.2)
	NORM.INV(probability, mean, standard dev.)	=NORM.INV(0.0668,16.3,0.2)	16.0000	P(X<**x?**) = 0.0668 N(16.3, 0.2)
Standard Normal, N(0,1)	NORM.S.DIST(z, cumulative)	=NORM.S.DIST(-1.75,TRUE)	0.0401	P(Z<-1.75)
	NORM.S.INV(probability)	=NORM.S.INV(0.04)	-1.7507	P(Z<**z?**)=0.04
Exponential	EXPON.DIST(x, lamda, cumulative)	=EXPON.DIST(1,1.33,TRUE)	0.7355	P(X<1), λ=4/3

Note that the commands here are for Excel 2013. These functions are available in earlier versions of Excel with similar commands (e.g., the pre-2010 Excel command for NORM.DIST was NORMDIST). In general, the functions ending in DIST will calculate a probability given a value from the distribution and the INV functions will calculate a value given a probability. When using the function bar or typing into a cell, Excel will search the functions to find what matches the typed characters, and this can be used to find the proper function.

XLSTAT

XLStat can make Normal probability plots (XLStat calls these Q-Q plots):

- Select **Visualizing data**, and then **Univariate plots**.
- On the General tab, click the **Quantitative** data box and then select the data on your worksheet.
- Click **OK**.
- If prompted, click **Continue**.

JMP

To make a "Normal Quantile Plot" in JMP,

- Make a histogram using **Distributions** from the **Analyze** menu.
- Click on the drop-down menu next to the variable name.
- Choose **Normal Quantile Plot** from the drop-down menu.
- JMP opens the plot next to the histogram.

Comments
JMP places the ordered data on the vertical axis and the Normal scores on the horizontal axis. The vertical axis aligns with the histogram's axis, a useful feature.

MINITAB

To make a "Normal Probability Plot" in MINITAB,

- Choose **Probability Plot** from the **Graph** menu.
- Select "Single" for the type of plot. Click **OK**.
- Enter the name of the variable in the "Graph variables" box. Click **OK**.

Comments
MINITAB places the ordered data on the horizontal axis and the Normal scores on the vertical axis.

R

To make a Normal probability (Q-Q) plot for X:

- **qqnorm(X)** will produce the plot.

To standardize a variable X:

- **Z = (X − mean(X))/sd(X)** will create a standardized variable **Z**.

Comments
By default, R places the ordered data on the vertical axis and the Normal scores on the horizontal axis, but that can be reversed by setting **datax = TRUE** inside qqnorm.

SPSS

To make a Normal "P-P plot" in SPSS,

- Choose **Descriptives > P-P Plots** from the **Analyze** menu.
- Select the variable to be displayed and add to "Variable".
- Make sure that "Normal" is selected under "Test Distribution". Leave all other defaults set.

Comments
SPSS places the ordered data on the horizontal axis and the Normal scores on the vertical axis.

Brief **Case**

Price/Earnings and Stock Value

The CAPE10 index is based on the Price/Earnings (P/E) ratios of stocks. We can examine the P/E ratios without applying the smoothing techniques used to find the CAPE10. The file **CAPE10** holds the data, giving dates, various economic variables, CAPE10 values, and P/E values.

Examine the P/E values. Split the data into two periods: 1870–1989 and 1990 to the present. Would you judge that a Normal model would be appropriate for those values from the 1880s through the 1980s? Explain (and show the plots you made.)

Now consider the more recent P/E values in this context. Do you think they have been extreme? What years, if any, appear to be particularly problematic? Explain.

EXERCISES

Normal model calculations can be performed using a variety of technology or with the tables in Appendix B. Different methods may yield slightly different results.

SECTION 7.1

1. An incoming MBA student took placement exams in economics and mathematics. In economics, she scored 82 and in math 86. The overall results on the economics exam had a mean of 72 and a standard deviation of 8, while the mean math score was 68, with a standard deviation of 12. On which exam did she do better compared with the other students?

2. The first Statistics exam had a mean of 65 and a standard deviation of 10 points; the second had a mean of 80 and a standard deviation of 5 points. Derrick scored an 80 on both tests. Julie scored a 70 on the first test and a 90 on the second. They both totaled 160 points on the two exams, but Julie claims that her total is better. Explain.

3. Your company's Human Resources department administers a test of "Executive Aptitude." They report test grades as z-scores, and you got a score of 2.20. What does this mean?

4. After examining a child at his 2-year checkup, the boy's pediatrician said that the z-score for his height relative to American 2-year-olds was −1.88. Write a sentence to explain to the parents what that means.

5. Your company will admit to the executive training program only people who score in the top 3% on the executive aptitude test discussed in Exercise 3.

a) With your z-score of 2.20, did you make the cut?
b) What do you need to assume about test scores to find your answer in part a?

6. The pediatrician in Exercise 4 explains to the parents that the most extreme 5% of cases often require special treatment or attention.

a) Does this child fall into that group?
b) What do you need to assume about the heights of 2-year-olds to find your answer to part a?

SECTION 7.2

7. The Environmental Protection Agency (EPA) fuel economy estimates for automobiles suggest a mean of 24.8 mpg and a standard deviation of 6.2 mpg for highway driving. Assume that a Normal model can be applied.

a) Draw the model for auto fuel economy. Clearly label it, showing what the 68–95–99.7 Rule predicts about miles per gallon.
b) In what interval would you expect the central 68% of autos to be found?

c) About what percent of autos should get more than 31 mpg?
d) About what percent of cars should get between 31 and 37.2 mpg?
e) Describe the gas mileage of the worst 2.5% of all cars.

8. Some IQ tests are standardized to a Normal model with a mean of 100 and a standard deviation of 16.

a) Draw the model for these IQ scores. Clearly label it, showing what the 68–95–99.7 Rule predicts about the scores.
b) In what interval would you expect the central 95% of IQ scores to be found?
c) About what percent of people should have IQ scores above 116?
d) About what percent of people should have IQ scores between 68 and 84?
e) About what percent of people should have IQ scores above 132?

9. What percent of a standard Normal model is found in each region? Be sure to draw a picture first.

a) $z > 1.5$
b) $z < 2.25$
c) $−1 < z < 1.15$
d) $|z| > 0.5$

10. What percent of a standard Normal model is found in each region? Draw a picture first.

a) $z > −2.05$
b) $z < −0.33$
c) $1.2 < z < 1.8$
d) $|z| < 1.28$

11. In a standard Normal model, what value(s) of z cut(s) off the region described? Don't forget to draw a picture.

a) the highest 20%
b) the highest 75%
c) the lowest 3%
d) the middle 90%

12. In a standard Normal model, what value(s) of z cut(s) off the region described? Remember to draw a picture first.

a) the lowest 12%
b) the highest 30%
c) the highest 7%
d) the middle 50%

SECTION 7.3

13. Speeds of cars were measured as they passed one point on a road to study whether traffic speed controls were needed. Here's a histogram and normal probability plot of the measured speeds. Is a Normal model appropriate for these data? Explain.

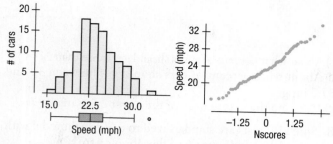

14. Has the Consumer Price Index (CPI) fluctuated around its mean according to a Normal model? Here are some displays. Is a Normal model appropriate for these data? Explain.

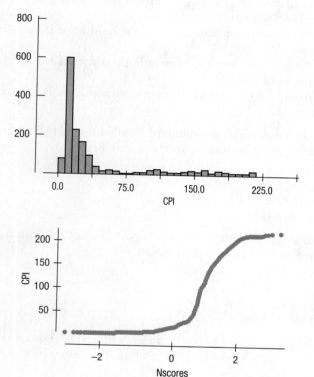

SECTION 7.4

15. For a new type of tire, a NASCAR team found the average distance a set of tires would run during a race is 168 miles, with a standard deviation of 14 miles. Assume that tire mileage is independent and follows a Normal model.

a) If the team plans to change tires twice during a 500-mile race, what is the expected value and standard deviation of miles remaining after two changes?

b) What is the probability they won't have to change tires a third time (and use a fourth set of tires) before the end of a 500-mile race?

16. In the 4 × 100 medley relay event, four swimmers swim 100 yards, each using a different stroke. A college

team preparing for the conference championship looks at the times their swimmers have posted and creates a model based on the following assumptions:

- The swimmers' performances are independent.
- Each swimmer's times follow a Normal model.
- The means and standard deviations of the times (in seconds) are as shown here.

Swimmer	Mean	SD
1 (backstroke)	50.72	0.24
2 (breaststroke)	55.51	0.22
3 (butterfly)	49.43	0.25
4 (freestyle)	44.91	0.21

a) What are the mean and standard deviation for the relay team's total time in this event?

b) The team's best time so far this season was 3:19.48. (That's 199.48 seconds.) What is the probability that they will beat that time in the next event?

SECTION 7.5

17. Because many passengers who make reservations do not show up, airlines often overbook flights (sell more tickets than there are seats). A Boeing 767-400ER holds 245 passengers. If the airline believes the rate of passenger no-shows is 5% and sells 255 tickets, is it likely they will not have enough seats and someone will get bumped?

a) Use the Normal model to approximate the Binomial to determine the probability of at least 246 passengers showing up.

b) Should the airline change the number of tickets they sell for this flight? Explain.

18. Shortly after the introduction of the Belgian euro coin, newspapers around the world published articles claiming the coin is biased. The stories were based on reports that someone had spun the coin 250 times and gotten 140 heads—that's 56% heads.

a) Use the Normal model to approximate the Binomial to determine the probability of spinning a fair coin 250 times and getting at least 140 heads.

b) Do you think this is evidence that spinning a Belgian euro is unfair? Would you be willing to use it at the beginning of a sports event? Explain.

SECTION 7.6

19. A cable provider wants to contact customers in a particular telephone exchange to see how satisfied they are with the new digital TV service the company has provided. All numbers are in the 452 exchange, so there are 10,000 possible numbers from 452-0000 to 452-9999. If they select the numbers with equal probability:

a) What distribution would they use to model the selection?

b) The new business "incubator" was assigned the 200 numbers between 452-2500 and 452-2699, but these businesses don't subscribe to digital TV. What is the probability that the randomly selected number will be for an incubator business?

c) Numbers above 9000 were only released for domestic use last year, so they went to newly constructed residences. What is the probability that a randomly selected number will be one of these?

20. In an effort to check the quality of their cell phones, a manufacturing manager decides to take a random sample of 10 cell phones from yesterday's production run, which produced cell phones with serial numbers ranging (according to when they were produced) from 43005000 to 43005999. If each of the 1000 phones is equally likely to be selected:

a) What distribution would they use to model the selection?

b) What is the probability that a randomly selected cell phone will be one of the last 100 to be produced?

c) What is the probability that the first cell phone selected is either from the last 200 to be produced or from the first 50 to be produced?

CHAPTER EXERCISES

For Exercises 21–28, use the 68–95–99.7 Rule to approximate the probabilities rather than using technology to find the values more precisely. Answers given for probabilities or percentages from Exercise 29 on assume that a calculator or software has been used. Answers found from using Z-tables may vary slightly.

21. Mutual fund returns 2013. In the first quarter of 2013, a group of domestic equity mutual funds had a mean return of 6.2% with a standard deviation of 1.8%. If a Normal model can be used to model them, what percent of the funds would you expect to be in each region?

Be sure to draw a picture first.

a) Returns of 8.0% or more

b) Returns of 6.2% or less

c) Returns between 2.6% and 9.8%

d) Returns of more than 11.6%

22. Human resource testing. Although controversial and the subject of some recent law suits (e.g., *Satchell et al. vs. FedEx Express*), some human resource departments administer standard IQ tests to all employees. The Stanford-Binet test scores are well modeled by a Normal model with mean 100 and standard deviation 16. If the applicant pool is well modeled by this distribution, a randomly selected applicant would have what probability of scoring in the following regions?

a) 100 or below

b) Above 148

c) Between 84 and 116

d) Above 132

23. Mutual funds, again. From the mutual funds in Exercise 21 with quarterly returns that are well modeled by a Normal model with a mean of 6.2% and a standard deviation of 1.8%, find the cutoff return value(s) that would separate the

a) highest 50%.

b) highest 16%.

c) lowest 2.5%.

d) middle 68%.

24. Human resource testing, again. For the IQ test administered by human resources and discussed in Exercise 22, what cutoff value would separate the

a) lowest 0.15% of all applicants?

b) lowest 16%?

c) middle 95%?

d) highest 2.5%?

25. Currency exchange rates. The daily exchange rates for the five-year period 2008 to 2013 between the euro (EUR) and the British pound (GBP) can be modeled by a Normal distribution with mean 1.19 euros (to pounds) and standard deviation 0.043 euros. Given this model, what is the probability that on a randomly selected day during this period, the pound was worth

a) less than 1.19 euros?

b) more than 1.233 euros?

c) less than 1.104 euros?

d) Which would be more unusual, a day on which the pound was worth less than 1.126 euros or more than 1.298 euros?

26. Stock prices. For the 300 trading days from January 11, 2012 to March 22, 2013, the daily closing price of IBM stock (in $) is well modeled by a Normal model with mean $197.92 and standard deviation $7.16. According to this model, what is the probability that on a randomly selected day in this period the stock price closed

a) above $205.08?

b) below $212.24?

c) between $183.60 and $205.08?

d) Which would be more unusual, a day on which the stock price closed above $206 or below $180?

27. Currency exchange rates, again. For the model of the EUR/GBP exchange rate discussed in Exercise 25, what would the cutoff rates be that would separate the

a) highest 16% of EUR/GBP rates?

b) lowest 50%?

c) middle 95%?

d) lowest 2.5%?

28. Stock prices, again. According to the model in Exercise 26, what cutoff value of price would separate the

a) lowest 16% of the days?

b) highest 0.15%?

c) middle 68%?

d) highest 50%?

29. Mutual fund probabilities. According to the Normal model $N(0.062, 0.018)$ describing mutual fund returns in the 1st quarter of 2013 in Exercise 21, what percent of this group of funds would you expect to have return

a) over 6.8%?
b) between 0% and 7.6%?
c) more than 1%?
d) less than 0%?

30. Normal IQs. Based on the Normal model $N(100, 16)$ describing IQ scores from Exercise 22, what percent of applicants would you expect to have scores

a) over 80?
b) under 90?
c) between 112 and 132?
d) over 125?

31. Mutual funds, once more. Based on the model $N(0.062, 0.018)$ for quarterly returns from Exercise 21, what are the cutoff values for the

a) highest 10% of these funds?
b) lowest 20%?
c) middle 40%?
d) highest 80%?

32. More IQs. In the Normal model $N(100, 16)$ for IQ scores from Exercise 22, what cutoff value bounds the

a) highest 5% of all IQs?
b) lowest 30% of the IQs?
c) middle 80% of the IQs?
d) lowest 90% of all IQs?

33. Mutual funds, finis. Consider the Normal model $N(0.062, 0.018)$ for returns of mutual funds in Exercise 21 one last time.

a) What value represents the 40th percentile of these returns?
b) What value represents the 99th percentile?
c) What's the IQR of the quarterly returns for this group of funds?

34. IQs, finis. Consider the IQ model $N(100, 16)$ one last time.

a) What IQ represents the 15th percentile?
b) What IQ represents the 98th percentile?
c) What's the IQR of the IQs?

35. Parameters. Every Normal model is defined by its parameters, the mean and the standard deviation. For each model described here, find the missing parameter. As always, start by drawing a picture.

a) $\mu = 20$, 45% above 30; $\sigma = ?$
b) $\mu = 88$, 2% below 50; $\sigma = ?$
c) $\sigma = 5$, 80% below 100; $\mu = ?$
d) $\sigma = 15.6$, 10% above 17.2; $\mu = ?$

36. Parameters, again. Every Normal model is defined by its parameters, the mean and the standard deviation. For each model described here, find the missing parameter. Don't forget to draw a picture.

a) $\mu = 1250$, 35% below 1200; $\sigma = ?$
b) $\mu = 0.64$, 12% above 0.70; $\sigma = ?$
c) $\sigma = 0.5$, 90% above 10.0; $\mu = ?$
d) $\sigma = 220$, 3% below 202; $\mu = ?$

37. SAT or ACT? Each year thousands of high school students take either the SAT or ACT, standardized tests used in the college admissions process. Combined SAT scores can go as high as 1600, while the maximum ACT composite score is 36. Since the two exams use very different scales, comparisons of performance are difficult. (A convenient rule of thumb is $SAT = 40 \times ACT + 150$; that is, multiply an ACT score by 40 and add 150 points to estimate the equivalent SAT score.) Assume that one year the combined SAT can be modeled by $N(1000, 200)$ and the ACT can be modeled by $N(27, 3)$. If an applicant to a university has taken the SAT and scored 1260 and another student has taken the ACT and scored 33, compare these students scores using z-values. Which one has a higher relative score? Explain.

38. Economics. Anna, a business major, took final exams in both Microeconomics and Macroeconomics and scored 83 on both. Her roommate Megan, also taking both courses, scored 77 on the Micro exam and 95 on the Macro exam. Overall, student scores on the Micro exam had a mean of 81 and a standard deviation of 5, and the Macro scores had a mean of 74 and a standard deviation of 15. Which student's overall performance was better? Explain.

39. Low job satisfaction. Suppose that job satisfaction scores can be modeled with $N(100, 12)$. Human resource departments of corporations are generally concerned if the job satisfaction drops below a certain score. What score would you consider to be unusually low? Explain.

40. Low return. Exercise 21 proposes modeling quarterly returns of a group of mutual funds with $N(0.062, 0.018)$. The manager of this group of funds would like to flag any fund whose return is unusually low for a quarter. What level of return would you consider to be unusually low? Explain.

41. Management survey. A survey of 200 middle managers showed a distribution of the number of hours of exercise they participated in per week with a mean of 3.66 hours and a standard deviation of 4.93 hours.

a) According to the Normal model, what percent of managers will exercise fewer than one standard deviation below the mean number of hours?
b) For these data, what does that mean? Explain.
c) Explain the problem in using the Normal model for these data.

42. Customer database. A large philanthropic organization keeps records on the people who have contributed to their cause. In addition to keeping records of past giving, the organization buys demographic data on neighborhoods from the U.S. Census Bureau. Eighteen of these variables concern the ethnicity of the neighborhood of the donor. Here is a histogram and summary statistics for the percentage of whites in the neighborhoods of 500 donors.

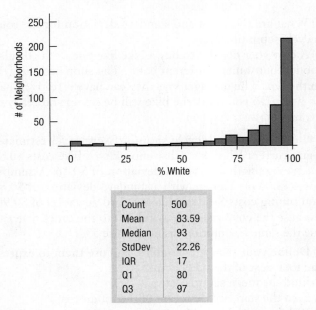

Count	500
Mean	83.59
Median	93
StdDev	22.26
IQR	17
Q1	80
Q3	97

a) Which is a better summary of the percentage of white residents in the neighborhoods, the mean or the median? Explain.
b) Which is a better summary of the spread, the IQR or the standard deviation? Explain.
c) From a Normal model, about what percentage of neighborhoods should have a percent white residents within one standard deviation of the mean?
d) What percentage of neighborhoods actually have a percent white within one standard deviation of the mean?
e) Explain the problem in using the Normal model for these data.

43. Drug company. Manufacturing and selling drugs that claim to reduce an individual's cholesterol level is big business. A company would like to market their drug to women if their cholesterol is in the top 15%. Assume the cholesterol levels of adult American women can be described by a Normal model with a mean of 188 mg/dL and a standard deviation of 24.

a) Draw and label the Normal model.
b) What percent of adult women do you expect to have cholesterol levels over 200 mg/dL?
c) What percent of adult women do you expect to have cholesterol levels between 150 and 170 mg/dL?
d) Estimate the interquartile range of the cholesterol levels.
e) Above what value are the highest 15% of women's cholesterol levels?

44. Tire company. A tire manufacturer believes that the tread life of its snow tires can be described by a Normal model with a mean of 32,000 miles and a standard deviation of 2500 miles.

a) If you buy a set of these tires, would it be reasonable for you to hope that they'll last 40,000 miles? Explain.
b) Approximately what fraction of these tires can be expected to last less than 30,000 miles?
c) Approximately what fraction of these tires can be expected to last between 30,000 and 35,000 miles?
d) Estimate the IQR for these data.
e) In planning a marketing strategy, a local tire dealer wants to offer a refund to any customer whose tires fail to last a certain number of miles. However, the dealer does not want to take too big a risk. If the dealer is willing to give refunds to no more than 1 of every 25 customers, for what mileage can he guarantee these tires to last?

45. Claims. Two companies make batteries for cell phone manufacturers. One company claims a mean life span of 2 years, while the other company claims a mean life span of 2.5 years (assuming average use of minutes/month for the cell phone).

a) Explain why you would also like to know the standard deviations of the battery life spans before deciding which brand to buy.
b) Suppose those standard deviations are 1.5 months for the first company and 9 months for the second company. Does this change your opinion of the batteries? Explain.

46. Car speeds. The police department of a major city needs to update its budget. For this purpose, they need to understand the variation in their fines collected from motorists for speeding. As a sample, they recorded the speeds of cars driving past a location with a 20 mph speed limit, a place that in the past has been known for producing fines. The mean of 100 readings was 23.84 mph, with a standard deviation of 3.56 mph. (The police actually recorded every car for a two-month period. These are 100 representative readings.)

a) How many standard deviations from the mean would a car going the speed limit be?
b) Which would be more unusual, a car traveling 34 mph or one going 10 mph?

47. CEOs. A business publication recently released a study on the total number of years of experience in industry among CEOs. The mean is provided in the article, but not the standard deviation. Is the standard deviation most likely to be 6 months, 6 years, or 16 years? Explain which standard deviation is correct and why.

48. Stocks. A newsletter for investors recently reported that the average stock price for a blue chip stock over the past 12 months was $72. No standard deviation was given. Is the standard deviation more likely to be $6, $16, or $60? Explain.

49. Cereal. The amount of cereal that can be poured into a small bowl varies with a mean of 1.5 ounces and a standard deviation of 0.3 ounces. A large bowl holds a mean of 2.5 ounces with a standard deviation of 0.4 ounces. You open a new box of cereal and pour one large and one small bowl.

a) How much more cereal do you expect to be in the large bowl?
b) What's the standard deviation of this difference?
c) If the difference follows a Normal model, what's the probability the small bowl contains more cereal than the large one?
d) What are the mean and standard deviation of the total amount of cereal in the two bowls?
e) If the total follows a Normal model, what's the probability you poured out more than 4.5 ounces of cereal in the two bowls together?
f) The amount of cereal the manufacturer puts in the boxes is a random variable with a mean of 16.3 ounces and a standard deviation of 0.2 ounces. Find the expected amount of cereal left in the box and the standard deviation.

50. Pets. The American Veterinary Association claims that the annual cost of medical care for dogs averages $100, with a standard deviation of $30, and for cats averages $120, with a standard deviation of $35.

a) What's the expected difference in the cost of medical care for dogs and cats?
b) What's the standard deviation of that difference?
c) If the costs can be described by Normal models, what's the probability that medical expenses are higher for someone's dog than for her cat?
d) What concerns do you have?

51. More cereal. In Exercise 49 we poured a large and a small bowl of cereal from a box. Suppose the amount of cereal that the manufacturer puts in the boxes is a random variable with mean 16.2 ounces and standard deviation 0.1 ounces.

a) Find the expected amount of cereal left in the box.
b) What's the standard deviation?
c) If the weight of the remaining cereal can be described by a Normal model, what's the probability that the box still contains more than 13 ounces?

52. More pets. You're thinking about getting two dogs and a cat. Assume that annual veterinary expenses are independent and have a Normal model with the means and standard deviations described in Exercise 50.

a) Define appropriate variables and express the total annual veterinary costs you may have.
b) Describe the model for this total cost. Be sure to specify its name, expected value, and standard deviation.
c) What's the probability that your total expenses will exceed $400?

53. Bikes. Bicycles arrive at a bike shop in boxes. Before they can be sold, they must be unpacked, assembled, and tuned (lubricated, adjusted, etc.). Based on past experience, the shop manager makes the following assumptions about how long this may take:

- The times for each setup phase are independent.
- The times for each phase follow a Normal model.
- The means and standard deviations of the times (in minutes) are as shown:

Phase	Mean	SD
Unpacking	3.5	0.7
Assembly	21.8	2.4
Tuning	12.3	2.7

a) What are the mean and standard deviation for the total bicycle setup time?
b) A customer decides to buy a bike like one of the display models but wants a different color. The shop has one, still in the box. The manager says they can have it ready in half an hour. Do you think the bike will be set up and ready to go as promised? Explain.

54. Bike sale. The bicycle shop in Exercise 53 estimates using current labor costs that unpacking a bike costs $0.82 on average with a standard deviation of $0.16. Assembly costs $8.00 on average with a standard deviation of $0.88 and tuning costs $4.10 with a standard deviation of $0.90. Because the costs are directly related to the times, you can use the same assumptions as in exercise 53.

a) Define your random variables, and use them to express the total cost of the bike set up.
b) Find the mean set up cost.
c) Find the standard deviation of the set up cost.
d) If the next shipment is 40 bikes, what is the probability that the total set up cost will be less than $500?

55. Coffee and doughnuts. At a certain coffee shop, all the customers buy a cup of coffee; some also buy a doughnut. The shop owner believes that the number of cups he sells each day is normally distributed with a mean of 320 cups and a standard deviation of 20 cups. He also believes that the number of doughnuts he sells each day is independent of the coffee sales and is normally distributed with a mean of 150 doughnuts and a standard deviation of 12.

a) The shop is open every day but Sunday. Assuming day-to-day sales are independent, what's the probability he'll sell more than 2000 cups of coffee in a week?
b) If he makes a profit of 50 cents on each cup of coffee and 40 cents on each doughnut, can he reasonably expect to have a day's profit of over $300? Explain.
c) What's the probability that on any given day he'll sell a doughnut to more than half of his coffee customers?

56. Weightlifting. The Atlas BodyBuilding Company (ABC) sells "starter sets" of barbells that consist of one bar, two 20-pound weights, and four 5-pound weights. The bars weigh an average of 10 pounds with a standard deviation of 0.25 pounds. The weights average the specified amounts, but the standard deviations are 0.2 pounds for the 20-pounders and 0.1 pounds for the 5-pounders. We can assume that all the weights are normally distributed.

a) ABC ships these starter sets to customers in two boxes: The bar goes in one box and the six weights go in another. What's the probability that the total weight in that second box exceeds 60.5 pounds? Define your variables clearly and state any assumptions you make.

b) It costs ABC $0.40 per pound to ship the box containing the weights. Because it's an odd-shaped package, though, shipping the bar costs $0.50 a pound plus a $6.00 surcharge. Find the mean and standard deviation of the company's total cost for shipping a starter set.

c) Suppose a customer puts a 20-pound weight at one end of the bar and the four 5-pound weights at the other end. Although he expects the two ends to weigh the same, they might differ slightly. What's the probability the difference is more than a quarter of a pound?

57. Lefties. A lecture hall has 200 seats with folding arm tablets, 30 of which are designed for left-handers. The typical size of classes that meet there is 188, and we can assume that about 13% of students are left-handed. Use a Normal approximation to find the probability that a right-handed student in one of these classes is forced to use a lefty arm tablet.

58. Seatbelts. Police estimate that 80% of drivers wear their seatbelts. They set up a safety roadblock, stopping cars to check for seatbelt use. If they stop 120 cars, what's the probability they find at least 20 drivers not wearing their seatbelt? Use a Normal approximation.

59. Rickets. Vitamin D is essential for strong, healthy bones. Although the bone disease rickets was largely eliminated in England during the 1950s, some people there are concerned that this generation of children is at increased risk because they are more likely to watch TV or play computer games than spend time outdoors. Recent research indicated that about 20% of British children are deficient in vitamin D. A company that sells vitamin D supplements tests 320 elementary school children in one area of the country. Use a Normal approximation to find the probability that no more than 50 of them have vitamin D deficiency.

60. Tennis. A tennis player has taken a special course to improve her serving. She thinks that individual serves are independent of each other. She has been able to make a successful first serve 70% of the time. Use a Normal approximation to find the probability she'll make at least 65 of her first serves out of the 80 she serves in her next match if her success percentage has not changed.

61. Wheel defects. Defects can occur anywhere on the wheel of a car during the manufacturing process. If X is the angle where the defect occurs, measured from a reference line, then X can be modeled as a uniform random variable on the interval from 0 to 360 degrees.

a) What is the probability that the defect is found between 0 and 180 degrees?

b) What is the probability that the defect is found between 0 and 45 degrees or between 315 and 360 degrees?

62. Quitting time. My employee seems to leave work anytime between 5PM and 6PM, uniformly.

a) What is the probability he will still be at work at 5:45 PM?

b) What is the probability he will still be at work at 5:45 PM every day this week (M–F)?

c) What did you assume to calculate b?

JUST CHECKING ANSWERS

1 a) On the first test, the mean is 88 and the SD is 4, so $z = (90 - 88)/4 = 0.5$. On the second test, the mean is 75 and the SD is 5, so $z = (80 - 75)/5 = 1.0$. The first test has the lower z-score, so it is the one that will be dropped.

  b) The second test is 1 standard deviation above the mean, farther away than the first test, so it's the better score relative to the class.

2 The mean is 184 centimeters, with a standard deviation of 8 centimeters. 2 meters is 200 centimeters, which is 2 standard deviations above the mean. We expect 2.28% of the men to be above 2 meters.

3 a) We know that 68% of the time we'll be within 1 standard deviation (2 min) of 20. So 32% of the time we'll arrive in less than 18 or more than 22 minutes. Half of those times (16%) will be greater than 22 minutes, so 84% will be less than 22 minutes.

  b) 24 minutes is 2 standard deviations above the mean. From Table Z we find that 2.28% of the times will be more than 24 minutes.

  c) Traffic incidents may occasionally increase the time it takes to get to school, so the driving times may be skewed to the right, and there may be outliers.

  d) If so, the Normal model would not be appropriate and the percentages we predict would not be accurate.

8

Surveys and Sampling

Roper Polls

Public opinion polls are a relatively new phenomenon. In 1948, as a result of telephone surveys of likely voters, all of the major organizations—Gallup, Roper, and Crossley—consistently predicted, throughout the summer and into the fall, that Thomas Dewey would defeat Harry Truman in the November presidential election. By October the results seemed so clear that *Fortune* magazine declared, "Due to the overwhelming evidence, *Fortune* and Mr. Roper plan no further detailed reports on change of opinion in the forthcoming presidential campaign . . . ."

Of course, Harry Truman went on to win the 1948 election, and the picture of Truman in the early morning after the election holding up the *Chicago Tribune* (printed the night before), with its headline declaring Dewey the winner, has become legend.

The public's faith in opinion polls plummeted after the election, but Elmo Roper vigorously defended the pollsters. Roper was a principal and founder of one of the first market research firms, Cherington, Wood, and Roper, and director of the *Fortune Survey*, which was the first national poll to use scientific sampling techniques. He argued that rather than abandoning polling, business leaders should learn what had gone wrong in the 1948 polls so that market research could be improved. His frank admission of the mistakes made in those polls helped to restore confidence in polling as a business tool.

For the rest of his career, Roper split his efforts between two projects, commercial polling and public opinion. He established the Roper Center for Public Opinion Research at Williams College as a place to house public opinion archives, convincing fellow polling leaders Gallup and Crossley to participate as well. Now located at the

University of Connecticut, the Roper Center is one of the world's leading archives of social science data. Roper's market research efforts started as Roper Research Associates and later became the Roper Organization, which was acquired in 2005 by GfK. Founded in Germany in 1934 as the Gesellschaft für Konsumforschung (literally, "Society for Consumption Research"), GfK now stands for "growth from knowledge." It is the fourth-largest international market research organization, with over 130 companies in 70 countries and more than 7700 employees worldwide.

GfK Roper Consulting conducts a yearly, global study to examine cultural, economic, and social information that may be crucial to companies doing business worldwide. These companies use the information provided by GfK Roper to help make marketing and advertising decisions in different markets around the world.

How do the researchers at GfK Roper know that the responses they get reflect the real attitudes of consumers? After all, they don't ask everyone, but they don't want to limit their conclusions to just the people they surveyed. Generalizing from the data at hand to the world at large is something that market researchers, investors, and pollsters do every day. To do it wisely, they need three fundamental ideas.

8.1   Three Ideas of Sampling

Idea 1: Sample—Examine a Part of the Whole

We'd like to know about an entire collection of individuals, called a **population**, but examining all of them is usually impractical, if not impossible. So we settle for examining a smaller group of individuals—a **sample**—selected from the population. For the Roper researchers the population of interest is the entire world, but it's not practical, cost-effective, or feasible to survey everyone. So they examine a sample selected from the population.

We take samples all the time. For example, if a restaurant chef wants to be sure that the vegetable soup she's cooking is up to her standards, she'll taste a spoonful. She doesn't need to consume the whole pot. She can trust that the taste will *represent* the flavor of the population—the entire pot. The idea of tasting is that a small sample, if selected properly, can represent the larger population. Sampling is common in many aspects of business practice. For example, auditors may sample some records rather than reading through all of them. Manufacturers monitor quality by testing a small sample off the line.

The GfK Roper Reports® Worldwide poll is an example of a **sample survey**, designed to ask questions of a small group of people in the hope of learning something about the entire population. Most likely, you've never been selected to be part of a national opinion poll. That's true of most people. So how can the pollsters claim that a sample represents the entire population? As we'll see, a representative sample can often provide a good idea of what the entire population is like. But the sample must be selected with care.

Selecting a sample to represent the population fairly is easy in theory, but in practice, it's more difficult than it sounds. For example, a sample may fail to represent part of the population. If a retail business samples customers as they come in the door, they may be missing an important part of their potential customer population—those who choose to shop elsewhere. Samples that over- or underemphasize some characteristics of the population are said to be biased. When a sample is

> **The W's and Sampling**
> The population we are interested in is usually determined by the *why* of our study. The participants or cases in the sample we draw will be the *who*. *When* and *how* we draw the sample may depend on what is practical.

biased, the summary characteristics of a sample differ from the corresponding characteristics of the population it is trying to represent, so they can produce misleading information. Conclusions based on biased samples are inherently flawed. There is usually no way to fix bias after the sample is drawn and no way to salvage useful information from it.

To make the sample as representative as possible, the best strategy is to select individuals for the sample *at random.* This may seem almost careless at first, but, as we will see, it is essential.

Idea 2: Randomize

Think back to our soup example. Suppose the chef adds some salt to the pot (the population). If she samples from the top before stirring, she'll get the misleading idea that the soup is salty. If she samples from the bottom, she'll get the equally misleading idea that it's bland. But by stirring the soup, she'll make each spoonful a more random sample of the soup, distributing the salt throughout the pot, and making each taste more typical of the saltiness of the whole pot. Deliberate randomization is one of the great tools of Statistics.

Randomization can also protect against factors that you aren't aware of. Suppose, while the chef isn't looking, an assistant adds a handful of peas to the soup. The peas sink to the bottom of the pot, mixing with the other vegetables. Stirring in the salt *also* randomizes the peas throughout the pot, making the sample taste more typical of the overall pot *even though the chef didn't know the peas were there.* So randomizing protects us by giving us a representative sample even for effects we were unaware of.

For a survey, we select participants at random, and this helps us represent *all* the features of our population, making sure that *on average* the sample looks like the rest of the population.

The essential feature of randomness is that the selection is "fair." We have discussed many facets of randomness in Chapter 5, and we can use some of those concepts here. What makes the sample fair is that each participant has an equal chance to be selected.

- **Why not match the sample to the population?** Rather than randomizing, we could try to design a sample to include every possible, relevant characteristic: income level, age, political affiliation, marital status, number of children, place of residence, etc. But we can't possibly think of all the things that might be important. Even if we could, we wouldn't be able to match our sample to the population for all these characteristics.

How well can a sample represent the population from which it was selected? Here's an example using the database of the Paralyzed Veterans of America, a philanthropic organization with a donor list of about 3.5 million people. We've taken two samples, each of 8000 individuals at random from the population. Table 8.1 shows how the means and proportions match up on seven variables.

	A	Age (yr)	White (%)	Female (%)	# of children	Income Bracket (1-7)	Wealth Bracket (1-9)	Homeowner? (% Yes)
1								
2	Sample 1	61.4	85.12	56.2	1.54	3.91	5.29	71.36
3	Sample 2	61.2	84.44	56.4	1.51	3.88	5.33	72.3

Table 8.1   Means and proportions for seven variables from two samples of size 8000 from the Paralyzed Veterans of America data. We drew these samples using Microsoft Excel's RAND function (Excel 2013), but you can use almost any statistics software to draw similar random samples. The fact that the summaries of the variables from these two samples are so similar gives us confidence that either one would be representative of the entire population.

The two samples match closely in every category. You can see how well randomizing has stirred the population. We didn't preselect the samples for these variables, but randomizing has matched the results closely. The two samples don't vary much from each other, so we can assume that they don't differ much from the rest of the population either.

Idea 3: The Sample Size Is What Matters

You probably weren't surprised by the idea that a sample can represent the whole. And the idea of sampling randomly to make the sample fair makes sense too. But the third important idea of sampling often surprises people. The third idea is that the *size of the sample* determines what we can conclude from the data *regardless of the size of the population*. Many people think that to provide a good representation of the population, the sample must be a large percentage, or *fraction*, of the population, but in fact all that matters is the size of the sample. The size of the *population* doesn't matter at all.[1] A random sample of 100 students in a college represents the student body just about as well as a random sample of 100 voters represents the entire electorate of the United States. This is perhaps the most surprising idea in designing surveys.

Think about the pot of soup again. The chef is probably making a large pot of soup. But she doesn't need a really big spoon to decide how the soup tastes. She'll get the same information from an ordinary spoonful no matter how large the pot—as long as the pot is sufficiently stirred. That's what randomness does for us. What *fraction* of the population you sample doesn't matter. It's the **sample size** itself that's important. This idea is of key importance to the design of any sample survey, because it determines the balance between how well the survey can measure the population and how much the survey costs.

How big a sample do you need? That depends on what you're estimating, but too small a sample won't be representative of the population. To get an idea of what's really in the soup, you need a large enough taste to be a *representative* sample from the pot, including, say, a selection of the vegetables. For a survey that tries to find the proportion of the population falling into a category, you'll usually need at least several hundred respondents.[2]

- **What do the professionals do?** How do professional polling and market research companies do their work? The most common polling method today is to contact respondents by telephone. Computers generate random telephone numbers for telephone exchanges known to include residential customers; so pollsters can contact people with unlisted phone numbers. The person who answers the phone will be invited to respond to the survey—if that person qualifies. (For example, only adults are usually surveyed, and the respondent usually must live at the residence phoned.) If the person answering doesn't qualify, the caller will ask for an appropriate alternative. When they conduct the interview, the pollsters often list possible responses (such as product names) in randomized orders to avoid biases that might favor the first name on the list.

Do these methods work? The Pew Research Center for the People and the Press reports on survey completion rates about every three years. Pew reports that by 2012 a telephone survey could contact about 62% of households whose

---

[1]Well, that's not exactly true. If the population is smaller than about 10 times the size of the sample it *can* matter. It doesn't matter whenever, as usual, our sample is a very small fraction of the population.

[2]Chapter 9 gives the details behind this statement and shows how to decide on a sample size for a survey.

phone numbers had been randomly generated. However, only 14% of those contacts yielded an interview, amounting to only 9% of the households originally sampled. Nevertheless, Pew concludes that "telephone surveys that include landlines and cell phones and are weighted to match the demographic composition of the population continue to provide accurate data on most political, social and economic measures." (www.people-press.org/2012/05/15 /assessing-the-representativeness-of-public-opinion-surveys/).

A Census—Does It Make Sense?

Why bother determining the right sample size? If you plan to open a store in a new community, why draw a sample of residents to understand their interests and needs? Wouldn't it be better to just include everyone and make the "sample" be the entire population? Such a special sample is called a **census**. Although a census would appear to provide the best possible information about the population, there are a number of reasons why it might not.

First, it can be difficult to complete a census. There always seem to be some individuals who are hard to locate or hard to measure. Do you really need to contact the folks away on vacation when you collect your data? How about those with no telephone or mailing address? The cost of locating the last few cases may far exceed the budget. It can also be just plain impractical to take a census. The quality control manager for Hostess® Twinkies® doesn't want to taste *all* the Twinkies on the production line to determine their quality. Aside from the fact that nobody could eat that many Twinkies, it would defeat their purpose: there would be none left to sell.

Second, the population you're studying may change. For example, in any human population, babies are born, people travel, and folks die during the time it takes to complete the census. News events and advertising campaigns can cause sudden shifts in opinions and preferences. A sample, surveyed in a shorter time frame, may actually generate more accurate information.

Finally, taking a census can be cumbersome. A census usually requires a team of pollsters and the cooperation of the population. Even with both, it's almost impossible to avoid errors. Because it tries to count everyone, the U.S. Census records too many college students. Many are included both by their families and in a report filed by their schools. Errors of this sort, of both under- and overcounting can be found throughout the U.S. Census.

FOR EXAMPLE Identifying sampling terms

A nonprofit organization has taken over the historic State Theater and hopes to preserve it with a combination of attractive shows and fundraising. The organization has asked a team of students to help them design a survey to better understand the customer base likely to purchase tickets. Fortunately, the theater's computerized ticket system records contact and some demographic information for ticket purchasers, and that database of 7345 customers is available.

QUESTIONS What is the population of interest? What would a census be in this case? Would it be practical?

ANSWERS The population is all potential ticket purchasers. A census would have to reach all potential purchasers. We don't know who they are or have any way to contact them.

8.2   Populations and Parameters

Statistic
Any quantity that we calculate from data could be called a "statistic." But in practice, we usually obtain a statistic from a sample and use it to estimate a population parameter.

Parameter
Population model parameters are not just unknown—usually they are unknowable. We take a sample and use the sample statistics to estimate them.

GfK Roper Reports Worldwide reports that 60.5% of people over 50 worry about food safety, but only 43.7% of teens do. What does this claim mean? We can be sure the Roper researchers didn't take a census. So they can't possibly know *exactly* what percentage of teenagers worry about food safety. So what does "43.7%" mean?

To generalize from a sample to the world at large, we need a model of reality. Such a model doesn't need to be complete or perfect. Just as a model of an airplane in a wind tunnel can tell engineers what they need to know about aerodynamics even though it doesn't include every rivet of the actual plane, models of data can give us summaries that we can learn from and use even though they don't fit each data value exactly. It's important to remember that they're only models of reality and not reality itself. But without models, what we can learn about the world at large is limited to only what we can say about the data we have at hand.

Models use mathematics to represent reality. We call the key numbers in those models **parameters**. Sometimes a parameter used in a model for a population is called (redundantly) a **population parameter**.

But let's not forget about the data. We use the data to try to estimate values for the population parameters. Any summary found from the data is a **statistic**. Those statistics that estimate population parameters are particularly interesting. Sometimes—and especially when we match statistics with the parameters they estimate—we use the term **sample statistic**.

We draw samples because we can't work with the entire population. We hope that the statistics we compute from the sample will estimate the corresponding parameters accurately. A sample that does this is said to be **representative**.

JUST CHECKING

1 Various claims are often made for surveys. Why is each of the following claims not correct?

 a) It is always better to take a census than to draw a sample.

 b) Stopping customers as they are leaving a restaurant is a good way to sample opinions about the quality of the food.

 c) We drew a sample of 100 from the 3000 students in a school. To get the same level of precision for a town of 30,000 residents, we'll need a sample of 1000.

 d) A poll taken at a popular website (www.statsisfun.org) garnered 12,357 responses. The majority of respondents said they enjoy doing Statistics. With a sample size that large, we can be sure that most Americans feel this way.

 e) The true percentage of all Americans who enjoy Statistics is called a "population statistic."

8.3   Common Sampling Designs

We've said that every individual in the population should have an equal chance of being selected in a sample. That makes the sample fair, but it's not quite enough to ensure that the sample is representative. Consider, for example, a market analyst who samples customers by drawing at random from product registration forms, half of which arrived by mail and half by online registration. She flips a coin. If it comes up heads, she'll draw 100 mail returns; tails, she'll draw 100 electronic returns. Each customer has an equal chance of being selected, but if tech-savvy customers are different, then the samples are hardly representative.

Simple Random Sample (SRS)

To make the sample representative, we must ensure that our sampling method gives each *combination* of individuals an equal chance as well. A sample drawn in this way is called a **simple random sample**, usually abbreviated **SRS**. An SRS is the sampling method on which the theory of working with sampled data is based and thus the standard against which we measure other sampling methods.

We'd like to select from the population, but often we don't have a list of all the individuals in the population. The list we actually draw from is called a **sampling frame**. A store may want to survey all its regular customers. But it can't draw a sample from the population of all regular customers, because it doesn't have such a list. The store may have a list of customers who have registered as "frequent shoppers." That list can be the sampling frame from which the store can draw its sample.

Of course, whenever the sampling frame and the population differ (as they almost always will), we must deal with the differences. Are the opinions of those who registered as frequent shoppers different from the rest of the regular shoppers? What about customers who used to be regulars but haven't shopped there recently? The answers to questions like these about the sampling frame may depend on the purpose of the survey and may impact the conclusions that one can draw.

Once we have a sampling frame, we need to *randomize* it so we can choose an SRS. Fortunately, random numbers are readily available these days in spreadsheets, statistics programs, and even on the Internet. Before this technology existed, people used to literally draw numbers out of a hat to randomize. But now, the easiest way to randomize your sampling frame is to match it with a parallel list of random numbers and then sort the random numbers, carrying along the cases so that they get "shuffled" into random order. Then you can just pick cases off the top of the randomized list until you have enough for your sample.

Samples drawn at random generally differ one from another. If we were to repeat the sampling process, a new draw of random numbers would select different people for our sample. These differences would lead to different values for the variables we measure. We call these sample-to-sample differences **sampling variability**. Sometimes they are called **sampling error** even though no error has taken place. Surprisingly, sampling variability isn't a problem; it's an opportunity. If different samples from a population vary little from each other, then most likely the underlying population harbors little variation. If the samples show much sampling variability, the underlying population probably varies a lot. In the coming chapters, we'll spend much time and attention working with sampling variability to better understand what we are trying to measure.

Sampling Errors vs. Bias

Referring to sample-to-sample variability as sampling error, makes it sound like it's some kind of mistake. It's not. We understand that samples will vary, so "sampling errors" are to be expected. It's bias we must strive to avoid. Bias means our sampling method distorts our view of the population. Of course, bias leads to mistakes. Even more insidious, bias introduces errors that we cannot correct with subsequent analysis.

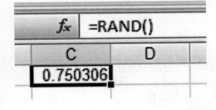

A Different Answer Every Time?

The RAND( ) function in Excel can take you by surprise. Every time the spreadsheet reopens, you get a new column of random numbers. But don't worry. Once you've shuffled the rows, you can ignore the new numbers. The order you got by shuffling won't keep changing. (Image created in Microsoft Excel 2013.)

FOR EXAMPLE   Choosing a random sample

Continuing the example on page 241, the student consultants select 200 ticket buyers at random from the database. First, the State Theater database is placed in a spreadsheet. Next, to draw random numbers, the students use the Excel command RAND( ). (They type =RAND( ) in the top cell of a column next to the data and then use *Fill Down* to populate the column down to the bottom.) They then sort the spreadsheet to put the random column in order and select ticket buyers from the top of the randomized spreadsheet until they complete 200 interviews. This makes it easy to select more respondents when (as always happens) some of the people they select can't be reached by telephone or decline to participate.

QUESTIONS   What is the sampling frame?

If the customer database held 30,000 records instead of 7345, how much larger a sample would we need to get the same information?

If we then draw a different sample of 200 customers and obtain different answers to the questions on the survey, how do we refer to these differences?

ANSWERS   The sampling frame is the customer database.

The size of the sample is all that matters, not the size of the population. We would need a sample of 200.

The differences in the responses from one sample to another are called sampling error, or sampling variability.

Simple random sampling is not the only fair way to sample. More complicated designs may save time or money or avert sampling problems. All statistical sampling designs have in common the idea that chance, rather than human choice, is used to select the sample.

Stratified Sampling

Designs that are used to sample from large populations—especially populations residing across large areas—are often more complicated than simple random samples. Sometimes we slice the population into homogeneous groups, called **strata,** and then use simple random sampling within each stratum, combining the results at the end. This is called **stratified random sampling**.

Why would we want to stratify? Suppose we want to survey how shoppers feel about a potential new anchor store at a large suburban mall. The shopper population is 60% women and 40% men, and we suspect that men and women have different views on their choice of anchor stores. If we use simple random sampling to select 100 people for the survey, we could end up with 70 men and 30 women or 35 men and 65 women. Our resulting estimates of the attractiveness of a new anchor store could vary widely. To help reduce this sampling variability, we can force a representative balance, selecting 40 men at random and 60 women at random. This would guarantee that the proportions of men and women within our sample match the proportions in the population, and that should make such samples more accurate in representing population opinion.

You can imagine that stratifying by race, income, age, and other characteristics can be helpful, depending on the purpose of the survey. When we use a sampling method that restricts by strata, additional samples are more like one another, so statistics calculated for the sampled values will vary less from one sample to another. This reduced sampling variability is the most important benefit of stratifying, but the analysis of data sampled with these designs is beyond the scope of our book.

Cluster and Multistage Sampling

Sometimes dividing the sample into homogeneous strata isn't practical, and even simple random sampling may be difficult. For example, suppose we wanted to assess the reading level of a product instruction manual based on the length of the sentences. Simple random sampling could be awkward; we'd have to number each sentence and then find, for example, the 576th sentence or the 2482nd sentence, and so on. Doesn't sound like much fun, does it?

We could make our task much easier by picking a few pages at random and then counting the lengths of the sentences on those pages. That's easier than picking individual sentences and works if we believe that the pages are all reasonably similar to one another in terms of reading level. Splitting the population in this way into parts or **clusters** that each represent the population can make sampling more practical. We select one or a few clusters at random and perform a census within each of them. This sampling design is called **cluster sampling.** If each cluster fairly represents the population, cluster sampling will generate an unbiased sample.

What's the difference between cluster sampling and stratified sampling? We stratify to ensure that our sample represents different groups in the population, and sample randomly within each stratum. This reduces the sample-to-sample variability. Strata are homogeneous, but differ from one another. By contrast, clusters are more or less alike, each heterogeneous and resembling the overall population. We cluster to save money or even to make the study practical.

Sometimes we use a variety of sampling methods together. In trying to assess the reading level of our instruction manual, we might worry that the "quick start" instructions are easy to read, but the "troubleshooting" chapter is more difficult.

Strata or Clusters?

We create strata by dividing the population into groups of similar individuals so that each stratum is different from the others. (For example, we often stratify by age, race, or sex.) By contrast, we create clusters that all look pretty much alike, each representing the wide variety of individuals seen in the population.

If so, we'd want to avoid samples that selected heavily from any one chapter. To guarantee a fair mix of sections, we could randomly choose one section from each chapter of the manual. Then we would randomly select a few pages from each of those sections. If altogether that made too many sentences, we might select a few sentences at random from each of the chosen pages. So, what is our sampling strategy? First we stratify by the chapter of the manual and randomly choose a section to represent each stratum. Within each selected section, we choose pages as clusters. Finally, we consider an SRS of sentences within each cluster. Sampling schemes that combine several methods are called **multistage samples**. Most surveys conducted by professional polling organizations and market research firms use some combination of stratified and cluster sampling as well as simple random samples.

FOR EXAMPLE Identifying more complex designs

The theater board wants to encourage people to come from out of town to attend theater events. They know that, in general, about 40% of ticket buyers are from out of town. These customers often purchase dinner at a local restaurant or stay overnight in a local inn, generating business for the town. The board hopes this information will encourage local businesses to advertise in the theater program, so they want to be sure out-of-town customers are represented in the samples. The database includes ZIP codes. The student consultants decide to sample 80 ticket buyers from ZIP codes outside the town and 120 from the town's ZIP code.

QUESTIONS What kind of sampling scheme are they using to replace the simple random sample?

What are the advantages of selecting 80 out of town and 120 local customers?

ANSWERS A stratified sample, consisting of a sample of 80 out-of-town customers and a sample of 120 local customers.

By stratifying, they can guarantee that 40% of the sample is from out of town, reflecting the overall proportions among ticket buyers. If out-of-town customers differ in important ways from local ticket buyers, a stratified sample will reduce the variation in the estimates for each group so that the combined estimates can be more precise.

Systematic Samples

Sometimes we draw a sample by selecting individuals systematically. For example, a **systematic sample** might select every tenth person on an alphabetical list of employees. To make sure our sample is random, we still must start the systematic selection with a randomly selected individual—not necessarily the first person on the list. When there is no reason to believe that the order of the list could be associated in any way with the responses measured, systematic sampling can give a representative sample. Systematic sampling can be much less expensive than true random sampling. When you use a systematic sample, you should justify the assumption that the systematic method is not associated with any of the measured variables.

Think about the reading level sampling example again. Suppose we have chosen a section of the manual at random, then three pages at random from that section, and now we want to select a sample of 10 sentences from the 73 sentences found on those pages. Instead of numbering each sentence so we can pick a simple random sample, it would be easier to sample systematically. A quick calculation shows $73/10 = 7.3$, so we can get our sample by picking every seventh sentence on the page. But where should you start? At random, of course. We've accounted for $10 \times 7 = 70$ of the sentences, so we'll throw the extra three into the starting group and choose a sentence at random from the first 10. Then we pick every seventh sentence after that and record its length.

JUST CHECKING

2 We need to survey a random sample of the 300 passengers on a flight from San Francisco to Tokyo. Name each sampling method described.

a) Pick every tenth passenger as people board the plane.

b) From the boarding list, randomly choose five people flying first class and 25 of the other passengers.

c) Randomly generate 30 seat numbers and survey the passengers who sit there.

d) Randomly select a seat position (right window, right center, right aisle, etc.) And survey all the passengers sitting in those seats.

GUIDED EXAMPLE   Market Demand Survey

In a course at a business school in the United States, the students form business teams, propose a new product, and use seed money to launch a business to sell the product on campus.

Before committing funds for the business, each team must complete the following assignment: "Conduct a survey to determine the potential market demand on campus for the product you are proposing to sell." Suppose your team's product is a 500-piece jigsaw puzzle of the map of your college campus. Design a marketing survey and discuss the important issues to consider.

PLAN	**Setup** State the goals and objectives of the survey.	Our team designed a study to find out how likely students at our school are to buy our proposed product—a 500-piece jigsaw puzzle of the map of our college campus.

Population and Parameters
Identify the population to be studied and the associated sampling frame. What are the parameters of interest?

The population studied will be students at our school. We have obtained a list of all students currently enrolled to use as the sampling frame. The parameter of interest is the proportion of students likely to buy this product. We'll also collect some demographic information about the respondents.

Sampling Plan Specify the sampling method and the planned sample size, n. Specify how the sample was actually drawn. What is the sampling frame?

The description should, if possible, be complete enough to allow someone to replicate the procedure, drawing another sample from the same population in the same manner. A good description of the procedure is essential, even if it could never practically be repeated. The question you ask is important, so state the

We will select a simple random sample of 200 students. The sampling frame is the master list of students we obtained from the registrar. We decided against stratifying by sex or class because we thought that students were all more or less alike in their likely interest in our product.

We will ask the students we contact:
Do you solve jigsaw puzzles for fun?
Then we will show them a prototype puzzle and ask:
If this puzzle sold for $10, would you purchase one?
We will also record the respondent's sex and class.

wording of the question clearly. Be sure that the question is useful in helping you with the overall goal of the survey.

 Sampling Practice Specify *when*, *where*, and *how* the sampling will be performed. Specify any other details of your survey, such as how respondents were contacted, any incentives that were offered to encourage them to respond, how nonrespondents were treated, and so on.

The survey will be administered in the middle of the fall semester during October. We have a master list of registered students, which we will randomize by matching it with random numbers from www .random.org and sorting on the random numbers, carrying the names. We will contact selected students by phone or e-mail and arrange to meet with them. If a student is unwilling to participate, the next name from the randomized list will be substituted until a sample of 200 participants is found.

We will meet with students in an office set aside for this purpose so that each will see the puzzle under similar conditions.

REPORT **Summary and Conclusion** This report should include a discussion of all the elements needed to design the study. It's good practice to discuss any special circumstances or other issues that may need attention.

MEMO

Re: Survey plans

Our team's plans for the puzzle market survey call for a simple random sample of students. Because subjects need to be shown the prototype puzzle, we must arrange to meet with selected participants. We have arranged an office for that purpose.

We will also collect demographic information so we can determine whether there is in fact a difference in interest level among classes or between men and women.

The Real Sample

We have been discussing sampling in a somewhat idealized setting. In the real world, things can be a bit messier. Here are some things to consider.

The population may not be as well-defined as it seems. For example, if a company wants the opinions of a typical mall "shopper," who should they sample? Should they only ask shoppers carrying a purchase? Should they include people eating at the food court? How about teenagers just hanging out in the mall? Even when the population is clear, it may not be possible to establish an appropriate sampling frame.

Usually, the practical sampling frame is not the group you *really* want to know about. For example, election polls want to sample from those who will actually vote in the next election—a group that is particularly tricky to identify before election day. The sampling frame limits what your survey can find out.

Then there's your target sample. These are the individuals selected according to your sample design for whom you *intend* to measure responses. You're not likely to get responses from all of them. ("I know it's dinner time, but I'm sure you wouldn't mind answering a few questions. It'll only take 20 minutes or so. Oh, you're busy?") Nonresponse is a problem in many surveys.

Sample designs are usually about the target sample. But in the real world, you won't get responses from everyone your design selects. So in reality, your sample consists of the actual respondents. These are the individuals about whom you *do*

What's the Sample?

The population we want to study is determined by asking *why*. When we design a survey, we use the term "sample" to refer to the individuals selected, from whom we hope to obtain responses. Unfortunately, the real sample is just those we can reach to obtain responses—the *who* of the study. These are slightly different uses of the same term sample. The context usually makes clear which we mean, but it's important to realize that the difference between the two samples could undermine even a well-designed study.

get data and can draw conclusions. Unfortunately, they might not be representative of either the sampling frame or the population.

At each step, the group we can study may be constrained further. The *who* of our study keeps changing, and each constraint can introduce biases. A careful study should address the question of how well each group matches the population of interest. The *who* in an SRS is the population of interest from which we've drawn a representative sample. That's not always true for other kinds of samples.

When people (or committees!) decide on a survey, they often fail to think through the important questions about who are the *who* of the study and whether they are the individuals from whom the answers would be interesting or have meaningful business consequences. This is a key step in performing a survey and should not be overlooked.

Calvin & Hobbes © 1993 Watterson. Distributed by Universal Uclick. Reprinted with permission. All rights reserved.

8.4   The Valid Survey

It isn't sufficient to draw a sample and start asking questions. You want to feel confident your survey can yield the information you need about the population you are interested in. We want a *valid survey*.

To help ensure a valid survey, you need to ask four questions:

- What do I want to know?
- Who are the right respondents?
- What are the right questions?
- What will be done with the results?

These questions may seem obvious, but there are a number of specific pitfalls to avoid:

Know what you want to know. Far too often, decision makers decide to perform a survey without any clear idea of what they hope to learn. Before considering a survey, you must be clear about what you hope to learn and what population you want to learn about. If you don't know that, you can't even judge whether you have a valid survey. The survey *instrument*—the questionnaire itself—can be a source of errors. Perhaps the most common error is to ask unnecessary questions. The longer the survey, the fewer people will complete it, leading to greater nonresponse bias. For each question on your survey, you should ask yourself whether you really want to know this and know what you would do with the responses if you had them. If you don't have a good use for the answer to a question, don't ask it.

Use the right sampling frame. A valid survey obtains responses from appropriate respondents. Be sure you have a suitable sampling frame. Have you identified the population of interest and sampled from it appropriately? A company looking to expand its base might survey customers who returned warrantee registration cards—after all, that's a readily available sampling frame—but if the company wants to know how to make its product more attractive, it needs to survey customers who rejected its product in favor of a competitor's product. This is the population that can tell the company what about its product needs to change to capture a larger market share. The errors in the presidential election polls of 1948 were likely due to the use of telephone samples in an era when telephones were not affordable by the less affluent—who were the folks most likely to vote for Truman.

It is equally important to be sure that your respondents actually know the information you hope to discover. Your customers may not know much about the competing products, so asking them to compare your product with others may not yield useful information.

Ask specific rather than general questions. It is better to be specific. "Do you usually recall TV commercials?" won't be as useful as "How many TV commercials can you recall from last night?" or better, yet, "Please describe for me all the TV commercials you can recall from your viewing last night."

Watch for biases. Even with the right sampling frame, you must beware of bias in your sample. If customers who purchase more expensive items are less likely to respond to your survey, this can lead to **nonresponse bias**. Although you can't expect all mailed surveys to be returned, if those individuals who don't respond have common characteristics, your sample will no longer represent the population you hope to learn about. Surveys in which respondents volunteer to participate, such as online surveys, suffer from **voluntary response bias**. Individuals with the strongest feelings on either side of an issue are more likely to respond; those who don't care may not bother.

Be careful with question phrasing. Questions must be carefully worded. A respondent may not understand the question—or may not understand the question the way the researcher intended it. For example, "Does anyone in your family own a Ford truck?" leaves the term "family" unclear. Does it include only spouses and children or parents and siblings, or do in-laws and second cousins count too? A question like "Was your Twinkie fresh?" might be interpreted quite differently by different people.

Be careful with answer phrasing. Respondents and survey-takers may also provide inaccurate responses, especially when questions are politically or sociologically sensitive. This also applies when the question does not take into account all possible answers, such as a true-false or multiple-choice question to which there may be other answers. Or the respondent may not know the correct answer to the question on the survey. In 1948, there were four major candidates for President,[3] but some survey respondents might not have been able to name them all. A survey question that just asked "Who do you plan to vote for?" might have underrepresented the less prominent candidates. And one that just asked "What do you think of Wallace?" might yield inaccurate results from voters who simply didn't know who he was. We refer to inaccurate responses (intentional or unintentional) as **measurement errors.** One way to cut down on measurement errors is to provide a range of possible responses. But be sure to phrase them in neutral terms.

The best way to protect a survey from measurement errors is to perform a pilot test. In a **pilot test,** a small sample is drawn from the sampling frame, and a draft form of the survey instrument is administered. A pilot test can point out flaws in the instrument. For example, during a staff cutback at one of our schools, a

---

[3] Harry Truman, Thomas Dewey, Strom Thurmond, and Henry Wallace.

researcher surveyed faculty members to ask how they felt about the reduction in staff support. The scale ran from "It's a good idea" to "I'm very unhappy." Fortunately, a pilot study showed that everyone was very unhappy or worse. The scale was re-tuned to run from "unhappy" to "ready to quit."

FOR EXAMPLE Survey design

A nonprofit organization has enlisted some student consultants to help design a fund-raising survey. The student consultants suggest to the board of directors that they may want to rethink their survey plans. They point out that there are differences among the population, the sampling frame, the target sample contacted by telephone, and the actual sample.

QUESTION How are the population, sampling frame, target sample, and sample likely to differ?

ANSWER The population is all potential ticket buyers.

The sampling frame consists of only those who have previously purchased tickets. Anyone who wasn't attracted to previous productions wouldn't be surveyed. That could keep the board from learning of ways to make the theater's offering more attractive to those who hadn't purchased tickets before.

The target sample is those selected from the database who can be contacted by telephone. Those with unlisted numbers or who had declined to give their phone number can't be contacted. It may be more difficult to contact those with caller ID.

The actual sample will be those previous customers selected at random from the database who can be reached by telephone and who agree to complete the survey.

8.5   How to Sample Badly

Bad sample designs yield worthless data. Many of the most convenient forms of sampling can be seriously biased. And there is no way to correct for the bias from a bad sample. So it's wise to pay attention to sample design—and to beware of reports based on poor samples.

Voluntary Response Sample

One of the most common dangerous sampling methods is the voluntary response sample. In a **voluntary response sample,** a large group of individuals is invited to respond, and all who do respond are counted. This method is used by call-in shows, 900 numbers, Internet polls, and letters written to members of Congress. Voluntary response samples are almost always biased, and so conclusions drawn from them are almost always wrong.

It's often hard to define the sampling frame of a voluntary response study. Practically, the frames are groups such as Internet users who frequent a particular website or viewers of a particular TV show. But those sampling frames don't correspond to the population you are likely to be interested in.

Even if the sampling frame is of interest, voluntary response samples are often biased toward those with strong opinions or those who are strongly motivated—and especially from those with strong negative opinions. A request that travelers who have used the local airport visit a survey site to report on their experiences is much more likely to hear from those who had long waits, cancelled flights, and lost luggage than from those whose flights were on time and carefree. The resulting voluntary response bias invalidates the survey.

Convenience Sampling

Another sampling method that doesn't work is convenience sampling. As the name suggests, in **convenience sampling** we simply include the individuals who are convenient. Unfortunately, this group may not be representative of the population. A survey of 437 potential home buyers in Orange County, California, found, among other things, that

> *all but 2 percent of the buyers have at least one computer at home, and 62 percent have two or more. Of those with a computer, 99 percent are connected to the Internet (Jennifer Hieger, "Portrait of Homebuyer Household: 2 Kids and a PC," Orange County Register, July 27, 2001).*

Later in the article, we learn that the survey was conducted via the Internet. That was a convenient way to collect data and surely easier than drawing a simple random sample, but perhaps home builders shouldn't conclude from this study that *every* family has a computer and an Internet connection.

Many surveys conducted at shopping malls suffer from the same problem. People in shopping malls are not necessarily representative of the population of interest. Mall shoppers tend to be more affluent and include a larger percentage of teenagers and retirees than the population at large. To make matters worse, survey interviewers tend to select individuals who look "safe," or easy to interview.

Convenience sampling is not just a problem for beginners. In fact, convenience sampling is a widespread problem in the business world. When a company wants to find out what people think about its products or services, it may turn to the easiest people to sample: its own customers. But the company will never learn how those who *don't* buy its product feel about it.

Do you use the Internet?
Click here ◯ for yes
Click here ◯ for no

Internet Surveys

Internet convenience surveys are often worthless. As voluntary response surveys, they have no well-defined sampling frame (all those who use the Internet and visit their site?) and thus report no useful information. Do not use them.

Bad Sampling Frame?

An SRS from an incomplete sampling frame introduces bias because the individuals included may differ from the ones not in the frame. It may be easier to sample workers from a single site, but if a company has many sites and they differ in worker satisfaction, training, or job descriptions, the resulting sample can be biased. There is serious concern among professional pollsters that the increasing numbers of people who can be reached only by cell phone may bias telephone-based market research and polling.

Undercoverage

Many survey designs suffer from **undercoverage,** in which some portion of the population is not sampled at all or has a smaller representation in the sample than it has in the population. Undercoverage can arise for a number of reasons, but it's always a potential source of bias. Are people who use answering machines to screen callers (and are thus less available to blind calls from market researchers) different from other customers in their purchasing preferences?

FOR EXAMPLE   Common mistakes in survey design

A board member proposes that rather than telephoning past customers, they simply post someone at the door to ask theater goers their opinions. Another suggests that it would be even easier to post a questionnaire on the theater website and invite responses there. A third suggests that rather than working with random numbers, they simply phone every 200th person on the list of past customers.

(continued)

> **QUESTION** Identify the three methods proposed and explain what strengths and weaknesses they have.
>
> **ANSWER** Questioning customers at the door would be a convenience sample. It would be cheap and fast but is likely to be biased by the nature and quality of the particular performance where the survey takes place.
>
> Inviting responses on the website would be a voluntary response sample. Only customers who frequented the website and decided to respond would be surveyed. This might, for example, underrepresent older customers or those without home Internet access.
>
> Sampling every 200th name from the customer list would be a systematic sample. It is slightly easier than randomizing. If the order of names on the list is unrelated to any questions asked, then this might be an acceptable method. But if, for example, the list is kept in the order of first purchases (when a customer's name and information were added to the database), then there might be a relationship between opinions and location on the list.

WHAT CAN GO WRONG?

- **Nonrespondents.** No survey succeeds in getting responses from everyone. The problem is that those who don't respond may differ from those who do. And if they differ on just the variables we care about, the lack of response will bias the results. Rather than sending out a large number of surveys for which the response rate will be low, it is often better to design a smaller, randomized survey for which you have the resources to ensure a high response rate.

- **Long, dull surveys.** Surveys that are too long are more likely to be refused, reducing the response rate and biasing all the results. Keep it short.

- **Response bias.** Response bias includes the tendency of respondents to tailor their responses to please the interviewer and the consequences of slanted question wording.

- **Push polls.** Push polls, which masquerade as surveys, present one side of an issue before asking a question. For example, a question like

  Would the fact that the new store that just opened by the mall sells mostly goods made overseas by workers in sweatshop conditions influence your decision to shop there rather than in the downtown store that features American-made products?

  is designed not to gather information, but to spread ill-will toward the new store.

THE WIZARD OF ID   **parker and hart**

The Wizard of Id © 2001 John L. Hart/Distributed by Creators Syndicate. Reprinted with permission. All rights reserved.

HOW TO THINK ABOUT BIASES

- **Look for biases in any survey.** If you design a survey of your own, ask someone else to help look for biases that may not be obvious to you. Do this *before* you collect your data. There's no way to recover from a biased sample or a survey that asks biased questions.

  A bigger sample size for a biased study just gives you a bigger useless study. A really big sample gives you a really big useless study.

- **Spend your time and resources reducing biases.** No other use of resources is as worthwhile as reducing the biases.

- **If you possibly can, pretest or pilot your survey.** Administer the survey in the exact form that you intend to use it to a small sample drawn from the population you intend to sample. Look for misunderstandings, misinterpretation, confusion, or other possible biases. Then redesign your survey instrument.

- **Always report your sampling methods in detail.** Others may be able to detect biases where you did not expect to find them.

ETHICS IN ACTION

The Lackawax River Group is interested in applying for state funds to continue their restoration and conservation of the Lackawax River, a river that has been polluted from years of industry and agricultural discharge. While they have managed to gain significant support for their cause through education and community involvement, the executive committee is now interested in presenting the state with more compelling evidence.

They decided to survey local residents regarding their attitudes toward the proposed expansion of the river restoration and conservation project. With limited time and money (the deadline for the grant application was fast approaching), the executive committee was delighted that one of its members, Harry Greentree, volunteered to undertake the project.

Harry owned a local organic food store and agreed to have a sample of his shoppers interviewed during the next one-week period. One committee member questioned whether a representative sample of residents could be found in this way, but the other members of the committee thought that the customers of Harry's store were likely to be just the kind of well-informed residents whose opinions they wanted to hear. The only instruction the committee decided to give was that the shoppers be selected in a systematic fashion, for instance, by interviewing every fifth person who entered the store. Harry had no problem with this request and was eager to help the Lackawax River Group.

- Identify the ethical dilemma in this scenario.
- What are the undesirable consequences?
- Propose an ethical solution that considers the welfare of all stakeholders.

WHAT HAVE WE LEARNED?

Learning Objectives

Know the three ideas of sampling.

- Examine a part of the whole: A sample can give information about the population.
- Randomize to make the sample representative.
- The sample size is what matters. It's the size of the sample—and not its fraction of the larger population—that determines the precision of the statistics it yields.

Be able to draw a Simple Random Sample (SRS) using a table of random digits or a list of random numbers from technology or an Internet site.

- In a **simple random sample** (SRS), every possible group of *n* individuals has an equal chance of being our sample.

Know the definitions of other sampling methods:

- **Stratified samples** can reduce sampling variability by identifying homogeneous subgroups and then randomly sampling within each.
- **Cluster samples** randomly select among heterogeneous subgroups that each resemble the population at large, making our sampling tasks more manageable.
- **Systematic samples** can work in some situations and are often the least expensive method of sampling. But we still want to start them randomly.
- **Multistage samples** combine several random sampling methods.

Identify and avoid causes of bias.

- **Nonresponse bias** can arise when sampled individuals will not or cannot respond.
- **Response bias** arises when respondents' answers might be affected by external influences, such as question wording or interviewer behavior.
- **Voluntary response samples** are almost always biased and should be avoided and distrusted.
- **Convenience samples** are likely to be flawed for similar reasons.
- **Undercoverage** occurs when individuals from a subgroup of the population are selected less often than they should be.

Terms

Bias	Any systematic failure of a sampling method to represent its population.
Census	An attempt to collect data on the entire population of interest.
Cluster	A subset of a population aggregated into larger sampling units. These units, chosen for reasons of cost or practicality are often natural groups thought to be representative of the population.
Cluster sampling	A sampling design in which groups, or clusters, representative of the population are chosen at random and a census is then taken of each.
Convenience sampling	A sample that consists of individuals who are conveniently available.
Measurement error	Any inaccuracy in a response, from any source, whether intentional or unintentional.
Multistage sample	A sampling scheme that combines several sampling methods.
Nonresponse bias	Bias introduced to a sample when a large fraction of those sampled fails to respond.
Parameter	A numerically valued attribute of a model for a population. We rarely expect to know the value of a parameter, but we do hope to estimate it from sampled data.
Pilot test	A small trial run of a study to check that the methods of the study are sound.
Population	The entire group of individuals or instances about whom we hope to learn.
Population parameter	A numerically valued attribute of a model for a population.
Randomization	A defense against bias in the sample selection process, in which each individual is given a fair, random chance of selection.
Representative sample	A sample from which the statistics computed accurately reflect the corresponding population parameters.
Response bias	Anything in a survey design that influences responses.
Sample	A subset of a population, examined in hope of learning about the population.
Sample size	The number of individuals in a sample.
Sample survey	A study that asks questions of a sample drawn from some population in the hope of learning something about the entire population.
Sampling frame	A list of individuals from which the sample is drawn. Individuals in the population of interest but who are not in the sampling frame cannot be included in any sample.
Sampling variability (or sampling error)	The natural tendency of randomly drawn samples to differ, one from another.

Simple random sample (SRS)	A sample in which each set of *n* elements in the population has an equal chance of selection.
Statistic, sample statistic	A value calculated for sampled data, particularly one that corresponds to, and thus estimates, a population parameter. The term "sample statistic" is sometimes used, usually to parallel the corresponding term "population parameter."
Strata	Subsets of a population that are internally homogeneous but may differ one from another.
Stratified random sample	A sampling design in which the population is divided into several homogeneous subpopulations, or strata, and random samples are then drawn from each stratum.
Systematic sample	A sample drawn by selecting individuals systematically from a sampling frame.
Voluntary response bias	Bias introduced to a sample when individuals can choose on their own whether to participate in the sample.
Voluntary response sample	A sample in which a large group of individuals are invited to respond and decide individually whether or not to participate. Voluntary response samples are generally worthless.
Undercoverage	A sampling scheme that biases the sample in a way that gives a part of the population less representation than it has in the population.

TECHNOLOGY HELP: Random Sampling

Computer-generated random numbers are usually quite good enough for drawing random samples. But there is little reason not to use the truly random values available on the Internet. Here's a convenient way to draw an SRS of a specified size using a computer-based sampling frame. The sampling frame can be a list of names or identification numbers arrayed, for example, as a column in a spreadsheet, statistics program, or database:

1 Generate random numbers of enough digits so that each exceeds the size of the sampling frame list by several digits. This makes duplication unlikely. (For example, in Excel, use the RAND function described in detail in Technology Help, Chapter 5 to fill a column with random numbers between 0 and 1. With many digits they will almost surely be unique.)

2 Assign the random numbers arbitrarily to individuals in the sampling frame list. For example, put them in an adjacent column.

3 Sort the list of random numbers, *carrying* along the sampling frame list.

4 Now the first *n* values in the sorted sampling frame column are an SRS of *n* values from the entire sampling frame.

Most statistics packages also offer commands to sample from your data, but you should be careful to see that they do what you intend.

EXCEL

To generate random numbers in Excel:

- Choose **Data > Data Analysis > Random Number Generation.** (Note: the Data Analysis add-in must be installed.)

- In the Random Number Generation window, fill in

  - Number of variables = number of columns of random numbers.

  - Number of random numbers = number of rows of random numbers.

- Select a distribution from the drop-down menu. Parameters for your selected distribution will appear below.

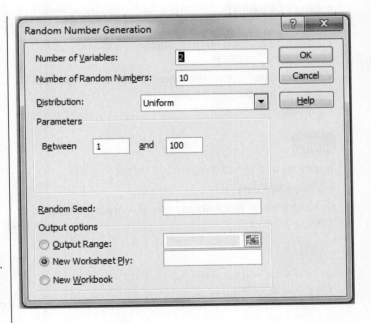

- Enter the minimum and maximum bounds for the random numbers. This will be the minimum and maximum of the random numbers generated.

- A list of random numbers will be generated in a new worksheet. The example shown here resulted from parameters of 1 to 100.

- Format cells to obtain values desired.

To sample from a column of data in Excel:

- Choose **Data > Data Analysis > Sampling.**

- Type in or select the cell range containing the data. If this column has a title, place a check in the box marked "Labels".

- Next to Random, indicate the "number of Samples" desired—this is actually the sample size, n.
- Finally, choose a location for the selected sample.

Warning: Excel samples with replacement. This is probably not the sampling method you want for drawing a sample from a population. The method given above using externally generated random numbers may be more appropriate.

MINITAB

To generate a list of random numbers in Minitab:

- Choose **Calc > Random Data > Uniform.**
- Enter the number of rows.
- Select the column where the random numbers will be stored.
- Click **OK.**

To sample from a variable in Minitab:

- Name a column in the data that will contain the sample; this column will be blank.
- Choose **Calc > Random Data > Sample From Columns.**
- Enter the number of rows to sample. This is the sample size, n.
- Indicate the column from which to select the data under "From Columns".
- Indicate the column in which the samples data should be placed under "Store Samples In".
- Minitab samples without replacement. To sample with replacement, check the box specifying that alternative.
- Click **OK.**

JMP

To generate a list of random numbers in JMP:

- Create a **New Data Table.**
- Choose **Rows > Add Rows** and enter the number of random numbers desired.

- Right-click on the top of Column 1.
- Choose **Formula . . .**
- Under **Functions (grouped)** choose **Random > Random Uniform.**
- Click **OK;** format data as desired.

To sample from a variable in JMP:

- Select the column of data to sample from by clicking the top of the column.
- Select **Tables > Subset.**
- Choose **Random – sample size:** and enter the desired sample size, n.
- Choose **Selected columns.**
- Fill in a name for the table where the sample will be stored.

A table will be created containing the random sample.

SPSS

To generate a list of random numbers in SPSS:

- Open a new dataset and assign a new variable with numbers 1 to n where n is the number of random numbers that will be generated.
- In the **Transform** menu, choose **Compute Variable . . .**
- Assign a name to the target variable.
- Under **Numeric Expression,** type **RV.UNIFORM(min,max),** where min = the lowest value of the variable and max = the highest value of the variable. For example, RV.UNIFORM (0, 1) will give you random numbers between 0 and 1.

To select a random sample in SPSS:

- From **Data** menu, choose **Select Cases.**
- Choose **Random sample of cases.**
- Click the **Sample** button to select either a percentage of cases or a number of cases.
- Select the desired output.

Brief **Case**

Market Survey Research

You are part of a marketing team that needs to research the potential of a new product. Your team decides to e-mail an interactive survey to a random sample of consumers. Write a short questionnaire that will generate the information you need about the new product. Select a sample of 200 using an SRS from your sampling frame. Discuss how you will collect the data and how the responses will help your market research.

The GfK Roper Reports Worldwide Survey

GfK Roper Consulting conducts market research for multinational companies who want to understand attitudes in different countries so they can market and advertise more effectively to different cultures. Every year they conduct a poll worldwide, which asks hundreds of questions of people in approximately 30 different countries. Respondents are asked a variety of questions about food. Some of the questions are simply yes/no (agree/disagree) questions: Please tell me whether you agree or disagree with each of these statements about your appearance: (Agree = 1; Disagree = 2; Don't know = 9).

The way you look affects the way you feel.

I am very interested in new skin care breakthroughs.

People who don't care about their appearance don't care about themselves.

Other questions are asked on a 5-point scale (Please tell me the extent to which you disagree or agree with it using the following scale: Disagree completely = 1; Disagree somewhat = 2; Neither disagree nor agree = 3; Agree somewhat = 4; Agree completely = 5; Don't know = 9).

Examples of such questions include:

I read labels carefully to find out about ingredients, fat content, and/or calories.

I try to avoid eating fast food.

When it comes to food I'm always on the lookout for something new.

Think about designing a survey on such a global scale:

- What is the population of interest?
- Why might it be difficult to select an SRS from this sampling frame?
- What are some potential sources of bias?
- Why might it be difficult to ensure a representative number of men and women and all age groups in some countries?
- What might be a reasonable sampling frame?

EXERCISES

SECTION 8.1

1. Indicate whether each statement below is true or false. If false, explain why.

a) We can eliminate sampling error by selecting an unbiased sample.

b) Randomization helps to ensure that our sample is representative.

c) Sampling error refers to sample-to-sample differences and is also known as sampling variability.

d) It is better to try to match the characteristics of the sample to the population rather than relying on randomization.

2. Indicate whether each statement below is true or false. If false, explain why.

a) To get a representative sample, you must sample a large fraction of the population.

b) Using modern methods, it is best to select a representative subset of a population systematically.

c) A census is the only true representative sample.

d) A random sample of 100 students from a school with 2000 students has the same precision as a random sample of 100 from a school with 20,000 students.

SECTION 8.2

3. An environmental advocacy group is interested in the perceptions of farmers about global climate change. Specifically, they wish to determine the percentage of organic farmers who are concerned that climate change will affect their crop yields. They use an alphabetized list of members of the Northeast Organic Farming Association (www.nofa .org), a nonprofit organization of over 5000 members with chapters in Connecticut, Massachusetts, New Hampshire, New Jersey, New York, Rhode Island, and Vermont. They use Excel to generate a randomly shuffled list of the members. They then select members to contact from this list until they have succeeded in contacting 150 members.

a) What is the population?
b) What is the sampling frame?
c) What is the population parameter of interest?
d) What sampling method is used?

4. An airline company is interested in the opinions of their frequent flyer customers about their proposed new routes. Specifically they want to know what proportion of them plan to use one of their new hubs in the next six months. They take a random sample of 10,000 from the database of all frequent flyers and send them an e-mail message with a request to fill out a survey in exchange for 1500 miles.

a) What is the population?
b) What is the sampling frame?
c) What is the population parameter of interest?
d) What sampling method is used?

SECTION 8.3

5. As discussed in the chapter, GfK Roper Consulting conducts a global consumer survey to help multinational companies understand different consumer attitudes throughout the world. In India, the researchers interviewed 1000 people aged 13–65 (www.gfkamerica.com). Their sample is designed so that they get 500 males and 500 females.

a) Are they using a simple random sample? How do you know?
b) What kind of design do you think they are using?

6. For their class project, a group of Business students decides to survey the student body to assess opinions about a proposed new student coffee shop to judge how successful it might be. Their sample of 200 contained 50 first-year students, 50 sophomores, 50 juniors, and 50 seniors.

a) Do you think the group was using an SRS? Why?
b) What kind of sampling design do you think they used?

7. The environmental advocacy group from Exercise 3 that was interested in gauging perceptions about climate change among organic farmers has decided to use a different method to sample. Instead of randomly selecting members from a shuffled list, they listed the members in alphabetical order and took every tenth member until they succeeded in contacting 150 members. What kind of sampling method have they used?

8. The airline company from Exercise 4, interested in the opinions of their frequent flyer customers about their proposed new routes, has decided that different types of customers might have different opinions. Of their customers, 50% are silver-level, 30% are blue, and 20% are red. They first compile separate lists of silver, blue, and red members and then randomly select 5000 silver members, 3000 blue members, and 2000 red members to e-mail. What kind of sampling method have they used?

For Exercises 9 and 10, identify the following if possible. (If not, say why.)

a) The population
b) The population parameter of interest
c) The sampling frame
d) The sample
e) The sampling method, including whether or not randomization was employed
f) Any potential sources of bias you can detect and any problems you see in generalizing to the population of interest

9. A business magazine mailed a questionnaire to the human resources directors of all Fortune 500 companies, and received responses from 23% of them. Those responding reported that they did not find that such surveys intruded significantly on their workday.

10. A question posted on the Lycos website asked visitors to the site to say whether they thought that businesses should be required to pay for their employees' health insurance.

SECTION 8.4

11. An intern for the environmental group in Exercise 3 has decided to make the survey process simpler by calling 150 of the members who attended the recent symposium on coping with climate change that was recently held in Burlington, VT. He has all the phone numbers, so it will be easy to contact them. He will start calling members from the top of the list, which was generated as the members enrolled for the symposium. He has written a script to read to them that follows,

"As we learned in Burlington, climate change is a serious problem for farmers. Given the evidence of impact on crops, do you agree that the government should be doing more to fight global warming?"

a) What is the population of interest?
b) What is the sampling frame?
c) Point out any problems you see either with the sampling procedure and/or the survey itself. What are the potential impacts of these problems?

12. The airline company in Exercise 4 has realized that some of its customers don't have e-mail or don't read it regularly. They decide to restrict the mailing only to customers who have recently registered for a "Win a trip to Miami" contest, figuring that those with Internet access are more likely to read and to respond to their e-mail. They send an e-mail with the following message:

"Did you know that National Airlines has just spent over $3 million refurbishing our brand new hub in Miami? By answering the following question, you may be eligible to win $1000 worth of coupons that can be spent in any of the fabulous restaurants or shops in the Miami airport. Might you possibly think of traveling to Miami in the next six months on your way to one of your destinations?"

a) What is the population?

b) What is the sampling frame?

c) Point out any problems you see either with the sampling procedure and/or the survey itself. What are the potential impacts of these problems?

13. An intern is working for Pacific TV (PTV), a small cable and Internet provider, and has proposed some questions that might be used in the survey to assess whether customers are willing to pay $50 for a new service.

Question 1: If PTV offered state-of-the-art, high-speed Internet service for $50 per month, would you subscribe to that service?

Question 2: Would you find $50 per month—less than the cost of a daily cappuccino—an appropriate price for high-speed Internet service?

a) Do you think these are appropriately worded questions? Why or why not?

b) Which one has more neutral wording? Explain.

14. Here are more proposed survey questions for the survey in Exercise 13:

Question 3: Do you find that the slow speed of DSL Internet access reduces your enjoyment of web services?

Question 4: Given the growing importance of high-speed Internet access for your children's education, would you subscribe to such a service if it were offered?

a) Do you think these are appropriately worded questions? Why or why not?

b) Suggest a question with better wording.

SECTION 8.5

15. Indicate whether each statement below is true or false. If false, explain why.

a) A local television news program that asks viewers to call in and give their opinion on an issue typically results in a biased voluntary response sample.

b) Convenience samples are generally representative of the population.

c) Measurement error is the same as sampling error.

d) A pilot test can be useful for identifying poorly worded questions on a survey.

16. Indicate whether each statement below is true or false. If false, explain why.

a) Asking viewers to call into an 800 number is a good way to produce a representative sample.

b) When writing a survey, it's a good idea to include as many questions as possible to ensure efficiency and to lower costs.

c) A recent poll on a website was valid because the sample size was over 1,000,000 respondents.

d) Malls are not necessarily good places to conduct surveys because people who frequent malls may not be representative of the population at large.

17. For your marketing class, you'd like to take a survey from a sample of all the Catholic Church members in your city to assess the market for a DVD about Pope Francis's first year as pope. A list of churches shows 17 Catholic churches within the city limits. Rather than try to obtain a list of all members of all these churches, you decide to pick 3 churches at random. For those churches, you'll ask to get a list of all current members and contact 100 members at random.

a) What kind of design have you used?

b) What could go wrong with the design that you have proposed?

18. The U.S. Fish and Wildlife Service plans to study the fishing industry around Saginaw Bay. To do that, they decide to randomly select five fishing boats at the end of a randomly chosen fishing day and count the numbers and types of all the fish on those boats.

a) What kind of design have they used?

b) What could go wrong with the design that they have proposed?

CHAPTER EXERCISES

19. Software licenses. The website asked, as their question of the day to which visitors to the site were invited to respond, *"Do you ever read the end-user license agreements when installing software or games?"* Of the 98,574 respondents, 63.47% said they never read those agreements—a fact that software manufacturers might find important.

a) What kind of sample was this?

b) How much confidence would you place in using 63.47% as an estimate of the fraction of people who don't read software licenses?

20. Drugs in baseball. Major League Baseball, responding to concerns about their "brand," tests players to see whether they are using performance-enhancing drugs. Officials select a team at random, and a drug-testing crew shows up unannounced to test all 40 players on the team. Each testing day can be considered a study of drug use in Major League Baseball.

a) What kind of sample is this?

b) Is that choice appropriate?

21. Pew. Pew Research Center publishes polls on issues important in the news and about American life at its website, www.pewinternet.org. At the end of a report about a survey you can find a paragraph such as this one:

These readings come from a national survey conducted between November 14 and December 9, 2012 of U.S. adults on landline and cell phones and in English and in Spanish. The results reported here come from the 1,802 respondents who are internet users and the margin of error is +/−2.6 percentage points.

a) For this survey, identify the population of interest.
b) Pew performs its surveys by phoning numbers generated at random by a computer program. What is the sampling frame? Does this seem representative of the population?

22. Defining the survey. At its website (www.gallup.com) the Gallup World Poll reports results of surveys conducted in various places around the world. At the end of one of these reports about the reliability of electric power in Africa, they describe their methods, including explanations such as the following:

Results are based on face-to-face interviews with 1,000 adults, aged 15 and older, conducted in 2010 in Botswana, Burkina Faso, Cameroon, Central African Republic, Chad, Ghana, Kenya, Liberia, Mali, Niger, Nigeria, Senegal, Sierra Leone, South Africa, Tanzania, Uganda, and Zimbabwe. For results based on the total sample of national adults, one can say with 95% confidence that the maximum margin of sampling error ranges from ±3.4 percentage points to ±4.0 percentage points. The margin of error reflects the influence of data weighting. In addition to sampling error, question wording and practical difficulties in conducting surveys can introduce error or bias into the findings of public opinion polls.[4]

a) Gallup is interested in the opinions of Africans. What kind of survey design are they using?
b) Some of the countries surveyed have large populations. (South Africa is estimated to have over 50 million people.) Some are quite small. (Zimbabwe has fewer than 13,000,000 people.) Nonetheless, Gallup sampled 1000 adults in each country. How does this affect the precision of its estimates for these countries?

23–30. Survey details. *For the following reports about statistical studies, identify the following items (if possible). If you can't tell, then say so—this often happens when we read about a survey.*

a) The population
b) The population parameter of interest
c) The sampling frame
d) The sample
e) The sampling method, including whether or not randomization was employed
f) Any potential sources of bias you can detect and any problems you see in generalizing to the population of interest

---

23. Teens and technology. Pew Internet & American Life Project surveyed 802 pairs of parents and teens (aged 12–17). They report that 93% of teens have access to a computer. 25% of teens access the Internet primarily on their cell phone rather than on a computer.

24. Global warming. The Gallup Poll interviewed 1022 randomly selected U.S. adults aged 18 and older, March 7–10, 2013. Gallup reports that when asked whether respondents thought that global warming was due primarily to human activities, 57% of respondents said it was.

25. At the bar. Researchers waited outside a bar they had randomly selected from a list of such establishments. They stopped every tenth person who came out of the bar and asked whether he or she thought drinking and driving was a serious problem.

26. Election poll. Hoping to learn what issues may resonate with voters in the coming election, the campaign director for a mayoral candidate selects one block at random from each of the city's election districts. Staff members go there and interview all the residents they can find.

27. Toxic waste. The Environmental Protection Agency took soil samples at 16 locations near a former industrial waste dump and checked each for evidence of toxic chemicals. They found no elevated levels of any harmful substances.

28. Housing discrimination. Inspectors send trained "renters" of various races and ethnic backgrounds, and of both sexes to inquire about renting randomly assigned advertised apartments. They look for evidence that landlords deny access illegally based on race, sex, or ethnic background.

29. Quality control. A company packaging snack foods maintains quality control by randomly selecting 10 cases from each day's production and weighing the bags. Then they open one bag from each case and inspect the contents.

30. Contaminated milk. Dairy inspectors visit farms unannounced and take samples of the milk to test for contamination. If the milk is found to contain dirt, antibiotics, or other foreign matter, the milk will be destroyed and the farm is considered to be contaminated pending further testing.

31. Instant poll. A local TV station conducted an "Instant Poll" to predict the winner in the upcoming mayoral election. Evening news viewers were invited to phone in their votes, with the results to be announced on the late-night news. Based on the phone calls, the station predicted that Amabo would win the election with 52% of the vote. They were wrong: Amabo lost, getting only 46% of the vote. Do you think the station's faulty prediction is more likely to be a result of bias or sampling error? Explain.

32. Paper poll. Prior to the mayoral election discussed in Exercise 31, the newspaper also conducted a poll. The paper surveyed a random sample of registered voters stratified by political party, age, sex, and area of residence. This poll predicted that Amabo would win the election with 52% of the vote. The newspaper was wrong: Amabo lost, getting only 46% of the vote. Do you think the newspaper's faulty prediction is more likely to be a result of bias or sampling error? Explain.

33. Cable company market research. A local cable TV company, Pacific TV (PTV), with customers in 15 towns is considering offering high-speed Internet service on its cable lines. Before launching the new service they want to find out whether customers would pay the $75 per month that they plan to charge. An intern has prepared several alternative plans for assessing customer demand. For each, indicate what kind of sampling strategy is involved and what (if any) biases might result.

a) Put a big ad in the newspaper asking people to log their opinions on the PTV website.
b) Randomly select one of the towns and contact every cable subscriber by phone.
c) Send a survey to each customer and ask them to fill it out and return it.
d) Randomly select 20 customers from each town. Send them a survey, and follow up with a phone call if they do not return the survey within a week.

34. Cable company market research, part 2. Four new sampling strategies have been proposed to help PTV determine whether enough cable subscribers are likely to purchase high-speed Internet service. For each, indicate what kind of sampling strategy is involved and what (if any) biases might result.

a) Run a poll on the local TV news, asking people to dial one of two phone numbers to indicate whether they would be interested.
b) Hold a meeting in each of the 15 towns, and tally the opinions expressed by those who attend the meetings.
c) Randomly select one street in each town and contact each of the households on that street.
d) Go through the company's customer records, selecting every 40th subscriber. Send employees to those homes to interview the people chosen.

35. Amusement park riders. An amusement park has opened a new roller coaster. It is so popular that people are waiting for up to three hours for a two-minute ride. Concerned about how patrons (who paid a large amount to enter the park and ride on the rides) feel about this, they survey every tenth person in line for the roller coaster, starting from a randomly selected individual.

a) What kind of sample is this?
b) Is it likely to be representative?
c) What is the sampling frame?

36. Playground. Some people have been complaining that the children's playground at a municipal park is too small and is in need of repair. Managers of the park decide to survey city residents to see if they believe the playground should be rebuilt. They hand out questionnaires to parents who bring children to the park. Describe possible biases in this sample.

37. Another ride. The survey of patrons waiting in line for the roller coaster in Exercise 35 asks whether they think it is worthwhile to wait a long time for the ride and whether they'd like the amusement park to install still more roller coasters. What biases might cause a problem for this survey?

38. Playground bias. The survey described in Exercise 36 asked,

Many people believe this playground is too small and in need of repair. Do you think the playground should be repaired and expanded even if that means imposing an entrance fee to the park?

Describe two ways this question may lead to response bias.

39. (Possibly) Biased questions. Examine each of the following questions for possible bias. If you think the question is biased, indicate how and propose a better question.

a) Should companies that pollute the environment be compelled to pay the costs of cleanup?
b) Should a company enforce a strict dress code?

40. More possibly biased questions. Examine each of the following questions for possible bias. If you think the question is biased, indicate how and propose a better question.

a) Do you think that price or quality is more important in selecting a tablet computer?
b) Given humanity's great tradition of exploration, do you favor continued funding for space flights?

41. Phone surveys. Anytime we conduct a survey, we must take care to avoid undercoverage. Suppose we plan to select 500 names from the city phone book, call their homes between noon and 4 p.m., and interview whoever answers, anticipating contacts with at least 200 people.

a) Why is it difficult to use a simple random sample here?
b) Describe a more convenient, but still random, sampling strategy.
c) What kinds of households are likely to be included in the eventual sample of opinion? Who will be excluded?
d) Suppose, instead, that we continue calling each number, perhaps in the morning or evening, until an adult is contacted and interviewed. How does this improve the sampling design?
e) Random-digit dialing machines can generate the phone calls for us. How would this improve our design? Is anyone still excluded?

42. Cell phone survey. What about drawing a random sample only from cell phone exchanges? Discuss the advantages and disadvantages of such a sampling method compared with surveying randomly generated telephone numbers from non–cell phone exchanges. Do you think these advantages and disadvantages have changed over time? How do you expect they'll change in the future?

43. Change. How much change do you have on you right now? Go ahead, count it.

a) How much change do you have?

b) Suppose you check on your change every day for a week as you head for lunch and average the results. What parameter would this average estimate?

c) Suppose you ask 10 friends to average *their* change every day for a week, and you average those 10 measurements. What is the population now? What parameter would this average estimate?

d) Do you think these 10 average change amounts are likely to be representative of the population of change amounts in your class? In your college? In the country? Why or why not?

44. Fuel economy. Occasionally, when I fill my car with gas, I figure out how many miles per gallon my car got. I wrote down those results after six fill-ups in the past few months. Overall, it appears my car gets 28.8 miles per gallon.

a) What statistic have I calculated?

b) What is the parameter I'm trying to estimate?

c) How might my results be biased?

d) When the Environmental Protection Agency (EPA) checks a car like mine to predict its fuel economy, what parameter is it trying to estimate?

45. Accounting. Between quarterly audits, a company likes to check on its accounting procedures to address any problems before they become serious. The accounting staff processes payments on about 120 orders each day. The next day, the supervisor rechecks 10 of the transactions to be sure they were processed properly.

a) Propose a sampling strategy for the supervisor.

b) How would you modify that strategy if the company makes both wholesale and retail sales, requiring different bookkeeping procedures?

46. Happy workers? A manufacturing company employs 14 project managers, 48 foremen, and 377 laborers. In an effort to keep informed about any possible sources of employee discontent, management wants to conduct job satisfaction interviews with a simple random sample of employees every month.

a) Do you see any danger of bias in the company's plan? Explain.

b) How might you select a simple random sample?

c) Why do you think a simple random sample might not provide the best estimate of the parameters the company wants to estimate?

d) Propose a better sampling strategy.

e) Listed below are the last names of the project managers. Use random numbers to select two people to be interviewed. Be sure to explain your method carefully.

Barrett	Bowman	Chen
DeLara	DeRoos	Grigorov
Maceli	Mulvaney	Pagliarulo
Rosica	Smithson	Tadros
Williams	Yamamoto	

47. Quality control. Sammy's Salsa, a small local company, produces 20 cases of salsa a day. Each case contains 12 jars and is imprinted with a code indicating the date and batch number. To help maintain consistency, at the end of each day, Sammy selects three bottles of salsa, weighs the contents, and tastes the product. Help Sammy select the sample jars. Today's cases are coded 07N61 through 07N80.

a) Carefully explain your sampling strategy.

b) Show how to use random numbers to pick the three jars for testing.

c) Did you use a simple random sample? Explain.

48. Fish quality. Concerned about reports of discolored scales on fish caught downstream from a newly sited chemical plant, scientists set up a field station in a shoreline public park. For one week they asked fishermen there to bring any fish they caught to the field station for a brief inspection. At the end of the week, the scientists said that 18% of the 234 fish that were submitted for inspection displayed the discoloration. From this information, can the researchers estimate what proportion of fish in the river have discolored scales? Explain.

49. Sampling methods. Consider each of these situations. Do you think the proposed sampling method is appropriate? Explain.

a) We want to know what percentage of local doctors accept Medicaid patients. We call the offices of 50 doctors randomly selected from local Yellow Pages listings.

b) We want to know what percentage of local businesses anticipate hiring additional employees in the upcoming month. We randomly select a page in the Yellow Pages and call every business listed there.

50. More sampling methods. Consider each of these situations. Do you think the proposed sampling method is appropriate? Explain.

a) We want to know if business leaders in the community support the development of an "incubator" site at a vacant lot on the edge of town. We spend a day phoning local businesses in the phone book to ask whether they'd sign a petition.

b) We want to know if travelers at the local airport are satisfied with the food available there. We go to the airport on a busy day and interview every tenth person in line in the food court.

JUST CHECKING ANSWERS

1 a) It can be hard to reach all members of a population, and it can take so long that circumstances change, affecting the responses. A well-designed sample is often a better choice.

b) This sample is probably biased—people who didn't like the food at the restaurant might not choose to eat there.

c) No, only the sample size matters, not the fraction of the overall population.

d) Students who frequent this website might be more enthusiastic about Statistics than the overall population of Statistics students. A large sample cannot compensate for bias.

e) It's the population "parameter." "Statistics" describe samples.

2 a) systematic

b) stratified

c) simple

d) cluster

9

Sampling Distributions and Confidence Intervals for Proportions

Marketing Credit Cards: The MBNA Story

When Delaware substantially raised its interest rate ceiling in 1981, banks and other lending institutions rushed to establish corporate headquarters there. One of these was the Maryland Bank National Association, which established a credit card branch in Delaware using the acronym MBNA. Starting in 1982 with 250 employees in a vacant supermarket in Ogletown, Delaware, MBNA grew explosively in the next two decades.

One of the reasons for this growth was MBNA's use of affinity groups—issuing cards endorsed by alumni associations, sports teams, interest groups, and labor unions, among others. MBNA sold the idea to these groups by letting them share a small percentage of the profit. By 2006, MBNA had become Delaware's largest private employer. At its peak, MBNA had more than 50 million cardholders and had outstanding credit card loans of $82.1 billion, making MBNA the third-largest U.S. credit card bank.

"In American corporate history, I doubt there are many companies that burned as brightly, for such a short period of time, as MBNA," said Rep. Mike Castle, R-Del.[1] MBNA was bought by Bank of America in 2005 for $35 billion. Bank of America kept the brand briefly before issuing all cards under its own name in 2007.

---

[1] Delaware *News Online*, January 1, 2006.

nlike the early days of the credit card industry when MBNA established itself, the environment today is intensely competitive, with companies constantly looking for ways to attract new customers and to maximize the profitability of the customers they already have. Many of the large companies have millions of customers, so instead of trying out a new idea with all their customers, they almost always conduct a pilot study or trial first, conducting a survey or an experiment on a sample of their customers.

Credit card companies make money on their cards in three ways: they earn a percentage of every transaction, they charge interest on balances that are not paid in full, and they collect fees (yearly fees, late fees, etc.). To generate all three types of revenue, the marketing departments of credit card banks constantly seek ways to encourage customers to increase the use of their cards.

A marketing specialist at one company has an idea of offering double air miles to their customers with an airline-affiliated card if they increase their spending by at least $800 in the month following the offer. Of course, offering double miles is not free. The company has to pay the airline for the added miles they give away. Her finance department tells her that if 20% of all customers increase spending by $800 then, based on past behavior, the double miles offer will be profitable. Unfortunately, she can't know what *all* customers will do until it's too late. So, she decides to send the offer to a random sample of 1000 customers. In that sample, she finds that 211 (21.1%) of the cardholders increase their spending by more than the required $800. Is that good enough? Could another sample of 1000 *different* people show 19.5%? If results vary from sample to sample how can we make good decisions? Variation like this is sometimes called **sampling error** even though no error has been committed. A better name for this variation that you'd expect to see from sample to sample might be **sampling variability**.

Even though we can't control this variability we can *predict* exactly how much different proportions will vary from sample to sample. This will enable us to make sound business decisions based on a single sample.

WHO	Cardholders of a bank's credit card
WHAT	Proportion of cardholders who increase their spending by at least $800 in the subsequent month
WHEN	Now
WHERE	United States
WHY	To predict costs and benefits of a program offer

9.1   The Distribution of Sample Proportions

NOTATION ALERT

We use p for the proportion in the population and $\hat{p}$ for the observed proportion in a sample. We'll also use q for the proportion of failures ($q = 1 - p$), and $\hat{q}$ for its observed value, just to simplify some formulas.

One way to understand sampling variability is to draw many samples of 1000 individuals using the same population proportion. And a computer is an ideal tool to help us *simulate* this random sampling. Figure 9.1 shows a histogram of the proportions of "success" (cardholders who increased spending by at least $800) in 2000 computer-generated samples where we set the true proportion, p, to be 0.21.

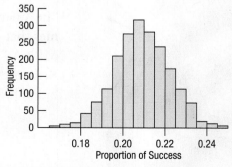

Figure 9.1   The distribution of 2000 sample values of $\hat{p}$, from simulated samples of size 1000 drawn from a population in which the true p is 0.21.

9.2 The Sampling Distribution for Proportions

Imagine

We see only the sample we actually drew. If we *imagine* the results of all the other possible samples we could have drawn (by modeling or simulating them), we can learn more.

The distribution of the sample proportion (shown in Figure 9.1), which we simulated for one situation is called the **sampling distribution** of the proportion. In general, the distribution of any statistic found over many independent samples is called a sampling distribution.

When we examine this distribution, we see that the middle 68% of the values extend from 0.188 to 0.233. That can give us some sense of the region in which the true proportion (which we don't know) is likely to be. But we can do better. If we knew the standard deviation of this distribution, we could make more precise statements about where the true proportion might be. By using some mathematics, we can be even more precise than by using a simulation, or a picture.

The Sampling Distribution for a Proportion

We have now answered the question raised at the start of the chapter. To discover how variable a sample proportion is, we need to know the proportion and the size of the sample. That's all.

The histogram in Figure 9.1 shows a simulation of the **sampling distribution** of $\hat{p}$. We know that the distribution of a Binomial random variable can be approximated by a Normal model when the sample size is large. Proportions are basically Binomials, so we can use the Normal to model the sampling distribution. The mean is naturally at the true proportion in the population and the standard deviation is:

$$SD(\hat{p}) = \sqrt{\frac{p(1-p)}{n}} = \sqrt{\frac{pq}{n}}.$$

So, we can say that the sampling distribution of a sample proportion from a sample of size n with true proportion p is Normal with mean p and standard deviation $\sqrt{\frac{pq}{n}}$. Here's a picture of that sampling distribution model:

Effect of Sample Size

Because n is in the denominator of $SD(\hat{p})$, the larger the sample, the smaller the standard deviation. We need a small standard deviation to make sound business decisions, but larger samples cost more. That tension is a fundamental issue in Statistics.

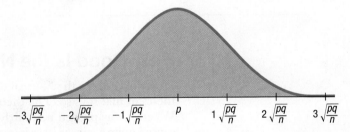

Figure 9.2 A Normal model centered at p with a standard deviation of $\sqrt{\frac{pq}{n}}$ is a good model for a collection of proportions found for many random samples of size n from a population with success probability p.

The Sampling Distribution Model for a Proportion

Provided that the sampled values are independent and the sample size is large enough, the sampling distribution of $\hat{p}$ is modeled by a Normal model with mean $\mu(\hat{p}) = p$ and standard deviation $SD(\hat{p}) = \sqrt{\frac{pq}{n}}.$

FOR EXAMPLE The distribution of a sample proportion

A supermarket has installed "self-checkout" stations that allow customers to scan and bag their own groceries. These are popular, but because customers occasionally encounter a problem, a staff member must be available to help out. The manager wants to estimate what proportion of customers need help so that he can optimize the number of self-check stations per staff member. He collects data from the stations for

(continued)

30 days, recording the proportion of customers on each day that need help and makes a histogram of the observed proportions.

QUESTIONS

1. If the proportion needing help is independent from day to day, what shape would you expect his histogram to follow?
2. Is the assumption of independence reasonable?

ANSWERS

1. Normal, centered at the true proportion.
2. Possibly not. For example, shoppers on weekends might be less experienced than regular weekday shoppers and would then need more help.

JUST CHECKING

1 You want to poll a random sample of 100 shopping mall customers about whether they like the proposed location for the new coffee shop on the third floor, with a panoramic view of the food court. Of course, you'll get just one number, your sample proportion, $\hat{p}$. But if you imagined all the possible samples of 100 customers you could draw and imagined the histogram of all the sample proportions from these samples, what shape would it have?

2 Where would the center of that histogram be?

3 If you think that about half the customers are in favor of the plan, what would the standard deviation of the sample proportions be?

How Good Is the Normal Model?

How do we know the Normal model works? It turns out that there is a mathematical theorem that says so. The theorem is called the **Central Limit Theorem** (CLT). When it was proved in a fairly general form in 1810 by Piere-Simon Laplace, it caused quite a stir (at least in mathematics circles) because it was so unexpected. The theorem applies to both proportions and means (so we'll discuss it at greater length in Chapter 11 when we deal with means). For our purposes here, the Central Limit Theorem says formally, just what we've observed: the distribution of proportions of many independent random samples gets closer and closer to a Normal model as the sample size grows.

One important consequence of the CLT is that we don't really need to draw repeated samples from the population or even simulate doing so to get an idea of the sampling distribution. The CLT tells us that we can simply use the Normal model.

There must be a catch. Suppose the samples were of size 2, for example. Then the only possible numbers of successes could be 0, 1, or 2, and the proportion values would be 0, 0.5, and 1. There's no way the histogram could ever look like a Normal model with only three possible values for the variable (Figure 9.3).

Well, there *is* a slight catch. The claim is only approximately true. But, the model becomes a better and better representation of the distribution of the sample proportions as the sample size gets bigger.[2] That's one reason we require np and nq to be at least 10. But the distributions of proportions from samples of the size you're likely to see in business do have histograms that are remarkably close to a Normal model.

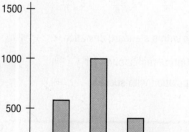

Figure 9.3 Proportions from samples of size 2 can take on only three possible values. A Normal model does not work well here.

---

[2]Formally, we say the claim is true in the limit as the sample size (n) grows.

FOR EXAMPLE   Sampling distribution for proportions

Time Warner provides cable, phone, and Internet services to customers, some of whom subscribe to "packages" including several services. Nationwide, suppose that 30% of their customers are "package subscribers" and subscribe to all three types of service. A local representative in Phoenix, Arizona, wonders if the proportion in his region is the same as the national proportion.

QUESTIONS If the same proportion holds in his region and he takes a survey of 100 customers at random from his subscriber list:

1. What proportion of customers would you expect to be package subscribers?
2. What is the standard deviation of the sample proportion?
3. What shape would you expect the sampling distribution of the proportion to have?
4. Would you be surprised to find out that in a sample of 100, 49 of the customers are package subscribers? Explain. What might account for this high percentage?

ANSWERS

1. Because 30% of customers nationwide are package subscribers, we would expect the same for the sample proportion.
2. The standard deviation is $SD(\hat{p}) = \sqrt{\dfrac{pq}{n}} = \sqrt{\dfrac{(0.3)(0.7)}{100}} = 0.046$.
3. Normal.
4. 49 customers results in a sample proportion of 0.49. The mean is 0.30 with a standard deviation of 0.046. This sample proportion is more than 4 standard deviations higher than the mean: $\dfrac{(0.49 - 0.30)}{0.046} = 4.13$. It would be very unusual to find such a large proportion in a random sample. Either it is a very unusual sample, or the proportion in his region is not the same as the national average.

Assumptions and Conditions

Most models are useful only when specific assumptions are true. For the Normal model to work as the CLT says it should, there are only two assumptions:

> **Independence Assumption:** The sampled values must be *independent* of each other.

> **Sample Size Assumption:** The sample size, *n*, must be *large* enough.

Of course, the best we can do with assumptions is to think about whether they are likely to be true, and we should do so. However, we often can check corresponding *conditions* that provide information about the assumptions as well. Think about the Independence Assumption and check the following corresponding conditions before using the Normal model to model the distribution of sample proportions:

> **Randomization Condition:** If your data come from an experiment, subjects should have been randomly assigned to treatments. If you have a survey, your sample should be a simple random sample of the population. If some other sampling design was used, be sure the sampling method was not biased and that the data are representative of the population.

> **10% Condition:** If the sample is a large fraction of the population, then the independence assumption won't be satisfied. Usually, we don't worry about this, but if you are sampling more than 10% of a population, you may want to use an adjusted formula for the standard deviation.

Success/Failure Condition: The Central Limit Theorem says that the sample size must be big enough for the approximation to work well. We'll take that to mean that both the number of "successes," np, and the number of "failures," nq, are expected to be at least 10.[3] Expressed without the symbols, this condition just says that we need to expect at least 10 successes and at least 10 failures to have enough data for sound conclusions. For the bank's credit card promotion example, we labeled as a "success" a cardholder who increases monthly spending by at least $800 during the trial. The bank observed 211 successes and 789 failures. Both are at least 10, so there are certainly enough successes and enough failures for the condition to be satisfied.[4]

FOR EXAMPLE   Assumptions and conditions for sample proportions

The analyst conducting the Time Warner survey says that, unfortunately, only 20 of the customers he tried to contact actually responded, but that of those 20, 8 are package subscribers.

QUESTIONS

1. If the proportion of package subscribers in his region is 0.30, how many package subscribers, on average, would you expect in a sample of 20?
2. Would you expect the shape of the sampling distribution of the proportion to be Normal? Explain.

ANSWERS

1. You would expect $0.30 \times 20 = 6$ package subscribers.
2. No. Because 6 is less than 10, we should be cautious in using the Normal as a model for the sampling distribution of proportions. (The number of *observed* successes, 8, is also less than 10.)

GUIDED EXAMPLE   Foreclosures

An analyst at a home loan lender was looking at a package of 90 mortgages that the company had recently purchased in central California. The analyst was aware that in that region about 13% of the homeowners with current mortgages will default on their loans in the next year and the houses will go into foreclosure. In deciding to buy the collection of mortgages, the finance department assumed that no more than 15 of the mortgages would go into default. Any amount above that will result in losses for the company. In the package of 90 mortgages, what's the probability that there will be more than 15 foreclosures?

PLAN	**Setup** State the objective of the study.	We want to find the probability that in a group of 90 mortgages, more than 15 will default. Since 15 out of 90 is 16.7%, we need the probability of finding more than 16.7% defaults out of a sample of 90, if the proportion of defaults is 13%.

---

[3] We saw where the 10 came from in the Math Box on page 222.
[4] The Success/Failure condition is about the number of successes and failures we *expect*, but if the number of successes and failures that *occurred* is ≥ 10, then you can use that.

We could simulate many samples of 90 from a population with $p = 0.130$, but it is more direct to trust the CLT and use the Normal model.

Model Check the conditions.

✓ **Independence Assumption.** If the mortgages come from a wide geographical area, one homeowner defaulting should not affect the probability that another does. However, if the mortgages come from the same neighborhood(s), the independence assumption may fail and our estimates of the default probabilities may be wrong.

✓ **Randomization Condition.** For the question asked, these 90 mortgages in the package can be considered as a random sample of mortgages in the region. If there are too many failures, we may doubt that they are a representative sample.

✓ **10% Condition.** The 90 mortgages are less than 10% of the population.

✓ **Success/Failure Condition**

$$np = 90(0.13) = 11.7 \geq 10$$
$$nq = 90(0.87) = 78.3 \geq 10$$

State the parameters and the sampling distribution model.

The population proportion is $p = 0.13$. The conditions are satisfied, so we'll model the sampling distribution of $\hat{p}$ with a Normal model, with mean 0.13 and standard deviation

$$SD(\hat{p}) = \sqrt{\frac{pq}{n}} = \sqrt{\frac{(0.13)(0.87)}{90}} \approx 0.035.$$

Our model for $\hat{p}$ is $N(0.13, 0.035)$. We want to find $P(\hat{p} > 0.167)$.

Plot Make a picture. Sketch the model and shade the area we're interested in, in this case the area to the right of 16.7%.

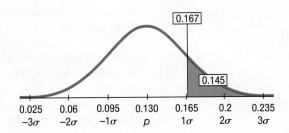

0.025	0.06	0.095	0.130	0.165	0.2	0.235
-3σ	-2σ	-1σ	p	1σ	2σ	3σ

DO

Mechanics Use the standard deviation as a ruler to find the z-score of the cutoff proportion. Find the resulting probability from a table, a computer program, or a calculator.

$$z = \frac{\hat{p} - p}{SD(\hat{p})} = \frac{0.167 - 0.13}{0.035} = 1.06$$

$$P(\hat{p} > 0.167) = P(z > 1.06) = 0.1446$$

REPORT

Conclusion Interpret the probability in the context of the question.

MEMO

Re: Mortgage defaults

Assuming that the 90 mortgages we recently purchased are a random sample of mortgages in this region, there is about a 14.5% chance that we will exceed the 15 foreclosures that Finance has determined as the break-even point.

9.3   A Confidence Interval for a Proportion

NOTATION ALERT

Remember that $\hat{p}$ is our sample estimate of the true proportion p. Recall also that q is just shorthand for $1 - p$, and $\hat{q} = 1 - \hat{p}$.

To plan their inventory and production needs, businesses use a variety of forecasts about the economy. One important attribute is consumer confidence in the overall economy. Tracking changes in consumer confidence over time can help businesses gauge whether the demand for their products is on an upswing or about to experience a downturn. The Gallup Poll periodically asks a random sample of U.S. adults whether they think economic conditions are getting better, getting worse, or staying about the same. When Gallup polled 3559 respondents in April 2013 (during the week ending April 21), only 1495 thought economic conditions in the United States were getting better—a sample proportion of $\hat{p} = 1495/3559 = 42\%$. We (and Gallup) hope that this observed proportion is close to the population proportion, p, but we know that a second sample of 3559 adults wouldn't have a sample proportion of exactly 42.0%. In fact, Gallup did sample another group of adults just a few days later and found a slightly different sample proportion.

What can we say about consumer confidence in the entire population when the proportion that we measure keeps bouncing around from sample to sample? That's where the sampling distribution model can help. By knowing how much they vary and the shape of their distribution, we'll get a clearer idea of where the true proportion might be and how much we know about it. So, what do we know about our sampling distribution model? We know that it's centered at the true proportion, p, of all U.S. adults who think the economy is improving. But we don't know p. It probably isn't 42.0%. That's the $\hat{p}$ from our sample. What we do know is that the sampling distribution model of $\hat{p}$ is centered at p, and we know that the standard deviation of the sampling distribution is $\sqrt{\dfrac{pq}{n}}$. We also know that the shape of the sampling distribution is approximately Normal, when the sample is large enough.

This is all fine in the model world, but we need to solve problems in the real world. In the real world, we don't know p. (If we did, we wouldn't have bothered to take a sample.) And so, we don't know $\sqrt{pq/n}$ either. But, we'll do the best we can and *estimate* it by using $\sqrt{\hat{p}\hat{q}/n}$. That may not seem like a big deal, but it gets a special name. Whenever we estimate the standard deviation of a sampling distribution, we call it a **standard error (SE)**. Using $\hat{p}$, we find the standard error:

$$SE(\hat{p}) = \sqrt{\frac{\hat{p}\hat{q}}{n}} = \sqrt{\frac{(0.42)(1 - 0.42)}{3559}} = 0.008$$

Now, we use that to draw our best guess of the sampling distribution for the true proportion who think the economy is getting better as shown in Figure 9.4.

Figure 9.4 The sampling distribution of sample proportions from samples of size 3559 is centered at the true proportion, p, with a standard deviation of 0.008.

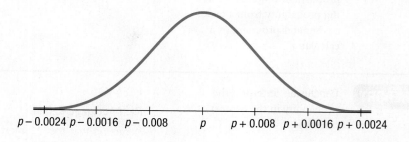

$$p - 0.0024 \quad p - 0.0016 \quad p - 0.008 \quad\quad p \quad\quad p + 0.008 \quad p + 0.0016 \quad p + 0.0024$$

Because the sampling distribution is Normal, we expect that about 68% of all samples of 3559 U.S. adults taken in April 2013 would have sample proportions within 1 standard deviation of p. And about 95% of all these samples will have proportions within $p \pm 2$ SEs. But where is *our* sample proportion in this picture? And what value does p really have? We still don't know!

We do know that for 95% of random samples, $\hat{p}$ will be no more than 2 SEs away from p. So here's the key to using sampling distributions. Let's reverse it and look at it from $\hat{p}$'s point of view. If I'm $\hat{p}$, there's a 95% chance that p is no more than 2 SEs away from me. If I reach out 2 SEs, or 2×0.008, away from me on both sides, I'm 95% sure that p will be within my grasp.

Figure 9.5 Reaching out 2 *SEs* on either side of $\hat{p}$ makes us 95% confident we'll trap the true proportion, p.

What Can We Say about a Proportion?

So what can we really say about p? Of course, I'm not *sure* that my interval catches p. And I don't know its true value, but I can state a probability that I've covered the true value in an interval. Here's a list of things we'd like to be able to say and the reasons we can't say most of them:

1. **"42.0% of *all* U.S. adults thought the economy was improving."** It would be nice to be able to make absolute statements about population values with certainty, but we just don't have enough information to do that. There's no way to be sure that the population proportion is the same as the sample proportion; in fact, it almost certainly isn't. Observations vary. Another sample would yield a different sample proportion.

2. **"It is *probably* true that 42.0% of all U.S. adults thought the economy was improving."** No. In fact, we can be pretty sure that whatever the true proportion is, it's not exactly 42.0%, so the statement is not true.

3. **"We don't know exactly what proportion of U.S. adults thought the economy was improving, but we know that it's within the interval 42.0% ± 2 × 0.8%. That is, it's between 40.4% and 43.6%."** This is getting closer, but we still can't be certain. We can't know for sure that the true proportion is in this interval—or in any particular interval.

4. **"We don't know exactly what proportion of U.S. adults thought the economy was improving, but the interval from 40.4% to 43.6% *probably* contains the true proportion."** Close! Now, we've fudged twice—first by giving an interval and second by admitting that we only think the interval "probably" contains the true value.

That last statement is true, but it's a bit wishy-washy. We can tighten it up by quantifying what we mean by "probably." We saw that 95% of the time when we reach out 2 SEs from $\hat{p}$, we capture p, *so we can be 95% confident that this is one of those times.* After putting a number on the probability that this interval covers the true proportion, we've given our best guess of where the parameter is and how certain we are that it's within some range.

5. **"We are 95% confident that between 40.4% and 43.6% of U.S. adults thought the economy was improving."** Statements like this are called **confidence intervals**. They don't tell us everything we might want to know, but they're the best we can do.

Each confidence interval discussed in this book has a name. You'll see many different kinds of confidence intervals in the following chapters. Some will be

"Far better an approximate answer to the right question, . . . than an exact answer to the wrong question."

—John W. Tukey

about more than *one* sample, some will be about statistics other than *proportions*, and some will use models other than the Normal. The interval calculated and interpreted here is an example of a **one-proportion z-interval**.[5] We'll lay out the formal definition in the next few pages.

FOR EXAMPLE   Finding a 95% confidence interval for a proportion

The Chamber of Commerce of a mid-sized city has supported a proposal to change the zoning laws for a new part of town. The new regulations would allow for mixed commercial and residential development. The vote on the measure is scheduled for three weeks from today, and the president of the Chamber of Commerce is concerned that they may not have the majority of votes that they will need to pass the measure. She commissions a survey that asks likely voters if they plan to vote for the measure. Of the 516 people selected at random from likely voters, 289 said they would likely vote for the measure.

QUESTIONS

a. Find a 95% confidence interval for the true proportion of voters who will vote for the measure. (Use the 68–95–99.7% Rule.)

b. What would you report to the president of the Chamber of Commerce?

ANSWERS

a. $\hat{p} = \dfrac{289}{516} = 0.56$ So, $SE(\hat{p}) = \sqrt{\dfrac{\hat{p}\hat{q}}{n}} = \sqrt{\dfrac{(0.56)(0.44)}{516}} = 0.022$

A 95% confidence interval for p can be found from $\hat{p} \pm 2\,SE(\hat{p}) = 0.56 \pm 2(0.022) = (0.516, 0.604)$ or 51.6% to 60.4%.

b. We are 95% confident that the true proportion of voters who plan to vote for the measure is between 51.6% and 60.4%. This assumes that the sample we have is representative of all likely voters.

What Does "95% Confidence" Really Mean?

What do we mean when we say we have 95% confidence that our interval contains the true proportion? Formally, what we mean is that "95% of samples of this size will produce confidence intervals that capture the true proportion." This is correct but a little long-winded, so we sometimes say "we are 95% confident that the true proportion lies in our interval." Our uncertainty is about whether the particular sample we have at hand is one of the successful ones or one of the 5% whose intervals fail to capture the true value. In this chapter, we have seen how proportions vary from sample to sample. If other pollsters had selected their own samples of adults, they would have found some who thought the economy was getting better, but each sample proportion would almost certainly differ from ours. When they each tried to estimate the true proportion, they'd center their confidence intervals at the proportions they observed in their own samples. Each would have ended up with a different interval.

Figure 9.6 shows the confidence intervals produced by simulating 20 samples. The purple dots are the simulated proportions of adults in each sample who thought the economy was improving, and the orange segments show the confidence intervals found for each simulated sample. The green line represents the true percentage of adults who thought the economy was improving. You can see that most of the simulated confidence intervals include the true value—but one

---

[5]In fact, this confidence interval is so standard for a single proportion that you may see it simply called a "confidence interval for the proportion."

Figure 9.6 The horizontal green line shows the true proportion of people in April 2013 who thought the economy was improving. Most of the 20 simulated samples shown here produced 95% confidence intervals that captured the true value, but one missed.

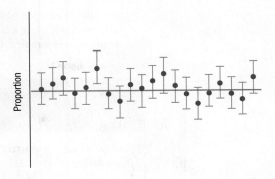

missed. (Note that it is the *intervals* that vary from sample to sample; the green line doesn't move.)

Of course, a huge number of possible samples *could* be drawn, each with its own sample proportion. This simulation approximates just some of them. Each sample can be used to make a confidence interval. That's a large pile of possible confidence intervals, and ours is just one of those in the pile. Did *our* confidence interval "work"? We can never be sure because we'll never know the true proportion of all U.S. adults who thought in April 2013 that the economy was improving. However, the Normal model assures us that 95% of the intervals in the pile are winners, covering the true value, and only 5%, on average, miss the target. That's why we're 95% *confident* that our interval is a winner.

The statements we made about what all U.S. adults thought about the economy were possible because we used a Normal model for the sampling distribution. But is that model appropriate?

As we've seen, all statistical models make assumptions. If those assumptions are not true, the model might be inappropriate, and our conclusions based on it may be wrong. Because the confidence interval is built on the Normal model for the sampling distribution, the assumptions and conditions are the same as those we discussed in Section 9.2. But, because they are so important, we'll go over them again.

You can never be certain that an assumption is true, but you can decide intelligently whether it is reasonable. When you have data, you can often decide whether an assumption is plausible by checking a related condition in the data. However, you'll want to make a statement about the world at large, not just about the data. So the assumptions you make are not just about how the data look, but about how representative they are.

Here are the assumptions and the corresponding conditions to check before creating (or believing) a confidence interval about a proportion.

Independence Assumption

You first need to think about whether the independence assumption is plausible. You can look for reasons to suspect that it fails. You might wonder whether there is any reason to believe that the data values somehow affect each other. (For example, might any of the adults in the sample be related?) This condition depends on your knowledge of the situation. It's not one you can check by looking at the data. However, now that you have data, there are two conditions that you can check:

- **Randomization Condition:** Were the data sampled at random or generated from a properly randomized experiment? Proper randomization can help ensure independence.
- **10% Condition:** Samples are almost always drawn without replacement. Usually, you'd like to have as large a sample as you can. But if you sample from a small population, the probability of success may be different for the last few individuals you draw than it was for the first few. For example, if most of the women have already been sampled, the chance of drawing a woman from the remaining population is lower. If

the sample exceeds 10% of the population, you will have to adjust the margin of error with methods more advanced than those found in this book. But if less than 10% of the population is sampled, it is safe to proceed without adjustment.

FOR EXAMPLE | Assumptions and conditions for a confidence interval for proportions

We previously reported a confidence interval to the president of the Chamber of Commerce.

QUESTION Were the assumptions and conditions for making this interval satisfied?

ANSWER Because the sample was randomized, we assume that the responses of the people surveyed were independent so the randomization condition is met. We assume that 516 people represent fewer than 10% of the likely voters in the town so the 10% condition is met. Because 289 people said they were likely to vote for the measure and thus 227 said they were not, both are much larger than 10 so the Success/Failure condition is also met.

All the conditions to make a confidence interval for the proportion appear to have been satisfied.

9.4 Margin of Error: Certainty vs. Precision

We've just claimed that at a certain confidence level we've captured the true proportion of all U.S. adults who thought the economy was improving in April 2013. Our confidence interval stretched out the same distance on either side of the estimated proportion with the form:

$$\hat{p} \pm 2\,SE(\hat{p}).$$

The *extent* of that interval on either side of $\hat{p}$ is called the **margin of error (ME)**. In general, confidence intervals look like this:

$$estimate \pm ME.$$

The margin of error for our 95% confidence interval was 2 SEs. What if we wanted to be more confident? To be more confident, we'd need to capture p more often, and to do that, we'd need to make the interval wider. For example, if we want to be 99.7% confident, the margin of error will have to be 3 SEs.

The more confident we want to be, the larger the margin of error must be. We can be 100% confident that any proportion is between 0% and 100%, but that's not very useful. Or we could give a narrow confidence interval, say, from 41.98% to 42.02%. But we couldn't be very confident about a statement this precise. Every confidence interval is a balance between certainty and precision.

> **Confidence Intervals**
> We'll see many confidence intervals in this book. All have the form:
>
> estimate $\pm$ ME.
>
> For proportions at 95% confidence:
>
> ME $\approx 2\,SE(\hat{p}).$

Figure 9.7 Reaching out 3 SEs on either side of $\hat{p}$ makes us 99.7% confident we'll trap the true proportion p. Compare the width of this interval with the interval in Figure 9.5.

NEW!!   ACME p-trap: Guaranteed* to capture p.   IMPROVED!!

*Now with 99.7% confidence!

$\hat{p} - 3\,SE$ $\hat{p}$ $\hat{p} + 3\,SE$

The tension between certainty and precision is always there. There is no simple answer to the conflict. Fortunately, in most cases we can be both sufficiently certain and sufficiently precise to make useful statements. The choice of confidence level is somewhat arbitrary, but you must choose the level yourself. The data can't do it for you. The most commonly chosen confidence levels are 90%, 95%, and 99%, but any percentage can be used. (In practice, though, using something like 92.9% or 97.2% might be viewed with suspicion.)

Garfield © 1999 Jim Davis/Distributed by Universal Uclick. Reprinted with permission. All rights reserved.

Critical Values

NOTATION ALERT

We put an asterisk on a letter to indicate a critical value. We usually use "z" when we talk about Normal models, so z^* is always a critical value from a Normal model.

Some common confidence levels and their associated critical values:

CI	z^*
90%	1.645
95%	1.960
99%	2.576

Figure 9.8 For a 90% confidence interval, the critical value is 1.645 because for a Normal model, 90% of the values fall within 1.645 standard deviations of the mean.

In our opening example, our margin of error was 2 SEs, which produced a 95% confidence interval. To change the confidence level, we'll need to change the *number* of SEs to correspond to the new level. A wider confidence interval means more confidence. For any confidence level, the number of SEs we must stretch out on either side of $\hat{p}$ is called the **critical value**. Because it is based on the Normal model, we denote it z^*. For any confidence level, we can find the corresponding critical value from a computer, a calculator, or a Normal probability table, such as Table Z in the back of the book.

For a 95% confidence interval, the precise critical value is $z^* = 1.96$. That is, 95% of a Normal model is found within ± 1.96 standard deviations of the mean. We've been using $z^* = 2$ from the 68–95–99.7 Rule because 2 is very close to 1.96 and is easier to remember. Usually, the difference is negligible, but if you want to be more precise, use 1.96.[6]

Suppose we could be satisfied with 90% confidence. What critical value would we need? We can use a smaller margin of error. Our greater precision is offset by our acceptance of being wrong more often (that is, having a confidence interval that misses the true value). Specifically, for a 90% confidence interval, the critical value is only 1.645 because for a Normal model, 90% of the values are within 1.645 standard deviations from the mean (Figure 9.8). By contrast, suppose your

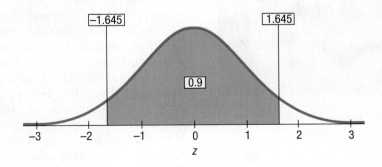

---

[6]It's been suggested that since 1.96 is both an unusual value and so important in Statistics, you can recognize someone who's had a Statistics course by just saying "1.96" and seeing whether they react.

boss demands more confidence. If she wants an interval in which she can have 99% confidence, she'll need to include values within 2.576 standard deviations, creating a wider confidence interval.

One-Proportion z-Interval

When the conditions are met, we are ready to find the confidence interval for the population proportion, p. The confidence interval is $\hat{p} \pm z^* \times SE(\hat{p})$, where the standard deviation of the proportion is estimated by $SE(\hat{p}) = \sqrt{\dfrac{\hat{p}\hat{q}}{n}}$.

FOR EXAMPLE   Finding confidence intervals for proportions with different levels of confidence

The president of the Chamber of Commerce is worried that 95% confidence is too low and wants a 99% confidence interval.

QUESTION   Find a 99% confidence interval. Would you reassure her that the measure will pass? Explain.

ANSWER   In the example on page 274, we used 2 as the value of z^* for 95% confidence. A more precise value would be 1.96 for 95% confidence. For 99% confidence, the critical z-value is 2.576. So, a 99% confidence interval for the true proportion is

$$\hat{p} \pm 2.576 \, SE(\hat{p}) = 0.56 \pm 2.576(0.022) = (0.503, 0.617)$$

The confidence interval is now wider: 50.3% to 61.7%.

The Chamber of Commerce needs at least 50% for the vote to pass. At a 99% confidence level, it looks as if the measure will pass. However, we must assume that the sample is representative of the voters in the actual election and that people vote in the election as they said they will when they took the survey.

GUIDED EXAMPLE   Public Opinion

In March of 2013, workers in the greeting card company Edit66, based in the southern French town of Cabestany, took their bosses hostage. Company chiefs Paul Denis and Merthus Bezemer had informed employees who were to be laid off that they would not receive severance pay that they are legally entitled to. The workers refused to allow their bosses to leave the premises. The town's mayor Jean Vila supported the action (www.english.rfi .fr/economy/20130329-greeetings-card-workers-kidnap-bosses-over-unpaid-layoff-pay). There had been a number of similar "bossnappings" in France in 2009. Incidents occurred at SONY, 3M, and Caterpillar plants in France. A poll taken by *Le Parisien* in April 2009 found 45% of the French "supportive" of such action. A similar poll taken by *Paris Match*, April 2–3, 2009, found 30% "approving" and 63% were "understanding" or "sympathetic" of the action. Only 7% condemned the practice of "bossnapping."

WHO	Adults in France
WHAT	Proportion who sympathize with the practice of bossnapping
WHEN	April 2–3, 2009
WHERE	France
HOW	1010 adults were randomly sampled by the French Institute of Public Opinion (l'Ifop) for the magazine *Paris Match*
WHY	To investigate public opinion of bossnapping

The *Paris Match* poll was based on a random representative sample of 1010 adults. What can we conclude about the proportion of all French adults who sympathize with (without supporting outright) the practice of bossnapping?

To answer this question, we'll build a confidence interval for the proportion of all French adults who sympathize with the practice of bossnapping. As with other procedures, there are three steps to building and summarizing a confidence interval for proportions: Plan, Do, and Report.

PLAN

Setup State the context of the question.

Identify the *parameter* you wish to estimate. Identify the *population* about which you wish to make statements.

Choose and state a confidence level.

Model Think about the assumptions and check the conditions to decide whether we can use the Normal model.

We want to find an interval that is likely with 95% confidence to contain the true proportion, p, of French adults who sympathize with the practice of bossnapping. We have a random sample of 1010 French adults, with a sample proportion of 63%.

✓ **Independence Assumption:** A French polling agency, l'Ifop, phoned a random sample of French adults. It is unlikely that any respondent influenced another.

✓ **Randomization Condition:** l'Ifop drew a random sample from all French adults. We don't have details of their randomization but assume that we can trust it.

✓ **10% Condition:** Although sampling was necessarily without replacement, there are many more French adults than were sampled. The sample is certainly less than 10% of the population.

✓ **Success/Failure Condition:**

$n\hat{p} = 1010 \times 0.63 = 636 \geq 10$ and
$n\hat{q} = 1010 \times 0.37 = 374 \geq 10$,

so the sample is large enough.

State the sampling distribution model for the statistic. Choose your method.

The conditions are satisfied, so I can use a Normal model to find a one-proportion z-interval.

DO

Mechanics Construct the confidence interval. First, find the standard error. (Remember: It's called the "standard error" because we don't know p and have to use $\hat{p}$ instead.)

$n = 1010, \hat{p} = 0.63$, so

$$SE(\hat{p}) = \sqrt{\frac{0.63 \times 0.37}{1010}} = 0.015$$

Next, find the margin of error. We could informally use 2 for our critical value, but 1.96 is more accurate.[7]

Because the sampling model is Normal, for a 95% confidence interval, the critical value $z^* = 1.96$. The margin of error is:

$$ME = z^* \times SE(\hat{p}) = 1.96 \times 0.015 = 0.029$$

Write the confidence interval.

So the 95% confidence interval is:

$$0.63 \pm 0.029 \text{ or } (0.601, 0.659).$$

(continued)

---

[7] If you are following along on your calculator and not rounding off (as we have done for this example), you'll get $SE = 0.0151918$ and an ME of 0.0297760.

Reality check	Check that the interval is plausible. We may not have a strong expectation for the center, but the width of the interval depends primarily on the sample size—especially when the estimated proportion is near 0.5.	The confidence interval covers a range of about plus or minus 3%. That's about the width we might expect for a sample size of about 1000 (when $\hat{p}$ is reasonably close to 0.5).
REPORT	**Conclusion** Interpret the confidence interval in the proper context. We're 95% confident that our interval captured the true proportion.	MEMO **Re: Bossnapping survey** The polling agency l'Ifop surveyed 1010 French adults and asked whether they approved, were sympathetic to, or disapproved of recent bossnapping actions. Although we can't know the true proportion of French adults who were sympathetic (without supporting outright), based on this survey we can be 95% confident that between 60.1% and 65.9% of all French adults were. Because this is an ongoing concern, we may want to repeat the survey to obtain more current data. We may also want to keep these results in mind for future corporate public relations.

JUST CHECKING

Think some more about the 95% confidence interval we just created in the guided example for the proportion of French adults who were sympathetic to bossnapping.

4 If we wanted to be 98% confident, would our confidence interval need to be wider or narrower?

5 Our margin of error was about ±3%. If we wanted to reduce it to ±2% without increasing the sample size, would our level of confidence be higher or lower?

6 If the organization had polled more people, would the interval's margin of error have likely been larger or smaller?

9.5　Choosing the Sample Size

Every confidence interval must balance precision—the width of the interval—against confidence. Although it is good to be precise and comforting to be confident, there is a trade-off between the two. A confidence interval that says that the percentage is between 10% and 90% wouldn't be of much use, although you could be quite confident that it covered the true proportion. An interval from 43% to 44% is reassuringly precise, but not if it carries a confidence level of 35%. It's a rare study that reports confidence levels lower than 80%. Levels of 95% or 99% are more common.

The time to decide whether the margin of error is small enough to be useful is when you design your study. Don't wait until you compute your confidence interval. To get a narrower interval without giving up confidence, you need to have less variability in your sample proportion. How can you do that? Choose a larger sample.

Consider a company planning to offer a new service to their customers. Product managers want to estimate the proportion of customers who are likely to purchase this new service to within 3% with 95% confidence. How large a sample do they need?

Let's look at the margin of error:

$$ME = z^* \sqrt{\frac{\hat{p}\hat{q}}{n}}$$

$$0.03 = 1.96 \sqrt{\frac{\hat{p}\hat{q}}{n}}.$$

They want to find n, the sample size. To find n, they need a value for $\hat{p}$. They don't know $\hat{p}$ because they don't have a sample yet, but they can probably guess a value. The worst case—the value that makes the SD (and therefore n) largest—is 0.50, so if they use that value for $\hat{p}$, they'll certainly be safe.

The company's equation, then, is:

$$0.03 = 1.96 \sqrt{\frac{(0.5)(0.5)}{n}}.$$

To solve for n, just multiply both sides of the equation by $\sqrt{n}$ and divide by 0.03:

$$0.03 \sqrt{n} = 1.96 \sqrt{(0.5)(0.5)}$$

$$\sqrt{n} = \frac{1.96 \sqrt{(0.5)(0.5)}}{0.03} \approx 32.67$$

Then square the result to find n:

$$n \approx (32.67)^2 \approx 1067.1$$

That method will probably give a value with a fraction. To be safe, always round up. The company will need at least 1068 respondents to keep the margin of error as small as 3% with a confidence level of 95%.

Unfortunately, bigger samples cost more money and require more effort. Because the standard error declines only with the *square root* of the sample size, to cut the standard error (and thus the ME) in half, you must *quadruple* the sample size.

Generally, a margin of error of 5% or less is acceptable, but different circumstances call for different standards. The size of the margin of error may be a marketing decision or one determined by the amount of financial risk you (or the company) are willing to accept. Drawing a large sample to get a smaller ME, however, can run into trouble. It takes time to survey 2400 people, and a survey that extends over a week or more may be trying to hit a target that moves during the time of the survey. A news event or new product announcement can change opinions in the middle of the survey process.

Keep in mind that the sample size for a survey is the number of respondents, not the number of people to whom questionnaires were sent or whose phone numbers were dialed. Also keep in mind that a low response rate turns any study essentially into a voluntary response study, which is of little value for inferring population values. It's almost always better to spend resources on increasing the response rate than on surveying a larger group. A complete or nearly complete response by a modest-size sample can yield useful results.

Surveys are not the only place where proportions pop up. Credit card banks sample huge mailing lists to estimate what proportion of people will accept a credit card offer. Even pilot studies may be mailed to 50,000 customers or more. Most of these customers don't respond. But in this case, that doesn't make the sample

---

What $\hat{p}$ Should We Use?

Often you'll have an estimate of the population proportion based on experience or perhaps on a previous study. If so, use that value as $\hat{p}$ in calculating what size sample you need. If not, the cautious approach is to use $\hat{p} = 0.5$. That will determine the largest sample necessary regardless of the true proportion. It's the *worst case* scenario.

Why 1000?

Public opinion polls often use a sample size of 1000, which gives an ME of about 3% (at 95% confidence) when p is near 0.5. But businesses and nonprofit organizations often use much larger samples to estimate the response to a direct mail campaign. Why? Because the proportion of people who respond to these mailings is very low, often 5% or even less. An ME of 3% may not be precise enough if the response rate is that low. Instead, an ME like 0.1% would be more useful, and that requires a very large sample size.

smaller. In fact, they did respond in a way—they just said "No thanks." To the bank, the response rate[8] is $\hat{p}$. With a typical success rate below 1%, the bank needs a very small margin of error—often as low as 0.1%—to make a sound business decision. That calls for a large sample, and the bank should take care when estimating the size needed. For our election poll example, we used $p = 0.5$, both because it's safe and because we honestly believed p to be near 0.5. If the bank used 0.5, they'd get an absurd answer. Instead they base their calculation on a value of p that they expect to find from their experience.

How Much of a Difference Can It Make?

A credit card company is about to send out a mailing to test the market for a new credit card. From that sample, they want to estimate the true proportion of people who will sign up for the card nationwide. To be within a tenth of a percentage point, or 0.001 of the true acquisition rate with 95% confidence, how big does the test mailing have to be? Similar mailings in the past lead them to expect that about 0.5% of the people receiving the offer will accept it. Using those values, they find:

$$ME = 0.001 = z^* \sqrt{\frac{pq}{n}} = 1.96 \sqrt{\frac{(0.005)(0.995)}{n}}$$

$$(0.001)^2 = 1.96^2 \frac{(0.005)(0.995)}{n} \Rightarrow n = \frac{1.96^2 (0.005)(0.995)}{(0.001)^2}$$

$$= 19{,}111.96 \; or \; 19{,}112$$

That's a perfectly reasonable size for a trial mailing. But if they had used 0.50 for their estimate of p they would have found:

$$ME = 0.001 = z^* \sqrt{\frac{pq}{n}} = 1.96 \sqrt{\frac{(0.5)(0.5)}{n}}$$

$$(0.001)^2 = 1.96^2 \frac{(0.5)(0.5)}{n} \Rightarrow n = \frac{1.96^2 (0.5)(0.5)}{(0.001)^2} = 960{,}400.$$

Quite a different result!

FOR EXAMPLE   Sample size calculations for a confidence interval for a proportion

The President of the Chamber of Commerce in the previous examples is worried that the 99% confidence interval is too wide. Recall that it was $(0.503, 0.617)$, which has a width of 0.114.

QUESTIONS   How large a sample would she need to take to have a 99% interval half as wide? One quarter as wide? What if she wanted a 99% confidence interval that was plus or minus 3 percentage points? How large a sample would she need?

---

[8]Be careful. In marketing studies like this *every* mailing yields a response—"yes" or "no"—and response rate means the success rate, the proportion of customers who accept the offer. That's a different use of the term response rate from the one used in survey response.

ANSWERS Because the formula for the confidence interval is dependent on the inverse of the square root of the sample size:

$$\hat{p} \pm z^* \sqrt{\frac{\hat{p}\hat{q}}{n}},$$

a sample size four times as large will produce a confidence interval *half* as wide. The original 99% confidence interval had a sample size of 516. If she wants it half as wide, she will need about $4 \times 516 = 2064$ respondents. To get it a quarter as wide she'd need $4^2 \times 516 = 8256$ respondents!

If she wants a 99% confidence interval that's plus or minus 3 percentage points, she must calculate

$$\hat{p} \pm z^* \sqrt{\frac{\hat{p}\hat{q}}{n}} = \hat{p} \pm 0.03$$

so

$$2.576 \sqrt{\frac{(0.5)(0.5)}{n}} = 0.03$$

which means that

$$n \approx \left(\frac{2.576}{0.03}\right)^2 (0.5)(0.5) = 1843.27$$

Rounding up, she'd need 1844 respondents. We used 0.5 because we didn't have any information about the election before taking the survey. Using $p = 0.56$ instead would give $n = 1817$.

WHAT CAN GO WRONG?

- **Don't confuse the sampling distribution with the distribution of the sample.** When you take a sample, you always look at the distribution of the values, usually with a histogram, and you may calculate summary statistics. Examining the distribution of the sample like this is wise. But that's not the sampling distribution. The sampling distribution is an imaginary collection of the values that a statistic might have taken for all the random samples—the one you got and the ones that you didn't get. Use the sampling distribution model to make statements about how the statistic varies.

- **Beware of observations that are not independent.** The CLT depends crucially on the assumption of independence. Unfortunately, this isn't something you can check in your data. You have to think about how the data were gathered. Good sampling practice and well-designed randomized experiments ensure independence.

- **Watch out for small samples.** The CLT tells us that the sampling distribution model is Normal if n is large enough. The Success/Failure condition assures us that if we have at least 10 successes and 10 failures, the Normal model will work well for modeling the sampling distribution of the sample proportion. If the population proportion is near 0.50, we could get by with even fewer successes and failures, but the Success/Failure condition is conservative and will protect us no matter how large or small the true underlying proportion happens to be.

Confidence intervals are powerful tools. Not only do they tell us what is known about the parameter value, but—more important—they also tell us what we *don't* know. In order to use confidence intervals effectively, you must be clear about what you say about them.

- **Be sure to use the right language to describe your confidence intervals.** Technically, you should say "I am 95% confident that the interval from 40.4% to 43.6% captures the true proportion of U.S. adults who thought the economy was improving in April 2013." That formal phrasing emphasizes that *your confidence (and your uncertainty) is about the interval, not the true proportion.* But you may choose a more casual phrasing like "I am 95% confident that between 40.4% and 43.6% of U.S. adults thought the economy was improving in April 2013." Because you've made it clear that the uncertainty is yours and you didn't suggest that the randomness is in the true proportion, this is OK. Keep in mind that it's the interval that's random. It's the focus of both our confidence and our doubt.

- **Don't suggest that the parameter varies.** A statement like "there is a 95% chance that the true proportion is between 40.4% and 43.6%" sounds as though you think the population proportion wanders around and sometimes happens to fall between 40.4% and 43.6%. When you interpret a confidence interval, make it clear that *you* know that the population parameter is fixed and that it is the interval that varies from sample to sample.

- **Don't claim that other samples will agree with yours.** Keep in mind that the confidence interval makes a statement about the true population proportion. An interpretation such as "in 95% of samples of U.S. adults the proportion who thought the economy was improving in April 2013 will be between 40.4% and 43.6%" is just wrong. The interval isn't about sample proportions but about the population proportion. There is nothing special about the sample we happen to have; it doesn't establish a standard for other samples.

- **Don't be certain about the parameter.** Saying "between 40.4% and 43.6% of U.S. adults thought the economy was improving in April 2013" asserts that the population proportion cannot be outside that interval. Of course, you can't be absolutely certain of that (just pretty sure).

- **Don't forget: It's about the parameter.** Don't say "I'm 95% confident that $\hat{p}$ is between 40.4% and 43.6%." Of course, you are—in fact, we calculated that our sample proportion was 42.0%. So we already *know* the sample proportion. The confidence interval is about the (unknown) population parameter, p.

- **Don't claim to know too much.** Don't say "I'm 95% confident that between 40.4% and 43.6% of all U.S. adults think the economy is improving." Gallup sampled adults during April 2013, and public opinion shifts over time.

- **Do take responsibility.** Confidence intervals are about *uncertainty*. *You* are the one who is uncertain, not the parameter. You have to accept the responsibility and consequences of the fact that not all the intervals you compute will capture the true value. In fact, about 5% of the 95% confidence intervals you find will fail to capture the true value of the parameter. You *can* say "I am 95% confident that between 40.4% and 43.6% of U.S. adults thought the economy was improving in April 2013."

Confidence intervals and margins of error depend crucially on the assumptions and conditions. When they're not true the results may be invalid. For your own surveys, follow the survey designs from Chapter 8. For surveys you read about, be sure to:

- **Watch out for biased sampling.** Just because we have more statistical machinery now doesn't mean we can forget what we've already learned. A questionnaire that finds that 85% of people enjoy filling out surveys still suffers from nonresponse bias even though now we're able to put confidence intervals around this (biased) estimate.

- **Think about independence.** The assumption that the values in a sample are mutually independent is one that you usually cannot check. It always pays to think about it, though.

- **Be careful of sample size.** The validity of the confidence interval for proportions may be affected by sample size. Avoid using the confidence interval on "small" samples.

ETHICS IN ACTION

Gold Key Agency is a regional real estate brokerage firm that features properties in northern Pennsylvania and southern New York. Ann Sheridan has been with the agency for about five years, working out of its Bradford County, PA, office. One of her current clients, Ben Rhodes, has been looking at a fairly large parcel of land with an old farmhouse to renovate. He seems very interested and she has met with him several times, but he has expressed some concern about gas drilling in the region.

Ann is well aware of how natural gas drilling in the area has affected the real estate business. Large reserves located in the Marcellus shale formation, now accessible as a result of advances in horizontal drilling and hydraulic fracturing or "fracking," have created an economic boom: new jobs, an influx of workers, and prosperity to landowners who have leased to gas drilling companies. At the same time, it has had undesirable consequences.

Because drilling companies are not required by law to disclose the chemicals used in fracking, many fear its potential negative effects on the surrounding environment. Indeed, there is evidence that some property values have actually decreased, particularly those depending on well water. Demonstrations in the media of how well water contaminated by fracking chemicals "ignites" have further heightened anxiety. Moreover, some banks and credit unions are reluctant to grant mortgages on properties leased for gas drilling.

Ann is getting ready to meet yet again with Ben. She really wants to close this deal, so she decides to gather some information to help persuade Ben to make the purchase. Ann collects both the selling price and appraised value for each of 20 properties recently sold by agents in her office. She finds that only one of them sold for below its appraised value. Based on these data, Ann constructs a 95% confidence interval and finds the upper limit to be 15%. That value is small enough to please her, so she doesn't bother to look more closely at whether her method is appropriate for her data. Assuming that one of Ben's concerns may be that property values in the region may decline in the future, she plans to use this figure to reassure him. She will tell him that she is 95% sure that no more than 15% of properties in the area run the risk of selling for less than the appraised value. She hopes this will convince Ben to finally make an offer on the property.

- Identify the ethical dilemma in this scenario.

- What are the undesirable consequences?

- Propose an ethical solution that considers the welfare of all stakeholders.

WHAT HAVE WE LEARNED?

Learning Objectives

Model the variation in statistics from sample to sample with a sampling distribution.

- The sampling distribution of the sample proportion is Normal as long as the sample size is large enough.

Understand that, usually, the mean of a sampling distribution is the value of the parameter estimated.

- For the sampling distribution of $\hat{p}$, the mean is p.

Interpret the standard deviation of a sampling distribution.

- The standard deviation of a sampling model is the most important information about it.

- The standard deviation of the sampling distribution of a proportion is $\sqrt{\dfrac{pq}{n}}$ where $q = 1 - p$.

Construct a confidence interval for a proportion, p, as the statistic, $\hat{p}$, plus and minus a margin of error.

- The margin of error consists of a **critical value** based on the sampling model times a **standard error** based on the sample.
- The critical value is found from the Normal model.
- The standard error of a sample proportion is calculated as $\sqrt{\dfrac{\hat{p}\hat{q}}{n}}$.

Interpret a confidence interval correctly.

- You can claim to have the specified level of confidence that the interval you have computed actually covers the true value.

Understand the importance of the sample size, n, in improving both the certainty (confidence level) and precision (margin of error).

- For the same sample size and proportion, more certainty requires less precision and more precision requires less certainty.

Know and check the assumptions and conditions for finding and interpreting confidence intervals.

- Independence Assumption or Randomization Condition
- 10% Condition
- Success/Failure Condition

Be able to invert the calculation of the margin of error to find the sample size required, given a proportion, a confidence level, and a desired margin of error.

Terms

Central Limit Theorem

The Central Limit Theorem (CLT) states that the sampling distribution model of the sample proportion is approximately Normal for large n, as long as the observations are independent. The theorem also applies to means (see Chapter 11).

Confidence interval

An interval of values usually of the form

$$estimate \pm margin\ of\ error$$

found from data in such a way that a particular percentage of all random samples can be expected to yield intervals that capture the true parameter value.

Critical value	The number of standard errors to move away from the estimate (mean of the sampling distribution) to correspond to the specified level of confidence. The critical value, denoted z^*, is usually found from a table or with technology.
Margin of error (ME)	In a confidence interval, the extent of the interval on either side of the estimate (the observed statistic value). A margin of error is typically the product of a critical value from the sampling distribution and a standard error from the data. A small margin of error corresponds to a confidence interval that pins down the parameter precisely. A large margin of error corresponds to a confidence interval that gives relatively little information about the estimated parameter.
One-proportion z-interval	A confidence interval for the true value of a proportion. The confidence interval is $$\hat{p} \pm z^*SE(\hat{p})$$ where z^* is a critical value from the Standard Normal model corresponding to the specified confidence level and $SE(\hat{p}) = \sqrt{\dfrac{\hat{p}\hat{q}}{n}}.$
Sampling distribution	The distribution of a statistic over many independent samples of the same size from the same population.
Sampling distribution model for a proportion	If the independence assumption and randomization condition are met and we expect at least 10 successes and 10 failures, then the sampling distribution of a proportion is well modeled by a Normal model with a mean equal to the true proportion value, p, and a standard deviation equal to $\sqrt{\dfrac{pq}{n}}.$
Sampling error Sampling variability	The variability we expect to see from sample to sample is often called the sampling error, although sampling variability is a better term.
Standard error (SE)	When the standard deviation of the sampling distribution of a statistic is estimated from the data, the resulting statistic is called a standard error (SE).

TECHNOLOGY HELP: Confidence Intervals for Proportions

Confidence intervals for proportions are so easy and natural that many statistics packages don't offer special commands for them. Most statistics programs want the "raw data" for computations. For proportions, the raw data are the "success" and "failure" status for each case. Usually, these are given as 1 or 0, but they might be category names like "yes" and "no." Often we just know the proportion of successes, $\hat{p}$, and the total count, n. Computer packages don't usually deal with summary data like this easily, but the statistics routines found on many graphing calculators allow you to create confidence intervals from summaries of the data—usually all you need to enter are the number of successes and the sample size.

In some programs you can reconstruct variables of 0's and 1's with the given proportions. But even when you have (or can reconstruct) the raw data values, you may not get *exactly* the same margin of error from a computer package as you would find working by hand. The reason is that some packages make approximations or use other methods. The result is very close but not exactly the same. Fortunately, Statistics means never having to say you're certain, so the approximate result is good enough.

EXCEL

Inference methods for proportions are not part of the standard Excel tool set, but you use Excel's equations to calculate a confidence interval for a proportion in Excel:

	A	B	C	D	E
1	z-Estimate of a Proportion				
2					
3	Sample proportion	0.63	Confidence Interval Estimate		
4	Sample size	1010	0.63	±	0.0298
5	Confidence level	0.95	Lower confidence limit		0.6002
6			Upper confidence limit		0.6598

- Enter the sample proportion in cell B3.
- Enter the sample size in cell B4.
- Enter the confidence level in cell B5.
- In cell C4, type: "=b3".
- In cell E4, type "=NORM.S.INV(0.5+B5/2)*(SQRT(B3*(1-B3)/B4))".

- Type "=C4−E4" in cell E5.
- Type "=C4+E4" in cell E6.

Comments

The method shown here will work for summarized data. When working with raw data, use the COUNTIF function in Excel to quickly count values to compute the sample proportion. You can also use the Pivot Table to quickly summarize the data.

JMP

For a categorical variable that holds category labels, the **Distribution** platform includes tests and intervals for proportions.

For raw data:

- Right-click on the column containing data.
- Identify "Modeling Type" as Nominal.
- Choose **Analyze > Distribution**.
- Select data column as **Y, Columns** and click **OK**.
- Expand menu next to variable name in output.
- Select **Confidence Interval** and choose confidence level.

Comments

JMP uses slightly different methods for proportion inferences than those discussed in this text. Your answers are likely to be slightly different, especially for small samples.

MINITAB

Choose **Basic Statistics** from the **Stat** menu.

- Choose **1Proportion** from the Basic Statistics submenu.
- If the data are category names in a variable, assign the variable from the variable list box to the **Samples in columns** box. If you have

summarized data, click the **Summarized Data** button and fill in the number of trials and the number of successes.

- Click the **Options** button and specify the remaining details. Leave the Alternative as ≠.
- If you have a large sample, check **Use test and interval based on normal distribution.** Click the **OK** button.

Comments

When working from a variable that names categories, MINITAB treats the last category as the "success" category. You can specify how the categories should be ordered.

R

The standard libraries in R do not contain a function for the confidence of a proportion, but a simple function can be written to do so. For example:

```
pconfint=function(phat,n,conf=.95)
    {
        se = sqrt(phat*(1-phat)/n)
        al2 = 1-(1-conf)/2
        zstar = qnorm(al2)
        ul = phat+zstar*se
        ll = phat·zstar*se
        return(c(ll,ul))
    }
```

For example, pconfint(0.3,500) will give a 95% confidence interval for *p* based on 150 successes out of 500.

SPSS

SPSS does not find confidence intervals for proportions.

Brief **Case**

Has Gold Lost Its Luster?

In 2011, when the Gallup organization polled investors, 34% rated gold the best long-term investment. But in April of 2013 Gallup surveyed a random sample of U.S. adults. Respondents were asked to select the best long-term investment from a list of possibilities. Only 241 of the 1005 respondents chose gold as the best long-term investment. By contrast, only 91 choose bonds.

Compute the standard error for each sample proportion. Compute and describe a 95% confidence interval in the context of the question.

Do you think opinions about the value of gold as a long-term investment have really changed from the old 34% favorability rate, or do you think this is just sample variability? Explain.

Forecasting Demand

Utilities must forecast the demand for energy use far into the future because it takes decades to plan and build new power plants. Ron Baker, who worked for New York State Electric and Gas (NYSEG), had the job of predicting the proportion of homes that would choose to use electricity to heat their homes. He was prepared to report a confidence interval for the true proportion, but after seeing his preliminary report, his management demanded a single number as his prediction. Help Ron explain to his management why a confidence interval for the desired proportion would be more useful for planning purposes. Explain how the precision of the interval and the confidence we can have in it are related to each other. Discuss the business consequences of an interval that is too narrow and the consequences of an interval with too low a confidence level.

EXERCISES

SECTION 9.1

1. An investment website can tell what devices are used to access the site. The site managers wonder whether they should enhance the facilities for trading via "smart phones" so they want to estimate the proportion of users who access the site that way (even if they also use their computers sometimes). They draw a random sample of 200 investors from their customers. Suppose that the true proportion of smart phone users is 36%.

a) What would you expect the shape of the sampling distribution for the sample proportion to be?
b) What would be the mean of this sampling distribution?
c) If the sample size were increased to 500, would your answers change? Explain.

2. The proportion of adult women in the United States is approximately 51%. A marketing survey telephones 400 people at random.

a) What proportion of women in the sample of 400 would you expect to see?
b) How many women, on average, would you expect to find in a sample of that size? (*Hint:* Multiply the expected proportion by the sample size.)

3. The investment website of Exercise 1 draws a random sample of 200 investors from their customers. Suppose that the true proportion of smart phone users is 36%.

a) What would the standard deviation of the sampling distribution of the proportion of smart phone users be?
b) What is the probability that the sample proportion of smart phone users is greater than 0.36?
c) What is the probability that the sample proportion is between 0.30 and 0.40?
d) What is the probability that the sample proportion is less than 0.28?
e) What is the probability that the sample proportion is greater than 0.42?

4. The proportion of adult women in the United States is approximately 51%. A marketing survey telephones 400 people at random.

a) What is the sampling distribution of the observed proportion that are women?
b) What is the standard deviation of that proportion?
c) Would you be surprised to find 53% women in a sample of size 400? Explain.
d) Would you be surprised to find 48% women in a sample of size 400? Explain.
e) Would you be surprised to find that there were fewer than 160 women in the sample? Explain.

5. A real estate agent wants to know how many owners of homes worth over $1,000,000 might be considering putting their home on the market in the next 12 months. He surveys 40 of them and finds that 10 of them are considering such a move. Are all the assumptions and conditions for finding the sampling distribution of the proportion satisfied? Explain briefly.

6. A tourist agency wants to know what proportion of visitors to the Eiffel Tower are from the Far East. To find out they survey 100 people in the line to purchase tickets to the top of the tower one Sunday afternoon in May. Are all the assumptions and conditions for finding the sampling distribution of the proportion satisfied? Explain briefly.

7. A marketing researcher for a phone company surveys 100 people and finds that that proportion of clients who are likely to switch providers when their contract expires is 0.15.

a) What is the standard deviation of the sampling distribution of the proportion?
b) If she wants to reduce the standard deviation by half, how large a sample would she need?

8. A market researcher for a provider of iPod accessories wants to know the proportion of customers who own cars to assess the market for a new iPod car charger. A survey of 500 customers indicates that 76% own cars.

a) What is the standard deviation of the sampling distribution of the proportion?
b) How large would the standard deviation have been if he had surveyed only 125 customers (assuming the proportion is about the same)?

SECTION 9.2

9. For each situation below identify the population and the sample and identify p and $\hat{p}$ if appropriate and what the value of $\hat{p}$ is. Would you trust a confidence interval for the true proportion based on these data? Explain briefly why or why not.

a) As concertgoers enter a stadium, a security guard randomly inspects their backpacks for alcoholic beverages. Of the 130 backpacks checked so far, 17 contained alcoholic beverages of some kind. The guards want to estimate the percentage of all backpacks of concertgoers at this concert that contain alcoholic beverages.

b) The website of the English newspaper *The Guardian* asked visitors to the site to say whether they approved of recent "bossnapping" actions by British workers who were outraged over being fired. Of those who responded, 49.2% said "Yes. Desperate times, desperate measures."

c) An airline wants to know the weight of carry-on baggage that customers take on their international routes, so they take a random sample of 50 bags and find that the average weight is 17.3 pounds.

10. For each situation below identify the population and the sample and explain what p and $\hat{p}$ represent and what the value of $\hat{p}$ is. Would you trust a confidence interval for the true proportion based on these data? Explain briefly why or why not.

a) A marketing analyst conducts a large survey of her customers to find out how much money they plan to spend at the company website in the next 6 months. The average amount reported from the 534 respondents is $145.34.

b) A campus survey on a large campus (40,000 students) is trying to find out whether students approve of a new parking policy allowing students to park in previously inaccessible parking lots, but for a small fee. Surveys are sent out by mail and e-mail to all students. Of the 243 surveys returned, 134 are in favor of the change.

c) The human resources department of a large Fortune 100 company wants to find out how many employees would take advantage of an on-site day care facility. They send out an e-mail to a random sample of 500 employees and receive responses from 450 of them. Of those responding, 75 say that they would take advantage of such a facility.

11. A survey of 200 students is selected randomly on a large university campus. They are asked if they use a laptop in class to take notes. The result of the survey is that 70 of the 200 students responded "yes."

a) What is the value of the sample proportion $\hat{p}$?
b) What is the standard error of the sample proportion?
c) Construct an approximate 95% confidence interval for the true proportion p by taking ± 2 SEs from the sample proportion.

12. From a survey of 250 coworkers you find that 155 would like the company to provide on-site day care.

a) What is the value of the sample proportion $\hat{p}$?

b) What is the standard error of the sample proportion?
c) Construct an approximate 95% confidence interval for the true proportion p by taking ± 2 SEs from the sample proportion.

13. From a survey of coworkers you find that 48% of 200 have already received this year's flu vaccine. An approximate 95% confidence interval is (0.409, 0.551). Which of the following are true? If not, explain briefly.

a) 95% of the coworkers fall in the interval (0.409, 0.551).
b) We are 95% confident that the proportion of coworkers who have received this year's flu vaccine is between 40.9% and 55.1%.
c) There is a 95% chance that a random selected coworker has received the vaccine.
d) There is a 48% chance that a random selected coworker has received the vaccine.
e) We are 95% confident that between 40.9% and 55.1% of the samples will have a proportion near 48%.

14. From the survey in Exercise 11, which of the following are true? If they are not true, explain briefly why not.

a) 95% of the 200 students are in the interval (0.283, 0.417).
b) The true proportion of students who use laptops to take notes is captured in the interval (0.283, 0.417) with probability 0.95.
c) There is a 35% chance that a student uses a laptop to take notes.
d) There is a 95% chance that the student uses a laptop to take notes 35% of the time.
e) We are 95% confident that the true proportion of students who use laptops to take notes is captured in the interval (0.283, 0.417).

SECTION 9.3

15. From the survey in Exercise 11,

a) How would the confidence interval change if the confidence level had been 90% instead of 95%?
b) How would the confidence interval change if the sample size had been 300 instead of 200? (Assume the same sample proportion.)
c) How would the confidence interval change if the confidence level had been 99% instead of 95%?
d) How large would the sample size have to be to make the margin of error half as big in the 95% confidence interval?

16. As in Exercise 13, from a survey of coworkers you find that 48% of 200 have already received this year's flu vaccine. An approximate 95% confidence interval is (0.409, 0.551).

a) How would the confidence interval change if the sample size had been 800 instead of 200?
b) How would the confidence interval change if the confidence level had been 90% instead of 95%?
c) How would the confidence interval change if the confidence level had been 99% instead of 95%?

SECTION 9.4

17. Suppose you want to estimate the proportion of traditional college students on your campus who own their own car. You have no preconceived idea of what that proportion might be.

a) What sample size is needed if you wish to be 95% confident that your estimate is within 0.02 of the true proportion?
b) What sample size is needed if you wish to be 99% confident that your estimate is within 0.02 of the true proportion?
c) What sample size is needed if you wish to be 95% confident that your estimate is within 0.05 of the true proportion?

18. As in Exercise 17, you want to estimate the proportion of traditional college students on your campus who own their own car. However, from some research on other college campuses, you believe the proportion will be near 20%.

a) What sample size is needed if you wish to be 95% confident that your estimate is within 0.02 of the true proportion?
b) What sample size is needed if you wish to be 99% confident that your estimate is within 0.02 of the true proportion?
c) What sample size is needed if you wish to be 95% confident that your estimate is within 0.05 of the true proportion?

19. It's believed that as many as 25% of adults over age 50 never graduated from high school. We wish to see if this percentage is the same among the 25 to 30 age group.

a) How many of this younger age group must we survey in order to estimate the proportion of nongrads to within 6% with 90% confidence?
b) Suppose we want to cut the margin of error to 4%. What's the necessary sample size?
c) What sample size would produce a margin of error of 3%?

20. In preparing a report on the economy, we need to estimate the percentage of businesses that plan to hire additional employees in the next 60 days.

a) How many randomly selected employers must we contact in order to create an estimate in which we are 98% confident with a margin of error of 5%?
b) Suppose we want to reduce the margin of error to 3%. What sample size will suffice?
c) Why might it not be worth the effort to try to get an interval with a margin of error of 1%?

CHAPTER EXERCISES

21. Send money. When they send out their fundraising letter, a philanthropic organization typically gets a return from about 5% of the people on their mailing list. To see what the response rate might be for future appeals, they did a simulation using samples of size 20, 50, 100, and 200. For each sample size, they simulated 1000 mailings with success rate $p = 0.05$ and constructed the histogram of the 1000 sample proportions, shown below. Explain what these histograms say about the sampling distribution model for sample proportions. Be sure to talk about shape, center, and spread.

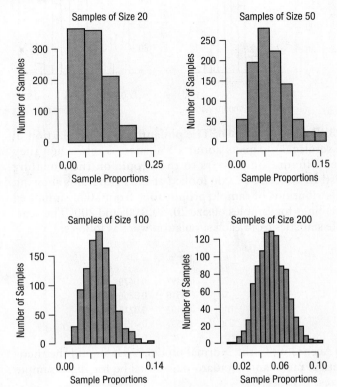

22. Character recognition. An automatic character recognition device can successfully read about 85% of handwritten credit card applications. To estimate what might happen when this device reads a stack of applications, the company did a simulation using samples of size 20, 50, 75, and 100. For each sample size, they simulated 1000 samples with success rate $p = 0.85$ and constructed the histogram of the 1000 sample proportions, shown here. Explain what these histograms say about the sampling distribution model for sample proportions. Be sure to talk about shape, center, and spread.

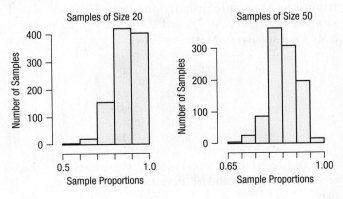

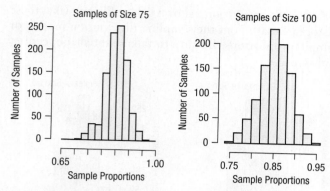

Samples of Size 75   Samples of Size 100

23. Send money, again. The philanthropic organization in Exercise 21 expects about a 5% success rate when they send fundraising letters to the people on their mailing list. In Exercise 21 you looked at the histograms showing distributions of sample proportions from 1000 simulated mailings for samples of size 20, 50, 100, and 200. The sample statistics from each simulation were as follows:

n	Mean	SD
20	0.0497	0.0479
50	0.0516	0.0309
100	0.0497	0.0215
200	0.0501	0.0152

a) According to the Normal model, what should the theoretical mean and standard deviations be for these sample sizes?
b) How close are those theoretical values to what was observed in these simulations?
c) Looking at the histograms in Exercise 21, at what sample size would you be comfortable using the Normal model as an approximation for the sampling distribution?
d) What does the Success/Failure Condition say about the choice you made in part c?

24. Character recognition, again. The automatic character recognition device discussed in Exercise 22 successfully reads about 85% of handwritten credit card applications. In Exercise 22 you looked at the histograms showing distributions of sample proportions from 1000 simulated samples of size 20, 50, 75, and 100. The sample statistics from each simulation were as follows:

n	Mean	SD
20	0.8481	0.0803
50	0.8507	0.0509
75	0.8481	0.0406
100	0.8488	0.0354

a) According to the Normal model, what should the theoretical mean and standard deviations be for these sample sizes?
b) How close are those theoretical values to what was observed in these simulations?

c) Looking at the histograms in Exercise 22, at what sample size would you be comfortable using the Normal model as an approximation for the sampling distribution?
d) What does the Success/Failure Condition say about the choice you made in part c?

25. Stock picking. In a large Business Statistics class, the professor has each person select stocks by throwing 16 darts at pages of the *Wall Street Journal*. They then check to see whether their stock picks rose or fell the next day and report their proportion of "successes." As a lesson, the professor has selected pages of the *Journal* for which exactly half the publicly traded stocks went up and half went down. The professor then makes a histogram of the reported proportions.

a) What shape would you expect this histogram to be? Why?
b) Where do you expect the histogram to be centered?
c) How much variability would you expect among these proportions?
d) Explain why a Normal model should not be used here.

26. Quality management. Manufacturing companies strive to maintain production consistency, but it is often difficult for outsiders to tell whether they have succeeded. Sometimes, however, we can find a simple example. The candy company that makes M&M's candies claims that 10% of the candies it produces are green and that bags are packed randomly. We can check on their production controls by sampling bags of candies. Suppose we open bags containing about 50 M&M's and record the proportion of green candies.

a) If we plot a histogram showing the proportions of green candies in the various bags, what shape would you expect it to have?
b) Can that histogram be approximated by a Normal model? Explain.
c) Where should the center of the histogram be?
d) What should the standard deviation of the proportion be?

27. Bigger portfolio. The class in Exercise 25 expands its stock-picking experiment.

a) The students use computer-generated random numbers to choose 25 stocks each. Use the 68–95–99.7 Rule to describe the sampling distribution model.
b) Confirm that you can use a Normal model here.
c) They increase the number of stocks picked to 64 each. Draw and label the appropriate sampling distribution model. Check the appropriate conditions to justify your model.
d) Explain how the sampling distribution model changes as the number of stocks picked increases.

28. More quality. Would a bigger sample help us to assess manufacturing consistency? Suppose instead of the 50-candy bags of Exercise 26, we work with bags that contain 200 M&M's each. Again we calculate the proportion of green candies found.

a) Explain why it's appropriate to use a Normal model to describe the distribution of the proportion of green M&M's they might expect.
b) Use the 68–95–99.7 Rule to describe how this proportion might vary from bag to bag.
c) How would this model change if the bags contained even more candies?

29. A winning investment strategy? One student in the class of Exercise 25 claims to have found a winning strategy. He watches a cable news show about investing and *during the show* throws his darts at the pages of the *Journal*. He claims that of 200 stocks picked in this manner, 58% were winners.
a) What do you think of his claim? Explain.
b) If there are 100 students in the class, are you surprised that one was this successful? Explain.

30. Even more quality. In a really large bag of M&M's, we found 12% of 500 candies were green. Is this evidence that the manufacturing process is out of control and has made too many greens? Explain.

31. Speeding. State police believe that 70% of the drivers traveling on a major interstate highway exceed the speed limit. They plan to set up a radar trap and check the speeds of 80 cars.
a) Using the 68–95–99.7 Rule, draw and label the distribution of the proportion of these cars the police will observe speeding.
b) Do you think the appropriate conditions necessary for your analysis are met? Explain.

32. Smoking, 2013. The most recent public health statistics available indicate that 19.0% of American adults smoke cigarettes. Using the 68–95–99.7 Rule, describe the sampling distribution model for the proportion of smokers among a randomly selected group of 50 adults. Be sure to discuss your assumptions and conditions.

33. Vision. It is generally believed that nearsightedness affects about 12% of all children. A school district has registered 170 incoming kindergarten children.
a) Can you use the Normal Model to describe the sampling distribution model for the sample proportion of children who are nearsighted? Check the conditions and discuss any assumptions you need to make.
b) Sketch and clearly label the sampling model, based on the 68–95–99.7 Rule.
c) How many of the incoming students might the school expect to be nearsighted? Explain.

34. Mortgages 2013. In early 2013 Realty Trac reported that foreclosures had settled down to 1 in 859 homes per month for a rate of 0.116%, far below the 1.6% seen during the financial crisis of 2007–2008. Suppose a large bank holds 9455 of these mortgages.
a) Can you use the Normal model to describe the sampling distribution model for the sample proportion of foreclosures? Check the conditions and discuss any assumptions you need to make.
b) Sketch and clearly label the sampling model, based on the 68–95–99.7 Rule.
c) How many of these homeowners might the bank expect will default on their mortgages? Explain.

35. Loans. Based on past experience, a bank believes that 7% of the people who receive loans will not make payments on time. The bank has recently approved 200 loans.
a) What are the mean and standard deviation of the proportion of clients in this group who may not make timely payments?
b) What assumptions underlie your model? Are the conditions met? Explain.
c) What's the probability that over 10% of these clients will not make timely payments?

36. Contacts. The campus representative for Lens.com wants to know what percentage of students at a university currently wear contact lens. Suppose the true proportion is 30%.
a) We randomly pick 100 students. Let $\hat{p}$ represent the proportion of students in this sample who wear contacts. What's the appropriate model for the distribution of $\hat{p}$? Specify the name of the distribution, the mean, and the standard deviation. Be sure to verify that the conditions are met.
b) What's the approximate probability that more than one third of this sample wear contacts?

37. Back to school? Best known for its testing program, ACT, Inc., also compiles data on a variety of issues in education. In 2012 the company reported that the national college freshman-to-sophomore retention rate at four-year colleges was about 80.0%. Consider colleges with freshman classes of 400 students. Use the 68–95–99.7 Rule to describe the sampling distribution model for the percentage of those students we expect to return to that school for their sophomore years. Do you think the appropriate conditions are met?

38. Binge drinking. A national study found that 44% of college students engage in binge drinking (5 drinks at a sitting for men, 4 for women). Use the 68–95–99.7 Rule to describe the sampling distribution model for the proportion of students in a randomly selected group of 200 college students who engage in binge drinking. Do you think the appropriate conditions are met?

39. Back to school, again. Based on the 80% national retention rate described in Exercise 37, does a college where 551 of the 603 freshmen returned the next year as sophomores have a right to brag that it has an unusually high retention rate? Explain.

40. Binge sample. After hearing of the national result that 44% of students engage in binge drinking (5 drinks at a sitting for men, 4 for women), a professor surveyed a random sample of 244 students at his college and found that 96 of them admitted to binge drinking in the past week. Should he be surprised at this result? Explain.

41. Polling. Just before a referendum on a school budget, a local newspaper polls 400 voters in an attempt to predict whether the budget will pass. Suppose that the budget actually has the support of 52% of the voters. What's the probability the newspaper's sample will lead them to predict defeat? Be sure to verify that the assumptions and conditions necessary for your analysis are met.

42. Seeds. Information on a packet of seeds claims that the germination rate is 92%. What's the probability that more than 95% of the 160 seeds in the packet will germinate? Be sure to discuss your assumptions and check the conditions that support your model.

43. Apples. When a truckload of apples arrives at a packing plant, a random sample of 150 is selected and examined for bruises, discoloration, and other defects. The whole truckload will be rejected if more than 5% of the sample is unsatisfactory. Suppose that in fact 8% of the apples on the truck do not meet the desired standard. What's the probability that the shipment will be accepted anyway?

44. Genetic defect. It's believed that 4% of children have a gene that may be linked to juvenile diabetes. Researchers hoping to track 20 of these children for several years test 732 newborns for the presence of this gene. What's the probability that they find enough subjects for their study?

45. Catalog sales. A catalog sales company promises to deliver orders placed on the Internet within 3 days. Follow-up calls to a few randomly selected customers show that a 95% confidence interval for the proportion of all orders that arrive on time is 88% ± 6%. What does this mean? Are the conclusions in parts a–e correct? Explain.

a) Between 82% and 94% of all orders arrive on time.
b) 95% of all random samples of customers will show that 88% of orders arrive on time.
c) 95% of all random samples of customers will show that 82% to 94% of orders arrive on time.
d) The company is 95% sure that between 82% and 94% of the orders placed by the customers in this sample arrived on time.
e) On 95% of the days, between 82% and 94% of the orders will arrive on time.

46. Belgian euro. Recently, two students made worldwide headlines by spinning a Belgian euro 250 times and getting 140 heads—that's 56%. That makes the 90% confidence interval (51%, 61%). What does this mean? Are the conclusions in parts a–e correct? Explain your answers.

a) Between 51% and 61% of all euros are unfair.
b) We are 90% sure that in this experiment this euro landed heads between 51% and 61% of the spins.
c) We are 90% sure that spun euros will land heads between 51% and 61% of the time.

d) If you spin a euro many times, you can be 90% sure of getting between 51% and 61% heads.
e) 90% of all spun euros will land heads between 51% and 61% of the time.

47. Confidence intervals. Several factors are involved in the creation of a confidence interval. Among them are the sample size, the level of confidence, and the margin of error. Which statements are true?

a) For a given sample size, higher confidence means a smaller margin of error.
b) For a specified confidence level, larger samples provide smaller margins of error.
c) For a fixed margin of error, larger samples provide greater confidence.
d) For a given confidence level, halving the margin of error requires a sample twice as large.

48. Confidence intervals, again. Several factors are involved in the creation of a confidence interval. Among them are the sample size, the level of confidence, and the margin of error. Which statements are true?

a) For a given sample size, reducing the margin of error will mean lower confidence.
b) For a certain confidence level, you can get a smaller margin of error by selecting a bigger sample.
c) For a fixed margin of error, smaller samples will mean lower confidence.
d) For a given confidence level, a sample 9 times as large will make a margin of error one third as big.

49. Cars. A student is considering publishing a new magazine aimed directly at owners of Japanese automobiles. He wanted to estimate the fraction of cars in the United States that are made in Japan. The computer output summarizes the results of a random sample of 50 autos. Explain carefully what it tells you.

```
z-interval for proportion
With 90.00% confidence
0.29938661 < p(Japan) < 0.46984416
```

50. Quality control. For quality control purposes, 900 ceramic tiles were inspected to determine the proportion of defective (e.g., cracked, uneven finish) tiles. Assuming that these tiles are representative of all tiles manufactured by an Italian tile company, what can you conclude based on the computer output?

```
z-interval for proportion
With 95.00% confidence
0.025 < p(defective) < 0.035
```

51. E-mail. A small company involved in e-commerce is interested in statistics concerning the use of e-mail. A poll found that 38% of a random sample of 1012 adults, who use a computer at their home, work, or school, said that they do not send or receive e-mail.

a) Find the margin of error for this poll if we want 90% confidence in our estimate of the percent of American adults who do not use e-mail.

b) Explain what that margin of error means.

c) If we want to be 99% confident, will the margin of error be larger or smaller? Explain.

d) Find that margin of error.

e) In general, if all other aspects of the situation remain the same, will smaller margins of error involve greater or less confidence in the interval?

52. Biotechnology. A biotechnology firm in Boston is planning its investment strategy for future products and research labs. A poll found that only 8% of a random sample of 1012 U.S. adults approved of attempts to clone a human.

a) Find the margin of error for this poll if we want 95% confidence in our estimate of the percent of American adults who approve of cloning humans.

b) Explain what that margin of error means.

c) If we only need to be 90% confident, will the margin of error be larger or smaller? Explain.

d) Find that margin of error.

e) In general, if all other aspects of the situation remain the same, would smaller samples produce smaller or larger margins of error?

53. Teenage drivers. An insurance company checks police records on 582 accidents selected at random and notes that teenagers were at the wheel in 91 of them.

a) Create a 95% confidence interval for the percentage of all auto accidents that involve teenage drivers.

b) Explain what your interval means.

c) Explain what "95% confidence" means.

d) A politician urging tighter restrictions on drivers' licenses issued to teens says, "In one of every five auto accidents, a teenager is behind the wheel." Does your confidence interval support or contradict this statement? Explain.

54. Advertisers. Direct mail advertisers send solicitations ("junk mail") to thousands of potential customers in the hope that some will buy the company's product. The response rate is usually quite low. Suppose a company wants to test the response to a new flyer and sends it to 1000 people randomly selected from their mailing list of over 200,000 people. They get orders from 123 of the recipients.

a) Create a 90% confidence interval for the percentage of people the company contacts who may buy something.

b) Explain what this interval means.

c) Explain what "90% confidence" means.

d) The company must decide whether to now do a mass mailing. The mailing won't be cost-effective unless it produces at least a 5% return. What does your confidence interval suggest? Explain.

55. Retailers. Some food retailers propose subjecting food to a low level of radiation in order to improve safety, but sale of such "irradiated" food is opposed by many people. Suppose a grocer wants to find out what his customers think. He has cashiers distribute surveys at checkout and ask customers to fill them out and drop them in a box near the front door. He gets responses from 122 customers, of whom 78 oppose the radiation treatments. What can the grocer conclude about the opinions of all his customers?

56. Local news. The mayor of a small city has suggested that the state locate a new prison there, arguing that the construction project and resulting jobs will be good for the local economy. A total of 183 residents show up for a public hearing on the proposal, and a show of hands finds 31 in favor of the prison project. What can the city council conclude about public support for the mayor's initiative?

57. Internet music. In a survey on downloading music, the Gallup Poll asked 703 Internet users if they "ever downloaded music from an Internet site that was not authorized by a record company, or not," and 18% responded "yes." Construct a 95% confidence interval for the true proportion of Internet users who have downloaded music from an Internet site that was not authorized.

58. Economy worries. During the week of April 15, 2013, a Gallup Poll asked 1500 U.S. adults, aged 18 or over, how they rated economic conditions. Only 17% rated the economy as Excellent/Good. Construct a 95% confidence interval for the true proportion of Americans who rated the U.S. economy as Excellent/Good.

59. International business. In Canada, the vast majority (90%) of companies in the chemical industry are ISO 14001 certified. The ISO 14001 is an international standard for environmental management systems. An environmental group wished to estimate the percentage of U.S. chemical companies that are ISO 14001 certified. Of the 550 chemical companies sampled, 385 are certified.

a) What proportion of the sample reported being certified?

b) Create a 95% confidence interval for the proportion of U.S. chemical companies with ISO 14001 certification. (Be sure to check conditions.) Compare to the Canadian proportion.

60. Worldwide survey. GfK Roper surveyed people worldwide asking them "how important is acquiring wealth to you." Of 1535 respondents in India, 1168 said that it was of more than average importance. In the United States of 1317 respondents, 596 said it was of more than average importance.

a) What proportion thought acquiring wealth was of more than average importance in each country's sample?

b) Create a 95% confidence interval for the proportion who thought it was of more than average importance in India. (Be sure to test conditions.) Compare that to a confidence interval for the U.S. population.

61. Business ethics. In a survey on corporate ethics, a poll split a sample at random, asking 538 faculty and corporate recruiters the question: "Generally speaking, do you believe that MBAs are more or less aware of ethical issues in business today than five years ago?" The other half were asked: "Generally speaking, do you believe that MBAs are less or more aware of ethical issues in business today than five years ago?" These may seem like the same questions, but sometimes the order of the choices matters. In response to the first question, 53% thought MBA graduates were more aware of ethical issues, but when the question was phrased differently, this proportion dropped to 44%.

a) What kind of bias may be present here?
b) Each group consisted of 538 respondents. If we combine them, considering the overall group to be one larger random sample, what is a 95% confidence interval for the proportion of the faculty and corporate recruiters that believe MBAs are more aware of ethical issues today?
c) How does the margin of error based on this pooled sample compare with the margins of error from the separate groups? Why?

62. Middle Eastern entrepreneurs. In 2012, Gallup published a report entitled "Qatar's Rising Entrepreneurial Spirit" in which they concluded that the 33% of 1057 Qatari youth they surveyed who responded that they plan to start their own business was the highest in the region. They conducted a variety of face to face and phone interviews with Qatari youth during their survey. They noted that the margin of error was between 6.6 and 7.8 percentage points, and that "in addition to sampling error, question wording and practical difficulties in conducting surveys can introduce error or bias into the findings of public opinion polls."

a) What kinds of bias might they be referring to?
b) Does their margin of error suggest that this was a simple random sample? Explain.

63. Pharmaceutical company. A pharmaceutical company is considering investing in a "new and improved" vitamin D supplement for children. Vitamin D, whether ingested as a dietary supplement or produced naturally when sunlight falls upon the skin, is essential for strong, healthy bones. The bone disease rickets was largely eliminated in England during the 1950s, but now there is concern that a generation of children more likely to watch TV or play computer games than spend time outdoors is at increased risk. A recent study of 2700 children randomly selected from all parts of England found 20% of them deficient in vitamin D.

a) Find a 98% confidence interval for the proportion of children in England who are deficient in vitamin D.
b) Explain carefully what your interval means.
c) Explain what "98% confidence" means.
d) Does the study show that computer games are a likely cause of rickets? Explain.

64. Real estate survey. A real estate agent looks over the 15 listings she has in a particular zip code in California and finds that 80% of them have swimming pools.

a) Check the assumptions and conditions for inference on proportions.
b) If it's appropriate, find a 90% confidence interval for the proportion of houses in this zip code that have swimming pools. If it's not appropriate, explain why.

65. Benefits survey. A paralegal at the Vermont State Attorney General's office wants to know how many companies in Vermont provide health insurance benefits to all employees. She chooses 12 companies at random and finds that all 12 offer benefits.

a) Check the assumptions and conditions for inference on proportions.
b) If conditions are met, find a 95% confidence interval for the true proportion of companies that provide health insurance benefits to all their employees. If conditions are not met, explain why.

66. Awareness survey. A telemarketer at a credit card company is instructed to ask the next 18 customers that call into the 800 number whether they are aware of the new Platinum card that the company is offering. Of the 18, 17 said they were aware of the program.

a) Check the assumptions and conditions for inference on proportions.
b) If conditions are met, find a 95% confidence interval for the true proportion of customers who are aware of the new card. If conditions are not met, explain why.

67. IRS. In a random survey of 226 self-employed individuals, 20 reported having had their tax returns audited by the IRS in the past year. Estimate the proportion of self-employed individuals nationwide who've been audited by the IRS in the past year.

a) Check the assumptions and conditions (to the extent you can) for constructing a confidence interval.
b) Construct a 95% confidence interval.
c) Interpret your interval.
d) Explain what "95% confidence" means in this context.

68. Internet music, again. A Gallup Poll (Exercise 57) asked Americans if the fact that they can make copies of songs on the Internet for free made them more likely—or less likely—to buy a performer's CD. Only 13% responded that it made them "less likely." The poll was based on a random sample of 703 Internet users.

a) Check that the assumptions and conditions are met for inference on proportions.
b) Find the 95% confidence interval for the true proportion of all U.S. Internet users who are "less likely" to buy CDs.

69. Politics. A recent poll of 1005 U.S. adults split the sample into four age groups: ages 18–29, 30–49, 50–64, and 65+. In the youngest age group, 62% said that they thought the United States was ready for a woman president, as opposed to 35% who said "no, the country was not ready" (3% were undecided). The sample included 250 18- to 29-year-olds.

a) Do you expect the 95% confidence interval for the true proportion of all 18- to 29-year-olds who think the United States is ready for a woman president to be wider or narrower than the 95% confidence interval for the true proportion of all U.S. adults? Explain.

b) Find the 95% confidence interval for the true proportion of all 18- to 29-year-olds who believe the United States is ready for a woman president.

70. More Internet music. A random sample of 168 students was asked how many songs were in their digital music library and what fraction of them was legally purchased. Overall, they reported having a total of 117,079 songs, of which 23.1% were legal. The music industry would like a good estimate of the proportion of songs in students' digital music libraries that are legal.

a) Think carefully. What is the parameter being estimated? What is the population? What is the sample size?
b) Check the conditions for making a confidence interval.
c) Construct a 95% confidence interval for the fraction of legal digital music.
d) Explain what this interval means. Do you believe that you can be this confident about your result? Why or why not?

71. CDs. A company manufacturing CDs is working on a new technology. A random sample of 703 Internet users were asked: "As you may know, some CDs are being manufactured so that you can only make one copy of the CD after you purchase it. Would you buy a CD with this technology, or would you refuse to buy it even if it was one you would normally buy?" Of these users, 64% responded that they would buy the CD.

a) Create a 90% confidence interval for this percentage.
b) If the company wants to cut the margin of error in half, how many users must they survey?

72. Internet music, last time. The research group that conducted the survey in Exercise 70 wants to provide the music industry with definitive information, but they believe that they could use a smaller sample next time. If the group is willing to have twice as big a margin of error, how many songs must be included?

73. Graduation. As in Exercise 19, we hope to estimate the percentage of adults aged 25 to 30 who never graduated from high school. What sample size would allow us to increase our confidence level to 95% while reducing the margin of error to only 2%?

74. Better hiring info. Editors of the business report in Exercise 20 are willing to accept a margin of error of 4% but want 99% confidence. How many randomly selected employers will they need to contact?

75. Pilot study. A state's environmental agency worries that a large percentage of cars may be violating clean air emissions standards. The agency hopes to check a sample of vehicles in order to estimate that percentage with a margin of error of 3% and 90% confidence. To gauge the size of the problem, the agency first picks 60 cars and finds 9 with faulty emissions systems. How many should be sampled for a full investigation?

76. Another pilot study. During routine conversations, the CEO of a new start-up reports that 22% of adults between the ages of 21 and 39 will purchase her new product. Hearing this, some investors decide to conduct a large-scale study, hoping to estimate the proportion to within 4% with 98% confidence. How many randomly selected adults between the ages of 21 and 39 must they survey?

77. Approval rating. A newspaper reports that the governor's approval rating stands at 65%. The article adds that the poll is based on a random sample of 972 adults and has a margin of error of 2.5%. What level of confidence did the pollsters use?

78. Amendment. The Board of Directors of a publicly traded company says that a proposed amendment to their bylaws is likely to win approval in the upcoming election because a poll of 1505 stock owners indicated that 52% would vote in favor. The Board goes on to say that the margin of error for this poll was 3%.

a) Explain why the poll is actually inconclusive.
b) What confidence level did the pollsters use?

T 79. Customer spending. The data set provided contains last month's credit card purchases of 500 customers randomly chosen from a segment of a major credit card issuer. The marketing department is considering a special offer for customers who spend more than $1000 per month on their card. From these data construct a 95% confidence interval for the proportion of customers in this segment who will qualify.

T 80. Advertising. A philanthropic organization knows that its donors have an average age near 60 and is considering taking out an ad in the *American Association of Retired People (AARP)* magazine. An analyst wonders what proportion of their donors are actually 50 years old or older. He takes a random sample of the records of 500 donors. From the data provided, construct a 95% confidence interval for the proportion of donors who are 50 years old or older.

JUST CHECKING ANSWERS

1 A Normal model (approximately).
2 At the actual proportion of all customers who like the new location.
3 $SD(\hat{p}) = \sqrt{\dfrac{(0.5)(0.5)}{100}} = 0.05$
4 Wider
5 Lower
6 Smaller

Case Study

Real Estate Simulation

Many variables important to the real estate market are skewed, limited to only a few values or considered as categorical variables. Yet, marketing and business decisions are often made based on means and proportions calculated over many homes.

Data on 1063 houses sold recently in the Saratoga, New York area, are in the file **Saratoga Real Estate**. Let's investigate how the sampling distribution of proportions approaches the Normal.

Part 1: Proportions

The variable *Fireplace* is a dichotomous variable where 1 = *has a fireplace* and 0 = *does not have a fireplace*.

- Calculate the proportion of homes that have fireplaces for all 1063 homes. Using this value, calculate what the standard error of the sample proportion would be for a sample of size 50.

- Using the software of your choice, draw 100 samples of size 50 from this population of homes, find the proportion of homes with fireplaces in each of these samples, and make a histogram of these proportions.

- Compare the mean and standard deviation of this (sampling) distribution to what you previously calculated.

- Examine the distribution of the sampled proportions. What do you expect it to look like? How closely does it match the theoretical distribution?

Part 2: Confidence Intervals for Proportion

Of the 1063 homes in the data set, 635, or 59.7% have fireplaces.

- Using appropriate software, draw 100 samples of size 25 from the data and compute 90% confidence intervals for the true proportion.

- How many of these contain 59.7%?

- Repeat this for 100 samples of size 100.

- Write up a short report explaining the main differences between the two sets of intervals.

10

Testing Hypotheses about Proportions

Dow Jones Industrial Average

More than a hundred years ago Charles Dow changed the way people look at the stock market. Surprisingly, he wasn't an investment wizard or a venture capitalist. He was a journalist who wanted to make investing understandable to ordinary people. Although he died at the relatively young age of 51 in 1902, his impact on how we track the stock market has been both long-lasting and far-reaching.

In the late 1800s, when Charles Dow reported on Wall Street, investors preferred bonds, not stocks. Bonds were reliable, backed by the real machinery and other hard assets the company owned. What's more, bonds were predictable; the bond owner knew when the bond would mature and so, knew when and how much the bond would pay. Stocks simply represented "shares" of ownership, which were risky and erratic. In May 1896, Dow and Edward Jones, whom he had known since their days as reporters for the *Providence Evening Press*, launched the now-famous Dow Jones Industrial Average (DJIA) to help the public understand stock market trends.

The original DJIA averaged 11 stock prices. Of those original industrial stocks, only General Electric is still in the DJIA.

Since then, the DJIA has become synonymous with overall market performance and is often referred to simply as the Dow. The index was expanded to 20 stocks in 1916 and to 30 in 1928 at the height of the roaring twenties bull market. That bull market peaked on September 3, 1929, when the Dow reached 381.17. On October 28 and 29, 1929, the Dow lost nearly 25% of its value. Then things got worse. Within four years, on July 8, 1932, the 30 industrials reached an all-time low of 40.65. The highs of September 1929 were not reached again until 1954.

Today the Dow is a weighted average of 30 stocks, with weights used to account for splits and other adjustments. The "Industrial" part of the name is largely historic. Today's DJIA includes the service industry and financial companies and is much broader than just heavy industry. And it is still a primary indicator of the volatility of the U.S. stock market and the global economy.

WHO	Days on which the stock market was open ("trading days")
WHAT	Closing price of the Dow Jones Industrial Average (*Close*)
UNITS	Points
WHEN	August 1982 to December 1986
WHY	To test theory of stock market behavior

How does the stock market move? Figure 10.1 shows the DJIA closing price for the bull market that ran from mid 1982 to the end of 1986.

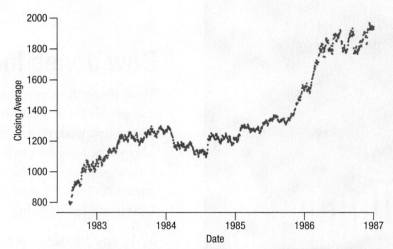

Figure 10.1   Daily closing prices of the Dow Jones Industrials from mid 1982 to the end of 1986.

The DJIA clearly increased during this famous bull market, more than doubling in value in less than five years. One common theory of market behavior says that on a given day, the market is just as likely to move up as down. Another way of phrasing this is that the daily behavior of the stock market is random. Can that be true during such periods of obvious increase? Let's investigate if the Dow is just as likely to move higher or lower on any given day. Out of the 1112 trading days in that period, the average increased on 573 days, a sample proportion of 0.5153 or 51.53%. That *is* more "up" days than "down" days, but is it far enough from 50% to cast doubt on the assumption of equally likely up or down movement?

10.1   Hypotheses

We've learned how to create a confidence interval for a proportion, but sometimes we have a question about a specific value that we need to answer. For the stock market, we'd like to know whether the market moves randomly up and down with equal probability ($p = 0.5$) or if there is some other underlying pattern. Tests like this are useful for many business questions. For example: Are our customers really more satisfied since the launch of our new website? Is the mean income of our Platinum customers higher than the income of our regular customers? Did our recent ad campaign really reach our target of 20% of all the adults in our region?

How can we state and test a hypothesis about daily changes in the stock market? In the logic of science, if we want to see whether a theory is true, we assume that it is and then gather evidence that either is consistent with, or seems to refute, that theory. So, to test whether the daily fluctuations in the Dow are equally likely

Hypothesis, *n.*;
pl. Hypotheses.
 A supposition; a proposition or principle which is supposed or taken for granted, in order to draw a conclusion or inference for proof of the point in question; something not proved, but assumed for the purpose of argument.
 —*Webster's Unabridged Dictionary, 1913*

to be up as down, we'll assume that they are and then see if the data are consistent with that hypothesis.

Our starting hypothesis, called the null hypothesis, is that the proportion of days on which the DJIA increases is 50%. The **null hypothesis**, H_0, is a working model that we adopt temporarily or for the sake of argument. It specifies a population model parameter (in this case, the true proportion of "up" days, p) and proposes a value for that parameter. We write $H_0: p = p_0$. This is a concise way to specify the two things we need. It names the parameter we hope to learn about and specifies a value for it. For our hypothesis about the DJIA, our null hypothesis is that $p = 0.5$ so we write $H_0: p = 0.5$.

The **alternative hypothesis**, which we denote H_A, contains all the values of the parameter that we will consider plausible if we reject the null hypothesis. For the stock market, our null hypothesis is $p = 0.5$. What's the alternative? If we are interested in deviations equally likely in either direction, then our alternative is $H_A: p \neq 0.5$.

Now that we have our null hypothesis, what evidence would convince you that the proportion of up days was not 50%? Suppose that the market closed up on 95% of the days you looked at. That would probably convince you that the hypothesis was wrong: it's just too unlikely an event. But if the sample proportion of up days is close to 50%, you might not be sure. After all, observations do vary, so you wouldn't be surprised to see some difference. So, how different from 50% must the sample proportion be before you would be convinced that the true proportion wasn't 50%? That's the crucial question in a hypothesis test. As usual in statistics, when we think about the size of a change, we think of using the standard deviation as the ruler to measure that change.

We learned in Chapter 7 how to estimate the standard deviation of a random proportion using the Binomial model. If we think the probability of an up day is 0.5, then the standard deviation of our random, observed proportion of up days would be

$$SD\,(\hat{p}) = \sqrt{\frac{pq}{n}} = \sqrt{\frac{(0.5)(1 - 0.5)}{1112}} = 0.015$$

NOTATION ALERT

Capital H is the standard letter for hypotheses. H_0 labels the null hypothesis, and H_A labels the alternative.

> ### Why Is This a Standard Deviation and Not a Standard Error?
>
> This is a standard deviation because we are using the model (hypothesized) value for p and *not* the estimated value, $\hat{p}$. Once we assume that the null hypothesis is true, it gives us a value for the model parameter p. With proportions, if we know p then we also automatically know its standard deviation. Because we find the standard deviation from the model parameter, this is a standard deviation and not a standard error. When we found a confidence interval for p, we could not assume that we knew its value, so we estimated the standard deviation from the sample value, $\hat{p}$.

The SD Comes from the Null Hypothesis

To remind us that the parameter value comes from the null hypothesis, it is sometimes written as p_0 and the standard deviation as

$$SD(\hat{p}) = \sqrt{\frac{p_0 q_0}{n}}.$$

Now, armed with that standard deviation, we want to model the behavior of the sample proportion to judge whether the proportion we've observed is unusually large. For that, we'll need the sampling distribution model. We saw in Chapter 9 how to model the distribution of random proportions with a Normal model. The null hypothesis provides all we need to find the mean and standard deviation. We adopt the null hypothesis temporarily to see whether it does actually describe the data, so we'll use a Normal model with a mean of .50 and a standard deviation of 0.015.

We want to know how likely it would be to see the observed value $\hat{p}$ as far away from 50% as the value of 51.53% that we actually have observed. Looking at Figure 10.2, we can see that 51.53% doesn't look very surprising. With a calculator, computer program, or the Normal table we can find that the probability of a proportion this far from 0.50 is about 0.308. This is the probability of observing more than

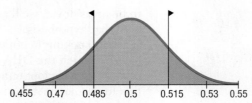

Figure 10.2   How likely is a proportion of more than 51.5% or less than 48.5% when the true mean is 50%? This is what it looks like. Each red area is 0.154 of the total area under the curve.

51.53% up days or more than 51.53% down days if the null model were true. In other words, if the chance of an up day for the Dow is 50%, we'd expect to see stretches of 1112 trading days with 51.53% or more up days about 15.4% of the time and with 51.53% or more down days about 15.4% of the time. That's not terribly unusual, so there's really no convincing evidence that the market did not act randomly.

It may surprise you that even during a bull market, the direction of daily movements is random. In fact, the probability that any given day will end up or down appears to be about 0.5 regardless of the longer-term trends. It may be that when the stock market has a long run-up (or possibly down, although we haven't checked that), it does so not by having more days of increasing or decreasing value, but by the actual amounts of the increases or decreases being unequal.

FOR EXAMPLE   Writing hypotheses

Many customers visit your site, and about 60% of all visitors "click-through" to the product page. You're happy with that 60% click-through rate, but one of your colleagues is proposing some major changes to your home page. You worry that the click-through rate might change, so you propose an experiment. To test the new design, you randomly send some visitors to the new site. Of the 90 visitors sent to the new site, 57 clicked through.

QUESTION   Does this suggest that the click-through proportion has changed with the new site? What are the right hypotheses for a test?

ANSWER   We'll assume that the click-through proportion is still 60% unless there is enough evidence to suggest that it's different.

$$H_0: p = 0.60$$
$$H_A: p \neq 0.60$$

10.2   A Trial as a Hypothesis Test

We started by assuming that the probability of an up day was 50%. Then we looked at the data and concluded that we couldn't say otherwise because the proportion that we actually observed wasn't far enough from 50%. Does this reasoning seem backward? That could be because we usually prefer to think about getting things right rather than getting them wrong. But, you've seen this reasoning before in a different context. This is the logic of jury trials.

Let's suppose a defendant has been accused of robbery. In British common law and those systems derived from it (including U.S. law), the null hypothesis is that the defendant is innocent. Instructions to juries are quite explicit about this.

The evidence takes the form of facts that seem to contradict the presumption of innocence. For us, this means collecting data. In the trial, the prosecutor presents evidence. ("If the defendant were innocent, wouldn't it be remarkable that the police found him at the scene of the crime with a bag full of money in his hand, a mask on his face, and a getaway car parked outside?") The next step is to judge the evidence. Evaluating the evidence is the responsibility of the jury in a trial, but it falls on your shoulders in hypothesis testing. The jury considers the evidence in light of the *presumption* of innocence and judges whether the evidence against the defendant would be plausible *if the defendant were in fact innocent.*

Like the jury, we ask: "Could these data plausibly have happened by chance if the null hypothesis were true?" If they are very unlikely to have occurred, then the evidence raises a reasonable doubt about the null hypothesis. Ultimately, *you* must make a decision. The standard of "beyond a reasonable doubt" is intentionally ambiguous because it leaves the jury to decide the degree to which the evidence contradicts the hypothesis of innocence. Juries don't explicitly use probability to help them decide whether to reject that hypothesis. But when you ask the same question of your null hypothesis, you have the advantage of being able to quantify exactly how surprising the evidence would be if the null hypothesis were true.

How unlikely is unlikely? Some people set rigid standards. Levels like 1 time out of 20 (0.05) or 1 time out of 100 (0.01) are common. But if *you* have to make the decision, you must judge for yourself in each situation whether the probability of observing your data is small enough to constitute "reasonable doubt."

10.3 P-Values

The fundamental step in our reasoning is the question: "Are the data surprising, given the null hypothesis?" And the key calculation is to determine exactly how likely the data we observed would be if the null hypothesis were the true model of the world. So we need a *probability*. Specifically, we want to find the probability of seeing data like these (or something even less likely) *given* the null hypothesis. This probability is called the **P-value**.

A low enough P-value says that the data we have observed would be very unlikely if our null hypothesis were true. We started with a model, and now that same model tells us that the data we have are unlikely to have happened. That's surprising. In this case, the model and data are at odds with each other, so we have to make a choice. Either the null hypothesis is correct and we've just seen something remarkable, or the null hypothesis is wrong (and, in fact, we were wrong to use it as the basis for computing our P-value). When you see a low P-value, you should reject the null hypothesis. There is no hard and fast rule about how low the P-value has to be. In fact, that decision is the subject of much of the rest of this chapter. Almost everyone would agree, however, that a P-value less than 0.001 indicates very strong evidence *against* the null hypothesis but a P-value greater than 0.05 provides very weak evidence.

When the P-value is *high* (or just not low *enough*), what do we conclude? In that case, we haven't seen anything unlikely or surprising at all. The data are consistent with the model from the null hypothesis, and we have no reason to reject the null hypothesis. Events that have a high probability of happening happen all the time. So, when the P-value is high does that mean we've proved the null hypothesis is true? No! We realize that many other similar hypotheses could also account for the data we've seen. The most we can say is that it doesn't appear to be false. Formally, we say that we "fail to reject" the null hypothesis. That may seem to be a pretty weak conclusion, but it's all we can say when the P-value is not low enough. All that means is that the data are consistent with the model that we started with.

Beyond a Reasonable Doubt

We ask whether the data were unlikely beyond a reasonable doubt. For the DJIA prices, we calculated how unlikely the data are if the null hypothesis is true. The probability that the observed statistic value (or an even more extreme value) could occur if the null model is true—in this case, 0.308—is the P-value. That probability is certainly not *beyond a reasonable doubt*, so we fail to reject the null hypothesis here.

How low is "low enough"? That's a judgment call that depends on the kind of decision you want to make. People typically select an arbitrary level such as .10, .05, or .01 and resolve to make a decision if the P-value falls below the level they selected. That arbitrary level is called an "alpha level" and (naturally enough) denoted α. We'll say more about alpha levels in Section 6.

What to Do with an "Innocent" Defendant

Let's see what that last statement means in a jury trial. If the evidence is not strong enough to reject the defendant's presumption of innocence, what verdict does the jury return? They do not say that the defendant is innocent. They say "not guilty." All they are saying is that they have not seen sufficient evidence to reject innocence and convict the defendant. The defendant may, in fact, be innocent, but the jury has no way to be sure.

Said statistically, the jury's null hypothesis is: innocent defendant. If the evidence is too unlikely (the P-value is low) then, given the assumption of innocence, the jury rejects the null hypothesis and finds the defendant guilty. But—and this is an important distinction—if there is *insufficient evidence* to convict the defendant (if the P-value is *not* low), the jury does not conclude that the null hypothesis is true and declare that the defendant is innocent. Juries can only *fail to reject* the null hypothesis and declare the defendant "not guilty."

In the same way, if the data are not particularly unlikely under the assumption that the null hypothesis is true, then the most we can do is to "fail to reject" our null hypothesis. We never declare the null hypothesis to be true. In fact, we simply do not know whether it's true or not. (After all, more evidence may come along later.)

Imagine a test of whether a company's new website design encourages a higher percentage of visitors to make a purchase (as compared to the site they've used for years). The null hypothesis is that the new site is no more effective at stimulating purchases than the old one. The test sends visitors randomly to one version of the website or the other. Of course, some will make a purchase, and others won't. If we compare the two websites on only 10 customers each, the results are likely *not to be clear*, and we'll be unable to reject the null hypothesis. Does this mean the new design is a complete bust? Not necessarily. It simply means that we don't have enough evidence to reject our null hypothesis. That's why we don't start by assuming that the new design is *more* effective. If we were to do that, then we could test just a few customers, find that the results aren't clear, and claim that since we've been unable to reject our original assumption the redesign must be effective. The Board of Directors is unlikely to be impressed by that argument.

> ### Don't We Want to Reject the Null?
>
> Often, people who collect data or perform an experiment hope to reject the null. They hope the new ad campaign is *better* than the old one, or they hope their candidate is *ahead* of the opponent. But, when we test a hypothesis, we must stay neutral. We can't let our hope bias our decision. As in a jury trial, we must stay with the null hypothesis until we are convinced otherwise. The burden of proof rests with the alternative hypothesis—innocent until proven guilty. When you test a hypothesis, you must act as judge and jury, but not as prosecutor.

> ### Conclusion
>
> If the P-value is "low," reject H_0 and conclude H_A.
>
> If the P-value is not "low enough," then fail to reject H_0 and the test is inconclusive.

JUST CHECKING

1. A pharmaceutical firm wants to know whether aspirin helps to thin blood. The null hypothesis says that it doesn't. The firm's researchers test 12 patients, observe the proportion with thinner blood, and get a P-value of 0.32. They proclaim that aspirin doesn't work. What would you say?

2. An allergy drug has been tested and found to give relief to 75% of the patients in a large clinical trial. Now the scientists want to see whether a new, "improved" version works even better. What would the null hypothesis be?

3. The new allergy drug is tested, and the P-value is 0.0001. What would you conclude about the new drug?

10.4   The Reasoning of Hypothesis Testing

Hypothesis tests follow a carefully structured path. To avoid getting lost, it helps to divide that path into four distinct sections: hypotheses, model, mechanics, and conclusion.

Hypotheses

First, state the null hypothesis. That's usually the skeptical claim that nothing's different. The null hypothesis assumes the default (often the status quo) is true (the defendant is innocent, the new method is no better than the old, customer preferences haven't changed since last year, etc.).

In statistical hypothesis testing, hypotheses are almost always about model parameters. To assess how unlikely our data may be, we need a null model. The null hypothesis specifies a particular parameter value to use in our model. In the usual notation, we write H_0: *parameter = hypothesized value*. The alternative hypothesis, H_A, contains the values of the parameter we consider plausible when we reject the null.

Model

To plan a statistical hypothesis test, specify the *model* for the sampling distribution of the statistic you will use to test the null hypothesis and the parameter of interest. For proportions, we use the Normal model for the sampling distribution. Of course, all models require assumptions, so you will need to state them and check any corresponding conditions. For a test of a proportion, the assumptions and conditions are the same as for a one-proportion z-interval. For the DJIA example, the sample size of 1112 is certainly big enough to satisfy the Success/Failure condition, but doesn't violate the 10% condition either. (We expect $0.50 \times 1112 = 556$ daily increases.) And, although the value of the DJIA from day to day is certainly not independent, it may be reasonable to assume that the daily price *changes* are random and independent.

Your model step should end with a statement such as: *Because the conditions are satisfied, we can model the sampling distribution of the proportion with a Normal model.* Watch out, though. Your Model step could end with: *Because the conditions are not satisfied, we can't proceed with the test.* (If that's the case, stop and reconsider.)

Each test we discuss in this book has a name that you should include in your report. We'll see many tests in the following chapters. Some will be about more than one sample, some will involve statistics other than proportions, and some will use models other than the Normal (and so will not use z-scores). The test about proportions is called a **one-proportion z-test**.[1]

> "The null hypothesis is never proved or established, but is possibly disproved, in the course of experimentation. Every experiment may be said to exist only in order to give the facts a chance of disproving the null hypothesis."
>
> —SIR RONALD FISHER, THE DESIGN OF EXPERIMENTS, 1931

When the Conditions Fail...

You might proceed with caution, explicitly stating your concerns. Or you may need to do the analysis with and without an outlier, or on different subgroups, or after re-expressing the response variable. Or you may not be able to proceed at all.

One-Proportion z-Test

The conditions for the one-proportion z-test are the same as for the one-proportion z-interval (except that we use the hypothesized values, p_0 and q_0, to check the Success/Failure condition). We test the hypothesis $H_0: p = p_0$ using the statistic

$$z = \frac{(\hat{p} - p_0)}{SD(\hat{p})}.$$

We also use p_0 to find the standard deviation: $SD(\hat{p}) = \sqrt{\dfrac{p_0 q_0}{n}}$. When the conditions are met and the null hypothesis is true, this statistic follows the standard Normal model, so we can use that model to obtain a P-value.

---

[1] It's also called the "one-sample test for a proportion."

> **FOR EXAMPLE**   Checking assumptions and conditions
>
> Consider the test of the new website from page 302.
>
> **QUESTION**   Are the conditions for inference satisfied?
>
> **ANSWER**
> **Independence assumption:** This is plausible because the **Randomization Condition** has been met through our decision to send a random selection of visitors to the new site.
>
> The population of potential visitors is large, so we don't have to worry about sampling more than 10% of it.
>
> **Success/Failure Condition:** We expect $np_0 = 90(.60) = 54$ successes and $nq_0 = 90(.40) = 36$ failures. Both are at least 10.
>
> So the assumptions and conditions appear to be satisfied.

Mechanics

Under "Mechanics" we perform the actual calculation of our test statistic from the data. Different tests we encounter will have different formulas and different test statistics. Usually, the mechanics are handled by a statistics program or calculator. The ultimate goal of the calculation is to obtain a P-value—the probability that the observed statistic value (or an even more extreme value) could occur if the null model were correct. If the P-value is small enough, we'll reject the null hypothesis.

Conclusions and Decisions

The primary conclusion in a formal hypothesis test is only a statement about the null hypothesis. It simply states whether we reject or fail to reject that hypothesis. As always, the conclusion should be stated in context, but your conclusion about the null hypothesis should never be the end of the process. You can't make a decision based solely on a P-value. Business decisions have consequences, with actions to take or policies to change. The conclusions of a hypothesis test can help *inform* your decision, but they shouldn't be the only basis for it.

Business decisions should always take into consideration three things: the statistical significance of the test, the *cost* of the proposed action, and the **effect size** (the difference between the hypothesized and observed value) of the statistic. For example, a cellular telephone provider finds that 30% of their customers switch providers (or *churn*) when their two-year subscription contract expires. They try a small experiment and offer a random sample of customers a free $350 top-of-the-line phone if they renew their contracts for another two years. Not surprisingly, they find that the new switching rate is lower by a statistically significant amount. Should they offer these free phones to all their customers? Obviously, the answer depends on more than the P-value of the hypothesis test. Even if the P-value is statistically significant, the correct business decision also depends on the cost of the free phones and by how much the churn rate is lowered (the effect size). It's rare that a hypothesis test alone is enough to make a sound business decision.

10.5   Alternative Hypotheses

In our example about the DJIA, we were equally interested in proportions that deviate from 50% in *either* direction. So we wrote our alternative hypothesis as $H_A: p \neq 0.5$. Such an alternative hypothesis is known as a **two-sided alternative** because we are equally interested in deviations on either side of the null hypothesis value. For two-sided alternatives, the P-value is the probability of deviating in *either* direction from the null hypothesis value. Figure 10.3 shows the P-value area when the alternative is two-sided.

Figure 10.3 The P-value for a two-sided alternative adds the probabilities in both tails of the sampling distribution model outside the value that corresponds to the test statistic.

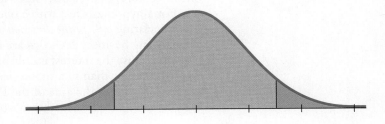

FOR EXAMPLE   Finding a P-value

We have sampled 90 visitors to our new website and 57 of them clicked through to another page (see page 302).

QUESTION   What's the P-value associated with a one-proportion z-test of our null hypothesis?

ANSWER   We have $n = 90$, $x = 57$, and a hypothesized $p = 0.60$

$$\hat{p} = \frac{57}{90} = 0.633$$

$$SD(\hat{p}) = \sqrt{\frac{p_0 q_0}{n}} = \sqrt{\frac{(0.6)(0.4)}{90}} = 0.0516$$

$$\text{So, } z = \frac{\hat{p} - p_0}{SD(\hat{p})} = \frac{0.633 - 0.60}{0.0516} = 0.64$$

P-value $= P(|z| > 0.64) = 0.52$

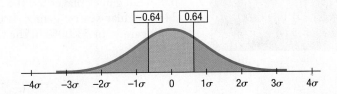

You can find from table Z, or other technology that the area in each tail is 0.26, so the P-value is $2 \times 0.26 = 0.52$.

Alternative Hypotheses

Proportions:

Two-sided

$H_0: p = p_0$

$H_A: p \neq p_0$

One-sided

$H_0: p = p_0$

$H_A: p < p_0$ or $p > p_0$

The alternative hypothesis gives the values of the parameter that we will consider plausible if we reject the null hypothesis. Sometimes values that deviate from the null in one direction are either not plausible at all or just of no interest to us in making a business decision. For example, suppose we want to test whether the proportion of customers returning merchandise has decreased under our new quality monitoring program. We're only going to continue the program if it has reduced returns, so we would only be interested in a sample proportion *smaller* than the null hypothesis value. We'd write our alternative hypothesis as $H_A: p < p_0$. An alternative hypothesis that focuses on deviations from the null hypothesis value in only one direction is called a **one-sided alternative**.

For a hypothesis test with a one-sided alternative, the P-value is the probability of deviating *only in the direction of the alternative* away from the null hypothesis value. Many business decisions are based on hypothesis tests with one-sided alternatives because the interest is only in increasing sales or profits or in reducing losses or errors rather than in a theoretical understanding of how things might change. Figure 10.4 shows the area of the P-value when the alternative is $H_A: p < p_0$. For $H_A: p > p_0$ the (red) shaded area would be on the right.

Figure 10.4 The P-value for a one-sided alternative considers only the probability of values beyond the test statistic value in the specified direction.

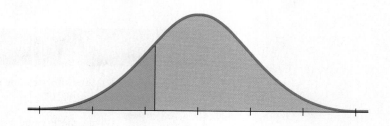

GUIDED EXAMPLE   Home Field Advantage

Major league sports are big business. And the fans are more likely to come out to root for the team if the home team has a good chance of winning. Anyone who plays or watches sports has heard of the "home field advantage." Teams tend to win more often when they play at home. Or do they?

If there were no home field advantage, the home teams would win about half of all games played. In the 2014 Major League Baseball season, there were 2430 regular-season games. It turns out that the home team won 1288 of the 2430 games, or 53.00% of the time.

Question: Could this deviation from 50% be explained just from natural sampling variability, or is it evidence to suggest that there really is a home field advantage, at least in professional baseball?

PLAN	**Setup** State what we want to know.	We want to know whether the home team in professional baseball is more likely to win. The data are all 2430 games from the 2014 Major League Baseball season. The variable is whether or not the home team won. The parameter of interest is the proportion of home team wins. If there's no advantage, I'd expect that proportion to be 0.50.
	Define the variables and discuss their context.	
	Hypotheses The null hypothesis makes the claim of no home field advantage.	$H_O: p = 0.50$ $H_A: p > 0.50$

We are interested only in a home field *advantage*, so the alternative hypothesis is one-sided.

Model Think about the assumptions and check the appropriate conditions.

✓ **Independence Assumption:** Generally, the outcome of one game has no effect on the outcome of another game. But this may not be strictly true. For example, if a key player is injured, the probability that the team will win in the next couple of games may decrease slightly, but independence is still roughly true. The data come from one entire season, but I expect other seasons to be similar.

This is not a random sample. But we can view the 2430 games here as a representative collection of games including those that might be played in the future.

✓ **Success/Failure Condition:** Both

$np_0 = 2430(0.50) = 1215.0$ and

$nq_0 = 2430(0.50) = 1215.0$ are at least 10.

Specify the sampling distribution model.

State what test you plan to use.

Because the conditions are satisfied, I'll use a Normal model for the sampling distribution of the proportion and do a **one-proportion z-test**.

DO

Mechanics The null model gives us the mean, and (because we are working with proportions) the mean gives us the standard deviation.

Next, we find the z-score for the observed proportion, to find out how many standard deviations it is from the hypothesized proportion.

From the z-score, we can find the P-value, which tells us the probability of observing a value that extreme (or more).

The probability of observing a sample proportion 2.96 or more standard deviations above the mean of a Normal model can be found by computer, calculator, or table to be about 0.0015.

The null model is a Normal distribution with a mean of 0.50 and a standard deviation of

$$SD(\hat{p}) = \sqrt{\frac{p_0 q_0}{n}} = \sqrt{\frac{(0.5)(1 - 0.5)}{2430}}$$

$$= 0.01014$$

The observed proportion, $\hat{p}$, is 0.5300.
So the z-value is

$$z = \frac{0.5300 - 0.5}{0.01014} = 2.96$$

The sample proportion lies 2.96 standard deviations above the mean.

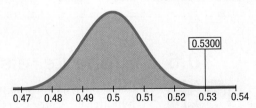

The corresponding P-value is about 0.0015

REPORT

Conclusion State your conclusion about the parameter—in context.

MEMO

Re: Home field advantage

Our analysis of games played during the 2014 season showed a statistically significant home field advantage. The P-value of about 0.0015 says that if the true proportion of home team wins were 0.50, then an observed value of 0.5300 (or larger) would occur about 1.5 times in 1000.

	The Hypothesis Testing Process
Setup	State your problem and identify the variables. Be sure that the identity of the cases and what the variables measure are both clear.
Hypotheses	State your hypotheses: • The null hypothesis, H_0, identifies a parameter and proposes a value for it. • The alternative hypothesis, H_A, specifies what kind of deviations from the null model would be of interest.
Model	Specify the statistic and its sampling distribution model. The sampling distribution model comes from theory and will usually be something you look up. It will likely require you to make assumptions about your data. You can often determine if an assumption is plausible by checking related conditions—usually with graphical displays. If the required assumptions are not plausible, STOP. You can't test your hypothesis with this method or with these data. Otherwise, state clearly what method you will use.
Mechanics	Calculate the test statistic. Usually, you'll use technology to do the work.
Find the P-value	Find the probability of observing the statistic value you found (or values even less likely). The sampling distribution model provides this probability. The hypothesized parameter value will be needed for the sampling distribution model and the alternative hypothesis will indicate what part of the distribution is of interest. Usually, we can use technology to find this probability. The probability is called the P-value. You should report it along with your decision.

Reasoning	Is the P-value small? Then **reject** the null hypothesis.
	    If the null hypothesis were true, then we have observed a rare event, and we don't want to base decisions on rare events.
	Is the P-value large? Then **fail to reject** the null hypothesis.
	    We don't know whether the null hypothesis is true, so we can't "accept" it. But we lack strong enough evidence to declare it false.

Judgment	You must decide for yourself what probability would be small enough that you would find an event with that probability rare. Although there are some common values (10%, 5%, and 1% are all used), your decision should be based on the circumstances of the test, including considerations of the costs should your test conclusion prove to be wrong.

JUST CHECKING

4 An accountant at an organic food chain is testing whether the customers prefer a new cheese supplier. She surveys 100 customers and finds out that the proportion preferring the new supplier is greater than the previous proportion with a P-value of 0.002. The new supplier costs more, but the P-value is so low. Should she switch to the new supplier?

5 A marketing analyst at a biotechnology firm has just learned that the P-value for testing whether the proportion of people who find pain relief using their new product Z-170 was 0.06. Should he conclude that the product doesn't work?

10.6   Alpha Levels and Significance

Sir Ronald Fisher (1890–1962) was one of the founders of modern Statistics.

Sometimes we need to make a firm decision about whether or not to reject the null hypothesis. A jury must *decide* whether the evidence reaches the level of "beyond a reasonable doubt." A business must *select* a Web design. You need to decide which section of a Statistics course to enroll in.

When the P-value is small, it tells us that our data are rare *given the null hypothesis*. As humans, we are suspicious of rare events. If the data are "rare enough," we just don't think that could have happened due to chance. Since the data *did* happen, something must be wrong. All we can do now is to reject the null hypothesis.

But how rare is "rare"? How low does the P-value have to be?

We can define "rare event" arbitrarily by setting a threshold for our P-value. If our P-value falls below that point, we'll reject the null hypothesis. We call such results *statistically significant*. The threshold is called an **alpha level**. Not surprisingly, it's labeled with the Greek letter α. Common α-levels are 0.10, 0.05, 0.01, and 0.001. You have the option—almost the *obligation*—to consider your alpha level carefully and choose an appropriate one for the situation. If you're assessing the safety of air bags, you'll want a low alpha level; even 0.01 might not be low enough. If you're just wondering whether folks prefer their pizza with or without pepperoni, you might be happy with $\alpha = 0.10$. It can be hard to justify your choice of α, though, so often we arbitrarily choose 0.05.

Where did the value 0.05 come from?

In 1931, in a famous book called *The Design of Experiments*, Sir Ronald Fisher discussed the amount of evidence needed to reject a null hypothesis. He said that it was *situation dependent*, but remarked, somewhat casually, that for many scientific applications, 1 out of 20 *might be* a reasonable value, especially in a *first* experiment—one that will be followed by confirmation. Since then, some people—indeed some entire disciplines—have acted as if the number 0.05 were sacrosanct.

The alpha level is also called the **significance level**. When we reject the null hypothesis, we say that the test is "significant at that level." For example, we might say that we reject the null hypothesis "at the 5% level of significance." You must select the alpha level *before* you look at the data. Otherwise you can be accused of finagling the conclusions by tuning the alpha level to the results after you've seen the data.

What can you say if the P-value does not fall below α? When you have not found sufficient evidence to reject the null according to the standard you have established, you should say: "The data have failed to provide sufficient evidence to reject the null hypothesis." Don't say: "We accept the null hypothesis." You certainly haven't proven or established the null hypothesis; it was assumed to begin with. You *could* say that you have *retained* the null hypothesis, but it's better to say that you've failed to reject it.

NOTATION ALERT

The first Greek letter, α, is used in Statistics for the threshold value of a hypothesis test. You'll hear it referred to as the alpha level. Common values are 0.10, 0.05, 0.01, and 0.001.

It could happen to you!

Of course, if the null hypothesis *is* true, no matter what alpha level you choose, you still have a probability α of rejecting the null hypothesis by mistake. When we do reject the null hypothesis, no one ever thinks that *this* is one of those rare times. As statistician Stu Hunter notes, "The statistician says 'rare events do happen—but not to me!' "

Conclusion

If the P-value $< \alpha$, then reject H_0.

If the P-value $\geq \alpha$, then fail to reject H_0.

Look again at the home field advantage example. The P-value was 0.0015. This is so much smaller than any reasonable alpha level that we can reject H_0. We concluded: "We reject the null hypothesis. There is sufficient evidence to conclude that there is a home field advantage over and above what we expect with random variation." On the other hand, when testing the proportion in the For Example (p. 307) the P-value was 0.52, a very high P-value. In this case we can say only that we have failed to reject the null hypothesis that $p = 0.60$. We certainly can't say that we've proved it, or even that we've accepted it.

The automatic nature of the reject/fail-to-reject decision when we use an alpha level may make you uncomfortable. If your P-value falls just slightly above your alpha level, you're not allowed to reject the null. Yet a P-value just barely below the alpha level leads to rejection. If this bothers you, you're in good company. Many statisticians think it better to report the P-value than to choose an alpha level and carry the decision through to a final reject/fail-to-reject verdict. So when you declare your decision, it's always a good idea to report the P-value as an indication of the strength of the evidence.

> **It's in the stars**
>
> Some disciplines carry the idea further and code P-values by their size. In this scheme, a P-value between 0.05 and 0.01 gets highlighted by a single asterisk (*). A P-value between 0.01 and 0.001 gets two asterisks (**), and a P-value less than 0.001 gets three (***). This can be a convenient summary of the weight of evidence against the null hypothesis, but it isn't wise to take the distinctions too seriously and make black-and-white decisions near the boundaries. The boundaries are a matter of tradition, not science; there is nothing special about 0.05. A P-value of 0.051 should be looked at seriously and not casually thrown away just because it's larger than 0.05, and one that's 0.009 is not very different from one that's 0.011.

> **Practical vs. Statistical Significance**
>
> A large insurance company mined its data and found a statistically significant ($P = 0.04$) difference between the proportion of policies renewed in 2011 and those renewed in 2012. The difference in the proportions was 2%. Even though it was statistically significant, management did not see this as an important difference when it factored in the expense of the effort to retain those customers. On the other hand, a marketable improvement of 10% in relief rate for a new pain medicine may not be statistically significant unless a large number of people are tested. The effect, which is economically (or medically) significant, might not be statistically significant.

Sometimes it's best to report that the conclusion is not yet clear and to suggest that more data be gathered. (In a trial, a jury may "hang" and be unable to return a verdict.) In such cases, it's an especially good idea to report the P-value, since it's the best summary we have of what the data say or fail to say about the null hypothesis.

What do we mean when we say that a test is statistically significant? All we mean is that the test statistic had a P-value lower than our alpha level. Don't be lulled into thinking that "statistical significance" necessarily carries with it any practical importance or impact.

For large samples, even small, unimportant ("insignificant") deviations from the null hypothesis can be statistically significant. On the other hand, if the sample is not large enough, even large, financially or scientifically important differences may not be statistically significant.

When you report your decision about the null hypothesis, it's good practice to report the effect size (the magnitude of the difference between the observed statistic value and the null hypothesis value in the data units) along with the P-value.

> ## FOR EXAMPLE   Decisions and the α level
>
> Not satisfied with the click through rate of 0.633 (For Example page 307), you hire Summit Projects in Hood River Oregon to redesign your website and improve your click-through rate. A new sample of 100 visitors shows 69 visitors clicking through. That has a one-sided P-value of 0.033, and you are prepared to reject the null hypothesis of no change at $\alpha = 0.05$.
>
> **QUESTION**   Your manager now insists on all hypotheses being tested using $\alpha = 0.01$. What should you tell your manager?
>
> **ANSWER**   Under the smaller α level you would not reject the null hypothesis and you might conclude that the new website did not improve your rate. However, if improving the click through rate to 0.69 is financially important you should note that there is some evidence of improvement so you can recommend that they consider running a larger test.

10.7   Critical Values

When building a confidence interval, we calculated the margin of error as the product of an estimated standard error for the statistic and a critical value. For proportions, we found a **critical value**, z^*, to correspond to our selected confidence level. Critical values can also be used as a shortcut for hypothesis tests. Before computers and calculators were common, P-values were hard to find. It was easier to select a few common alpha levels (0.05, 0.01, 0.001, for example) and learn the corresponding critical values for the Normal model (that is, the critical values corresponding to confidence levels 0.95, 0.99, and 0.999, respectively). Rather than find the probability that corresponded to your observed statistic, you'd just calculate how many standard deviations it was away from the hypothesized value and compare that value directly against these z^* values. (Remember that whenever we measure the distance of a value from the mean in standard deviations, we are finding a z-score.) Any z-score larger in magnitude (that is, more extreme) than a particular critical value has to be less likely, so it will have a P-value smaller than the corresponding alpha.

If we were willing to settle for a flat reject/fail-to-reject decision, comparing an observed z-score with the critical value for a specified alpha level would give a shortcut path to that decision. For the home field advantage example, if we choose $\alpha = 0.05$, then in order to reject H_0, our z-score has to be larger than the one-sided critical value of 1.645. The observed proportion was 4.74 standard deviations above 0.5, so we clearly reject the null hypothesis. This is perfectly correct and does give us a yes/no decision, but it gives us less information about the hypothesis because we don't have the P-value to think about. With technology, P-values are easy to find. And since they give more information about the strength of the evidence, you should report them.

Here are the traditional z^* critical values from the Normal model:[2]

α	1-Sided	2-Sided
0.05	1.645	1.96
0.01	2.33	2.576
0.001	3.09	3.29

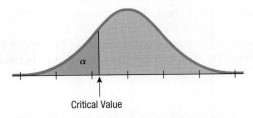

Figure 10.5 When the alternative is one-sided, the critical value puts all of α on one side.

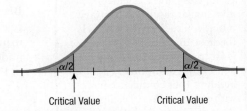

Figure 10.6 When the alternative is two-sided, the critical value splits α equally into two tails.

When testing means, you'll need to know both the α level and the degrees of freedom to find the t^* critical value. With large n, the t^* critical values will be close to the z^* critical values (above) that you use for testing proportions.

10.8 Confidence Intervals and Hypothesis Tests

Confidence intervals and hypothesis tests are built from the same calculations. They have the same assumptions and conditions. As we have just seen, you can approximate a hypothesis test by examining the confidence interval. Just ask whether the null hypothesis value is consistent with a confidence interval for the parameter at the corresponding confidence level. Because confidence intervals are naturally two-sided, they correspond to two-sided tests. For example, a 95% confidence interval corresponds to a two-sided hypothesis test at $\alpha = 5\%$. In general, a confidence

---

[2] In a sense, these are the flip side of the 68–95–99.7 Rule. There we chose simple statistical distances from the mean and recalled the areas of the tails. Here we select convenient tail areas (0.05, 0.01, and 0.001, either on one side or adding both together), and record the corresponding statistical distances.

NOTATION ALERT

We've attached symbols to many of the *p*'s. Let's keep them straight.

p is a population parameter—the true proportion in the population.

p_0 is a hypothesized value of p.

$\hat{p}$ is an observed proportion.

p^* is a critical value of a proportion corresponding to a specified α (see page 317).

A Technical Note

As we have seen (page 301), it is not exactly true that hypothesis tests and confidence intervals are equivalent for proportions. For a confidence interval, we estimate the standard deviation of $\hat{p}$ from $\hat{p}$ itself, making it a *standard error*. For the corresponding hypothesis test, we use the model's *standard deviation* for $\hat{p}$ based on the null hypothesis value p_0. When $\hat{p}$ and p_0 are close, these calculations give similar results. When they differ, you're likely to reject H_0 (because the observed proportion is far from your hypothesized value). In that case, you're better off building your confidence interval with a standard error estimated from the data rather than rely on the model you just rejected.

Figure 10.7 The one-sided 95% confidence interval (top) leaves 5% on one side (in this case the left), but leaves the other side unbounded. The 90% confidence interval is symmetric and matches the one-sided interval on the side of interest.

---

interval with a confidence level of C% corresponds to a two-sided hypothesis test with an α level of $100 - C\%$.

The relationship between confidence intervals and one-sided hypothesis tests gives us a choice. For a one-sided test with $\alpha = 5\%$, you could construct a one-sided confidence level of 95%, leaving 5% in one tail.

A one-sided confidence interval leaves one side unbounded. For example, in the home field example, we wondered whether the home field gave the home team an *advantage*, so our test was naturally one-sided. A 95% one-sided confidence interval for a proportion would be constructed from one side of the associated two-sided confidence interval:

$$0.5593 - 1.645 \times 0.01014 = 0.543$$

In order to leave 5% on one side, we used the z^* value 1.645 that leaves 5% in one tail. Writing the one-sided interval as $(0.543, \infty)$ allows us to say with 95% confidence that we know the home team will win, on average, at least 54.3% of the time. To test the hypothesis $H_0: p = 0.50$ we note that the value 0.50 is not in this interval. The lower bound of 0.543 is clearly above 0.50, showing the connection between hypothesis testing and confidence intervals.

For convenience, and to provide more information, however, we sometimes report a two-sided confidence interval even though we are interested in a one-sided test. For the home field example, we could report a 90% two-sided confidence interval:

$$0.5593 \pm 1.645 \times 0.01014 = (0.543, 0.576).$$

Notice that we *matched* the left endpoint by leaving α in *both* sides, which made the corresponding confidence level 90%. We can still see the correspondence. Because the 90% (two-sided) confidence interval for $\hat{p}$ doesn't contain 0.50, we reject the null hypothesis. Using the two-sided interval also tells us that the home team winning percentage is unlikely to be greater than 57.6%, an added benefit to understanding. You can see the relationship between the two confidence intervals in Figure 10.7.

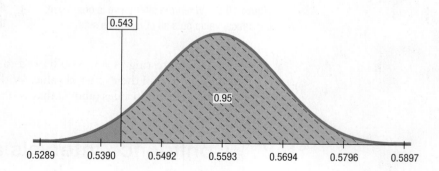

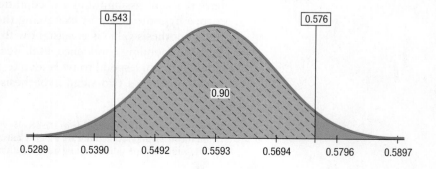

There's another good reason for finding a confidence interval along with a hypothesis test. Although the test can tell us whether the observed statistic differs from the hypothesized value, it doesn't say by how much. Often, business decisions depend not only on whether there is a statistically significant difference, but also on whether the difference is meaningful. For the home field advantage, the corresponding confidence interval shows that over a full season, home field advantage adds an average of about two to six extra victories for a team. That could make a meaningful difference in both the team's standing and in the size of the crowd.

JUST CHECKING

6 A bank is testing a new method for getting delinquent customers to pay their past-due credit card bills. The standard way was to send a letter (costing about $0.60 each) asking the customer to pay. That worked 30% of the time. The bank wants to test a new method that involves sending a DVD to the customer encouraging them to contact the bank and set up a payment plan. Developing and sending the DVD costs about $10.00 per customer. What is the parameter of interest? What are the null and alternative hypotheses?

7 The bank sets up an experiment to test the effectiveness of the DVD. The DVD is mailed to several randomly selected delinquent customers, and employees keep track of how many customers then contact the bank to arrange payments. The bank just got back the results on their test of the DVD strategy. A 90% confidence interval for the success rate is (0.29, 0.45). Their old send-a-letter method had worked 30% of the time. Can you reject the null hypothesis and conclude that the method increases the proportion at $\alpha = 0.05$? Explain.

8 Given the confidence interval the bank found in the trial of the DVD mailing, what would you recommend be done? Should the bank scrap the DVD strategy?

GUIDED EXAMPLE   Credit Card Promotion

A credit card company plans to offer a special incentive program to customers who charge at least $500 next month. The marketing department has pulled a sample of 500 customers from the same month last year and noted that the mean amount charged was $478.19 and the median amount was $216.48. The finance department says that the only relevant quantity is the proportion of customers who spend more than $500. If that proportion is not more than 25%, the program will lose money.

Among the 500 customers, 148 or 29.6% of them charged $500 or more. Can we use a confidence interval to test whether the goal of 25% for all customers was met? (Data in **Credit Card Charges**).

PLAN	**Setup** State the problem and discuss the variables and the context.	We want to know whether 25% or more of the customers will spend $500 or more in the next month and qualify for the special program. We will use the data from the same month a year ago to estimate the proportion and see whether the proportion was at least 25%.
	Hypotheses The null hypothesis is that the proportion qualifying is 25%. The alternative is that it is higher. It's clearly a one-sided test, so if we use a confidence interval, we'll have to be careful about what level we use.	The statistic is $\hat{p} = 0.296$, the proportion of customers who charged $500 or more. $$H_O: p = 0.25$$ $$H_A: p > 0.25$$

(continued)

Model Check the conditions. (Because this is a confidence interval, we use the observed successes and failures to check the Success/Failure condition.)

✓ **Independence Assumption.** Customers are not likely to influence one another when it comes to spending on their credit cards.
✓ **Randomization Condition.** This is a random sample from the company's database.
✓ **10% Condition.** The sample is less than 10% of all customers.
✓ **Success/Failure Condition.** There were 148 successes and 352 failures, both at least 10. The sample is large enough.

State your method. Here we are using a confidence interval to test a hypothesis.

Under these conditions, the sampling model is Normal. We'll create a one-proportion z-interval.

DO

Mechanics Write down the given information and determine the sample proportion.

To use a confidence interval, we need a confidence level that corresponds to the alpha level of the test. If we use $\alpha = 0.05$, we should construct a 90% confidence interval because this is a one-sided test. That will leave 5% on *each* side of the observed proportion. Determine the standard error of the sample proportion and the margin of error. The critical value is $z^* = 1.645$.

$n = 500$, so

$$\hat{p} = \frac{148}{500} = 0.296 \text{ and}$$

$$SE(\hat{p}) = \sqrt{\frac{\hat{p}\hat{q}}{n}} = \sqrt{\frac{(0.296)(0.704)}{500}} = 0.020$$

$$ME = z^* \times SE(\hat{p})$$

$$= 1.645(0.020) = 0.033$$

The confidence interval is estimate ± margin of error.

The 90% confidence interval is 0.296 ± 0.033 or $(0.263, 0.329)$.

REPORT

Conclusion Link the confidence interval to your decision about the null hypothesis, then state your conclusion in context.

MEMO

Re: Credit card promotion

Our study of a sample of customer records indicates that between 26.3% and 32.9% of customers charge $500 or more. We are 90% confident that this interval includes the true value. Because the minimum suitable value of 25% is below this interval, we conclude that it is not a plausible value, and so we reject the null hypothesis that only 25% of the customers charge more than $500 a month. The goal appears to have been met assuming that the month we studied is typical.

FOR EXAMPLE   Confidence intervals and hypothesis tests

QUESTION   Construct appropriate confidence intervals for testing the hypothesis that $p = 0.60$ from the new website design on page 312 and see what these confidence intervals says about testing at $\alpha = 0.05$ and $\alpha = 0.01$.

ANSWER   For a one-sided test at $\alpha = 0.05$ we construct a 90% confidence interval and find: $\hat{p} \pm 1.645 \times SE(\hat{p}) = 0.69 \pm 1.645 \times \sqrt{\frac{0.69 \times 0.31}{100}} = (0.614, 0.766)$. We would reject the hypothesis at $\alpha = 0.05$. But a 98% confidence interval shows: $0.69 \pm 2.054 \times \sqrt{\frac{0.69 \times 0.31}{100}} = (0.595, 0.785)$, which does not reject the null hypothesis at $\alpha = 0.01$.

10.9   Two Types of Errors

Nobody's perfect. Even with lots of evidence, we can still make the wrong decision. In fact, when we perform a hypothesis test, we can make mistakes in *two* ways:

I. The null hypothesis is true, but we mistakenly reject it.
II. The null hypothesis is false, but we fail to reject it.

These two types of errors are known as **Type I** and **Type II errors**, respectively. One way to keep the names straight is to remember that we start by assuming the null hypothesis is true, so a Type I error is the first kind of error we could make.

In medical disease testing, the null hypothesis is usually the assumption that a person is healthy. The alternative is that he or she has the disease we're testing for. So a Type I error is a *false positive*—a healthy person is diagnosed with the disease. A Type II error, in which an infected person is diagnosed as disease free, is a *false negative*. These errors have other names, depending on the particular discipline and context.

Which type of error is more serious depends on the situation. In a jury trial, a Type I error occurs if the jury convicts an innocent person. A Type II error occurs if the jury fails to convict a guilty person. Which seems more serious? In medical diagnosis, a false negative could mean that a sick patient goes untreated. A false positive might mean that the person receives unnecessary treatments or even surgery.

In business planning, a false positive result could mean that money will be invested in a project that turns out not to be profitable. A false negative result might mean that money won't be invested in a project that would have been profitable. Which error is worse, the lost investment or the lost opportunity? The answer always depends on the situation, the cost, and your point of view.

Here's an illustration of the situations:

Figure 10.8 The two types of errors occur on the diagonal where the truth and decision don't match. Remember that we *start* by assuming H_0 to be true, so an error made (rejecting it) when H_0 is true is called a Type I error. A Type II error is made when H_0 is false (and we fail to reject it).

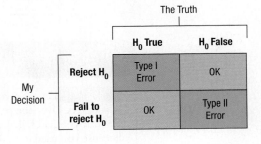

How often will a Type I error occur? It happens when the null hypothesis is true but we've had the bad luck to draw an unusual sample. To reject H_0, the P-value must fall below α. When H_0 is true, that happens *exactly* with probability α. So when you choose level α, you're setting the probability of a Type I error to α.

What if H_0 is not true? Then we can't possibly make a Type I error. You can't get a false positive from a sick person. A Type I error can happen only when H_0 is true.

When H_0 is false and we reject it, we have done the right thing. A test's ability to detect a false hypothesis is called the **power** of the test. In a jury trial, power is a measure of the ability of the criminal justice system to convict people who are guilty. We'll have a lot more to say about power soon.

When H_0 is false but we fail to reject it, we have made a Type II error. We assign the letter β to the probability of this mistake. What's the value of β? That's harder to assess than α because we don't know what the value of the parameter really is. When H_0 is true, it specifies a single parameter value. But when H_0 is false, we don't have a specific one; we have many possible values. We can compute the probability β for any parameter value in H_A, but the choice of which one to pick is not always clear.

NOTATION ALERT

In Statistics, α is the probability of a Type I error and β is the probability of a Type II error.

One way to focus our attention is by thinking about the *effect size*. That is, ask: "How big a difference would matter?" Suppose a charity wants to test whether placing personalized address labels in the envelope along with a request for a donation increases the response rate above the baseline of 5%. If the minimum response that would pay for the address labels is 6%, they would calculate β for the alternative $p = 0.06$.

Of course, we could reduce β for *all* alternative parameter values by increasing α. By making it easier to reject the null, we'd be more likely to reject it whether it's true or not. The only way to reduce *both* types of error is to collect more evidence or, in statistical terms, to collect more data. Otherwise, we just wind up trading off one kind of error against the other. Whenever you design a survey or experiment, it's a good idea to calculate β (for a reasonable α level). Use a parameter value in the alternative that corresponds to an effect size that you want to be able to detect. Too often, studies fail because their sample sizes are too small to detect the change they are looking for.

JUST CHECKING

9 Remember our bank that's sending out DVDs to try to get customers to make payments on delinquent loans? It is looking for evidence that the costlier DVD strategy produces a higher success rate than the letters it has been sending. Explain what a Type I error is in this context and what the consequences would be to the bank.

10 What's a Type II error in the bank experiment context and what would the consequences be?

11 If the DVD strategy *really* works well—actually getting 60% of the people to pay their balances—would the power of the test be higher or lower than if the rate were 32%? Explain briefly.

FOR EXAMPLE   Type I and Type II errors

QUESTION   After several retests of the clickthrough rate (page 312), the new design is adopted. Over the past year, on all customers, the clickthrough rate has been over 70%. If you had failed to reject the null hypothesis (and didn't adopt the new deign) because you used $\alpha = 0.01$, what error would you have made?

ANSWER   If we had failed to reject the null hypothesis, we would have made a Type II error.

*10.10   Power

Remember, we can never prove a null hypothesis true. We can only fail to reject it. But when we fail to reject a null hypothesis, it's natural to wonder whether we looked hard enough. Might the null hypothesis actually be false and our test too weak to tell?

When the null hypothesis actually *is* false, we hope our test is strong enough to reject it. We'd like to know how likely we are to succeed. The power of the test gives us a way to think about that. The power of a test is the probability that it correctly rejects a false null hypothesis. When the power is high, we can be confident that we've looked hard enough. We know that β is the probability that a test *fails* to reject a false null hypothesis, so the power of the test is $1 - \beta$. We might have just written $1 - \beta$, but power is such an important concept that it gets its own name.

Power and Effect Size

When planning a study, it's wise to think about the size of the effect we're looking for. We've called the effect size the difference between the null hypothesis and the observed statistic. In planning, it's the difference between the null hypothesis and a particular alternative we're interested in. It's easier to see a larger effect size, so the power of the study will increase with the effect size.

Once the study has been completed we'll base our business decision on the observed effect size, the difference between the null hypothesis and the observed value.

Whenever a study fails to reject its null hypothesis, the test's power comes into question. Was the sample size big enough to detect an effect had there been one? Might we have missed an effect large enough to be interesting just because we failed to gather sufficient data or because there was too much variability in the data we could gather? Might the problem be that the experiment simply lacked adequate power to detect their ability?

When we calculate power, we base our calculation on the smallest effect that might influence our business decision. The value of the power depends on how large this effect is. For proportions, that effect size is $p - p_0$; for means it's $\mu - \mu_0$. The power depends directly on the effect size. It's easier to see larger effects, so the further p_0 is from p the greater the power.

How can we decide what power we need? Choice of power is more a financial or scientific decision than a statistical one because to calculate the power, we need to specify the alternative parameter value we're interested in. In other words, power is calculated for a particular effect size, and it changes depending on the size of the effect we want to detect.

Graph It!

It makes intuitive sense that the larger the effect size, the easier it should be to see it. Obtaining a larger sample size decreases the probability of a Type II error, so it increases the power. It also makes sense that the more we're willing to accept a Type I error, the less likely we will be to make a Type II error.

Figure 10.9 may help you visualize the relationships among these concepts. Although we'll use proportions to show the ideas, a similar picture and similar statements also hold true for means as we'll see in the next chapter. Suppose we are testing $H_0: p = p_0$ against the alternative $H_A: p > p_0$. We'll reject the null if

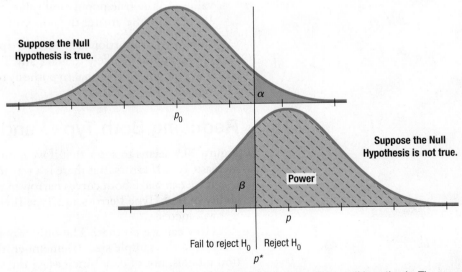

Figure 10.9   The power of a test is the probability that it rejects a false null hypothesis. The upper figure shows the null hypothesis model. We'd reject the null in a one-sided test if we observed a value in the red region to the right of the critical value, p^*. The lower figure shows the model if we assume that the true value is p. If the true value of p is greater than p_0, then we're more likely to observe a value that exceeds the critical value and make the correct decision to reject the null hypothesis. The power of the test is the green region on the right of the lower figure. Of course, even drawing samples whose observed proportions are distributed around p, we'll sometimes get a value in the red region on the left and make a Type II error of failing to reject the null.

the observed proportion, $\hat{p}$, is big enough. By *big enough*, we mean $\hat{p} > p^*$ for some critical value p^* (shown as the red region in the right tail of the upper curve). The upper model shows a picture of the sampling distribution model for the proportion when the null hypothesis is true. If the null were true, then this would be a picture of that truth. We'd make a Type I error whenever the sample gave us $\hat{p} > p^*$ because we would reject the (true) null hypothesis. Unusual samples like that would happen only with probability α.

In reality, though, the null hypothesis is rarely *exactly* true. The lower probability model supposes that H_0 is not true. In particular, it supposes that the true value is p, not p_0. It shows a distribution of possible observed $\hat{p}$ values around this true value. Because of sampling variability, sometimes $\hat{p} < p^*$ and we fail to reject the (false) null hypothesis. Then we'd make a Type II error. The area under the curve to the left of p^* in the bottom model represents how often this happens. The probability is β. In this picture, β is less than half, so most of the time we *do* make the right decision. The *power* of the test—the probability that we make the right decision—is shown as the region to the right of p^*. It's $1 - \beta$.

We calculate p^* based on the upper model because p^* depends only on the null model and the alpha level. No matter what the true proportion, p^* doesn't change. After all, we don't *know* the truth, so we can't use it to determine the critical value. But we always reject H_0 when $\hat{p} > p^*$.

How often we reject H_0 when it's *false* depends on the effect size. We can see from the picture that if the true proportion were further from the hypothesized value, the bottom curve would shift to the right, making the power greater.

We can see several important relationships from this figure:

- Power = $1 - \beta$.
- Moving the critical value (p^* in the case of proportions) to the right, reduces α, the probability of a Type I error, but increases β, the probability of a Type II error. It correspondingly reduces the power.
- The larger the true effect size, the real difference between the hypothesized value and the true population value, the smaller the chance of making a Type II error and the greater the power of the test.

If the two proportions are very far apart, the two models will barely overlap, and we would not be likely to make any Type II errors at all—but then, we are unlikely to really need a formal hypothesis testing procedure to see such an obvious difference.

Reducing Both Type I and Type II Errors

Figure 10.9 seems to show that if we reduce Type I errors, we automatically must increase Type II errors. But there is a way to reduce both. Can you think of it?

If we can make both curves narrower, as shown in Figure 10.10, then the probability of both Type I errors and Type II errors will decrease, and the power of the test will increase.

How can we do that? The only way is to reduce the standard deviations by increasing the sample size. (Remember, these are pictures of sampling distribution models, not of data.) Increasing the sample size works regardless of the true population parameters. But recall the curse of diminishing returns. The standard deviation of the sampling distribution model decreases only as the *square root* of the sample size, so to halve the standard deviations, we must *quadruple* the sample size.

Figure 10.10 Making the standard deviations smaller increases the power without changing the alpha level or the corresponding *z*-critical value. The proportions are just as far apart as in Figure 10.9, but the error rates are reduced.

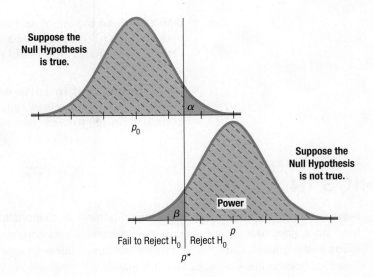

WHAT CAN GO WRONG?

- **Don't base your null hypotheses on what you see in the data.** You are not allowed to look at the data first and then adjust your null hypothesis so that it will be rejected. If your sample value turns out to be $\hat{p} = 51.8\%$ with a standard deviation of 1%, don't form a null hypothesis just big enough so you'll be able to reject it like $H_0: \hat{p} = 49.8\%$. The null hypothesis should not be based on the data you collect. It should describe the "nothing interesting" or "nothing has changed" situation.

- **Don't base your alternative hypothesis on the data either.** You should always think about the situation you are investigating and base your alternative hypothesis on that. Are you interested only in knowing whether something has *increased*? Then write a one-tail (upper tail) alternative. Or would you be equally interested in a change in either direction? Then you want a two-tailed alternative. You should decide whether to do a one- or two-tailed test based on what results would be of interest to you, not on what you might see in the data.

- **Don't make your null hypothesis what you want to show to be true.** Remember, the null hypothesis is the status quo, the nothing-is-strange-here position a skeptic would take. You wonder whether the data cast doubt on that. You can reject the null hypothesis, but you can never "accept" or "prove" the null.

- **Don't forget to check the conditions.** The reasoning of inference depends on randomization. No amount of care in calculating a test result can save you from a biased sample. The probabilities you compute depend on the independence assumption. And your sample must be large enough to justify your use of a Normal model.

- **Don't believe too strongly in arbitrary alpha levels.** There's not really much difference between a P-value of 0.051 and a P-value of 0.049, but sometimes it's regarded as the difference between night (having to retain H_0) and day (being able to shout to the world that your results are "statistically significant"). It may just be better to report the P-value and a confidence interval and let the world (perhaps your manager or client) decide along with you.

- **Don't confuse practical and statistical significance.** A large sample size can make it easy to discern even a trivial change from the null hypothesis value. On the other hand, you could miss an important difference if your test lacks sufficient power.

- **Don't forget that in spite of all your care, you might make a wrong decision.** No one can ever reduce the probability of a Type I error (α) or of a Type II error (β) to zero (but increasing the sample size helps).

ETHICS IN ACTION

Shellie Cooper, longtime owner of a small organic food store, specializes in locally produced organic foods and products. Over the years Shellie's customer base has been quite stable, consisting mainly of health-conscious individuals who tend not to be very price sensitive, opting to pay higher prices for better-quality local, organic products. However, faced with increasing competition from grocery chains offering more organic choices, Shellie is now thinking of offering coupons. She needs to decide between the newspaper and the Internet. She recently read that the percentage of consumers who use printable Internet coupons is on the rise but, at 15%, is still much less than the 40% who clip and redeem newspaper coupons. Nonetheless, she is interested in learning more about the Internet and sets up a meeting with Jack Kasor, a Web consultant. She discovers that for an initial investment and continuing monthly fee, Jack would design Shellie's website, host it on his server, and broadcast e-coupons to her customers at regular intervals. While she was concerned about the difference in redemption rates for e-coupons vs. newspaper coupons, Jack assured her that e-coupon redemptions are continuing to rise and that she should expect between 15% and 40% of her customers to redeem them. Shellie agreed to give it a try. After the first six months, Jack informed Shellie that the proportion of her customers who redeemed e-coupons was significantly greater than 15%. He determined this by selecting several broadcasts at random and found the number redeemed (483) out of the total number sent (3000). Shellie thought that this was positive and made up her mind to continue the use of e-coupons.

- Identify the ethical dilemma in this scenario.

- What are the undesirable consequences?

- Propose an ethical solution that considers the welfare of all stakeholders.

WHAT HAVE WE LEARNED?

Learning Objectives

Know how to formulate a null and alternative hypothesis for a question of interest.

- The null hypothesis specifies a parameter and a (null) value for that parameter.
- The alternative hypothesis specifies a range of plausible values should we fail to reject the null.

Be able to perform a hypothesis test for a proportion.

- The null hypothesis has the form $H_0: p = p_0$.
- We estimate the standard deviation of the sampling distribution of the sample proportion by assuming that the null hypothesis is true:

$$SD(\hat{p}) = \sqrt{\frac{p_0 q_0}{n}}$$

- We refer the statistic $z = \dfrac{\hat{p} - p_0}{SD(\hat{p})}$ to the standard Normal model.

Understand P-values.

- A P-value is the estimated probability of observing a statistic value at least as far from the (null) hypothesized value as the one we have actually observed.
- A small P-value indicates that the statistic we have observed would be unlikely were the null hypothesis true. That leads us to doubt the null.
- A large P-value just tells us that we have insufficient evidence to doubt the null hypothesis. In particular, it does not prove the null to be true.

Know the reasoning of hypothesis testing.

- State the **hypotheses**.
- Determine (and check assumptions for) the sampling distribution **model**.
- Calculate the test statistic—the **mechanics**.
- State your **conclusions and decisions**.

Be able to decide on a two-sided or one-sided alternative hypothesis, and justify your decision.

Compare P-values to a pre-determined α-level to decide whether to reject the null hypothesis.

Know the value of estimating and reporting the effect size.

- A test may be statistically significant, but practically meaningless if the estimated effect is of trivial importance.

Be aware of the risks of making errors when testing hypotheses.

- A Type I error occurs when rejecting a null hypothesis if that hypothesis is, in fact, true.
- A Type II error occurs when failing to reject a null hypothesis if that hypothesis is, in fact, false.

Understand the concept of the power of a test.

- We are particularly concerned with power when we fail to reject a null hypothesis.
- The power of a test reports, for a specified effect size, the probability that the test would reject a false null hypothesis.
- Remember that increasing the sample size will generally improve the power of any test.

Terms

Alpha level
The threshold P-value that determines when we reject a null hypothesis. Using an alpha level of α, if we observe a statistic whose P-value based on the null hypothesis is less than α, we reject that null hypothesis.

Alternative hypothesis
The hypothesis that proposes what we should conclude if we find the null hypothesis to be unlikely.

Critical value
The value in the sampling distribution model of the statistic whose P-value is equal to the alpha level. Any statistic value further from the null hypothesis value than the critical value will have a smaller P-value than α and will lead to rejecting the null hypothesis. The critical value is often denoted with an asterisk, as z^*, for example.

Effect size
The difference between the null hypothesis and an observed (or proposed) value.

Null hypothesis
The claim being assessed in a hypothesis test. Usually, the null hypothesis is a statement of "no change from the traditional value," "no effect," "no difference," or "no relationship." For a claim to be a testable null hypothesis, it must specify a value for some population parameter that can form the basis for assuming a sampling distribution for a test statistic.

One-proportion z-test
A test of the null hypothesis that the proportion of a single sample equals a specified value ($H_0: p = p_0$) by comparing the statistic $z = \dfrac{\hat{p} - p_0}{SD(\hat{p})}$ to a standard Normal model.

One-sided alternative	An alternative hypothesis is one-sided (e.g., $H_A: p > p_0$ or $H_A: p < p_0$) when we are interested in deviations in *only one* direction away from the hypothesized parameter value.
P-value	The probability of observing a value for a test statistic at least as far from the hypothesized value as the statistic value actually observed if the null hypothesis is true. A small P-value indicates that the observation obtained is improbable given the null hypothesis and thus provides evidence against the null hypothesis.
*Power	The probability that a hypothesis test will correctly reject a false null hypothesis. To find the power of a test, we must specify a particular alternative parameter value as the "true" value. For any specific value in the alternative, the power is $1 - \beta$.
Significance level	Another term for the alpha level, used most often in a phrase such as "at the 5% significance level."
Two-sided alternative	An alternative hypothesis is two-sided (e.g., $H_A: p \neq p_0$) when we are interested in deviations in *either* direction away from the hypothesized parameter value.
Type I error	The error of rejecting a null hypothesis when in fact it is true (also called a "false positive"). The probability of a Type I error is α.
Type II error	The error of failing to reject a null hypothesis when in fact it is false (also called a "false negative"). The probability of a Type II error is commonly denoted β and depends on the effect size.

TECHNOLOGY HELP: Hypothesis Tests

Hypothesis tests for proportions are so easy and natural that many statistics packages don't offer special commands for them. Most statistics programs want to know the "success" and "failure" status for each case. Usually these are given as 1 or 0, but they might be category names like "yes" and "no." Often we just know the proportion of successes, $\hat{p}$, and the total count, n. Computer packages don't usually deal naturally with summary data like this, but see below for a couple of important exceptions (Minitab and JMP).

In some programs you can reconstruct the original values. But even when you have reconstructed (or can reconstruct) the raw data values, often you won't get *exactly* the same test statistic from a computer package as you would find working by hand. The reason is that many packages make approximations. The result is very close, but not exactly the same. If you use a computer package, you may notice slight discrepancies between your answers and the answers in the back of the book, but they're not important.

Reports about hypothesis tests generated by technologies don't follow a standard form. Most will name the test and provide the test statistic value, its standard deviation, and the P-value. But these elements may not be labeled clearly. For example, the expression "Prob > $|z|$ means the probability (the "Prob") of observing a test statistic whose magnitude (the absolute value tells us this) is larger than that

of the one (the "z") found in the data (which, because it is written as "z," we know follows a Normal model). That is a fancy (and not very clear) way of saying two-sided P-value for a test based on the Normal model. In some packages, you can specify that the test be one-sided. Others might report three P-values, covering the ground for both one-sided tests and two-sided tests.

Sometimes a confidence interval and hypothesis test are automatically given together.

Often, the standard deviation of the statistic is called the "standard error," and usually that's appropriate because we've had to estimate its value from the data. That's not the case for proportions, however: We get the standard deviation for a proportion from the null hypothesis value. Nevertheless, you may see the standard deviation called a "standard error" even for tests with proportions.

It's common for statistics packages and calculators to report more digits of "precision" than could possibly have been found from the data. You can safely ignore them. Round values such as the standard deviation to one digit more than the number of digits reported in your data.

Here are the kind of results you might see in typical computer output.

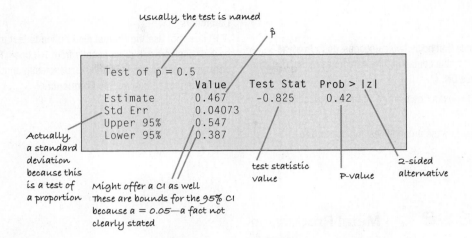

usually, the test is named

$\hat{p}$

```
Test of p = 0.5
                 Value   Test Stat   Prob > |z|
Estimate         0.467    -0.825        0.42
Std Err          0.04073
Upper 95%        0.547
Lower 95%        0.387
```

Actually, a standard deviation because this is a test of a proportion

Might offer a CI as well These are bounds for the 95% CI because a = 0.05—a fact not clearly stated

test statistic value

P-value

2-sided alternative

EXCEL

Inference methods for proportions are not part of the standard Excel tool set.

To calculate a z-test of a proportion in Excel:

	A	B	C	D	E
1	z-Test of a Proportion				
2					(Expression for Column D)
3	Sample proportion	0.467	z Stat	-0.81	=(B3-B5)/((B5*(1-B5)/B4)^0.5)
4	Sample size	150	P(Z<=z) one-tail	0.2094	=(1-NORM.S.DIST(ABS(D3),TRUE))
5	Hypothesized proportion	0.5	z Critical one-tail	1.6449	=NORM.S.INV(1-B6)
6	Alpha	0.05	P(Z<=z) two-tail	0.4189	=2*D4
7			z Critical two-tail	1.9600	=NORM.S.INV(1-B6/2)

- To calculate the values in cells D3–D7, type the formulas in cells E3 through E7.

- After the formulas are set in column D, enter information for any z-test for a proportion in cells B3 through B6, and the results will show in column D. Note that the one- and two-tail critical values given are for the upper end of the distribution. If a lower one-tail test is being conducted, the critical value will be negative, and for a two-tail test the lower tail critical value will be negative.

Note: In pre-2010 versions of Microsoft Excel, the functions used in these calculations are "=NORMSINV" and "=NORMSDIST".

XLSTAT

To find a one-proportion z interval:

- Select **Parametric Tests,** and then choose **Tests for one proportion**.

- Under the **General** tab, choose the **Data format** to be either **Frequency** or **Proportion** depending on the form of your data.

- Enter the **frequency** of your variable (or **proportion**) and the **sample size**.

- Enter the **test proportion**.

- Under **Data format** choose the appropriate button.

- Under the **Options** tab choose the **Alternative hypothesis** of **Proportion – Test proportion ≠ D**. Enter **0** for the **Hypothesized difference (D)**.

- Enter 5 under **Significance Level.** The output will show the 95% confidence interval.

MINITAB

Choose **Basic Statistics** from the **Stat** menu.

- Choose **1 Proportion** from the Basic Statistics submenu.

- If the data are category names in a variable, assign the variable from the variable list box to the **Samples in columns** box.

- If you have summarized data, click the **Summarized Data** button and fill in the number of trials and the number of successes.

- Mark a check next to **Perform hypothesis test** and enter the hypothesized proportion.

- Click the **Options** button and specify the remaining details.

- If you have a large sample, check **Use test and interval based on Normal distribution.**

- Click the **OK** button.

Comments

When working from a variable that names categories, Minitab treats the last category as the "success" category. You can specify how the categories should be ordered.

R

In library(stats):

- prop.test (X, n, p = NULL, alternative = c("two.sided," "less," "greater"), conf.level = 0.95, correct=FALSE)

- will test the hypothesis that $p = p_0$ against various alternatives. For example with 260 successes out of 500, to test that $p = 0.5$ vs. $p \neq 0.5$, use:

- prop.test(260,500,0.5,"two.sided," correct=FALSE)

SPSS

SPSS does not perform the hypothesis test for one proportion using the Normal approximation.

For a categorical variable that holds two categories, go to **Analyze >
Distribution**. Put the variable containing the categories in the **Y,
Columns** box and press **OK**.

- Click the red triangle next to the column name and select **Test
  Probabilities**.

- Put both p_0 and $1-p_0$ into the probabilities and click **Done**.

Comments

JMP does not use the Normal distribution to test the proportions and
so your answers will differ slightly from the book. For summarized
data, you will need to have the frequencies in another column and
designate that column as the **Frequencies**.

Brief **Case**

Metal Production

Ingots are huge pieces of metal, often weighing in excess of 20,000 pounds,
made in a giant mold. They must be cast in one large piece for use in fabricat-
ing large structural parts for cars and planes. If they crack while being made,
the crack may propagate into the zone required for the part, compromising its
integrity. Airplane manufacturers insist that metal for their planes be defect-
free, so the ingot must be made over if any cracking is detected.

Even though the metal from the cracked ingot is recycled, the scrap cost
runs into the tens of thousands of dollars. Metal manufacturers would like to
avoid cracking if at all possible, but only about 75% of the ingots have been free
of cracks. The data from 5000 ingots produced after some changes were made
to the process are found in the file **Ingots**. The variable *Crack* indicates whether
a crack was found (1) or not (0). The variable *Impurities* shows the amount (in
ppm) of impurities found in a sample from each ingot. Select a random sample
of 100 ingots and test the claim that the cracking rate has decreased from 25%.
Find a confidence interval for the cracking rate as well. Now select a random
sample of 500 ingots and test the claim and find the confidence interval again.
Prepare a short report about your findings including any differences you see
between the conclusions from the two samples.

Loyalty Program

A marketing manager has sent out 10,000 mail pieces to a random sample of
customers to test a new web-based loyalty program and see its impact on customer spending. The customers either received nothing (No Offer), a free
companion airline ticket (Free Flight), or free flight insurance on their next
flight (Free Insurance). The analyst who designed the test used a stratified random sample to ensure that all the market segments are represented, but the
manager has several concerns. First, she worries that the *Travel* segment (which
comprises 25% of all customers) is underrepresented (variable *Segment*). In addition, she worries that fewer than 1/3 of the customers in that segment were
held out as controls and received no offer. Using the data found in the file
Loyalty Program write a short report to the manager testing the appropriate
hypotheses and summarizing your findings. Include in your report 95% confidence intervals for the proportion of customers who responded to the offer
by signing up for the loyalty program in the various segments. (The variable
Response indicates a 1 for responders and 0 for nonresponders.)

EXERCISES

SECTION 10.1

1. For each of the following situations, write the null and alternative hypotheses in terms of parameter values. Example: We want to know if the proportion of up days in the stock market is 50%. Answer: Let p = the proportion of up days. $H_0: p = 0.5$ vs. $H_A: p \neq 0.5$.

a) A casino wants to know if their slot machine really delivers the 1 in 100 win rate that it claims.

b) A pharmaceutical company wonders if their new drug has a cure rate different from the 30% reported by the placebo.

c) A bank wants to know if the percentage of customers using their website has changed from the 40% that used it before their system crashed last week.

2. As in Exercise 1, for each of the following situations, write the null and alternative hypotheses in terms of parameter values.

a) A recent UMass-Amherst study found that seat-belt compliance in Massachusetts was 73% in 2011. The state wants to know if it has changed.

b) Last year, a survey found that 45% of the employees were willing to pay for on-site day care. The company wants to know if that has changed.

c) Regular card customers have a default rate of 6.7%. A credit card bank wants to know if that rate is different for their Gold card customers.

SECTION 10.3

3. Which of the following are true? If false, explain briefly.

a) A very high P-value is strong evidence that the null hypothesis is false.

b) A very low P-value proves that the null hypothesis is false.

c) A high P-value shows that the null hypothesis is true.

d) A P-value below 0.05 is always considered sufficient evidence to reject a null hypothesis.

4. Which of the following are true? If false, explain briefly.

a) A very low P-value provides evidence against the null hypothesis.

b) A high P-value is strong evidence in favor of the null hypothesis.

c) A P-value above 0.10 shows that the null hypothesis is true.

d) If the null hypothesis is true, you can't get a P-value below 0.01.

SECTION 10.4

5. A consulting firm had predicted that 35% of the employees at a large firm would take advantage of a new company Credit Union, but management is skeptical. They doubt the rate is that high. A survey of 300 employees shows that 138 of them are currently taking advantage of the Credit Union. From the sample proportion

a) Find the standard deviation of the sample proportion based on the null hypothesis.

b) Find the z-statistic.

c) Does the z-statistic seem like a particularly large or small value?

6. A survey of 100 CEOs finds that 60 think the economy will improve next year. Is there evidence that the rate is higher among all CEOs than the 55% reported by the public at large?

a) Find the standard deviation of the sample proportion based on the null hypothesis.

b) Find the z-statistic.

c) Does the z-statistic seem like a particularly large or small value?

SECTION 10.5

7. For each of the following, write out the null and alternative hypothesis, being sure to state whether it is one-sided or two-sided.

a) A company reports that last year 40% of their reports in accounting were on time. From a random sample this year, they want to know if that proportion has changed.

b) Last year, 42% of the employees enrolled in at least one wellness class at the company's site. Using a survey, they want to know if a greater percentage is planning to take a wellness class this year.

c) A political candidate wants to know from recent polls if she's going to garner a majority of votes in next week's election.

8. For each of the following, write out the alternative hypothesis, being sure to indicate whether it is one-sided or two-sided.

a) *Consumer Reports* discovered that 20% of a certain computer model had warranty problems over the first three months. From a random sample, the manufacturer wants to know if a new model has improved that rate.

b) The last time a philanthropic agency requested donations, 4.75% of people responded. From a recent pilot mailing, they wonder if that rate has increased.

c) A student wants to know if other students on her campus prefer Coke or Pepsi.

SECTION 10.6

9. Which of the following statements are true? If false, explain briefly.

a) Using an alpha level of 0.05, a P-value of 0.04 results in rejecting the null hypothesis.
b) The alpha level depends on the sample size.
c) With an alpha level of 0.01, a P-value of 0.10 results in rejecting the null hypothesis.
d) Using an alpha level of 0.05, a P-value of 0.06 means the null hypothesis is true.

10. Which of the following statements are true? If false, explain briefly.

a) It is better to use an alpha level of 0.05 than an alpha level of 0.01.
b) If we use an alpha level of 0.01, then a P-value of 0.001 is statistically significant.
c) If we use an alpha level of 0.01, then we reject the null hypothesis if the P-value is 0.001.
d) If the P-value is 0.01, we reject the null hypothesis for any alpha level greater than 0.01.

SECTION 10.7

11. For each of the following situations, find the critical value(s) for z.

a) $H_0: p = 0.5$ vs. $H_A: p \neq 0.5$ at $\alpha = 0.05$.
b) $H_0: p = 0.4$ vs. $H_A: p > 0.4$ at $\alpha = 0.05$.
c) $H_0: p = 0.5$ vs. $H_A: p > 0.5$ at $\alpha = 0.01; n = 345$.

12. For each of the following situations, find the critical value for z.

a) $H_0: p = 0.05$ vs. $H_A: p > 0.05$ at $\alpha = 0.05$.
b) $H_0: p = 0.6$ vs. $H_A: p \neq 0.6$ at $\alpha = 0.01$.
c) $H_0: p = 0.5$ vs. $H_A: p < 0.5$ at $\alpha = 0.01; n = 500$.
d) $H_0: p = 0.2$ vs. $H_A: p < 0.2$ at $\alpha = 0.01$.

SECTION 10.8

13. Suppose that you are testing the hypotheses $H_0: p = 0.20$ vs. $H_A: p \neq 0.20$. A sample of size 250 results in a sample proportion of 0.25.

a) Construct a 95% confidence interval for p.
b) Based on the confidence interval, at $\alpha = .05$ can you reject H_0? Explain.
c) What is the difference between the standard error and standard deviation of the sample proportion?
d) Which is used in computing the confidence interval?

14. Suppose that you are testing the hypotheses $H_0: p = 0.40$ vs. $H_A: p > 0.40$. A sample of size 200 results in a sample proportion of 0.55.

a) Construct a 90% confidence interval for p.
b) Based on the confidence interval, at $\alpha = .05$ can you reject H_0? Explain.

c) What is the difference between the standard error and standard deviation of the sample proportion?
d) Which is used in computing the confidence interval?

SECTION 10.9

15. For each of the following situations, state whether a Type I, a Type II, or neither error has been made. Explain briefly.

a) A bank wants to know if the enrollment on their website is above 30% based on a small sample of customers. They test $H_0: p = 0.3$ vs. $H_A: p > 0.3$ and reject the null hypothesis. Later they find out that actually 28% of all customers enrolled.
b) A student tests 100 students to determine whether other students on her campus prefer Coke or Pepsi and finds no evidence that preference for Coke is not 0.5. Later, a marketing company tests all students on campus and finds no difference.
c) A pharmaceutical company tests whether a drug lifts the headache relief rate from the 25% achieved by the placebo. They fail to reject the null hypothesis because the P-value is 0.465. Further testing shows that the drug actually relieves headaches in 38% of people.

16. For each of the following situations, state whether a Type I, a Type II, or neither error has been made.

a) A test of $H_0: p = 0.8$ vs. $H_A: p < 0.8$ fails to reject the null hypothesis. Later it is discovered that $p = 0.9$.
b) A test of $H_0: p = 0.5$ vs. $H_A: p \neq 0.5$ rejects the null hypothesis. Later is it discovered that $p = 0.65$.
c) A test of $H_0: p = 0.7$ vs. $H_A: p < 0.7$ fails to reject the null hypothesis. Later is it discovered that $p = 0.6$.

CHAPTER EXERCISES

17. Hypotheses. Write the null and alternative hypotheses to test each of the following situations.

a) An online clothing company is concerned about the timeliness of their deliveries. The VP of Operations and Marketing recently stated that she wanted the percentage of products delivered on time to be greater than 90%, and she wants to know if the company has succeeded.
b) A realty company recently announced that the proportion of houses taking more than three months to sell is now greater than 50%.
c) A financial firm's accounting reports that after improvements in their system, they now have an error rate below 2%.

18. More hypotheses. Write the null and alternative hypotheses to test each of the following situations.

a) A 2010 *Harvard Business Review* article looked at 1109 CEOs from global companies and found that 32% had MBAs. Has the percentage changed?

b) Recently, 20% of cars of a certain model have needed costly transmission work after being driven between 50,000 and 100,000 miles. The car manufacturer hopes that the redesign of a transmission component has solved this problem.

c) A market researcher for a cola company decides to field test a new flavor soft drink, planning to market it only if he is sure that over 60% of the people like the flavor.

19. Deliveries. The clothing company in Exercise 17a looks at a sample of delivery reports. They test the hypothesis that 90% of the deliveries are on time against the alternative that greater than 90% are on time and find a P-value of 0.22. Which of these conclusions is appropriate?

a) There's a 22% chance that 90% of the deliveries are on time.

b) There's a 78% chance that 90% of the deliveries are on time.

c) There's a 22% chance that the sample they drew shows the correct percentage of on-time deliveries

d) There's a 22% chance that natural sampling variation could produce a sample with an observed proportion of on-time deliveries such as the one they obtained if, in fact, 90% of deliveries are on time.

20. House sales. The realty company in Exercise 17b looks at a recent sample of houses that have sold. On testing the null hypothesis that 50% of the houses take more than three months to sell against the hypothesis that more than 50% of the houses take more than three months to sell, they find a P-value of 0.034. Which of these conclusions is appropriate?

a) There's a 3.4% chance that 50% of the houses take more than three months to sell.

b) If 50% of the houses take more than three months to sell, there's a 3.4% chance that a random sample would produce a sample proportion as high as the one they obtained.

c) There's a 3.4% chance that the null hypothesis is correct.

d) There's a 96.6% chance that 50% of the houses take more than three months to sell.

21. P-value. Have harsher penalties and ad campaigns increased seat-belt use among drivers and passengers? Observations of commuter traffic have failed to find evidence of a significant change compared with three years ago. Explain what the study's P-value of 0.17 means in this context.

22. Another P-value. A company developing scanners to search for hidden weapons at airports has concluded that a new device is significantly better than the current scanner. The company made this decision based on a P-value of 0.03. Explain the meaning of the P-value in this context.

23. Ad campaign. An information technology analyst believes that they are losing customers on their website who find the checkout and purchase system too complicated.

She adds a one-click feature to the website, to make it easier but finds that only about 10% of the customers are using it. She decides to launch an ad awareness campaign to tell customers about the new feature in the hope of increasing the percentage. She doesn't see much of a difference, so she hires a consultant to help her. The consultant selects a random sample of recent purchases, tests the hypothesis that the ads produced no change against the alternative that the percent who use the one-click feature is now greater than 10%, and finds a P-value of 0.22. Which conclusion is appropriate? Explain.

a) There's a 22% chance that the ads worked.

b) There's a 78% chance that the ads worked.

c) There's a 22% chance that the null hypothesis is true.

d) There's a 22% chance that natural sampling variation could produce poll results like these if the use of the one-click feature has increased.

e) There's a 22% chance that natural sampling variation could produce poll results like these if there's really no change in website use.

24. Mutual funds. A mutual fund manager claims that at least 70% of the stocks she selects will increase in price over the next year. We examined a sample of 200 of her selections over the past three years. Our P-value turns out to be 0.03. Test an appropriate hypothesis. Which conclusion is appropriate? Explain.

a) There's a 3% chance that the fund manager is correct.

b) There's a 97% chance that the fund manager is correct.

c) There's a 3% chance that a random sample could produce the results we observed, so it's reasonable to conclude that the fund manager is correct.

d) There's a 3% chance that a random sample could produce the results we observed if $p = .7$, so it's reasonable to conclude that the fund manager is not correct.

e) There's a 3% chance that the null hypothesis is correct.

25. Product effectiveness. A pharmaceutical company's old antacid formula provided relief for 70% of the people who used it. The company tests a new formula to see if it is better and gets a P-value of 0.27. Is it reasonable to conclude that the new formula and the old one are equally effective? Explain.

26. Car sales. A German automobile company is counting on selling more cars to the younger market segment—drivers under the age of 20. The company's market researchers survey to investigate whether or not the proportion of today's high school seniors who own their own cars is higher than it was a decade ago. They find a P-value of 0.017. Is it reasonable to conclude that more high school seniors have cars? Explain.

27. False claims? A candy company claims that in a large bag of holiday M&M's® half the candies are red and half the candies are green. You pick candies at random from a bag and discover that of the first 20 you eat, 12 are red.

a) If it were true that half are red and half are green, what is the probability you would have found that at least 12 out of 20 were red?

b) Do you think that half of the M&M's® candies in the bag are really red? Explain.

28. Scratch off. A retail company offers a "scratch off" promotion. Upon entering the store, you are given a card. When you pay, you may scratch off the coating. The company advertises that half the cards are winners and have immediate cash-back savings of $5 (the others offer $1 off any future purchase of coffee in the cafe). You aren't sure the percentage is really 50% winners.

a) The first time you shop there, you get the coffee coupon. You try again and again get the coffee coupon. Do two failures in a row convince you that the true fraction of winners isn't 50%? Explain.

b) You try a third time. You get coffee again! What's the probability of not getting a cash savings three times in a row if half the cards really do offer cash savings?

c) Would three losses in a row convince you that the store is cheating?

d) How many times in a row would you have to get the coffee coupon instead of cash savings to be pretty sure that the company isn't living up to its advertised percentage of winners? Justify your answer by calculating a probability and explaining what it means.

29. MBAs. A 2010 *Harvard Business Review* article looked at the success of CEOs with MBAs. Of the 1109 CEOs who responded, only 32% said that they had earned an MBA.

a) Estimate the percentage of all CEOs who have earned an MBA. Use a 98% confidence interval. Don't forget to check the conditions first.

b) Some believe that fewer contemporary CEOs are earning MBAs. Suppose we wished to conduct a hypothesis test to see if the fraction has fallen below the 35% mark. What does your confidence interval indicate? Explain.

c) What is the significance level of this test? Explain.

30. Stocks. A young investor in the stock market is concerned that investing in the stock market is actually gambling, since the chance of the stock market going up on any given day is 50%. She decides to track her favorite stock for 250 days and finds that on 140 days the stock was "up."

a) Find a 95% confidence interval for the proportion of days the stock is "up." Don't forget to check the conditions first.

b) Does your confidence interval provide any evidence that the market is not random? Explain.

c) What is the significance level of this test? Explain.

31. Economy. Gallup reported in 2012 that 53% of American investors are likely to say the price of energy (including gas and oil) is hurting the U.S. investment climate "a lot," according to a Wells Fargo/Gallup Investor and Retirement Optimism Index survey. The survey results are based on questions asked February 3–12, 2012, of a random sample of 1022 U.S. adults having investable assets of $10,000 or more. The percentage reported to this same question in September 2011 was 62%.

a) Is the percentage in 2012 different from that in 2011? Test the appropriate hypothesis. Find a 95% confidence interval for the sample Economic Confidence Index. Check conditions.

b) Does your confidence interval provide evidence that the percentage has changed from 2011?

c) What is the significance level of the test in b? Explain.

32. Gallup Poll. Gallup tracks daily the percentage of Americans who approve or disapprove of the job Barack Obama is doing as President. Daily results are based on telephone interviews with approximately 1500 national adults. The Gallup Poll conducted March 12–15, 2015, reported that 49% of adults approve of Obama. A media outlet claimed the true proportion to be 51%. Does the Gallup Poll contradict this claim?

a) Test the appropriate hypothesis. Find a 95% confidence interval for the sample proportion of U.S. adults who approved of the President's performance. Check conditions.

b) Does your confidence interval provide evidence to contradict the claim?

c) What is the significance level of the test in b? Explain.

33. Convenient alpha. An enthusiastic junior executive has run a test of his new marketing program. He reports that it resulted in a "significant" increase in sales (percentage of customers who make a purchase). A footnote on his report explains that he used an alpha level of 7.2% for his test. Presumably, he performed a hypothesis test against the null hypothesis of no change in sales.

a) If instead he had used an alpha level of 5%, is it more or less likely that he would have rejected his null hypothesis? Explain.

b) If he chose the alpha level 7.2% so that he could claim statistical significance, explain why this is not an ethical use of statistics.

34. Safety. The manufacturer of a new sleeping pill suspects that it may increase the risk of sleepwalking (percentage of patients who sleepwalk while taking pill), which could be dangerous. A test of the drug fails to reject the null hypothesis of increased sleepwalking when tested at alpha = 0.01.

a) If the test had been performed at alpha = 0.05, would the test have been more or less likely to reject the null hypothesis of no increase in sleepwalking?

b) Which alpha level do you think the company should use? Why?

35. Product testing. Since many people have trouble programming their VCRs, an electronics company has developed what it hopes will be easier instructions. The goal is to have at least 96% of customers succeed at being able to program their VCRs. The company tests the new system on 200 people, 188 of whom were successful. Is this strong evidence that the new system fails to meet the company's goal? A student's test of this hypothesis is shown here. How many mistakes can you find?

$$H_0: \hat{p} = 0.96$$
$$H_A: \hat{p} \neq 0.96$$
$$SRS, 0.96(200) > 10$$

$$\frac{188}{200} = 0.94; SD(\hat{p}) = \sqrt{\frac{(0.94)(0.06)}{200}} = 0.017$$

$$z = \frac{0.96 - 0.94}{0.017} = 1.18$$

$$P = P(z > 1.18) = 0.12$$

There is strong evidence that the new system does not work.

36. Gallup Poll, part 2. The Gallup-Healthways Well-Being Index tracks daily how Americans evaluate their lives, both now and in five years, on the Cantril Self-Anchoring Striving Scale, where "0" represents the worst possible life and "10" represents the best possible life. Respondents are classified by Gallup as "thriving" if they rate their current life a 7 or higher and their future life an 8 or higher. Daily results are based on a three-day rolling average, based on telephone interviews with approximately 1000 national adults. As of March 1, 2012, Gallup reported that 530 of 1000 adults were thriving. Do these responses provide strong evidence that more than half of adults are thriving? Correct the mistakes you find in the following student's attempt to test an appropriate hypothesis.

$$H_0: \hat{p} = 0.5$$
$$H_A: \hat{p} > 0.5$$
$$SRS, 1000 > 10$$

$$\frac{530}{1000} = 0.530; SD(\hat{p}) = \sqrt{\frac{(0.53)(0.47)}{1000}} = 0.016$$

Since the 90% confidence interval is $0.530 + /-1.645$ (0.016) or $(0.504, 0.556)$, we can be confident that the true percentage of those thriving is above 50%.

37. Equal opportunity? A company is sued for job discrimination because only 19% of the newly hired candidates were minorities when 27% of all applicants were minorities. Is this strong evidence that the company's hiring practices are discriminatory?

a) Is this a one-tailed or a two-tailed test? Why?
b) In this context, what would a Type I error be?
c) In this context, what would a Type II error be?
d) In this context, what is meant by the power of the test?
e) If the hypothesis is tested at the 5% level of significance instead of 1%, how would this affect the power of the test?

f) The lawsuit is based on the hiring of 37 employees. Is the power of the test higher than, lower than, or the same as it would be if it were based on 87 hires?

38. Stop signs. Highway safety engineers test new road signs, hoping that increased reflectivity will make them more visible to drivers. Volunteers drive through a test course with several of the new- and old-style signs and rate which kind shows up the best.

a) Is this a one-tailed or a two-tailed test? Why?
b) In this context, what would a Type I error be?
c) In this context, what would a Type II error be?
d) In this context, what is meant by the power of the test?
e) If the hypothesis is tested at the 1% level of significance instead of 5%, how would this affect the power of the test?
f) The engineers hoped to base their decision on the reactions of 50 drivers, but time and budget constraints may force them to cut back to 20. How would this affect the power of the test? Explain.

39. Environment. In the 1980s, it was generally believed that congenital abnormalities affected about 5% of the nation's children. Some people believe that the increase in the number of chemicals in the environment has led to an increase in the incidence of abnormalities. A recent study examined 384 children and found that 46 of them showed signs of an abnormality. Is this strong evidence that the risk has increased? (We consider a P-value of around 5% to represent reasonable evidence.)

a) Write appropriate hypotheses.
b) Check the necessary assumptions.
c) Perform the mechanics of the test. What is the P-value?
d) Explain carefully what the P-value means in this context.
e) What's your conclusion?
f) Do environmental chemicals cause congenital abnormalities?

40. Billing company. A billing company that collects bills for doctors' offices in the area is concerned that the percentage of bills being paid by Medicare has risen. Historically, that percentage has been 31%. An examination of 8368 recent bills reveals that 32% of these bills are being paid by Medicare. Is this evidence of a change in the percent of bills being paid by Medicare?

a) Write appropriate hypotheses.
b) Check the assumptions and conditions.
c) Perform the test and find the P-value.
d) State your conclusion.
e) Do you think this difference is meaningful? Explain.

41. Global warming. Gallup found in 2012 that 55% of Americans worry a great deal or a fair amount about global warming, up from 51% in 2011. Results for this 2012 Gallup poll were based on telephone interviews conducted March 8–11, 2012, with a random sample of 1024 adults, aged 18 and older, living in the United States. Using the

51% from 2011 as a baseline, does the increase from 2011 to 2012 show a significant increase in concern?

a) Write appropriate hypotheses.
b) Check the assumptions and conditions.
c) Perform the test and find the P-value.
d) State your conclusion.
e) Do you think this difference is meaningful? Explain.

42. Global warming, part 2. Gallup previously found that the percent of Americans who worry a great deal or a fair amount about global warming was 60% in 2009. Assume they asked the same number of Americans in 2009. Do the responses in this problem and Exercise 41 give evidence that concerns for global warming have decreased since 2009?

a) Write appropriate hypotheses.
b) Check the assumptions and conditions.
c) Perform the test and find the P-value.
d) State your conclusion.
e) Do you think this difference is meaningful? Explain.

43. Retirement. A survey of 1000 workers indicated that approximately 520 have invested in an individual retirement account. National data suggests that 44% of workers invest in individual retirement accounts.

a) Create a 95% confidence interval for the proportion of workers who have invested in individual retirement accounts based on the survey.
b) Does this provide evidence of a change in behavior among workers? Using your confidence interval, test an appropriate hypothesis and state your conclusion.

44. Customer satisfaction. A company hopes to improve customer satisfaction, setting as a goal no more than 5% negative comments. A random survey of 350 customers found only 10 with complaints.

a) Create a 95% confidence interval for the true level of dissatisfaction among customers.
b) Does this provide evidence that the company has reached its goal? Using your confidence interval, test an appropriate hypothesis and state your conclusion.

45. Maintenance costs. A limousine company is concerned with increasing costs of maintaining their fleet of 150 cars. After testing, the company found that the emissions systems of 7 out of the 22 cars they tested failed to meet pollution control guidelines. They had forecasted costs assuming that a total of 30 cars would need updating to meet the latest guidelines. Is this strong evidence that more than 20% of the fleet might be out of compliance? Test an appropriate hypothesis and state your conclusion. Be sure the appropriate assumptions and conditions are satisfied before you proceed.

46. Damaged goods. An appliance manufacturer stockpiles washers and dryers in a large warehouse for shipment to retail stores. Sometimes in handling them the appliances get damaged. Even though the damage may be minor, the company must sell those machines at drastically reduced prices. The company goal is to keep the proportion of damaged machines below 2%. One day an inspector randomly checks 60 washers and finds that 5 of them have scratches or dents. Is this strong evidence that the warehouse is failing to meet the company goal? Test an appropriate hypothesis and state your conclusion. Be sure the appropriate assumptions and conditions are satisfied before you proceed.

47. WebZine. A magazine called *WebZine* is considering the launch of an online edition. The magazine plans to go ahead only if it's convinced that more than 25% of current readers would subscribe. The magazine contacts a simple random sample of 500 current subscribers, and 137 of those surveyed expressed interest. What should the magazine do? Test an appropriate hypothesis and state your conclusion. Be sure the appropriate assumptions and conditions are satisfied before you proceed.

48. Truth in advertising. A garden center wants to store leftover packets of vegetable seeds for sale the following spring, but the center is concerned that the seeds may not germinate at the same rate a year later. The manager finds a packet of last year's green bean seeds and plants them as a test. Although the packet claims a germination rate of 92%, only 171 of 200 test seeds sprout. Is this evidence that the seeds have lost viability during a year in storage? Test an appropriate hypothesis and state your conclusion. Be sure the appropriate assumptions and conditions are satisfied before you proceed.

49. Women executives. A company is criticized because only 13 of 43 people in executive-level positions are women. The company explains that although this proportion is lower than it might wish, it's not surprising given that only 40% of their employees are women. What do you think? Test an appropriate hypothesis and state your conclusion. Be sure the appropriate assumptions and conditions are satisfied before you proceed.

50. Jury. Census data for a certain county shows that 19% of the adult residents are Hispanic. Suppose 72 people are called for jury duty, and only 9 of them are Hispanic. Does this apparent underrepresentation of Hispanics call into question the fairness of the jury selection system? Explain.

51. Real estate. A national real estate magazine advertised that 15% of first-time home buyers had a family income below $40,000. A national real estate firm believes this percentage is too low and samples 100 of its records. The firm finds that 25 of its first-time home buyers did have a family income below $40,000. Does the sample suggest that the proportion of first-time home buyers with an income less than $40,000 is more than 15%? Comment and write up your own conclusions based on an appropriate confidence interval as well as a hypothesis test. Include any assumptions you made about the data.

52. TV ads. A start-up company is about to market a new computer printer. It decides to gamble by running commercials during the Super Bowl. The company hopes that name recognition will be worth the high cost of the ads. The goal of the company is that over 40% of the public recognize its brand name and associate it with computer equipment. The day after the game, a pollster contacts 420 randomly chosen adults and finds that 181 of them know that this company manufactures printers. Would you recommend that the company continue to advertise during the Super Bowl? Explain.

53. Business ethics. One study reports that 30% of newly hired MBAs are confronted with unethical business practices during their first year of employment. One business school dean wondered if her MBA graduates had similar experiences. She surveyed recent graduates from her school's MBA program and found that 27% of the 120 graduates from the previous year claim to have encountered unethical business practices in the workplace. Can she conclude that her graduates' experiences are different?

54. Stocks, part 2. A young investor believes that he can beat the market by picking stocks that will increase in value. Assume that on average 50% of the stocks selected by a portfolio manager will increase over 12 months. Of the 25 stocks that the young investor bought over the last 12 months, 14 have increased. Can he claim that he is better at predicting increases than the typical portfolio manager?

55. U.S. politics. Gallup reported in March 2012 that 73% of American investors say that a politically divided federal government is hurting the U.S. investment climate "a lot". Is there any evidence that the percentage has changed significantly from the 74% reported in September 2011? (These results are based on questions asked of a random sample of 1022 U.S. adults having investable assets of $10,000 or more.)

a) Find the z-score of the observed proportion.
b) Compare the z-score to the critical value for a 0.1% significance level using a two-sided alternative.
c) Explain your conclusion.

56. U.S. politics, part 2. In the same 2012 survey conducted in Exercise 55, Gallup reported that 66% of American investors say that the federal budget deficit is hurting the U.S. investment climate "a lot." Is there any evidence that the percentage has decreased from the 79% reported in September 2011?

a) Find the z-score of the observed proportion.
b) Compare the z-score to the critical value for a 0.1% significance level using a one-sided alternative.
c) Explain your conclusion.

57. Testing cars. A clean air standard requires that vehicle exhaust emissions not exceed specified limits for various pollutants. Many states require that cars be tested annually to be sure they meet these standards. Suppose state regulators double-check a random sample of cars that a suspect repair shop has certified as okay. They will revoke the shop's license if they find significant evidence that the shop is certifying vehicles that do not meet standards.

a) In this context, what is a Type I error?
b) In this context, what is a Type II error?
c) Which type of error would the shop's owner consider more serious?
d) Which type of error might environmentalists consider more serious?

58. Quality control. Production managers on an assembly line must monitor the output to be sure that the level of defective products remains small. They periodically inspect a random sample of the items produced. If they find a significant increase in the proportion of items that must be rejected, they will halt the assembly process until the problem can be identified and repaired.

a) Write null and alternative hypotheses for this problem.
b) What is the Type I and Type II error in this context?
c) Which type of error would the factory owner consider more serious?
d) Which type of error might customers consider more serious?

59. Testing cars, again. As in Exercise 57, state regulators are checking up on repair shops to see if they are certifying vehicles that do not meet pollution standards.

a) In this context, what is meant by the power of the test the regulators are conducting?
b) Will the power be greater if they test 20 or 40 cars? Why?
c) Will the power be greater if they use a 5% or a 10% level of significance? Why?
d) Will the power be greater if the repair shop's inspectors are only a little out of compliance or a lot? Why?

60. Quality control, part 2. Consider again the task of the quality control inspectors in Exercise 58.

a) In this context, what is meant by the power of the test the inspectors conduct?
b) They are currently testing 5 items each hour. Someone has proposed they test 10 items each hour instead. What are the advantages and disadvantages of such a change?
c) Their test currently uses a 5% level of significance. What are the advantages and disadvantages of changing to a significance level of 1%?
d) Suppose that as a day passes one of the machines on the assembly line produces more and more items that are defective. How will this affect the power of the test?

61. Statistics software. A Statistics professor has observed that for several years about 13% of the students who initially enroll in his Introductory Statistics course withdraw before the end of the semester. A salesperson suggests that he try a statistics software package that gets students more involved with computers, predicting that it will cut the dropout rate. The software is expensive, and the salesperson offers to let the professor use it for a semester to see if the dropout rate goes down significantly. The professor will have to pay for the software only if he chooses to continue using it.

a) Is this a one-tailed or two-tailed test? Explain.
b) Write the null and alternative hypotheses.
c) In this context, explain what would happen if the professor makes a Type I error.
d) In this context, explain what would happen if the professor makes a Type II error.
e) What is meant by the power of this test?

62. Radio ads. A company is willing to renew its advertising contract with a local radio station only if the station can prove that more than 20% of the residents of the city have heard the ad and recognize the company's product. The radio station conducts a random phone survey of 400 people.

a) What are the hypotheses?
b) The station plans to conduct this test using a 10% level of significance, but the company wants the significance level lowered to 5%. Why?
c) What is meant by the power of this test?
d) For which level of significance will the power of this test be higher? Why?
e) They finally agree to use $\alpha = 0.05$, but the company proposes that the station call 600 people instead of the 400 initially proposed. Will that make the risk of Type II error higher or lower? Explain.

63. Statistics software, part 2. Initially, 203 students signed up for the Statistics course in Exercise 59. They used the software suggested by the salesperson, and only 11 dropped out of the course.

a) Should the professor spend the money for this software? Support your recommendation with an appropriate test.
b) Explain what your P-value means in this context.

64. Radio ads, part 2. The company in Exercise 62 contacts 600 people selected at random, and 133 can remember the ad.

a) Should the company renew the contract? Support your recommendation with an appropriate test.
b) Explain carefully what your P-value means in this context.

JUST CHECKING ANSWERS

1 You can't conclude that the null hypothesis is true. You can conclude only that the experiment was unable to reject the null hypothesis. They were unable, on the basis of 12 patients, to show that aspirin was effective.

2 The null hypothesis is $H_0: p = 0.75$.

3 With a P-value of 0.0001, this is very strong evidence against the null hypothesis. We can reject H_0 and conclude that the improved version of the drug gives relief to a higher proportion of patients.

4 Not necessarily. The difference is statistically significant, but she should look at how big the difference actually is using a confidence interval and whether it is economically justified.

5 No. He can conclude only that the test failed to find a difference. He should construct a confidence interval to see if the upper end of the difference is financially interesting. If so, they may need a larger sample size to see if the difference is repeatable.

6 The parameter of interest is the proportion, p, of all delinquent customers who will pay their bills. $H_0: p = 0.30$ and $H_A: p > 0.30$.

7 At $\alpha = 0.05$, you can't reject the null hypothesis because 0.30 is contained in the 90% confidence interval—it's plausible that sending the DVDs is no more effective than sending letters.

8 The confidence interval is from 29% to 45%. The DVD strategy is more expensive and may not be worth it. We can't distinguish the success rate from 30% given the results of this experiment, but 45% would represent a large improvement. The bank should consider another trial, increasing the sample size to get a narrower confidence interval.

9 A Type I error would mean deciding that the DVD success rate is higher than 30%, when it isn't. The bank would adopt a more expensive method for collecting payments that's no better than its original, less expensive strategy.

10 A Type II error would mean deciding that there's not enough evidence to say the DVD strategy works when in fact it does. The bank would fail to discover an effective method for increasing revenue from delinquent accounts.

11 Higher; the larger the effect size, the greater the power. It's easier to detect an improvement to a 60% success rate than to a 32% rate.

11

Confidence Intervals and Hypothesis Tests for Means

Guinness & Co.

In 1759, when Arthur Guinness was 34 years old, he took an incredible gamble, signing a 9000-year lease on a run-down, abandoned brewery in Dublin. The brewery covered four acres and consisted of a mill, two malt houses, stabling for 12 horses, and a loft that could hold 200 tons of hay. At the time, brewing was a difficult and competitive market. Gin, whiskey, and the traditional London porter were the drinks of choice.

In addition to the lighter ales that Dublin was known for, Guinness began to brew dark porters to compete directly with those of the English brewers. Forty years later, Guinness stopped brewing light Dublin ales altogether to concentrate on his stouts and porters. Upon his death in 1803, his son Arthur Guinness II took over the business, and a few years later the company began to export Guinness stout to other parts of Europe. By the 1830s, the Guinness St. James's Gate Brewery had become the largest in Ireland. In 1886, the Guinness Brewery, with an annual production of 1.2 million barrels, was the first major brewery to be incorporated as a public company on the London Stock Exchange. During the 1890s, the company began to employ scientists. One of those, William S. Gosset, was hired as a chemist to test the quality of the brewing process. Gosset was not only an early pioneer of quality control methods in industry but a statistician whose work made modern statistical inference possible.[1]

---

[1] Source: Guinness & Co., www.guinness.com/global/story/history.

A s a chemist at the Guinness Brewery in Dublin, William S. Gosset was in charge of quality control. His job was to make sure that the stout (a thick, dark beer) leaving the brewery was of high enough quality to meet the standards of the brewery's many discerning customers. It's easy to imagine, when testing stout, why testing a large amount of stout might be undesirable, not to mention dangerous to one's health. So to test for quality, Gosset often used a sample of only 3 or 4 observations per batch. But he noticed that with samples of this size, his tests for quality weren't quite right. He knew this because when the batches that he rejected were sent back to the laboratory for more extensive testing, too often the test results turned out to be wrong. As a practicing statistician, Gosset knew he had to be wrong *some* of the time, but he hated being wrong more often than the theory predicted. One result of Gosset's frustrations was the development of a test to handle small samples, the main subject of this chapter.

11.1   The Central Limit Theorem

We've learned a lot about proportions. We know that when we sample at random, the proportions we get will vary from sample to sample. We know that a mathematical theorem called the Central Limit Theorem (CLT) says that the Normal model does a remarkably good job at summarizing all that variation. The CLT also applies to means. A good way to see why this theorem works for means is to look at simulations.

Simulating the Sampling Distribution of a Mean

Here's a simple simulation with a quantitative variable. Let's start with one fair die. If we toss this die 10,000 times, what should the histogram of the numbers on the face of the die look like? Here are the results of a simulated 10,000 tosses:

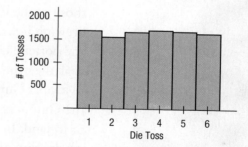

That's the *uniform distribution*—it's certainly not Normal. Now let's toss a *pair* of dice and record the average of the two. If we repeat this (or at least simulate repeating it) 10,000 times, recording the average of each pair, what will the histogram of these 10,000 averages look like? Before you look, think a minute. Is getting an average of 1 on *two* dice as likely as getting an average of 3 or 3.5? Let's see:

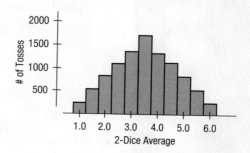

We're much more likely to get an average near 3.5 than we are to get one near 1 or 6. Without calculating those probabilities exactly, it's fairly easy to see that the *only* way to get an average of 1 is to get two 1s. To get a total of 7 (for an average of 3.5), though, there are many more possibilities. This distribution even has a name—the *triangular distribution*.

What if we average three dice? We'll simulate 10,000 tosses of three dice and take their average.

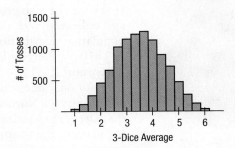

First notice that it's getting harder to have averages near the ends. Getting an average of 1 or 6 with three dice requires all three to come up 1 or 6, respectively. That's less likely than for two dice to come up both 1 or both 6. The distribution is being pushed toward the middle. But what's happening to the shape?

Let's continue this simulation to see what happens with larger samples. Here's a histogram of the averages for 10,000 tosses of five dice.

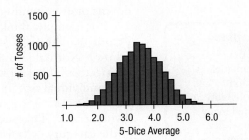

The pattern is becoming clearer. Two things are happening. The first fact we knew already from the Law of Large Numbers, which we saw in Chapter 5. It says that as the sample size (number of dice) gets larger, each sample average tends to become closer to the population mean. So we see the shape continuing to tighten around 3.5. But the shape of the distribution is the surprising part. It's becoming bell-shaped. In fact, it's approaching the Normal model.

Let's skip ahead and try 20 dice. The histogram of averages for 10,000 tosses of 20 dice looks like this.

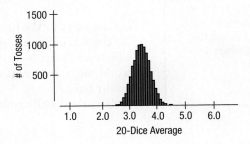

Now we see the Normal shape again (and notice how much smaller the spread is). But can we count on this happening for situations other than dice throws?

Pierre-Simon Laplace, 1749–1827.

"The theory of probabilities is at bottom nothing but common sense reduced to calculus."
—LAPLACE, IN *THÉORIE ANALYTIQUE DES PROBABILITIÉS*, 1812

> ### Pierre-Simon Laplace
>
> Laplace was one of the greatest scientists and mathematicians of his time. In addition to his contributions to probability and statistics, he published many new results in mathematics, physics, and astronomy (where his nebular theory was one of the first to describe the formation of the solar system in much the way it is understood today). He also played a leading role in establishing the metric system of measurement.
>
> His brilliance, though, sometimes got him into trouble. A visitor to the Académie des Sciences in Paris reported that Laplace let it be known widely that he considered himself the best mathematician in France. The effect of this on his colleagues was not eased by the fact that Laplace was right.

The Central Limit Theorem

The dice simulation may look like a special situation. But it turns out that what we saw with dice is true for means of repeated samples for almost every situation. When we looked at the sampling distribution of a proportion, we had to check only a few conditions. For means, the result is even more remarkable. There are almost no conditions at all.

Let's say that again: The sampling distribution of *any* mean becomes Normal as the sample size grows. All we need is for the observations to be independent and collected with randomization. We don't even care about the shape of the population distribution![2] This surprising fact was proved in a fairly general form in 1810 by Pierre-Simon Laplace. Laplace's result is called the **Central Limit Theorem**[3] (CLT).

Not only does the distribution of means of many random samples get closer and closer to a Normal model as the sample size grows, but *this is true regardless of the shape of the population distribution!* Even if we sample from a skewed or bimodal population, the Central Limit Theorem tells us that means of repeated random samples will tend to follow a Normal model as the sample size grows. You probably won't be surprised to learn that it works better and faster the closer the population distribution is to a Normal model. And it works better for larger samples. If the data come from a population that's exactly Normal to start with, then the observations themselves are Normal. If we take samples of size 1, their "means" are just the observations—so, of course, they have a Normal sampling distribution. But now suppose the population distribution is very skewed (like the CEO data from Chapter 3, for example). The CLT works, although it may take a sample size of dozens or even hundreds of observations for the Normal model to work well.

For example, think about a very bimodal population, one that consists of only 0s and 1s. The CLT says that even means of samples from this population will follow a Normal sampling distribution model. But wait. Suppose we have a categorical variable and we assign a 1 to each individual in the category and a 0 to each individual not in the category. Then we find the mean of these 0s and 1s. That's the same as counting the number of individuals who are in the category and dividing by *n*. That mean will be the *sample proportion*, $\hat{p}$, of individuals who are in the category (a "success"). That's why the CLT works for proportions; they are actually just a special case of Laplace's remarkable theorem. Of course, for such an extremely bimodal population, we need a reasonably large sample size—and that's where the Success/Failure condition for proportions comes in.

> ### The Central Limit Theorem (CLT)
> The mean of a random sample has a sampling distribution whose shape can be approximated by a Normal model. The larger the sample, the better the approximation will be.

Be careful. We have been slipping smoothly between the real world, in which we draw random samples of data, and a magical mathematical-model world, in which we describe how the sample means and proportions we observe in the real world might behave if we could see the results from every random sample that we might have drawn. Now we have *two* distributions to deal with. The first is the real-world distribution of the sample, which we might display with a histogram

---

[2] Technically, the data must come from a population with a finite variance.

[3] The word "central" in the name of the theorem means "fundamental." It doesn't refer to the center of a distribution.

(for quantitative data) or with a bar chart or table (for categorical data). The second is the math-world *sampling distribution* of the statistic, which we model with a Normal model based on the Central Limit Theorem. Don't confuse the two.

For example, don't mistakenly think the CLT says that the *data* are Normally distributed as long as the sample is large enough. In fact, as samples get larger, we expect the distribution of the data to look more and more like the distribution of the population from which it is drawn—skewed, bimodal, whatever—but not necessarily Normal. You can collect a sample of CEO salaries for the next 1000 years, but the histogram will never look Normal. It will be skewed to the right. The Central Limit Theorem doesn't talk about the distribution of the data from the sample. It talks about the sample *means* and sample *proportions* of many different random samples drawn from the same population. Of course, we never actually draw all those samples, so the CLT is talking about an imaginary distribution—the sampling distribution model.

When the population shape is not unimodal and symmetric it takes longer for the sampling distribution to resemble the Normal. But with a large enough sample, the CLT applies to means and proportions from almost any data set.

JUST CHECKING

> **The Central Limit Theorem**
>
> A supermarket manager examines the amount spent by customers using a self-checkout station. He finds that the distribution of these amounts is unimodal but skewed to the high end because some customers make unusually expensive purchases. He finds the mean spent on each of the 30 days studied and makes a histogram of those values.
>
> 1 What shape would you expect for this histogram?
>
> 2 If, instead of averaging all customers on each day, he selects the first 10 for each day and just averages those, how would you expect his histogram of the means to differ from the one in (1)?

11.2 The Sampling Distribution of the Mean

The CLT says that the sampling distribution of any mean or proportion is approximately Normal. But which Normal? We know that any Normal model is specified by its mean and standard deviation. For proportions, the sampling distribution is centered at the population proportion. For means, it's centered at the population mean. What else would we expect?

What about the standard deviations? We noticed in our dice simulation that the histograms got narrower as the number of dice we averaged increased. This shouldn't be surprising. Means vary less than the individual observations. Think about it for a minute. Which would be more surprising, having *one* person in your Statistics class who is over 6′9″ tall or having the *mean* of 100 students taking the course be over 6′9″? The first event is fairly rare.[4] You may have seen somebody this tall in one of your classes sometime. But finding a class of 100 whose mean height is over 6′9″ tall just won't happen. Why? *Means have smaller standard deviations than individuals.*

"The n's justify the means."
—APOCRYPHAL
STATISTICAL SAYING

---

[4] If students are a random sample of adults, fewer than 1 out of 10,000 should be taller than 6′9″. Why might college students not really be a random sample with respect to height? Even if they're not a perfectly random sample, a college student over 6′9″ tall is still rare.

That is, the Normal model for the sampling distribution of the mean has a standard deviation equal to $SD(\bar{y}) = \dfrac{\sigma}{\sqrt{n}}$ where σ is the standard deviation of the population. To emphasize that this standard deviation is a *parameter* of the sampling distribution model for the sample mean, $\bar{y}$, we write $SD(\bar{y})$ or $\sigma(\bar{y})$.

The Sampling Distribution Model for a Mean

When a random sample is drawn from any population with mean μ and standard deviation σ, its sample mean, $\bar{y}$, has a sampling distribution with the same mean μ but whose standard deviation is $\dfrac{\sigma}{\sqrt{n}}$, and we write $\sigma(\bar{y}) = SD(\bar{y}) = \dfrac{\sigma}{\sqrt{n}}$. No matter what population the random sample comes from, the shape of the sampling distribution is approximately Normal as long as the sample size is large enough. The larger the sample used, the more closely the Normal approximates the sampling distribution model for the mean.

FOR EXAMPLE   Working with the sampling distribution of the mean

Suppose that the weights of boxes shipped by a company follow a unimodal, symmetric distribution with a mean of 12 lbs and a standard deviation of 4 lbs. Boxes are shipped in palettes of 10 boxes. The shipper has a limit of 150 lbs for such shipments.

QUESTION   What's the probability that a palette will exceed that limit?

ANSWER   Asking the probability that the total weight of a sample of 10 boxes exceeds 150 lbs is the same as asking the probability that the *mean* weight exceeds 15 lbs. First we'll check the conditions. We will assume that the 10 boxes on the palette are a random sample from the population of boxes and that their weights are mutually independent. We are told that the underlying distribution of weights is unimodal and symmetric, so a sample of 10 boxes should be large enough.

Under these conditions, the CLT says that the sampling distribution of $\bar{y}$ has a Normal model with mean 12 and standard deviation

$$SD(\bar{y}) = \frac{\sigma}{\sqrt{n}} = \frac{4}{\sqrt{10}} = 1.26 \text{ and } z = \frac{\bar{y} - \mu}{SD(\bar{y})} = \frac{15 - 12}{1.26} = 2.38$$

$$P(\bar{y} > 15) = P(z > 2.38) = 0.0087$$

So the chance that the shipper will reject a palette is only .0087—less than 1%.

We now have two closely related sampling distribution models. Which one we use depends on which kind of data we have.

- When we have categorical data, we calculate a sample proportion, $\hat{p}$. Its sampling distribution follows a Normal model with a mean at the population proportion, p, and a standard deviation $SD(\hat{p}) = \sqrt{\dfrac{pq}{n}} = \dfrac{\sqrt{pq}}{\sqrt{n}}$.

- When we have quantitative data, we calculate a sample mean, $\bar{y}$. Its sampling distribution has a Normal model with a mean at the population mean, μ, and a standard deviation $SD(\bar{y}) = \dfrac{\sigma}{\sqrt{n}}$.

The means of these models are easy to remember, so all you need to be careful about is the standard deviations. Remember that these are standard deviations of the *statistics* $\hat{p}$ and $\bar{y}$. They both have a square root of n in the denominator. That tells us that the larger the sample, the less either statistic will vary. The only difference is in the numerator. If you just start by writing $SD(\bar{y})$ for quantitative data and $SD(\hat{p})$ for categorical data, you'll be able to remember which formula to use.

But both of these expressions for the standard deviations use facts about the population that we usually don't know. Often we know only the observed proportion, $\hat{p}$, or the sample standard deviation, s. Naturally, we just use what we know, and we estimate. That may not seem like a big deal, but it gets a special name. Whenever we *estimate* the standard deviation of a sampling distribution, we call it a **standard error (SE)**.

For a sample proportion, $\hat{p}$, the standard error is:

$$SE(\hat{p}) = \sqrt{\frac{\hat{p}\hat{q}}{n}}.$$

For the sample mean, $\bar{y}$, the standard error is:

$$SE(\bar{y}) = \frac{s}{\sqrt{n}}.$$

You may see a "standard error" reported by a computer program in a summary or offered by a calculator. It's safe to assume that if no statistic is specified, what was meant is $SE(\bar{y})$, the standard error of the mean.

JUST CHECKING

3 The entrance exam for business schools, the GMAT, given to 100 students had a mean of 520 and a standard deviation of 120. What was the standard error for the mean of this sample of students?

4 As the sample size increases, what happens to the standard error, assuming the standard deviation remains constant?

5 If the sample size is doubled, what is the impact on the standard error?

11.3 How Sampling Distribution Models Work

To keep track of how the concepts we've seen combine, we can draw a diagram relating them. At the heart is the idea that *the statistic itself (the sample proportion or the sample mean) is a random quantity.* We can't know what our statistic will be because it comes from a random sample. A different random sample would have given a different result. This sample-to-sample variability is what generates the sampling distribution, the distribution of all the possible values that the statistic could have had.

We could simulate that distribution by pretending to take lots of samples. Fortunately, for the mean and the proportion, the CLT tells us that we can model their sampling distribution directly with a Normal model.

The two basic truths about sampling distributions are:

1. Sampling distributions arise because samples vary. Each random sample will contain different cases and, so, a different value of the statistic.
2. Although we can always simulate a sampling distribution, the Central Limit Theorem saves us the trouble for means and proportions.

When we don't know σ, we estimate it with the standard deviation of the one real sample. That gives us the standard error, $SE(\bar{y}) = \frac{s}{\sqrt{n}}$.

Figure 11.1 diagrams the process.

Figure 11.1   We start with a population model, which can have any shape. It can even be bimodal or skewed (as this one is). We label the mean of this model μ and its standard deviation, σ.

We draw one real sample (solid line) of size n and show its histogram and summary statistics. We *imagine* (or simulate) drawing many other samples (dotted lines), which have their own histograms and summary statistics.

We (imagine) gathering all the means into a histogram.

The CLT tells us we can model the shape of this histogram with a Normal model. The mean of this Normal is μ, and the standard deviation is $SD(\bar{y}) = \dfrac{\sigma}{\sqrt{n}}$.

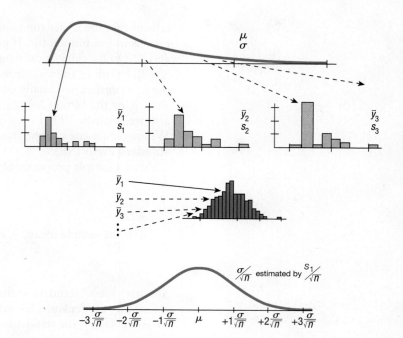

11.4   Gosset and the *t*-Distribution

Now that we have a sampling distribution model for means, we can find confidence intervals. For proportions we found the confidence interval as

$$\hat{p} \pm ME.$$

The *ME* was equal to a critical value, z^*, times $SE(\hat{p})$. Our confidence interval for means will look very similar:

$$\bar{y} \pm ME.$$

And our *ME* will be a critical value times $SE(\bar{y})$. So let's put the pieces together.

The Central Limit Theorem gives us a sampling distribution and a standard deviation for the mean. All we need is a random sample of quantitative data and the true value of the population standard deviation σ.

But wait. That could be a problem. To compute $\sigma/\sqrt{n}$ we need to know σ. How are we supposed to know σ? Suppose we told you that for 25 young executives the mean value of their stock portfolios is \$125,672. Would that tell you the value of σ? No, the standard deviation depends on how similarly the executives invest, not on how well they invested (the mean tells us that). Because we don't know

σ we have to estimate it and use the standard error: $SE(\bar{y}) = \dfrac{s}{\sqrt{n}}$.

A century ago, people just plugged the standard error into the Normal model, assuming it would work. And for large sample sizes it *did* work pretty well. But they began to notice problems with smaller samples. Because s is a statistic, it also varies from sample to sample. The extra variation in the standard error due to using s was wreaking havoc with the margins of error.

Gosset was the first to investigate this phenomenon. He realized that we need a new sampling distribution model. In fact, we need a whole *family* of models, depending on the sample size, n. These models are unimodal, symmetric, and bell-shaped, but the smaller our sample, the more we must stretch out the tails. Gosset's work transformed Statistics, but most people who use his work don't even know his name.

To find the sampling distribution of $\frac{\bar{y}}{s/\sqrt{n}}$, Gosset simulated it *by hand.*

He drew paper slips of small samples from a hat *hundreds of times* and computed the means and standard deviations with a mechanically cranked calculator. Today, you could repeat in seconds on a computer the experiment that took him over a year. Gosset's work was so meticulous that not only did he get the shape of the new histogram approximately right, but he even figured out the exact formula for it from his sample. The formula was not confirmed mathematically until years later by Sir Ronald Aylmer Fisher.

NOTATION ALERT

Ever since Gosset, the letter *t* has been reserved in Statistics for his distribution.

Gosset's *t*

Gosset made decisions about the stout's quality by using statistical inference. He knew that if he used a 95% confidence interval, he would fail to capture the true quality of the batch about 5% of the time. However, the lab told him that he was in fact rejecting about 15% of the good batches. Gosset knew something was wrong, and it bugged him.

Gosset took time off from his job to study the problem and earn a graduate degree in the emerging field of Statistics. He figured out that when he used the standard error $\frac{s}{\sqrt{n}}$, the shape of the sampling model was no longer Normal. He even figured out what the new model was and called it a *t*-distribution.

The Guinness Company didn't give Gosset a lot of support for his work. In fact, it had a policy against publishing results. Gosset had to convince the company that he was not publishing an industrial secret and (as part of getting permission to publish) had to use a pseudonym. The pseudonym he chose was "Student," and ever since, the model he found has been known as **Student's *t*.**

Using a Known Standard Deviation

Variation is inherent in manufacturing, even under the most tightly controlled processes. To ensure that parts do not vary too much, however, quality professionals monitor the processes by selecting samples at regular intervals. The mean performance of these samples is measured, and if it lies too far from the desired target mean, the process may be stopped until the underlying cause of the problem can be determined. In silicon wafer manufacturing, the thickness of the film is a crucial measurement. To assess a sample of wafers, quality engineers compare the mean thickness of the sample to the target mean. But, they don't estimate the standard deviation of the mean by using the standard error derived from the same sample. Instead they base the standard deviation of the mean on the historical process standard deviation, estimated from a vast collection of similar parts. In this case, the standard deviation can be treated as "known" and the Normal model can be used for the sampling distribution instead of the *t* distribution.

Student's *t*-models form a family of related distributions that depend on a parameter known as **degrees of freedom.** We often denote degrees of freedom as df and the model as t_{df}, with the numerical value of the degrees of freedom as a subscript.

Student's *t*-models are unimodal, symmetric, and bell-shaped, just like the Normal model. But *t*-models with only a few degrees of freedom have a narrower peak than the Normal model and have much fatter tails. (See Figure 11.2.) (That's what makes the margin of error bigger.) As the degrees of freedom increase, the

Figure 11.2 The *t*-model (solid curve) with 2 degrees of freedom has fatter tails than the Normal model (dashed curve). So the 68–95–99.7 Rule doesn't work for *t*-models with only a few degrees of freedom.

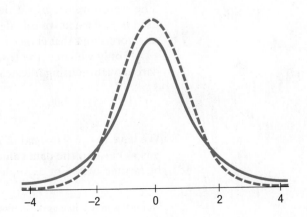

t-models look more and more like the Normal model. In fact, the t-model with infinite degrees of freedom is exactly Normal.[5] This is great news if you happen to have an infinite number of data values. Unfortunately, that's not practical. Fortunately, above a few hundred degrees of freedom it's very hard to tell the difference. Of course, in the rare situation that we *know* σ, it would be foolish not to use that information. If we don't have to estimate σ, we can use the Normal model. Typically that value of σ would be based on (lots of) experience, or on a theoretical model. Usually, however, we estimate σ by s from the data and use the t-model.

Practical Sampling Distribution Model for Means

When certain conditions are met, the standardized sample mean,

$$t = \frac{\bar{y} - \mu}{SE(\bar{y})}$$

follows a Student's t-model with $n - 1$ degrees of freedom. We find the standard error from:

$$SE(\bar{y}) = \frac{s}{\sqrt{n}}.$$

11.5   A Confidence Interval for Means

To make confidence intervals, we need to use Gosset's model. Which one? Well, for means, it turns out the right value for degrees of freedom is df $= n - 1$.

One-Sample t-Interval

When the assumptions and conditions are met, we are ready to find the **one-sample t-interval for the population mean**, μ. The confidence interval is:

$$\bar{y} \pm t^*_{n-1} \times SE(\bar{y})$$

where the standard error of the mean is:

$$SE(\bar{y}) = \frac{s}{\sqrt{n}}.$$

The critical value t^*_{n-1} depends on the particular confidence level, C, that you specify and on the number of degrees of freedom, $n - 1$, which we get from the sample size.

Degrees of Freedom—Why $n - 1$?

The reason we use $n - 1$ degrees of freedom for the t is closely related to why we used $n - 1$ when we calculated the standard deviation. We promised back then to say more about that choice later, and this seems like a good time to bring it up.

If only we knew the true population mean, μ, we would find the sample standard deviation using n instead of $n - 1$ as:

$$s = \sqrt{\frac{\sum (y - \mu)^2}{n}} \text{ and we'd call it } s.$$

We have to use $\bar{y}$ instead of μ, though, and that causes a problem. For any sample, $\bar{y}$ is as close to the data values as possible. Generally the population mean, μ, will be farther away. For example, GMAT scores have a population mean of 525. If you

---

[5] Formally, in the limit as the number of degrees of freedom goes to infinity.

took a random sample of 5 students who took the test, their sample mean wouldn't be 525. The five data values will be closer to their own $\bar{y}$ than to 525. So if we use $\Sigma (y - \bar{y})^2$ instead of $\Sigma (y - \mu)^2$ in the equation to calculate s, our standard deviation estimate will be too small. The amazing mathematical fact is that we can compensate for the fact that $\Sigma (y - \bar{y})^2$ is too small just by dividing by $n - 1$ instead of by n. So that's all the $n - 1$ is doing in the denominator of s. We call $n - 1$ the degrees of freedom.

Finding t^*-Values

The Student's t-model is different for each value of degrees of freedom. Statistics books usually have one table of t-model critical values for a selected set of confidence levels. This one does too; see Table T in Appendix B. (You can also find tables on the Internet. Search for "Students t calculator," for example.)

The t-tables run down the page for as many degrees of freedom as can fit, and, as you can see from Figure 11.3, they are much easier to use than the Normal tables. But there is only room on the page for a limited number of degrees of freedom. Of course, for *enough* degrees of freedom, the t-model gets closer and closer to the Normal, so the tables give a final row with the critical values from the Normal model and label it "∞ df." Online calculators and apps for your smartphone can work with any number of degrees of freedom.

Figure 11.3 Part of Table T in Appendix B.

Two tail probability One tail probability		0.20 0.10	0.10 0.05	0.05 0.025
Table T Values of t_α	**df**			
	1	3.078	6.314	12.706
	2	1.886	2.920	4.303
	3	1.638	2.353	3.182
	4	1.533	2.132	2.776
	5	1.476	2.015	2.571
	6	1.440	1.943	2.447
	7	1.415	1.895	2.365
	8	1.397	1.860	2.306
	9	1.383	1.833	2.262
	10	1.372	1.812	2.228
	11	1.363	1.796	2.201
	12	1.356	1.782	2.179
	13	1.350	1.771	2.160
	14	1.345	1.761	2.145
	15	1.341	1.753	2.131
	16	1.337	1.746	2.120
	17	1.333	1.740	2.110
	18	1.330	1.734	2.101
	19	1.328	1.729	2.093
	$\vdots$	$\vdots$	$\vdots$	$\vdots$
	∞	1.282	1.645	1.960
Confidence levels		80%	90%	95%

Two tails: $\frac{\alpha}{2}$, $-t_{\alpha/2}$, 0, $t_{\alpha/2}$

One tail: 0, t_α, α

FOR EXAMPLE Finding a confidence interval for the mean

According to the Environmental Defense Fund, "Americans are eating more and more salmon, drawn to its rich taste and health benefits. Increasingly they are choosing *farmed* salmon because of its wide availability and low price. But in the last few years, farmed salmon has been surrounded by controversy over its health risks and the ecological impacts of salmon aquaculture operations. Studies have shown that some farmed salmon

(continued)

is relatively higher in contaminants like PCBs than wild salmon, and there is mounting concern over the industry's impact on wild salmon populations."

In a widely cited study of contaminants in farmed salmon, fish from many sources were analyzed for 14 organic contaminants.[6] One of those was the insecticide mirex, which has been shown to be carcinogenic and is suspected of being toxic to the liver, kidneys, and endocrine system. Summaries for 150 mirex concentrations (in parts per million) from a variety of farmed salmon sources were reported as:

$$n = 150; \quad \bar{y} = 0.0913 \text{ ppm}; \quad s = 0.0495 \text{ ppm}$$

QUESTION The Environmental Protection Agency (EPA) recommends to recreational fishers as a "screening value" that mirex concentrations be no larger than 0.08 ppm. What does the 95% confidence interval say about that value?

ANSWER Because $n = 150$, there are 149 df. There isn't an entry in Table T for 149 df, so if we're using the table, we'll use the next *smaller* value to be conservative. We find $t^*_{140, 0.025} = 1.977$ (from technology, $t^*_{149, 0.025} = 1.976$), so a 95% confidence interval is

$$\bar{y} \pm t^* \times SE(\bar{y}) = \bar{y} \pm 1.977 \times \frac{s}{\sqrt{n}} = 0.0913 \pm 1.977 \frac{0.0495}{\sqrt{150}} = (0.0833, 0.0993)$$

If this sample is representative (as the authors claim it is), we can be 95% confident that it contains the true value of the mean mirex concentration. Because the interval from 0.0834 to 0.0992 ppm is entirely above the recommended value set by the EPA, we have reason to believe that the true mirex concentration exceeds the EPA guidelines.

11.6   Assumptions and Conditions

Gosset found the *t*-model by simulation. Years later, when Sir Ronald Fisher showed mathematically that Gosset was right, he needed to make some assumptions to make the proof work. These are the assumptions we need in order to use the Student's *t*-models.

Independence Assumption

The data values should be independent. There's really no way to check independence of the data by looking at the sample, but we should think about whether the assumption is reasonable.

Randomization Condition: The data arise from a random sample or suitably randomized experiment. Randomly sampled data—and especially data from a Simple Random Sample (SRS)—are ideal.

When a sample is drawn without replacement, technically we ought to confirm that we haven't sampled a large fraction of the population, which would threaten the independence of our selections. In that case, we can check the following condition.

10% Condition: The sample size should be no more than 10% of the population. In practice, though, we often don't mention the 10% Condition when estimating means. Why not? When we made inferences about proportions, this condition was a greater concern because we usually had large samples. But for means, our samples are generally smaller, so this problem arises only if we're sampling from a small population. If you are concerned that you may have sampled too large a fraction of your population, consult a Statistician to learn how to correct for that problem.

> ### We Don't *Want* to Stop
> We check conditions hoping that we can make a meaningful analysis of our data. The conditions serve as *disqualifiers*—we keep going unless there's a serious problem. If we find minor issues, we note them and express caution about our results. If the sample is not an SRS, but we believe it's representative of some populations, we limit our conclusions accordingly. If there are outliers, rather than stop, we perform the analysis both with and without them. If the sample looks bimodal, we try to analyze subgroups separately. Only when there's major trouble—like a strongly skewed small sample or an obviously nonrepresentative sample—are we unable to proceed at all.

---

[6] Ronald A. Hites, Jeffery A. Foran, David O. Carpenter, M. Coreen Hamilton, Barbara A. Knuth, and Steven J. Schwager, "Global Assessment of Organic Contaminants in Farmed Salmon," *Science* 9, January 2004: Vol. 303, no. 5655, pp. 226–229.

Normal Population Assumption

Student's *t*-models assume that the data are from a population that follows a Normal model. Practically speaking, there's no way to be certain this is true.

And it's almost certainly *not* true. Models are idealized; real data are, well, real. The good news, however, is that even for small samples, it's sufficient to check a condition.

Nearly Normal Condition. The data come from a distribution that is unimodal and symmetric. This is a much more practical condition and one we can check by making a histogram.[7]

For very small samples (*n* < 15 or so), the data should follow a Normal model pretty closely. Of course, with so little data, it's rather hard to tell. But if you do find outliers or strong skewness, don't use these methods.[8]

For moderate sample sizes (*n* between 15 and 40 or so), the *t* methods will work well as long as the data are unimodal and reasonably symmetric. Make a histogram to check.

When the sample size is larger than 40 or 50, the *t* methods are safe to use unless the data are extremely skewed. Make a histogram anyway. If you find outliers in the data and they aren't errors that are easy to fix, it's always a good idea to perform the analysis twice, once with and once without the outliers, even for large samples. The outliers may well hold additional information about the data, so they deserve special attention. If you find multiple modes, you may well have different groups that should be analyzed and understood separately.

If the data are extremely skewed, the mean may not be the most appropriate summary. But when our data consist of a collection of instances whose *total* is the business consequence—as when we add up the profits (or losses) from many transactions or the costs of many supplies—then the mean is just that total divided by *n*. And that's the value with a business consequence. Fortunately, in this instance, the Central Limit Theorem comes to our rescue. Even when we must sample from a very skewed distribution, the sampling distribution of our sample mean will be close to Normal, so we can use Student's *t* methods without much worry as long as the sample size is *large enough*.

How large is large enough? Figure 11.4 shows a histogram of CEO compensations ($000) for Fortune 500 companies.

Figure 11.4 It's hard to imagine a distribution more skewed than these annual compensations from the Fortune 500 CEOs.

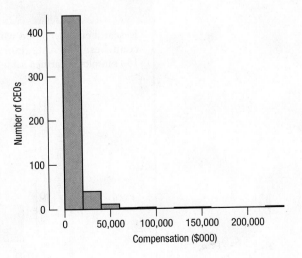

---

[7] Or we could check a Normal probability plot.

[8] If the outlier is an error, then correct it and go ahead with Student's *t*.

Although this distribution is very skewed, the Central Limit Theorem will make the sampling distribution of the means of samples from this distribution more and more Normal as the sample size grows. Figure 11.5 is a histogram of the means of many samples of 100 CEOs:

Figure 11.5 Even samples as small as 100 from the CEO data set produce means whose sampling distribution is nearly Normal. Larger samples will have sampling distributions even more Normal.

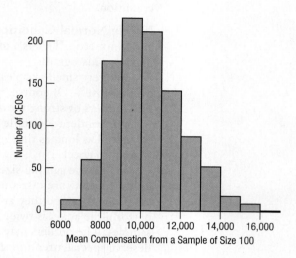

Often, in modern business applications, even if we have a sample of many hundreds, or thousands, we should still be on guard for outliers and multiple modes and we should think about whether the observations are independent. But if the mean is of interest, the Central Limit Theorem works quite well in ensuring that the sampling distribution of the mean will be close to the Normal for samples of this size.

FOR EXAMPLE   Checking the assumptions and conditions for a confidence interval for means

Researchers purchased whole farmed salmon from 51 farms in eight regions in six countries. The histogram shows the concentrations of the insecticide mirex in the 150 samples of farmed salmon we examined in the previous example.

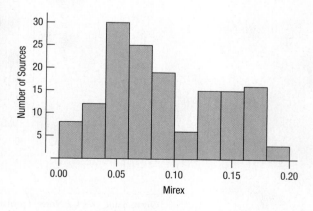

QUESTION   Are the assumptions and conditions for making a confidence interval for the mean mirex concentration satisfied?

ANSWER

✓ **Independence Assumption:** The fish were raised in many different places, and samples were purchased independently from several sources.
✓ **Randomization Condition:** The fish were selected randomly from those available for sale.
✓ **Nearly Normal Condition:** The histogram of the data looks bimodal. While it might be interesting to learn the reason for that and possibly identify the subsets, we can proceed because the sample size is large.

It's okay to use these data about farm-raised salmon to make a confidence interval for the mean.

JUST CHECKING

The alumni organization of your university is trying to get information on the success of their recent graduates, so they have surveyed a sample of graduates. We know that incomes are likely to be skewed to the right (although perhaps not as much as the salaries of CEOs!).

6 If they are successful at obtaining the salaries of only 20 recent graduates, what should they be concerned about when reporting a confidence interval for mean salaries of all recent grads?

7 Why do they need to base these confidence intervals on *t*-models?

8 If they manage to get 60 salaries instead of 20, how will this affect their confidence interval?

GUIDED EXAMPLE   Insurance Profits

Profit (in $) from 30 Policies		
222.80	463.35	2089.40
1756.23	−66.20	2692.75
1100.85	57.90	2495.70
3340.66	833.95	2172.70
1006.50	1390.70	3249.65
445.50	2447.50	−397.10
3255.60	1847.50	−397.31
3701.85	865.40	186.25
−803.35	1415.65	590.85
3865.90	2756.94	578.95

Insurance companies take risks. When they insure a property or a life, they must price the policy in such a way that their expected profit enables them to survive. They can base their projections on actuarial tables, but the reality of the insurance business often demands that they discount policies to a variety of customers and situations. Managing this risk is made even more difficult by the fact that until the policy expires, the company won't know if they've made a profit, no matter what premium they charge.

A manager wanted to see how well one of her sales representatives was doing, so she selected 30 matured policies that had been sold by the sales rep and computed the (net) profit (premium charged minus paid claims), for each of the 30 policies.

The manager would like you, as a consultant, to construct a 95% confidence interval for the mean profit of the policies sold by this sales rep.

(continued)

PLAN

Setup State what we want to know. Identify the variables and their context.

Make a picture. Check the distribution shape and look for skewness, multiple modes, and outliers.

We wish to find a 95% confidence interval for the mean profit of policies sold by this sales rep. We have data for 30 matured policies.

Here's a boxplot and histogram of these values.

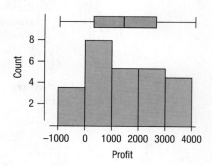

The sample appears to be unimodal and fairly symmetric with profit values between −$1000 and $4000 and no outliers.

Model Think about the assumptions and check the conditions.

✓ **Independence Assumption**
This is a random sample so observations should be independent.

✓ **Randomization Condition**
This sample was selected randomly from the matured policies sold by the sales representative of the company.

✓ **Nearly Normal Condition**
The distribution of profits is unimodal and fairly symmetric without strong skewness.

State the sampling distribution model for the statistic.

We will use a Student's t-model with $n - 1 = 30 - 1 = 29$ degrees of freedom and find a one-sample t-interval for the mean.

DO

Mechanics Compute basic statistics and construct the confidence interval.

Using software, we obtain the following basic statistics:

$$n = 30$$
$$\bar{y} = \$1438.90$$
$$s = \$1329.60$$

Remember that the standard error of the mean is equal to the standard deviation divided by the square root of n.

The standard error of the mean is:

$$SE(\bar{y}) = \frac{s}{\sqrt{n}} = \frac{1329.60}{\sqrt{30}} = \$242.75$$

The critical value we need to make a 95% confidence interval comes from a Student's t-table, a computer program, or a calculator. We have $30 - 1 = 29$ degrees of freedom. So we look up the corresponding t^* value in Table T.

There are $30 - 1 = 29$ degrees of freedom. The manager has specified a 95% level of confidence, so the critical value (from Table T) is 2.045.

The margin of error is:

$$ME = 2.045 \times SE(\bar{y})$$
$$= 2.045 \times 242.75$$
$$= \$496.42$$

The 95% confidence interval for the mean profit is:

$$\$1438.90 \pm \$496.42$$
$$= (\$942.48, \$1935.32)$$

REPORT

Conclusion Interpret the confidence interval in the proper context.

When we construct confidence intervals in this way, we expect 95% of them to cover the true mean and 5% to miss the true value. That's what "95% confident" means.

MEMO

Re: Profit from policies

From our analysis of the selected policies, we are 95% confident that the true mean profit of policies sold by this sales rep is contained in the interval from $942.48 to $1935.32.

Caveat: Insurance losses are notoriously subject to outliers. One very large loss could influence the average profit substantially. However, there were no such cases in this data set.

Cautions about Interpreting Confidence Intervals

Confidence intervals for means offer new, tempting, wrong interpretations. Here are some ways to keep from going astray:

- **Don't say,** "*95% of all the policies* sold by this sales rep have profits between $942.48 and $1935.32." The confidence interval is about the *mean*, not about the measurements of individual policies.
- **Don't say,** "We are 95% confident that *a randomly selected policy* will have a net profit between $942.48 and $1935.32." This false interpretation is also about individual policies rather than about the *mean* of the policies. We are 95% confident that the *mean* profit of all (similar) policies sold by this sales rep is between $942.48 and $1935.32.
- **Don't say,** "The mean profit is $1438.90 *95% of the time*." That's about means, but still wrong. It implies that the true mean varies, when in fact it is the confidence interval that would have been different had we gotten a different sample.
- **Finally, don't say,** "*95% of all samples* will have mean profits between $942.48 and $1935.32." That statement suggests that *this* interval somehow sets a standard for every other interval. In fact, this interval is no more (or less) likely to be correct than any other. You could say that 95% of all possible samples would produce intervals that contain the true mean profit. (The problem is that because we'll never know what the true mean profit is, we can't know if our sample was one of those 95%.)

So, what *should* you say? Since 95% of random samples yield an interval that captures the true mean, you should say:

- "I am 95% confident that the interval from $942.48 to $1935.32 contains the mean profit of all policies sold by this sales representative." It's also okay to make this a little less formal by saying something like:
- "I am 95% confident that the mean profit for all policies sold by this sales rep is between $942.48 and $1935.32."

Remember: Your uncertainty is about the interval, not the true mean. The interval varies randomly. The true mean profit is neither variable nor random—just unknown.

JUST CHECKING

9 The alumni organization finds a confidence interval for the mean starting salary of their recent graduates to be ($55,000, $62,000). They release a statement to prospective students saying that 95% of them will earn between $55,000 and $62,000 their first year on the job. Why is this wrong?

11.7   Testing Hypotheses about Means— the One-Sample *t*-Test

The manager we met in the Guided Example above has a more specific concern. Company policy states that if a sales rep's mean profit is below $1500, the sales rep has been discounting too much and will have to adjust his pricing strategy. Is there evidence from this sample that the mean is really less than $1500? This question calls for a hypothesis test called the **one-sample *t*-test for the mean**.

When testing a hypothesis, it's natural to compare the difference between the observed statistic and a hypothesized value to the standard error. For means that looks like: $\dfrac{\bar{y} - \mu_0}{SE(\bar{y})}$. We already know that the appropriate probability model to use is Student's *t* with $n - 1$ degrees of freedom.

> **One-Sample *t*-Test for the Mean**
>
> The conditions for the one-sample *t*-test for the mean are the same as for the one-sample *t*-interval. We test the hypothesis $H_0: \mu = \mu_0$ using the statistic
>
> $$t_{n-1} = \frac{\bar{y} - \mu_0}{SE(\bar{y})},$$
>
> where the standard error of $\bar{y}$ is: $SE(\bar{y}) = \dfrac{s}{\sqrt{n}}.$
>
> When the conditions are met and the null hypothesis is true, this statistic follows a Student's *t*-model with $n - 1$ degrees of freedom. We use that model to obtain a P-value.

GUIDED EXAMPLE   Insurance Profits Revisited

Let's apply the one-sample *t*-test to the 30 mature policies sampled by the manager. From these 30 policies, the management would like to know if there's evidence that the mean profit of policies sold by this sales rep is less than $1500.

PLAN	**Setup** State what we want to know. Make clear what the population and parameter are. Identify the variables and context. **Hypotheses** We give benefit of the doubt to the sales rep. The null hypothesis is that the true mean profit is equal to $1500. Because we're interested in whether the profit is less, the alternative is one-sided.	We want to test whether the mean profit of the sales rep's policies is less than $1500. We have a random sample of 30 mature policies from which to judge. $H_O: \mu = \$1500$ $H_A: \mu < \$1500$

Make a graph. Check the distribution for skewness, multiple modes, and outliers.	We checked the histogram of these data in the previous Guided Example and saw that it had a unimodal, symmetric distribution.
Model Check the conditions.	We checked the Randomization and Nearly Normal Conditions in the previous Guided Example.
State the sampling distribution model. Choose your method.	The conditions are satisfied, so we'll use a Student's t-model with $n - 1 = 29$ degrees of freedom and a one-sample t-test for the mean.

DO

| **Mechanics** Compute the sample statistics. Be sure to include the units when you write down what you know from the data.

The t-statistic calculation is just a standardized value. We subtract the hypothesized mean and divide by the standard error.

We assume the null model is true to find the P-value. Make a picture of the t-model, centered at μ_0. Since this is a lower-tail test, shade the region to the left of the observed average profit.

The P-value is the probability of observing a sample mean as small as $1438.90 (or smaller) *if* the true mean were $1500, as the null hypothesis states. We can find this P-value from a table, calculator, or computer program. | Using software, we obtain the following basic statistics:

$$n = 30$$
$$Mean = \$1438.90$$
$$s = \$1329.60$$
$$t = \frac{1438.90 - 1500}{1329.60/\sqrt{30}} = -0.2517$$

(The observed mean is less than one standard error below the hypothesized value.)

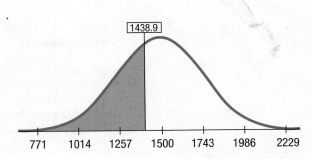

P-value $= P(t_{29} < -0.2517) = 0.4015$ (or from a table, we know at least that $P > 0.10$) |

REPORT

| **Conclusion** Link the P-value to your decision about H_0, and state your conclusion in context. | MEMO
Re: Sales performance
The mean profit on 30 sampled contracts closed by the sales rep in question has fallen below our standard of $1500, but there is not enough evidence in this sample of policies to indicate that the true mean is below $1500. If the mean were $1500, we would expect a sample of size 30 to have a mean this low about 40.15% of the time. |

Notice that the way this hypothesis was set up, the sales rep's mean profit would have to be well below $1500 to reject the null hypothesis. Because the null hypothesis was that the mean was $1500 and the alternative was that it was less, this setup gave some benefit of the doubt to the sales rep. There's nothing intrinsically wrong with that, but keep in mind that it's always a good idea to make sure that the hypotheses are stated in ways that will guide you to make the right business decision.

FOR EXAMPLE   Testing a mean

Summit Projects is a full-service interactive agency based in Hood River, Oregon, that offers companies a variety of website services. Summit recently redesigned a client company's website and wants to show that sales have increased. An analyst selects 58 sales records at random from the client's 2000 access logs and she finds a mean amount spent of $26.05 with a standard deviation of $10.20.

QUESTION   Test the hypothesis that the mean is $24.85 (as it was before the redesign) against the alternative that it has increased.

ANSWER   We can write: $H_0: \mu = \$24.85$ *vs.* $H_A: \mu > \$24.85$. Then

$$t = \frac{(26.05 - 24.85)}{10.2/\sqrt{58}} = 0.896.$$

Because the alternative is *one-sided*, we find $P(t > 0.896)$ with 57 degrees of freedom. From technology, $P(t > 0.896) = 0.1870$, a large P-value. This would not be a surprising value if the hypothesized mean of $24.85 were the true value. Therefore we *fail to reject* the null hypothesis and conclude that there is not sufficient evidence to suggest that the mean has increased from 24.85. Had we used a two-sided alternative, the P-value would have been twice 0.1870 or 0.3740.

Sample Size

How large a sample do you need? More information is always better, but acquiring more observations costs money, effort, and time. So how much data is enough?

As you make plans to collect data, you should have some idea of how small a margin of error is required to be able to draw a conclusion or detect a difference we want to see. If the size of the effect you're studying is large, then you may be able to tolerate a larger ME. If you need greater precision, however, you'll want a smaller ME, and, of course, that means a larger sample size. Armed with the ME and confidence level, you can find the sample size we'll need. Almost.

We know that for a mean, $ME = t^*_{n-1} \times SE(\bar{y})$ and that $SE(\bar{y}) = \dfrac{s}{\sqrt{n}}$, so we can determine the sample size by solving this equation for n:

$$ME = t^*_{n-1} \times \frac{s}{\sqrt{n}}.$$

The good news is that we have an equation; the bad news is that we won't know most of the values we need to compute it. When we thought about sample size for proportions, we ran into a similar problem. There we had to guess a working value for p to compute a sample size. Here, we need to know s, and if we're thinking about a very small sample, we need to know how many degrees of freedom to use. As an approximation for the critical value of t for 95% confidence (for the number of df we are likely to see), we can use 2.0 instead of t^*. That's much simpler to calculate, and it's a pretty good approximation. We don't know s until we get some data, but we want to calculate the sample size *before* collecting the data. We might be able to make a good guess, and that is often good enough for this purpose. If we have no idea what the standard deviation might be or if the sample size really matters (for example, because each additional individual is very expensive to sample or experiment on), it might be a good idea to run a small *pilot study* to get some feeling for the size of the standard deviation.

There are software packages available that can compute sample sizes for both confidence intervals and tests. It's important to keep in mind that all sample size calculations are approximate. You won't know the *actual* margin of error or the power of the test until after you've collected the data.

JUST CHECKING

You've seen an ad for some software that claims to lower the time it takes to download movies. It costs $49.95, so you're thinking about testing it on a few movies to see if it really works before you buy. Right now it takes you about 20 minutes on average to download a 2-hour movie. The standard deviation is about 5 minutes.

10 For which situation would you need a larger sample size to see if the software works: the software really reduces the time by 2 minutes or by 10 minutes on average? Why?

11 Suppose the standard deviation of the time it takes this software is only 2 minutes instead of 5. Will this widen or narrow your confidence interval for the mean time it takes to download a movie? Why?

FOR EXAMPLE   Finding the sample size for a confidence interval for means

In the 150 samples of farmed salmon (see page 346), the mean concentration of mirex was 0.0913 ppm with a standard deviation of 0.0495 ppm. A 95% confidence interval for the mean mirex concentration was found to be (0.0833, 0.0993).

QUESTION How large a sample would be needed to produce a 95% confidence interval with a margin of error of 0.004?

ANSWER We will assume that the standard deviation is 0.0495 ppm. The margin of error is equal to the critical value times the standard error. Using 2 for t^*, we find:

$$0.004 = 2 \times \frac{0.0495}{\sqrt{n}}$$

Solving for n, we find:

$$\sqrt{n} = 2 \times \frac{0.0495}{0.004}$$

or

$$n = \left(2 \times \frac{0.0495}{0.004} \right)^2 = 612.56$$

You should use a sample of at least 613 to have a good chance of getting a margin of error of 0.004.

WHAT CAN GO WRONG?

First, you must decide when to use Student's *t* methods.

- **Don't confuse proportions and means.** When you treat your data as categorical, counting successes and summarizing with a sample proportion, make inferences using the Normal model methods. When you treat your data as quantitative, summarizing with a sample mean, make your inferences using Student's *t* methods.

Student's *t* methods work only when the Normal Population Assumption is true. Naturally, many of the ways things can go wrong turn out to be ways that the Normal Population Assumption can fail. It's always a good idea to look for the most common kinds of failure. It turns out that you can even fix some of them.

- **Beware of multimodality.** The Nearly Normal Condition clearly fails if a histogram of the data has two or more modes. When you see this, look for the possibility that your data come from two groups. If so, your best bet is to try to separate the data into groups. (Use the variables to help distinguish the modes, if possible. For example, if the modes seem to be composed mostly of men in one and women in the other, split the data according to the person's sex.) Then you can analyze each group separately.

- **Beware of skewed data.** Make a histogram of the data. If the data are severely skewed, you might try re-expressing the variable. Re-expressing may yield a distribution that is unimodal and symmetric, making it more appropriate for the inference methods for means. Re-expression cannot help if the sample distribution is not unimodal.

> **What to Do with Outliers**
>
> As tempting as it is to get rid of annoying values, you can't just throw away outliers and not discuss them. It is not appropriate to lop off the highest or lowest values just to improve your results. The best strategy is to report the analysis with *and* without the outliers and comment on any differences.

- **Investigate outliers.** The Nearly Normal Condition also fails if the data have outliers. If you find outliers in the data, you need to investigate them. Sometimes, it's obvious that a data value is wrong and the justification for removing or correcting it is clear. When there's no clear justification for removing an outlier, you might want to run the analysis both with and without the outlier and note any differences in your conclusions. Any time data values are set aside, you *must* report on them individually. Often they will turn out to be the most informative part of your report on the data.[9]

Of course, Normality issues aren't the only risks you face when doing inferences about means.

- **Watch out for bias.** Measurements of all kinds can be biased. If your observations differ from the true mean in a systematic way, your confidence interval may not capture the true mean. And there is no sample size that will save you. A bathroom scale that's 5 pounds off will be 5 pounds off even if you weigh yourself 100 times and take the average. We've seen several sources of bias in surveys, but measurements can be biased, too. Be sure to think about possible sources of bias in your measurements.

- **Make sure data are independent.** Student's *t* methods also require the sampled values to be mutually independent. We check for random sampling. You should also think hard about whether there are likely violations of independence in the data collection method. If there are, be very cautious about using these methods.

---

[9] This suggestion may be controversial in some disciplines. Setting aside outliers is seen by some as unethical because the result is likely to be a narrower confidence interval or a smaller P-value. But an analysis of data with outliers left in place is *always* wrong. The outliers violate the Nearly Normal Condition and also the implicit assumption of a homogeneous population, so they invalidate inference procedures. An analysis of the nonoutlying points, along with a separate discussion of the outliers, is often much more informative, and can reveal important aspects of the data.

ETHICS IN ACTION

It has been three years since Mohammed Al-Tamimi opened his computer repair business, Mo's Mending Station. Unlike the well-known Nerd Squad of the big electronics retailer, Mo's Mending Station fixes only computers and does not deal with any other electronics such as TVs, phones, cameras, or appliances. Nor does Mo's provide any in-home services, such as networking or computer setup. Mo's main objective is clear: to provide standard repair services for computers and laptops, virus and spyware removal, and data recovery, each at a fixed low price. He charges the competitive rate of $45 per hour for more complicated computer issues.

Mo's slogan is *"Get twice the nerd at half the cost!"* This strategy has worked well, allowing Mohammed to grow his business to include six repair technicians and one office manager. However, recent monthly receipts indicate that the demand for Mo's services may be slowing down. Worried that the Mending Station might be losing its competitive price advantage, Mohammed gathers his staff together for a brainstorming session. Ed Ramsey, who has been with Mo's since it opened, mentions the possibility that the Mending Station's low prices may give some potential customers the impression that it offers poor quality service. He suggests hiring a local advertising firm to help brand Mo's Mending Station as affordable AND high quality by emphasizing its team of professional, experienced, and friendly repair technicians. In other words, put the focus on *"twice the nerd"* rather than *"half the cost."*

Mohammed thinks that this is a great idea and wonders if they can also prepare some statistics to strengthen the message. Because customer receipts include both when a computer is brought to Mo's (date and time) as well as when the repair is finished, he asks his office manager to select a sample so they can estimate the average service time. Based on 36 receipts, she finds a mean service time of 2 hours and 10 minutes with a standard deviation of 30 minutes. Further statistical analysis yielded a 95% confidence interval for the mean service time of 1.99 to 2.33 hours. Mohammed plans to advertise that 95% of his customers can expect to wait between 1.99 and 2.33 hours to get their computers back from repair! He is anxious to include this claim in all of the Mending Station's future marketing communication materials.

- Identify the ethical dilemma in this scenario.
- Has Mohammed interpreted the confidence interval correctly?
- What are the undesirable consequences?
- Propose an ethical solution that considers the welfare of all stakeholders.

WHAT HAVE WE LEARNED?

Learning Objectives

Know the sampling distribution of the mean.
- To apply the Central Limit Theorem for the mean in practical applications, we must estimate the standard deviation. This *standard error* is

$$SE(\bar{y}) = \frac{s}{\sqrt{n}}$$

- When we use the SE, the sampling distribution that allows for the additional uncertainty is Student's *t*.

Construct confidence intervals for the true mean, μ.
- A confidence interval for the mean has the form $\bar{y} \pm ME$.
- The Margin of Error is $ME = t^*_{df} \times SE(\bar{y})$.

Find *t** values by technology or from tables.
- When constructing confidence intervals for means, the correct degrees of freedom is $n - 1$.

Check the Assumptions and Conditions before using any sampling distribution for inference.

Write clear summaries to interpret a confidence interval.

Be able to perform a hypothesis test for a mean.

- The null hypothesis has the form $H_0: \mu = \mu_0$.
- We refer the test statistic $t = \dfrac{\bar{y} - \mu_0}{SE(\bar{y})}$ to the Student's t distribution with $n - 1$ degrees of freedom.

Terms	
Central Limit Theorem	The Central Limit Theorem (CLT) states that the sampling distribution model of the sample mean (and proportion) from a random sample is approximately Normal for large n, regardless of the distribution of the population, as long as the observations are independent.
Degrees of freedom (df)	A parameter of the Student's t-distribution that depends upon the sample size. Typically, more degrees of freedom reflects increasing information from the sample.
One-sample t-interval for the mean	A one-sample t-interval for the population mean is: $$\bar{y} \pm t^*_{n-1} \times SE(\bar{y}) \text{ where } SE(\bar{y}) = \frac{s}{\sqrt{n}}.$$ The critical value t^*_{n-1} depends on the particular confidence level, C, that you specify and on the number of degrees of freedom, $n - 1$.
One-sample t-test for the mean	The one-sample t-test for the mean tests the hypothesis $H_0: \mu = \mu_0$ using the statistic $$t_{n-1} = \frac{\bar{y} - \mu_0}{SE(\bar{y})}, \text{ where } SE(\bar{y}) = \frac{s}{\sqrt{n}}.$$
Sampling distribution model for a mean	If the independence assumption and randomization condition are met and the sample size is large enough, the sampling distribution of the sample mean is well modeled by a Normal model with a mean equal to the population mean, and a standard deviation equal to $\dfrac{\sigma}{\sqrt{n}}$.
Standard error	An estimate of the standard deviation of a statistic's sampling distribution based on the data
Student's t	A family of distributions indexed by its degrees of freedom. The t-models are unimodal, symmetric, and bell-shaped, but generally have fatter tails and a narrower center than the Normal model. As the degrees of freedom increase, t-distributions approach the Normal model.

TECHNOLOGY HELP: Inference for Means

Statistics packages offer convenient ways to make histograms of the data. That means you have no excuse for skipping the check that the data are nearly Normal.

Any standard statistics package can compute a hypothesis test. Here's what the package output might look like in general (although no package we know gives the results in exactly this form).

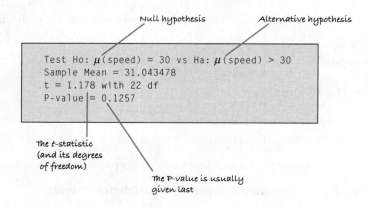

The package computes the sample mean and sample standard deviation of the variable and finds the P-value from the *t*-distribution based on the appropriate number of degrees of freedom. All modern statistics packages report P-values. The package may also provide additional information such as the sample mean, sample standard deviation, *t*-statistic value, and degrees of freedom. These are useful for interpreting the resulting P-value and telling the difference between a meaningful result and one that is merely statistically significant.

Statistics packages that report the estimated standard deviation of the sampling distribution usually label it "standard error" or "SE."

Inference results are also sometimes reported in a table. You may have to read carefully to find the values you need. Often, test results and the corresponding confidence interval bounds are given together. And often you must read carefully to find the alternative hypotheses. The example below shows some of those combinations.

The commands to do inference for means on common statistics programs and calculators may vary. The way that data is set up differs from program to program. Consult the help file for the program.

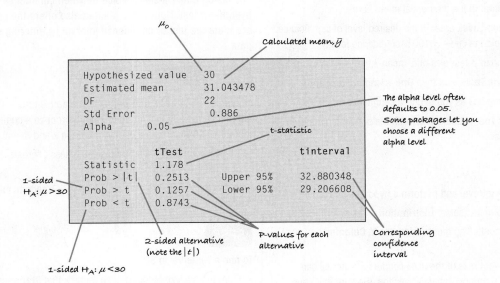

EXCEL

To find a one-sample *t*-interval and one-sample *t*-test for the mean using Excel, follow these examples:

	A	B	C	D	E	F
1	t-Estimate of a Mean					
2						
3	Sample mean	15.02	Confidence Interval Estimate			Syntax for Column E
4	Sample standard deviation	8.31	15.02	±	1.81	=ABS(T.INV((1-B6)/2,(B5-1))*(B4/B5^0.5))
5	Sample size	83	Lower confidence limit		13.21	=B3-E4
6	Confidence level	0.95	Upper confidence limit		16.83	=B3+E4

	A	B	C	D	E
1	t-Test of a Mean				
2					Syntax for Column D
3	Sample mean	460.38	t Stat	1.89	=(B3-B6)/(B4/B5^0.5)
4	Sample standard deviation	38.83	P(T<=t) one-tail	0.0323	=1-(T.DIST(ABS(D3),(B5-1),1))
5	Sample size	50	t Critical lower one-tail	-1.6766	=T.INV(B7,B5-1)
6	Hypothesized mean	450	t Critical upper one tail	1.6766	=ABS(T.INV(B7,B5-1))
7	Alpha	0.05	P(T<=t) two-tail	0.0646	=2*D4
8			t Critical two-tail	± 2.0096	=T.INV(B7/2,B5-1)

XLSTAT

To find a one-sample *z*-interval or a one-sample *t*-interval:

- Choose **Parametric Tests**, and then **One-sample t-test and z-test**.
- Under the **General** tab, enter your data cell range and choose either **z-test** or **Student's t test**.
- On the **Options** tab, choose the **Alternative hypothesis** of **Mean 1 ≠ Theoretical mean**.
- For calculating just a confidence interval, you can leave the **Theoretical mean** field blank. If you are also conducting a hypothesis test, enter in the theoretical mean here.
- Under **Significance Level**, enter in the desired level of significance. The output will yield the $(1 - \alpha)100\%$ confidence level.

To conduct a one-mean *z*-test or a one-mean *t*-test:

- Choose **Parametric Tests**, and then **One-sample t-test and z-test**.
- Complete the dialog box as you did for a confidence interval.
- Fill in the field for **Theoretical mean** with the population mean from your null hypothesis.

JMP

To find a confidence interval and perform a hypothesis test using JMP:

- From the **Analyze** menu, select **Distribution**.
- Drag the column containing the data into the **Y, Columns** box and press **OK**.
- The upper and lower limits of the 95% confidence interval can be found in the "Summary Statistics" section. (Be sure that your variables are "Continuous" type so that this section will be available.)
- To perform a *t*-test for the mean, click the arrow next to the variable name and choose **Test Mean**.

Comments

Choosing **Confidence Interval** in the expanded menu under the arrow next to the variable name allows for confidence levels other than 95%. In **Test Mean**, one can also run a *z*-test for the mean by providing a known standard deviation.

MINITAB

To perform a *z*-test and *z*-interval or *t*-test and *t*-interval in Minitab:

- From the Stat menu, choose the **Basic Statistics** submenu.

- From that menu, choose:
  - **1-sample Z…** for a z-interval or z-test
  - **1-sample t…** for a *t*-interval or *t*-test

Then fill in the dialog

- If a hypothesis test is desired, check **Perform hypothesis test** and enter the population mean.
- Click **Options** and set confidence level and alternative hypothesis.

Comments

The dialog offers a clear choice between confidence interval and hypothesis test. Notice that Minitab also offers the opportunity to calculate the test or confidence interval by entering summarized data.

R

To test the hypothesis that $\mu = mu$ (default is mu = 0) against an alternative (default is two-sided) and to produce a confidence interval (default is 95%), create a vector of data in x and then:

- **t.test**(x, alternative = c("two.sided", "less", "greater"), mu = 0, conf.level = 0.95)

provides the *t*-statistic, P-value, degrees of freedom, and the confidence interval for a specified alternative.

SPSS

To find a *t*-interval in SPSS:

- From the Analyze Menu, choose **Descriptive Statistics**.
- Choose **Explore**.
- Click the **Statistics** button to change confidence level and choose **Statistics** only if no plots are desired.

To perform a *t*-test in SPSS:

- From the Analyze menu, choose the **Compare Means** submenu.
- Choose the **One-Sample t-test** command.
- Select the column containing data and enter the test value for the mean.

Brief **Case**

Real Estate

A real estate agent is trying to understand the pricing of homes in her area, a region comprised of small to midsize towns and a small city. For each of 1200 homes recently sold in the region, the file **Real Estate sample 1200** holds the following variables:

- *Sale Price* (in $)
- *Lot size* (size of the lot in acres)
- *Waterfront* (Yes, No)
- *Age* (in years)
- *Central Air* (Yes, No)
- *Fuel Type* (Wood, Oil, Gas, Electric, Propane, Solar, Other)
- *Condition* (1 to 5, 1 = Poor, 5 = Excellent)
- *Living Area* (living area in square feet)
- *Pct College* (% in ZIP code who attend a four-year college)
- *Full Baths* (number of full bathrooms)
- *Half Baths* (number of half bathrooms)
- *Bedrooms* (number of bedrooms)
- *Fireplace* (Yes, No)

The agent has a family interested in a four-bedroom house. Using confidence intervals, how should she advise the family on what the average price of a four-bedroom house might be in this area? Compare that to a confidence interval for two-bedroom homes. How much more, on average, does a house that has central air conditioning sell for? Restrict your attention to two-bedroom houses and answer the question again. What about four-bedroom houses? Repeat this investigation for houses with and without fireplaces. What cautions might you give the agent before coming to any overall conclusions?

Explore other questions that might be useful for the real estate agent in knowing how different categorical factors affect the sale price and write up a short report on your findings.

Donor Profiles

A philanthropic organization collects and buys data on their donor base. The full database contains about 4.5 million donors and over 400 variables collected on each, but the data set **Donor Profiles** is a sample of 916 donors and includes the variables:

- *Age* (in years)
- *Homeowner* (H = Yes, U = Unknown)
- *Gender* (F = Female, M = Male, U = Unknown)
- *Wealth* (Ordered categories of total household wealth from 1 = Lowest to 9 = Highest)
- *Children* (Number of children)
- *Donated Last* (0 = Did not donate to last campaign, 1 = Did donate to last campaign)
- *Amt Donated Last* ($ amount of contribution to last campaign)

The analysts at the organization want to know how much people donate on average to campaigns, and what factors might influence that amount. Compare the confidence intervals for the mean *Amt Donated Last* by those known to own their homes with those whose homeowner status is unknown. Perform similar comparisons for *Gender* and two of the *Wealth* categories. Write up a short

(*continued*)

report using graphics and confidence intervals for what you have found. (Be careful not to make inferences directly about the differences between groups. We'll discuss that in Chapter 13. Your inference should be about single groups.)

(The distribution of *Amt Donated Last* is highly skewed to the right, and so the median might be thought to be the appropriate summary. But the median is $0.00 so the analysts must use the mean. From simulations, they have ascertained that the sampling distribution for the mean is unimodal and symmetric for samples larger than 250 or so. Note that small differences in the mean could result in millions of dollars of added revenue nationwide. The average cost of their solicitation is $0.67 per person to produce and mail.)

EXERCISES

SECTION 11.1

1. Games for the iPad have a distribution of prices that is skewed to the high end.

a) Explain why this is what you would expect.
b) Members of the iPad gamers club each own about 50 games. Pat is one such member. What would you expect the shape of the distribution of game prices on her iPad to be?
c) Each club member computes the average price of his or her games. What shape would you expect the distribution of these averages to have?

2. For a sample of 36 houses, what would you expect the distribution of the sale prices to be? A real-estate agent has been assigned 10 houses at random to sell this month. She wants to know whether the mean price of those houses is typical. What, if anything, does she need to assume about the distribution of prices to be able to use the Central Limit Theorem? Are those assumptions reasonable?

3. According to the Gallup Poll, 27% of U.S. adults have high levels of cholesterol. Gallup reports that such elevated levels "could be financially devastating to the U.S. healthcare system" and are a major concern to health insurance providers. According to recent studies, cholesterol levels in healthy U.S. adults average about 215 mg/dL with a standard deviation of about 30 mg/dL and are roughly Normally distributed. If the cholesterol levels of a sample of 42 healthy U.S. adults is taken,

a) What shape should the sampling distribution of the mean have?
b) What would the mean of the sampling distribution be?
c) What would its standard deviation be?
d) If the sample size were increased to 100, how would your answers to parts a–c change?

4. As in Exercise 3, cholesterol levels in healthy U.S. adults average about 215 mg/dL with a standard deviation of about 30 mg/dL and are roughly Normally distributed. If the cholesterol levels of a sample of 42 healthy U.S. adults is taken, what is the probability that the mean cholesterol level of the sample

a) Will be no more than 215?
b) Will be between 205 and 225?
c) Will be less than 200?
d) Will be greater than 220?

SECTION 11.3

5. Organizers of a fishing tournament believe that the lake holds a sizable population of largemouth bass. They assume that the weights of these fish have a model that is skewed to the right with a mean of 3.5 pounds and a standard deviation of 2.32 pounds.

a) Explain why a skewed model makes sense here.
b) Explain why you cannot use a Normal model to determine the probability that a largemouth bass randomly selected ("caught") from the lake weighs over 3 pounds.
c) Each contestant catches 5 fish each day. Can you determine the probability that someone's catch averages over 3 pounds? Explain.
d) The 12 contestants competing each caught the limit of 5 fish. What's the standard deviation of the mean weight of the 60 fish caught?
e) Would you be surprised if the mean weight of the 60 fish caught in the competition was more than 4.5 pounds? Use the 68–95–99.7 Rule.

6. In 2008 and 2009, Systemax bought two failing electronics stores, Circuit City and CompUSA. They have kept both the names active and customers can purchase products from either website. If they take a random sample of a mixture of recent purchases from the two websites, the distribution of the amounts purchased will be bimodal.

a) As their sample size increases, what's the expected shape of the distribution of amounts purchased in the sample?

b) As the sample size increases, what's the expected shape of the sampling model for the mean amount purchased of the sample?

SECTION 11.4

7. A survey of 25 randomly selected customers found the following ages (in years):

20	32	34	29	30
30	30	14	29	11
38	22	44	48	26
25	22	32	35	32
35	42	44	44	48

The mean was 31.84 years and the standard deviation was 9.84 years.

a) What is the standard error of the mean?
b) How would the standard error change if the sample size had been 100 instead of 25? (Assume that the sample standard deviation didn't change.)

8. A random sample of 20 purchases showed the following amounts (in $):

39.05	2.73	32.92	47.51
37.91	34.35	64.48	51.96
56.95	81.58	47.80	11.72
21.57	40.83	38.24	32.98
75.16	74.30	47.54	65.62

The mean was $45.26 and the standard deviation was $20.67.

a) What is the standard error of the mean?
b) How would the standard error change if the sample size had been 5 instead of 20? (Assume that the sample standard deviation didn't change.)

9. For the data in Exercise 7:

a) How many degrees of freedom does the t-statistic have?
b) How many degrees of freedom would the t-statistic have if the sample size had been 100?

10. For the data in Exercise 8:

a) How many degrees of freedom does the t-statistic have?
b) How many degrees of freedom would the t-statistic have if the sample size had been 5?

SECTION 11.5

11. Find the critical value t^* for:

a) a 95% confidence interval based on 24 df.
b) a 95% confidence interval based on 99 df.

12. Find the critical value t^* for:

a) a 90% confidence interval based on 19 df.
b) a 90% confidence interval based on 4 df.

13. For the ages in Exercise 7:

a) Construct a 95% confidence interval for the mean age of all customers, assuming that the assumptions and conditions for the confidence interval have been met.
b) How large is the margin of error?
c) How would the confidence interval change if you had assumed that the standard deviation was known to be 10.0 years?

14. For the purchase amounts in Exercise 8:

a) Construct a 90% confidence interval for the mean purchases of all customers, assuming that the assumptions and conditions for the confidence interval have been met.
b) How large is the margin of error?
c) How would the confidence interval change if you had assumed that the standard deviation was known to be $20?

SECTION 11.6

15. For the confidence intervals of Exercise 13, a histogram of the data looks like this:

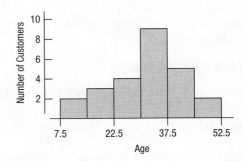

Check the assumptions and conditions for your inference.

16. For the confidence intervals of Exercise 14, a histogram of the data looks like this:

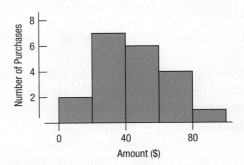

Check the assumptions and conditions for your inference.

SECTION 11.7

17. The owner of the store from Exercise 7 wants to know if the mean age of all customers is 25 years old.

a) What is the null hypothesis?
b) Is the alternative one- or two-sided?
c) What is the value of the test statistic?
d) What is the P-value of the test statistic?
e) What do you conclude at alpha = 0.05?

18. The analyst in Exercise 8 wants to know if the mean purchase amount of all transactions is at least $40.

a) What is the null hypothesis?
b) Is the alternative one- or two-sided?
c) What is the value of the test statistic?
d) What is the P-value of the test statistic?
e) What do you conclude at alpha = 0.05?

19. For the confidence interval in Exercise 13:

a) How large would the sample size have to be to cut the margin of error in half?
b) About how large would the sample size have to be to cut the margin of error by a factor of 10?

20. For the confidence interval in Exercise 14 part a:

a) To reduce the margin of error to about $4, how large would the sample size have to be?
b) How large would the sample size have to be to reduce the margin of error to $0.80?

CHAPTER EXERCISES

21. *t*-models. Using the *t*-tables, software, or a calculator, estimate:

a) the critical value of *t* for a 90% confidence interval with df = 17.
b) the critical value of *t* for a 98% confidence interval with df = 88.

22. *t*-models, part 2. Using the *t*-tables, software, or a calculator, estimate:

a) the critical value of *t* for a 95% confidence interval with df = 7.
b) the critical value of *t* for a 99% confidence interval with df = 102.

23. Confidence intervals. Describe how the width of a 95% confidence interval for a mean changes as the standard deviation (s) of a sample increases, assuming sample size remains the same.

24. Confidence intervals, part 2. Describe how the width of a 95% confidence interval for a mean changes as the sample size (n) increases, assuming the standard deviation remains the same.

25. Confidence intervals and sample size. A confidence interval for the price of gasoline from a random sample of 30 gas stations in a region gives the following statistics:

$$\bar{y} = \$4.49 \quad s = \$0.29$$

a) Find a 95% confidence interval for the mean price of regular gasoline in that region.
b) Find the 90% confidence interval for the mean.
c) If we had the same statistics from a sample of 60 stations, what would the 95% confidence interval be now?

26. Confidence intervals and sample size, part 2. A confidence interval for the price of gasoline from a random sample of 30 gas stations in a region gives the following statistics:

$$\bar{y} = \$4.49 \quad SE(\bar{y}) = \$0.06$$

a) Find a 95% confidence interval for the mean price of regular gasoline in that region.
b) Find the 90% confidence interval for the mean.
c) If we had the same statistics from a sample of 60 stations, what would the 95% confidence interval be now?

27. Marketing livestock feed. A feed supply company has developed a special feed supplement to see if it will promote weight gain in livestock. Their researchers report that the 77 cows studied gained an average of 56 pounds and that a 95% confidence interval for the mean weight gain this supplement produces has a margin of error of ±11 pounds. Staff in their marketing department wrote the following conclusions. Did anyone interpret the interval correctly? Explain any misinterpretations.

a) 95% of the cows studied gained between 45 and 67 pounds.
b) We're 95% sure that a cow fed this supplement will gain between 45 and 67 pounds.
c) We're 95% sure that the average weight gain among the cows in this study was between 45 and 67 pounds.
d) The average weight gain of cows fed this supplement is between 45 and 67 pounds 95% of the time.
e) If this supplement is tested on another sample of cows, there is a 95% chance that their average weight gain will be between 45 and 67 pounds.

28. Meal costs. A company is interested in estimating the costs of lunch in their cafeteria. After surveying employees, the staff calculated that a 95% confidence interval for the mean amount of money spent for lunch over a period of six months is ($780, $920). Now the organization is trying to write its report and considering the following interpretations. Comment on each.

a) 95% of all employees pay between $780 and $920 for lunch.
b) 95% of the sampled employees paid between $780 and $920 for lunch.
c) We're 95% sure that employees in this sample averaged between $780 and $920 for lunch.
d) 95% of all samples of employees will have average lunch costs between $780 and $920.
e) We're 95% sure that the average amount all employees pay for lunch is between $780 and $920.

29. CEO compensation. A sample of 20 CEOs from the Forbes 500 shows total annual compensations ranging from a minimum of $0.1 to $62.24 million. The average for these 20 CEOs is $7.946 million. The histogram and boxplot are as follows:

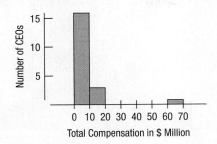

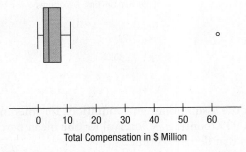

Based on these data, a computer program found that a confidence interval for the mean annual compensation of all Forbes 500 CEOs is (1.69, 14.20) $M. Why should you be hesitant to trust this confidence interval?

30. Credit card charges. A credit card company takes a random sample of 100 cardholders to see how much they charged on their card last month. A histogram and boxplot are as follows:

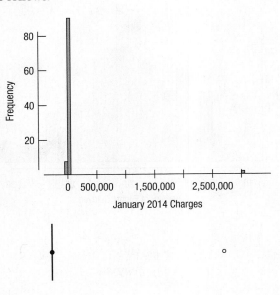

A computer program found that the 95% confidence interval for the mean amount spent in January 2014 is (−$28,366.84, $90,691.49). Explain why the analysts didn't find the confidence interval useful, and explain what went wrong.

31. Parking. Hoping to lure more shoppers downtown, a city builds a new public parking garage in the central business district. The city plans to pay for the structure through parking fees. For a random sample of 44 weekdays, daily fees collected averaged $126, with a standard deviation of $15.

a) What assumptions must you make in order to use these statistics for inference?
b) Find a 90% confidence interval for the mean daily income this parking garage will generate.
c) Explain in context what this confidence interval means.
d) Explain what 90% confidence means in this context.
e) The consultant who advised the city on this project predicted that parking revenues would average $128 per day. Based on your confidence interval, what do you think of the consultant's prediction? Why?

32. Housing. In 2012, a large number of foreclosed homes in the Washington, DC, metro area were sold. In one community, a sample of 30 foreclosed homes sold for an average of $443,705 with a standard deviation of $196,196.

a) What assumptions and conditions must be checked before finding a confidence interval for the mean? How would you check them?
b) Find a 95% confidence interval for the mean value per home.
c) Interpret this interval and explain what 95% confidence means.
d) Suppose nationally, the average foreclosed home sold for $350,000. Do you think the average sale price in the sampled community differs significantly from the national average? Explain.

33. Parking, part 2. Suppose that for budget planning purposes the city in Exercise 31 needs a better estimate of the mean daily income from parking fees.

a) Someone suggests that the city use its data to create a 95% confidence interval instead of the 90% interval first created. How would this interval be better for the city? (You need not actually create the new interval.)
b) How would the 95% confidence interval be worse for the planners?
c) How could they achieve a confidence interval estimate that would better serve their planning needs?

34. Housing, part 2. In Exercise 32, we found a 95% confidence interval to estimate the average value of foreclosed homes.

a) Suppose the standard deviation of the values was $300,000 instead of the $196,196 used for that interval. What would the larger standard deviation do to the width of the confidence interval (assuming the same level of confidence)?

b) Your classmate suggests that the margin of error in the interval could be reduced if the confidence level were changed to 90% instead of 95%. Do you agree with this statement? Why or why not?

c) Instead of changing the level of confidence, would it be more statistically appropriate to draw a bigger sample?

35. State budgets. States that rely on sales tax for revenue to fund education, public safety, and other programs often end up with budget surpluses during economic growth periods (when people spend more on consumer goods) and budget deficits during recessions (when people spend less on consumer goods). Fifty-one small retailers in a state with a growing economy were recently sampled. The sample showed a mean increase of $2350 in additional sales tax revenue collected per retailer compared to the previous quarter. The sample standard deviation is $425.

a) Find a 95% confidence interval for the mean increase in sales tax revenue.

b) What assumptions have you made in this inference? Do you think the appropriate conditions have been satisfied?

c) Explain what your interval means and provide an example of what it does not mean.

36. State budgets, part 2. Suppose the state in Exercise 35 sampled 16 small retailers instead of 51, and for the sample of 16, the sample mean increase again equaled $2350 in additional sales tax revenue collected per retailer compared to the previous quarter. Also assume the sample standard deviation is $425.

a) What is the standard error of the mean increase in sales tax revenue collected?

b) What happens to the accuracy of the estimate when the interval is constructed using the smaller sample size?

c) Find and interpret a 95% confidence interval.

d) How does the margin of error for the interval constructed in Exercise 35 compare with the margin of error constructed in this exercise? Explain statistically how sample size changes the accuracy of the constructed interval. Which sample would you prefer if you were a state budget planner? Why?

T 37. Departures 2015. What are the chances your flight will leave on time? The U.S. Bureau of Transportation Statistics of the Department of Transportation publishes information about airline performance. Here are a histogram and summary statistics for the percentage of flights departing on time each month from January 1995 through January 2015:

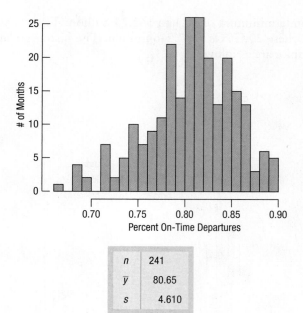

n	241
$\bar{y}$	80.65
s	4.610

a) Check the assumptions and conditions for inference.

b) Find a 90% confidence interval for the mean percentage of flights that depart on time.

c) Interpret this interval for a traveler planning to fly.

T 38. Late arrivals 2015. Will your flight get you to your destination on time? The U.S. Bureau of Transportation Statistics also reports arrival statistics each month. Here's a histogram, along with some summary statistics of the percentage of delayed arrivals each month from January 1995 through January 2015:

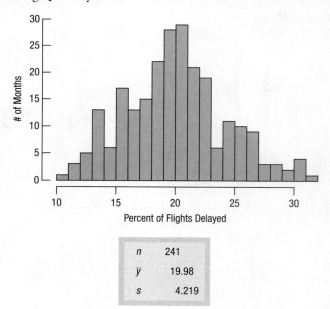

n	241
$\bar{y}$	19.98
s	4.219

We can consider these data to be a representative sample of all months. There is no evidence of a time trend.

a) Check the assumptions and conditions for inference about the mean.

b) Find a 99% confidence interval for the true percentage of flights that arrive late.

c) Interpret this interval for a traveler planning to fly.

39. Computer lab fees. The technology committee has stated that the average time spent by students per lab visit has increased, and the increase supports the need for increased lab fees. To substantiate this claim, the committee randomly samples 12 student lab visits and notes the amount of time spent using the computer. The times in minutes are as follows:

Time	
52	74
57	53
54	136
76	73
62	8
52	62

a) Plot the data. Are any of the observations outliers? Explain.

b) The previous mean amount of time spent using the lab computer was 55 minutes. Find a 95% confidence interval for the true mean. What do you conclude about the claim? If there are outliers, find intervals with and without the outliers present.

40. Cell phone batteries. A company that produces cell phones claims its standard phone battery lasts longer on average than other batteries in the market. To support this claim, the company publishes an ad reporting the results of a recent experiment showing that under normal usage, their batteries last at least 35 hours. To investigate this claim, a consumer advocacy group asked the company for the raw data. The company sends the group the following results:

35, 34, 32, 31, 34, 34, 32, 33, 35, 55, 32, 31

Find a 95% confidence interval and state your conclusion. Explain how you dealt with the outlier, and why.

41. Growth and air pollution. Government officials have difficulty attracting new business to communities with troubled reputations. Nevada has been one of the fastest growing states in the country for a number of years. Accompanying the rapid growth are massive new construction projects. Since Nevada has a dry climate, the construction creates visible dust pollution. High pollution levels may paint a less than attractive picture of the area, and can also result in fines levied by the federal government. As required by government regulation, researchers continually monitor pollution levels. In the most recent test of pollution levels, 121 air samples were collected. The dust particulate levels must be reported to the federal regulatory agencies. In the report sent to the federal agency, it was noted that the mean particulate level = 57.6 micrograms/cubic liter of air, and the 95% confidence interval estimate is 52.06 mg to 63.07 mg. A graph of the distribution of the particulate amounts was also included and is shown below.

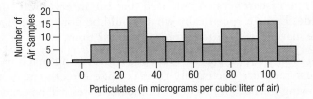

a) Discuss the assumptions and conditions for using Student's t inference methods with these data.

b) Do you think the confidence interval noted in the report is valid? Briefly explain why or why not.

42. Convention revenues. At one time, Nevada was the only U.S. state that allowed gambling. Although gambling continues to be one of the major industries in Nevada, the proliferation of legalized gambling in other areas of the country has required state and local governments to look at other growth possibilities. The convention and visitor's authorities in many Nevada cities actively recruit national conventions that bring thousands of visitors to the state. Various demographic and economic data are collected from surveys given to convention attendees. One statistic of interest is the amount visitors spend on slot machine gambling. Nevada often reports the slot machine expenditure as amount spent per hotel guest room. A recent survey of 500 visitors asked how much they spent on gambling. The average expenditure per room was $180.

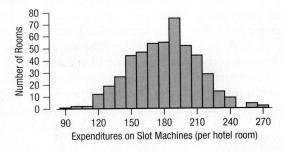

Casinos will use the information reported in the survey to estimate slot machine expenditure per hotel room. Do you think the estimates produced by the survey will accurately represent expenditures? Explain using the statistics reported and graph shown.

43. Traffic speed. Police departments often try to control traffic speed by placing speed-measuring machines on roads that tell motorists how fast they are driving. Traffic safety experts must determine where machines should be placed. In one recent test, police recorded the average speed clocked by cars driving on one busy street close to an elementary school. For a sample of 25 speeds, it was determined that the average amount over the speed limit for the 25 clocked speeds was 11.6 mph with a standard deviation of 8 mph. The 95% confidence interval estimate for this sample is 8.30 mph to 14.90 mph.

a) What is the margin of error for this problem?

b) The researchers commented that the interval was too wide. Explain specifically what should be done to reduce the margin of error to no more than ±2 mph.

44. Traffic speed, part 2. The speed-measuring machines must measure accurately to maximize effectiveness in slowing traffic. The accuracy of the machines will be tested before placement on city streets. To ensure that error rates are estimated accurately, the researchers want to take a large enough sample to ensure usable and accurate interval estimates of how much the machines may be off in measuring actual speeds. Specifically, the researchers want the margin of error for a single speed measurement to be no more than ±1.5 mph.

a) Discuss how the researchers may obtain a reasonable estimate of the standard deviation of error in the measured speeds.

b) Suppose the standard deviation for the error in the measured speeds equals 4 mph. At 95% confidence, what sample size should be taken to ensure that the margin of error is no larger than ±1.0 mph?

45. Tax audits. Certified public accountants are often required to appear with clients if the IRS audits the client's tax return. Some accounting firms give the client an option to pay a fee when the tax return is completed that guarantees tax advice and support from the accountant if the client were audited. The fee is charged up front like an insurance premium and is less than the amount that would be charged if the client were later audited and then decided to ask the firm for assistance during the audit. A large accounting firm is trying to determine what fee to charge for next year's returns. In previous years, the actual mean cost to the firm for attending a client audit session was $650. To determine if this cost has changed, the firm randomly samples 32 client audit fees. The sample mean audit cost was $680 with a standard deviation of $75.

a) Develop a 95% confidence interval estimate for the mean audit cost.

b) Based on your confidence interval, what do you think of the claim that the mean cost has changed?

46. Tax audits, part 2. While reviewing the sample of audit fees, a senior accountant for the firm notes that the fee charged by the firm's accountants depends on the complexity of the return. A comparison of actual charges therefore might not provide the information needed to set next year's fees. To better understand the fee structure, the senior accountant requests a new sample that measures the time the accountants spent on the audit. Last year, the

average hours charged per client audit was 3.25 hours. A new sample of 10 audit times shows the following times in hours:

$$4.2, 3.7, 4.8, 2.9, 3.1, 4.5, 4.2, 4.1, 5.0, 3.4$$

a) Assume the conditions necessary for inference are met. Find a 90% confidence interval estimate for the mean audit time.

b) Based on your answer to part a, do you think that the audit times have, in fact, increased?

47. Wind power. Should you generate electricity with your own personal wind turbine? That depends on whether you have enough wind on your site. To produce enough energy, your site should have an annual average wind speed of at least 8 miles per hour, according to the Wind Energy Association. One candidate site was monitored for a year, with wind speeds recorded every 6 hours. A total of 1114 readings of wind speed averaged 8.019 mph with a standard deviation of 3.813 mph. You've been asked to make a statistical report to help the landowner decide whether to place a wind turbine at this site.

a) Discuss the assumptions and conditions for using Student's t inference methods with these data. Here are some plots that may help you decide whether the methods can be used:

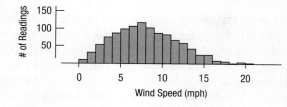

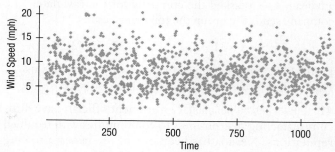

b) What would you tell the landowner about whether this site is suitable for a small wind turbine? Explain

48. Real estate crash? After the sub-prime crisis of late 2007, real estate prices fell almost everywhere in the U.S. In 2006–2007 before the crisis, the average selling price of homes in a region in upstate New York was $191,300. A real estate agency wants to know how much the prices have fallen since then. They collect a sample of 1231 homes in the region in mid-2013 and find the average asking price to be $178,613.50 with a standard deviation of $92,701.56. You have been retained by the real estate agency to report on the current situation.

a) Discuss the assumptions and conditions for using *t*-methods for inference with these data. Here are some plots that may help you decide what to do.

b) What would you report to the real estate agency about the current situation?

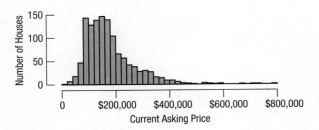

49. E-commerce. A market researcher at a major clothing company that has traditionally relied on catalog mail-order sales decides to investigate whether the amount of monthly online sales has changed. She compares the mean monthly online sales of the past several months with a historical figure for mean monthly sales for online purchases. She gets a P-value of 0.01. Explain in this context what the 1% means.

50. Performance standards. The U.S. Golf Association (USGA) sets performance standards for golf balls. For example, the initial velocity of the ball may not exceed 250 feet per second when measured by an apparatus approved by the USGA. Suppose a manufacturer introduces a new kind of ball and provides a randomly selected sample of balls for testing. Based on the mean speed in the sample, the USGA comes up with a P-value of 0.34. Explain in this context what the 34% represents.

51. E-commerce, part 2. The average age of online consumers a few years ago was 23.3 years. As older individuals gain confidence with the Internet, it is believed that the average age has increased. We would like to test this belief.

a) Write appropriate hypotheses.

b) We plan to test the null hypothesis by selecting a random sample of 40 individuals who have made an online purchase this year. Do you think the necessary assumptions for inference are satisfied? Explain.

c) The online shoppers in our sample had an average age of 24.2 years, with a standard deviation of 5.3 years. What's the P-value for this result?

d) Explain (in context) what this P-value means.

e) What's your conclusion?

52. Fuel economy. A company with a large fleet of cars hopes to keep gasoline costs down and sets a goal of attaining a fleet average of at least 26 miles per gallon. To see if the goal is being met, they check the gasoline usage for 50 company trips chosen at random, finding a mean of 25.02 mpg and a standard deviation of 4.83 mpg. Is this strong evidence that they have failed to attain their fuel economy goal?

a) Write appropriate hypotheses.

b) Are the necessary assumptions to perform inference satisfied?

c) Test the hypothesis and find the P-value.

d) Explain what the P-value means in this context.

e) State an appropriate conclusion.

T 53. Yogurt. *Consumer Reports* tested 11 brands of vanilla yogurt and found these numbers of calories per serving:

130 160 150 120 120 110 170 160 110 130 90

a) Check the assumptions and conditions for inference.

b) Create a 95% confidence interval for the average calorie content of vanilla yogurt.

c) A diet guide claims that you will get an average of 120 calories from a serving of vanilla yogurt. What does this evidence indicate? Use your confidence interval to test an appropriate hypothesis and state your conclusion.

T 54. Driving distance 2013. How far do professional golfers drive a ball? (For nongolfers, the drive is the shot hit from a tee at the start of a hole and is typically the longest shot.) Here's a histogram of the average driving distances of the 190 leading professional golfers by May 5, 2013, along with summary statistics (www.pgatour.com).

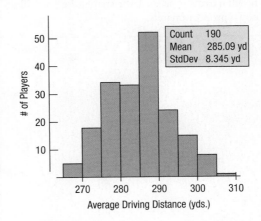

a) Find a 95% confidence interval for the mean drive distance.

b) Interpreting this interval raises some problems. Discuss.

c) The data are the mean driving distance for each golfer. Is that another concern in interpreting the interval?

JUST CHECKING ANSWERS

1 Normal. It doesn't matter that the sample is drawn from a skewed distribution; the CLT tells us that the means can be modeled with a Normal model as long as the sample size is large enough.

2 I would expect the histogram to be skewed. A sample of 10 is fairly small. With a skewed population the CLT requires a larger sample size before the sampling distribution of the mean will appear to be Normal.

3 $SE(\bar{y}) = 120/\sqrt{100} = 12$

4 Decreases.

5 The standard error decreases by $1/\sqrt{2}$.

6 Because the distribution is right skewed, the confidence interval based on only 20 observations is not likely to be accurate.

7 They don't know the population standard deviation, so they must use the sample SD as an estimate. The additional uncertainty is taken into account by t-models.

8 With a sample size of 60, the conditions are more nearly satisfied because the sampling distribution of the mean is closer to the Normal model, so the confidence interval should be accurate. It will also be narrower than the one based on 20.

9 The confidence interval is about the *mean* starting salary, not about individuals.

10 2 minutes. You'll need a larger sample size to see a smaller effect size because the sample mean is likely to be closer to the hypothesized mean if the effect size is smaller.

11 It will narrow it because the standard error will go down proportionately.

12

Comparing Two Means

Visa Global Organization

Today, more than one billion people and 24 million merchants use the Visa card in 170 countries worldwide. But back in the early 1950s when the idea of cashless transactions first took hold, only Diners Club and some retailers, notably oil companies, issued charge cards. The vast majority of purchases were made by cash or personal check. Bank of America pioneered its BankAmericard program in Fresno, California, in 1958, and American Express issued the first plastic card in 1959. The idea of a credit "card" really gained momentum a decade later when a group of banks formed a joint venture to create a centralized system of payment. National BankAmericard, Inc. (NBI) took ownership of the credit card system in 1970 and for simplicity and marketability changed its name to Visa in 1976. (The name Visa is pronounced nearly the same way in every language.) That year, Visa processed 679,000 transactions—a volume that is processed on average every four minutes today. As technology changed, so did the credit card industry. By 1986, cardholders were able to use their Visa cards to get cash from ATMs.

Now a global organization, Visa is divided into several regional entities including: Visa Asia Pacific, Visa Canada, Visa Europe, and Visa USA. In the fiscal year ending September 2014, Visa's global network processed 64.9 billion transactions for a total volume of $7.3 trillion.[1]

---

[1] Visa Inc. Annual Report 2014.

Champion vs. Challenger

In the credit card industry, among others, it is common to run experiments testing a new idea (offer, promotion, incentive, etc.) against the current, or "business as usual" practice. The new idea is called the *challenger* and the previously tested offer is the *champion*. Much of the direct mail that you see consists of marketing experiments whose goal is to determine if the challenger's performance warrants dethroning the champion and using the challenger going forward.

For credit card promotions, the typical measure used to judge success of the challenger offer is the amount charged on the card, from which the credit card issuer derives revenue. Because customers vary so much in the amount they charge on their cards, it is more efficient to compare the amount charged after the offer is received to the amount charged before the offer is received. This change in amount charged is often referred to as the *spend lift*. The word *lift* connotes the optimistic sentiment that the offer will increase the amount a customer charges, although it can be and often is negative.

The credit card business can be extremely profitable. The average American household has eight cards and a total of $7073 in credit card debt. But, since many households have no debt, that average is misleading. Among households that carry credit card debt, the average debt was $15,162 in 2012. Not surprisingly, the credit card business is also intensely competitive. Rival banks and lending agencies are constantly trying to create new products and offers to win new customers, keep current customers, and provide incentives for current customers to charge more on their cards.

Are some credit card promotions more effective than others? For example, do customers spend more using their credit card if they know they will be given "double miles" or "double coupons" toward flights, hotel stays, or store purchases? To answer questions such as this, credit card issuers often perform experiments on a sample of customers, making some of them an *offer* of an incentive, while other customers receive no offer. Promotions cost the company money, so the company needs to estimate the size of any increased revenue to judge whether it is sufficient to cover their expenses. By comparing the performance of the two offers on the sample, they can decide whether the new offer would provide enough potential profit if they were to "roll it out" and offer it to their entire customer base.

Experiments that compare two groups are common throughout both science and industry. Other applications include comparing the effects of a new drug with the traditional therapy, the fuel efficiency of two car engine designs, or the sales of new products on two different customer segments. Usually the experiment is carried out on a subset of the population, often a much smaller subset. Using statistics, we can make statements about whether the means of the two groups differ in the population at large, and how large that difference might be.

12.1   Comparing Two Means

The natural display for comparing the distributions of two groups is side-by-side boxplots (see Figure 12.1). For the credit card promotion, the company judges performance by comparing the *mean* spend lift (the change in spending from before receiving the promotion to after receiving it) for the two samples. If the difference in spend lift between the group that received the promotion and the group that didn't is high enough, this will be viewed as evidence that the promotion worked. Looking at the two boxplots, it's not obvious that there's much of a difference. Can we conclude that the slight increase seen for those who received the promotion is more than just random fluctuation? We'll need statistical inference.

For two groups, the statistic of interest is the difference in the observed means of the offer and no offer groups: $\bar{y}_{Offer} - \bar{y}_{No\ Offer}$. We've offered the promotion to a random sample of cardholders and used another sample of cardholders, who got no special offer, as a control group. We know what happened in our samples, but what we'd really like to know is the difference of the means in the population at large: $\mu_{Offer} - \mu_{No\ Offer}$.

We compare two means in much the same way as we compared a single mean to a hypothesized value. But now both means are statistics, each with its own standard deviation, and the population model parameter of interest is the *difference* between the means. In our example, it's the true difference between the mean spend lift for customers offered the promotion and for customers for whom no offer was

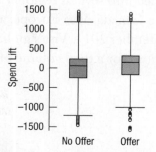

Figure 12.1   Side-by-side boxplots show a small increase in spending for the group that received the promotion.

made. We estimate the difference with $\bar{y}_{Offer} - \bar{y}_{No\ Offer}$. How can we tell if a difference we observe in the sample means indicates a real difference in the underlying population means? We'll need to know the sampling distribution model and standard deviation of the difference. Once we know those, we can build a confidence interval and test a hypothesis just as we did for a single mean.

We have data on 500 randomly selected customers who were offered the promotion and another randomly selected 500 who were not. It's easy to find the mean and standard deviation of the spend lift for each of these groups. From these, we can find the standard deviation of each mean, but that's not what we want. We need the standard deviation of the *difference* in the means. For that, we can use a simple rule: *If the sample means come from independent samples, the variance of their sum or difference is the sum of their variances.*

> ### Variances Add for Sums *and* Differences
> At first, it may seem that this can't be true for differences as well as for sums. Here's some intuition about why variation increases even when we subtract two random quantities. Grab a full box of cereal. The label claims that it contains 16 ounces of cereal. We know that's not exact. There's a random quantity of cereal in the box with a mean (presumably) of 16 ounces and some variation from box to box. Now pour a 2-ounce serving of cereal into a bowl. Of course, your serving isn't exactly 2 ounces. There's some variation there, too. How much cereal would you guess was left in the box? Can you guess as accurately as you could for the full box? *After* you pour your bowl, the amount of cereal in the box is still a random quantity (with a smaller mean than before), but you've made it *more variable* because of the uncertainty in the amount you poured. Notice that we don't add the *standard deviations* of these two random quantities. It's the *variance* of the amount of cereal left in the box that's the sum of the two variances.

As long as the two groups are independent, we find the standard deviation of the *difference* between the two sample means by adding their variances and then taking the square root:

$$SD(\bar{y}_1 - \bar{y}_2) = \sqrt{Var(\bar{y}_1) + Var(\bar{y}_2)}$$

$$= \sqrt{\left(\frac{\sigma_1}{\sqrt{n_1}}\right)^2 + \left(\frac{\sigma_2}{\sqrt{n_2}}\right)^2}$$

$$= \sqrt{\frac{\sigma_1^2}{n_1} + \frac{\sigma_2^2}{n_2}}$$

Of course, usually we don't know the true standard deviations of the two groups, σ_1 and σ_2, so we substitute the estimates, s_1 and s_2, and find a *standard error*:

$$SE(\bar{y}_1 - \bar{y}_2) = \sqrt{\frac{s_1^2}{n_1} + \frac{s_2^2}{n_2}}.$$

Just as we did for one mean, we'll use the standard error to judge how big the difference really is. You shouldn't be surprised that, just as for a single mean, the ratio of the difference in the means to the standard error of that difference has a sampling model that follows a Student's t distribution.

To do inference, we'll need the degrees of freedom for the Student's t-model. Unfortunately, that formula isn't as simple as $n - 1$. The formula is straightforward but doesn't help our understanding much, so we leave it to the computer or calculator. (If you are curious and really want to see the formula, look in the footnote.[2])

> ### An Easier Rule?
>
> The formula for the degrees of freedom of the sampling distribution of the difference between two means is complicated. So some books teach an easier rule: The number of degrees of freedom is always at *least* the smaller of $n_1 - 1$ and $n_2 - 1$ and at most $n_1 + n_2 - 2$. The problem is that if you need to use this rule, you'll have to be conservative and use the lower value. And *that* approximation can be a poor choice because it can give less than *half* the degrees of freedom you're entitled to from the correct formula. So, we don't recommend using it unless you have lots of data.

A Sampling Distribution for the Difference Between Two Means

We denote the difference in the means of two groups by

$$\mu_1 - \mu_2 = \Delta$$

When the conditions are met (see Section 12.3), the standardized sample difference between the means of two independent groups,

$$t = \frac{(\bar{y}_1 - \bar{y}_2) - \Delta}{SE(\bar{y}_1 - \bar{y}_2)},$$

can be modeled by a Student's *t*-model with a number of degrees of freedom found with a special formula. We estimate the standard error with

$$SE(\bar{y}_1 - \bar{y}_2) = \sqrt{\frac{s_1^2}{n_1} + \frac{s_2^2}{n_2}}.$$

FOR EXAMPLE   Sampling distribution of the difference of two means

The owner of a large car dealership wants to understand the negotiation process for buying a new car. Cars are given a "sticker price," but that a potential buyer may negotiate a better price is well known. He wonders if there is a difference in how men and women negotiate and who, if either, obtains the larger discount.

The owner takes a random sample of 100 customers from the last six months' sales and finds that 54 were men and 46 were women. On average the 54 men received an average discount of $962.96 with a standard deviation of $458.95; the 46 women received an average discount of $1262.61 with a standard deviation of $399.70.

QUESTION   What is the mean difference of the discounts received by men and women? What is its standard error? If there is no difference between them, does this seem like an unusually large value?

ANSWER   The mean difference is $1262.61 − $962.96 = $299.65. The women received, on average, a discount that was larger by $299.65. The standard error is

$$SE(\bar{y}_{Women} - \bar{y}_{Men}) = \sqrt{\frac{s_{Women}^2}{n_{Women}} + \frac{s_{Men}^2}{n_{Men}}} = \sqrt{\frac{(399.70)^2}{46} + \frac{(458.95)^2}{54}} = \$85.87$$

So, the difference is $299.65/85.87 = 3.49 standard errors away from 0. That sounds like a reasonably large number of standard errors for a Student's *t* statistic with at least 45 degrees of freedom. (The approximation formula gives 97.94 degrees of freedom.)

---

[2] The result is due to Satterthwaite and Welch.

Satterthwaite, F. E. (1946). "An Approximate Distribution of Estimates of Variance Components," *Biometrics Bulletin* 2: 110–114.

Welch, B. L. (1947). "The Generalization of Student's Problem when Several Different Population Variances are Involved," *Biometrika* 34: 28–35.

$$df = \frac{\left(\dfrac{s_1^2}{n_1} + \dfrac{s_2^2}{n_2}\right)^2}{\dfrac{1}{n_1 - 1}\left(\dfrac{s_1^2}{n_1}\right)^2 + \dfrac{1}{n_2 - 1}\left(\dfrac{s_2^2}{n_2}\right)^2}$$

This approximation formula usually doesn't even give a whole number. If you are using a table, you'll need a whole number, so round down to be safe. If you are using technology, the approximation formulas that computers and calculators use for the Student's *t*-distribution can deal with fractional degrees of freedom.

12.2   The Two-Sample *t*-Test

Now we've got everything we need to construct the hypothesis test, and you already know how to do it. It's the same idea we used when testing one mean against a hypothesized value. Here, we start by hypothesizing a value for the true difference of the means. We'll call that hypothesized difference Δ_0. (It's so common for that hypothesized difference to be zero that we often just assume $\Delta_0 = 0$.) We then take the ratio of the difference in the means from our samples to its standard error and use that ratio to find a P-value from a Student's *t*-model. The test is called the **two-sample *t*-test**.

Two-Sample *t*-Test

When the appropriate assumptions and conditions are met, we test the hypothesis:

$$H_0 : \mu_1 - \mu_2 = \Delta_0$$

where the hypothesized difference Δ_0 is almost always 0. We use the statistic:

$$t = \frac{(\bar{y}_1 - \bar{y}_2) - \Delta_0}{SE(\bar{y}_1 - \bar{y}_2)}.$$

The standard error of $\bar{y}_1 - \bar{y}_2$ is:

$$SE(\bar{y}_1 - \bar{y}_2) = \sqrt{\frac{s_1^2}{n_1} + \frac{s_2^2}{n_2}}.$$

When the null hypothesis is true, the statistic can be closely modeled by a Student's *t*-distribution with a number of degrees of freedom given by a special formula. We use that model to compare our *t* ratio with a critical value for *t* or to obtain a P-value.

The Two-Sample *t*-Methods

Two-sample *t*-methods assume that the two groups are independent. This is a crucial assumption. If it's not met, it is not safe to use these methods. One common way for groups to fail independence is when each observation in one group is related to one (and only one) observation in the other group. For example, if we test the *same* subjects before and after an event, or if we measure a variable on both husbands and wives. In that case, the observations are said to be **paired** and you'll need to use the paired *t*-methods discussed in Section 12.7.

Sometimes you may see the two-sample *t*-methods referred to as the two-sample **independent** *t*-methods for emphasis. In this book, however, when we say two-sample, we'll always assume that the groups are independent.

FOR EXAMPLE   The *t*-test for the difference of two means

QUESTION   We saw (on page 374) that the difference between the average discount obtained by men and women appeared to be large if we assume that there is no true difference. Test the hypothesis, find the P-value, and state your conclusions.

ANSWER   The null hypothesis is: $H_0 : \mu_{Women} - \mu_{Men} = 0$ vs.

$$H_A : \mu_{Women} - \mu_{Men} \neq 0.$$

The difference in the sample means, $\bar{y}_{Women} - \bar{y}_{Men}$ is \$299.65 with a standard error of 85.87. The *t*-statistic is that difference divided by the standard error:

$$t = \frac{\bar{y}_{Women} - \bar{y}_{Men}}{SE(\bar{y}_{Women} - \bar{y}_{Men})} = \frac{299.65}{85.87} = 3.49.$$ The approximation formula gives 97.94

degrees of freedom (which is close to the maximum possible of $n_1 + n_2 - 2 = 98$). The P-value (from technology) for $t = 3.49$ with 97.94 df is 0.00073. We reject the null hypothesis. There is strong evidence to suggest that the difference in mean discount received by men and women is not 0.

12.3   Assumptions and Conditions

Before we can perform a two-sample *t*-test, we have to check the assumptions and conditions.

Independence Assumption

We can't *assume* that the data, taken as one big group, come from a homogeneous population because that would mean assuming the means were equal—and we're trying to test that. So, the data in each group must be drawn independently and at random from each group's own homogeneous population separately or generated by a randomized comparative experiment. We should think about whether the independence assumption is reasonable. We can also check two conditions:

Randomization Condition: Data collected with suitable randomization are likely to be independent. For surveys, are the data a representative random sample? For experiments, was the experiment randomized?

10% Condition: We usually check this condition for differences of means only if we have a very small population or an extremely large sample. We needn't worry about it at all for randomized experiments.

Normal Population Assumption

With Student's *t*-models, we need the assumption that the underlying populations are *each* Normally distributed. So we check one condition.

Nearly Normal Condition: We must check normality for *both* groups; a violation by either one violates the condition. As we saw for single sample means, the Normality Assumption matters most when sample sizes are small. When either group is small ($n < 15$), you should not use these methods if the histogram or Normal probability plot shows skewness. For *n*'s closer to 40, a mildly skewed histogram is OK, but you should remark on any outliers you find and not work with severely skewed data. When both groups are bigger than that, the Central Limit Theorem starts to work, so the Nearly Normal Condition for the data matters less. Even in large samples, however, you should still be on the lookout for outliers, extreme skewness, and multiple modes.

Independent Groups Assumption

To use the two-sample *t*-methods, the two groups we are comparing must be independent of each other. In fact, the test is sometimes called the two *independent samples t*-test. No statistical test can verify that the groups are independent. You have to think about how the data were collected. The assumption would be violated, for example, if one group were comprised of husbands and the other group, their wives. Similarly, if we compared subjects' performances before some treatment with their performances afterward, we'd expect a relationship of each "before" measurement with its corresponding "after" measurement. When the observational units in the two groups are related or matched, we need the methods of Section 12.7.

JUST CHECKING

Many office "coffee stations" collect voluntary payments for the food consumed. Researchers at the University of Newcastle upon Tyne performed an experiment to see whether the image of eyes watching would change employee behavior.[3] Each week on the cupboard behind the "honesty box," they alternated pictures of eyes looking at the viewer with pictures of flowers. They measured the consumption of milk to approximate the amount of food consumed and recorded the contributions (in £) each week per liter of milk. The table summarizes their results.

	Eyes	**Flowers**
n (# weeks)	5	5
$\bar{y}$	0.417 £/liter	0.151 £/liter
s	0.1811	0.067

1 What null hypothesis were the researchers testing?
2 Check the assumptions and conditions needed to test whether there really is a difference in behavior due to the difference in pictures.
3 What alternative hypothesis would you test?
4 The P-value of the test was less than 0.05. State a brief conclusion.

[3] Melissa Bateson, Daniel Nettle, and Gilbert Roberts, "Cues of Being Watched Enhance Cooperation in a Real-World Setting," *Biol. Lett.* Doi:10.1098/rsbl.2006.0509.

FOR EXAMPLE   Checking assumptions and conditions for a two-sample *t*-test

QUESTION In the example on page 375, we rejected the null hypothesis that the mean discount received by men and women is the same. Here are the histograms of the discounts for both women and men. Check the assumptions and conditions and state any concerns that you might have about the conclusion we reached.

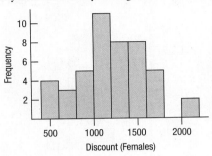
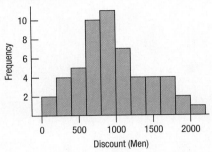

ANSWER We were told that the sample was random, so the Randomization Condition is satisfied. There is no reason to think that the men's and women's responses are related (as they might be if they were husband and wife pairs) so the Independent Group Assumption is plausible. Because both groups have more than 40 observations, the discounts can be mildly skewed, which is the case. There are no obvious outliers (there is a small gap in the women's distribution, but the observations are not far from the center), so all the assumptions and conditions seem to be satisfied. We have no real concerns about the conclusion we reached that the mean difference is not 0.

GUIDED EXAMPLE   Credit Card Promotions and Spending

Our preliminary market research has suggested that a new incentive may increase customer spending. However, before we invest in this promotion on the entire population of cardholders, let's test a hypothesis on a sample. To judge whether the incentive works, we will examine the change in spending (called the *spend lift*) over a six-month period. We will see whether the *spend lift* for the group that received the offer was greater than the *spend lift* for the group that received no offer. If we observe differences, how can we decide whether these differences are important (or real) enough to justify our costs?

PLAN	**Setup** State what we want to know.	We want to know if cardholders who are offered a promotion spend more on their credit card. We have the spend lift (in $) for a random sample of 500 cardholders who were offered the promotion and for a random sample of 500 customers who were not.
	Identify the *parameter* we wish to estimate. Here our parameter is the difference in the means, not the individual group means.	H_O: The mean spend lift for the group who received the offer is the same as for the group who did not:
	Identify the *population(s)* about which we wish to make statements.	H_O: $\mu_{Offer} - \mu_{No\ Offer} = O$ H_A: The mean spend lift for the group who received the offer is higher:
	Identify the variables and context.	H_A: $\mu_{Offer} - \mu_{No\ Offer} > O$

(continued)

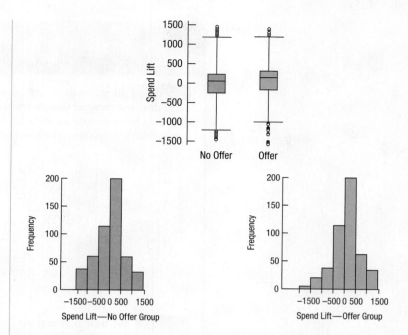

Make a graph to compare the two groups and check the distribution of each group. For completeness, we should report any outliers. If any outliers are extreme enough, we should consider performing the test both with and without the outliers and reporting the difference.

The boxplots and histograms show the distribution of both groups. It looks like the distribution for each group is unimodal and fairly symmetric.

The boxplots indicate several outliers in each group, but we have no reason to delete them, and their impact is minimal.

Model Check the assumptions and conditions.

✓ **Independence Assumption.** We have no reason to believe that the spending behavior of one customer would influence the spending behavior of another customer in the same group. The data report the "spend lift" for each customer for the same time period.

✓ **Randomization Condition.** The customers who were offered the promotion were selected at random.

✓ **Nearly Normal Condition.** The samples are large, so we are not overly concerned with this condition, and the boxplots and histograms show unimodal, symmetric distributions for both groups.

✓ **Independent Groups Assumption.** Customers were assigned to groups at random. There's no reason to think that those in one group can affect the spending behavior of those in the other group.

State the sampling distribution model for the statistic. Here the degrees of freedom will come from the approximation formula in footnote 2.

Under these conditions, it's appropriate to use a Student's *t*-model.

Specify your method.

We will use a two-sample *t*-test.

DO   **Mechanics** List the summary statistics. Be sure to include the units along with the statistics. Use meaningful subscripts to identify the groups.

We know $n_{No\ Offer} = 500$ and $n_{Offer} = 500$.
From technology, we find:

$$\bar{y}_{No\ Offer} = \$7.69 \qquad \bar{y}_{Offer} = \$127.61$$
$$s_{No\ Offer} = \$611.62 \qquad s_{Offer} = \$566.05$$

The observed difference in the two means is:

$$\bar{y}_{Offer} - \bar{y}_{No\ Offer} = \$127.61 - \$7.69 = \$119.92$$

The groups are independent, so:

$$SE(\bar{y}_{Offer} - \bar{y}_{No\ Offer}) = \sqrt{\frac{(566.05)^2}{500} + \frac{(611.62)^2}{500}}$$
$$= \$37.27$$

The observed *t*-value is:

$$t = 119.92/37.27 = 3.218$$

with 992 df (from technology).

(To use critical values, we could find that the one-sided 0.01 critical value for a *t* with 992 df is $t^* = 2.33$.

Our observed *t*-value is larger than this, so we could reject the null hypothesis at the 0.01 level.)

Using software to obtain the P-value, we get:

Promotional Group	N	Mean	StDev
No	500	7.69	611.62
Yes	500	127.61	566.05

```
Difference = mu(1) - mu(0)
Estimate for difference: 119.9231
t = 3.2178, df = 992.007
One-sided P-value = 0.0006669
```

Use the sample standard deviations to find the standard error of the sampling distribution.

The best alternative is to let the computer use the approximation formula for the degrees of freedom and find the P-value.

REPORT

Conclusion Interpret the test results in the proper context.

MEMO

Re: Credit card promotion

Our analysis of the credit card promotion experiment found that customers offered the promotion spent more than those not offered the promotion. The difference was statistically significant, with a P-value < 0.001. So we conclude that this promotion will increase spending. The difference in spend lift averaged $119.92, but our analyses so far have not determined how much income this will generate for the company and thus whether the estimated increase in spending is worth the cost of the offer.

12.4 A Confidence Interval for the Difference Between Two Means

We rejected the null hypothesis that customers' mean spending would not change when offered a promotion. Because the company took a random sample of customers for each group, and our P-value was convincingly small, we concluded this difference is not zero for the population. Does this mean that we should offer the promotion to all customers?

A hypothesis test really says nothing about the size of the difference. All it says is that the observed difference is large enough that we can be confident it isn't zero. That's what the term "statistically significant" means. It doesn't say that the difference is important, financially significant, or interesting. Rejecting a null hypothesis simply says that a value as extreme as the observed statistic is unlikely to have been observed if the null hypothesis were true.

So, what recommendations can we make to the company? Almost every business decision will depend on looking at a range of likely scenarios—precisely the kind of information a confidence interval gives. We construct the confidence interval for the difference in means in the usual way, starting with our observed statistic, in this case $(\bar{y}_1 - \bar{y}_2)$. We then add and subtract a multiple of the standard error $SE(\bar{y}_1 - \bar{y}_2)$ where the multiple is based on the Student's t distribution with the same df formula we saw before.

Confidence Interval for the Difference Between Two Means

When the conditions are met, we are ready to find a **two-sample t-interval** for the difference between means of two independent groups, $\mu_1 - \mu_2$. The confidence interval is:

$$(\bar{y}_1 - \bar{y}_2) \pm t^*_{df} \times SE(\bar{y}_1 - \bar{y}_2),$$

where the standard error of the difference of the means is:

$$SE(\bar{y}_1 - \bar{y}_2) = \sqrt{\frac{s_1^2}{n_1} + \frac{s_2^2}{n_2}},$$

and the degrees of freedom is found by the special formula (see p. 374).

The critical value t^*_{df} depends on the particular confidence level, and on the number of degrees of freedom.

FOR EXAMPLE   A confidence interval for the difference between two means

QUESTION   We've concluded on page 375 that, on average, women receive a larger discount than men at this car dealership. How big is the difference, on average? Find a 95% confidence interval for the difference.

ANSWER   We've seen that the difference from our sample is $299.65 with a standard error of $85.87 and that it has 97.94 degrees of freedom. The 95% critical value for a t with 97.94 degrees of freedom is 1.984.

$$\bar{y}_{Women} - \bar{y}_{Men} \pm t^*_{97.94} \times SE(\bar{y}_{Women} - \bar{y}_{Men}) = 299.65 \pm 1.984 \times 85.87$$
$$= (\$129.28, \$470.02)$$

We are 95% confident that, at this dealership, the average discount received by women exceeds the average discount received by men by between $129.28 and $470.02.

GUIDED EXAMPLE   Confidence Interval for Credit Card Spending

We rejected the null hypothesis that the mean spending in the two groups was equal. But, to find out whether we should consider offering the promotion nationwide, we need to estimate the magnitude of the spend lift.

PLAN	**Setup** State what we want to know.	We want to find a 95% confidence interval for the mean difference in spending between those who are offered a promotion and those who aren't.
	Identify the *parameter* we wish to estimate. Here our parameter is the difference in the means, not the individual group means.	
	Identify the *population(s)* about which we wish to make statements.	We looked at the boxplots and histograms of the groups and checked the conditions before. The same assumptions and conditions are appropriate here, so we can proceed directly to the confidence interval.
	Identify the variables and context.	
	Specify the method.	We will use a two-sample *t*-interval.

DO

Mechanics Construct the confidence interval. Be sure to include the units along with the statistics. Use meaningful subscripts to identify the groups.

Use the sample standard deviations to find the standard error of the sampling distribution.

The best alternative is to let the computer use the approximation formula for the degrees of freedom and find the confidence interval.

Ordinarily, we rely on technology for the calculations. In our hand calculations, we rounded values at intermediate steps to show the steps more clearly. The computer keeps full precision and is the one you should report. The difference between the hand and computer calculations is about $0.08.

In our previous analysis, we found:

$$\bar{y}_{No\ Offer} = \$7.69 \qquad \bar{y}_{Offer} = \$127.61$$
$$s_{No\ Offer} = \$611.62 \qquad s_{Offer} = \$566.05$$

The observed difference in the two means is:

$$\bar{y}_{Offer} - \bar{y}_{No\ Offer} = \$127.61 - \$7.69 = \$119.92,$$

and the standard error is:

$$SE(\bar{y}_{Offer} - \bar{y}_{No\ Offer}) = \$37.27$$

From technology, the df is 992, and the one-sided 0.025 critical value for *t* with 992 df is 1.96. So the 95% confidence interval is:

$$\$119.92 \pm 1.96(\$37.27) = (\$46.87, \$192.97)$$

Using software to obtain these computations, we get:

```
95% confidence interval:
  (46.78784, 193.05837)
sample means:
No Offer      Offer
7.690882      127.613987
```

REPORT

Conclusion Interpret the confidence interval in the proper context.

MEMO
Re: Credit card promotion experiment

In our experiment, the promotion resulted in an increased spend lift of $119.92 on average. Further analysis gives a 95% confidence interval of ($46.79, $193.06). In other words, we expect with 95% confidence that under similar conditions, the mean spend lift that we achieve when we roll out the offer to all similar customers will be in this interval. We recommend that the company consider whether the values in this interval will justify the cost of the promotion program.

12.5   The Pooled *t*-Test

If you bought a used camera in good condition from a friend, would you pay the same as you would if you bought the same item from a stranger? A researcher at Cornell University[4] wanted to know how friendship might affect simple sales such as this. She randomly divided subjects into two groups and gave each group descriptions of items they might want to buy. One group was told to imagine buying from a friend whom they expected to see again. The other group was told to imagine buying from a stranger.

Here are the prices they offered to pay for a used camera in good condition.

Price Offered for a Used Camera ($)	
Buying from a Friend	**Buying from a Stranger**
275	260
300	250
260	175
300	130
255	200
275	225
290	240
300	

WHO	University students
WHAT	Prices offered to pay for a used camera ($)
WHEN	1990s
WHERE	Cornell University
WHY	To study the effects of friendship on transactions

The researcher who designed the friendship study was interested in testing the impact of friendship on negotiations. Previous theories had doubted that friendship had a measurable effect on pricing, but she hoped to find such an effect. The usual null hypothesis is that there's no difference in means and that's what we'll use for the camera purchase prices.

When we performed *t*-tests earlier in the chapter, we used an approximation formula that adjusts the degrees of freedom downward. Because this is an experiment, we might be willing to make another assumption. The null hypothesis says that whether you buy from a friend or a stranger should have no effect on the mean amount you're willing to pay for a camera. If it has no effect on the means, should it affect the variance of the transactions?

If we're willing to *assume* that the variances of the groups are equal (at least when the null hypothesis is true), then we can use a slightly more powerful method. We **pool** the data to estimate a common variance:

$$s_{pooled}^2 = \frac{(n_1 - 1)\, s_1^2 + (n_2 - 1)\, s_2^2}{(n_1 - 1) + (n_2 - 1)}.$$

Then we substitute this pooled variance in place of each of the variances in the standard error formula, simplifying it:

$$SE_{pooled}(\bar{y}_1 - \bar{y}_2) = \sqrt{\frac{s_{pooled}^2}{n_1} + \frac{s_{pooled}^2}{n_2}} = s_{pooled}\sqrt{\frac{1}{n_1} + \frac{1}{n_2}}.$$

The formula for degrees of freedom for the Student's *t*-model is simpler, too:

$$df = (n_1 - 1) + (n_2 - 1).$$

---

[4] J. J. Halpern (1997). "The Transaction Index: A Method for Standardizing Comparisons of Transaction Characteristics Across Different Contexts," *Group Decision and Negotiation*, 6, no. 6: 557–572.

Substitute the pooled-*t* estimate of the standard error and its degrees of freedom into the steps of the confidence interval or hypothesis test and you'll be using pooled-*t* methods.

To use the pooled-*t* methods, also check the **Equal Variance Assumption** that the variances of the two populations from which the samples have been drawn are equal. An easy way to do that is to compare their boxplots to check whether the boxes are of roughly equal size and that there are no extreme outliers in need of special attention.

Pooled *t*-Test and Confidence Interval for the Difference Between Means

The conditions for the **pooled *t*-test** for the difference between the means of two independent groups are the same as for the two-sample *t*-test with the additional assumption that the variances of the two groups are the same. We test the hypothesis:

$$H_0: \mu_1 - \mu_2 = \Delta_0,$$

where the hypothesized difference Δ_0 is almost always 0, using the statistic

$$t = \frac{(\bar{y}_1 - \bar{y}_2) - \Delta_0}{SE_{pooled}(\bar{y}_1 - \bar{y}_2)}.$$

The standard error of $\bar{y}_1 - \bar{y}_2$ is:

$$SE_{pooled}(\bar{y}_1 - \bar{y}_2) = s_{pooled}\sqrt{\frac{1}{n_1} + \frac{1}{n_2}},$$

where the pooled variance is:

$$s_{pooled}^2 = \frac{(n_1 - 1)\, s_1^2 + (n_2 - 1)\, s_2^2}{(n_1 - 1) + (n_2 - 1)}.$$

When the conditions are met and the null hypothesis is true, we can model this statistic's sampling distribution with a Student's *t*-model with $(n_1 - 1) + (n_2 - 1)$ degrees of freedom. We use that model to obtain a P-value for a test or a margin of error for a confidence interval.

The corresponding **pooled-*t* confidence interval** is:

$$(\bar{y}_1 - \bar{y}_2) \pm t_{df}^* \times SE_{pooled}(\bar{y}_1 - \bar{y}_2),$$

where the critical value t^* depends on the confidence level and is found with $(n_1 - 1) + (n_2 - 1)$ degrees of freedom.

GUIDED EXAMPLE   Role of Friendship in Negotiations

The usual null hypothesis in a pooled *t*-test is that there's no difference in means and that's what we'll use for the camera purchase prices.

(continued)

PLAN

Setup State what we want to know.

Identify the *parameter* we wish to estimate. Here our parameter is the difference in the means, not the individual group means. Identify the variables and context.

We want to know whether people are likely to offer a different amount for a used camera when buying from a friend than when buying from a stranger. We wonder whether the difference between mean amounts is zero. We have bid prices from 8 subjects buying from a friend and 7 subjects buying from a stranger, found in a randomized experiment.

Hypotheses State the null and alternative hypotheses.

The research claim is that friendship changes what people are willing to pay. The natural null hypothesis is that friendship makes no difference.

We didn't start with any knowledge of whether friendship might increase or decrease the price, so we choose a two-sided alternative.

H_O: The difference in mean price offered to friends and the mean price offered to strangers is zero:

$$\mu_F - \mu_S = 0,$$

H_A: The difference in mean prices is not zero:

$$\mu_F - \mu_S \neq 0.$$

Make a graph. Boxplots are the display of choice for comparing groups. We'll also want to check the distribution of each group. Histograms may do a better job.

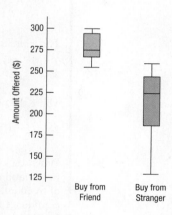

REALITY CHECK ▶ Looks like the prices are higher if you buy from a friend. The two ranges barely overlap, so we'll be pretty surprised if we don't reject the null hypothesis.

Model Think about the assumptions and check the conditions. (Because this is a randomized experiment, we haven't sampled at all, so the 10% Condition doesn't apply.)

✓ **Independence Assumption.** There is no reason to think that the behavior of one subject influenced the behavior of another.

✓ **Randomization Condition.** The experiment was randomized. Subjects were assigned to treatment groups at random.

✓ **Independent Groups Assumption.** Randomizing the experiment gives independent groups.

✓ **Nearly Normal Condition.** Histograms of the two sets of prices show no evidence of skewness or extreme outliers.

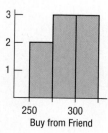

Buy from Friend

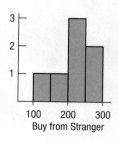

Buy from Stranger

State the sampling distribution model. Specify the method.	Because this is a randomized experiment with a null hypothesis of no difference in means, we can make the Equal Variance Assumption. If, as we are assuming from the null hypothesis, the treatment doesn't change the means, then it is reasonable to assume that it also doesn't change the variances. Under these assumptions and conditions, we can use a Student's t-model to perform a pooled t-test.

DO

Mechanics List the summary statistics. Be sure to use proper notation.

Use the null model to find the P-value. First determine the standard error of the difference between sample means.

From the data:

$$n_F = 8 \qquad n_S = 7$$
$$\bar{y}_F = \$281.88 \qquad \bar{y}_S = \$211.43$$
$$s_F = \$18.31 \qquad s_S = \$46.43$$

The pooled variance estimate is:

$$s_p^2 = \frac{(n_F - 1)s_F^2 + (n_S - 1)s_S^2}{n_F + n_S - 2}$$
$$= \frac{(8 - 1)(18.31)^2 + (7 - 1)(46.43)^2}{8 + 7 - 2}$$
$$= 1175.48$$

The standard error of the difference becomes:

$$SE_{pooled}(\bar{y}_F - \bar{y}_S) = \sqrt{\frac{s_p^2}{n_F} + \frac{s_p^2}{n_S}}$$
$$= 17.744$$

The observed difference in means is:

$$(\bar{y}_F - \bar{y}_S) = 281.88 - 211.43 = \$70.45$$

which results in a t-ratio

Find the t-value.

Make a graph. Sketch the t-model centered at the hypothesized difference of zero. Because this is a two-tailed test, shade the region to the right of the observed difference and the corresponding region in the other tail.

A statistics program can find the P-value.

$$t = \frac{(\bar{y}_F - \bar{y}_S) - (0)}{SE_{pooled}(\bar{y}_F - \bar{y}_S)} = \frac{70.45}{17.744} = 3.97$$

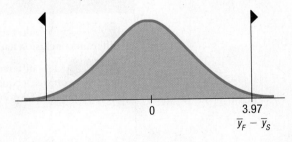

$$\begin{array}{cc} 0 & \begin{array}{c} 3.97 \\ \bar{y}_F - \bar{y}_S \end{array} \end{array}$$

The computer output for a pooled t-test appears here.

```
Pooled T Test for friend vs. stranger
             N     Mean     StDev
Friend       8     281.9    18.3
Stranger     7     211.4    46.4
t = 3.97, df = 13,P-value = 0.0016
Alternative hypothesis: true difference in means
is not equal to 0
95% confidence interval:
(32.11, 108.78)
```

(continued)

REPORT

Conclusion Link the P-value to your decision about the null hypothesis and state the conclusion in context.

Be cautious about generalizing to items whose prices are outside the range of those in this study. The confidence interval can reveal more detailed information about the size of the difference. In the original article (referenced in footnote 4 in this chapter), the researcher tested several items and proposed a model relating the size of the difference to the price of the items.

MEMO

Re: Role of friendship in negotiations

Results of a small experiment show that people are likely to offer a different amount for a used camera when bargaining with a friend than when bargaining with a stranger. The difference in mean offers was statistically significant ($P = .0016$).

The confidence interval suggests that people tend to offer more to a friend than they would to a stranger. For the camera, the 95% confidence interval for the mean difference in price was $32.11 to $108.78, but we suspect that the actual difference may vary with the price of the item purchased.

When Should You Use the Pooled *t*-Test?

> Because the advantages of pooling are small, and you are allowed to pool only rarely (when the equal variances assumption is met), our advice is: **don't**.
> **It's never wrong *not* to pool.**

When the variances of the two groups are in fact equal, the two methods give pretty much the same result. On average, pooled methods have a small advantage (slightly narrower confidence intervals, slightly more powerful tests), but the advantage is slight. When the variances are *not* equal, the pooled methods are just not valid and can give poor results. You have to use the two-sample methods instead. Pooled *t*-methods are most appropriate for experimental data, where the equal variance assumption is most plausible. But for most other data, we recommend using two-sample *t*.

FOR EXAMPLE   The pooled *t*-test

QUESTION   Would the dealership owner of the example on page 380 have reached a different conclusion had he used the pooled *t*-test for the difference in mean discounts?

ANSWER   The difference in the pooled *t*-test is that it assumes that the variances of the two groups are equal, and pools those two estimates to find the standard error of the difference. The pooled estimate of the common standard deviation is:

$$s_{pooled} = \sqrt{\frac{(n_{Women} - 1)s^2_{Women} + (n_{Men} - 1)s^2_{Men}}{n_{Women} + n_{Men} - 2}}$$

$$= \sqrt{\frac{(46 - 1)(399.70)^2 + (54 - 1)(458.95)^2}{46 + 54 - 2}}$$

$$= 432.75$$

We use that to find the *SE* of the difference:

$$SE_{pooled}(\bar{y}_{Women} - \bar{y}_{Men}) = s_{pooled}\sqrt{\frac{1}{n_{Women}} + \frac{1}{n_{Men}}} = 432.75\sqrt{\frac{1}{46} + \frac{1}{54}} = \$86.83$$

(Without pooling, our estimate was $85.87.) This pooled *t* has $46 + 54 - 2 = 98$ degrees of freedom.

$$t_{98} = \frac{\bar{y}_{Women} - \bar{y}_{Men}}{SE_{pooled}(\bar{y}_{Women} - \bar{y}_{Men})} = \frac{299.65}{86.83} = 3.45$$

The P-value for a t of 3.45 with 98 degrees of freedom is (from technology) 0.0008. The two-sample t-test value was 0.0007. There is no practical difference between these, and we reach the same conclusion that the difference is not 0. The assumption of equal variances did not affect the conclusion.

Tukey's Quick Test

If you think that the t-test is a lot of work for what seemed like an easy comparison, you're not alone. The famous statistician John Tukey[5] was once challenged to come up with a simpler alternative to the two-sample t-test that, like the 68–95–99.7 Rule, had critical values that could be remembered easily. The test he came up with asks you only to count and to remember three numbers: 7, 10, and 13.

When you first looked at the boxplots of the friendship data, you might have noticed that they didn't overlap very much. That's the basis for Tukey's test. To use Tukey's test, one group must have the highest value, and the other must have the lowest. We just count how many values in the high group are higher than *all* the values of the lower group. Add to this the number of values in the low group that are lower than *all* the values of the higher group. (You can count ties as $\frac{1}{2}$.) If the total of these exceedances is 7 or more, we can reject the null hypothesis of equal means at $\alpha = 0.05$. The "critical values" of 10 and 13 correspond to α's of 0.01 and 0.001.

Let's try it. The "Friend" group has the highest value ($300), and the "Stranger" group has the lowest value ($130). Six of the values in the Friend group are higher than the highest value of the Stranger group ($260), and one is a tie. Six of the Stranger values are lower than the lowest value for the Friend group. That's a total of $12\frac{1}{2}$ exceedances. That's more than 10, but less than 13. So the P-value is between 0.01 and 0.001—just what we found with the pooled t.

This is a remarkably good test. The only assumption it requires is that the two samples be independent. It's so simple to do that there's no reason not to do one to check your two-sample t results. If they disagree, check the assumptions. Tukey's quick test, however, is not as widely known or accepted as the two-sample t-test, so you still need to know and use the two-sample t.

FOR EXAMPLE   Tukey's Quick Test

Here are the discounts (in order) received by the men (see page 374):
130, 158, 303, 340, 353, 390, 415, 423, 536, 566, 588 . . . 1606, 1616, 1658, 1763, 1840, 1881, 2030

Here are the discounts received by the women:
503, 526, 574, 579, 603, 630, 794, 831 . . . 1727, 1742, 1748, 2142, 2192

QUESTION   Is Tukey's Quick Test applicable? If so, find the number of exceedances and report the P-value and conclusion. How does the test compare to the two-sample t-test?

ANSWER   Tukey's Quick Test is applicable because the groups are independent; also, one group (Men) has the smallest value ($130) and the other group (Women) has the largest value ($2192).

(continued)

---

[5] Famous seems like an understatement for John Tukey. The *New York Times* called Tukey "one of the most influential statisticians" of the 20th century. The *Times* also noted that he is credited with inventing the term "bit" (in its computer sense) and with the first printed use of the word "software." Tukey also invented both the stem-and-leaf display and the boxplot.

There were 8 discounts received by men that were smaller than all the women's discounts and 2 discounts received by women that were larger than all the men's discounts for a total of 10 exceedances. That corresponds to a P-value of about 0.01 (but larger than 0.001). This gives strong evidence to reject the null hypothesis that the two distributions are the same. We conclude that the mean difference is not 0. The two-sample *t*-test gave a P-value of 0.0028, which is in line with this test.

12.6   Paired Data

The two-sample *t*-test depends crucially on the assumption that the two groups are independent of each other. One common violation of this assumption is when we have data on the *same* cases in two different circumstances. For example, we might want to compare the same customers' spending at our website last January to this January, or we might have each participant in a focus group rate two different product designs. Data such as these are said to be **paired**. When pairs arise from an experiment, we compare measurements before and after a treatment, and the pairing is a type of *blocking*. When they arise from an observational study, it is called *matching*.

When the data are paired, the groups are not independent, so you *should not* use the two-sample (or pooled two-sample) method. You must decide, from the way the data were collected whether the data are paired. Be careful. There is no statistical test to determine whether the data are paired. You must decide whether the data are paired from understanding how they were collected and what they mean (check the W's).

Once we recognize that our data are matched pairs, it makes sense to concentrate on the *difference* between the two measurements in each pair. That is, we look at the collection of pairwise differences in the measured variable. For example, if studying customer spending, we would analyze the *difference* between this January's and last January's spending for each customer. Because it is the *differences* we care about, we can analyze the data as a single variable holding those differences. With only one variable to consider, we can use a simple one-sample *t*-test. A **paired *t*-test** is just a one-sample *t*-test for the mean of the pairwise differences. The sample size is the number of pairs.

The assumptions and conditions for the test are exactly the same as the ones we used for the one-sample *t*-test. But now we have the additional assumption that the data are in fact paired.

> **Are the Data Paired?**
>
> To decide whether your data are paired you'll need to understand the business context of the data and how the data were collected. Pairing is not a choice, but a fact about the data. If the data are paired, you can't treat the groups as independent. But to be paired, the data must be related between the two groups in a way that you can justify. That justification may involve knowledge about the data not present in the variables themselves.

Paired Data Assumption

The data must be paired. When you have two groups with the same number of observations, it may be tempting to match them up, but that's not valid. Nor can you pair data just because they "seem to go together." To use paired methods you must determine from knowing how the data were collected that the individuals are paired. Usually the context will make it clear.

Be sure to recognize paired data when you have it. Remember, two-sample *t*-methods aren't valid unless the groups are independent, and paired groups aren't independent.

Independence Assumption

For these methods, it's the *differences* that must be independent of each other. This is just the one-sample *t*-test assumption of independence, now applied to the differences. As always, randomization helps to ensure independence.

Randomization Condition. Randomness can arise in many ways. The *pairs* may be a random sample. For example, we may be comparing opinions of husbands and wives from a random selection of couples. In an experiment, the order of the two treatments may be randomly assigned, or the treatments may be randomly assigned to one member of each pair. In a before-and-after study, we may believe that the observed differences are a representative sample from a population of interest. What we want to know usually focuses our attention on where the randomness should be.

10% Condition. When we sample from a finite population, we should be careful not to sample more than 10% of that population. Sampling too large a fraction of the population calls the independence assumption into question. We don't usually check the 10% condition, but it is a good idea to think about it.

Normal Population Assumption

The population of *differences* follow a Normal model. There is no need to check the distribution of the two individual groups. In fact, each group can be quite skewed and yet the differences can still be unimodal and symmetric.

Nearly Normal Condition. Make a histogram of the differences. The Normal population assumption matters less as we have more pairs to consider. You may be pleasantly surprised when you check this condition. Even if your original measurements are skewed or bimodal, the *differences* may be nearly Normal. After all, the individual who was way out in the tail on an initial measurement is likely to still be out there on the second one, giving a perfectly ordinary difference.

FOR EXAMPLE Paired data

QUESTION The owner of the car dealership (see page 374) is curious about the maximum discount his salesmen are willing to give to customers. In particular, two of his salespeople, Frank and Ray, seem to have very different ideas about how much discount to allow. To test his suspicion, he selects 30 cars from the lot and asks each to say how much discount they would allow a customer. If we want to test whether the mean discount given by them is the same, what test would he use? Explain.

ANSWER A paired *t*-test. The responses (maximum discount in \$) are not independent between the two salespeople because they are evaluating the same 30 cars.

12.7   Paired *t*-Methods

The paired *t*-test is mechanically a one-sample *t*-test applied to the paired differences. Compare the mean difference to its standard error. If the ratio is large enough, reject the null hypothesis.

Paired *t*-Test

When the conditions are met, we are ready to test whether the mean paired difference is significantly different from a hypothesized value (called Δ_0). We test the hypothesis:

$$H_0\colon \mu_d = \Delta_0,$$

where the *d*'s are the pairwise differences and Δ_0 is almost always 0.
    We use the statistic:

$$t = \frac{\bar{d} - \Delta_0}{SE(\bar{d})},$$

(continued)

where $\bar{d}$ is the mean of the pairwise differences, n is the number of *pairs*, and

$$SE(\bar{d}) = \frac{s_d}{\sqrt{n}},$$

where s_d is the standard deviation of the pairwise differences.

When the conditions are met and the null hypothesis is true, the sampling distribution of this statistic is a Student's t-model with $n - 1$ degrees of freedom and we use that model to obtain the P-value.

Similarly, we can construct a confidence interval for the true mean difference. As in a one-sample t-interval, we center our estimate at the mean difference in our data. The margin of error on either side is the standard error multiplied by a critical t-value (based on our confidence level and the number of pairs we have).

> ### Paired t-Interval
>
> When the conditions are met, we are ready to find the confidence interval for the mean of the paired differences. The confidence interval is:
>
> $$\bar{d} \pm t^*_{n-1} \times SE(\bar{d}),$$
>
> where the standard error of the mean difference is $SE(\bar{d}) = \dfrac{s_d}{\sqrt{n}}$.
>
> The critical value t^* from the Student's t-model depends on the particular confidence level that you specify and on the degrees of freedom, $n - 1$, based on the number of pairs, n.

JUST CHECKING

Think about each of the following situations. Would you use a two-sample t or paired t-method (or neither)? Why?

5   Random samples of 50 men and 50 women are surveyed on the amount they invest on average in the stock market on an annual basis. We want to estimate any gender difference in how much they invest.

6   Random samples of students were surveyed on their perception of ethical and community service issues both in their first year and fourth year at a university. The university wants to know whether their required programs in ethical decision-making and service learning change student perceptions.

7   A random sample of work groups within a company was identified. Within each work group, one male and one female worker were selected at random. Each was asked to rate the secretarial support that their work group received. When rating the same support staff, do men and women rate them equally on average?

8   A total of 50 companies are surveyed about business practices. Some are privately held and others are publicly traded. We wish to investigate differences between these two kinds of companies.

9   These same 50 companies are surveyed again one year later to see if their perceptions, business practices, and R&D investment have changed.

GUIDED EXAMPLE   Seasonal Spending

Economists and credit card banks know that people tend to spend more near the holidays in December. In fact, sales in the few days after Thanksgiving (the fourth Thursday of November in the United States) provide an indication of the strength of the holiday season sales and an early look at the strength of the economy in general. After the holidays, spending decreases substantially. Because credit card banks receive a percentage of each transaction, they need to forecast how much the average spending will increase or decrease from month to month. How much less do people tend to spend in January than December? For any particular segment of cardholders, a credit card bank could select two random samples—one for each month—and simply compare the average amount spent in January with that in December. A more sensible approach might be to select a single random sample and compare the spending between the two months for *each cardholder*. Designing the study in this way and examining the paired differences gives a more precise estimate of the actual change in spending.

Here we have a sample of 911 cardholders from a particular market segment and the amount they charged on their credit card in both December and the following January. We can test whether the mean difference in spending is 0 by using a paired *t*-test and create a **paired *t*-confidence interval** to estimate the true mean difference in spending between the two months.

WHO	Cardholders in a particular market segment of a major credit card issuer
WHAT	Amount charged on their credit card in December and January
WHERE	United States
WHY	To estimate the amount of decrease in spending one could expect after the holiday shopping season

PLAN

Setup State what we want to know.

Identify the *parameter* we wish to estimate and the sample size.

We want to know how much we can expect credit card charges to change, on average, from December to January for this market segment. We have the total amount charged in December and January for $n = 911$ cardholders in this segment. We want to test whether the mean spending is the same for customers in the two months and find a confidence interval for the true mean difference in charges between these two months for all cardholders in this segment. Because we know that people tend to spend more in December, we will look at the difference:

$$\text{December spend} - \text{January spend}$$

and use a one-sided test. A positive difference will mean a decrease in spending.

Hypotheses State the null and alternative hypotheses.

H_O: Mean spending was the same in December and January; the mean difference is zero: $\mu_d = O$.

H_A: Mean spending was greater in December than January; the mean difference was greater than zero: $\mu_d > O$.

Model Check the conditions.

State why the data are paired. Simply having the same number of individuals in each group or displaying them in side-by-side columns, doesn't make them paired.

Think about what we hope to learn and where the randomization comes from.

✓ **Paired Data Assumption:** The data are paired because they are measurements on the same cardholders in two different months.

✓ **Independence Assumption:** The behavior of any individual is independent of the behavior of the others, so the differences are mutually independent.

✓ **Randomization Condition:** This was a random sample from a large market segment.

✓ **Nearly Normal Condition:** The distribution of the differences is unimodal and symmetric. Although the tails of the distribution are long, the distributions are symmetric. (This is typical of the behavior of credit card spending.) There are no isolated cases that would unduly dominate the mean difference, and our sample size is large so the Central Limit Theorem will protect us.

(continued)

Make a picture of the differences. Don't plot separate distributions of the two groups—that would entirely miss the pairing. For paired data, it's the Normality of the differences that we care about. Treat those paired differences as you would a single variable, and check the Nearly Normal condition.

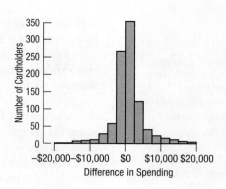

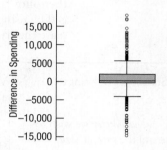

Specify the sampling distribution model.

Choose the method.

The conditions are met, so we'll use a Student's *t*-model with $(n - 1) = 910$ degrees of freedom, perform a paired *t*-test, and find a paired *t*-confidence interval.

DO   **Mechanics** *n* is the number of *pairs*, in this case, the number of cardholders.

$\bar{d}$ is the mean difference.

s_d is the standard deviation of the differences.

Make a picture. Sketch a *t*-model centered at the observed mean of 788.18.

Find the standard error and the *t*-score of the observed mean difference. There is nothing new in the mechanics of the paired *t*-methods. These are the mechanics of the *t*-test and *t*-interval for a mean applied to the differences.

The computer output tells us:

$$n = 911 \ pairs$$
$$\bar{d} = \$788.18$$
$$s_d = \$3740.22$$

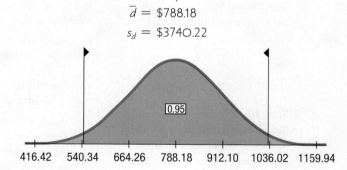

We estimate the standard error of $\bar{d}$ using:

$$SE(\bar{d}) = \frac{s_d}{\sqrt{n}} = \frac{3740.22}{\sqrt{911}} = \$123.919$$

We find $t = \dfrac{\bar{d} - 0}{SE(\bar{d})} = \dfrac{788.180}{123.919} = 6.36$

A *t*-statistic with 910 degrees of freedom and a value of 6.36 has a one-sided P-value <0.001.

The critical value for a 95% confidence interval is:

$$t^*_{910} = 1.96$$

The margin of error, $ME = t^*_{910} \times SE(\bar{d})$
$$= 1.96 \times 123.919 = 242.88$$

So a 95% CI is $\bar{d} \pm ME = (\$545.30, \$1031.06)$.

REPORT

Conclusion Link the results of the confidence interval to the context of the problem.

MEMO

Re: Credit card expenditure changes

In the sample of cardholders studied, the change in expenditures between December and January averaged $788.18, which means that, on average, cardholders spend $788.18 less in January than the month before. There is strong evidence to suggest that the mean difference is not zero and that customers really do spend more in December than January. Although we didn't measure the change for all cardholders in the segment, we can be 95% confident that the true mean decrease in spending from December to January was between $545.30 and $1031.06.

FOR EXAMPLE   The paired *t*-test

Here are some summary statistics from the study undertaken by the car dealership owner (see page 389):

$$\bar{y}_{Frank} = \$414.48 \qquad \bar{y}_{Ray} = \$478.88$$

$$SD_{Frank} = \$87.33 \qquad SD_{Ray} = \$175.12$$

$$\bar{y}_{Diff} = \$64.40 \qquad SD_{Diff} = \$146.74$$

QUESTION   Test the hypothesis that the mean maximum discount that Frank and Ray would give is the same. Give a 95% confidence interval for the mean difference.

ANSWER   We use a paired *t*-test because Frank and Ray were asked to give opinions about the same 30 cars.

$$t_{n-1} = \frac{\bar{d}}{SE(\bar{d})}$$

$$SE(\bar{d}) = \frac{s_d}{\sqrt{n}} = \frac{146.74}{\sqrt{30}} = \$26.79$$

$$t_{29} = \frac{64.40}{26.79} = 2.404$$

which has a (two-sided) P-value of 0.0228. We reject the null hypothesis that the mean difference is 0 and conclude that there is strong evidence to suggest that they are not the same.

A 95% confidence interval:

$$t^*_{29} = 2.045 \text{ at } 95\% \text{ confidence}$$

$$\bar{d} \pm t^*_{29} \times SE(\bar{d}) = 64.40 \pm 2.045 \times 26.79$$

$$= (\$9.61, \$119.19)$$

shows that Ray gives, on average, somewhere between $9.61 and $119.19 more for his maximum discount than Frank does.

WHAT CAN GO WRONG?

- **Watch out for paired data when using the two-sample *t*-test.** The Independent Groups Assumption deserves special attention. Some researcher designs *deliberately* violate the Independent Groups Assumption. For example, suppose you wanted to test a diet program. You select 10 people at random to take part in your diet. You measure their weights at the beginning of the diet and after 10 weeks of the diet. So, you have two columns of weights, one for *before* and one for *after*. But the data are related; each "after" weight goes naturally with the "before" weight for the *same* person. If the samples are *not* independent, you can't use two-sample methods. This is probably the main thing that can go wrong when using two-sample methods.

- **Don't use individual confidence intervals for each group to test the difference between their means.** If you make 95% confidence intervals for the means of each group separately and you find that the intervals don't overlap, you can reject the hypothesis that the means are equal (at the corresponding α level). But, if the intervals do overlap, that doesn't mean that you *can't* reject the null hypothesis. The margin of error for the difference between the means is smaller than the sum of the individual confidence interval margins of error. Comparing the individual confidence intervals is like adding the standard deviations. But we know that it's the variances that we add, and when we do it right, we actually get a more powerful test. So, don't test the difference between group means by looking at separate confidence intervals. Always make a two-sample *t*-interval or perform a two-sample *t*-test.

- **Look at the plots.** The usual (by now) cautions about checking for outliers and non-Normal distributions apply. The simple defense is to make and examine plots (side-by-side boxplots and histograms for two independent samples, histograms of the differences for paired data). You may be surprised how often this simple step saves you from the wrong or even absurd conclusions that can be generated by a single undetected outlier. You don't want to conclude that two methods have very different means just because one observation is atypical.

- **Don't use a paired *t*-method when the samples aren't paired.** When two groups don't have the same number of values, it's easy to see that they can't be paired. But just because two groups have the same number of observations doesn't mean they can be paired, even if they are shown side by side in a table. We might have 25 men and 25 women in our study, but they might be completely independent of one another. If they were siblings or spouses, we might consider them paired. Remember that you cannot *choose* which method to use based on your preferences. Only if the data are from an experiment or study in which observations were paired, can you use a paired method.

- **Don't forget to look for outliers when using paired methods.** For two-sample *t*-methods watch out for outliers in *either* group. For paired *t*-methods, the outliers we care about now are in the differences. A subject who is extraordinary both before and after a treatment may still have a perfectly typical difference. But one outlying difference can completely distort your conclusions. Be sure to plot the differences (even if you also plot the data) when using paired methods.

> **Do What We Say, Not What We Do**
>
> Precision machines used in industry often have a bewildering number of parameters that have to be set, so experiments are performed in an attempt to try to find the best settings. Such was the case for a hole-punching machine used by a well-known computer manufacturer to make printed circuit boards. The data were analyzed by one of the authors, but because he was in a hurry, he didn't look at the boxplots first and just performed *t*-tests on the experimental factors. When he found extremely small P-values even for factors that made no sense, he plotted the data. Sure enough, there was one observation 1,000,000 times bigger than the others. It turns out that it had been recorded in microns (millionths of an inch), while all the rest were in inches.

ETHICS IN ACTION

Joan Martinez is just beginning her third year as dean of the business school at a small regional university in the Midwest. When she joined, the university's president and academic provost expressed concern over dwindling enrollments in the MBA program, and made it clear to her that increasing MBA enrollment was to be her highest priority. She spent her first year evaluating the situation. She discovered that the main reason for the business school's declining enrollments in the MBA program is its location. The region had experienced an economic downturn that forced companies and businesses, many of which had tuition reimbursement programs for their employees, to close or relocate. Given the lack of prospects for increasing MBA enrollments in the traditional regional market, Joan made plans to start an online MBA program.

The university's administration was very enthusiastic about the initiative and approved funding for a new position, Online MBA Director. Joan hired Raj Patel to fill it. Raj not only had considerable experience with online MBA programs, but a proven track record for getting faculty support, something that Joan really needed. Faculty who teach the more quantitative, technical courses had already begun criticizing the initiative. They argued that it would be difficult to achieve desired student learning outcomes in these types of courses using the online mode of delivery. Therefore, Raj's first task as the new Online MBA Director was to change their beliefs.

He remembered that a faculty colleague from his former institution, Rob Lowry, taught finance in a hybrid format: the first half in a traditional classroom and the other half online. Raj contacted him to ask if he would share his test scores (without identifying students) for the mid-term (based on traditional in-class instruction) and the final (based upon online teaching). Rob agreed. Based on test scores for 40 students, Raj calculated the mid-term average as 78.65 and the final exam average as 72.55. Although students' test scores after online instruction were lower, the two-sample *t*-test revealed that the averages on the two exams were not significantly different ($t = 1.82$, P-value $= 0.073$). Raj felt that these results help make the case that online and in-class instruction are equivalent when it comes to student learning in quantitative business courses, and Joan agreed. She was happy to have these results in time for the next faculty meeting.

- Identify the ethical dilemma in this scenario.

- What are the undesirable consequences?

- Propose an ethical solution that considers the welfare of all stakeholders.

WHAT HAVE WE LEARNED?

Learning Objectives

Know how to test whether the difference in the means of two independent groups is equal to some hypothesized value.

- The two-sample *t*-test is appropriate for independent groups. It uses a special formula for degrees of freedom.
- The Assumptions and Conditions are the same as for one-sample inferences for means with the addition of assuming that the groups are independent of each other.
- The most common null hypothesis is that the means are equal.

Be able to construct and interpret a confidence interval for the difference between the means of two independent groups.

- The confidence interval inverts the *t*-test in the natural way.

Know how and when to use pooled *t* inference methods.

- There is an additional assumption that the variances of the two groups are equal.
- This may be a plausible assumption in a randomized experiment.

Recognize when you have paired or matched samples and use an appropriate inference method.

- Paired t-methods are the same as one-sample t-methods applied to the pairwise differences.
- If data are paired they cannot be independent, so two-sample t and pooled-t methods would not be applicable.

Terms

Paired data

Data are paired when the observations are collected in pairs or the observations in one group are naturally related to observations in the other. The simplest form of pairing is to measure each subject twice—often before and after a treatment is applied. Pairing in experiments is a form of blocking and arises in other contexts. Pairing in observational and survey data is a form of matching.

Paired t-test

A hypothesis test for the mean of the pairwise differences of two groups. It tests the null hypothesis $H_0: \mu_d = \Delta_0$, where the hypothesized difference is almost always 0, using the statistic $t = \dfrac{\bar{d} - \Delta_0}{SE(\bar{d})}$ with $n - 1$ degrees of freedom, where $SE(\bar{d}) = \dfrac{s_d}{\sqrt{n}}$ and n is the number of pairs.

Paired t-confidence interval

A confidence interval for the mean of the pairwise differences between paired groups found as $\bar{d} \pm t^*_{n-1} \times SE(\bar{d})$, where $SE(\bar{d}) = \dfrac{s_d}{\sqrt{n}}$ and n is the number of pairs.

Pooled t-interval

A confidence interval for the difference in the means of two independent groups used when we are willing and able to make the additional assumption that the variances of the groups are equal. It is found as:

$$(\bar{y}_1 - \bar{y}_2) \pm t^*_{df} \times SE_{pooled}(\bar{y}_1 - \bar{y}_2),$$

where $SE_{pooled}(\bar{y}_1 - \bar{y}_2) = s_{pooled} \sqrt{\dfrac{1}{n_1} + \dfrac{1}{n_2}}$,

and the pooled variance is

$$s^2_{pooled} = \dfrac{(n_1 - 1) s^2_1 + (n_2 - 1) s^2_2}{(n_1 - 1) + (n_2 - 1)}.$$

The number of degrees of freedom is $(n_1 - 1) + (n_2 - 1)$.

Pooled t-test

A hypothesis test for the difference in the means of two independent groups when we are willing and able to assume that the variances of the groups are equal. It tests the null hypothesis

$$H_0: \mu_1 - \mu_2 = \Delta_0,$$

where the hypothesized difference Δ_0 is almost always 0, using the statistic

$$t_{df} = \dfrac{(\bar{y}_1 - \bar{y}_2) - \Delta_0}{SE_{pooled}(\bar{y}_1 - \bar{y}_2)},$$

where the pooled standard error is defined as for the pooled interval and the degrees of freedom is $(n_1 - 1) + (n_2 - 1)$.

Pooling

Data from two or more populations may sometimes be combined, or *pooled*, to estimate a statistic (typically a pooled variance) when we are willing to assume that the estimated value is the same in both populations. The resulting larger sample size may lead to an estimate with lower sample variance. However, pooled estimates are appropriate only when the required assumptions are true.

Two-sample t-interval

A confidence interval for the difference in the means of two independent groups found as

$$(\bar{y}_1 - \bar{y}_2) \pm t^*_{df} \times SE(\bar{y}_1 - \bar{y}_2), \text{ where}$$

$$SE(\bar{y}_1 - \bar{y}_2) = \sqrt{\dfrac{s^2_1}{n_1} + \dfrac{s^2_2}{n_2}}$$

and the number of degrees of freedom is given by the approximation formula in footnote 2 of this chapter, or with technology.

Two-sample t-test

A hypothesis test for the difference in the means of two independent groups. It tests the null hypothesis $H_0: \mu_1 - \mu_2 = \Delta_0,$

where the hypothesized difference Δ_0 is almost always 0, using the statistic

$$t_{df} = \dfrac{(\bar{y}_1 - \bar{y}_2) - \Delta_0}{SE(\bar{y}_1 - \bar{y}_2)},$$

with the number of degrees of freedom given by the approximation formula in footnote 2 of this chapter, or with technology.

TECHNOLOGY HELP: Comparing Two Groups

Two-Sample Methods

Here's some typical computer package output with comments:

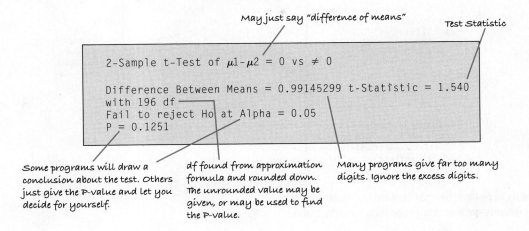

May just say "difference of means"

Test Statistic

```
2-Sample t-Test of μ1-μ2 = 0 vs ≠ 0

Difference Between Means = 0.99145299  t-Statistic = 1.540
with 196 df
Fail to reject Ho at Alpha = 0.05
P = 0.1251
```

Some programs will draw a conclusion about the test. Others just give the P-value and let you decide for yourself.

df found from approximation formula and rounded down. The unrounded value may be given, or may be used to find the P-value.

Many programs give far too many digits. Ignore the excess digits.

Some statistics software automatically tries to test whether the variances of the two groups are equal. Some automatically offer both the two-sample-*t* and pooled-*t* results. Ignore the test for the variances; it has little power in any situation in which its results could matter. If the pooled and two-sample methods differ in any important way, you should stick with the two-sample method. Most likely, the Equal Variance Assumption needed for the pooled method has failed.

The degrees of freedom approximation usually gives a fractional value. Most packages seem to round the approximate value down to the next smallest integer (although they may actually compute the P-value with the fractional value, gaining a tiny amount of power).

There are two ways to organize data when we want to compare two independent groups. The first, called **unstacked data**, lists the data in two columns, one for each group. Each list can be thought of as a variable. In this method, the variables in the credit card example would be "Offer" and "No Offer." Graphing calculators usually prefer this form, and some computer programs can use it as well.

The alternative way to organize the data is as **stacked data**. What is the response variable for the credit card experiment? It's the "Spend Lift"—the amount by which customers increased their spending. But the values of this variable in the unstacked lists are in both columns, and actually there's an experiment factor here, too—namely, whether the customer was offered the promotion or not. So we could put the data into two different columns, one with the "Spend Lifts" in it and one with a "Yes" for those who were offered the promotion and a "No" for those who weren't. The stacked data would look like this:

Spend Lift	Offer
969.74	Yes
915.04	Yes
197.57	No
77.31	No
196.27	Yes
…	…

This way of organizing the data makes sense as well. Now the factor and the response variables are clearly visible. You'll have to see which method your program requires. Some packages even allow you to structure the data either way.

The commands to do inference for two independent groups on common statistics technology are not always found in obvious places. The step-by-step instructions contain some starting guidelines.

Paired Methods

Most statistics programs can compute paired-*t* analyses. Some may want you to find the differences yourself and use the one-sample *t*-methods. Those that perform the entire procedure will need to know the two variables to compare. The computer, of course, cannot verify that the variables are naturally paired. Most programs will check whether the two variables have the same number of observations, but some stop there, and that can cause trouble. Most programs will automatically omit any pair that is missing a value for either variable. You must look carefully to see whether that has happened.

As we've seen with other inference results, some packages pack a lot of information into a simple table, but you must locate what you want for yourself. Here's a generic example with comments.

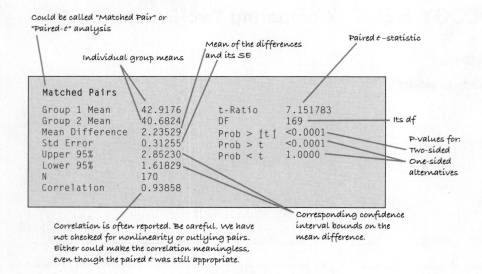

Could be called "Matched Pair" or "Paired-t" analysis

Individual group means

Mean of the differences and its SE

Paired t-statistic

```
Matched Pairs
Group 1 Mean      42.9176       t-Ratio    7.151783
Group 2 Mean      40.6824       DF         169
Mean Difference    2.23529      Prob > │t│ <0.0001
Std Error          0.31255      Prob > t   <0.0001
Upper 95%          2.85230      Prob < t   1.0000
Lower 95%          1.61829
N                 170
Correlation        0.93858
```

Its df

P-values for:
Two-sided
One-sided
alternatives

Correlation is often reported. Be careful. We have not checked for nonlinearity or outlying pairs. Either could make the correlation meaningless, even though the paired t was still appropriate.

Corresponding confidence interval bounds on the mean difference.

Other packages try to be more descriptive. It may be easier to find the results, but you may get less information from the output table.

Groups may have missing values. Only cases with both values present are used in a paired-t analysis. You may not learn that from some packages.

Even simple tables can have superfluous numbers such as these.

SD (differences)

SE($\bar{d}$)

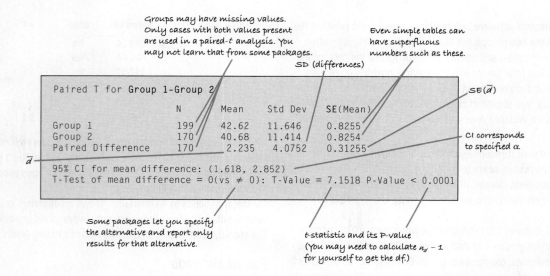

```
Paired T for Group 1-Group 2
                    N      Mean    Std Dev    SE(Mean)
Group 1            199    42.62    11.646     0.8255
Group 2            170    40.68    11.414     0.8254
Paired Difference  170     2.235    4.0752    0.31255

95% CI for mean difference: (1.618, 2.852)
T-Test of mean difference = 0(vs ≠ 0): T-Value = 7.1518 P-Value < 0.0001
```

$\bar{d}$

CI corresponds to specified α.

Some packages let you specify the alternative and report only results for that alternative.

t-statistic and its P-value (You may need to calculate $n_d - 1$ for yourself to get the df.)

Computers make it easy to examine the boxplots of the two groups and the histogram of the differences—both important steps. Some programs offer a scatterplot of the two variables. That can be helpful. In terms of the scatterplot, a paired t-test is about whether the points tend to be above or below the 45° line $y = x$. (Note that pairing says nothing about whether the scatterplot should be straight. That doesn't matter for our t-methods.)

EXCEL

To perform a two-sample t-test in Excel using Data Analysis Tool Pack:

- In the **Data** menu, Choose **Data Analysis** (found under Analysis).
- Choose the appropriate t-test from the menu.
- Identify the following:
  - The cell range for the data in the two samples (for testing independent groups) or the difference data (for a paired t-test).
  - The hypothesized difference.
- Check the box next to Labels if the first row of the data contains labels, and identify where you would like the results to be placed.

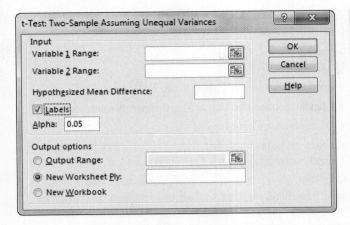

To perform a two-sample *t*-test using built-in formulas:

- From Formulas, choose **More functions** > **Statistical** > **T.TEST**, and specify:

  - Array 1 and Array 2 = Location in the spreadsheet of the two groups to compare.

  - Tails of the distribution = 1 for one-tailed test and 2 for two-tailed test.

- Type = 1 for a paired test, 2 for equal variance *t*-test, and 3 for unequal variance *t*-test.

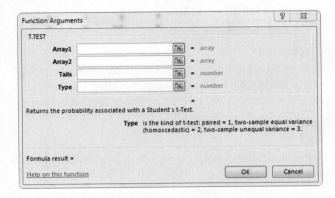

To perform a paired *t*-test in Excel:

- Enter paired data into two columns with one pair in each row.

- In another column, calculate the difference of the values in the two columns.

- Analyze the difference column using a one-sample *t*-test. An example is shown below:

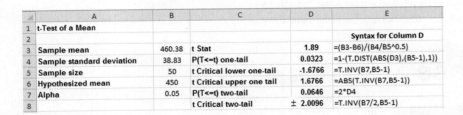

	A	B	C	D	E
1	t-Test of a Mean				
2					**Syntax for Column D**
3	Sample mean	460.38	t Stat	1.89	=(B3-B6)/(B4/B5^0.5)
4	Sample standard deviation	38.83	P(T<=t) one-tail	0.0323	=1-(T.DIST(ABS(D3),(B5-1),1))
5	Sample size	50	t Critical lower one-tail	-1.6766	=T.INV(B7,B5-1)
6	Hypothesized mean	450	t Critical upper one tail	1.6766	=ABS(T.INV(B7,B5-1))
7	Alpha	0.05	P(T<=t) two-tail	0.0646	=2*D4
8			t Critical two-tail	± 2.0096	=T.INV(B7/2,B5-1)

XLSTAT

To conduct a *t* test for the difference between two means:

- Choose **Parametric tests,** and then **Two-sample t-test and z-test.**

- If the two samples are in two separate columns in your workbook, choose the option **One column per sample.**

- If your data are *stacked* (one column lists the data, and the other the group identifier) choose the option **One column per variable.**

- If the data are paired, choose the option **Paired samples.**

- Enter the range of cell data.

- On the **Options** tab, for a test that is not pooled uncheck the box next to **Assume equality.**

The output yields the results of a hypothesis test and a confidence interval for the difference between means.

JMP

To conduct a two-sample *t*-test in JMP:

- From the **Analyze** menu, select **Fit X by Y.**

- Select variables: a **Y, Response** variable that holds the data and an **X, Factor** variable that holds the group names. Click OK, and JMP

will make a dotplot. Be sure that the variable type for the **X, Factor** variable is set to nominal or ordinal.

- Click the Red triangle next to "Oneway Analysis…" and select:

  - *t* **Test** to perform a two-sample *t*-test.

To compute a paired *t*-test using JMP:

- From the **Analyze** menu, select **Matched Pairs.**

- Place two variables containing matched data into **Y, Column.**

Comments

To perform a pooled *t*-test, select **Means/ANOVA/Pooled *t*** under the red triangle next to "Oneway Analysis…"

MINITAB

To perform a two-sample *t*-test in Minitab:

- From the **Stat** menu, choose the **Basic Statistics** submenu.

- Choose **2t 2-sample t…** and fill in the dialog for:

  - Data in "one column" (unstacked data)

  - Data "two columns" (stacked data) or

  - Summarized data

To conduct a paired *t*-test using Minitab:

- From the **Stat** menu, choose the **Basic Statistics** submenu.
- Choose **t-t Paired t** and fill in the dialog for:
  - Data in "two columns" or
  - Summarized data

Note: For two-sample *t*-tests, the **Graphs** button creates boxplots of your two samples and the **Options** button allows you to conduct either a one-sided or two-sided test or change the confidence level.

To test the hypothesis that $\mu_1 = \mu_2$ against an alternative (default is two-sided):

- Create a vector for the data of each group, say *X* and *Y*, and produce the confidence interval (default is 95%):
- t.test(X,Y, alternative = c("two.sided", "less", "greater"), conf.level = 0.95)

will produce the *t*-statistic, degrees of freedom, P-value, and confidence interval for a specified alternative. For paired data, add the statement paired=TRUE in the t.test function.

Comments

This is the same function as for the one-sample *t*-test, but with two vectors. In R, the default *t*-test does not assume equal variances in the two groups (default is **var.equal = FALSE**). To get the "pooled" t-test, add **var.equal = TRUE** to the function call.

SPSS

To conduct a two-sample *t*-test in SPSS:

- From the **Analyze** menu, choose the **Compare Means** submenu.
- Choose **Independent-Samples t-test.**
- Specify the data variable and grouping variable, click the button under the grouping variable, and identify numerical codes for groups.

Note: SPSS offers both the two-sample and pooled-*t* results in the same table. SPSS expects the data in one variable and group names in the other. If there are more than two group names in the group variable, only the two that are named in the dialog box will be compared.

To perform a paired *t*-test in SPSS:

- From the **Analyze** menu, choose the **Compare Means** submenu.
- Choose **Paired Samples t-test.**
- Highlight the two variables containing the paired data and add to the list.

Brief **Case**

Real Estate

In Chapter 4, we examined the regression of the sales price of a home on its size and saw that larger homes generally fetch a higher price. How much can we learn about a house from the fact that it has a fireplace or more than the average number of bedrooms? Data for a random sample of 1063 homes from the upstate New York area can be found in the file **Saratoga Real Estate**. There are 6 quantitative variables: *Price($)*, *Living Area(sq.ft)*, *Bathrooms(#)*, *Bedrooms(#)*, *Lot Size(Acres)*, and *Age(years)*, and one categorical variable, *Fireplace?* (1 = *Yes*; 0 = *No*) denoting whether the house has at least one fireplace. We can use *t*-methods to see, for example, whether homes with fireplaces sell for more, on average, and by how much. For the quantitative variables, create new categorical variables by splitting them at the median or some other splitting point of your choice, and compare home prices above and below this value. For example, the median number of *Bedrooms* of these homes is 2. You might compare the prices of homes with 1 or 2 bedrooms to those with more than 2. Write up a short report summarizing the differences in mean price based on the categorical variables that you created.

Consumer Spending Patterns (Data Analysis)

You are on the financial planning team for monitoring a high spending segment of a credit card. You know that customers tend to spend more during December before the holidays, but you're not sure about the pattern of spending in the months after the holidays. Look at the data set **Consumer spending post holiday**. It contains the monthly credit card spending of 1200 customers during the months December, January, February, March, and April. Report on the spending differences between the months. If you had failed to realize that these are paired data, what difference would that have made in your reported confidence intervals and tests?

EXERCISES

SECTION 12.1

1. A developer wants to know if the houses in two different neighborhoods were built at roughly the same time. She takes a random sample of six houses from each neighborhood and finds their ages from local records. The table shows the data for each sample (in years).

Neighborhood 1	Neighborhood 2
57	50
60	43
46	35
62	53
67	46
56	55

a) Find the sample means for each neighborhood.
b) Find the estimated difference of the mean ages of the two neighborhoods.
c) Find the sample variances for each neighborhood.
d) Find the sample standard deviations for each neighborhood.
e) Find the standard error of the difference of the two sample means.

2. A market analyst wants to know if the new website he designed is showing increased page views per visit. A customer is randomly sent to one of two different websites, offering the same products, but with different designs. Here are the page views from five randomly chosen customers from each website:

Website A	Website B
9	10
4	13
14	2
7	3
2	7

a) Find the sample mean page views for each website.
b) Find the estimated difference of the sample mean page views of the two websites.
c) Find the sample variances for each website.
d) Find the sample standard deviations for each website.
e) Find the standard error of the difference of the sample means.

3. The developer in Exercise 1 hires an assistant to collect a random sample of houses from each neighborhood and finds that the summary statistics for the two neighborhoods look as follows:

Neighborhood 1	Neighborhood 2
$n_1 = 30$	$n_2 = 35$
$\bar{y}_1 = 57.2$ yrs	$\bar{y}_2 = 47.6$ yrs
$s_1 = 7.51$ yrs	$s_2 = 7.85$ yrs

a) Find the estimated mean age difference between the two neighborhoods.
b) Find the standard error of the estimated mean difference.
c) Calculate the t-statistic for the observed difference in mean ages assuming that the true mean difference is 0.

4. Not happy with the previous results, the analyst in Exercise 2 takes a much larger random sample of customers from each website and records their page views. Here are the data:

Website A	Website B
$n_1 = 80$	$n_2 = 95$
$\bar{y}_1 = 7.7$ pages	$\bar{y}_2 = 7.3$ pages
$s_1 = 4.6$ pages	$s_2 = 4.3$ pages

a) Find the estimated mean difference in page visits between the two websites.
b) Find the standard error of the estimated mean difference.
c) Calculate the t-statistic for the observed difference in mean page visits assuming that the true mean difference is 0.

SECTION 12.2

5. For the data in Exercise 1, we want to test the null hypothesis that the mean age of houses in the two neighborhoods is the same. Assume that the data come from a population that is Normally distributed.

a) Using the values you found in Exercise 1, find the value of the t-statistic for the difference in mean ages for the null hypothesis.
b) Calculate the degrees of freedom from the formula in the footnote of page 374.
c) Calculate the degrees of freedom using the rule that $df = \min(n_1 - 1, n_2 - 1)$.
d) Find the P-value using the degrees of freedom from part b. (You can either round the number of df and use a table or use technology or a website).
e) Find the P-value using the degrees of freedom from part d.
f) What do you conclude at $\alpha = 0.05$?

6. For the data in Exercise 2, we want to test the null hypothesis that the mean number of page visits is the same for the two websites. Assume that the data come from a population that is Normally distributed.

a) Using the values you found in Exercise 2, find the value of the t-statistic for the difference in mean ages for the null hypothesis.
b) Calculate the degrees of freedom from the formula in the footnote of page 374.
c) Calculate the degrees of freedom using the rule that $df = \min(n_1 - 1, n_2 - 1)$.
d) Find the P-value using the degrees of freedom from part b. (You can either round the number of df and use a table or use technology or a website).
e) Find the P-value using the degrees of freedom from part c.
f) What do you conclude at $\alpha = 0.05$?

7. Using the data in Exercise 3, test the hypothesis that the mean age of houses in the two neighborhoods is the same. You may assume that the ages of houses in each neighborhood follow a Normal distribution.

a) Calculate the P-value of the statistic knowing that the approximation formula gives 62.2 df (you will need to round or use technology).
b) Calculate the P-value of the statistic using the rule that df is at least $min(n_1 - 1, n_2 - 1)$.
c) What do you conclude at $\alpha = 0.05$?

8. Using the data in Exercise 4, test the hypothesis that the mean number of page views from the two websites is the same. You may assume that the number of page views from each website follow a Normal distribution.

a) Calculate the P-value of the statistic knowing that the approximation formula gives 163.6 df.
b) Calculate the P-value of the statistic using the rule that df is at least $\min(n_1 - 1, n_2 - 1)$.
c) What do you conclude at $\alpha = 0.05$?

SECTION 12.4

9. Using the data in Exercise 1, and assuming that the data come from a distribution that is Normally distributed,

a) Find a 95% confidence interval for the mean difference in ages of houses in the two neighborhoods.
b) Is 0 within the confidence interval?
c) What does it say about the null hypothesis that the mean difference is 0?

10. Using the data in Exercise 2, and assuming that the data come from a distribution that is Normally distributed,

a) Find a 95% confidence interval for the mean difference in page views from the two websites.
b) Is 0 within the confidence interval?
c) What does it say about the null hypothesis that the mean difference is 0?

11. Using the summary statistics in Exercise 3, and assuming that the data come from a distribution that is Normally distributed,

a) Find a 95% confidence interval for the mean difference in ages of houses in the two neighborhoods using the df given in Exercise 7.
b) Why is the confidence interval narrower than the one you found in Exercise 9?
c) Is 0 within the confidence interval?
d) What does it say about the null hypothesis that the mean difference is 0?

12. Using the summary statistics in Exercise 4, and assuming that the data come from a distribution that is Normally distributed,

a) Find a 95% confidence interval for the mean difference in page views from the two websites.
b) Why is the confidence interval narrower than the one you found in Exercise 10?
c) Is 0 within the confidence interval?
d) What does it say about the null hypothesis that the mean difference is 0?

SECTION 12.5

13. For the data in Exercise 1,

a) Test the null hypothesis at $\alpha = 0.05$ using the pooled t-test. (Show the t-statistic, P-value, and conclusion.)
b) Find a 95% confidence interval using the pooled degrees of freedom.
c) Are your answers different from what you previously found in Exercise 9? Explain briefly why or why not.

14. For the data in Exercise 2,

a) Test the null hypothesis at $\alpha = 0.05$ using the pooled t-test. (Show the t-statistic, P-value, and conclusion.)
b) Find a 95% confidence interval using the pooled degrees of freedom.
c) Are your answers different from what you previously found in Exercise 10? Explain briefly why or why not.

15. For the data in Exercise 3,

a) Test the null hypothesis at $\alpha = 0.05$ using the pooled t-test. (Show the t-statistic, P-value, and conclusion.)
b) Find a 95% confidence interval using the pooled degrees of freedom.
c) Are your answers different from what you previously found in Exercise 11? Explain briefly why or why not.

16. For the data in Exercise 4,

a) Test the null hypothesis at $\alpha = 0.05$ using the pooled t-test. (Show the t-statistic, P-value, and conclusion.)
b) Find a 95% confidence interval using the pooled degrees of freedom.

c) Are your answers different from what you previously found in Exercise 12? Explain briefly why or why not.

SECTION 12.6

17. For each of the following scenarios, say whether the data should be treated as independent or paired samples. Explain briefly. If paired, explain what the pairing involves.

a) An efficiency expert claims that a new ergonomic desk chair makes typing at a computer terminal easier and faster. To test it, 15 volunteers are selected. Using both the new chair and their old chair, each volunteer types a randomly selected passage for 2 minutes and the number of correct words typed is recorded.

b) A developer wants to know if the houses in two different neighborhoods have the same mean price. She selects 10 houses from each neighborhood at random and tests the null hypothesis that the means are equal.

c) A manager wants to know if the mean productivity of two workers is the same. For a random selection of 30 hours in the past month, he compares the number of items produced by each worker in that hour.

18. For each of the following scenarios, say whether the data should be treated as independent or paired samples. Explain briefly. If paired, explain what the pairing involves.

a) An efficiency expert claims that a new ergonomic desk chair makes typing at a computer terminal easier and faster. To test it, 30 volunteers are selected. Half of the volunteers will use the new chair and half will use their old chairs. Each volunteer types a randomly selected passage for 2 minutes and the number of correct words typed is recorded.

b) A real estate agent wants to know how much extra a fireplace adds to the price of a house. She selects 25 city blocks. In each block, she randomly chooses a house with a fireplace and one without and records the assessment value.

c) A manager wants to know if the mean productivity of two workers is the same. For each worker, he randomly selects 30 hours in the past month and compares the number of items produced.

SECTION 12.7

19. A supermarket chain wants to know if their "buy one, get one free" campaign increases customer traffic enough to justify the cost of the program. For each of 10 stores they select two days at random to run the test. For one of those days (selected by a coin flip), the program will be in effect. They want to test the hypothesis that there is no mean difference in traffic against the alternative that the program increases the mean traffic. Here are the results in number of customer visits to the 10 stores:

Store #	With Program	Without Program
1	140	136
2	233	235
3	110	108
4	42	35
5	332	328
6	135	135
7	151	144
8	33	39
9	178	170
10	147	141

a) Are the data paired? Explain.
b) Compute the mean difference.
c) Compute the standard deviation of the differences.
d) Compute the standard error of the mean difference.
e) Find the value of the *t*-statistic.
f) How many degrees of freedom does the *t*-statistic have?
g) Is the alternative one- or two-sided? Explain.
h) What is the P-value associated with this *t*-statistic? (Assume that the other assumptions and conditions for inference are met.)
i) At $\alpha = 0.05$, what do you conclude?

20. A city wants to know if a new advertising campaign to make citizens aware of the dangers of driving after drinking has been effective. They count the number of drivers who have been stopped with more alcohol in their systems than the law allows for each day of the week in the week before and the week a month after the campaign starts. Here are the results:

Day of Week	Before	After
M	5	2
T	4	0
W	2	2
Th	4	1
F	6	8
S	14	7
Su	6	7

a) Are the data paired? Explain.
b) Compute the mean difference.
c) Compute the standard deviation of the differences.
d) Compute the standard error of the mean difference.
e) Find the value of the *t*-statistic.
f) How many degrees of freedom does the *t*-statistic have?
g) Is the alternative one- or two-sided? Explain.
h) What is the P-value associated with this *t*-statistic? (Assume that the other assumptions and conditions for inference are met.)
i) At $\alpha = 0.05$, what do you conclude?

21. In order to judge whether the program is successful, the manager of the supermarket chain in Exercise 19 wants to know the plausible range of values for the mean increase in customers using the program. Construct a 90% confidence interval.

22. A new operating system is installed in every workstation at a large company. The claim of the operating system manufacturer is that the time to shut down and turn on the machine will be much faster. To test it an employee selects 36 machines and tests the combined shut down and restart time of each machine before and after the new operating system has been installed. The mean and standard deviation of the differences (before – after) is 23.5 seconds with a standard deviation of 40 seconds.

a) What is the standard error of the mean difference?
b) How many degrees of freedom does the *t*-statistic have?
c) What is the 90% confidence interval for the mean difference?
d) What do you conclude at $\alpha = 0.05$?

CHAPTER EXERCISES

23. Hot dogs and calories. Consumers increasingly make food purchases based on nutrition values. *Consumer Reports* examined the calorie content of two kinds of hot dogs: meat (usually a mixture of pork, turkey, and chicken) and all beef. The researchers purchased samples of several different brands. The meat hot dogs averaged 111.7 calories, compared to 135.4 for the beef hot dogs. A test of the null hypothesis that there's no difference in mean calorie content yields a P-value of 0.124. What would you conclude?

24. Hot dogs and sodium. The *Consumer Reports* article described in Exercise 23 also listed the sodium content (in mg) for the various hot dogs tested. A test of the null hypothesis that beef hot dogs and meat hot dogs don't differ in the mean amounts of sodium yields a P-value of 0.110. What would you conclude?

25. Learning math. The Core Plus Mathematics Project (CPMP) is an innovative approach to teaching mathematics that engages students in group investigations and mathematical modeling. After field tests in 36 high schools over a three-year period, researchers compared the performances of CPMP students with those taught using a traditional curriculum. In one test, students had to solve applied algebra problems using calculators. Scores for 320 CPMP students were compared with those of a control group of 273 students in a traditional math program. Computer software was used to create a confidence interval for the difference in mean scores (*Journal for Research in Mathematics Education*, 31, no. 3, 2000).

```
Conf. level: 95%
Variable: μ(CPMP) − μ(Ctrl)
Interval: (5.573, 11.427)
```

a) What is the margin of error for this confidence interval?
b) If we had created a 98% confidence interval, would the margin of error be larger or smaller?
c) Explain what the calculated interval means in this context.
d) Does this result suggest that students who learn mathematics with CPMP will have (statistically) significantly higher mean scores in applied algebra than those in traditional programs? Explain.

26. Sales performance. A chain that specializes in healthy and organic food would like to compare the sales performance of two of its primary stores in the state of Massachusetts. These stores are both in urban, residential areas with similar demographics. A comparison of the weekly sales randomly sampled over a period of nearly two years for these two stores yields the following information:

Store	N	Mean	StDev	Minimum	Median	Maximum
Store#1	9	242170	23937	211225	232901	292381
Store#2	9	235338	29690	187475	232070	287838

a) Create a 95% confidence interval for the difference in the mean store weekly sales. (df from technology is 15.31)
b) Interpret your interval in context.
c) Does it appear that one store sells more on average than the other store?
d) What is the margin of error for this interval?
e) Would you expect a 99% confidence interval to be wider or narrower? Explain.
f) If you computed a 99% confidence interval, would your conclusion in part c change? Explain.

27. CPMP, again. During the study described in Exercise 25, students in both CPMP and traditional classes took another algebra test that did not allow them to use calculators. The table shows the results. Are the mean scores of the two groups significantly different? Assume that the assumptions for inference are satisfied.

Math Program	*n*	Mean	SD
CPMP	312	29.0	18.8
Traditional	265	38.4	16.2

a) Write an appropriate hypothesis.
b) Here is computer output for this hypothesis test. Explain what the P-value means in this context.

```
2-Sample t-Test of μ₁ − μ₂ ≠ 0
t-Statistic = −6.451 w/574.8761 df
P < 0.0001
```

c) State a conclusion about the CPMP program.

28. IT training costs. An accounting firm is trying to decide between IT training conducted in-house and the use of third party consultants. To get some preliminary cost data, each type of training was implemented at two of the firm's offices located in different cities. The table below shows the average annual training cost per employee at each location. Are the mean costs significantly different? Assume that the assumptions for inference are satisfied.

IT Training	n	Mean	SD
In-House	210	$490.00	$32.00
Consultants	180	$500.00	$48.00

a) Write the appropriate hypotheses.
b) Below is computer output for this hypothesis test. Explain what the P-value means in this context.

$$\text{2-Sample t-Test of } \mu_1 - \mu_2 \neq 0$$
$$\text{t-Statistic} = -2.38 \text{ w/303 df}$$
$$P = .018$$

c) State a conclusion about IT training costs.

29. CPMP and word problems. The study of the new CPMP mathematics methodology described in Exercise 25 also tested students' abilities to solve word problems. This table shows how the CPMP and traditional groups performed. What do you conclude? (Assume that the assumptions for inference are met.)

Math Program	n	Mean	SD
CPMP	320	57.4	32.1
Traditional	273	53.9	28.5

30. Statistical training. The accounting firm described in Exercise 28 is interested in providing opportunities for its auditors to gain more expertise in statistical sampling methods. They wish to compare traditional classroom instruction with online self-paced tutorials. Auditors were assigned at random to one type of instruction, and the auditors were then given an exam. The table shows how the two groups performed. What do you conclude? (Assume the assumptions for inference are met.)

Program	n	Mean	SD
Traditional	296	74.5	11.2
Online	275	72.9	12.3

31. Trucking company. A trucking company would like to compare two different routes for efficiency. Truckers are randomly assigned to two different routes. Twenty truckers following Route A report an average of 40 minutes, with a standard deviation of 3 minutes. Twenty truckers following Route B report an average of 43 minutes, with a standard deviation of 2 minutes. Histograms of travel times for the routes are roughly symmetric and show no outliers.

a) Find a 95% confidence interval for the difference in average time for the two routes.
b) Will the company save time by always driving one of the routes? Explain.

32. Change in sales. Suppose the specialty food chain from Exercise 26 wants to now compare the change in sales across different regions. An examination of the difference in sales over a 37-week period in a recent year for 8 stores in the state of Massachusetts compared to 12 stores in nearby states reveals the following descriptive statistics for relative increase in sales. (If these means are multiplied by 100, they show % increase in sales.)

State	N	Mean	StDev
MA	8	0.0738	0.0666
Other	12	0.0559	0.0503

a) Find the 90% confidence interval for the difference in relative increase in sales over this time period.
b) Is there a significant difference in increase in sales between these two groups of stores? Explain.
c) What would you like to see to check the conditions?

33. Cereal company. A food company is concerned about recent criticism of the sugar content of their children's cereals. The data show the sugar content (as a percentage of weight) of several national brands of children's and adults' cereals.

Children's cereals: 40.3, 55, 45.7, 43.3, 50.3, 45.9, 53.5, 43, 44.2, 44, 47.4, 44, 33.6, 55.1, 48.8, 50.4, 37.8, 60.3, 46.6

Adults' cereals: 20, 30.2, 2.2, 7.5, 4.4, 22.2, 16.6, 14.5, 21.4, 3.3, 6.6, 7.8, 10.6, 16.2, 14.5, 4.1, 15.8, 4.1, 2.4, 3.5, 8.5, 10, 1, 4.4, 1.3, 8.1, 4.7, 18.4

a) Write the null and alternative hypotheses.
b) Check the conditions.
c) Find the 95% confidence interval for the difference in means.
d) Is there a significant difference in mean sugar content between these two types of cereals? Explain.

34. Italian wines. Chemical analyses of 1599 red and 4898 white Italian wines revealed the following summary statistics for pH (a measure of acidity):

Type	Count	Mean	St Dev
Red	1599	3.311	0.154
White	4898	3.188	0.151

Is there a difference in pH between red and white wines?

a) Write the null and alternative hypotheses
b) What conditions would you check?
c) Test the hypothesis and find the P-value.
d) Is there a significant difference in pH?

35. Bond funds. Morningstar (www.morningstar.com) selects mutual funds as "Medalist" funds expected to perform well over the long term. You have decided to invest in a bond fund and plan to limit your choice of funds to Morningstar "medalist" funds. But now you must choose between a taxable fund and a municipal bond fund that is at least partially tax-free. Which is better? Here are the % returns for the three-year period leading up to spring of 2013:

Taxable bond funds

10.83, 6.45, 8.52, 10.9, 4.16, 10.48, 6.07, 2.69, 1.24, 1.58, 4.02, 5.64, 6.29, 12.36

Municipal bond funds

8.34, 7.3, 6.07, 6.46, 5.77, 5.89, 5.76, 5.81, 5.12, 5.63 4.71, 5.22, 5.21, 3.12, 4.77, 2.2

a) Write the null and alternative hypotheses.
b) Check the conditions.
c) Test the hypothesis and find the P-value.
d) Is there a significant difference in 3-year returns between these two kinds of funds?

36. Technology adoption. The Pew Internet & American Life Project (www.pewinternet.org/) conducts surveys to gauge how the Internet and technology impact daily life of individuals, families, and communities. In a recent survey Pew asked respondents if they thought that computers and technology give people more or less control over their lives. Companies that are involved in innovative technologies use the survey results to better understand their target market. One might suspect that younger and older respondents might differ in their opinions of whether computers and technology give them more control over their lives. A subset of the data from this survey shows the mean ages of two groups of respondents, those who reported that they believed that computers and technology give them "more" control and those that reported "less" control.

Group	N	Mean	StDev	Min	Q1	Med	Q3	Max
More	74	54.42	19.65	18	41.5	53.5	68.5	99.0
Less	29	54.34	18.57	20	41.0	58.0	70.0	84.0

a) Write the null and alternative hypotheses.
b) Find the 95% confidence interval for the difference in mean age between the two groups of respondents.
c) Is there a significant difference in the mean ages between these two groups? Explain.

37. Product testing. A company is producing and marketing new reading activities for elementary school children that it believes will improve reading comprehension scores. A researcher randomly assigns third graders to an eight-week program in which some will use these activities and others will experience traditional teaching methods. At the end of the experiment, both groups take a reading comprehension exam. Their scores are shown in the back-to-back stem-and-leaf display. Do these results suggest that the new activities are better? Test an appropriate hypothesis and state your conclusion.

New Activities		Control
	1	07
4	2	068
3	3	377
96333	4	12222238
9876432	5	355
721	6	02
1	7	
	8	5

38. Product placement. The owner of a small organic food store was concerned about her sales of a specialty yogurt manufactured in Greece. As a result of increasing fuel costs, she recently had to increase its price. To help boost sales, she decided to place the product on a different shelf (near eye level for most consumers) and in a location near other popular international products. She kept track of sales (number of containers sold per week) for six months after she made the change. These values are shown below, along with the sales numbers for the six months prior to making the change, in stem-and-leaf displays.

After Change			Before Change	
3	2		2	0
3	9		2	899
4	23		3	224
4	589		3	7789
5	0012		4	0000223
5	55558		4	5567
6	00123		5	0
6	67		5	6
7	0			

Do these results suggest that sales are better after the change in product placement? Test an appropriate hypothesis and state your conclusion. Be sure to check assumptions and conditions.

39. Named tropical cyclones 2014. It has been suggested that global climate change may be affecting the frequency of tropical storms. The data here show the number of tropical cyclones (including hurricanes) assigned official names by the National Hurricane Center. Is there evidence of a change?

1995–2004	2005–2014
19, 13, 7, 14, 12, 14, 15, 12, 16, 15	27, 9, 15, 16, 9, 19, 18, 19, 14, 9

a) Write the null and alternative hypotheses.
b) Are the conditions for hypothesis testing satisfied?
c) If so, test the hypothesis.

40. Hurricanes 2014. Exercise 39 considered possible changes in the numbers of tropical cyclones that have grown large enough to be officially named. Regardless of the *number* of storms, has the *fraction* of named storms that grew to hurricane strength changed? Here are those fractions for the same periods as those considered in the previous exercise.

1995–2004	2005–2014
0.579, 0.692, 0.429, 0.714, 0.667, 0.571, 0.600, 0.333, 0.438, 0.600	0.556, 0.556, 0.400, 0.500, 0.333, 0.632, 0.333, 0.526, 0.667, 0.142

a) Write the null and alternative hypotheses.
b) Are the conditions for hypothesis testing satisfied?
c) If so, test the hypothesis.

T 41. Gingko test. A pharmaceutical company is producing and marketing a ginkgo biloba supplement to enhance memory. In an experiment to test the product, subjects were assigned randomly to take ginkgo biloba supplements or a placebo. Their memory was tested to see whether it improved. Here are boxplots comparing the two groups and some computer output from a two-sample *t*-test computed for the data.

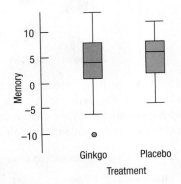

```
2-Sample t-Test of μ_G − μ_P > 0
Difference Between Means = −0.9914
t-Statistic = −1.540 w/196 df
P = 0.9374
```

a) Explain in this context what the P-value means.
b) State your conclusion about the effectiveness of ginkgo biloba.
c) Proponents of ginkgo biloba continue to insist that it works. What type of error do they claim your conclusion makes?

T 42. Designated hitter 2014. American League baseball teams play their games with the designated hitter rule, meaning that pitchers do not bat. The league believes that replacing the pitcher, traditionally a weak hitter, with another player in the batting order produces more runs and generates more interest among fans. The data provided in the file on the CD include the average numbers of runs scored per game (*Runs per game*) by American League and National League teams for the 2014 season (www.baseball-reference.com).

American League			National League		
Team	**R/G**	**HR**	**Team**	**R/G**	**HR**
BAL	4.35	211	ARI	3.8	118
BOS	3.91	123	ATL	3.54	123
CHW	4.07	155	CHC	3.79	157
CLE	4.13	142	CIN	3.67	131
DET	4.67	155	COL	4.66	186
KCR	4.02	95	HOU	3.88	163
LAA	4.77	155	LAD	4.43	134
MIN	4.41	128	MIA	3.98	122
NYY	3.91	147	MIL	4.01	150
OAK	4.5	146	NYM	3.88	125
SEA	3.91	136	PHI	3.82	125
TBR	3.78	117	PIT	4.21	156
TEX	3.93	111	SDP	3.3	109
TOR	4.46	177	SFG	4.1	132
			STL	3.82	105
			WSN	4.23	152

a) Create an appropriate display of these data. What do you see?
b) With a 95% confidence interval, estimate the mean number of runs scored by American League teams.
c) With a 95% confidence interval, estimate the mean number of runs scored by National League teams.
d) Explain why you should not use two separate confidence intervals to decide whether the two leagues differ in average number of runs scored.

43. Productivity. A factory hiring people to work on an assembly line gives job applicants a test of manual agility. This test counts how many strangely shaped pegs the applicant can fit into matching holes in a one-minute period. The table summarizes the data by gender of the job applicant. Assume that all conditions necessary for inference are met.

	Male	**Female**
Number of Subjects	50	50
Pegs Placed:		
Mean	19.39	17.91
SD	2.52	3.39

a) Find 95% confidence intervals for the average number of pegs that males and females can each place.
b) Those intervals overlap. What does this suggest about any gender-based difference in manual agility?
c) Find a 95% confidence interval for the difference in the mean number of pegs that could be placed by men and women.
d) What does this interval suggest about any gender-based difference in manual agility?

e) The two results seem contradictory. Which method is correct: doing two-sample inference, or doing one-sample inference twice?
f) Why don't the results agree?

T **44. Designated hitter 2014, part 2.** Do the data in Exercise 42 suggest that the American League's designated hitter rule may lead to more runs per game scored?

a) Write the null and alternative hypotheses.
b) Find a 95% confidence interval for the difference in mean runs per game, and interpret your interval.
c) Test the hypothesis stated above in part a and find the P-value.
d) Interpret the P-value and state your conclusion. Does the test suggest that the American League scores more runs on average?

T **45. Water hardness.** In an investigation of environmental causes of disease, data were collected on the annual mortality rate (deaths per 100,000) for males in 61 large towns in England and Wales. In addition, the water hardness was recorded as the calcium concentration (parts per million, ppm) in the drinking water. The data set also notes for each town whether it was south or north of Derby. Is there a significant difference in mortality rates in the two regions? Here are the summary statistics.

```
Summary of:              mortality
For categories in:       Derby

Group    Count    Mean      Median    StdDev
North    34       1631.59   1631      138.470
South    27       1388.85   1369      151.114
```

a) Test appropriate hypotheses and state your conclusion.
b) The boxplots of the two distributions show a possible outlier among the data north of Derby. What effect might that have had on your test?

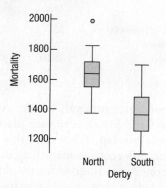

46. Technology investment. The Price-to-Book-value ratio is often used by investors to indicate whether a stock's price is particularly high or low relative to the value of the company. But different market sectors expect different Price/Book values. Here are data on technology companies (biz.yahoo.com/p/8conameu.html accessed in May 2013).

We'll compare applications software companies with manufacturers of peripheral equipment. For both, we have taken the logarithm of the Price/Book ratio to make the distributions more nearly symmetric.

```
Group        Count   Mean       Median     StdDev
Peripherals  19      0.339035   0.264818   0.342258
Software     102     0.566206   0.516535   0.527435
```

a) State and test appropriate hypotheses and state your conclusion. (So you don't need to compute that strange degrees of freedom formula, the correct df is 36.)
b) Here are boxplots. What more do they tell you about the distributions? Do you think this changes your conclusions in part a?

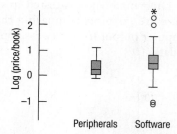

T **47. Job satisfaction.** A company institutes an exercise break for its workers to see if this will improve job satisfaction, as measured by a questionnaire that was given to a random sample of workers to assess their satisfaction.

Worker Number	Job Satisfaction Index	
	Before	After
1	34	33
2	28	36
3	29	50
4	45	41
5	26	37
6	27	41
7	24	39
8	15	21
9	15	20
10	27	37

a) Identify the procedure you would use to assess the effectiveness of the exercise program and check to see if the conditions allow for the use of that procedure.
b) Test an appropriate hypothesis and state your conclusion.

T **48. ERP effectiveness.** When implementing a packaged Enterprise Resource Planning (ERP) system, many companies report that the module they first install is Financial Accounting. Among the measures used to gauge the effectiveness of their ERP system implementation is acceleration of the financial close process. Below is a sample of 8 companies that report their average time (in weeks) to financial close before and after the implementation of their ERP system.

Company	Before	After
1	6.5	4.2
2	7.0	5.9
3	8.0	8.0
4	4.5	4.0
5	5.2	3.8
6	4.9	4.1
7	5.2	6.0
8	6.5	4.2

a) Identify the procedure you would use to assess the effectiveness of the ERP system and check to see if the conditions allow for the use of that procedure.
b) Test an appropriate hypothesis and state your conclusion.

49. Delivery time. A small appliance company is interested in comparing delivery times of their product during two months. They are concerned that the summer slow-downs in August cause delivery times to lag during this month. Given the following delivery times (in days) of their appliances to the customer for a random sample of 6 orders each month, test if delivery times differ across these two months.

June	54	49	68	66	62	62
August	50	65	74	64	68	72

50. Branding. The *Journal of Applied Psychology* reported on a study that examined whether the content of TV shows influenced the ability of viewers to recall brand names of items featured in the commercials. The researchers randomly assigned volunteers to watch one of three programs, each containing the same nine commercials. One of the programs had violent content, another sexual content, and the third neutral content. After the shows ended, the subjects were asked to recall the brands of products that were advertised. The table shows summaries for how many brands were recalled.

	Program Type		
	Violent	Sexual	Neutral
n	108	108	108
Mean	2.08	1.71	3.17
SD	1.87	1.76	1.77

a) Do these results indicate that viewer memory for ads may differ depending on program content? Test the hypothesis that there is no difference in ad memory between programs with sexual content and those with violent content. State your conclusion.
b) Is there evidence that viewer memory for ads may differ between programs with sexual content and those with neutral content? Test an appropriate hypothesis and state your conclusion.

51. Ad campaign. You are a consultant to the marketing department of a business preparing to launch an ad campaign for a new product. The company can afford to run ads during one TV show, and has decided not to sponsor a show with sexual content (see Exercise 50). You create a confidence interval for the difference in mean number of brand names remembered between the groups watching violent shows and those watching neutral shows.

```
Two-Sample t
95% CI for μ_viol − μ_neut: (−1.578, −0.602)
```

a) At the meeting of the marketing staff, you have to explain what this output means. What will you say?
b) What advice would you give the company about the upcoming ad campaign?

52. Branding, part 2. In the study described in Exercise 50, the researchers also contacted the subjects again, 24 hours later, and asked them to recall the brands advertised. Results for the number of brands recalled are summarized in the table.

	Program Type		
	Violent	Sexual	Neutral
No. of Subjects	101	106	103
Mean	3.02	2.72	4.65
SD	1.61	1.85	1.62

a) Is there a significant difference in viewers' abilities to remember brands advertised in shows with violent vs. neutral content?
b) Find a 95% confidence interval for the difference in mean number of brand names remembered between the groups watching shows with sexual content and those watching neutral shows. Interpret your interval in this context.

53. Ad recall. In Exercises 50 and 52, we see the number of advertised brand names people recalled immediately after watching TV shows and 24 hours later. Strangely enough, it appears that they remembered more about the ads the next day. Should we conclude this is true in general about people's memory of TV ads?

a) Suppose one analyst conducts a two-sample hypothesis test to see if memory of brands advertised during violent TV shows is higher 24 hours later. The P-value is 0.00013. What might she conclude?
b) Explain why her procedure was inappropriate. Which of the assumptions for inference was violated?
c) How might the design of this experiment have tainted these results?
d) Suggest a design that could compare immediate brand name recall with recall one day later.

54. Hybrid SUVs. The Chevy Tahoe Hybrid got a lot of attention when it first appeared. It is a relatively high-priced hybrid SUV that makes use of the latest technologies for fuel efficiency. One of the more popular hybrid SUVs on the market is the modestly priced Ford Escape Hybrid. A consumer group was interested in comparing the gas mileage of these two models. In order to do so, each vehicle was driven on the same 10 routes that combined both highway and city streets. The results showed that the mean mileage for the Chevy Tahoe was 29 mpg and for the Ford Escape it was 31 mpg. The standard deviations were 3.2 mpg and 2.5 mpg, respectively.

a) An analyst for the consumer group computed a two-sample 95% t-interval for the difference between the two means as $(-0.71, 4.71)$. What conclusion would he reach based on this analysis?

b) Why is this procedure inappropriate? What assumption is violated?

c) In what way do you think this may have impacted the results?

55. The Internet. The National Assessment in Education Program compared science scores for students who had home Internet access with the scores of those who did not, as shown in the graph. They report that the differences are statistically significant.

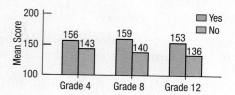

a) Explain what "statistically significant" means in this context.

b) If their conclusion is incorrect, which type of error did the researchers commit?

c) Does this prove that using the Internet at home can improve a student's performance in science?

d) What companies might be interested in this information?

56. Credit card debt public or private. The average credit card debt carried by college students was compared at public versus private universities. It was reported that a significant difference existed between the two types of institutions and that students at private universities carried higher credit card debt.

a) Explain what "statistically significant" means in this context.

b) If this conclusion is incorrect, which type of error was committed?

c) Does this prove that students who choose to attend public institutions will carry lower credit card debt?

57. Pizza sales. A national food product company believes that it sells more frozen pizza during the winter months than during the summer months. Average weekly sales for a sample of stores in the Baltimore area over a three-year period provided the following data for sales volume (in pounds) during the two seasons.

Season	N	Mean	StDev	Minimum	Maximum
Winter	38	31234	13500	15312	73841
Summer	40	22475	8442	12743	54706

a) How much difference is there between the mean amount of this brand of frozen pizza sold (in pounds) between the two seasons? (Assume that this time frame represents typical sales in the Baltimore area.)

b) Construct and interpret a 95% confidence interval for the difference between weekly sales during the winter and summer months.

c) Suggest factors that might have influenced the sales of the frozen pizza during the winter months.

58. More pizza sales. Here's some additional information about the pizza sales data presented in Exercise 57. It is generally thought that sales spike during the weeks leading up to AFC and NFC football championship games, as well as leading up to the Super Bowl at the end of January each year. If we omit those 6 weeks of sales from this three-year period of weekly sales, the summary statistics look like this.

Season	N	Mean	StDev	Minimum	Maximum
Winter	32	28995	9913	15312	48354
Summer	40	22475	8442	12743	54706

Do sales appear to be higher during the winter months after omitting those weeks most influenced by football championship games?

a) Write the null and alternative hypotheses.

b) Test the null hypotheses and state your conclusion.

c) Suggest additional factors that may influence pizza sales not accounted for in this exercise.

T 59. Olympic heats 2012. In Olympic running events, preliminary heats are determined by random draw, so we should expect the ability level of runners in the various heats to be about the same, on average. The table gives the times (in seconds) for the 800-m men's run in the 2012 Olympics in London for preliminary heats 1 and 3. Is there any evidence that the mean time to finish is different for randomized heats? Explain. Be sure to include a discussion of assumptions and conditions for your analysis.

Country	Name	Time	Heat
BOT	Nigel Amos	105.9	1
BRA	Fabiano Pecanha	106.3	1
ESP	Luis Alberto Marco	106.9	1
USA	Khadevis Robinson	107.2	1
POL	Marcin Lewandowski	107.6	1
RUS	Ivan Tukhtachev	109.8	1
GUM	Derek Mandell	118.9	1
SUD	Abubaker Kaki	105.5	3
KEN	Timothy Kitum	105.7	3
KSA	Abdulaziz Ladan Mohammed	106.1	3
CUB	Andy González	106.2	3
GBR	Gareth Warburton	107.0	3
HUN	Tamás Kazi	107.1	3
GER	Sören Ludolph	108.6	3
VAN	Arnold Sorina	114.3	3

60. Swimming heats 2012. In Exercise 59, we looked at the times in two different heats for the 800-m men's run from the 2012 Olympics. Unlike track events, swimming heats are *not* determined at random. Instead, swimmers are seeded so that better swimmers are placed in later heats. Here are the times (in seconds) for the women's 400-m freestyle from heats 3 and 5.

Country	Name	Time (secs)	Heat
GBR	Rebecca Adlington	245.7	3
USA	Chloe Sutton	247.1	3
ESP	Mireia Belmonte	248.2	3
HUN	Eva Risztov	249.1	3
CHN	Li Xuanxu	250.9	3
ROU	Camelia Potec	251.4	3
GBR	Joanne Jackson	251.5	3
RSA	Wendy Trott	251.6	3
FRA	Camille Muffat	243.3	5
USA	Allison Schmitt	243.3	5
NZL	Lauren Boyle	243.6	5
DEN	Lotte Friis	244.2	5
ESP	Melanie Costa	246.8	5
AUS	Bronte Barratt	248.0	5
HUN	Boglarka Kapas	250.0	5
RUS	Elena Sokolova	252.2	5

Do these results suggest that the mean times of heat 5 are faster than heat 3? Explain. Include a discussion of assumptions and conditions for your analysis.

61. Tee tests. Does it matter what kind of tee a golfer places the ball on? The company that manufactures "Stinger" tees claims that the thinner shaft and smaller head will lessen resistance and drag, reducing spin and allowing the ball to travel farther. Golf Laboratories, Inc. compared the distance traveled by golf balls hit off regular wooden tees to those hit off Stinger tees. All the balls were struck by the same golf club using a robotic device set to swing the club head at approximately 95 miles per hour. Summary statistics from the test are shown in the table. Assume that 6 balls were hit off each tee and that the data were suitable for inference. Is there evidence that balls hit off the Stinger tees would have a higher initial velocity?

		Total Distance (yards)	Ball Velocity (mph)	Club Velocity (mph)
Regular Tee	Mean	227.17	127.00	96.17
	SD	2.14	0.89	0.41
Stinger Tee	Mean	241.00	128.83	96.17
	SD	2.76	0.41	0.52

62. Tee tests, part 2. Given the test results on golf tees described in Exercise 61, is there evidence that balls hit off Stinger tees travel farther? Assume that 6 balls were hit off each tee and that the data are suitable for inference.

63. Marketing slogan. A company is considering marketing their classical music as "music to study by." Is this a valid slogan? In a study conducted by some Statistics students, 62 people were randomly assigned to listen to rap music, music by Mozart, or no music while attempting to memorize objects pictured on a page. They were then asked to list all the objects they could remember. Here are summary statistics for each group.

	Rap	Mozart	No Music
Count	29	20	13
Mean	10.72	10.00	12.77
SD	3.99	3.19	4.73

a) Does it appear that it is better to study while listening to Mozart than to rap music? Test an appropriate hypothesis and state your conclusion.
b) Create a 90% confidence interval for the mean difference in memory score between students who study to Mozart and those who listen to no music at all. Interpret your interval.

64. Marketing slogan, part 2. Using the results of the experiment described in Exercise 63, does it matter whether one listens to rap music while studying, or is it better to study without music at all?

a) Test an appropriate hypothesis and state your conclusion.
b) If you concluded there is a difference, estimate the size of that difference with a 90% confidence interval and explain what your interval means.

65. Mutual fund returns 2013. You have heard that if you leave your money in mutual funds for a longer period of time, you will see a greater return. So you would like to compare the 3-year and 5-year returns of a random sample of mutual funds to see if indeed, your return is expected to be greater if you leave your money in the funds for 5 years.

a) Using the data provided, check the conditions for this test.
b) Write the null and alternative hypotheses for this test.
c) Test the hypothesis and find the P-value if appropriate.
d) Find a 95% confidence interval for the mean difference.

66. Mutual fund returns 2013, part 2. An investor now tells you that if you leave your money in as long as 10 years, you will see an even greater return, so you would like to compare the 5-year and 10-year returns of a random sample of mutual funds to see if your return is expected to be greater if you leave your money in the funds for 10 years.

a) Using the data provided, check the conditions for this test.
b) Write the null and alternative hypotheses for this test.
c) Test the hypothesis and find the P-value if appropriate.
d) Find a 95% confidence interval for the mean difference.

67. Real estate, two towns. Residents of neighboring towns in a state in the United States have an ongoing disagreement over who lays claim to the higher average price of a single-family home. Since you live in one of these towns, you decide to obtain a random sample of homes listed for sale with a major local realtor to investigate if there is actually any difference in the average home price.

a) Using the data provided, check the conditions for this test.
b) Write the null and alternative hypotheses for this test.
c) Test the hypothesis and find the P-value.
d) What is your conclusion?

68. Real estate, two towns, bigger sample. Residents of one of the towns discussed in Exercise 67 claim that since their town is much smaller, the sample size should be increased. Instead of random sampling 30 homes, you decide to sample 42 homes from the database to test the difference in the mean price of single-family homes in these two towns.

a) Using the data provided on the CD, check the conditions for this test.
b) Write the null and alternative hypotheses for this test.
c) Test the hypothesis and find the P-value.
d) What is your conclusion? Did the sample size make a difference?
e) Use the pooled *t*-test to test the hypothesis and compare your answer to parts c and d.

69. Designated hitter 2014, part 3. For the same reasons identified in Exercise 42, a friend of yours claims that the average number of home runs hit per game is higher in the American League than in the National League. Using the same 2014 data as in Exercises 42 and 44, you decide to test your friend's theory.

a) Using the data provided, check the conditions for this test.

b) Write the null and alternative hypotheses for this test.
c) Test the hypothesis and find the P-value.
d) What is your conclusion?
e) Use the pooled *t*-test to test the hypothesis and compare your answer to parts c and d.

70. Cloud seeding. It has long been a dream of farmers to summon rain when it is needed for their crops. Crop losses to drought have significant economic impact. One possibility is cloud seeding in which chemicals are dropped into clouds in an attempt to induce rain. Simpson, Alsen, and Eden (*Technometrics*, 1975) report the results of trials in which clouds were seeded and the amount of rainfall recorded. The authors report on 26 seeded (Group 2) and 26 unseeded (Group 1) clouds. Each group has been sorted in order of the amount of rainfall, largest amount first. Here are two possible tests to study the question of whether cloud seeding works.

```
Paired t-Test of μ(1 - 2)
Mean of Paired Differences = -277.4
t-Statistic = -3.641 w/25 df  p = 0.0012
2-Sample t-test of μ1 - μ2
Difference Between Means = -277.4
t-Statistic = -1.998 w/33 df  p = 0.0538
```

a) Which of these tests is appropriate for these data? Explain.
b) Using the test you selected, state your conclusion.

71. Online insurance. After seeing countless commercials claiming one can get cheaper car insurance from an online company, a local insurance agent was concerned that he might lose some customers. To investigate, he randomly selected profiles (type of car, coverage, driving record, etc.) for 10 of his clients and checked online price quotes for their policies. The comparisons are shown in the table.

Local	Online	PriceDiff
568	391	177
872	602	270
451	488	-37
1229	903	326
605	677	-72
1021	1270	-249
783	703	80
844	789	55
907	1008	-101
712	702	10

His statistical software produced the following summaries (where *PriceDiff = Local − Online*):

Variable	Count	Mean	StdDev
Local	10	799.200	229.281
Online	10	753.300	256.267
PriceDiff	10	45.900	175.663

At first, the insurance agent wondered whether there was some kind of mistake in this output. He thought the Pythagorean Theorem of Statistics should work for finding the standard deviation of the price differences—in other words, that

$$SD(Local - Online) = \sqrt{SD^2(Local) + SD^2(Online)}.$$

But when he checked, he found that

$$\sqrt{(229.281)^2 + (256.267)^2} = 343.864,$$

not 175.663 as given by the software. Tell him where his mistake is.

72. Windy. Alternative sources of energy are of increasing interest throughout the energy industry. Wind energy has great potential. But appropriate sites must be found for the turbines. To select the site for an electricity-generating wind turbine, wind speeds were recorded at several potential sites every 6 hours for a year. Two sites not far from each other looked good. Each had a mean wind speed high enough to qualify, but we should choose the site with a higher average daily wind speed. Because the sites are near each other and the wind speeds were recorded at the same times, we should view the speeds as paired. Here are the summaries of the speeds (in miles per hour):

Variable	Count	Mean	StdDev
site2	1114	7.452	3.586
site4	1114	7.248	3.421
site2 − site4	1114	0.204	2.551

Is there a mistake in this output? Why doesn't the Pythagorean Theorem of Statistics work here? In other words, shouldn't

$$SD(site2 - site4) = \sqrt{SD^2(site2) + SD^2(site4)}?$$

But $\sqrt{(3.586)^2 + (3.421)^2} = 4.956$, not 2.551 as given by the software. Explain why this happened.

73. Online insurance, part 2. In Exercise 71, we saw summary statistics for 10 drivers' car insurance premiums quoted by a local agent and an online company. Here are displays for each company's quotes and for the difference (*Local − Online*):

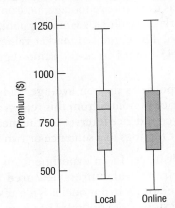

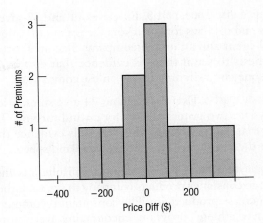

a) Which of the summaries would help you decide whether the online company offers cheaper insurance? Why?
b) The standard deviation of *PriceDiff* is quite a bit smaller than the standard deviation of prices quoted by either the local or online companies. Discuss why.
c) Using the information you have, discuss the assumptions and conditions for inference with these data.

74. Windy, part 2. In Exercise 72, we saw summary statistics for wind speeds at two sites near each other, both being considered as locations for an electricity-generating wind turbine. The data, recorded every 6 hours for a year, showed each of the sites had a mean wind speed high enough to qualify, but how can we tell which site is best? Here are some displays:

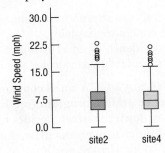

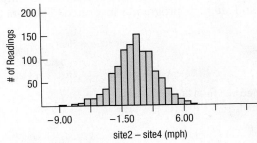

a) The boxplots show outliers for each site, yet the histogram shows none. Discuss why.
b) Which of the summaries would you use to select between these sites? Why?
c) Using the information you have, discuss the assumptions and conditions for paired *t* inference for these data. (*Hint:* Think hard about the independence assumption in particular.)

75. Online insurance, part 3. Exercises 71 and 73 give summaries and displays for car insurance premiums quoted by a local agent and an online company. Test an appropriate hypothesis to see if there is evidence that drivers might save money by switching to the online company.

T 76. Windy, part 3. Exercises 72 and 74 give summaries and displays for two potential sites for a wind turbine. Test an appropriate hypothesis to see if there is evidence that either of these sites has a higher average wind speed.

77. Employee athletes. An ergonomics consultant is engaged by a large consumer products company to see what they can do to increase productivity. The consultant recommends an "employee athlete" program, encouraging every employee to devote 5 minutes an hour to physical activity. The company worries that the gains in productivity will be offset by the loss in time on the job. They'd like to know if the program increases or decreases productivity. To measure it, they monitor a random sample of 145 employees who word process, measuring their hourly key strokes both before and after the program is instituted. Here are the data:

	Keystrokes per Hour		
	Before	After	Difference (After − Before)
Mean	1534.2	1556.9	22.7
SD	168.5	149.5	113.6
N	145	145	145

a) What are the null and alternative hypotheses?
b) What can you conclude? Explain.
c) Give a 95% confidence interval for the mean change in productivity (as measured by keystrokes per hour).

78. Employee athletes, part 2. A small company, on hearing about the employee athlete program (see Exercise 77) at the large company down the street, decides to try it as well. To measure the difference in productivity, they measure the average number of keystrokes per hour of 23 employees before and after the program is instituted. The data follow:

	Keystrokes per Hour		
	Before	After	Difference (After − Before)
Mean	1497.3	1544.8	47.5
SD	155.4	136.7	122.8
N	23	23	23

a) Is there evidence to suggest that the program increases productivity?
b) Give a 95% confidence interval for the mean change in productivity (as measured by keystrokes per hour).
c) Given this information and the results of Exercise 77, what recommendations would you make to the company about the effectiveness of the program?

T 79. Exercise equipment. A leading manufacturer of exercise equipment wanted to collect data on the effectiveness of their equipment. An August 2001 article in the journal *Medicine and Science in Sports and Exercise* compared how long it would take men and women to burn 200 calories during light or heavy workouts on various kinds of exercise equipment. The results summarized in the table are the average times for a group of physically active young men and women whose performances were measured on a representative sample of exercise equipment.

		Average Minutes to Burn 200 Calories			
		Hard Exertion		Light Exertion	
		Men	Women	Men	Women
Machine Type	Treadmill	12	17	14	22
	X-C Skier	12	16	16	23
	Stair Climber	13	18	20	37
	Rowing Machine	14	16	21	25
	Exercise Rider	22	24	27	36
	Exercise Bike	16	20	29	44

a) On average, how many minutes longer than a man must a woman exercise at a light exertion rate in order to burn 200 calories? Find a 95% confidence interval.
b) Estimate the average number of minutes longer a woman must work out at light exertion than at heavy exertion to get the same benefit. Find a 95% confidence interval.
c) These data are actually averages rather than individual times. How might this affect the margins of error in these confidence intervals?

T 80. Market value. Real estate agents want to set correctly the price of a house that's about to go on the real estate market. They must choose a price that strikes a balance between one that is so high that the house takes too long to sell and one that's so low that not enough value will go to the homeowner. One appraisal method is the "Comparative Market Analysis" approach by which the market value of a house is based on recent sales of similar homes in the neighborhood. Because no two houses are exactly the same, appraisers have to adjust comparable homes for such features as extra square footage, bedrooms, fireplaces, upgrading, parking facilities, swimming pool, lot size, location, and so on. The appraised market values and the selling prices of 45 homes from the same region are in the data file.

a) Test the hypothesis that on average, the market value and the sale price of homes from this region are the same.
b) Find a 95% confidence interval for the mean difference.
c) Explain your findings in a sentence or two in context.

T 81. Stopping distance. In an experiment on braking performance, a tire manufacturer measured the stopping distance for one of its tire models. On a test track, a car made repeated stops from 60 miles per hour. Twenty tests were run, 10 each on both dry and wet pavement, with results shown in the table. (Note that actual *braking distance*, which takes into account the driver's reaction time, is much longer, typically nearly 300 feet at 60 mph!)

Stopping Distance (ft)	
Dry Pavement	**Wet Pavement**
145	211
152	191
141	220
143	207
131	198
148	208
126	206
140	177
135	183
133	223

a) Find a 95% confidence interval for the mean dry pavement stopping distance. Be sure to check the appropriate assumptions and conditions, and explain what your interval means.

b) Find a 95% confidence interval for the mean increase in stopping distance on wet pavement. Be sure to check the appropriate assumptions and conditions, and explain what your interval means.

82. Stopping distances, again. For another test of the tires in Exercise 81, the company tried them on 10 different cars, recording the stopping distance for each car on both wet and dry pavement. Results are shown in the following table.

	Stopping Distance (ft)	
Car #	**Dry Pavement**	**Wet Pavement**
1	150	201
2	147	220
3	136	192
4	134	146
5	130	182
6	134	173
7	134	202
8	128	180
9	136	192
10	158	206

a) Find a 95% confidence interval for the mean dry pavement stopping distance. Be sure to check the appropriate assumptions and conditions, and explain what your interval means.

b) Find a 95% confidence interval for the mean increase in stopping distance on wet pavement. Be sure to check the appropriate assumptions and conditions, and explain what your interval means.

83. Airline "bumping" 2014. Commercial airlines overbook flights, selling more tickets than they have seats, because a sizeable number of reservation holders don't show up in time for their flights. But sometimes, there are more passengers wishing to board than there are seats. Most airlines try to entice travelers to voluntarily give up their seats in return for free travel or other awards, but they do have to "bump" some travelers involuntarily. Of course, they don't like to offend passengers by bumping, so they are constantly trying to improve their systems for predicting how many passengers will show up. Have the rates of "bumping" changed? Here are data on the number of passengers involuntarily denied boarding ("bumping" is not the approved term) per 10,000 passengers during the periods of January to June in 2013 and 2014 by airline.

Airline	2013	2014
Hawaiian Airlines	0.18	0.00
Jetblue Airways	0.00	0.04
Delta Air Lines	0.52	0.07
Virgin America	0.01	0.10
Alaska Airlines	0.29	0.28
American	0.36	0.38
US Airways	0.55	0.53
United Airlines	1.37	0.52
Southwest	0.66	0.57
Airtran	1.31	0.75
Frontier Airlines	1.20	1.41
Envoy Air	1.34	1.68
Expressjet Airlines	2.46	1.79
Skywest Airlines	3.38	2.00

*2014 is the final year that American and US Airways are recorded separately before their merger.

2014 is the final year than Southwest and Airtran are recorded separately before their merger.

Envoy Air was formerly American Eagle.

(Source: http://www.dot.gov/sites/dot.gov/files/docs/2015MarchATCR_1.pdf)

a) Are these paired data? Why or why not?

b) Was there a statistically significant change in the number of passengers involuntarily denied boarding per 10,000 passengers?

84. Grocery prices. WinCo Foods, a large discount grocery retailer in the western United States, promotes itself as the lowest priced grocery retailer. In newspaper ads WinCo Foods published a price comparison for products between WinCo and several competing grocery retailers. One of the retailers compared against WinCo was Walmart, also known as a low price competitor. WinCo selected a variety of products, listed the price of the product charges at each retailer, and showed the sales receipt to prove the prices at WinCo were the lowest in the area. A sample of the products and their price comparison at both WinCo and Walmart are shown in the following table:

Item	WinCo Price	Walmart Price
Bananas (lb)	0.42	0.56
Red Onions (lb)	0.58	0.98
Mini Peeled Carrots (1 lb bag)	0.98	1.48
Roma Tomatoes (lb)	0.98	2.67
Deli Tater Wedges (lb)	1.18	1.78
Beef Cube Steak (lb)	3.83	4.11
Beef Top Round London Broil (lb)	3.48	4.12
Pillsbury Devils Food Cake Mix (18.25 oz)	0.88	0.88
Lipton Rice and Sauce Mix (5.6 oz)	0.88	1.06
Sierra Nevada Pale Ale (12 – 12 oz bottles)	12.68	12.84
GM Cheerios Oat Clusters (11.3 oz)	1.98	2.74
Charmin Bathroom Tissue (12 roll)	5.98	7.48
Bumble Bee Pink Salmon (14.75 oz)	1.58	1.98
Pace Thick & Chunky Salsa, Mild (24 oz)	2.28	2.78
Nalley Chili, Regular w/Beans (15 oz)	0.78	0.78
Challenge Butter (lb quarters)	2.18	2.58
Kraft American Singles (12 oz)	2.27	2.27
Yuban Coffee FAC (36 oz)	5.98	7.56
Totino's Pizza Rolls, Pepperoni (19.8 oz)	2.38	2.42
Rosarita Refried Beans, Original (16 oz)	0.68	0.73
Barilla Spaghetti (16 oz)	0.78	1.23
Sun-Maid Mini Raisins (14 – .5 oz)	1.18	1.36
Jif Peanut Butter, Creamy (28 oz)	2.54	2.72
Dole Fruit Bowl, Mixed Fruit (4 – 4 oz)	1.68	1.98
Progresso Chicken Noodle Soup (19 oz)	1.28	1.38
Precious Mozzarella Ball, Part Skim (16 oz)	3.28	4.23
Mrs. Cubbison Seasoned Croutons (6 oz)	0.88	1.12
Kellogg's Raisin Bran (20 oz)	1.98	2.50
Campbell's Soup at Hand, Cream of Tomato (10.75 oz)	1.18	1.26

a) Do the prices listed indicate that, on average, prices at WinCo are lower than prices at Walmart?

b) At the bottom of the price list, the following statement appears: "Though this list is not intended to represent a typical weekly grocery order or a random list of grocery items, WinCo continues to be the area's low price leader." Why do you think WinCo added this statement?

c) What other comments could be made about the statistical validity of the test on price comparisons given in the ad?

JUST CHECKING ANSWERS

1 H_0: $\mu_{eyes} - \mu_{flowers} = 0$

2 ✓ **Independence Assumption:** The amount paid by one person should be independent of the amount paid by others.

  ✓ **Randomization Condition:** This study was observational. Treatments alternated a week at a time and were applied to the same group of office workers.

  ✓ **Nearly Normal Condition:** We don't have the data to check, but it seems unlikely there would be outliers in either group.

  ✓ **Independent Groups Assumptions:** The same workers were recorded each week, but week-to-week independence is plausible.

3 H_A: $\mu_{eyes} - \mu_{flowers} \neq 0$. An argument could be made for a one-sided test because the research hypothesis was that eyes would improve honest compliance.

4 Office workers' compliance in leaving money to pay for food at an office coffee station was different when a picture of eyes was placed behind the "honesty box" than when the picture was one of flowers.

5 These are independent groups sampled at random, so use a two-sample t confidence interval to estimate the size of the difference.

6 If the same random sample of students was sampled both in the first year and again in the fourth year of their university experience, then this would be a paired t-test.

7 A male and female are selected from each work group. The question calls for a paired t-test.

8 Since we have no reason to believe that public and private companies are not independent, we should use a two-sample test.

9 Since the same 50 companies are surveyed twice to examine a change in variables over time, this would be a paired t-test.

13

Inference for Counts: Chi-Square Tests

SAC Capital

Hedge funds, like mutual funds and pension funds, pool investors' money in an attempt to make profits. Unlike these other funds, however, hedge funds are not required to register with the U.S. Securities and Exchange Commission (SEC) because they issue securities in "private offerings" only to "qualified investors" (investors with either $1 million in assets or annual income of at least $200,000).

Hedge funds don't necessarily "hedge" their investments against market moves. But typically these funds use multiple, often complex, strategies to exploit inefficiencies in the market. For these reasons, hedge fund managers have the reputation for being obsessive traders, and the SEC has recently begun to investigate their practices.

One of the most successful hedge funds is SAC Capital, which was founded by Steven (Stevie) A. Cohen in 1992 with nine employees and $25 million in assets under management (AUM). SAC Capital returned annual gains of 40% or more through much of the 1990s and is now reported to have more than 1000 employees and nearly $14 billion in assets under management. According to *Forbes*, Cohen's $9.3 billion fortune ranks him as the 41st wealthiest American and 117th richest person in the world.

Cohen, a legendary figure on Wall Street, is known for taking advantage of any information he can find and for turning that information into profit. Unlike most hedge funds, which trade by computer and pay fractions of a cent per trade, SAC still trades the "old fashioned way," paying 3 to 5¢. It is generally agreed to be the largest payer of fees on Wall Street, a largesse that may earn it tips from traders.

In March of 2013, SAC agreed to pay (without admitting guilt) a fine of $616 million—the largest fine in the history of the Securities and Exchange Commission—to settle charges of insider trading in just two trades. In October of 2013, SAC pleaded guilty and agreed to pay fines of $1.2B. The government also forced SAC to stop managing money for outside investors. But don't feel too sorry for Mr. Cohen. The firm will continue to manage his fortune of over $9B. He is still the 44th richest person in the United States and 109th richest in the world, with an estimated net worth of over $11 billion as of April 2015.

I n a business as competitive as hedge fund management, information is gold. Being the first to have information and knowing how to act on it can mean the difference between success and failure. Hedge fund managers look for small advantages everywhere, hoping to exploit inefficiencies in the market and to turn those inefficiencies into profit.

Wall Street has plenty of "wisdom" about market patterns. For example, investors are advised to watch for "calendar effects," certain times of year or days of the week that are particularly good or bad: "As goes January, so goes the year" and "Sell in May and go away." Some analysts claim that the "bad period" for holding stocks is from the sixth trading day of June to the fifth-to-last trading day of October. Of course, there is also Mark Twain's advice:

> October. This is one of the peculiarly dangerous months to speculate in stocks. The others are July, January, September, April, November, May, March, June, December, August, and February.
>
> —Mark Twain, Pudd'nhead Wilson, 1894

One common claim is that stocks show a weekly pattern. For example, some argue that there is a *weekend effect* in which stock returns on Mondays are often lower than those of the immediately preceding Friday. Are patterns such as this real? We have the data, so we can check. Between October 1, 1928 and February 25, 2013, there were 21,186 trading sessions. Let's first see how many trading days fell on each day of the week. It's not exactly 20% for each day because of holidays. The distribution of days is shown in Table 13.1.

Of these 21,186 trading sessions, 11,113, or about 52% of the days, saw a gain in the Dow Jones Industrial Average (DJIA). To test for a pattern, we need a model. The model comes from the supposition that any day is as likely to show a gain as any other. In any sample of positive or "up" days, we should expect to see the same distribution of days as in Table 13.1—in other words, about 19.31% of "up" days would be Mondays, 20.26% would be Tuesdays, and so on. Here is the distribution of days in one such random sample of 1110 "up" days.

Of course, we expect some variation. We wouldn't expect the proportions of days in the two tables to match exactly. In our sample, the percentage of Mondays in Table 13.2 is slightly lower than in Table 13.1, and the proportion of Fridays is a little higher. Are these deviations enough for us to declare that there is a recognizable pattern?

Day of the Week	Count	% of Day
Monday	4090	19.305
Tuesday	4293	20.263
Wednesday	4317	20.377
Thursday	4253	20.075
Friday	4233	19.980

Table 13.1 The distribution of days of the week among the 21,186 trading days from October 1, 1928 to February 25, 2013. We expect about 20% to fall in each day, with minor variations due to holidays and other events.

Data in **Stock Market Patterns**

13.1   Goodness-of-Fit Tests

To address this question, we test the table's **goodness-of-fit**, where *fit* refers to the null model proposed. Here, the null model is that there is no pattern, that the distribution of *up* days should be the same as the distribution of trading days overall. (If there were no holidays or other closings, that would just be 20% for each day of the week.)

Day of the Week	Count	% of Days in the Sample of "Up" days
Monday	195	17.568
Tuesday	223	20.090
Wednesday	239	21.532
Thursday	223	20.090
Friday	230	20.721

Table 13.2 The distribution of days of the week for a sample of 1110 "up" trading days selected at random from October 1, 1928 to February 25, 2013. If there is no pattern, we would expect the proportions here to match fairly closely the proportions observed among all trading days in Table 13.1.

Assumptions and Conditions

Data for a goodness-of-fit test are organized in tables, and the assumptions and conditions reflect that. Rather than having an observation for each individual, we typically work with summary counts in categories. Here, the individuals are trading days, but rather than list all 1110 trading days in the sample, we have totals for each weekday.

Counted Data Condition. The data must be counts for the categories of a categorical variable. This might seem a silly condition to check. But many kinds of values can be assigned to categories, and it is unfortunately common to find the methods of this chapter applied incorrectly (even by business professionals) to proportions or quantities just because they happen to be organized in a two-way table. So check to be sure that you really have counts.

Independence Assumption

Independence Assumption. The counts in the cells should be independent of each other. You should think about whether that's reasonable. If the data are a random sample you can simply check the randomization condition.

Randomization Condition. The individuals counted in the table should be a random sample from some population. We need this condition if we want to generalize our conclusions to that population. We took a random sample of 1110 trading days on which the DJIA rose. That lets us assume that the market's performance on any one day is independent of performance on another. If we had selected 1110 consecutive trading days, there would be a risk that market performance on one day could affect performance on the next, or that an external event could affect performance for several consecutive days.

Sample Size Assumption

Sample Size Assumption. We must have enough data for the methods to work. We usually just check the following condition:

Expected Cell Frequency Condition. We should expect to see at least 5 individuals in each cell. The expected cell frequency condition should remind you of—and is, in fact, quite similar to—the condition that np and nq be at least 10 when we test proportions.

Chi-Square Model

We have observed a count in each category (weekday). We can compute the number of up days we'd *expect* to see for each weekday if the null model were true. For the

> **Expected Cell Frequencies**
> Companies often want to assess the relative successes of their products in different regions. However, a company whose sales regions had 100, 200, 300, and 400 representatives might not expect equal sales in all regions. They might expect observed sales to be proportional to the size of the sales force. The null hypothesis in that case would be that the proportions of sales were 1/10, 2/10, 3/10, and 4/10, respectively. With 500 total sales, their expected counts would be 50, 100, 150, and 200.

trading days example, the expected counts come from the null hypothesis that the up days are distributed among weekdays just as trading days are. Of course, we could imagine almost any kind of model and base a null hypothesis on that model.

To decide whether the null model is plausible, we look at the differences between the expected values from the model and the counts we observe. We wonder: Are these differences so large that they call the model into question, or could they have arisen from natural sampling variability? We denote the *differences* between these observed and expected counts, (*Obs* − *Exp*). As we did with variance, we square them. That gives us positive values and focuses attention on any cells with large differences. Because the differences between observed and expected counts generally get larger the more data we have, we also need to get an idea of the *relative* sizes of the differences. To do that, we divide each squared difference by the expected count for that cell.

The test statistic, called the **chi-square (or chi-squared) statistic**, is found by adding up the sum of the squares of the deviations between the observed and expected counts divided by the expected counts:

$$\chi^2 = \sum_{all\ cells} \frac{(Obs - Exp)^2}{Exp}.$$

The chi-square statistic is denoted χ^2, where χ is the Greek letter chi (pronounced kī). The resulting family of sampling distribution models is called the **chi-square models**.

The members of this family of models differ in the number of degrees of freedom. The number of degrees of freedom for a goodness-of-fit test is $k - 1$, where k is the number of cells—in this example, 5 weekdays.

We will use the chi-square statistic only for testing hypotheses, not for constructing confidence intervals. A small chi-square statistic means that our model fits the data well, so a small value gives us no reason to doubt the null hypothesis. If the observed counts don't match the expected counts, the statistic will be large. If the calculated statistic value is large enough, we'll reject the null hypothesis. So the chi-square test is always one-sided. What could be simpler? Let's see how it works.

NOTATION ALERT

We compare the counts *observed* in each cell with the counts we *expect* to find. The usual notation uses *Obs* and *Exp* as we've used here. The expected counts are found from the null model.

NOTATION ALERT

The only use of the Greek letter χ in Statistics is to represent the chi-square statistic and the associated sampling distribution. This violates the general rule that Greek letters represent population parameters. Here we are using a Greek letter simply to name a family of distribution models and a statistic.

FOR EXAMPLE   Goodness-of-fit test

Atara manages 8 call center operators at a telecommunications company. To develop new business, she gives each operator a list of randomly selected phone numbers of rival phone company customers. She also provides the operators with a script that tries to convince the customers to switch providers. Atara notices that some operators have found more than twice as many new customers as others, so she suspects that some of the operators are performing better than others.

The 120 new customer acquisitions are distributed as follows:

Operator	1	2	3	4	5	6	7	8
New Customers	11	17	9	12	19	18	13	21

QUESTION   Is there evidence to suggest that some of the operators are more successful than others?

ANSWER   Atara has randomized the potential new customers to the operators so the Randomization Condition is satisfied. The data are counts and there are at least 5 expected counts in each cell, so we can apply a chi-square goodness-of-fit test to the null hypothesis that the operator performance is uniform and that each of the operators will convince the same number of customers. Specifically we expect each operator to have converted 1/8 of the 120 customers that switched providers.

Operator	1	2	3	4	5	6	7	8
Observed	11	17	9	12	19	18	13	21
Expected	15	15	15	15	15	15	15	15
Observed-Expected	−4	2	−6	−3	4	3	−2	6
(Obs-Exp)^2	16	4	36	9	16	9	4	36
(Obs-Exp)^2/Exp	16/15 = 1.07	4/15 = 0.27	36/15 = 2.40	9/15 = 0.60	16/15 = 1.07	9/15 = 0.60	4/15 = 0.27	36/15 = 2.40

$$\sum \frac{(Obs - Exp)^2}{Exp} = 1.07 + 0.27 + 2.40 + \cdots + 2.40 = 8.68$$

The number of degrees of freedom is $k - 1 = 7$.

$$P(\chi_7^2 > 8.68) = 0.2765$$

8.68 is not a surprising value for a chi-square statistic with 7 degrees of freedom. So, we fail to reject the null hypothesis that the operators actually find new customers at different rates.

BY HAND The chi-square calculation

Here are the steps to calculate the chi-square statistic:

1. **Find the expected values.** These come from the null hypothesis model. Every null model gives a hypothesized proportion for each cell. The expected value is the product of the total number of observations times this proportion. (The result need not be an integer.)

2. **Compute the residuals.** Once you have expected values for each cell, find the residuals, $Obs - Exp$.

3. **Square the residuals.** $(Obs - Exp)^2$

4. **Compute the components.** Find $\dfrac{(Obs - Exp)^2}{Exp}$ for each cell.

5. **Find the sum of the components.** That's the chi-square statistic,

$$\chi^2 = \sum_{all\ cells} \frac{(Obs - Exp)^2}{Exp}.$$

6. **Find the degrees of freedom.** It's equal to the number of cells minus one.

7. **Test the hypothesis.** Large chi-square values mean lots of deviation from the hypothesized model, so they give small P-values. Look up the critical value from a table of chi-square values such as Table X in Appendix B, or use technology to find the P-value directly.

(continued)

The steps of the chi-square calculations are often laid out in tables. Use one row for each category, and columns for observed counts, expected counts, residuals, squared residuals, and the contributions to the chi-square total:

	A	B	C	D	E	F	G
1	Day	Pop%	Obs	Exp	Resid	Resid^2	Component
2	Monday	0.19305	195	214.28774	-19.28774	372.016806	1.736062
3	Tuesday	0.20263	223	224.92353	-1.92353	3.6999846	0.01645
4	Wednesday	0.20377	239	226.18097	12.81903	164.327567	0.7265312
5	Thursday	0.20075	223	222.82781	-0.17219	0.0296491	0.0001331
6	Friday	0.1998	230	221.77995	8.22005	67.5692381	0.3046679

Table 13.3 Calculations for the chi-square statistic in the trading days example can be performed conveniently in Microsoft Excel 2013. Set up the calculation in the first row and Fill Down, and then find the sum of the rightmost column. The CHIDIST function looks up the chi-square total to find the P-value.

GUIDED EXAMPLE Stock Market Patterns

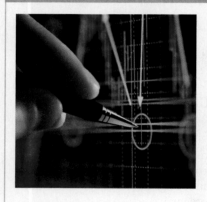

We have counts of the "up" days for each day of the week. The economic theory we want to investigate is whether there is a pattern in "up" days. So, our null hypothesis is that across all days in which the DJIA rose, the days of the week are distributed as they are across all trading days. (As we saw, the trading days are not quite *evenly* distributed because of holidays, so we use the *trading days* percentages as the null model.) The alternative hypothesis is that the observed percentages are *not* uniform. The test statistic looks at how closely the observed data match this idealized situation.

PLAN **Setup** State what you want to know. Identify the variables and context.	We want to know whether the distribution for "up" days differs from the null model (the trading days distribution). We have the number of times each weekday appeared among a random sample of 1110 "up" days.
Hypotheses State the null and alternative hypotheses. For χ^2 tests, it's usually easier to state the hypotheses in words than in symbols.	H_O: The days of the workweek are distributed among the up days as they are among all trading days. H_A: The trading days model does not fit the up days distribution.
Model Think about the assumptions and check the conditions.	✓ **Counted Data Condition** We have counts of the days of the week for all trading days and for the "up" days. ✓ **Independence Assumption** We have no reason to expect that one day's performance will affect another's, but to be safe we've taken a random sample of days. The randomization should make them far enough apart to alleviate any concerns about dependence. ✓ **Randomization Condition** We have a random sample of 1110 days from the time period. ✓ **Expected Cell Frequency Condition** All the expected cell frequencies are much larger than 5.

Specify the sampling distribution model.

Name the test you will use.

The conditions are satisfied, so we'll use a χ^2 model with $5 - 1 = 4$ degrees of freedom and do a **chi-square goodness-of-fit test.**

DO

Mechanics To find the expected number of days, we take the fraction of each weekday from all days and multiply by the number of "up" days.

For example, there were 4090 Mondays out of 21,186 trading days.

So, we'd expect there would be $1110 \times 4090/21186$ or 214.2877 Mondays among the 1110 "up" days.

Each cell contributes a value equal to $\dfrac{(Obs - Exp)^2}{Exp}$ to the chi-square sum.

Add up these components. If you do it by hand, it can be helpful to arrange the calculation in a table or spreadsheet.

The P-value is the probability in the upper tail of the χ^2 model. It can be found using software or a table (see Table X in Appendix B).

Large χ^2 statistic values correspond to small P-values, which would lead us to reject the null hypothesis, but the value here is not particularly large.

The expected values are:

Monday: 214.28774
Tuesday: 224.92353
Wednesday: 226.18097
Thursday: 222.82781
Friday: 221.77995

And we observe:

Monday: 195
Tuesday: 223
Wednesday: 239
Thursday: 223
Friday: 230

$$\chi^2 = \frac{(195 - 214.28774)^2}{214.28774} + \cdots + \frac{(230 - 221.77995)^2}{221.77995}$$
$$= 2.7838$$

Using Table X in Appendix B, we find that for a significance level of 5% and 4 degrees of freedom, we'd need a value of 9.488 or more to have a P-value less than .05. Our value of 2.7838 is less than that.

Using a computer to generate the P-value, we find:

$$P\text{-}value = P(\chi^2_4 > 2.7838) = 0.5946$$

REPORT

Conclusion Link the P-value to your decision. Be sure to say more than a fact about the distribution of counts. State your conclusion in terms of what the data mean.

MEMO

Re: Stock market patterns

Our investigation of whether there are day-of-the-week patterns in the behavior of the DJIA in which one day or another is more likely to be an "up" day found no evidence of such a pattern. Our statistical test indicated that a pattern such as the one found in our sample of trading days would happen by chance about 60% of the time.

We conclude that there is, unfortunately, no evidence of a pattern that could be used to guide investment in the market. We were unable to detect a "weekend" or other day-of-the-week effect in the market.

13.2   Interpreting Chi-Square Values

When we calculated χ^2 for the trading days example, we got 2.7838. That value was not large for 4 degrees of freedom, so we were unable to reject the null hypothesis. In general, what *is* big for a χ^2 statistic?

Think about how χ^2 is calculated. In every cell, any deviation from the expected count contributes to the sum. Large deviations generally contribute more, but if there are a lot of cells, even small deviations can add up, making the χ^2 value larger. So the more cells there are, the higher the value of χ^2 has to be before it becomes significant. For χ^2, the decision about how big is big depends on the number of degrees of freedom.

Unlike the Normal and t families, χ^2 models are skewed. Curves in the χ^2 family change both shape and center as the number of degrees of freedom grows. For example, Figure 13.1 shows the χ^2 curves for 5 and for 9 degrees of freedom.

Figure 13.1 The χ^2 curves for 5 and 9 degrees of freedom.

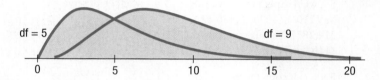

Notice that the value $\chi^2 = 10$ might seem somewhat extreme when there are 5 degrees of freedom, but appears to be rather ordinary for 9 degrees of freedom. Here are two simple facts to help you think about χ^2 models:

- The mode is at $\chi^2 = $ df $- 2$. (Look at the curves; their peaks are at 3 and 7.)
- The expected value (mean) of a χ^2 model is its number of degrees of freedom. That's a bit to the right of the mode—as we would expect for a skewed distribution.

Goodness-of-fit tests are often performed by people who have a theory of what the proportions *should* be in each category and who believe their theory to be true. In some cases, unlike our market example, there isn't an obvious null hypothesis against which to test the proposed model. So, unfortunately, in those cases, the only null hypothesis available is that the proposed theory is true. And as we know, the hypothesis testing procedure allows us only to reject the null or fail to reject it. We can never confirm that a theory is in fact true; we can never confirm the null hypothesis.

At best, we can point out that the data are consistent with the proposed theory. But this doesn't prove the theory. The data *could* be consistent with the model even if the theory were wrong. In that case, we fail to reject the null hypothesis but can't conclude anything for sure about whether the theory is true.

Why Can't We Prove the Null?

A student claims that it really makes no difference to your starting salary how well you do in your Statistics class. He surveys recent graduates, categorizes them according to whether they earned an A, B, or C in Statistics, and according to whether their starting salary is above or below the median for their class. He calculates the proportion above the median salary for each grade. His null model is that in each grade category, 50% of students are above the median. With 40 respondents, he gets a P-value of 0.07 and declares that Statistics grades don't matter. But then more questionnaires are returned, and he finds that with a sample size of 70, his P-value is 0.04. Can he ignore the second batch of data? Of course not. If he could do that, he could claim almost any null model was true just by having too little data to refute it.

13.3   Examining the Residuals

Chi-square tests are always one-sided. The chi-square statistic is always positive, and a large value provides evidence against the null hypothesis (because it shows that the fit to the model is *not* good), while small values provide little evidence that the model doesn't fit. In another sense, however, chi-square tests are really many-sided; a large statistic doesn't tell us *how* the null model doesn't fit. In our market theory example, if we had rejected the uniform model, we wouldn't have known *how* it failed. Was it because there were not enough Mondays represented, or was it that all five days showed some deviation from the uniform?

When we reject a null hypothesis in a goodness-of-fit test, we can examine the residuals in each cell to learn more. In fact, whenever we reject a null hypothesis, it's a good idea to examine the residuals. (We don't need to do that when we fail to reject because when the χ^2 value is small, all of its components must have been small.) Because we want to compare residuals for cells that may have very different counts, we standardize the residuals. We know the mean residual is zero,[1] but we need to know each residual's standard deviation. When we tested proportions, we saw a link between the expected proportion and its standard deviation. For counts, there's a similar link. To standardize a cell's residual, we divide by the square root of its expected value:[2]

$$\frac{(Obs - Exp)}{\sqrt{Exp}}.$$

Notice that these **standardized residuals** are the square roots of the components we calculated for each cell, with the plus (+) or the minus (−) sign indicating whether we observed more or fewer cases than we expected.

The standardized residuals give us a chance to think about the underlying patterns and to consider how the distribution differs from the model. Now that we've divided each residual by its standard deviation, they are z-scores. If the null hypothesis were true, we could even use the 68–95–99.7 Rule to judge how extraordinary the large ones are.

Ordinarily, you should only look at the residuals if the chi-square value is significantly large. But for illustration, we'll take a peek at the standardized residuals for the trading days data:

	Standardized Residual $= \dfrac{(Obs - Exp)}{\sqrt{Exp}}$
Monday	−1.3176
Tuesday	−0.1283
Wednesday	0.8524
Thursday	0.0115
Friday	0.5520

Table 13.4   Standardized residuals.

As we would expect from the small chi-square value, none of these values is remarkable.

---

[1] Residual = Observed − Expected. Because the total of the expected values is the same as the observed total, the residuals must sum to zero.

[2] It can be shown mathematically that the square root of the expected value estimates the appropriate standard deviation.

FOR EXAMPLE Examining residuals from a chi-square test

QUESTION In the call center example (see page 420), examine the residuals to see if any operators stand out as having especially strong or weak performance.

ANSWER Because we failed to reject the null hypothesis, we don't expect any of the standardized residuals to be large, but we will examine them nonetheless.

The standardized residuals are the square roots of the components (from the bottom row of the table in the Example on page 421).

Standardized Residuals	−1.03	0.52	−1.55	−0.77	1.03	0.77	−0.52	1.55

As we expected, none of the residuals are large. Even though Atara notices that some of the operators enrolled more than twice the number of new customers as others, the variation is typical (within two standard deviations) of what we would expect if all their performances were, in fact, equal.

13.4 The Chi-Square Test of Homogeneity

In Chapter 2 we saw data from a Pew Research survey about social networking in several countries. The question of interest to businesses asked whether respondents had access to and used social networking. Responses were "yes" (use social networking), "no", and "not available". We organized the responses into a contingency table that looked like this:

Country	No	Yes	N/A	Total
Britain	336	529	153	**1018**
Egypt	70	300	630	**1000**
Germany	460	340	200	**1000**
Russia	90	500	420	**1010**
U.S.	293	506	212	**1011**
Total	**1249**	**2175**	**1615**	**5039**

Table 13.5 Responses to a question about use and availability of social networking. ("N/A" = not available)

The natural question to ask is whether there are any differences among the countries (and, if so, what those are). We can start with a stacked bar chart.

Figure 13.2 The data of Table 13.5 in a stacked bar chart. The bars are easy to compare because they have roughly the same total counts.

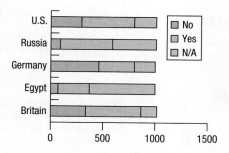

In Figure 13.2, Egypt stands out for having less availability of social networking. Germany may have an unusually large number of people saying they don't use social networking despite its availability.

But are the observed differences in the percentages real or just natural sampling variation? Our null hypothesis is that the proportions choosing each alternative are the same for each country. To test that hypothesis, we use a **chi-square test of homogeneity**. This is just another chi-square test. It turns out that the mechanics of the test of this hypothesis are nearly identical to the chi-square goodness-of-fit test we just saw in Section 13.1. The difference is that the goodness-of-fit test compared our observed counts to the expected counts from a *given* model. The test of homogeneity, by contrast, has a null hypothesis that the distributions are the same for all the groups. The test examines the differences between the observed counts and what we'd expect under that assumption of homogeneity.

For example, 529/1018 or 52% of British respondents said that they use social networking. If the distributions were homogeneous across the five countries (as the null hypothesis asserts), then that proportion should be the same for all five countries. So 52% of the 1010 Russian respondents, or 525, would have said that they use social networking. That's the number we'd *expect* under the rule hypothesis.

Working in this way, we (or, more likely, the computer) can fill in expected values for each cell. The following table shows these expected values for each response and each country.

	No	Yes	N/A	Total
Britain	252.328	439.403	326.269	**1018**
Egypt	247.867	431.633	320.500	**1000**
Germany	247.867	431.633	320.500	**1000**
Russia	250.345	435.950	323.705	**1010**
U.S.	250.593	436.381	324.026	**1011**
Total	**1249**	**2175**	**1615**	**5039**

Table 13.6   Expected values for the responses. Because these are theoretical values, they don't have to be integers.

The term *homogeneity* refers to the hypothesis that things are the same. Here, we ask whether the distribution of responses is the same across the five countries. The chi-square test looks for differences large enough to step beyond what we might expect from random sample-to-sample variation. It can reveal a large deviation in a single category or small but persistent differences over all the categories—or anything in between.

Assumptions and Conditions

The assumptions and conditions are the same as for the chi-square test for goodness-of-fit. The **Counted Data Condition** says that these data must be counts. You can never perform a chi-square test on a quantitative variable. For example, if Pew had asked how many hours in a typical day respondents spent on social networking, we wouldn't be able to use a chi-square test to determine whether the mean time expenditures in the five countries were the same.[3]

Independence Assumption. The counts must be independent of each other. We can check the **Randomization Condition**. Here, we have random samples, so we *can* assume that the observations are independent.

---
[3] To do that, you'd use a method called Analysis of Variance, which we'll see in Chapter 20.

> **Large Samples and Chi-Square Tests**
>
> Whenever we test any hypothesis, a very large sample size means that small effects have a greater chance of being statistically significant. This is especially true for chi-square tests. So it's important to look at the effect sizes when the null hypothesis is rejected to see if the differences are practically significant. Don't rely only on the P-value when making a business decision. This applies to many of the examples in this chapter which have large sample sizes typical of those seen in today's business environment.

The **Sample Size Assumption** can be checked with the **Expected Cell Frequency Condition**, which says that the expected count in each cell must be at least 5. Here, our samples are certainly large enough.

Following the pattern of the goodness-of-fit test, we compute the component for each cell of the table:

$$\text{Component} = \frac{(Obs - Exp)^2}{Exp}.$$

Summing these components across all cells gives the chi-square value:

$$\chi^2 = \sum_{all\ cells} \frac{(Obs - Exp)^2}{Exp}.$$

The degrees of freedom are different than they were for the goodness-of-fit test. For a test of homogeneity, there are $(R - 1) \times (C - 1)$ degrees of freedom, where R is the number of rows and C is the number of columns.

In our example, we have $4 \times 2 = 8$ degrees of freedom. We'll need the degrees of freedom to find a P-value for the chi-square statistic.

BY HAND How to find expected values

In a contingency table, to test for homogeneity, we need to find the expected values when the null hypothesis is true. To find the expected value for row i and column j, we take:

$$Exp_{ij} = \frac{Total_{Row\ i} \times Total_{Col\ j}}{Table\ Total}.$$

Here's an example:

Suppose we ask 100 people, 40 men and 60 women, to name their magazine preference: *Sports Illustrated*, *Cosmopolitan*, or *The Economist*, with the following result, shown in Microsoft Excel 2013 below.

	A	B	C	D	E
1	Actual	SI	Cosmo	Economist	Total
2	Men	25	5	10	40
3	Women	10	45	5	60
4	Total	35	50	15	100

Then, for example, the expected value under homogeneity for Men (row 1) who prefer *The Economist* (column 3) would be:

$$Exp_{13} = \frac{40 \times 15}{100} = 6$$

Performing similar calculations for all cells gives the expected values:

6	Expected	SI	Cosmo	Economist	Total
7	Men	14	20	6	40
8	Women	21	30	9	60
9	Total	35	50	15	100

It is a simple matter to do these calculations in a spreadsheet (as we have done here).

For the social networking data, the chi-square value is 1049, with 8 degrees of freedom. We don't need a table to know that this is large enough to be statistically significant. So we can conclude that these five countries are not homogeneous in the use of social networking.

If you find that simply rejecting the hypothesis of homogeneity is a bit unsatisfying, you're in good company. It's hardly a shock that responses to this question differ from country to country especially with sample sizes this large. What we'd really like to know is where the differences were and how big they were. The test for homogeneity doesn't answer these interesting questions, but it does provide some evidence that can help us. A look at the standardized residuals can help identify cells that don't match the homogeneity pattern.

FOR EXAMPLE   Testing homogeneity

QUESTION Although annual inflation in the United States has been low for several years, many Americans fear that inflation may return. In May 2010, a Gallup poll asked 680 adults nationwide, "Are you very concerned, somewhat concerned, or not at all concerned that inflation will climb?" Does the distribution of responses appear to be the same for Conservatives as Liberals?

Ideology	Very Concerned	Somewhat Concerned	Not At All Concerned	Total
Conservative	232	83	25	340
Liberal	143	126	71	340
Total	375 (55.15%)	209 (30.74%)	96 (14.12%)	680

ANSWER This is a test of homogeneity, testing whether the distribution of responses is the same for the two ideological groups. The data are counts, the Gallup poll selected adults randomly (stratified by ideology), and all expected cell frequencies are much greater than 5 (see table below).

There are $(3 - 1) \times (2 - 1)$ or 2 degrees of freedom.

If the distributions were the same, we would expect each cell to have expected values that are 55.15%, 30.74%, and 14.12% of the row totals for Very Concerned, Somewhat Concerned, and Not at all Concerned, respectively. These values can be computed explicitly from:

$$Exp_{ij} = \frac{TotalRow_i \times TotalCol_j}{Table\ Total}$$

So, in the first cell (Conservative, Very Concerned):

$$Exp_{11} = \frac{TotalRow_1 \times TotalCol_1}{Table\ Total} = \frac{340 \times 375}{680} = 187.5$$

Expected counts for all cells are:

Expected Numbers	Very Concerned	Somewhat Concerned	Not At All Concerned
Conservative	187.5	104.5	48.0
Liberal	187.5	104.5	48.0

(continued)

The components $\dfrac{(Obs - Exp)^2}{Exp}$ are:

Components	Very Concerned	Somewhat Concerned	Not At All Concerned
Conservative	10.56	4.42	11.02
Liberal	10.56	4.42	11.02

Summing these gives $\chi^2 = 10.56 + 4.42 + \cdots + 11.02 = 52.00$, which, with 2 df, has a P-value of < 0.0001.

We, therefore, reject the hypothesis that the distribution of responses is the same for Conservatives and Liberals.

13.5   Comparing Two Proportions

In Chapter 2 we saw data from a survey comparing the reasons that men and women watch the Super Bowl. Advertisers spend huge amounts on Super Bowl ads and, consequently, care deeply about who watches them. If we extract the data on the 708 survey respondents who answered and said they would be watching the Super Bowl, we get the data in Table 13.7.

A chi-square test of homogeneity gives 37.26 with 1 degree of freedom. That has a P-value ≤ 0.0001, so we can reject the null hypothesis of no difference.

When we have a 2×2 table like this, we are really just comparing two proportions. In this example, $154/352 = 43.75\%$ of the women and $79/356 = 22.19\%$ of the men say they watch the Super Bowl primarily for the commercials. In this special case of testing the equality of two proportions, there is a z-test, which is equivalent and gives the same P-value.

	Female	Male	Total
Game	198	277	475
Commercials	154	79	233
Total	352	356	708

Table 13.7   Counts of respondents categorized by sex and whether they watch the Super Bowl primarily for the game or for the commercials.

Confidence Interval for the Difference of Two Proportions

The approach of working directly with proportions offers an advantage. As we saw, 43.75% of the women and 22.19% of the men watch the Super Bowl primarily for the commercials. That's a difference of 21.56%.

If we knew the standard error of that quantity, we could use a z-statistic to construct a confidence interval for the true difference in the population. It's not hard to find the standard error. All we need is the formula:[4]

$$SE(\hat{p}_1 - \hat{p}_2) = \sqrt{\dfrac{\hat{p}_1\hat{q}_1}{n_1} + \dfrac{\hat{p}_2\hat{q}_2}{n_2}}$$

> **Confidence Interval for the Difference of Two Proportions**
>
> When the conditions are met, we can find the confidence interval for the difference of two proportions, $p_1 - p_2$. The confidence interval is
>
> $$(\hat{p}_1 - \hat{p}_2) \pm z^*SE(\hat{p}_1 - \hat{p}_2),$$
>
> where we find the standard error of the difference as
>
> $$SE(\hat{p}_1 - \hat{p}_2) = \sqrt{\dfrac{\hat{p}_1\hat{q}_1}{n_1} + \dfrac{\hat{p}_2\hat{q}_2}{n_2}}$$
>
> from the observed proportions.
>
> The critical value z^* depends on the particular confidence level that you specify.

---

[4] The standard error of the difference is found from the general fact that the variance of a difference of two independent quantities is the *sum* of their variances. See Chapter 6 for details.

The confidence interval has the same form as the confidence interval for a single proportion, with this new standard error:

$$(\hat{p}_1 - \hat{p}_2) \pm z^* SE(\hat{p}_1 - \hat{p}_2).$$

For the Super Bowl survey, a 95% confidence interval for the true difference between the proportion of women and the proportion of men preferring the commercials is:

$$(0.4375 - 0.2219) \pm 1.96 \times \sqrt{\frac{0.4375 \times 0.5625}{352} + \frac{0.2219 \times 0.7781}{356}}$$

$$= 0.2156 \pm 0.0674$$

or from 14.82% to 28.30%.

We can be 95% confident that between 14.82% and 28.30% more women than men watch the Super Bowl for the commercials more than for the game.

This confidence interval helps us to address the question of whether the difference matters. That depends on the reason we are asking the question. The confidence interval shows the *effect size*—or at least the interval of plausible values for the effect size. A company planning to advertise on the Super Bowl might consider this difference large enough to consider in designing their ads. Be sure to consider the effect size whenever you make a business decision based on rejecting a null hypothesis.

FOR EXAMPLE A confidence interval for the difference of proportions

QUESTION In the Gallup poll on inflation (see page 429), 68.2% (232 of 340) of those identifying themselves as Conservative were very concerned about the rise of inflation, but only 42.1% (143 of 340) of Liberals responded the same way. That's a difference of 26.2% in this sample of 680 adults. Find a 95% confidence interval for the true difference.

ANSWER The confidence interval can be found from:

$$(\hat{p}_C - \hat{p}_L) \pm z^* SE(\hat{p}_C - \hat{p}_L) \text{ where } SE(\hat{p}_C - \hat{p}_L) = \sqrt{\frac{\hat{p}_C \hat{q}_C}{n_C} + \frac{\hat{p}_L \hat{q}_L}{n_L}}$$

$$= \sqrt{\frac{(0.682)(0.338)}{340} + \frac{(0.421)(0.579)}{340}} = 0.037.$$

Since we know the 95% confidence critical value for z is 1.96, we have:

$$0.262 \pm 1.96\,(0.037) = (0.189, 0.335).$$

In other words, we are 95% confident that the proportion of Conservatives who are very concerned about inflation is between 18.9% and 33.5% higher than the same proportion of Liberals.

13.6 Chi-Square Test of Independence

If the importance people place on their personal appearance varies a great deal by age, that might be crucial for the marketing department of a global cosmetics company. Fortunately, we have data. The GfK Roper Reports® Worldwide Survey asked 30,000 consumers in 23 countries about their attitudes on health, beauty, and other personal values. One question participants were asked was how important their personal appearance is to them.

We might think of the different age groups as categories to be compared with a test of homogeneity. But it makes more sense to think of *Age* as a second variable

				Age				
		13–19	**20–29**	**30–39**	**40–49**	**50–59**	**60+**	**Total**
Appearance	**7—Extremely Important**	396	337	300	252	142	93	**1520**
	6	325	326	307	254	123	86	**1421**
	5	318	312	317	270	150	106	**1473**
	4—Average Importance	397	376	403	423	224	210	**2033**
	3	83	83	88	93	54	45	**446**
	2	37	43	53	58	37	45	**273**
	1—Not At All Important	40	37	53	56	36	52	**274**
	Total	**1596**	**1514**	**1521**	**1406**	**766**	**637**	**7440**

Table 13.8   Responses to the question about personal appearance by age group.

whose value has been measured for each respondent along with his or her response to the appearance question. Asking whether the distribution of responses changes with *Age* now raises the question of whether the variables personal *Appearance* and *Age* are independent.

Whenever we have two variables in a contingency table like this, the natural test is a **chi-square test of independence**. Mechanically, this chi-square test is identical to a test of homogeneity. The difference between the two tests is in how we think of the data and, thus, what conclusion we draw.

Here we ask whether the response to the personal appearance question is independent of age. Remember, that for any two events, **A** and **B**, to be independent, the probability of event **A** given that event **B** occurred must be the same as the probability of event **A**. Here, this means the probability that a randomly selected respondent thinks personal appearance is extremely important doesn't depend on his or her age group. Of course, from a table based on data, the probabilities will never be exactly the same. But to tell whether they are different enough, we use a chi-square test of independence.

Here we have two categorical variables measured on a single population. For the homogeneity test, we had a single categorical variable measured independently on two or more populations. Now we ask a different question: "Are the variables independent?" rather than "Are the groups homogeneous?" These are subtle differences, but they are important when we draw conclusions.

> **Homogeneity or Independence?**
>
> The only difference between the test for homogeneity and the test for independence is in the decision you need to make.

Assumptions and Conditions

Of course, we still need counts and enough data so that the expected counts are at least five in each cell.

If we're interested in the independence of variables, we usually want to generalize from the data to some population. In that case, we'll need to check that the data are a representative random sample from that population.

GUIDED EXAMPLE   Personal Appearance and Age

We want to help marketers discover whether a person's age influences how they respond to the question: "How important is seeking the utmost attractive appearance to you?" We have data from a randomized survey conducted in 30 countries. We'll look at just five of these countries. We have the values of *Age* in six age categories. We will view *Age* as a variable, and ask whether the variables *Age* and *Appearance* are independent.

| | PLAN | **Setup** State what you want to know.

Identify the variables and context. | We want to know whether the categorical variables personal *Appearance* and *Age* are statistically independent. We have a contingency table of 7440 respondents from a sample of five countries. |

Setup State what you want to know.

Identify the variables and context.

We want to know whether the categorical variables personal *Appearance* and *Age* are statistically independent. We have a contingency table of 7440 respondents from a sample of five countries.

Hypotheses State the null and alternative hypotheses.

We perform a test of independence when we suspect the variables may not be independent. We are making the claim that knowing the respondents' age will change the distribution of their response to the question about personal *Appearance*, and testing the null hypothesis that it is *not* true.

H_O: Personal *Appearance* and *Age* are independent.[5]

H_A: Personal *Appearance* and *Age* are not independent.

Model Check the conditions.

✓ **Counted Data Condition** We have counts of individuals categorized on two categorical variables.

✓ **Randomization Condition** These data are from a randomized survey conducted in 30 countries. We have data from five of them. Although they are not an SRS, the samples within each country were selected to avoid biases.

✓ **Expected Cell Frequency Condition** The expected values are all much larger than 5.

The table below shows the expected counts for each cell. The expected counts are calculated exactly as they were for a test of homogeneity; in the first cell, for example, we expect $\frac{1520}{7440} = 20.43\%$ of 1596, which is 326.065.

		Expected Values					
		Age					
		13–19	**20–29**	**30–39**	**40–49**	**50–59**	**60+**
	7—Extremely Important	326.065	309.312	310.742	287.247	156.495	130.140
	6	304.827	289.166	290.503	268.538	146.302	121.664
Appearance	5	315.982	299.748	301.133	278.365	151.656	126.116
	4—Average Importance	436.111	413.705	415.617	384.193	209.312	174.062
	3	95.674	90.759	91.178	84.284	45.919	38.186
	2	58.563	55.554	55.811	51.591	28.107	23.374
	1—Not At All Important	58.777	55.758	56.015	51.780	28.210	23.459

(continued)

---

[5] As in other chi-square tests, the hypotheses are usually expressed in words, without parameters. The hypothesis of independence itself tells us how to find expected values for each cell of the contingency table. That's all we need.

The stacked bar graph shows that the response seems to be dependent on *Age*. Older people tend to think personal appearance is less important than younger people.

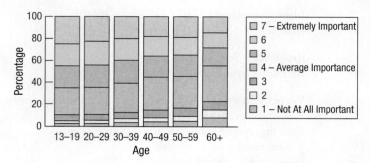

	Specify the model. Name the test you will use.	(The counts are shown in Table 13.8.) We'll use a χ^2 model with $(7-1) \times (6-1) = 30$ df and do a **chi-square test of independence**.
DO	**Mechanics** Calculate χ^2 and find the P-value using software. The shape of a chi-square model depends on its degrees of freedom. Even with 30 df, this chi-square statistic is extremely large, so the resulting P-value is small.	$$\chi^2 = \sum_{all\ cells} \frac{(Obs - Exp)^2}{Exp} = 170.7762$$ $P\text{-value} = P(\chi^2_{30} > 170.7762) < 0.001$
REPORT	**Conclusion** Link the P-value to your decision. State your conclusion.	MEMO **Re: Investigation of the relationship between age of consumer and attitudes about personal appearance** It appears from our analysis of the Roper survey that attitudes on personal *Appearance* are not independent of *Age*. It seems that older people find personal appearance less important than younger people do (on average in the five countries selected).

We rejected the null hypothesis of independence between *Age* and attitudes about personal *Appearance*. With a sample size this large, we can detect very small deviations from independence, so it's almost guaranteed that the chi-square test will reject the null hypothesis. Examining the residuals can help you see the cells that deviate farthest from independence. To make a meaningful business decision, you'll have to look at effect sizes as well as the P-value. We should also look at each country's data individually since country-to-country differences could affect marketing decisions.

Suppose the company was specifically interested in deciding how to split advertising resources between the teen market and the 30–39-year-old market. How much of a difference is there between the proportions of those in each age group that rated personal *Appearance* as very important (responding either 6 or 7)?

For that we'll need to construct a confidence interval on the difference. From Table 13.8, we find that the percentages of those answering 6 and 7 are 45.18%

and 39.91% for the teen and 30–39-year-old groups, respectively. The 95% confidence interval is:

$$(\hat{p}_1 - \hat{p}_2) \pm z^*SE(\hat{p}_1 - \hat{p}_2)$$

$$= (0.4518 - 0.3991) \pm 1.96 \times \sqrt{\frac{(0.4518)(0.5482)}{1596} + \frac{(0.3991)(0.6009)}{1521}}$$

$$= (0.018, 0.087), \text{ or } (1.8\% \text{ to } 8.7\%)$$

This is a statistically significant difference, but now we can see that the difference may be as small as 1.8%. When deciding how to allocate advertising expenditures, it is important to keep these estimates of the effect size in mind.

FOR EXAMPLE   A chi-square test of independence

QUESTION   In May 2010, the Gallup poll asked U.S. adults their opinion on whether they are in favor of or opposed to using profiling to identify potential terrorists at airports, a practice used routinely in Israel, but not in the United States. Does opinion depend on age? Or are opinion and age independent? Here are numbers similar to the ones Gallup found (the percentages are the same, but the totals have been changed to make the calculations easier).

	Age				
	18–29	**30–49**	**50–64**	**65+**	**Total**
Favor	57	66	77	87	**287**
Oppose	43	34	23	13	**113**
Total	**100**	**100**	**100**	**100**	**400**

ANSWER   The null hypothesis is that *Opinion* and *Age* are independent. We can view this as a test of independence as opposed to a test of homogeneity if we view *Age* and *Opinion* as variables whose relationship we want to understand. This was a random sample and there are at least 5 expected responses in every cell. The expected values are calculated using the formula:

$$Exp_{ij} = \frac{TotalRow_i \times TotalCol_j}{Table\ Total} \Longrightarrow$$

$$Exp_{11} = \frac{TotalRow_1 \times TotalCol_1}{Table\ Total} = \frac{287 \times 100}{400} = 71.75$$

Expected Values	Age				
	18–29	**30–49**	**50–64**	**65+**	**Total**
Favor	71.75	71.75	71.75	71.75	**287**
Oppose	28.25	28.25	28.25	28.25	**113**
Total	**100**	**100**	**100**	**100**	**400**

The components are:

Components	Age			
	18–29	**30–49**	**50–64**	**65+**
Favor	3.03	0.46	0.38	3.24
Oppose	7.70	1.17	0.98	8.23

There are $(r - 1) \times (c - 1) = 1 \times 3 = 3$ degrees of freedom. Summing all the components gives:

$$\chi_3^2 = 3.03 + 0.46 + \cdots + 8.23 = 25.19,$$

(continued)

which has a P-value < 0.0001.

Thus, we reject the null hypothesis and conclude that *Age* and *Opinion* about Profiling are not independent. Looking at the residuals,

Residuals	Age			
	18–29	**30–49**	**50–64**	**65 +**
Favor	−1.74	−0.68	0.62	1.80
Oppose	2.78	1.08	−0.99	−2.87

we see a pattern. These two variables fail to be independent because increasing age is associated with more favorable attitudes toward profiling.

Bar charts arranged in *Age* order make the pattern clear:

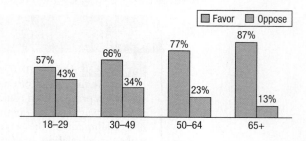

JUST CHECKING

Which of the three chi-square tests would you use in each of the following situations—goodness-of-fit, homogeneity, or independence?

1 A restaurant manager wonders whether customers who dine on Friday nights have the same preferences among the chef's four special entrées as those who dine on Saturday nights. One weekend he has the wait staff record which entrées were ordered each night. Assuming these customers to be typical of all weekend diners, he'll compare the distributions of meals chosen Friday and Saturday.

2 Company policy calls for parking spaces to be assigned to everyone at random, but you suspect that may not be

so. There are three lots of equal size: lot A, next to the building; lot B, a bit farther away; and lot C on the other side of the highway. You gather data about employees at middle management level and above to see how many were assigned parking in each lot.

3 Is a student's social life affected by where the student lives? A campus survey asked a random sample of students whether they lived in a dormitory, in off-campus housing, or at home and whether they had been out on a date 0, 1–2, 3–4, or 5 or more times in the past two weeks.

Chi-Square Tests and Causation

Chi-square tests are common. Tests for independence are especially widespread. Unfortunately, many people interpret a small P-value as proof of causation. We know better. Just as correlation between quantitative variables does not demonstrate causation, a failure of independence between two categorical variables does not show a cause-and-effect relationship between them, nor should we say that one variable *depends* on the other.

The chi-square test for independence treats the two variables symmetrically. There is no way to differentiate the direction of any possible causation from one variable to the other. While we can see that attitudes on personal *Appearance* and *Age* are related, we can't say that getting older *causes* you to change attitudes. And certainly it's not correct to say that changing attitudes on personal appearance makes you older.

Of course, there's never any way to eliminate the possibility that a lurking variable is responsible for the observed lack of independence. In some sense, a failure of independence between two categorical variables is less impressive than a strong, consistent association between quantitative variables. Two categorical variables can fail the test of independence in many ways, including ways that show no consistent pattern of failure. Examination of the chi-square standardized residuals can help you think about the underlying patterns.

WHAT CAN GO WRONG?

- **Don't use chi-square methods unless you have counts.** All three of the chi-square tests apply only to counts. Other kinds of data can be arrayed in two-way tables. Just because numbers are in a two-way table doesn't make them suitable for chi-square analysis. Data reported as proportions or percentages can be suitable for chi-square procedures, *but only after they are converted to counts*. If you try to do the calculations without first finding the counts, your results will be wrong.

- **Beware large samples.** Beware *large* samples? That's not the advice you're used to hearing. The chi-square tests, however, are unusual. You should be wary of chi-square tests performed on very large samples. No hypothesized distribution fits perfectly, no two groups are exactly homogeneous, and two variables are rarely perfectly independent. The degrees of freedom for chi-square tests don't grow with the sample size. With a sufficiently large sample size, a chi-square test can always reject the null hypothesis. But we have no measure of how far the data are from the null model. There are no confidence intervals to help us judge the effect size except in the case of two proportions.

- **Don't say that one variable "depends" on the other just because they're not independent.** "Depend" can suggest a model or a pattern, but variables can fail to be independent in many different ways. When variables fail the test for independence, it may be better to say they are "associated."

ETHICS IN ACTION

Forever Youthful, Inc. specializes in products that contain biologically active ingredients with anti-aging properties often referred to as "cosmeceuticals." Forever Youthful, Inc. has been exclusively focused on skin care. Its line of luxurious creams, masks, and eye treatments, based on various combinations of antioxidants, peptides, and even growth factors, claim to reverse the signs of aging. Michele Kaplan, Vice President for Sales and Marketing, and her team have been very successful in partnering with high-end spas, resorts, and boutiques to deliver Forever Youthful skin care products to its target customer: older wealthy women. Consequently, Michele was the only top level executive to survive the recent restructuring at Forever Youthful after a new president and CEO, J. C. Burnett, was hired.

Under Burnett's leadership the company has begun to move in several different directions. An emerging trend in the anti-aging market is hair care, and Forever Youthful is about to launch a line of expensive shampoos, conditioners, and scalp treatments designed to make hair shiny, full, and younger looking. Michele is excited about the prospects for sales in this new market segment, and she believes the current strategy of targeting high-end distributors, in this case fashionable salons, would work best. However, the new leadership team is proposing that the hair care line of products be sold online direct to customers, arguing that this is more cost-effective because it doesn't require establishing relationships with hair salons. Michele disagrees. She is concerned about the possible negative consequences of such a move, particularly on the brand's "elite" image, not to mention on her division.

To prove her point, Michele directed her research team to design some questions for gathering data from current and potential customers. Working with some of the businesses that sell Forever Youthful skin care products, clients were interviewed about their online purchasing behaviors and perceptions of brand image as they checked in for appointments. Demographic data were also collected. Based on a sample of 3500 clients, chi-square tests for independence were performed to determine if responses about online shopping and brand image were independent of age category. The results revealed what Michele had expected. Online purchasing behavior is dependent on age. Moreover, level of agreement with the statement "hair products sold online are of inferior quality to those sold in salons" also showed dependence on age. Because Forever Youthful targets older customers, Michele is hopeful that these findings will help convince the team that her strategy is best. She is now looking forward to the next leadership team meeting.

- Identify the ethical dilemma in this scenario.

- What are the undesirable consequences?

- Propose an ethical solution that considers the welfare of all stakeholders.

WHAT HAVE WE LEARNED?

Learning Objectives

Recognize when a chi-square test of goodness-of-fit, homogeneity, or independence is appropriate.

For each test, find the expected cell frequencies.

For each test, check the assumptions and corresponding conditions and know how to complete the test.
- Counted data condition.
- Independence assumption; randomization makes independence more plausible.
- Sample size assumption with the expected cell frequency condition; expect at least 5 observations in each cell.

Interpret a chi-square test.
- Even though we might believe the model, we cannot prove that the data fit the model with a chi-square test because that would mean confirming the null hypothesis.

Examine the standardized residuals to understand what cells were responsible for rejecting a null hypothesis.

Compare two proportions.

State the null hypothesis for a test of independence and understand how that is different from the null hypothesis for a test of homogeneity.
- Both are computed the same way. You may not find both offered by your technology. You can use either one as long as you interpret your result correctly.

Terms

Chi-square models

Chi-square models are skewed to the right. They are parameterized by their degrees of freedom and become less skewed with increasing degrees of freedom.

Chi-square (or chi-squared) statistic

The chi-square statistic is found by summing the chi-square components. Chi-square tests can be used to test goodness-of-fit, homogeneity, or independence.

Chi-square goodness-of-fit test

A test of whether the distribution of counts in one categorical variable matches the distribution predicted by a model. A chi-square test of goodness-of-fit finds

$$\chi^2 = \sum_{all\ cells} \frac{(Obs - Exp)^2}{Exp},$$

where the expected counts come from the predicting model. It finds a P-value from a chi-square model with $n - 1$ degrees of freedom, where n is the number of categories in the categorical variable.

Chi-square test of homogeneity

A test comparing the distribution of counts for two or more groups on the same categorical variable. A chi-square test of homogeneity finds

$$\chi^2 = \sum_{all\ cells} \frac{(Obs - Exp)^2}{Exp},$$

where the expected counts are based on the overall frequencies, adjusted for the totals in each group. We find a P-value from a chi-square distribution with $(R - 1) \times (C - 1)$ degrees of freedom, where R gives the number of categories (rows) and C gives the number of independent groups (columns).

Chi-square test of independence

A test of whether two categorical variables are independent. It examines the distribution of counts for one group of individuals classified according to both variables. A chi-square test of *independence* uses the same calculation as a test of homogeneity. We find a P-value from a chi-square distribution with $(R - 1) \times (C - 1)$ degrees of freedom, where R gives the number of categories in one variable and C gives the number of categories in the other.

Standardized residual	In each cell of a two-way table, a standardized residual is the square root of the chi-square component for that cell with the sign of the *Observed − Expected* difference:

$$\frac{(Obs - Exp)}{\sqrt{Exp}}$$

When we reject a chi-square test, an examination of the standardized residuals can sometimes reveal more about how the data deviate from the null model.

TECHNOLOGY HELP: Chi-Square

Most statistics packages associate chi-square tests with contingency tables. Often chi-square is available as an option only when you make a contingency table. This organization can make it hard to locate the chi-square test and may confuse the three different roles that the chi-square test can take. In particular, chi-square tests for goodness-of-fit may be hard to find or missing entirely. Chi-square tests for homogeneity are computationally the same as chi-square tests for independence, so you may have to perform the mechanics as if they were tests of independence and interpret them afterward as tests of homogeneity.

Most statistics packages work with data on individuals rather than with the summary counts. If the only information you have is the table of counts, you may find it more difficult to get a statistics package to compute chi-square. Some packages offer a way to reconstruct the data from the summary counts so that they can then be passed back through the chi-square calculation, finding the cell counts again. Many packages offer chi-square standardized residuals (although they may be called something else).

EXCEL

To perform a chi-square goodness-of-fit test using Excel:

If you have two parallel columns, one holding observed counts and the other expected counts, follow these steps:

- In columns D through F, you can use spreadsheet calculations and built-in functions to calculate (Obs-Exp), (Obs-Exp)^2, and (Obs-Exp)^2/Exp. Type the following into cells D2, E2, and F2 respectively:
- =(B2-C2)
- =D2^2
- =E2/C2
- Copy the formulas from D2, E2, and F2 and paste into the remainder of the rows in columns D, E, and F.
- Using the =SUM function, add the values contained in the cells in column F to get the χ^2 value.
- The CHISQ.INV.RT and CHISQ.DIST.RT functions can then be used to calculate the critical value and P-value of the chi-square statistic that is calculated. See the image below for commands. Note that the function used to calculate degrees of freedom (=COUNT) should reflect rows containing data.

	A	B	C	D	E	F	G
1	Operator	Observed	Expected	(Obs-Exp)	(Obs-Exp)^2	(Obs-Exp)^2/Exp	**Excel Syntax**
2	1	11	15	-4	16	1.0667	
3	2	17	15	2	4	0.2667	
4	3	9	15	-6	36	2.4000	
5	4	12	15	-3	9	0.6000	
6	5	19	15	4	16	1.0667	
7	6	18	15	3	9	0.6000	
8	7	13	15	-2	4	0.2667	
9	8	21	15	6	36	2.4000	
10					*Chi-Square=*	8.6667	=SUM(F2:F9)
11					*Alpha=*	0.0500	[enter value]
12					*df=*	7	=COUNT(A2:A9)-1
13					χ^2 *critical=*	14.0671	=CHISQ.INV.RT(F11,F12)
14					*p-value=*	0.2775	=CHISQ.DIST.RT(F10,F12)

To perform a chi-square test of independence using Excel:

- Summarize raw data into a row by column pivot table—this will be the observed table. It should have row and column totals and a grand total.

- In the same spreadsheet, create a row by column table that mirrors the observed table without row and colum totals—this will be the expected table.

- In each corresponding cell in the expected table, calculate the expected value. In the example shown below, the formula in cell B11 is =(D5*B7)/D7; repeat this for all cells in the expected table.

- Create a third row by column table that mirrors the observed table without row and column totals—this will be the table that holds the chi-square calculation for each cell.

- In each corresponding cell in the chi-square table, calculate the chi-square value. In the example shown here, the code for cell B15 =((POWER((B5-B11),2)/B11)); repeat this for all cells in the expected table.

- Below the table, designate cells to hold the values for alpha and degrees of freedom (recall that degrees of freedom for this statistic are ([number of rows] -1)*([number of columns] -1), and then type in the code to calculate the chi-square value, critical chi-square value, and P-value.

	A	B	C	D
1				
2				
3	*Pivot Table*	**Column Labels**		
4	**Row Labels**	**No AC**	**AC**	**Grand Total**
5	Old	1055	592	1647
6	New	38	43	81
7	**Grand Total**	**1093**	**635**	**1728**
8				
9				
10	*Expected Values*	**No AC**	**AC**	
11	**Old**	1041.765625	605.234375	
12	**New**	51.234375	29.765625	
13				
14	*(O-E)²/E Values*	**No AC**	**AC**	
15	**Old**	0.168126762	0.28938984	
16	**New**	3.418577501	5.88426017	
17				
18			**Excel Syntax**	
19	*Chi-Square=*	9.7604	=SUM(B15,B16,C15,C16)	
20	*Alpha=*	0.0500	[enter value]	
21	*df=*	1	[enter value]	
22	χ^2 *critical=*	3.8415	=CHISQ.INV.RT(B20,B21)	
23	*p-value=*	0.0018	=CHISQ.DIST.RT(B19,B21)	

Comments

Excel offers the function CHISQ.TEST (actual_range, expected_range), which computes a chi-square P-value for independence. However, this command will only provide the P-value and no other values. Both ranges are of the form UpperLeftCell: LowerRightCell, specifying two rectangular tables. The two tables must be of the same size and shape. The function is called CHITEST in Excel versions earlier than 2010.

XLSTAT

To perform a chi-square goodness-of-fit test:

- Choose **Parametric Tests**, and then select **Multinomial goodness of fit test**.

- In one column you should have the observed frequencies of the categories of the variable. Enter this under **Frequencies**.

- In another column you should have either the expected frequencies or the proportions. Enter this under either **Expected frequencies** (or expected proportions) and choose the appropriate **data format**.

- Be sure to check the box that says **Chi-square test**.

- Enter in the desired **Significance level**.

- Select **OK**.

- Select **Continue** (if prompted).

To perform a chi-square test for homogeneity:

- Choose **Correlation/Association tests**, and then select **Tests on contingency tables**.

- If your data already are in a contingency table, choose the **Data format** option **Contingency table**.

- If your data are not in a contingency table, choose **Qualitative variables**.

- Enter the cell range of your data.

- On the **Options** tab, check **Chi-square test**.

- On the **Outputs** tab, choose **Proportions/Row** or **Proportions/Column** if you wish to see the conditional distributions.

JMP

To perform a chi-square goodness-of-fit test:

- Select **Analyze > Distribution**.

- Choose a categorical variable for the **Y, Columns** dialog box.

- From the red triangle next to the variable, name, select **Test Probabilities** and enter the probabilities you want to test.

To perform other chi-square tests:

- Select **Analyze > Fit Y by X**.

- Choose one variable as the Y, response variable, and the other as the X, factor variable. Both selected variables must be Nominal or Ordinal.

- JMP will make a plot and a contingency table. Below the contingency table, **JMP** offers a **Tests** panel. In that panel, the Chi Square for independence is called **Pearson.** The table also offers the P-value.
- Click on the contingency Table title bar to drop down a menu that offers to include a **Deviation** and Cell **Chi square** in each cell of the table.

Comments

JMP will perform a chi-square analysis for a **Fit Y by X** if both variables are nominal or ordinal (marked with an N or O), but not otherwise. Be sure the variables have the right type. Deviations are the observed—expected differences in counts. Cell chi-squares are the squares of the standardized residuals. Refer to the deviations for the sign of the difference.

MINITAB

From the **Start** menu,

- Choose the **Tables** submenu.
- From that menu, choose **Chi Square Test**....
- In the dialog, identify the columns that make up the table. Minitab will display the table and print the chi-square value and its P-value.

Comments

Alternatively, select the **Cross Tabulation** ... command to see more options for the table, including expected counts and standardized residuals.

R

Goodness-of-fit test:

To test the probabilities in a vector prob with observed values in a vector x,

- **chisq.test**$(x, p = prob)$

Test of Independence or homogeneity:

With counts in a contingency table (a matrix called, say, con.table), the test of independence (or homogeneity) is found by

- **chisq.test**(con.table)

Comments

Using the function xtabs you can create a contingency table from two variables x and y in a data frame called mydata by

- con.table $=$ xtabs$(\sim x + y, data = mydata)$ then
- **chisq.test**(con.table)

SPSS

From the **Analyze** menu,

- Choose the **Descriptive Statistics** submenu.
- From that submenu, choose **Crosstabs**....
- In the Crosstabs dialog, assign the row and column variables from the variable list. Both variables must be categorical.
- Click the **Cells** button to specify that standardized residuals should be displayed.
- Click the **Statistics** button to specify a chi-square test.

Comments

SPSS offers only variables that it knows to be categorical in the variable list for the Crosstabs dialog. If the variables you want are missing, check that they have the right type.

Brief **Case**

Health Insurance

In 2010, the U.S. Congress passed the historic Affordable Care Act that provides coverage for most of the 32 million Americans who were without health care insurance. Just how widespread was the lack of medical coverage? The media claim that the segments of the population most at risk were women, children, the elderly, and the poor. The tables give the number of uninsured (in thousands) by sex, by age, and by household income in 2011.[6] Using the appropriate summary statistics, graphical displays, statistical tests, and confidence intervals, investigate the accuracy of the media's statement using these data. Be sure to discuss your assumptions, methods, results, and conclusions. (Note: some totals between tables may not match exactly due to rounding.)

(continued)

---

[6] Source: U.S. Census Bureau, Current Population Survey, Annual Social and Economic Supplement, 2011 (www.census.gov/hhes/www/cpstables/032012/health/h01_000.htm).

	Sex		
	Male	**Female**	**Total**
Uninsured	25,571	23,043	**48,614**
Insured	125,604	134,610	**260,214**
Total	**151,175**	**157,653**	**308,828**

Age		**Uninsured**	**Insured**	**Total**
	0–17	6,964	67, 143	**74,108**
	18–64	40,959	152,253	**193,212**
	65 +	690	40,817	**41,507**
	Total	**48,613**	**260,213**	**308,827**

Household Income		**Uninsured**	**Insured**	**Total**
	Less than $25,000	19,166	51,255	**70,421**
	$25,000 to $49,999	14,714	57,965	**72,679**
	$50,000 to $74,999	7,286	46,445	**53,731**
	$75,000 or more	7,448	104,549	**111,997**
	Total	**48,614**	**260,214**	**308,828**

Loyalty Program

A marketing executive tested two incentives to see what percentage of customers would enroll in a new Web-based loyalty program. The customers were asked to log on to their accounts on the Web and provide some demographic and spending information. As an incentive, they were offered either Nothing (No Offer), free flight insurance on their next flight (Free Insurance), or a free companion Airline ticket (Free Flight). The customers were segmented according to their past year's spending patterns as spending primarily in one of five areas: *Travel, Entertainment, Dining, Household,* or *Balanced.* The executive wanted to know whether the incentives resulted in different enrollment rates (*Response*). Specifically, she wanted to know how much higher the enrollment rate for the free flight was compared to the free insurance. She also wanted to see whether *Spending Pattern* was associated with *Response.* Using the data **Loyalty Program,** write up a report for the marketing executive using appropriate graphics, summary statistics, statistical tests, and confidence intervals.

EXERCISES

SECTION 13.1

1. If there is no seasonal effect on human births, we would expect equal numbers of children to be born in each season (winter, spring, summer, and fall). A student takes a census of her statistics class and finds that of the 120 students in the class, 25 were born in winter, 35 in spring, 32 in summer, and 28 in fall. She wonders if the excess in the spring is an indication that births are not uniform throughout the year.

a) What is the expected number of births in each season if there is no "seasonal effect" on births?

b) Compute the χ^2 statistic.

c) How many degrees of freedom does the χ^2 statistic have?

2. At a major credit card bank, the percentages of people who historically apply for the Silver, Gold, and Platinum cards are 60%, 30%, and 10%, respectively. In a recent sample of customers responding to a promotion, of 200 customers, 110 applied for Silver, 55 for Gold, and 35 for Platinum. Is there evidence to suggest that the percentages for this promotion may be different from the historical proportions?

a) What is the expected number of customers applying for each type of card in this sample if the historical proportions are still true?
b) Compute the χ^2 statistic.
c) How many degrees of freedom does the χ^2 statistic have?

SECTION 13.2

3. For the births in Exercise 1,

a) If there is no seasonal effect, about how big, on average, would you expect the χ^2 statistic to be (what is the mean of the χ^2 distribution)?
b) Does the statistic you computed in Exercise 1 seem large in comparison to this mean? Explain briefly.
c) What does that say about the null hypothesis?
d) Find the $\alpha = 0.05$ critical value for the χ^2 distribution with the appropriate number of df.
e) Using the critical value, what do you conclude about the null hypothesis at $\alpha = 0.05$?

4. For the customers in Exercise 2,

a) If the customers apply for the three cards according to the historical proportions, about how big, on average, would you expect the χ^2 statistic to be (what is the mean of the χ^2 distribution)?
b) Does the statistic you computed in Exercise 2 seem large in comparison to this mean? Explain briefly.
c) What does that say about the null hypothesis?
d) Find the $\alpha = 0.05$ critical value for the χ^2 distribution with the appropriate number of df.
e) Using the critical value, what do you conclude about the null hypothesis at $\alpha = 0.05$?

SECTION 13.3

5. For the data in Exercise 1,

a) Compute the standardized residual for each season.
b) Are any of these particularly large? (Compared to what?)
c) Why should you have anticipated the answer to part b?

6. For the data in Exercise 2,

a) Compute the standardized residual for each type of card.
b) Are any of these particularly large? (Compared to what?)
c) What does the answer to part b say about this new group of customers?

SECTION 13.4

7. An analyst at a local bank wonders if the age distribution of customers coming for service at his branch in town is the same as at the branch located near the mall. He selects 100 transactions at random from each branch and researches the age information for the associated customer. Here are the data:

	Age			
	Less than 30	30–55	56 or Older	Total
In-Town Branch	20	40	40	100
Mall Branch	30	50	20	100
Total	50	90	60	200

a) What is the null hypothesis?
b) What type of test is this?
c) What are the expected numbers for each cell if the null hypothesis is true?
d) Find the χ^2 statistic.
e) How many degrees of freedom does it have?
f) Find the critical value at $\alpha = 0.05$.
g) What do you conclude?

8. A market researcher working for the bank in Exercise 2 wants to know if the distribution of applications by card is the same for the past three mailings. She takes a random sample of 200 from each mailing and counts the number applying for Silver, Gold, and Platinum. The data follow:

	Type of Card			
	Silver	Gold	Platinum	Total
Mailing 1	120	50	30	200
Mailing 2	115	50	35	200
Mailing 3	105	55	40	200
Total	340	155	105	600

a) What is the null hypothesis?
b) What type of test is this?
c) What are the expected numbers for each cell if the null hypothesis is true?
d) Find the χ^2 statistic.
e) How many degrees of freedom does it have?
f) Find the critical value at $\alpha = 0.05$.
g) What do you conclude?

SECTION 13.5

9. Markets have become interested in the potential of social networking sites. But they need to understand the demographics of social networking users. Pew Research conducted a survey in late 2012 that addressed these questions (www.pewinternet.org/Reports/2013/Social-media-users/Social-Networking-Site-Users). That survey found that 525 of 846 surveyed male Internet users use social networking. By contrast 679 of 956 female Internet users use social networking.

a) Find the proportions of male and female Internet users who said they use social networking.
b) What is the difference in proportions?
c) What is the standard error of the difference?
d) Find a 95% confidence interval for the difference between the proportions.

10. From the same survey as in Exercise 9, 294 of the 409 respondents who reported earning less than $30,000 per year said they were social networking users. At the other end of the income scale, 333 of the 504 respondents reporting earnings of $75,000 or more were social networking users.

a) Find the proportions of each income group who are social networking users.
b) What is the difference in proportions?
c) What is the standard error of the difference?
d) Find a 95% confidence interval for the difference between these proportions.

SECTION 13.6

11. The same poll as in Exercise 9 has asked about social networking over several years and for different age groups. Here is a table of responses:

		May-2010	Aug-2011	Aug-2012	Total
Age	18–29	273	277	293	843
	30–49	325	362	388	1075
	50–64	259	270	314	843
	Total	857	909	995	2761

a) Under the usual null hypothesis, what are the expected values?
b) Compute the χ^2 statistic.
c) How many degrees of freedom does it have?
d) What do you conclude?

12. To complete the poll reported in Exercise 9, Pew Research surveyed respondents by telephone, drawing a random sample of landlines and another random sample of cell phones. For those numbers that were valid, they report the following:

	Land	Cell	Total
No Answer/Busy	552	42	594
Voicemail	3347	2843	6190
Contact	8399	8612	17,011
Total	12,298	11,497	23,795

Are the results they find independent of the telephone type?

a) Under the usual null hypothesis, what are the expected values?
b) Compute the χ^2 statistic.
c) How many degrees of freedom does it have?
d) What do you conclude?

CHAPTER EXERCISES

13. Concepts. For each of the following situations, state whether you'd use a chi-square goodness-of-fit test, chi-square test of homogeneity, chi-square test of independence, or some other statistical test.

a) A brokerage firm wants to see whether the type of account a customer has (Silver, Gold, or Platinum) affects the type of trades that customer makes (in person, by phone, or on the Internet). It collects a random sample of trades made for its customers over the past year and performs a test.

b) That brokerage firm also wants to know if the type of account affects the size of the account (in dollars). It performs a test to see if the mean size of the account is the same for the three account types.

c) The academic research office at a large community college wants to see whether the distribution of courses chosen (Humanities, Social Science, or Science) is different for its residential and nonresidential students. It assembles last semester's data and performs a test.

14. Concepts, part 2. For each of the following situations, state whether you'd use a chi-square goodness-of-fit test, a chi-square test of homogeneity, a chi-square test of independence, or some other statistical test.

a) Is the quality of a car affected by what day it was built? A car manufacturer examines a random sample of the warranty claims filed over the past two years to test whether defects are randomly distributed across days of the workweek.

b) A researcher for the American Booksellers Association wants to know if retail sales/sq. ft. is related to serving coffee or snacks on the premises. She examines a database of 10,000 independently owned bookstores testing whether retail sales (dollars/sq. ft.) is related to whether or not the store has a coffee bar.

c) A researcher wants to find out whether education level (some high school, high school graduate, college graduate, advanced degree) is related to the type of transaction most likely to be conducted using the Internet (shopping, banking, travel reservations, auctions). He surveys 500 randomly chosen adults and performs a test.

15. Dice. After getting trounced by your little brother in a children's game, you suspect that the die he gave you is unfair. To check, you roll it 60 times, recording the number of times each face appears. Do these results cast doubt on the die's fairness?

Face	Count
1	11
2	7
3	9
4	15
5	12
6	6

a) If the die is fair, how many times would you expect each face to show?
b) To see if these results are unusual, will you test goodness-of-fit, homogeneity, or independence?
c) State your hypotheses.
d) Check the conditions.
e) How many degrees of freedom are there?

f) Find χ^2 and the P-value.
g) State your conclusion.

16. Online Mating. According to recent research (www.nas .org) married couples who met their spouse through an on-line dating service may have a different divorce rate than those who met "off-line." (From "Marital satisfaction and break-ups differ across on-line and off-line meeting venues" by Cacioppo et al., Proceedings of the National Academy of Sciences of the United States of America. Copyright © 2013 by National Academy of Sciences. Used by permission of National Academy of Sciences.) The survey polled couples who married between 2005 and 2012. The baseline divorce rate (by 2013) for off-line marriages in this cohort is 7.73%. The report gives the following divorce statistics according to the on-line dating service where the couple met:

Service	Couples (*n*)	Divorces/Separations
eHarmony	791	29
Match	775	58
Plenty of Fish	201	20
Yahoo	227	12
Small Sites	777	37

a) If the divorce rate were the same for these couples as for those who met offline, how many divorces would you expect for each group of couples?
b) To test whether these couples are different from offline couples, will you perform a goodness-of-fit test, a test of homogeneity, or a test of independence?
c) State the hypotheses.
d) Check the conditions.
e) Find the standardized residuals and the chi-square components. (Hint: use a spreadsheet to perform the calculations.)
f) State the number of degrees of freedom and find χ^2 and the P-value.
g) State your conclusion.
h) Online dating services are a billion-dollar business in the United States. Does it change your conclusion to know that the study was funded by eHarmony?

17. Quality control. A company advertises that its premium mixture of nuts contains 10% Brazil nuts, 20% cashews, 20% almonds, 10% hazelnuts, and that the rest are peanuts. You buy a large can and separate the various kinds of nuts. Upon weighing them, you find there are 112 grams of Brazil nuts, 183 grams of cashews, 207 grams of almonds, 71 grams of hazelnuts, and 446 grams of peanuts. You wonder whether your mix is significantly different from what the company advertises.

a) Explain why the chi-square goodness-of-fit test is not an appropriate way to find out.
b) What might you do instead of weighing the nuts in order to use a χ^2 test?

18. Sales rep travel. A sales representative who is on the road visiting clients thinks that, on average, he drives the same distance each day of the week. He keeps track of his mileage for several weeks and discovers that he averages 122 miles on Mondays, 203 miles on Tuesdays, 176 miles on Wednesdays, 181 miles on Thursdays, and 108 miles on Fridays. He wonders if this evidence contradicts his belief in a uniform distribution of miles across the days of the week. Is it appropriate to test his hypothesis using the chi-square goodness-of-fit test? Explain.

19. Maryland lottery. For a lottery to be successful, the public must have confidence in its fairness. One of the lotteries in Maryland is Pick-3 Lottery, where 3 random digits are drawn each day.[7] A fair game depends on every value (0 to 9) being equally likely at each of the three positions. If not, then someone detecting a pattern could take advantage of that and beat the lottery. To investigate the randomness, we'll look at data collected over a recent 32-week period. Although the winning numbers look like three-digit numbers, in fact, each digit is a randomly drawn numeral. We have 654 random digits in all. Are each of the digits from 0 to 9 equally likely? Here is a table of the frequencies (Maryland State Lottery Agency, www.mdlottery.com).

Group	Count	%
0	62	9.480
1	55	8.410
2	66	10.092
3	64	9.786
4	75	11.468
5	57	8.716
6	71	10.856
7	74	11.315
8	69	10.550
9	61	9.327

a) Select the appropriate procedure.
b) Check the assumptions.
c) State the hypotheses.
d) Test an appropriate hypothesis and state your results.
e) Interpret the meaning of the results and state a conclusion.

20. Employment discrimination? Census data for New York City indicate that 29.2% of the under-18 population is white, 28.2% black, 31.5% Latino, 9.1% Asian, and 2% are of other ethnicities. The New York Civil Liberties Union points out that of 26,181 police officers, 64.8% are white, 14.5% black, 19.1% Hispanic, and 1.4% Asian. Do the police officers reflect the ethnic composition of the city's youth? (Equate Latino with Hispanic.)

a) Select the appropriate procedure.
b) Check the assumptions.
c) State the hypotheses.
d) Test an appropriate hypothesis and state your results.
e) Interpret the meaning of the results and state a conclusion.

[7] Source: Maryland State Lottery Agency, www.mdlottery.com.

T **21.** *Titanic.* Recently, the *Encyclopedia Titanica*[8] has updated the data on who survived the sinking of the *Titanic* based on whether they were crew members or passengers booked in first-, second-, or third-class staterooms.

	Crew	First	Second	Third	Total
Alive	212	201	119	180	712
Dead	677	123	166	530	1496
Total	889	324	285	710	2208

a) If we draw an individual at random from this table, what's the probability that we will draw a member of the crew?
b) What's the probability of randomly selecting a third-class passenger who survived?
c) What's the probability of a randomly selected passenger surviving, given that the passenger was in a first-class stateroom?
d) If someone's chances of surviving were the same regardless of their status on the ship, how many members of the crew would you expect to have lived?
e) State the null and alternative hypotheses we would test here (and the name of the test).
f) Give the degrees of freedom for the test.
g) The chi-square value for the table is 187.56, and the corresponding P-value is barely greater than 0. State your conclusions about the hypotheses.

T **22.** Promotion discrimination? The table shows the rank attained by male and female officers in the New York City Police Department (NYPD). Do these data indicate that men and women are equitably represented at all levels of the department? (All possible ranks in the NYPD are shown.)

Rank		Male	Female
	Officer	21,900	4281
	Detective	4058	806
	Sergeant	3898	415
	Lieutenant	1333	89
	Captain	359	12
	Higher Ranks	218	10

a) What's the probability that a person selected at random from the NYPD is a female?
b) What's the probability that a person selected at random from the NYPD is a detective?
c) Assuming no bias in promotions, how many female detectives would you expect the NYPD to have?
d) To see if there is evidence of differences in ranks attained by males and females, will you test goodness-of-fit, homogeneity, or independence?
e) State the hypotheses.
f) Test the conditions.
g) How many degrees of freedom are there?
h) Find the chi-square value and the associated P-value.

i) State your conclusion.
j) If you concluded that the distributions are not the same, analyze the differences using the standardized residuals of your calculations.

23. Birth order and college choice. Students in an Introductory Statistics class at a large university were classified by birth order and by the college they attend.

College	Birth Order (1 = oldest or only child)				
	1	**2**	**3**	**4 or More**	**Total**
Arts and Sciences	34	14	6	3	57
Agriculture	52	27	5	9	93
Social Science	15	17	8	3	43
Professional	13	11	1	6	31
Total	114	69	20	21	224

College	Expected Values Birth Order (1 = oldest or only child)			
	1	**2**	**3**	**4 or More**
Arts and Sciences	29.0089	17.5580	5.0893	5.3438
Agriculture	47.3304	28.6473	8.3036	8.7188
Social Science	21.8839	13.2455	3.8393	4.0313
Professional	15.7768	9.5491	2.7679	2.9063

a) What kind of chi-square test is appropriate—goodness-of-fit, homogeneity, or independence?
b) State your hypotheses.
c) State and check the conditions.
d) How many degrees of freedom are there?
e) The calculation yields $\chi^2 = 17.78$, with $P = 0.0378$. State your conclusion.
f) Examine and comment on the standardized residuals. Do they challenge your conclusion? Explain.

College	Standardized Residuals Birth Order (1 = oldest or only child)			
	1	**2**	**3**	**4 or More**
Arts and Sciences	0.92667	−0.84913	0.40370	−1.01388
Agriculture	0.67876	−0.30778	−1.14640	0.09525
Social Science	−1.47155	1.03160	2.12350	−0.51362
Professional	−0.69909	0.46952	−1.06261	1.81476

24. Automobile manufacturers. *Consumer Reports* uses surveys given to subscribers of its magazine and website (www.ConsumerReports.org) to measure reliability in automobiles. This annual survey asks about problems that consumers have had with their cars, vans, SUVs, or trucks during the previous 12 months. Each analysis is based on the number of problems per 100 vehicles.

	Origin of Manufacturer			
	Asia	Europe	U.S.	Total
No Problems	88	79	83	250
Problems	12	21	17	50
Total	100	100	100	300

	Expected Values		
	Asia	**Europe**	**U.S.**
No Problems	83.33	83.33	83.33
Problems	16.67	16.67	16.67

a) State your hypotheses.
b) State and check the conditions.
c) How many degrees of freedom are there?
d) The calculation yields $\chi^2 = 2.928$, with P = 0.231. State your conclusion.
e) Would you expect that a larger sample might find statistical significance? Explain.

T 25. Cranberry juice. It's common folk wisdom that cranberries can help prevent urinary tract infections in women. A leading producer of cranberry juice would like to use this information in their next ad campaign, so they need evidence of this claim. In 2001, the *British Medical Journal* reported the results of a Finnish study in which three groups of 50 women were monitored for these infections over 6 months. One group drank cranberry juice daily, another group drank a lactobacillus drink, and the third group drank neither of those beverages, serving as a control group. In the control group, 18 women developed at least one infection compared with 20 of those who consumed the lactobacillus drink and only 8 of those who drank cranberry juice. Does this study provide supporting evidence for the value of cranberry juice in warding off urinary tract infections in women?

a) Select the appropriate procedure.
b) Check the assumptions.
c) State the hypotheses.
d) Test an appropriate hypothesis and state your results.
e) Interpret the meaning of the results and state a conclusion.
f) If you concluded that the groups are not the same, analyze the differences using the standardized residuals of your calculations.

26. College value? In March and April of 2011, the Pew Research Center asked 2142 U.S. adults and 1055 college presidents whether they would "rate the job the higher education system is doing in providing value for the money spent by students and their families as" Excellent, Good, Only Fair, or Poor.

	Poor	**Only Fair**	**Good**	**Excellent**	**No Answer/ Don't Know**
U.S. Adults	321	900	750	107	64
College Presidents	32	222	622	179	0

Is there a difference in the distribution of responses between U.S. adults and college presidents?

a) Is this a test of independence or homogeneity?
b) Write appropriate hypotheses.

c) Check the necessary assumptions and conditions.
d) Find the P-value of your test.
e) State your conclusion and analysis.

27. Eating out. The Cornell National Social Survey[9] polls 1000 U.S. adults each year throughout that time. One question asked in each survey was "How much does your family spend weekly on eating out?" The table shows the results for 2008 and 2010. Was there a change?

	2008	**2010**	**Total**
<$50/Wk	662	691	1353
$51–100	185	179	364
$101–150	71	57	128
$151–200	40	25	65
>$200	35	20	55
Total	993	972	1965

a) Would you interpret a chi-square test as a test of homogeneity or a test of independence?
b) Write appropriate hypotheses.
c) Are the conditions for inference satisfied?
d) A calculation gives $\chi^2 = 9.581$, P = 0.0481. What conclusion do you draw?
e) Here are the standardized residuals. State a more complete description of what the data say.

	2008	**2010**
<$50/Wk	−0.8310	0.83995
$51–100	0.0778	−0.0786
$101–150	0.7853	−0.7938
$151–200	1.2480	−1.261
>$200	1.3669	−1.3815

28. Seafood company. A large company in the northeastern United States that buys fish from local fishermen and distributes them to major companies and restaurants is considering launching a new ad campaign on the health benefits of fish. As evidence, they would like to cite the following study. Medical researchers followed 6272 Swedish men for 30 years to see if there was any association between the amount of fish in their diet and prostate cancer ("Fatty Fish Consumption and Risk of Prostate Cancer," *Lancet*, June 2001).

		Prostate Cancer	
		No	**Yes**
Fish Consumption	**Never/Seldom**	110	14
	Small Part of Diet	2420	201
	Moderate Part	2769	209
	Large Part	507	42

a) Is this a survey, a retrospective study, a prospective study, or an experiment? Explain.
b) Is this a test of homogeneity or independence?

[9] Cornell Survey Research Institute, 2009 and 2011.

c) Do you see evidence of an association between the amount of fish in a man's diet and his risk of developing prostate cancer?

d) Does this study prove that eating fish does not prevent prostate cancer? Explain.

29. Shopping. A survey of 430 randomly chosen adults finds that 47 of 222 men and 37 of 208 women had purchased books online.

a) Is there evidence that the sex of the person and whether they buy books online are associated?

b) If your conclusion in fact proves to be wrong, did you make a Type I or Type II error?

c) Give a 95% confidence interval for the difference in proportions of buying online for men and women.

30. Information technology. A recent report suggests that Chief Information Officers (CIOs) who report directly to Chief Financial Officers (CFOs) rather than Chief Executive Officers (CEOs) are more likely to have IT agendas that deal with cost cutting and compliance (SearchCIO.com, March 14, 2006). In a random sample of 535 companies, it was found that CIOs reported directly to CFOs in 173 out of 335 service firms and in 95 out of 200 manufacturing companies.

a) Is there evidence that type of business (service versus manufacturing) and whether or not the CIO reports directly to the CFO are associated?

b) If your conclusion proves to be wrong, did you make a Type I or Type II error?

c) Give a 95% confidence interval for the difference in proportions of companies in which the CIO reports directly to the CFO between service and manufacturing firms.

31. Fast food. GfK Roper Consulting gathers information on consumer preferences around the world to help companies monitor attitudes about health, food, and health care products. They asked people in many different cultures how they felt about the following statement: *I try to avoid eating fast foods.*

In a random sample of 800 respondents, 411 people were 35 years old or younger, and, of those, 197 agreed (completely or somewhat) with the statement. Of the 389 people over 35 years old, 246 people agreed with the statement.

a) Is there evidence that the percentage of people avoiding fast food is different in the two age groups?

b) Give a 90% confidence interval for the difference in proportions.

32. Libraries. Public libraries are not run for profit, but they must know their customers. A Pew Research survey in November 2012 found that 426 of 584 surveyed parents of young children had current library cards. 967 of 1668 adults without young children had library cards.

a) Is there evidence that the percentage of adults with library cards is different between these two groups?

b) Give a 95% confidence interval for the difference.

c) Do you think the difference should affect how a library markets its services?

33. Under water. In early 2012, the proportion of mortgages that were "under water"—a negative equity position in which the homeowner owes more than the value of the home—was highest in Nevada and Arizona (s.wsj .net/public/resources/documents/info-NEGATIVE_ EQUITY_0911.html). Lenders are also interested in homes at risk—those within 5% of being in a negative equity position. A sample of mortgages from these two states suggest that the problem may be less severe in Nevada. The sample found that 62 of 1369 Arizona mortgages were in a "near negative equity" state. A sample of 604 Nevada mortgages found 22 in a near negative equity state.

a) Is there evidence that the percentage of near negative equity mortgages is different in the two states?

b) Give a 90% confidence interval for the difference in proportions.

34. Labor force. Immigration reform has focused on dividing illegal immigrants into two groups: long-term and short-term. In a random sample of 958 construction workers from the Northeast, 66 are illegal short-term immigrants. In the Midwest, 42 out of a sample of 1070 are illegal short-term immigrants.

a) Is there evidence that the percentage of construction workers who are illegal short-term immigrants differs in the two regions?

b) Give a 90% confidence interval for the difference in proportions.

35. Groceries. The surveys discussed in Exercise 27 also asked about weekly household expenditures for groceries. Here is a table with the results for the years 2008, 2009, and 2010:

	Groceries 2008	Groceries 2009	Groceries 2010	Total
< $50/Wk	138	153	129	420
$51–100	336	327	345	1008
$101–150	226	270	248	744
$151–200	134	100	130	364
> $200	146	104	95	345
Total	980	954	947	2881

State and test an appropriate hypothesis about household expenditures for groceries in these three years.

36. Investment options. A full service brokerage firm surveyed a random sample of 1200 clients asking them to indicate the likelihood that they would add inflation-linked annuities and bonds to their portfolios within the next year. The table below shows the distribution of responses by the investors' tolerance for risk. Test an appropriate hypothesis for the relationship between risk tolerance and the likelihood of investing in inflation-linked options.

		Risk Tolerance			
		Averse	Neutral	Seeking	Total
Likelihood of Investing in Inflation-Linked Options	Certain Will Invest	191	93	40	324
	Likely to Invest	82	106	123	311
	Not Likely to Invest	64	110	101	275
	Certain Will Not Invest	63	91	136	290
	Total	400	400	400	1200

37. Pew Research surveyed U.S., adults in December 2011. They asked how important it is "to you personally" to be successful in a high-paying career or profession. Among 18–34 year-old respondents, do men and women have the same ideas about the importance of this kind of success? Here's a table of the responses:

	F	M	Total
Most important	110	77	187
Very important	293	331	624
Somewhat important	158	218	376
Not important	49	77	126
Total	610	703	1313

a) Select the appropriate procedure.
b) Check the assumptions.
c) State the hypotheses.
d) Test an appropriate hypothesis and state your results.
e) Interpret the meaning of the results and state a conclusion.

38. Entrepreneurial executives. A leading CEO mentoring organization offers a program for chief executives, presidents, and business owners with a focus on developing entrepreneurial skills. Women and men executives that recently completed the program rated its value. Are perceptions of the program's value the same for men and women?

		Men	Women
Perceived Value	Excellent	3	9
	Good	11	12
	Average	14	8
	Marginal	9	2
	Poor	3	1

a) Will you test goodness-of-fit, homogeneity, or independence?
b) Write appropriate hypotheses.
c) Find the expected counts for each cell, and explain why the chi-square procedures are not appropriate for this table.

39. The Cornell National Social Survey asked 1000 U.S. adults about their employment status and whether they owned stocks. This table gives the counts of the 938 respondents:

		Stocks		
		No	Yes	Total
Employed	Yes	330	248	578
	No	123	45	168
	Retired	120	72	192
	Total	573	365	938

Is there a relationship between employment status and stock ownership?

a) Select an appropriate procedure.
b) Check the assumptions.
c) Test the hypothesis.
d) Examine the residuals.
e) Discuss what you find.

40. Online shopping. A recent report concludes that while Internet users like the convenience of online shopping, they do have concerns about privacy and security (*Online Shopping*, Washington, DC, Pew Internet & American Life Project, February 2008). A random sample of adults were asked to indicate their level of agreement with the statement "I don't like giving my credit card number or personal information online." The table gives a subset of responses. Test an appropriate hypothesis for the relationship between age and level of concern about privacy and security online.

		Strongly Agree	Agree	Disagree	Strongly Disagree	Total
Age Category	Ages 18–29	127	147	138	10	422
	Ages 30–49	141	129	78	55	403
	Ages 50–64	178	102	64	51	395
	Ages 65 +	180	132	54	14	380
	Total	626	510	334	130	1600

a) Select the appropriate procedure.
b) Check the assumptions.
c) State the hypotheses.
d) Test an appropriate hypothesis and state your results.
e) Interpret the meaning of the results and state a conclusion.

41. Entrepreneurial executives again. In some situations where the expected counts are too small, as in Exercise 38, we can complete an analysis anyway. We can often proceed after combining cells in some way that makes sense and also produces a table in which the conditions are satisfied. Here is a new table displaying the same data, but combining "Marginal" and "Poor" into a new category called "Below Average."

		Men	Women
Perceived Value	Excellent	3	9
	Good	11	12
	Average	14	8
	Below Average	12	3

a) Find the expected counts for each cell in this new table, and explain why a chi-square procedure is now appropriate.

b) With this change in the table, what has happened to the number of degrees of freedom?

c) Test your hypothesis about the two groups and state an appropriate conclusion.

42. Small business. The director of a small business development center located in a mid-sized city is reviewing data about its clients. In particular, she is interested in examining if the distribution of business owners across the various stages of the business life cycle is the same for white-owned and Hispanic-owned businesses. The data are shown below.

Stage in Business	White-Owned	Hispanic-Owned
Planning	11	9
Starting	14	11
Managing	20	2
Getting Out	15	1

a) Will you test goodness-of-fit, homogeneity, or independence?

b) Write the appropriate hypotheses.

c) Find the expected counts for each cell and explain why chi-square procedures are not appropriate for this table.

d) Create a new table by combining categories so that a chi-square procedure can be used.

e) With this change in the table, what has happened to the number of degrees of freedom?

f) Test your hypothesis about the two groups and state an appropriate conclusion.

T 43. Racial steering. A subtle form of racial discrimination in housing is "racial steering." Racial steering occurs when real estate agents show prospective buyers only homes in neighborhoods already dominated by that family's race. This violates the Fair Housing Act of 1968. Tenants at a large apartment complex filed a lawsuit alleging racial steering. The complex is divided into two parts: Section A and Section B. The plaintiffs claimed that white potential renters were steered to Section A, while African-Americans were steered to Section B. The following table displays the data that were presented in court to show the locations of recently rented apartments. Do you think there is evidence of racial steering?

	New Renters		
	White	Black	Total
Section A	87	8	95
Section B	83	34	117
Total	170	42	212

44. Titanic, again. Newspaper headlines at the time and traditional wisdom in the succeeding decades have held that women and children escaped the *Titanic* in greater proportion than men. Here's a table with the relevant data. Do you think that survival was independent of whether the person was male or female? Defend your conclusion.

	Female	Male	Total
Alive	343	367	710
Dead	127	1364	1491
Total	470	1731	2201

45. Racial steering, revisited. Find a 95% confidence interval for the difference in the proportions of Black renters in the two sections for the data in Exercise 43.

46. Titanic, one more time. Find a 95% confidence interval for the difference in the proportion of women who survived and the proportion of men who survived for the data in Exercise 44. (Assume the passengers on the *Titanic* were representative of others who might have taken the trip.)

47. Industry sector and outsourcing. Many companies have chosen to outsource segments of their business to external providers in order to cut costs and improve quality and/or efficiencies. Common business segments that are outsourced include Information Technology (IT) and Human Resources (HR). The data below show the types of outsourcing decisions made (no outsourcing, IT only, HR only, both IT and HR) by a sample of companies from various industry sectors.

Industry Sector	No Outsourcing	IT Only	HR Only	Both IT and HR
Health Care	810	6429	4725	1127
Financial	263	1598	549	117
Industrial Goods	1031	1269	412	99
Consumer Goods	66	341	305	197

Do these data highlight significant differences in outsourcing by industry sector?

a) Select the appropriate procedure.

b) Check the assumptions.

c) State the hypotheses.

d) Test an appropriate hypothesis and state your results.

e) Interpret the meaning of the results and state a conclusion.

48. Industry sector and outsourcing, part 2. Consider only the companies that have outsourced their IT and HR business segments. Do these data suggest significant differences between companies in the financial and industrial goods sectors with regard to their outsourcing decisions?

Industry Sector	IT Only	HR Only	Both IT and HR
Financial	1598	549	117
Industrial Goods	1269	412	99

a) Select the appropriate procedure.

b) Check the assumptions.

c) State the hypotheses.

d) Test an appropriate hypothesis and state your results.

e) Interpret the meaning of the results and state the conclusion.

49. Management styles. Use the survey results in the table below to investigate differences in employee job satisfaction among organizations in the United States with different management styles.

	Employee Job Satisfaction			
Management Styles	Very Satisfied	Satisfied	Somewhat Satisfied	Not Satisfied
Exploitative Authoritarian	27	82	43	48
Benevolent Authoritarian	50	19	56	75
Laissez Faire	52	88	26	34
Consultative	71	83	20	26
Participative	101	59	20	20

a) Select the appropriate procedure.
b) Check the assumptions.
c) State the hypotheses.
d) Test an appropriate hypothesis and state your results.
e) Interpret the meaning of the results and state a conclusion.

50. Ranking companies. Every year, *Fortune* magazine lists the 100 best companies to work for, based on criteria such as pay, benefits, turnover rate, and diversity. In 2013, the top three were Google, SAS, and CHG Healthcare. Of the top 30, 11 experienced double-digit job growth (10% or more), 16, single-digit growth (1% to 10%), 2 had no growth, and Google did not report. Of the bottom 30, only 5 experienced double-digit job growth, 16 had single-digit growth, and 9 had no growth or job loss. Ignoring Google, is job growth in the best of the best places to work different from job growth in the bottom of that elite list?

a) Select the appropriate procedure.
b) Check the assumptions.
c) State the hypotheses.
d) Test an appropriate hypothesis and state your results.
e) Interpret the meaning of the results and state a conclusion.

51. Businesses and blogs. The Pew Internet & American Life Project routinely conducts surveys to gauge the impact of the Internet and technology on daily life. A recent survey asked respondents if they read online journals or blogs, an Internet activity of potential interest to many businesses. A subset of the data from this survey (*February–March 2007 Tracking Data Set*) shows responses to this question. Test whether reading online journals or blogs is independent of generation.

	Read Online Journal or Blog			
Generation	Yes, Yesterday	Yes, But Not Yesterday	No	Total
Gen-Y (18–30)	29	35	62	126
Gen X (31–42)	12	34	137	183
Trailing Boomers (43–52)	15	34	132	181
Leading Boomers (53–61)	7	22	83	112
Matures (62+)	6	21	111	138
Total	69	146	525	740

52. CyberShopping. It has become more common for shoppers to "comparison shop" using the Internet. Respondents to a Pew survey in 2013 who owned cell phones were asked whether they had, in the past 30 days, looked up the price of a product while they were in a store to see if they could get a better price somewhere else. Here is a table of their responses by income level.

	< $30K	$30K–$49.9K	$50K–$74.9K	> $75K
Yes	207	115	134	204
No	625	406	260	417

a) Is the frequency of comparison shopping on the Internet independent of the income level of the respondent? Perform an appropriate chi-square test and state your conclusion.
b) Calculate and examine the standardized residuals. What pattern (if any, do they show that would be of interest to retailers concerned about cybershopping comparisons?

53. Information systems. In a recent study of enterprise resource planning (ERP) system effectiveness, researchers asked companies about how they assessed the success of their ERP systems. Out of 335 manufacturing companies surveyed, they found that 201 used return on investment (ROI), 100 used reductions in inventory levels, 28 used improved data quality, and 6 used on-time delivery. In a survey of 200 service firms, 40 used ROI, 40 used inventory levels, 100 used improved data quality, and 20 used on-time delivery. Is there evidence that the measures used to assess ERP system effectiveness differ between service and manufacturing firms? Perform the appropriate test and state your conclusion.

54. U.S. Gross Domestic Product. The U.S. Bureau of Economic Analysis provides information on the Gross Domestic Product (GDP) in the United States by state (www.bea .gov). The Bureau recently released figures that showed the real GDP by state for 2007. Using the data in the table below examine if GDP and *Region* of the country are independent. (Alaska and Hawaii are part of the West Region. D.C. is included in the Mideast Region.)

	GDP		
	Top 40%	Bottom 60%	Total
West (Far West, Southwest, and Rocky Mtn.)	5	10	15
Midwest (Great Lakes and Plains States)	5	7	12
Southeast	5	7	12
Northeast (Mideast and New England States)	5	7	12
Total	20	31	51

55. Economic growth. The U.S. Bureau of Economic Analysis also provides information on the growth of the U.S. economy (www.bea.gov). The Bureau recently released figures that they claimed showed a growth spurt in the western region of the United States. Using the table and map below, determine if the percent change in real GDP by state for 2005–2006 was independent of region of the country. (Alaska and Hawaii are part of the West Region. D.C. is included in the Mideast Region.)

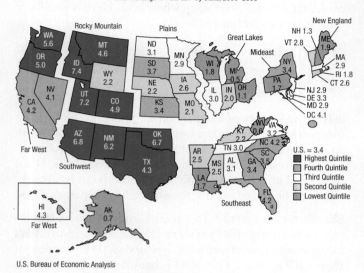

Percent Change in Real GDP by State, 2005–2006

U.S. Bureau of Economic Analysis

	GDP % Change		
	Top 40%	Bottom 60%	Total
West (Far West, Southwest, and Rocky Mtn.)	13	2	15
Midwest (Great Lakes and Plains States)	2	10	12
Southeast	4	8	12
Northeast (Mideast and New England States)	2	10	12
Total	21	30	51

56. Economic growth, revisited. The U.S. Bureau of Economic Analysis provides information on the GDP in the United States by metropolitan area (www.bea.gov). The Bureau recently released figures that showed the percent change in real GDP by metropolitan area for 2004–2005. Using the data in the following table, examine if there is independence of the growth in metropolitan GDP and region of the country. (Alaska and Hawaii are part of the West Region. Some of the metropolitan areas may have been combined for this analysis.)

	GDP Growth		
	Top Two Quintiles (top 40%)	Bottom Three Quintiles (bottom 60%)	Total
West (Far West, Southwest, and Rocky Mtn.)	62	46	108
Midwest (Great Lakes and Plains States)	9	87	96
Southeast	38	58	96
Northeast (Mideast and New England States)	12	36	48
Total	121	227	348

Percent Change in Real GDP by Metropolitan Area, 2004–2005

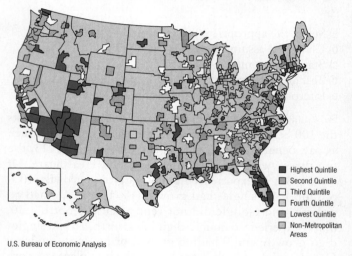

U.S. Bureau of Economic Analysis

JUST CHECKING ANSWERS

1 This is a test of homogeneity. The clue is that the question asks whether the distributions are alike.

2 This is a test of goodness-of-fit. We want to test the model of equal assignment to all lots against what actually happened.

3 This is a test of independence. We have responses on two variables for the same individuals.

Investment Strategy Segmentation

I n the aftermath of the financial crisis of 2008, brokerage firms struggled to get individual investors back into the stock market. To gain competitive advantage, market analysts in nearly every industry group their customers into different segments by placing advertisements where they can have the greatest impact. The brokerage business is no different. Because different groups of people invest differently, customizing advertising to the needs of these groups leads to more efficient advertising placement and response. Brokerage firms get their information about the investment practices of individuals from a variety of sources, among which the U.S. Census Bureau figures prominently.

The U.S. Census Bureau, with the help of the Bureau of Labor Statistics (BLS) and the Internal Revenue Service (IRS), monitors the incomes and expenditures of Americans. A random sample of Americans are surveyed periodically about their investment practices. In the file **Investment Strategy Segmentation** you'll find a random sample of 1000 people from the 48,842 records found in the file Census Income Data set on the University of California at Irvine machine learning repository. This subset was sampled from those that showed some level of investment in the stock market as evidenced by claiming *Capital Gains ($)*, *Capital Losses ($)*, or *Dividends ($)*. Included as well are the demographic variables for these people: *Age (years)*, *Sex (male/female)*, *Union Member (Yes/No)*, *Citizenship (several categories)*, *College (No College/Some)*, *Married (Married/Single)*, *Filer Status (Joint/Single)*.

To support her segmentation efforts, a market analyst at an online brokerage firm wants to study differences in investment behaviors. If she can find meaningful differences in the types and amounts of investing that various groups engage in, she can use that information to inform the advertising and marketing departments in their strategies to attract new investors. Using the techniques of Part III, including confidence intervals and hypothesis tests, what differences in investment behaviors can you find among the various demographic groups?

Some specific questions to consider:

1. Do men and women invest similarly? Construct confidence intervals for the differences in mean *Capital Gains*, *Capital Losses* and *Dividends*.

   Be sure to make a suitable display to check assumptions and conditions. If you find outlier(s), consider the analysis with and without the outlier(s).

2. Make a suitable display to compare investment results for the various levels of *Citizenship*.

3. Do those who file singly have the same investment results as those who file jointly? Select, perform, and interpret an appropriate test.

4. Compare differences in investment results for the other demographic variables, being careful to check assumptions and conditions.

   Once again, make suitable displays. Are there outliers to be concerned with? Discuss.

Summarize your findings and conclusions about the investment practices of various groups. Write a short report in order to help the market analyst.

14

Inference for Regression

Nambé Mills

Nambé (nam 'bei) Mills, Inc. was founded in 1951 near the tiny village of Nambé Pueblo, about 10 miles north of Santa Fe, New Mexico. Known for its elegant, functional cooking and tableware, Nambé Mills now sells its products in luxury stores throughout the world. Many of its products are made from an eight-metal alloy created at the Los Alamos National Laboratory (where the atomic bomb was developed during World War II) and now used exclusively by Nambé Mills. The alloy has the luster of silver and the solidity of iron, but its main component is aluminum. In fact, it does not contain silver, lead, or pewter (a tin and copper alloy), and it does not tarnish. Because it's a trade secret, Nambé Mills does not divulge the rest of the formula. Up to 15 craftsmen may be involved in the production process of an item, which includes molding, pouring, grinding, polishing, and buffing.

Because Nambé Mills's metal products are sand-cast, they must go through a lengthy production process. To rationalize its production schedule, management examined the total polishing times of 59 tableware items. Here's a scatterplot showing the retail price of the items and the amount of time (in minutes) spent in the polishing phase (Figure 14.1).

Figure 14.1 A scatterplot of *Price* ($) against polishing *Time* (minutes) for Nambé tableware products shows that the items that take longer to polish cost more, on average.

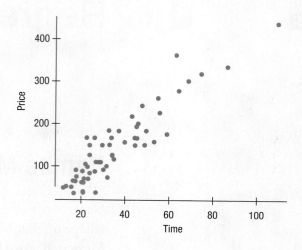

The equation of the least squares line for these data is:

$$\widehat{Price} = -4.871 + 4.200 \times Time$$

The slope says that, on average, the price increases by \$4.20 for every extra minute of polishing time.

In Chapter 4, we used a regression line just as a description of the data at hand. Now we'd like to know what the regression model can tell us beyond the sample. To do that, we'll make confidence intervals and test hypotheses about the slope and intercept of the regression line as we've done for the mean and proportion of a sample.

> **The Regression Model**
> Remember that we find the least squares line as $\hat{y} = b_0 + b_1 x$ where $b_1 = r s_y / s_x$ and $b_0 = \bar{y} - b_1 \bar{x}$.

14.1 A Hypothesis Test and Confidence Interval for the Slope

Our data are a sample of 59 items. If we take another sample, we hope the regression line will be similar to the one we found here, but we know it won't be exactly the same. Observations vary from sample to sample. But we can imagine a true idealized line that models the relationship between *Price* and *Time*. Following our usual conventions, we write the model using Greek letters and consider the coefficients (slope and intercept) to be parameters: β_0 is the intercept, and β_1 is the slope. Corresponding to our fitted line of $\hat{y} = b_0 + b_1 x$, we write $\mu_y = \beta_0 + \beta_1 x$. We write μ_y instead of y because the regression model assumes that the *means* of the y values for each value of x fall exactly on the line. We can picture the model as in Figure 14.2. The means are on the line, and the y values at each x are distributed around them.

Now, if only we had all the values in the population, we could find the slope and intercept of this *idealized regression line* explicitly by using least squares.

Figure 14.2 There's a distribution of *Prices* for each value of polishing *Time*. The regression model assumes that the means line up perfectly like this, and that the *y*-values are distributed with a Normal model around the line at each *x*-value.

Of course, the individual *y*'s are not at these means. In fact, the line will miss most—and usually all—of the plotted points. Some *y*'s lie above the line and some below the line, so like all models, this one makes errors. To account for each individual value of *y* in our model, we can include these errors, which we denote by ε:

$$y = \beta_0 + \beta_1 x + \varepsilon.$$

This equation has an ε to soak up the deviation at each point, so the model gives a value of *y* for each value of *x*. To do hypothesis tests and make confidence intervals, we assume that the errors have a Normal distribution around the line at each *x*-value, as shown in Figure 14.2.

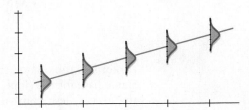

We estimate the β's by finding a regression line, $\hat{y} = b_0 + b_1 x$, as we did in Chapter 4. The residuals, $e = y - \hat{y}$ are the sample-based versions of the errors, ε. We'll use them to help us assess the regression model.

We know that least squares regression will give us reasonable estimates of the parameters of this model from a random sample of data. We also know that our estimates won't be equal to the parameters in the idealized or "true" model. Our challenge is to account for the uncertainty in our estimates by making confidence intervals as we've done for means and proportions. For that, we need to find the standard errors of the slope and intercept.

We expect the estimated slope for any sample, b_1, to be close to—but not actually equal to—the model slope, β_1. If we could see the collection of slopes from many samples (imagined or real) we would see a distribution of values around the true slope. That's the sampling distribution of the slope.

What is the standard deviation of this distribution? That depends on several aspects of our situation: how much the data vary, how solid a basis we have for our estimate, and how much data we have.

- **Spread around the line.** Figure 14.3 shows samples from two populations. Which underlying population would give rise to the more consistent slopes?

Figure 14.3 Which of these scatterplots would give the more consistent regression slope estimate if we were to sample repeatedly from its underlying population?

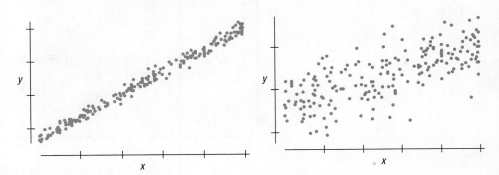

Less scatter around the line means the slope will be more consistent from sample to sample. In Chapter 4, we measured the spread around the line with the **residual standard deviation:**

$$s_e = \sqrt{\frac{\Sigma (y - \hat{y})^2}{n - 2}}.$$

The less scatter around the line, the smaller the residual standard deviation and the stronger the relationship between x and y.

- **Spread of the x's:** Here are samples from two more populations (Figure 14.4). Which of these would yield more consistent slopes?

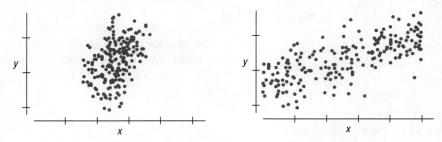

Figure 14.4   Which of these scatterplots would give the more consistent regression slope estimate if we were to sample repeatedly from the underlying population?

A plot like the one on the right has a broader range of x-values, so it gives a more stable base for the slope. We might expect the slopes of samples from situations like that to vary less from sample to sample. A large standard deviation of x, s_x, as in the figure on the right, provides a more stable regression.

- **Sample size.** What about the two scatterplots in Figure 14.5?

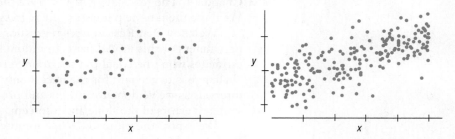

Figure 14.5   Which of these scatterplots would give the more consistent regression slope estimate if we were to sample repeatedly from the underlying population?

It shouldn't shock you that a larger sample size (scatterplot on the right) gives more consistent estimates from sample to sample.

Let's summarize what we've seen in these three figures:

The Standard Error of the Regression Slope

Three aspects of the scatterplot that affect the standard error of the regression slope are:

- Spread around the line: s_e
- Spread of x values: s_x
- Sample size: n

These are in fact the *only* things that affect the standard error of the slope. The formula for the standard error of the slope is:

$$SE(b_1) = \frac{s_e}{s_x \sqrt{n-1}}.$$

The error standard deviation, s_e, is in the *numerator*, since a larger spread around the line *increases* the slope's standard error. On the other hand, the *denominator* has both a sample size term $(\sqrt{n-1})$ and s_x because increasing either of these *decreases* the slope's standard error.

Our goal is a hypothesis test or confidence interval for the slope. The standard deviation of the sampling distribution of the slope is one part, but we need to know the shape of the sampling distribution as well. Here the Central Limit

Theorem and Gosset come to the rescue again. We do just what we've done with the mean and proportion. We standardize the estimated slope by subtracting the model value and dividing by the estimated standard error. And just as we did with the mean, we get a Student's t-model, this time with $n - 2$ degrees of freedom:

$$\frac{b_1 - \beta_1}{SE(b_1)} \sim t_{n-2}.$$

> ## The Sampling Distribution for the Regression Slope
> When the conditions are met, the standardized estimated regression slope,
> $$t = \frac{b_1 - \beta_1}{SE(b_1)},$$
> follows a Student's t-model with $n - 2$ degrees of freedom. We estimate the standard error with $SE(b_1) = \dfrac{s_e}{s_x \sqrt{n - 1}}$, where $s_e = \sqrt{\dfrac{\sum (y - \hat{y})^2}{n - 2}}$, n is the number of data values, and s_x is the standard deviation of the x-values.

> ### What If the Slope Is 0?
> If $b_1 = 0$, our prediction is $\hat{y} = b_0 + 0x$, and the equation collapses to just $\hat{y} = b_0$. Now x is nowhere in sight, so y doesn't depend on x at all.
>
> In this case, b_0 would turn out to be $\bar{y}$. Why? Because we know that $b_0 = \bar{y} - b_1\bar{x}$, and when $b_1 = 0$, that becomes simply $b_0 = \bar{y}$. It turns out, that when the slope is 0, the entire regression equation is just $\hat{y} = \bar{y}$, so for every value of x, we predict the mean value ($\bar{y}$) for y.

Now that we have the standard error of the slope and its sampling distribution, we can test a hypothesis about it and make confidence intervals. The usual null hypothesis about the slope is that it's equal to 0. Why? Well, a slope of zero would say that y doesn't tend to change linearly when x changes—in other words, that there is no linear association between the two variables. If the slope were zero, there wouldn't be much left of our regression equation.

A null hypothesis of a zero slope questions the entire claim of a linear relationship between the two variables, and often that's just what we want to know. In fact, every software package or calculator that does regression simply assumes that you want to test the null hypothesis that the slope is really zero.

> ## The t-Test for the Regression Slope
> When the assumptions and conditions are met, we can test the hypothesis $H_0: \beta_1 = 0$ vs. $H_A: \beta_1 \neq 0$ (or a one-sided alternative hypothesis) using the standardized estimated regression slope,
> $$t = \frac{b_1 - \beta_1}{SE(b_1)},$$
> which follows a Student's t-model with $n - 2$ degrees of freedom. We can use the t-model to find the P-value of the test.

This is just like every other t-test we've seen: a difference between the statistic and its hypothesized value divided by its standard error. This test is the t-test that the regression slope is 0, usually referred to as the ***t*-test for the regression slope**.

Another use of these values might be to make a confidence interval for the slope. We can build a confidence interval in the usual way, as an estimate plus or minus a margin of error. As always, the margin of error is just the product of the standard error and a critical value.

> ## The Confidence Interval for the Regression Slope
> When the assumptions and conditions are met, we can find a confidence interval for β_1 from
> $$b_1 \pm t^*_{n-2} \times SE(b_1),$$
> where the critical value t^* depends on the confidence level and has $n - 2$ degrees of freedom.

The same reasoning applies for the intercept. We write:

$$\frac{b_0 - \beta_0}{SE(b_0)} \sim t_{n-2}.$$

We could use this statistic to construct confidence intervals and test hypotheses about the intercept, but often the value of the intercept isn't interesting. Most hypothesis tests and confidence intervals for regression are about the slope. But in case you really want to see the formula for the standard error of the intercept, we've parked it in a footnote.[1]

Regression models are almost always found with a computer or calculator. The calculations are too long to do conveniently by hand for data sets of any reasonable size. No matter how the regression is computed, the results are usually presented in a table that has a standard form. Figure 14.6 shows a portion of a typical regression results table, along with annotations showing where the numbers come from.

Figure 14.6 A typical computer regression table presents regression results in a standard format.

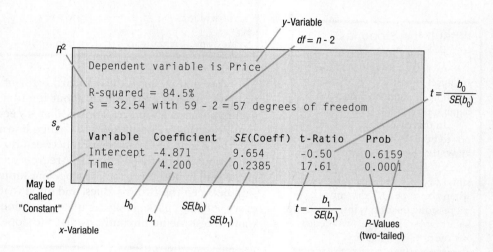

14.2   Assumptions and Conditions

Back in Chapter 4 when we fit lines to data, we needed both the **Linearity** and the **Equal Variance Assumptions**. Now, to make inferences about the coefficients of the line, we'll have to assume that the errors are independent and Normally distributed as well.

It is important to check the assumptions and their associated conditions in the right order, so we number the assumptions, and check conditions for each in that order: (1) Linearity Assumption, (2) Independence Assumption, (3) Equal Variance Assumption, and (4) Normal Population Assumption.

1. Linearity Assumption

If the true relationship of two quantitative variables is far from linear and we use a straight line to fit the data, our entire analysis will be useless, so we always check linearity first (and we check the **Quantitative Variable Condition** for both variables as well).

The **Linearity Condition** is satisfied if a scatterplot looks straight. It's generally not a good idea to draw a line through the scatterplot when checking. That can fool your eye into seeing the plot as straighter than it really is. Sometimes it's easier to see

---

[1] $SE(b_0) = s_e \sqrt{\dfrac{1}{n} + \dfrac{\bar{x}^2}{\sum (x - \bar{x})^2}}$

violations of this condition by looking at a scatterplot of the residuals against x or against the predicted values, $\hat{y}$. That plot should have no pattern if the condition is satisfied.

If the scatterplot shows a reasonably straight relationship, we can go on to some assumptions about the errors. If not, we stop here, or consider transforming the variables to make the scatterplot more linear.

2. Independence Assumption

The errors in the true underlying regression model (the ε's) must be independent of each other. As usual, there's no way to be sure that the Independence Assumption is true.

When we care about inference for the regression parameters, it's often because we think our regression model might apply to a larger population. In such cases, we can check the **Randomization Condition** that the individuals are a random sample from that population.

We can also check displays of the regression residuals for evidence of patterns, trends, or clumping, any of which would suggest a failure of independence. In the special case when we have a time series, a common violation of the Independence Assumption is for successive errors to be correlated (autocorrelated). (The error our model makes today may be similar to the one it made yesterday.) We can check this violation by plotting the residuals against time (usually the x-variable for a time series) and looking for patterns.

3. Equal Variance Assumption

The variability of y should be about the same for all values of x. In Chapter 4, we found the standard deviation of the residuals (s_e). Now we need this standard deviation to estimate the standard errors of the coefficients but the residual standard deviation only makes sense if the scatter of the residuals is the same everywhere (else what are we estimating?)

We check the **Equal Spread Condition** by looking at a scatterplot of residuals against either x or $\hat{y}$. Make sure the spread around the line is nearly constant. Be alert for a "fan" shape or other tendency for the variation to grow or shrink in one part of the scatterplot.

If the plot is reasonably straight, the data are independent, and the spread doesn't change, we can move on to the final assumption and its associated condition.

4. Normal Population Assumption

To use a Student's t-model for inference, we must assume the errors around the idealized regression line at each value of x follow a Normal model. As we did before when we used Student's t, we'll settle for the residuals satisfying the **Nearly Normal Condition**.[2] As we have noted before, the Normality Assumption becomes less important as the sample size grows because the model is about means, and the Central Limit Theorem takes over. You can check a histogram of the residuals or look at a **Normal probability plot** (see Section 7.3), which finds deviations from the Normal model more efficiently. Another common failure of Normality is the presence of an outlier. So, we still check the **Outlier Condition**.

Summary of Assumptions and Conditions

We don't expect the assumptions to be exactly true. As George Box said, "all models are wrong." But the linear model is often close enough to be useful as long as the assumptions of the model are reasonably met.

Before we compute the regression, we should check the linearity condition. (You can also look for outliers here.) Then, after we fit the model we should check the rest.

---

[2] *This* is why we check the conditions in order. We check that the residuals are independent and that the variation is the same for all x's before we can lump all the residuals together to check the Normal Condition.

"Truth will emerge more readily from error than from confusion."
—FRANCIS BACON (1561–1626),
ENGLISH PHILOSOPHER

So we work in this order:

1. **Make a scatterplot of the data** to check the Linearity Condition and to look for outliers (and always check that the variables are quantitative as well). (This checks the **Linearity Assumption**.)
2. If the data show a reasonably straight relationship, **fit a regression and find the residuals, *e*, and predicted values, *ŷ*.**
3. If you know when the measurements were made, **plot the residuals against time** to check for evidence of patterns that suggest they may not be independent (**Independence Assumption**).
4. **Make a scatterplot of the residuals against *x* or the predicted values.** This plot should have no pattern. Check in particular for any bend (which would suggest that the data weren't that straight after all), for any thickening (or thinning), and, of course, for any unusual observations. (If you discover any errors, correct them or omit those points, and go back to step 1. Otherwise, consider performing two regressions—one with and one without the unusual observations.) (**Equal Variance Assumption**)
5. If the scatterplots look OK, then **make a histogram and Normal probability plot of the residuals** to check the **Nearly Normal** and **Outlier Conditions** (**Normal Population Assumption**).

GUIDED EXAMPLE   Nambé Mills

Now that we have a method to draw inferences from our regression equation, let's try it out on the Nambé Mills data. The slope of the regression gives the impact of *Time* on *Price*. Let's test the hypothesis that the slope is different from zero.

PLAN **Setup** State the objectives.	We want to test the null hypothesis that the price of Nambé Mills items is not related to the time it takes to polish them. We have data for 59 items sold by Nambé Mills. The slope of this relationship will model the relationship between *Time* and *Price*. Our null hypothesis is that the slope of the regression is 0.

Identify the parameter you wish to estimate. Here our parameter is the slope.

Identify the variables and their context.

Hypotheses   Write the null and alternative hypotheses.

H_O: The *Price* of an item is not linearly related to the polishing *Time*: $\beta_1 = 0$.
H_A: The *Price* is related to the *Time*: $\beta_1 \neq 0$.

Model   Check the assumptions and conditions.

Make graphs. Because our scatterplot of *y* versus *x* looks reasonably straight, we can find the least squares regression and plot the residuals.

✓ **Linearity Condition:** There is no obvious curve in the scatterplot of *y* versus *x*.

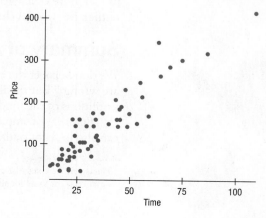

If it is appropriate, we check for suggestions that the Independence Assumption fails by plotting the residuals against time. Patterns or trends in that plot raise our suspicions.

✓ **Independence Assumption:** These data are on 59 different items manufactured by the company. There is no reason to suggest that the error in modeling the price of one item should be influenced by another.

✓ **Randomization Condition:** The data are *not* a random sample, but we assume they are representative of the prices and polishing times of Nambé Mills items.

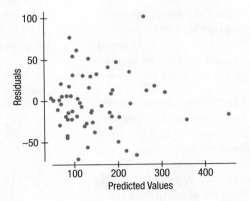

✓ **Equal Spread Condition:** The plot of residuals against the predicted values shows no obvious patterns. The spread is about the same for all predicted values, and the scatter appears random.

✓ **Nearly Normal Condition:** A histogram of the residuals is unimodal and symmetric, and the normal probability plot is reasonably straight.

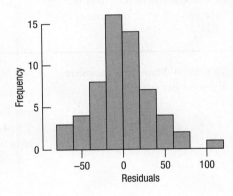

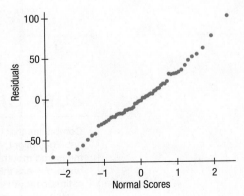

State the sampling distribution model.

Choose the method.

Under these conditions, the sampling distribution of the regression slope can be modeled by a Student's t-model with $(n - 2) = 59 - 2 = 57$ degrees of freedom, so we'll proceed with a regression slope t-test.

(*continued*)

DO

Mechanics The regression equation can be found from the formulas in Chapter 4, but regressions are almost always found from a computer program or calculator.

The P-values given in the regression output table are from the Student's *t*-distribution on $(n - 2) = 57$ degrees of freedom. They are appropriate for two-sided alternatives.

Create a confidence interval for the true slope. To obtain the *t*-value for 57 degrees of freedom, use the *t*-table at the back of your textbook. The estimated slope and SE for the slope are obtained from the regression output.

Interpret the interval.

Simply rejecting the standard null hypothesis doesn't guarantee that the size of the effect is large enough to be important.

Here's the computer output for this regression.

Variable	Coefficient	SE(coeff)	t-Ratio	P-Value
Intercept	−4.871	9.654	−0.50	0.6159
Time	4.200	0.2385	17.61	< 0.0001

$s = 32.54$ R-Sq = 84.5%

The P-value<0.0001 means that the association we see in the data is unlikely to have occurred by chance. Therefore, we reject the null hypothesis and conclude that there is strong evidence that the *Price* is linearly related to the polishing *Time*.

A 95% confidence interval for β_1 is:

$$b_1 \pm t^*_{n-2} \times SE(b_1) = (3.72, 4.68)\$/minute$$

We can be 95% confident that the price is higher on average, between $3.72 and $4.68 for each additional minute of polishing time. (Technically: we are 95% confident that the interval from $3.72 to $4.68 per minute captures the true rate at which the *Price* increases with polishing *Time*.)

REPORT

Conclusion State the conclusion in the proper context.

MEMO

Re: Nambé Mills pricing

We investigated the relationship between polishing time and pricing of 59 Nambé Mills items. The regression analysis showed that, on average, the price is higher by $4.20 for every additional minute of polishing time. Assuming that these items are representative, we are 95% confident that the actual price of a metal item manufactured by Nambé Mills is higher by between $3.72 and $4.68 on average for each additional minute of polishing work required.

JUST CHECKING

Companies that market food items conduct research into how and how much people eat. They might, for example, study how big people's mouths tend to be. Researchers measured mouth volume by pouring water into the mouths of subjects who lay on their backs. Unless this is your idea of a good time, it would be helpful to have a model to estimate mouth volume more simply. Fortunately, mouth volume is related to height. (Mouth volume is measured in cubic centimeters and height in meters.)

The data were checked and deemed suitable for regression. Take a look at the computer output below.

```
Summary of Mouth Volume
Mean            60.2704
StdDev          16.8777
```

```
Dependent variable is Mouth Volume
R-squared = 15.3%
s = 15.66 with 61 - 2 = 59 degrees of freedom

Variable      Coefficient   SE(coeff)   t-Ratio   P-Value
Intercept      -44.7113       32.16      -1.39     0.1697
Height          61.3787       18.77       3.27     0.0018
```

1 What does the t-ratio of 3.27 for the slope tell about this relationship? How does the P-value help your understanding?

2 Would you say that measuring a person's height could reliably be used as a substitute for the wetter method of determining how big a person's mouth is? What numbers in the output helped you reach that conclusion?

3 What does the value of s_e add to this discussion?

*A Hypothesis Test for Correlation

We just tested whether the slope, β_1, was 0. To test it, we estimated the slope from the data and then, using its standard error and the t-distribution, measured how far the slope was from 0: $t = \dfrac{b_1 - 0}{SE(b_1)}$. What if we wanted to test whether the *correlation* between x and y is 0? We write ρ for the parameter (true population value) of the correlation, so we're testing $H_0: \rho = 0$. Remember that the regression slope estimate is $b_1 = r\dfrac{s_y}{s_x}$. The same is true for the parameter versions of these statistics: $\beta_1 = \rho\dfrac{\sigma_y}{\sigma_x}$. That means that if the slope is really 0, then the correlation has to be 0, too. So if we test $H_0: \beta_1 = 0$, that's really the same as testing $H_0: \rho = 0$. Sometimes a researcher, however, might want to test correlation without fitting a regression, so you'll see the test of correlation as a separate test (it's also slightly more general), but the results are mathematically the same even though the form looks a little different. Here's the t-test for the correlation coefficient.

The t-Test for the Correlation Coefficient

When the conditions are met, we can test the hypothesis $H_0: \rho = 0$ vs. $H_A: \rho \neq 0$ using the test statistic:

$$t = r\sqrt{\frac{n - 2}{1 - r^2}},$$

which follows a Student's t-model with $n - 2$ degrees of freedom. We can use the t-model to find the P-value of the test.

A test that rejects the null hypothesis for the correlation tells us that there is a real linear relationship between the variables, but it doesn't tell enough. In the For Example, we conclude that the correlation isn't 0, but where do we go from here? For business decisions, we need to know more. We need the information provided by the full regression model. The test that the correlation is zero is equivalent to the test that the slope is zero, but the regression model helps us to understand what's going on and to make real decisions.

14.3   Standard Errors for Predicted Values

Often a business decision will depend not on the value of a slope coefficient, but on a predicted value of the response variable, *y*, for some given value of *x*. We saw how to find a predicted value for any value of *x* back in Chapter 4. This predicted value would be our best estimate, but it's still just an informed guess. Now that we have standard errors, we can use them to construct confidence intervals for the predictions. That makes it possible to report our uncertainty about those predictions honestly—something we'd want to know before relying on them for a decision.

From our model of Nambé Mills items, we can use polishing *Time* to get a reasonable estimate of *Price*. Suppose a manager wanted to estimate the selling *Price* of an item that takes 40 minutes of *Time* to polish. A confidence interval can indicate how precise that prediction is. The precision depends on the question asked, however, and there are two different questions we could ask:

Do we want to know the mean *Price* for *all items* that have a polishing *Time* of 40 minutes?

or,

Do we want to estimate the *Price* for a *particular* item whose polishing *Time* is 40 minutes?

What's the difference between the two questions? The manufacturer might be more interested in the first question which asks about the *mean Price* of all items that take a certain *Time* to polish. But a customer or sales manager might be more interested in the second question—estimating an *individual* item's *Price*. The predicted *Price* value is the same for both, but one question leads to a much more precise interval than the other. If your intuition says that it's easier to be more precise about the mean than about an individual, you're on the right track. Because, as we have seen, means vary much less than individuals, we can predict the *mean Price* for all items with the same polishing *Time*, more precisely than we can predict the *Price* of a particular item with that polishing *Time*.

Let's start by predicting the mean *Price* for a new *Time*, one that was not necessarily part of the original data set. To emphasize this, we'll call this *x*-value "*x* sub new" and write it x_ν.[3] As an example, we'll take x_ν to be 40 minutes. The regression equation predicts *Price* by $\hat{y}_\nu = b_0 + b_1 x_\nu$. Now that we have the predicted value, we can construct a confidence interval around this number. It has the usual form:

$$\hat{y}_\nu \pm t^*_{n-2} \times SE.$$

The t^* value is the critical value (from Table T or technology) for $n - 2$ degrees of freedom and the specified confidence level.

The Confidence Interval for the Predicted Mean Value

When the conditions are met, we find the confidence interval for the predicted mean value μ_ν at a value x_ν as

$$\hat{y}_\nu \pm t^*_{n-2} \times SE,$$

where the standard error is

$$SE(\hat{\mu}_\nu) = \sqrt{SE^2(b_1) \times (x_\nu - \bar{x}^2) + \frac{s_e^2}{n}}.$$

Figure 14.7 shows the confidence intervals for the mean predictions. In this plot, the intervals for all the mean *Prices* at all values of *Time* are shown together as confidence bands. Notice that the bands get wider as we attempt to predict values that lie farther away from the mean *Time*. (That's due to the $(x_\nu - \bar{x})^2$ term in the SE formula.) As we move away from the mean *x* value, there is more uncertainty associated with our prediction.

Figure 14.7 The confidence intervals for the mean *Price* at a given polishing *Time* are shown as the green dotted lines. Near the mean *Time* (35.8 minutes) our confidence interval for the mean *Price* is narrower than for values far from the mean, like 100 minutes.

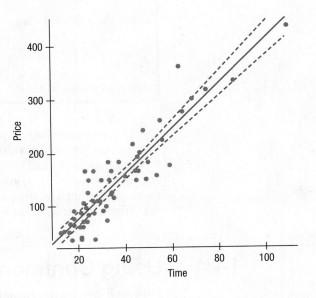

---

[3] Yes, this is a bilingual pun. The Greek letter ν is called "nu." Don't blame me; my co-author suggested this.

Like all confidence intervals, the width of these confidence intervals varies with the sample size. A larger sample would result in narrower intervals.

The last factor affecting our confidence intervals is the spread of the data around the line. If there were more spread around the line, predictions would be less certain, and the confidence interval bands would be wider.

From Figure 14.7, it's easy to see that most *points* don't fall within the confidence interval bands—and we shouldn't expect them to. These bands show confidence intervals for the *mean*. An even larger sample would have given even narrower bands. Then we'd expect an even smaller percentage of the points to fall within them.

If we want to capture an *individual* price, we need to use a wider interval, called a **prediction interval**. Figure 14.8 shows the prediction intervals for the Nambé Mills data. Prediction intervals are based on the same quantities as the confidence intervals, but the standard error includes an extra term for the spread around the line. As Figure 14.8 shows, these bands also widen as we move from the mean of x.

Figure 14.8 Prediction intervals (in red) estimate the interval that contains say, 95% of the distribution of the y values that might be observed at a given value of x. If the assumptions and conditions hold, then there's about a 95% chance that a particular y-value at x_ν will be covered by the interval.

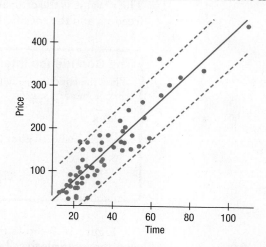

The Prediction Interval for an Individual Value

When the conditions are met, we can find the prediction interval for all values of y at a value x_ν as

$$\hat{y}_\nu \pm t^*_{n-2} \times SE,$$

where the standard error is

$$SE(\hat{y}_\nu) = \sqrt{SE^2(b_1) \times (x_\nu - \bar{x})^2 + \frac{s_e^2}{n} + s_e^2}.$$

The critical value t^* depends on the confidence level that you specify.

The good news about confidence intervals for prediction is that they are not hard to find and are provided by most statistics software. The bad news is that intervals for an individual predicted value can be quite wide—often so wide that they are not very useful for a business decision. By contrast, confidence intervals for the mean prediction are often narrow enough to be useful. But you must remember the difference between the two kinds of prediction intervals.

14.4 Using Confidence and Prediction Intervals

How well can our regression model predict the mean price for objects that take 25 minutes to polish? The regression output table provides most of the numbers we need.

```
Variable    Coefficient   SE(coeff)   t-Ratio   P-Value
Intercept     -4.871        9.654      -0.50     0.6159
Time           4.200        0.2385     17.61    < 0.0001

s = 32.54  R-Sq = 84.5%
```

The regression model gives a predicted value at x_ν = 25 minutes of:

$$-4.871 + 4.200(25) = \$100.13$$

Many statistics programs can also provide prediction confidence intervals. The 95% confidence interval for the mean price of items with a 25-minute polishing time is ($90.20, $110.06). That may be precise enough to support decisions about where to advertise these products, for example.

But the 95% confidence interval for a particular item with a 25-minute polishing time is ($34.22, $166.04). That's so wide that it serves more as a warning not to believe that the predicted value of $100.13 provides that much information about a single item. Figure 14.9 shows both intervals.

For the details of how the formulas lead to these intervals, see the Math Box.

Figure 14.9 A scatterplot of *Price* versus *Time* with a least squares regression line. The inner lines (green) near the regression line show the extent of the 95% confidence intervals, and the outer lines (red) show the prediction intervals. Most of the points are contained within the prediction intervals (as they should be), but not within the confidence interval for the means.

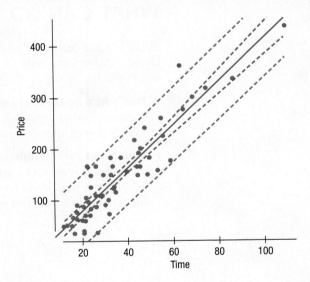

MATH BOX Finding Confidence and Prediction Intervals

We find the standard errors from the formula using the values in the regression output and associated values. (The mean *Polishing Time* in these data is 35.82 minutes.)

$$SE(\hat{\mu}_\nu) = \sqrt{(SE^2(b_1))\,(x_\nu - \bar{x})^2 + \left(\frac{s_e}{\sqrt{n}}\right)^2}$$

$$= \sqrt{(0.2385)^2\,(25 - 35.82)^2 + \left(\frac{32.54}{\sqrt{59}}\right)^2} = \$4.96$$

The t^* value that excludes 2.5% in either tail with $59 - 2 = 57$ df is (according to the tables) 2.002.

Putting it all together, the margin of error is:

$$ME = 2.002(4.96) = \$9.93$$

So, the 95% confidence interval is

$$\$100.13 \pm 9.93 = (\$90.20, \$110.06)$$

(continued)

To make a prediction interval for an *individual* item's price with a polishing time of 25 minutes, we use the formula

$$SE(\hat{y}_\nu) = \sqrt{(SE^2(b_1))(x_\nu - \bar{x})^2 + \frac{s_e^2}{n} + s_e^2} = \$32.92,$$

and find the ME to be

$$ME = t^* SE(\hat{y}_\nu) = 2.002 \times 32.92 = \$65.91,$$

and so the prediction interval is

$$\hat{y} \pm ME = 100.13 \pm 65.91 = (\$34.22, \$166.04).$$

WHAT CAN GO WRONG?

With inference, we've put numbers on our estimates and predictions, but these numbers are only as good as the model. Here are the main things to watch out for:

- **Don't fit a linear regression to data that aren't straight.** This is the most fundamental assumption. If the relationship between x and y isn't approximately linear, there's no sense in fitting a straight line to it.

- **Watch out for changing spread.** The common part of confidence and prediction intervals is the estimate of the error standard deviation, the spread around the line. If it changes with x, the estimate won't make sense. Imagine making a prediction interval for these data:

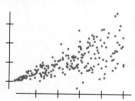

When x is small, we can predict y precisely, but as x gets larger, it's much harder to pin y down. Unfortunately, if the spread changes, the single value of s_e won't pick that up. The prediction interval will use the average spread around the line, with the result that we'll be too pessimistic about our precision for low x-values and too optimistic for high x-values. A re-expression of y is often a good fix for changing spread.

- **Watch out for non-Normal errors.** When we make a prediction interval for an individual y-value, the Central Limit Theorem can't come to our rescue. For us to believe the prediction interval, the errors must follow the Normal model. Check the histogram and Normal probability plot of the residuals to see if this assumption looks reasonable.

- **Watch out for one-tailed tests.** Because tests of hypotheses about regression coefficients are usually two-tailed, software packages report two-tailed P-values. If you are using that type of software to conduct a one-tailed test about the slope, you'll need to divide the reported P-value by two.

ETHICS IN ACTION

The need for senior care businesses that offer companionship and nonmedical home services is increasing as the U.S. population continues to age. One such franchise, Independent Senior Care, tries to set itself apart from its competitors by offering an additional service to prospective franchisees. In addition to standard information packets that provide tools, training, and mentorship opportunities, Independent Senior Care has an analyst on staff, Allen Ackman, to help prospective franchisees evaluate the feasibility of opening an elder care business in their area.

Allen was contacted recently by Kyle Sennefeld, a recent business school graduate with a minor in gerontology, who is interested in starting a senior care franchise in northeastern Pennsylvania. Allen decides to use a regression model that relates annual profit to the number of residents over the age of 65 that live within a 100-mile radius of a franchise location. Even though the R^2 for this model is small, the slope is statistically significant, and the model is easy to explain to prospective franchisees. Allen sends Kyle a report that estimates the annual profit at Kyle's proposed location. Kyle was excited to see that opening an Independent Senior Care franchise in northeastern Pennsylvania would be a good business decision.

- Identify the ethical dilemma in this scenario.

- What are the undesirable consequences?

- Propose an ethical solution that considers the welfare of all stakeholders.

WHAT HAVE WE LEARNED?

Learning Objectives

Apply your understanding of inference for means using Student's *t* to make inference about regression coefficients.

Know the Assumptions and Conditions for inference about regression coefficients and how to check them, in this order:

- **Linearity Assumption,** checked with the Linearity Condition by examining a scatterplot of *y* vs. *x* or a scatterplot of the residuals plotted against the predicted values.
- **Independence Assumption,** which can't be checked, but is more plausible if the data were collected with appropriate randomization—the Randomization Condition.
- **Equal Variance Assumption,** which requires that the spread around the regression model be the same everywhere. We check it with the Equal Spread Condition, assessed with a scatterplot of the residuals versus the predicted values.
- **Normal Population Assumption,** which is required to use Student's *t*-models unless the sample size is large. Check it with the Nearly Normal Condition by making a histogram or normal probability plot of the residuals.

Know the components of the standard error of the slope coefficient:

- The standard deviation of the residuals, $s_e = \sqrt{\dfrac{\sum (y - \hat{y})^2}{n - 2}}$

- The standard deviation of *x*, $s_x = \sqrt{\dfrac{\sum (x - \bar{x})^2}{n - 1}}$

- The sample size, *n*

Be able to find and interpret the standard error of the slope.

- $SE(b_1) = \dfrac{s_e}{s_x \sqrt{n - 1}}$

- The standard error of the slope is the estimated standard deviation of the sampling distribution of the slope.

State and test the standard null hypothesis on the slope.

- $H_0: \beta_1 = 0$. This would mean that *x* and *y* are not linearly related.

- We test this null hypothesis using the t-statistic $t = \dfrac{b_1 - 0}{SE(b_1)}$.

Construct and interpret a confidence interval for the predicted mean value corresponding to a specified value, x_ν.

- $\hat{y}_\nu \pm t^*_{n-2} \times SE(\hat{\mu}_\nu)$, where $SE(\hat{\mu}_\nu) = \sqrt{SE^2(b_1) \times (x_\nu - \overline{x})^2 + \dfrac{s_e^2}{n}}$.

Construct and interpret a confidence interval for an individual predicted value corresponding to a specified value, x_ν.

- $\hat{y}_\nu \pm t^*_{n-2} \times SE(\hat{y}_\nu)$, where $SE(\hat{y}_\nu) = \sqrt{SE^2(b_1) \times (x_\nu - \overline{x})^2 + \dfrac{s_e^2}{n} + s_e^2}$.

Terms

Confidence interval for the predicted mean value

Different samples will give different estimates of the regression model and, so, different predicted values for the same value of x. We find a confidence interval for the mean of these predicted values at a specified x-value, x_ν, as

$$\hat{y}_\nu \pm t^*_{n-2} \times SE(\hat{\mu}_\nu),$$

where

$$SE(\hat{\mu}_\nu) = \sqrt{SE^2(b_1) \times (x_\nu - \overline{x})^2 + \dfrac{s_e^2}{n}}.$$

The critical value, t^*_{n-2}, depends on the specified confidence level and the Student's t-model with $n-2$ degrees of freedom.

Confidence interval for the regression slope

When the assumptions are satisfied, we can find a confidence interval for the slope parameter from $b_1 \pm t^*_{n-2} \times SE(b_1)$. The critical value, t^*_{n-2}, depends on the confidence interval specified and on the Student's t-model with $n-2$ degrees of freedom.

Prediction interval for a future observation

A confidence interval for individual values. Prediction intervals are to observations as confidence intervals are to parameters. They predict the distribution of individual values, while confidence intervals specify likely values for a true parameter. When the assumptions are satisfied, the prediction interval takes the form

$$\hat{y}_\nu \pm t^*_{n-2} \times SE(\hat{y}_\nu),$$

where

$$SE(\hat{y}_\nu) = \sqrt{SE^2(b_1) \times (x_\nu - \overline{x})^2 + \dfrac{s_e^2}{n} + s_e^2}.$$

The critical value, t^*_{n-2}, depends on the specified confidence level and the Student's t-model with $n-2$ degrees of freedom. The extra s_e^2 in $SE(\hat{y}_\nu)$ makes the interval wider than the corresponding confidence interval for the mean.

Residual standard deviation

The measure, denoted s_e, of the spread of the data around the regression line:

$$s_e = \sqrt{\frac{\sum (y - \hat{y})^2}{n - 2}} = \sqrt{\frac{\sum e^2}{n - 2}}.$$

t-test for the regression slope

The usual null hypothesis is that the true value of the slope is zero. The alternative is that it is not. A slope of zero indicates a complete lack of linear relationship between y and x.

To test $H_0: \beta_1 = 0$ we find

$$t = \frac{b_1 - 0}{SE(b_1)},$$

where $SE(b_1) = \dfrac{s_e}{s_x \sqrt{n-1}}$, $s_e = \sqrt{\dfrac{\sum (y - \hat{y})^2}{n - 2}}$, n is the number of cases, and s_x is the standard deviation of the x-values. We find the P-value from the Student's t-model with $n-2$ degrees of freedom.

TECHNOLOGY HELP: Regression Analysis

All statistics packages make a table of results for a regression. These tables differ slightly from one package to another, but all are essentially the same.

All packages offer analyses of the residuals. With some, you must request plots of the residuals as you request the regression. Others let you find the regression first and then analyze the residuals afterward. Either way, your analysis is not complete if you don't check the residuals with a histogram or Normal probability plot and a scatterplot of the residuals against x or the predicted values.

You should, of course, always look at the scatterplot of your two variables before computing a regression.

Regressions are almost always found with a computer or calculator. The calculations are too long to do conveniently by hand for data sets of any reasonable size. No matter how the regression is computed, the results are usually presented in a table that has a standard form. Here's a portion of a typical regression results table, along with annotations showing where the numbers come from.

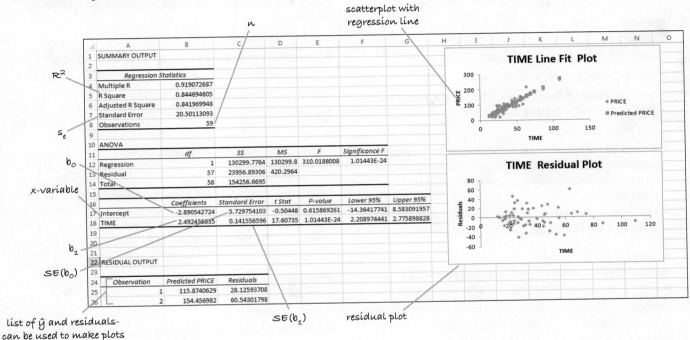

The regression table gives the coefficients (once you find them in the middle of all this other information). This regression (for different items than in the example in the text) predicts Price from Time. The regression equation is

$$\widehat{Price} = -2.891 + 2.492\ Time$$

and the R^2 for the regression is 84.5%.

The column of t-ratios gives the test statistics for the respective null hypotheses that the true values of the coefficients are zero. The corresponding P-values are also usually reported.

EXCEL

To perform a regression analysis in Excel:

- From **Data**, select **Data Analysis** and select **Regression.**
- Enter the data range holding the y-variable in the box labeled "Input Y range".
- Enter the range of cells holding the x-variable in the box labeled "Input X range".
- Select the **New Worksheet Ply** option to report results in a new worksheet (or identify output range in current worksheet or new

workbook) and **Labels** if the first row of the data holds the variable labels.

- Select **Residuals**, **Residual Plots**, and **Line Fit Plots** options.

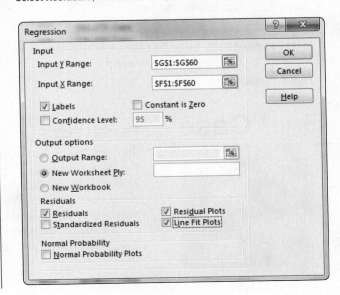

- After the plots are generated, you can delete the predicted values and/or add the least squares line, called a Trendline in Excel, on the Line Fit Plot.
- To obtain a histogram of the residuals, use residuals listed in Excel output to create a histogram using **Data > Data Analysis > Histogram**.

Comments

The *Y* and *X* ranges do not need to be in the same rows of the spreadsheet, although they must cover the same number of cells. But it is a good idea to arrange your data in parallel columns as in a data table to reduce the chance of error. Although the dialog offers a Normal probability plot of the residuals, the data analysis add-in does not make a correct probability plot, so don't use this option.

JMP

To perform a regression analysis in JMP,

- From the **Analyze** menu, select **Fit Y by X.**
- Select variables: a **Y, Response** variable, and an **X, Factor** variable. Both must be continuous (quantitative).
- JMP makes a scatterplot.
- Click on the red triangle beside the heading labeled **Bivariate Fit . . .** and choose **Fit Line.** JMP draws the least squares regression line on the scatterplot and displays the results of the regression in tables below the plot.
- The portion of the table labeled "Parameter Estimates" gives the coefficients and their standard errors, *t*-ratios, and P-values.

Comments

JMP chooses a regression analysis when both variables are "Continuous." If you get a different analysis, check the variable types.

The Parameter table does not include the residual standard deviation s_e. You can find that as Root Mean Square Error in the Summary of Fit panel of the output.

MINITAB

To compute a regression analysis in Minitab,

- Choose **Regression** from the **Stat** menu.
- Choose **Regression . . .** from the **Regression** submenu.
- In the Regression dialog, assign the Y-variable to the Response box and assign the X-variable to the Predictors box.
- Click the **Graphs** button.

- In the Regression-Graphs dialog, select **Standardized residuals,** and check **Normal plot of residuals**, **Residuals versus fits**, and **Residuals versus order.**
- Click the **OK** button to return to the Regression dialog.
- Click the **OK** button to compute the regression.

Comments

You can also start by choosing a Fitted Line plot from the **Regression** submenu to see the scatterplot first—usually good practice.

R

From a data frame called mydata with variables x and y, to find the linear model:

- mylm=lm(*y*~*x*,mydata)
- summary(mylm) # shows a summary of the model including estimates, SEs, and the F-statistic

To get confidence or prediction intervals use:

- predict(mylm, interval="confidence")

or

- predict(mylm, interval="prediction")

Comments

Predictions on points not found in the original data frame can be found from predict as well. In the new data frame (called, say, mynewdata), there must be a predictor variable with the same name as the original "x" variable. Then predict(mylm, newdata=mynewdata) will produce predictions at all the "x" values of mynewdata.

SPSS

To find a regression in SPSS,

- Choose **Regression** from the **Analyze** menu.
- Choose **Linear** from the **Regression** submenu.
- In the Linear Regression dialog that appears, select the Y-variable and move it to the dependent target. Then move the X-variable to the independent target.
- Click the **Plots** button.
- In the Linear Regression Plots dialog, choose to plot the *SRESIDs against the *ZPRED values.
- Click the **Continue** button to return to the Linear Regression dialog.
- Click the **OK** button to compute the regression.

Brief **Case**

Frozen Pizza

The product manager at a subsidiary of Kraft Foods, Inc. is interested in learning how sensitive sales are to changes in the unit price of a frozen pizza in Dallas, Denver, Baltimore, and Chicago. The product manager has been provided data on both *Price* and *Sales* volume every fourth week over a period of nearly four years for the four cities (**Frozen Pizza**).

Examine the relationship between *Price* and *Sales* for each city. Be sure to discuss the nature and validity of this relationship. Is it linear? Is it negative? Is it significant? Are the conditions of regression met? Some individuals in the product manager's division suspect that frozen pizza sales are more sensitive to price in some cities than in others. Is there any evidence to suggest that? Write up a short report on what you find. Include 95% confidence intervals for the mean *Sales* if the *Price* is $2.50 and discuss how that interval changes if the *Price* is $3.50.

Global Warming?

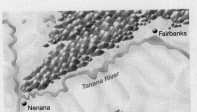

Every spring, Nenana, Alaska, hosts a contest in which participants try to guess the exact minute that a wooden tripod placed on the frozen Tanana River will fall through the breaking ice. The contest started in 1917 as a diversion for railroad engineers, with a jackpot of $800 for the closest guess. It has grown into an event in which hundreds of thousands of entrants enter their guesses on the Internet and vie for more than $300,000.

Because so much money and interest depends on the time of the ice breakup, it has been recorded to the nearest minute with great accuracy ever since 1917 (**Nenana** 2014). And because a standard measure of breakup has been used throughout this time, the data are consistent. An article in *Science* ("Climate Change in Nontraditional Data Sets," *Science* 294, October 2001) used the data to investigate global warming. Researchers are interested in the following questions. What is the rate of change in the date of breakup over time (if any)? If the ice is breaking up earlier, what is your conclusion? Does this necessarily suggest global warming? What could be other reasons for this trend? What is the predicted breakup date for the year 2020? (Be sure to include an appropriate prediction or confidence interval.) Write up a short report with your answers.

EXERCISES

SECTION 14.1

1. A website that rents movies online recorded the age and the number of movies rented during the past month for some of their customers. Here are their data:

Age	Rentals
35	9
40	8
50	4
65	3
40	10
30	12

Make a scatterplot for these data. What does it tell you about the relationship between these two variables? From computer output, the regression line has $b_0 = 18.9$ and $b_1 = -0.260$.

a) Use the estimated regression equation to predict *Rentals* for all six values of *Age*.
b) Find the residuals e_i.
c) Calculate the residual standard deviation, s_e.

2. A training center, wishing to demonstrate the effectiveness of their methods, tests some of their clients after different numbers of days of training, recording their scores on a sample test. Their data are:

Training Days	Correct Responses
1	4
4	6
8	7
10	9
12	10

The regression model they calculate is

$$\widehat{Correct\ responses} = 3.525 + 0.525\ Training\ days.$$

a) Use the model to predict the correct responses for each number of training days.
b) Find the residuals, e_i.
c) Calculate the residual standard deviation, s_e.

3. For the regression of Exercise 1, find the standard error of the regression slope. Show all three values that go into the calculation.

4. For the regression of Exercise 2, find the standard error of the regression slope. Show all three values that go into the calculation.

5. A data set of 5 observations for *Concession Sales per person* ($) at a theater and *Minutes before the movie begins* results in the following estimated regression model:

$$\widehat{Sales} = 4.3 + 0.265\ Minutes.$$

The standard error of the regression slope is 0.0454.

a) Compute the value of the *t*-statistic to test if there is a significant relationship between *Sales* and *Minutes*.
b) What are the degrees of freedom associated with the *t*-statistic?
c) What is the P-value associated with the *t*-statistic?
d) At $\alpha = .05$, can you reject the standard null hypothesis for the slope? Explain.

6. A soap manufacturer tested a standard bar of soap to see how long it would last. A test subject showered with the soap each day for 15 days and recorded the *Weight* (in grams) of the soap after the shower. The resulting regression computer output looks, in part, like this:

```
Dependent variable is: Weight
R-squared = 99.5%
s = 2.949

Variable   Coefficient  SE(Coeff)  t-Ratio  P-Value
Intercept  123.141      1.382       89.1    <0.0001
Day        −5.57476     0.1068     −52.2    <0.0001
```

Find the following facts in this output, or determine them from what you know.

a) The standard deviation of the residuals
b) The slope of the regression line
c) The standard error of b_1
d) The P-value appropriate for testing $H_0: \beta_1 = 0$ versus $H_A: \beta_1 \neq 0$
e) Is the null hypothesis rejected at $\alpha = 0.05$?

SECTION 14.2

7. For the data from Exercise 1, which of the following conditions can you check from the scatterplot? Are satisfied?

a) Linearity
b) Independence
c) Equal spread
d) Normal population

8. Here's a scatterplot of the % of income spent on food versus household income for respondents to the Cornell National Social Survey:

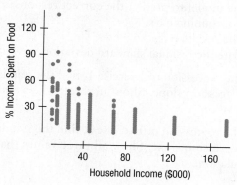

For each of the regression assumptions, state whether it is satisfied, not satisfied, or can't be determined from this plot.

a) Linearity
b) Independence
c) Equal spread
d) Normal population

SECTION 14.3

9. Here are data from a small bookstore.

Number of Salespeople Working	Sales (in $1000's)
2	10
3	11
7	13
9	14
10	18
10	20
12	20
15	22
16	22
20	26
$\bar{x} = 10.4$	$\bar{y} = 17.6$
$SD(x) = 5.64$	$SD(y) = 5.34$

The regression line is:

$$\widehat{Sales} = 8.10 + 0.9134\ Number\ of\ Salespeople\ Working.$$

The assumptions and conditions for regression are met, and from technology we learn that

$$SE(b_1) = 0.0873 \qquad s_e = 1.477$$

a) Find the predicted *Sales* on a day with 12 employees working.
b) Find a 95% confidence interval for the mean *Sales* on days that have 12 employees working.
c) Find the 95% prediction interval for *Sales* on a day with 12 employees working.

10. The study of external disk drives from Chapter 4, exercise 2 (with the outlier removed) finds the following:

	Capacity (TB)	Price ($)
	0.15	35
	0.25	39.95
	0.32	49.95
	1	75
	2	110
	3	140
	4	325
Mean	1.53	110.7
SD	1.51	102.05

The least squares line was found to be: $\widehat{Price} = 15.112 + 62.417\ Capacity$ with $s_e = 42.037$ and $SE(b_1) = 11.328$.

a) Find the predicted *Price* of a 2 TB hard drive.

b) Find a 95% confidence interval for the mean *Price* of 2 TB disk drives.

c) Find the 95% prediction interval for the *Price* of a 2 TB hard drive.

SECTION 14.4

11. A survey designed to study how much households spend on eating out finds the following regression model,

$$\widehat{EatOut \ \$/wk} = 17.28 + 0.354 \ HHIncome$$

relating the amount respondents said they spent individually to eat out each week to their household income in $1000's.

a) A 95% prediction interval for a customer with a household income of $80,000 is ($35.60, $55.60). Explain to the restaurant owner how she should interpret this interval.

b) A 95% confidence interval for the mean amount spent weekly to eat out by people with a household income of $80,000 is ($40.60, $50.60). Explain to a restaurant owner how to interpret this interval.

c) Now explain to her why these intervals are different.

12. In Exercise 5, we saw a regression to predict the sales per person at a movie theater in terms of the time (in minutes) before the show. The model was:

$$\widehat{Sales} = 4.3 + 0.265 \ Minutes.$$

a) A 90% prediction interval for sales to a concessions customer 10 minutes before the movie starts is ($4.60, $9.30). Explain how to interpret this interval.

b) A 90% confidence interval for the mean of sales per person 10 minutes before the movie starts is ($6.65, $7.25). Explain how to interpret this interval.

c) Which interval is of particular interest to the concessions manager? Which one is of particular interest to you, the moviegoer?

SECTION 14.5

13. Recall the small bookstore we saw in Exercise 9. The regression line is:

$$\widehat{Sales} = 8.10 + 0.9134 \ Number \ of \ Sales \ People \ Working$$

and the assumptions and conditions for regression are met. Calculations with technology find that

$$s_e = 1.477.$$

a) Find the predicted sales on a day with 500 employees working.

b) Find a 95% prediction interval for the sales on a day with 500 employees working.

c) Are these predictions likely to be useful? Explain.

14. Look back at the prices for the external disk drives we saw in Exercise 10.

The least squares line is $\widehat{Price} = 15.112 + 62.417 \ Capacity$.

The assumptions and conditions for regression are met.

$$SE(b_1) = 11.328$$

a) Disk drives keep growing in capacity. Some tech experts now talk about *Petabyte* (PB = 1000 TB = 1,000,000 GB) drives. What does this model predict that a Petabyte-capacity drive will cost?

b) Find a 95% prediction interval for the price of a 1 PB drive.

c) Are these predictions likely to be useful? Explain.

CHAPTER EXERCISES

T 15. Online shopping. Several studies have found that the frequency with which shoppers browse Internet retailers is related to the frequency with which they actually purchase products and/or services online. Here are data showing the age of respondents and their answer to the question "How many minutes do you browse online retailers per week?"

Age	Browsing Time (min/wk)
22	492
50	186
44	180
32	384
55	120
60	120
38	276
22	480
21	510
45	252
52	126
33	360
19	570
17	588
21	498

a) Make a scatterplot for these data.

b) Do you think a linear model is appropriate? Explain.

c) Find the equation of the regression line.

d) Check the residuals to see if the conditions for inference are met.

T 16. Climate change 2014. The Earth has been getting warmer. Most climate scientists agree that one important cause of the warming is the increase in atmospheric levels of carbon dioxide (CO_2), a greenhouse gas. Here is part of a regression analysis of the mean annual CO_2 concentration in the atmosphere, measured in parts per thousand (ppt), at the top of Mauna Loa in Hawaii and the mean annual air temperature over both land and sea across the globe, in degrees Celsius for the years 1959–2014. The scatterplots and residuals plots indicated that the data were appropriate for inference and the response variable is *Temp*.

Variable	Coeff	SE(Coeff)
Intercept	10.952	0.1733
CO2	0.009427	0.000493

R-squared = 87.7%
s = 0.091 with 56 − 2 = 54 degrees of freedom

a) Write the equation of the regression line.

b) Find the value of the correlation.

c) Find the *t*-value and P-value for the slope. Is there evidence of an association between CO_2 level and global temperature? What do you know from the slope and *t*-test that you might not have known from the correlation?

d) Do you think predictions made by this regression will be very accurate? Explain.

17. Movie budgets. How does the cost of a movie depend on its length? Data on the cost (millions of dollars) and the running time (minutes) for major release films in one recent year are summarized in these plots and computer output:

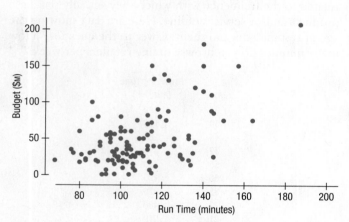

```
Dependent variable is: Budget($million)
R-squared = 27.3%
s = 32.95 with 120 - 2 = 118 degrees of freedom

Variable    Coefficient  SE(Coeff)  t-Ratio  P-Value
Intercept   -63.9981     17.12      -3.74    0.0003
Run time     1.02648      0.1540     6.66    ≤0.0001
```

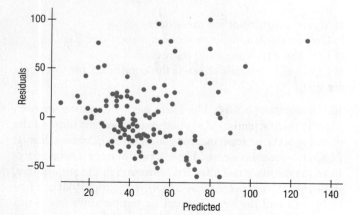

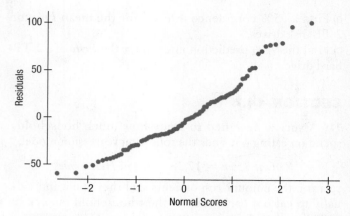

a) Explain in words and numbers what the regression says.

b) The intercept is negative. Discuss its value.

c) The output reports $s = 32.95$. Explain what that means in this context.

d) What's the value of the standard error of the slope of the regression line?

e) Explain what that means in this context.

18. House prices. How does the price of a house depend on its size? Data from Saratoga, New York, on 1064 randomly selected houses that had been sold include data on price ($1000s) and size (1000 ft^2), producing the following graphs and computer output:

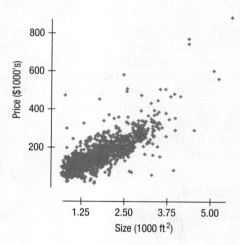

```
Dependent variable is: Price
R-squared = 59.5%
s = 53.79 with 1064 - 2 = 1062 degrees of freedom

Variable    Coefficient  SE(Coeff)  t-Ratio  P-Value
Intercept   -3.1169      4.688      -0.665   0.5063
Size        94.4539      2.393      39.465   <0.0001
```

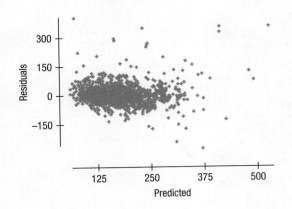

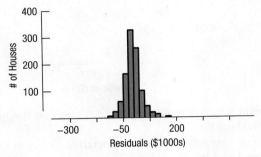

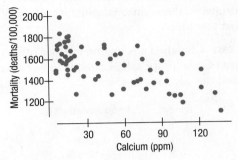

R-squared = 42.9%
s = 143.0 with 61 − 2 = 59 degrees of freedom

a) Explain in words and numbers what the regression says.
b) The intercept is negative. Discuss its value, taking note of its P-value.
c) The output reports $s = 53.79$. Explain what that means in this context.
d) What's the value of the standard error of the slope of the regression line?
e) Explain what that means in this context.

19. Movie budgets, part 2. Exercise 17 shows computer output examining the association between the length of a movie and its cost.

a) Check the assumptions and conditions for inference.
b) Find a 95% confidence interval for the slope and interpret it.

20. House prices, part 2. Exercise 18 shows computer output examining the association between the sizes of houses and their sale prices.

a) Check the assumptions and conditions for inference.
b) Find a 95% confidence interval for the slope and interpret it.

21. Water hardness. In an investigation of environmental causes of disease, data were collected on the annual mortality rate (deaths per 100,000) for males in 61 large towns in England and Wales. In addition, the water hardness was recorded as the calcium concentration (parts per million, or ppm) in the drinking water. Here are the scatterplot and regression analysis of the relationship between mortality and calcium concentration, where the dependent variable is *Mortality*.

a) Is there an association between the hardness of the water and the mortality rate? Write the appropriate hypothesis.
b) Assuming the assumptions for regression inference are met, what do you conclude?
c) Create a 95% confidence interval for the slope of the true line relating calcium concentration and mortality.
d) Interpret your interval in context.

22. Mutual fund returns 2013. The brief case for Chapter 4 listed the rate of return for 92 mutual funds over the previous 3-year and 5-year periods. It's common for advertisements to carry the disclaimer that "past returns may not be indicative of future performance." Do these data indicate that there was an association between 3-year and 5-year rates of return?

23. Male labor force participation rate 2014. The International Labor Organization (ILO) reports the Labor Force Participation Rate (LFPR)—the percentage of the relevant population who are either employed or actively seeking work—worldwide. The datafile holds this data for the years 1990 to 2014 for men and women.

a) Find a regression model to describe any trend in the LFPR for men over this time period. State in simple language what the model says.
b) Examine the residuals to determine if a linear regression is appropriate. Are there other concerns about the data to consider?
c) Test an appropriate hypothesis to determine if the association is statistically significant.
d) What percentage of the variability in the LFPR can be accounted for by the regression model?

24. Female labor force participation rate 2014. The International Labor Organization (ILO) reports the Labor Force Participation Rate (LFPR)—the percentage of the relevant population who are either employed or actively seeking work—worldwide. The datafile holds this data for the years 1990 to 2014 for men and women.

a) Find a regression model to describe any trend in the LFPR for women over this time period. State in simple language what the model says.

b) Test an appropriate hypothesis to determine if the association is statistically significant.

c) What percentage of the variability in the LFPR can be accounted for by the regression model?

d) Examine the residuals to determine if a linear regression is appropriate. Make additional plots if necessary and describe what you find.

T 25. Used cars. Classified ads in a newspaper offered several used Toyota Corollas for sale. Listed below are the ages of the cars and the advertised prices.

Age (yr)	Prices Advertised ($)
1	13990
1	13495
3	12999
4	9500
4	10495
5	8995
5	9495
6	6999
7	6950
7	7850
8	6999
8	5995
10	4950
10	4495
13	2850

a) Make a scatterplot for these data.

b) Do you think a linear model is appropriate? Explain.

c) Find the equation of the regression line.

d) Check the residuals to see if the conditions for inference are met.

T 26. Property assessments. The following software results provide information about the size (in square feet) of 18 homes in Ithaca, New York, and the city's assessed value of those homes, where the response variable is *Assessment*.

Predictor	Coeff	SE(Coeff)	t-Ratio	P-Value
Intercept	37108.85	8664.33	4.28	0.0006
Size	11.90	4.29	2.77	0.0136

s = 4682.10 R-Sq = 32.5%

Variable	Mean	StdDev
Assessment	60946.7	5527.62
Size	2003.39	264.727

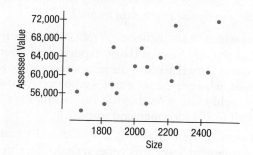

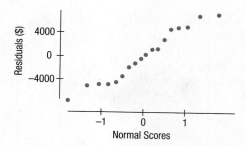

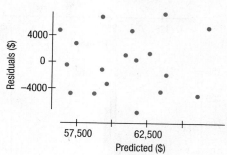

a) Explain why inference for linear regression is appropriate with these data.

b) Is there a significant linear association between the *Size* of a home and its *Assessment*? Test an appropriate hypothesis and state your conclusion.

c) What percentage of the variability in assessed value is accounted for by this regression?

d) Give a 90% confidence interval for the slope of the true regression line, and explain its meaning in the proper context.

e) From this analysis, can we conclude that adding a room to your house will increase its assessed value? Why or why not?

f) The owner of a home measuring 2100 square feet files an appeal, claiming that the $70,200 assessed value is too high. Do you agree? Explain your reasoning.

T 27. Used cars, part 2. Based on the analysis of used car prices you did for Exercise 25, if appropriate, create a 95% confidence interval for the slope of the regression line and explain what your interval means in context.

T 28. Assets and sales. A business analyst is looking at a company's assets and sales to determine the relationship (if any) between the two measures. She has data (in $million) from a random sample of 79 Fortune 500 companies, and obtained the linear regression below:

Predictor	Coeff	SE(Coeff)	t-Ratio	P-Value
Intercept	1867.4	804.5	2.32	0.0230
Sales	0.975	0.099	9.84	<0.0001

s = 6132.59 R-Sq = 55.7%

Use the data provided to find a 95% confidence interval for the slope of the regression line and interpret your interval in context.

T 29. Fuel economy and weight. A consumer organization has reported test data for 50 car models. We will examine the association between the weight of the car (in thousands of pounds) and the fuel efficiency (in miles per gallon). Use the data provided on the disk to answer the following questions, where the response variable is *Fuel Efficiency* (mpg).

a) Create the scatterplot and obtain the regression equation.
b) Are the assumptions for regression satisfied?
c) Write the appropriate hypotheses for the slope.
d) Test the hypotheses and state your conclusion.

30. Auto batteries. *Consumer Reports* listed the price (in dollars) and power (in cold cranking amps) of auto batteries. We want to know if more expensive batteries are generally better in terms of starting power. Here are the regression and residual output, where the response variable is *Power*.

```
Dependent variable is: Power
R-squared = 25.2%
s = 116.0 with 33 - 2 = 31 degrees of freedom
```

Variable	Coefficient	SE(Coeff)	t-Ratio	P-Value
Intercept	384.594	93.55	4.11	0.0003
Price	4.146	1.282	3.23	0.0029

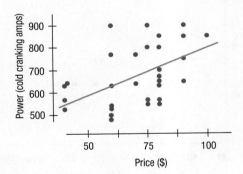

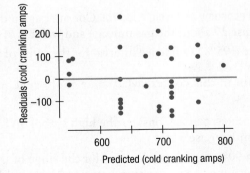

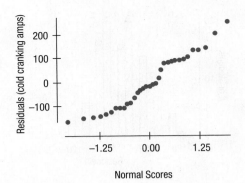

a) How many batteries were tested?
b) Are the conditions for inference satisfied? Explain.
c) Is there evidence of a linear association between the price and cranking power of auto batteries? Test an appropriate hypothesis and state your conclusion.
d) Is the association strong? Explain.
e) What is the equation of the regression line?
f) Create a 90% confidence interval for the slope of the true line.
g) Interpret your interval in this context.

T 31. SAT scores. How strong was the association between student scores on the Math and Verbal sections of the old SAT? Scores on this exam ranged from 200 to 800 and were widely used by college admissions offices. Here are summary statistics, regression analysis, and plots of the scores for a graduating class of 162 students at Ithaca High School, where the response variable is *Math Score*.

Predictor	Coeff	SE(Coeff)	t-Ratio	P-Value
Intercept	209.55	34.35	6.10	<0.0001
Verbal	0.675	0.057	11.88	<0.0001

```
s = 71.75   R-Sq = 46.9%
```

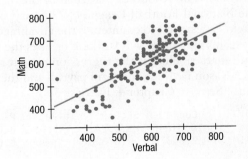

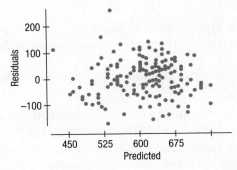

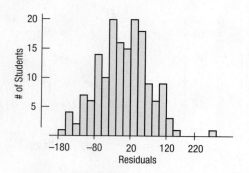

a) Is there evidence of a linear association between *Math* and *Verbal* scores? Write an appropriate hypothesis.
b) Discuss the assumptions for inference.
c) Test your hypothesis and state an appropriate conclusion.

T **32. Productivity.** How strong is the association between labor productivity and labor costs? Data from the Bureau of Labor Statistics for labor productivity, as measured by *Output per Hour*, and *Unit Labor Costs* across 124 industries, are used to examine this relationship (ftp://ftp.bls.gov; accessed June 2013).

a) From a scatterplot, is there evidence of a linear association between *Labor Productivity* and *Unit Labor Costs*? Plot the reciprocal, *Hours per output (000s)*, against *Unit Labor Costs*. Why did the analysts prefer this measure of productivity?
b) Using the reciprocal measure, *Hours per Output (000s)*, test an appropriate null hypothesis and state an appropriate conclusion (assume that assumptions and conditions are now met).

T **33. Football salaries 2013.** Football owners are constantly in competition for good players. The more wins, the more likely that the team will provide good business returns for the owners. The resources that each of the 32 teams has in the National Football League (NFL) vary, but the draft system is designed to counteract the advantages that wealthier teams may have. Is it working or does the size of the payroll matter? Here is the regression output for the 2012/2013 season between the team payroll and the number of wins. (See also Chapter 4, Exercise 33.)

Predictor	Coeff	SE(Coeff)	t-Ratio	P-Value
Intercept	-16.31846	8.49246	-1.922	0.06421
Payroll($M)	0.21875	0.07636	2.865	0.00755

s = 2.779 R-Sq = 21.48%

a) State the hypotheses about the slope.
b) Perform the hypothesis test and state your conclusion in context.
c) Using a statistics program, check the assumptions and conditions.

T **34. Female president.** The Gallup organization has, over six decades, periodically asked the following question:

If your party nominated a generally well-qualified person for president who happened to be a woman, would you vote for that person?

We wonder if the proportion of the public who have "no opinion" on this issue has changed over the years. Here is a regression for the proportion of those respondents whose response to this question about voting for a woman president was "no opinion." Assume that the conditions for inference are satisfied and that the response variable is proportion responding *No Opinion*.

Predictor	Coeff	SE(Coeff)	t-Ratio	P-Value
Intercept	7.693	2.445	3.15	0.0071
Year	-0.043	0.035	-1.21	0.2460

s = 2.28 R-Sq = 9.5%

a) State the hypotheses about the slope (both numerically and in words) that describes how voters' thoughts have changed about voting for a woman.
b) Assuming that the conditions for inference are satisfied, perform the hypothesis test and state your conclusion.
c) Examine the scatterplot corresponding to the regression for No Opinion. How does it change your opinion of the trend in "no opinion" responses? Do you think the true slope is negative as shown in the regression output?

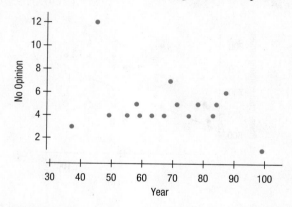

T **35. Fuel economy and weight, part 2.** Consider again the data in Exercise 29 about the gas mileage and weights of cars.

a) Create a 95% confidence interval for the slope of the regression line.
b) Explain in this context what your confidence interval means.

T **36. SAT scores, part 2.** Consider the high school SAT scores data from Exercise 31.

a) Find a 90% confidence interval for the slope of the true line describing the association between Math and Verbal scores.
b) Explain in this context what your confidence interval means.

T **37. Ozone 2013.** The Environmental Protection Agency is examining the relationship between the ozone level (in parts per million) and the population density (in million people per square km) of U.S. cities. Part of the regression analysis is shown.

```
Dependent variable is Ozone
R-squared = 23.5%
s = 0.0090 with 19 - 2 = 17 df

Variable        Coeff      SE(Coeff)
Intercept       0.0698       0.0033
Pop density    -2.8713       1256
```

a) Is the relationship statistically significant at $\alpha = 0.05$? Assuming the conditions for inference are satisfied, test an appropriate hypothesis and state your conclusion in context.

b) Do you think that the population density of a city is a useful predictor of ozone level? Use the values of both R^2 and s in your explanation.

38. Sales and profits. A business analyst was interested in the relationship between a company's sales and its profits. She collected data (in millions of dollars) from a random sample of Fortune 500 companies and created the regression analysis and summary statistics shown. The assumptions for regression inference appeared to be satisfied.

```
           Profits    Sales    | Dependent variable is Profits
Count      79         79       | R-squared = 66.2% s = 466.2
Mean       209.839    4178.29  | Variable     Coefficient  SE(Coeff)
Variance   635,172    49,163,000| Intercept   -176.644     61.16
Std Dev    796.977    7011.63  | Sales        0.092498     0.0075
```

a) Is there a statistically significant association between sales and profits? Test an appropriate hypothesis and state your conclusion in context.

b) Do you think that a company's sales serve as a useful predictor of its profits? Use the values of both R^2 and s in your explanation.

39. Ozone, again. Using a statistics program, consider again the relationship between the population density and ozone level of U.S. cities that you analyzed in Exercise 37.

a) Give a 90% confidence interval for the approximate change in ozone level associated with differences in population density.

b) For the cities studied, the mean population density was 0.0020 million people per square km. The population of Lincoln, NB is approximately .0011 million people per square km. Predict the mean ozone level for cities of that population density with an interval in which you have 90% confidence.

40. More sales and profits. Using a statistics program, consider again the relationship between the sales and profits of Fortune 500 companies that you analyzed in Exercise 38.

a) Find a 95% confidence interval for the slope of the regression line. Interpret your interval in context.

b) Last year, the drug manufacturer Eli Lilly, Inc., reported gross sales of $23 billion (that's $23,000 million). Create a 95% prediction interval for the company's profits, and interpret your interval in context.

41. Tablet computers. In July 2013, cnet.com listed the battery life (in hours) and luminous intensity (i.e., screen brightness, in cd/m²) for a sample of tablet computers. We want to know if screen brightness is associated with battery life. (reviews.cnet.com/8301-19736_7-20080768-251/cnet-updates-tablet-test-results/?tag=contentBody; contentHighlights)

```
Dependent variable is Video battery life (in hours)
R-squared = 4.82%
s = 1.946 with 69 - 2 = 67 degrees of freedom
Variable      Coeff     SE(Coeff)   t-Ratio   P-Value
Intercept     6.21108   0.9609      6.46      6.0001
Brightness    0.00486   0.0026      1.84      0.0699
```

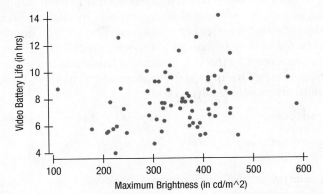

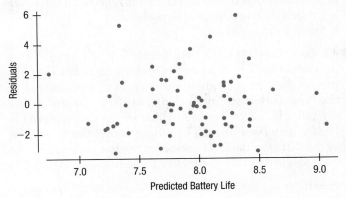

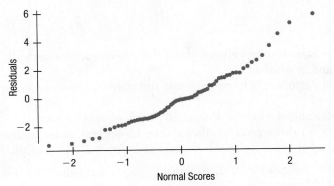

a) How many tablet computers were tested?

b) Are the conditions for inference satisfied? Explain.

c) Is there evidence of an association between maximum brightness of the screen and battery life? Test an appropriate hypothesis and state your conclusion.

d) Is the association strong? Explain.

e) What is the equation of the regression line?

f) Create a 95% confidence interval for the slope of the true line.

g) Interpret your interval in this context.

T 42. Marketing managers. Are wages for various marketing managerial positions related? One way to determine this is to examine the relationship between the mean hourly wages for two managerial occupations in marketing: sales managers and advertising managers. The average hourly wage for both occupations is reported for all U.S. states and territories by the U.S. Bureau of Labor Statistics (data .bls.gov/oes; Occupational Employment Statistics). Here are the regression analysis results.

```
Predictor        Coeff    SE(Coeff)  t-Ratio  P-Value
Intercept        10.317   4.382      2.35     0.0227
Sales Mgr Avg    0.56349  0.09786    5.76     <0.0001
  Hourly Wage
```

a) State the null and alternative hypothesis under investigation.

b) Assuming that the assumptions for regression inference are reasonable, test the null hypothesis.

c) State your conclusion.

T 43. Cost index 2011. The *Worldwide Cost of Living Survey* published by *The Economist* provides an index that expresses the cost of living in other cities as a percentage of the New York cost. For example, in 2011, the cost of living index in Tokyo was 161, which means that it was 61% higher than New York. The output shows the regression of the 2011 on the 2010 index for the ten most expensive cities in 2011, where *Index 2011* is the response variable.

```
Predictor    Coeff    SE(Coeff)  t-Ratio  P-Value
Intercept    46.73    15.45      3.02     0.0164
Index 2010   0.744    0.115      6.49     0.0002
s = 3.474  R-Sq = 84.0%
```

a) State the hypotheses about the slope (both numerically and in words).

b) Perform the hypothesis test and state your conclusion in context.

c) Explain what the *R*-squared in this regression means.

d) Do these results indicate that, in general, cities with a higher index in 2010 had a higher index in 2011? Explain.

44. Job growth 2012. *Fortune magazine* publishes the top 100 companies to work for every year. Among the information listed is the percentage growth in jobs at each company. The period from 2009 to 2011 was a difficult one for job growth in the United States. The output below shows the regression of the 2012 job growth (%) on the 2010 job

growth for those companies that appear on both years' lists. (One outlier has been omitted.) *Job Growth 2012* is the response variable (money.cnn.com/magazines/fortune/ bestcompanies/; accessed March 2012).

```
Dependent variable is: Job Growth 2012
R-squared = 5.7%
s = 0.0738 with 68 − 2 = 6 degrees of freedom
Variable          Coeff    SE(Coeff)  t-Ratio  P-Value
Intercept         0.0628   0.009      7.00     <0.0001
Job Growth 2010   0.2      0.100      2.00     0.0498
```

a) State the hypotheses about the slope (both numerically and in words).

b) Assuming that the assumptions for inference are satisfied, perform the hypothesis test and state your conclusion in context.

c) Explain what the *R*-squared in this regression means.

d) Do these results indicate that, in general, companies with a higher job growth in 2010 had a higher job growth in 2012? Explain.

T 45. Cost index again. In Exercise 43, we examined the *Worldwide Cost of Living Survey* cost of living index for the ten most expensive cities, whose 2011 indices range from Singapore's 137 to Tokyo's 161. Let's add the ten *least* expensive cities to the data. Their 2011 indices range from 46 for Karachi (the least expensive city in 2011) to Dhaka and Manila at 62. Here is the resulting regression:

```
Dependent variable is: Index 2011
R-squared = 99.3%
s = 4.061 with 20 − 2 = 18 degrees of freedom
Predictor    Coeff    SE(Coeff)  t-Ratio  P-Value
Intercept    −2.6787  2.272      −1.18    0.2537
Index 2010   1.1082   0.022      50.1     <0.0001
```

a) Sketch what a scatterplot of *Index2011* vs. *Index2010* is likely to look like. You do not need to see the data.

b) Explain why the R^2 of this regression is higher than the R^2 of the regression in Exercise 43.

T 46. Job growth again. In Exercise 44, the company Zappos was omitted. Here is a scatterplot of the data with Zappos plotted as an *x*:

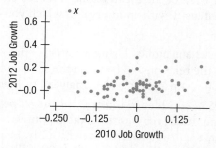

a) In words, what does the outlying point say about Zappos?

b) What effect would this point have on the regression, had it been left with the rest of the data?

c) Using the data on the DVD, find the regression with and without the outlier.

47. Old pitchers. Many factors may affect fans' decision to go to a ball game. Is it possible that fans prefer teams with an older pitching staff?

a) Examine a scatterplot of Attend/Game and PitchAge. Check the conditions for regression.
b) Do you think there is a linear association between Attendance and Pitcher Age?
c) Compute and discuss the regression model.

48. Little League product testing. Ads for a Little League instructional video claimed that the techniques would improve the performances of Little League pitchers. To test this claim, 20 Little Leaguers threw 50 pitches each, and we recorded the number of strikes. After the players participated in the training program, we repeated the test. The following table shows the number of strikes each player threw before and after the training. A test of paired differences failed to show that this training was effective in improving a player's ability to throw strikes. Is there any evidence that the *Effectiveness* (*After* − *Before*) of the video depends on the player's *Initial Ability* (*Before*) to throw strikes? Test an appropriate hypothesis and state your conclusion. Propose an explanation for what you find.

Number of Strikes (out of 50)			
Before	After	Before	After
28	35	33	33
29	36	33	35
30	32	34	32
32	28	34	30
32	30	34	33
32	31	35	34
32	32	36	37
32	34	36	33
32	35	37	35
33	36	37	32

49. Fuel economy and weight, part 3. Consider again the data in Exercise 29 about the fuel economy and weights of cars.

a) Create a 95% confidence interval for the average fuel efficiency among cars weighing 2500 pounds, and explain what your interval means.
b) Create a 95% prediction interval for the gas mileage you might get driving your new 3450-pound SUV, and explain what that interval means.

50. SAT scores, part 3. Consider the high school SAT scores data from Exercise 31 once more. The mean Verbal score was 596.30 and the standard deviation was 99.52.

a) Find a 90% confidence interval for the mean SAT Math score for all students with an SAT Verbal score of 500.
b) Find a 90% prediction interval for the Math score of the senior class president, if you know she scored 710 on the Verbal section.

51. Little League product testing, part 2. Using the same data provided in Exercise 48, answer the following questions.

a) Find the 95% prediction interval for the effectiveness of the video on a pitcher with an initial ability of 33 strikes.
b) Do you think predictions made by this regression will be very accurate? Explain.

52. Assets and sales, part 2. The analyst in Exercise 28 realized the data were in need of transformation because of the nonlinearity between the variables. Economists commonly take the logarithm of these variables to make the relationship more nearly linear, and she did too. (These are base 10 logs.) The dependent variable is *LogSales*. The conditions for regression inference now appear to be satisfied.

```
Dependent variable is: LogSales
R-squared = 33.9%
s = 0.4278 with 79 - 2 = 77 degrees of freedom

Variable    Coefficient  SE(Coeff)  t-Ratio  P-Value
Intercept   1.303        0.3211     4.06     0.0001
LogAssets   0.578        0.0919     6.28     <0.0001
```

a) Is there a significant linear association between *LogAssets* and *LogSales*? Find the *t*-value and P-value to test an appropriate hypothesis and state your conclusion in context.
b) Do you think that a company's assets serve as a useful predictor of their sales?

53. All the efficiency money can buy 2013. A sample of 61 model-2013 cars from an online information service was examined to see how fuel efficiency (as highway mpg) relates to the cost (Manufacturer's Suggested Retail Price in dollars) of cars. Here are displays and computer output:

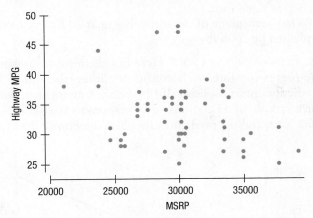

```
Dependent variable is: Highway MPG
R-squared = 10.36%
s = 4,870 with 61 - 2 = 59 degrees of freedom

Variable    Coefficient  SE(Coeff)  t-Ratio  P-Value
Intercept   45.6898      4.849      9.42     <0.0001
MSRP        -0.000416    0.000159   -2.61    0.0114
```

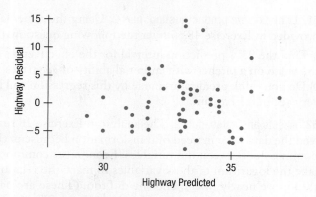

a) State what you want to know, identify the variables, and give the appropriate hypotheses.
b) Check the assumptions and conditions.
c) If the conditions are met, complete the analysis.

T 54. Energy use and recession. The great recession of 2008 changed spending and energy use habits worldwide. Based on data collected from the United Nations Millennium Indicators Database related to measuring the goal of *ensuring environmental sustainability*, investigate the association between energy use (kg oil equivalent per $1000 GDP) before (2006) and after (2010) the crisis, for a sample of 33 countries (unstats.un.org/unsd/mi/mi_goals.asp; accessed June 2013).

a) Find a regression model showing the relationship between *2010 Energy Use* (response variable) and *2006 Energy Use* (predictor variable).
b) Examine the residuals to determine if a linear regression is appropriate.
c) Test an appropriate hypothesis to determine if the association is significant.
d) What percentage of the variability in *2010 Energy Use* is explained by *2006 Energy Use*?

T 55. Youth employment 2012. Here is a scatterplot showing the regression line, 95% confidence intervals, and 95% prediction intervals, using 2012 youth unemployment data for a sample of 33 nations. The response variable is the *Male Rate*, and the predictor variable is the *Female Rate*.

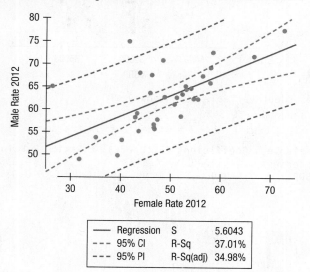

	Regression	S	5.6043
---	95% CI	R-Sq	37.01%
---	95% PI	R-Sq(adj)	34.98%

a) Explain the meaning of the 95% prediction intervals in this context.
b) Explain the meaning of the 95% confidence intervals in this context.
c) Using a statistics program, identify any unusual observations, and discuss their potential impact on the regression.

T 56. Male unemployment 2012. Here is a scatterplot showing the regression line, 95% confidence intervals, and 95% prediction intervals, using 2011 and 2012 male unemployment data for a sample of 33 nations. The response variable is the *2012-Male Rate*, and the predictor variable is the *2011-Male Rate*.

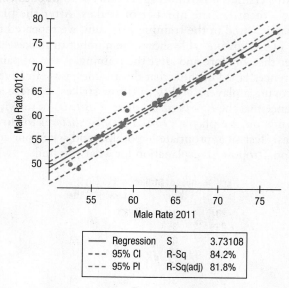

	Regression	S	3.73108
---	95% CI	R-Sq	84.2%
---	95% PI	R-Sq(adj)	81.8%

a) Explain the meaning of the 95% prediction intervals in this context.
b) Explain the meaning of the 95% confidence intervals in this context.
c) Using a statistics program, identify any unusual observations, and discuss their potential impact on the regression.

T 57. Energy use and recession, part 2. Examine the regression and scatterplot showing the regression line, 95% confidence intervals, and 95% prediction intervals using *2006* and *2010 energy use* (kg oil equivalent per $1000 GDP) for a sample of 31 countries first examined in Exercise 54. The response variable is *2010 Energy Use*.

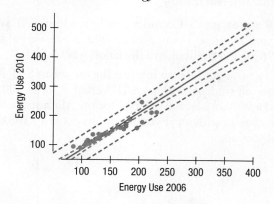

a) Explain the meaning of the 95% prediction intervals in this context.

b) Explain the meaning of the 95% confidence intervals in this context.

You can see the regression model in Exercise 54. The extraordinary point is Iceland. If we set it aside, the model looks like this

```
Response variable is: Energy Use 2010
R squared = 88.6%
s = 13.10 with 32 − 2 = 30 degrees of freedom

Variable     Coefficient  SE(Coeff)  t-Ratio  P-Value
Intercept    5.74436      9.061      0.634    0.5309
Energy 2006  0.921540     0.0605     15.2     <0.0001
```

c) How has setting aside Iceland changed the regression model? How is it likely to affect the intervals discussed in a) and b)?

58. Global reach 2012. The Internet has revolutionized business and offers unprecedented opportunities for globalization. However, the ability to access the Internet varies greatly among different regions of the world. One of the variables the United Nations collects data on each year is *Personal Computers per 100 Population* (unstats.un.org/unsd/cdb/cdb_help/cdb_quick_start.asp) for various countries. Below is a scatterplot showing the regression line, 95% confidence intervals, and 95% prediction intervals using 2000 and 2012 computer adoption (personal computers per 100 population) for a sample of 85 countries. The response variable is *PC/100 2012*.

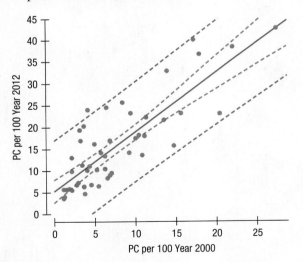

a) Working with the data set on the DVD with the same name as this exercise, find a regression model showing the relationship between personal computer adoption in 2012 *PC/100 2012* (the response variable) and personal computer adoption in 2000 *PC/100 2000* (the predictor variable).

b) Explain the meaning of the 95% prediction intervals in this context.

c) Explain the meaning of the 95% confidence intervals in this context.

59. Seasonal spending. Spending on credit cards decreases after the Christmas spending season (as measured by amount charged on a credit card in December). The data set on the DVD with the same name as this exercise contains the monthly credit card charges of a random sample of 99 cardholders.

a) Build a regression model to predict January spending from December's spending.

b) How much, on average, will cardholders who charged $2000 in December charge in January?

c) Give a 95% confidence interval for the average January charges of cardholders who charged $2000 in December.

d) From part c, give a 95% confidence interval for the average decrease in the charges of cardholders who charged $2000 in December.

e) What reservations, if any, do you have about the confidence intervals you made in parts c and d?

60. Seasonal spending, part 2. Financial analysts know that January credit card charges will generally be much lower than those of the month before. What about the difference between January and the next month? Does the trend continue? The data set on the DVD contains the monthly credit card charges of a random sample of 99 cardholders.

a) Build a regression model to predict February charges from January charges.

b) How much, on average, will cardholders who charged $2000 in January charge in February?

c) Give a 95% confidence interval for the average February charges of cardholders who charged $2000 in January.

d) From part c, give a 95% confidence interval for the average decrease in the charges of cardholders who charged $2000 in January.

e) What reservations, if any, do you have about the confidence intervals you made in parts c and d?

JUST CHECKING ANSWERS

1 A high *t*-ratio of 3.27 indicates that the slope is different from zero—that is, that there is a linear relationship between height and mouth size. The small P-value says that a slope this large would be very unlikely to occur by chance if, in fact, there was no linear relationship between the variables.

2 Not really. The R^2 for this regression is only 15.3%, so height doesn't account for very much of the variability in mouth size.

3 The value of s_e tells the standard deviation of the residuals. Mouth sizes have a mean of 60.3 cubic centimeters. A standard deviation of 15.7 in the residuals indicates that the errors made by this regression model can be quite large relative to what we are estimating. Errors of 15 to 30 cubic centimeters would be common.

15

Multiple Regression

Zillow.com

Zillow.com is a real estate research site, founded in 2005 by Rich Barton and Lloyd Frink. Both are former Microsoft executives and founders of Expedia.com, the Internet-based travel agency. Zillow collects publicly available data and provides an estimate (called a Zestimate®) of a home's worth. The estimate is based on a model of the data that Zillow has been able to collect on a variety of predictor variables, including the past history of the home's sales, the location of the home, and characteristics of the house such as its size and number of bedrooms and bathrooms.

The site is enormously popular among both potential buyers and sellers of homes. According to Rismedia.com, Zillow is one of the most-visited U.S. real estate sites on the Web, with approximately 5 million unique users each month. These users include more than one-third of all mortgage professionals in the United States— or approximately 125,000—in any given month. Additionally, 90% of Zillow users are homeowners, and two-thirds are either buying and selling now, or plan to in the near future.

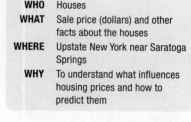

WHO	Houses
WHAT	Sale price (dollars) and other facts about the houses
WHERE	Upstate New York near Saratoga Springs
WHY	To understand what influences housing prices and how to predict them

H ow exactly does Zillow figure the worth of a house? According to the Zillow.com site, "We compute this figure by taking zillions of data points—much of this data is public—and entering them into a formula. This formula is built using what our statisticians call 'a proprietary algorithm'—big words for 'secret formula.' When our statisticians developed the model to determine home values, they explored how homes in certain areas were similar (i.e., number of bedrooms and baths, and a myriad of other details) and then looked at the relationships between actual sale prices and those home details." These relationships form a pattern, and they use that pattern to develop a model to estimate a market value for a home. In other words, the Zillow statisticians use a model, most likely a regression model, to predict home value from the characteristics of the house. We've seen how to predict a response variable based on a single predictor. That's been useful, but the types of business decisions we'll want to make are often too complex for simple regression.[1] In this chapter, we'll expand the power of the regression model to take into account many predictor variables into what's called a multiple regression model. With our understanding of simple regression as a base, getting to multiple regression isn't a big step, but it's an important and worthwhile one. Multiple regression is probably the most powerful and widely used statistical tool today.

As anyone who's ever looked at house prices knows, house prices depend on the local market. To control for that, we will restrict our attention to a single market. We have a random sample of 1057 home sales from the public records of sales in upstate New York, in the region around the city of Saratoga Springs. The first thing often mentioned in describing a house for sale is the number of bedrooms. Let's start with just one predictor variable. Can we use *Bedrooms* to predict home *Price*?

Figure 15.1 Side-by-side boxplots of *Price* against *Bedrooms* show that price increases, on average, with more bedrooms.

(Data in **Housing Prices GE15**).

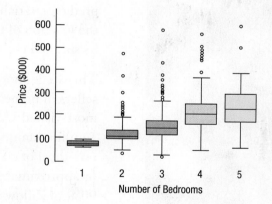

The number of *Bedrooms* is a quantitative variable, but it holds only a few values (from 1 to 5 in this data set). So a scatterplot may not be the best way to examine the relationship between *Bedrooms* and *Price*. In fact, at each value for *Bedrooms* there is a whole distribution of prices. Side-by-side boxplots of *Price* against *Bedrooms* (Figure 15.1) show a general increase in price with more bedrooms, and an approximately linear growth.

Figure 15.1 also shows a clearly increasing spread from left to right, violating the Equal Spread Condition, and that's a possible sign of trouble. For now, we'll proceed cautiously. We'll fit the regression model, but we will be cautious about using inference methods for the model. Later we'll add more variables to increase the power and usefulness of the model.

---

[1] When we need to note the difference, a regression with a single predictor is called a **simple regression**.

SUMMARY OUTPUT				
Response Variable is Price				
Regression Statistics				
R Square	0.214			
Standard Error	68432.21			
Observations	1057			
	Coefficients	Standard Error	t Stat	P-value
Intercept	14349.48	9297.69	1.54	0.1230
Bedrooms	48218.91	2843.88	16.96	0.0000

A linear regression model of *Price* on *Bedrooms* gives the model

$$\widehat{Price} = 14349.48 + 48218.91 \times Bedrooms.$$

The model tells us that, on average, we'd expect the price to be higher by almost $50,000 for each bedroom in the house, as we can see from the slope value of $48,218.91. But the R^2 for this regression is only 21.4%. So the regression model accounts for about 21.4% of the variation in house prices. Perhaps some of the other facts about these houses can account for portions of the remaining variation.

The standard deviation of the residuals is $s = 68,432$, which tells us that the model only does a modestly good job of accounting for the price of a home. Approximating with the 68–95–99.7 Rule, we'd guess that only about 68% of home prices predicted by this model would be within $68,432 of the actual price. That's not likely to be close enough to be useful for a home buyer.

15.1   The Multiple Regression Model

For simple regression, we wrote the predicted values in terms of one predictor variable:

$$\hat{y} = b_0 + b_1 x.$$

To include more predictors in the model, we simply write the same regression model with more predictor variables. The resulting **multiple regression** looks like this:

$$\hat{y} = b_0 + b_1 x_1 + b_2 x_2 + \cdots + b_k x_k$$

where b_0 is still the intercept and each b_k is the estimated coefficient of its corresponding predictor x_k. Although the model doesn't look much more complicated than a simple regression, it isn't practical to determine a multiple regression by hand. This is a job for a statistics program on a computer. Remember that for simple regression, we found the coefficients for the model using the least squares solution, the one whose coefficients made the sum of the squared residuals as small as possible. For multiple regression, a statistics package does the same thing and can find the coefficients of the least squares model easily.

If you know how to find the regression of *Price* on *Bedrooms* using a statistics package, you can probably just add another variable to the list of predictors in your program to compute a multiple regression. A multiple regression of *Price* on the two variables *Bedrooms* and *Living Area* generates a multiple regression table like this one.

```
Response variable: Price
R² = 57.8%
s = 50142.4 with 1057 − 3 = 1054 degrees of freedom
```

Variable	Coeff	SE(Coeff)	t-ratio	P-value
Intercept	20986.09	6816.3	3.08	0.0021
Bedrooms	−7483.10	2783.5	−2.69	0.0073
Living Area	93.84	3.11	30.18	<0.0001

Table 15.1   Multiple regression output for the linear model predicting *Price* from *Bedrooms* and *Living Area*.

You should recognize most of the numbers in this table, and most of them mean what you expect them to. The value of R^2 for a regression on two variables gives the fraction of the variability of *Price* accounted for by both predictor variables together. With *Bedrooms* alone predicting *Price*, the R^2 value was 21.4%, but this model accounts for 57.8% of the variability in *Price* and the standard deviation of the residuals is now $50,142.40. We shouldn't be surprised that the variability explained by the model has gone up. It was for this reason—the hope of accounting for some of that leftover variability—that we introduced a second predictor. We also shouldn't be surprised that the size of the house, as measured by *Living Area*, also contributes to a good prediction of house prices. Collecting the coefficients of the multiple regression of *Price* on *Bedrooms* and *Living Area* from Table 15.1, we can write the estimated regression as:

$$\widehat{Price} = 20{,}986.09 - 7{,}483.10 \ Bedrooms + 93.84 \ Living\,Area.$$

As before, we define the residuals as:

$$e = y - \hat{y}.$$

The standard deviation of the residuals is still denoted s (or also sometimes s_e to distinguish it from the standard deviation of y). The degrees of freedom is the number of observations ($n = 1057$) minus one for each coefficient estimated:

$$\text{df} = n - k - 1,$$

where k is the number of predictor variables and n is the number of cases. For this model, subtract 3 (the two coefficients and the intercept). To find the standard deviation of the residuals, use that number of degrees of freedom in the denominator:

$$s_e = \sqrt{\frac{\sum (y - \hat{y})^2}{n - k - 1}}.$$

For each predictor, the regression output shows a coefficient, its standard error, a *t*-ratio, and the corresponding P-value. As with simple regression, the **t-ratio** measures how many standard errors the coefficient is away from 0. Using a Student's *t*-model, we can find its P-value and use that to test the null hypothesis that the true value of the coefficient is 0.

What's different? With so much of the multiple regression looking just like simple regression, why devote an entire chapter to the subject?

There are several answers to this question. First, and most important, is that the meaning of the coefficients in the regression model has changed in a subtle, but important, way. Because that change is not obvious, multiple regression coefficients are often misinterpreted. And that can lead to dangerously wrong decisions.

Second, multiple regression is an extraordinarily versatile model, underlying many widely used statistical methods. A sound understanding of the multiple regression model will help you to understand these other applications as well.

Third, multiple regression offers you a first glimpse into statistical models that use more than two quantitative variables. The real world is complex. Simple models are a great start, but they're not detailed enough to be useful for understanding, predicting, and making business decisions in many real-world situations. Models that use several variables can be a big step toward realistic and useful modeling of complex phenomena and relationships.

FOR EXAMPLE The multiple regression model

A large clothing store has recently sent out a special catalog of Fall clothes, and Leiram, its marketing analyst, wants to find out which customers responded to it and which ones bought the most. She plans to conduct an RFM (Recency, Frequency, Monetary) analysis. The RFM method is based on the principle that 80% of your business comes from the best 20% of your customers, and states that the following attributes will be useful predictors of who the best customers will be:

- How *recently* the customer has purchased (Recency)
- How *frequently* the customer shops (Frequency)
- How *much* the customer spends (Monetary)

For each customer, Leiram has information for the past 5 years on: *Date of Last Purchase, Number of Purchases,* and *Total Amount Spent.* In addition, she has demographic information including *Age, Marital Status, Sex, Income,* and *Number of Children.* She chooses a random sample of 149 customers who bought something from the catalog and who have purchased at least 3 times in the past 5 years (*Number of Purchases*). She wants to model how much they bought from the new catalog (*Respond Amount*).

Leiram fits the following multiple regression model to the response variable *Respond Amount*:

```
Response Variable: Respond Amount
R² = 91.48% Adjusted R² = 91.31%
s = 18.183 with 149 − 4 = 145 degrees of freedom
```

Variable	Coeff	SE(Coeff)	t-ratio	P-value
Intercept	111.01	6.459	17.187	<0.0001
Income	0.00091	0.00012	7.643	<0.0001
Total Amount Spent	0.154	0.00852	18.042	<0.0001
Number of Purchases	−14.81	0.716	−20.695	<0.0001

QUESTIONS How much of the variation in *Respond Amount* is explained by this model? What does the term $s = 18.183$ mean? Which variables seem important in the model?

ANSWERS This model has an R^2 of 91.48% which means that the model has explained 91.48% of the variation in *Respond Amount* using the three predictors: *Income, Total Amount Spent,* and *Number of Purchases.* The term $s = 18.183$ means that the standard deviation of the residuals is about $18.18. Using the 68–95–99.7 Rule we know that most prediction errors will be no larger than about $36.36. All terms seem important in this model since all three have very large *t*-ratios and correspondingly small P-values.

15.2 Interpreting Multiple Regression Coefficients

It makes sense that both the number of bedrooms and the size of the living area would influence the price of a house. We'd expect both variables to have a positive effect on price—houses with more bedrooms typically sell for more money, as do larger houses. But look at the coefficient for *Bedrooms* in the multiple regression equation. It's negative: −7483.10. How can it be that the coefficient of *Bedrooms* in the multiple regression is negative? And not just slightly negative, its *t*-ratio is large enough for us to be quite confident that the true value is really negative. Yet we saw the coefficient was clearly positive when *Bedrooms* was the sole predictor in the model (Figure 15.2).

The explanation of this apparent paradox is that in a multiple regression, coefficients have a more subtle meaning. Each coefficient takes into account the other predictor(s) in the model.

To see how variables can interact, let's look at a group of similarly sized homes and examine the relationship between *Bedrooms* and *Price* just for houses with 2500 to

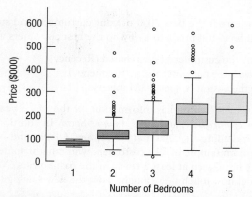

Figure 15.2 The slope of *Bedrooms* is positive. Each is worth about $48,000 in the price of a house as the simple regression model estimated.

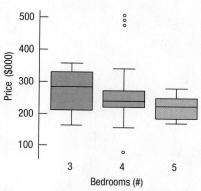

Figure 15.3 For the 96 houses with *Living Area* between 2500 and 3000 square feet, the slope of *Price* on *Bedrooms* is negative. For each additional bedroom, restricting data to homes of this size, we would predict that the house's *Price* was about $17,800 lower.

3000 square feet of living area (Figure 15.3). For houses of this size, those with *fewer* bedrooms have a higher price, on average, than those with more bedrooms. This makes sense when we think about houses in terms of *both*. A 2500-square-foot house with five bedrooms would have either relatively small, cramped bedrooms or not much common living space. The same size house with only three bedrooms could have larger, more attractive bedrooms and still have adequate common living space. What the coefficient of *Bedrooms* says in the multiple regression is that, after accounting for living area, houses with more bedrooms tend to sell for a *lower* price—that the pattern for moderate-size homes is generally true across all sizes.

JUST CHECKING

Body fat percentage is an important health indicator, but it is difficult to measure accurately. One way to do so is via dual energy X-ray absorptiometry (DEXA) at a cost of several hundred dollars per image. Insurance companies want to know if body fat percentage can be estimated from easier-to-measure characteristics such as *Height* and *Weight*. A scatterplot of *Percent Body Fat* against *Height* for 250 adult men shows no pattern, and the correlation is −0.03, which is not statistically significant. A multiple regression using *Height (inches), Age (years), and Weight (pounds)* finds the following model: (Data in **Bodyfat**)

	Coeff	SE(Coeff)	t-ratio	P-value
Intercept	57.27217	10.39897	5.507	<0.0001
Height	−1.27416	0.15801	−8.064	<0.0001
Weight	0.25366	0.01483	17.110	<0.0001
Age	0.13732	0.02806	4.895	<0.0001

$s = 5.382$ on 246 degrees of freedom
Multiple R-squared: 0.584,
F-statistic: 115.1 on 3 and 246 DF, P-value: <0.0001

1 Interpret the R^2 of this regression model.

2 Interpret the coefficient of *Age*.

3 How can the coefficient of *Height* have such a small P-value in the multiple regression when the correlation between *Height* and *Percent Body Fat* was not statistically distinguishable from zero?

Of course, without taking *Living Area* into account, *Price* tends to go *up* with more bedrooms. But that's because *Living Area* and *Bedrooms* are related. Multiple regression coefficients must always be interpreted in terms of the other predictors in the model. That can make their interpretation more subtle, more complex, and more challenging, but it is also what makes multiple regression so versatile and effective. The more sophisticated interpretations are more appropriate.

People often make another error when interpreting coefficients. Regression coefficients should not be interpreted causally. A regression model describes the world as measured by the data, but it can't say very much about what would happen if things were changed. For example, this analysis cannot tell a homeowner how much the price of his home will change if he combines two of his four bedrooms into a new master bedroom. And it can't be used to predict whether adding a 100-square-foot child's bedroom onto the house would increase or decrease its value. The model simply reports the relationship between the number of *Bedrooms* and *Living Area* and *Price* for existing houses. As always with regression, you should be careful not to assume causation between the predictor variables and the response.

FOR EXAMPLE   Interpreting multiple regression coefficients

QUESTION   In the regression model of *Respond Amount*, interpret the intercept and the regression coefficients of the three predictors.

ANSWER   The model says that from a base of \$111.01 of spending, customers (who have purchased at least 3 times in the last 12 months), on average, spent \$0.91 for every \$1000 of *Income* (after accounting for *Number of Purchases* and *Total Amount Spent*), \$0.154 for every dollar they have spent in the past 5 years (after accounting for *Income* and *Number of Purchases*), but \$14.81 less for every additional purchase that they've made in the last 5 years (after accounting for *Income* and *Total Amount Spent*). It is important to note especially that the coefficient in *Number of Purchases* is negative, but only after accounting for both *Income* and *Total Amount Spent*.

15.3   Assumptions and Conditions for the Multiple Regression Model

We can write the multiple regression model, numbering the predictors arbitrarily (the order doesn't matter), writing betas for the model coefficients (which we will estimate from the data), and including the errors in the model:

$$y = \beta_0 + \beta_1 x_1 + \beta_2 x_2 + \cdots + \beta_k x_k + \varepsilon.$$

The assumptions and conditions for the multiple regression model are nearly the same as for simple regression, but with more variables in the model, we'll have to make a few changes, as described in the following sections.

Linearity Assumption

We are fitting a linear model.[2] For that to be the right kind of model for this analysis, we need to verify an underlying linear relationship. But now we're thinking about

---

[2] By *linear* we mean that each *x* appears simply multiplied by its coefficient and added to the model, and that no *x* appears in an exponent or some other more complicated function. That ensures that as we move along any *x*-variable, our prediction for *y* will change at a constant rate (given by the coefficient) if nothing else changes.

several predictors. To confirm that the assumption is reasonable, we'll check the Linearity Condition for *each* of the predictors.

Linearity Condition. Scatterplots of *y* against each of the predictors are reasonably straight. The scatterplots need not show a strong (or any) slope; just check to make sure that there isn't a bend or other nonlinearity. For the real estate data, the scatterplot is linear in both *Bedrooms* and *Living Area*.

As in simple regression, it's a good idea to check the residual plot for any violations of the linearity condition. Fit the regression and plot the residuals against the predicted values (Figure 15.4), checking to make sure there are no patterns—especially bends or other nonlinearities.

Figure 15.4 A scatterplot of residuals against the predicted values shows no obvious pattern.

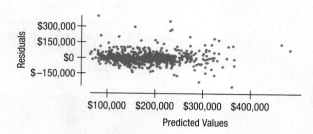

Independence Assumption

As with simple regression, the errors in the true underlying regression model must be independent of each other. As usual, there's no way to be sure that the Independence Assumption is true, but you should think about how the data were collected to see if that assumption is reasonable. You should check the randomization condition as well.

Randomization Condition. Ideally, the data should arise from a random sample or randomized experiment. Randomization assures us that the data are representative of some identifiable population. If you can't identify the population, you can interpret the regression model as a description of the data you have, but you can't interpret the hypothesis tests at all because such tests are about a regression model for a specific population. Regression methods are often applied to data that were not collected with randomization. Regression models fit to such data may still do a good job of modeling the data at hand, but without some reason to believe that the data are representative of a particular population, you should be reluctant to believe that the model generalizes to other situations.

We also check the regression residuals for evidence of patterns, trends, or clumping, any of which would suggest a failure of independence. In the special case when one of the *x*-variables is related to time (or is itself *Time*), be sure that the residuals do not have a pattern when plotted against that variable. In addition to checking the plot of residuals against the predicted values, we recommend that you check the individual plots of the residuals against each of the explanatory, or *x*, variables in the model. These individual plots can yield important information on necessary transformations, or re-expressions, for the predictor variables.

The real estate data were sampled from a larger set of public records for sales during a limited period of time. The error the model makes in predicting one is not likely to be related to the error it makes in predicting another.

Equal Variance Assumption

The variability of the errors should be about the same for all values of *each* predictor. To see whether this assumption is valid, look at scatterplots and check the Equal Spread Condition.

Equal Spread Condition. The same scatterplot of residuals against the predicted values (Figure 15.4) is a good check of the consistency of the spread. We saw what appeared to be a violation of the equal spread condition when *Price* was plotted against *Bedrooms* (Figure 15.2). But here in the multiple regression, the problem has dissipated when we look at the residuals. Apparently, much of the tendency of houses with more bedrooms to have greater variability in prices was accounted for in the model when we included *Living Area* as a predictor.

If residual plots show no pattern, if the errors are plausibly independent, and if the plots of residuals against each *x*-variable don't show changes in variation, you can feel good about interpreting the regression model. Before testing hypotheses, however, you must check one final assumption: the normality assumption.

Normality Assumption

We assume that the errors around the idealized regression model at any specified values of the *x*-variables follow a Normal model. We need this assumption so that we can use a Student's *t*-model for inference. As with other times when we've used Student's *t*, we'll settle for the residuals satisfying the Nearly Normal Condition. As with means, the assumption is less important as the sample size grows. For large data sets, you have to be careful only of extreme skewness or large outliers. These inference methods will work well even when the residuals are moderately skewed, if the sample size is large. If the distribution of residuals is unimodal and symmetric, there is little to worry about.[3]

Nearly Normal Condition. Because there is only one set of residuals, this is the same set of conditions we had for simple regression. Look at a histogram or Normal probability plot of the residuals.

Figure 15.5 A histogram of the residuals shows a unimodal, symmetric distribution, but the tails seem a bit longer than one would expect from a Normal model. The Normal probability plot confirms that.

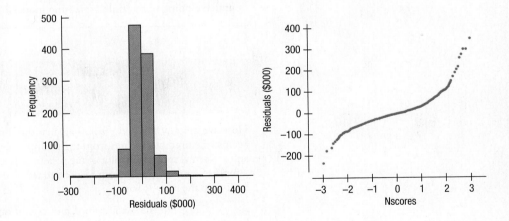

The histogram of residuals in the real estate example certainly looks unimodal and symmetric. The Normal probability plot has some bend on both sides, which indicates that the residuals in the tails straggle away from the center more than Normally distributed data would. However, as we have said before, the Normality Assumption becomes less important as the sample size grows, and here we have no skewness and more than 1000 cases. (The Central Limit Theorem helps our confidence intervals and tests based on the *t*-statistic to be valid when we have large samples.)

Let's summarize all the checks of conditions that we've made and the order in which we've made them.

---

[3] The only time we need strict adherence to the Normality Assumption of the errors is when finding prediction intervals for individuals in multiple regression. Because they are based on individual Normal probabilities and not the Central Limit Theorem, the errors must closely follow a Normal model.

1. Check the Linearity Condition with scatterplots of the *y*-variable against each *x*-variable.
2. If the scatterplots are straight enough, fit a multiple regression model to the data. (Otherwise, either stop or consider re-expressing an *x*-variable or the *y*-variable.)
3. Find the residuals and predicted values.
4. Make a scatterplot of the residuals against the predicted values (and ideally against each predictor variable separately). These plots should look patternless. Check, in particular, for any bend (which would suggest that the data weren't all that straight after all) and for any changes in variation. If there's a bend, consider re-expressing the *y* and/or the *x*-variables. If the variation in the plot grows from one side to the other, consider re-expressing the *y*-variable. If you re-express a variable, start the model fitting over.
5. Think about how the data were collected. Was suitable randomization used? Are the data representative of some identifiable population? If the data are measured over time, check for evidence of patterns that might suggest they're not independent by plotting the residuals against time to look for patterns.
6. If the conditions check out this far, feel free to interpret the regression model and use it for prediction.
7. Make a histogram and Normal probability plot of the residuals to check the Nearly Normal Condition. If the sample size is large, the Normality is less important for inference, but always be on the lookout for skewness or outliers.

JUST CHECKING

> **4** Give two ways that we use histograms to support the construction, inference, and understanding of multiple regression models.
>
> **5** Give two ways that we use scatterplots to support the construction, inference, and understanding of multiple regression models.
>
> **6** What role does the Normal model play in the construction, inference, and understanding of multiple regression models?

FOR EXAMPLE   **Assumptions and conditions for multiple regression**

Here are plots of *Respond Amount* against the three predictors, a plot of the residuals against the predicted values from the multiple regression, a histogram of the residuals, and a Normal probability plot of the residuals. Recall that the data come from a random sample of customers who both responded to the catalog and who had purchased at least 3 times in the past 5 years.

QUESTION   Do the assumptions and conditions for multiple regression appear to have been met?

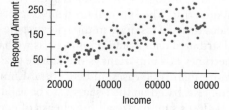

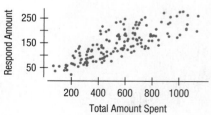

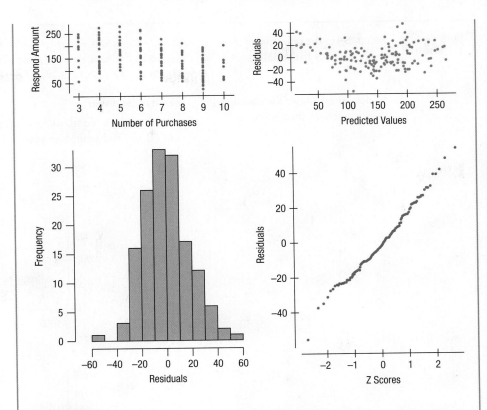

ANSWER Because the sample is random, the randomization condition is satisfied and we assume the responses are independent. The plots of *Respond Amount* against each predictor are reasonably linear with the possible exception of *Number of Purchases*. There may also be some curvature and increasing spread in the residual plot. The histogram of residuals is unimodal and symmetric with no extreme outliers. The Normal probability plot shows that the distribution is fairly Normally distributed. The conditions are not completely satisfied and we should proceed somewhat cautiously especially with regard to *Number of Purchases*.

GUIDED EXAMPLE   Housing Prices

Zillow.com attracts millions of users each month who are interested in finding out how much their house is worth. Let's see how well a multiple regression model can do. The variables available include: (Data in **Housing Prices GE15**).

Price	The price of the house as sold in 2002
Living Area	The size of the living area of the house in square feet
Bedrooms	The number of bedrooms
Bathrooms	The number of bathrooms (a half bath is a toilet and sink only)
Age	Age of the house in years
Fireplaces	Number of fireplaces in the house

(continued)

PLAN

Setup State the objective of the study. Identify the variables.

We want to build a model to predict house prices for a region in upstate New York. We have data on *Price* ($), *Living Area* (sq ft), *Bedrooms* (#), *Bathrooms* (#), *Fireplaces* (#), and *Age* (in years).

Model Think about the assumptions and check the conditions.

Linearity Condition

To fit a regression model, we first require linearity. Scatterplots (or side-by-side boxplots) of *Price* against all potential predictor variables are shown.

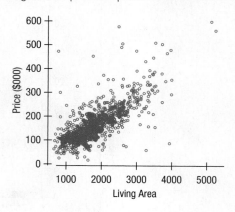

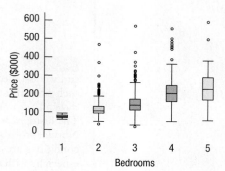

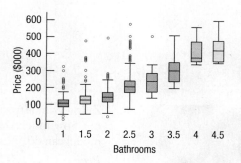

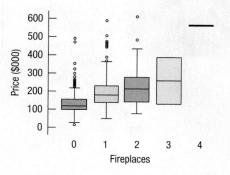

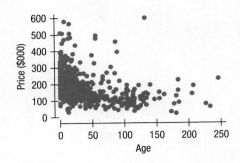

Remarks

There are a few anomalies in the plots that deserve discussion. The plot of *Price* against *Bathrooms* shows a positive relationship, but it is not quite linear. There seem to be two slopes, one from 1 to 2 bathrooms and then a steeper one from 2 to 4. For now, we'll proceed cautiously, realizing that any slope we find will average these two. The plot of *Price* against *Fireplaces* shows an outlier—an expensive home with four fireplaces. We tried setting this home aside and running the regression without it, but its influence on the coefficients was not large, so we decided to include it in the model. The plot of *Price* against *Age* shows that there may be some curvature. We may want to consider re-expressing *Age* to improve the linearity of the relationship.

✓ **Independence Assumption.** We can regard the house prices as being independent of one another. The error the regression model makes in predicting one isn't likely to be related to the error it will make in predicting another.

✓ **Randomization Condition.** These 1057 houses are a random sample of a much larger set.

✓ **Equal Spread Condition.** A scatterplot of residuals vs. predicted values shows no evidence of changing spread. There is a group of homes whose residuals are larger (both negative and positive) than the vast majority. This is also seen in the long tails of the histogram of residuals.

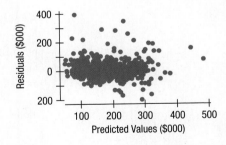

We need the Nearly Normal Condition only if we want to do inference and the sample size is not large. If the sample size is large, as it is here, we need the distribution to be Normal only if we plan to produce prediction intervals.

✓ **Nearly Normal Condition, Outlier Condition.** The histogram of residuals is unimodal and symmetric, but long tailed. The Normal probability plot supports that.

(continued)

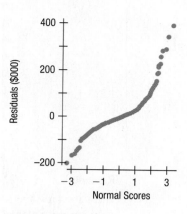
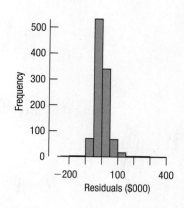

Under these conditions, we can proceed with caution to a multiple regression analysis. We will return to some of our concerns about curvature in some plots in the discussion.

DO

Mechanics We always fit multiple regression models with computer software. An output table like this one isn't exactly what any of the major packages produce, but it is enough like all of them to look familiar.

Here is some computer output for the multiple regression, using all five predictors.

	Coeff	SE(Coeff)	t-ratio	P-value
Intercept	15712.702	7311.427	2.149	0.03186
Living Area	73.446	4.009	18.321	<0.0001
Bedrooms	-6361.311	2749.503	-2.314	0.02088
Bathrooms	19236.678	3669.080	5.243	<0.0001
Fireplaces	9162.791	3194.233	2.869	0.00421
Age	-142.740	48.276	-2.957	0.00318

Residual standard error: 48615.95 on 1051 degrees of freedom
Multiple R-squared: 0.6049.
F-statistic: 321.8 on 5 and 1051 DF, P-value: <0.0001

The estimated equation is:

$$\widehat{Price} = 15{,}712.70 + 73.45\ Living\ Area - 6361.31\ Bedrooms + 19{,}236.68\ Bathrooms + 9162.79\ Fireplaces - 142.74\ Age$$

All of the P-values are small which indicates that even with five predictors in the model, all are contributing. The R^2 value of 60.49% indicates that more than 60% of the overall variation in house prices has been accounted for by this model. The residual standard error of $48,620 gives us a rough indication that we can predict the price of a typical home to within about $2 \times \$48{,}620 = \$97{,}240$. That seems too large to be useful, but the model does give some idea of the price of a home.

REPORT

Summary and Conclusions
Summarize your results and state any limitations of your model in the context of your original objectives.

MEMO

Re: Regression analysis of home price predictions

A regression model of *Price* on *Living Area, Bedrooms, Bathrooms, Fireplaces,* and *Age* accounts for 60.5% of the variation in the price of homes in upstate New York. The residual standard deviation is $48,620. A statistical test of each coefficient shows that each one is almost certainly not zero, so each of these variables appears to be a contributor to the price of a house.

This model reflects the common wisdom in real estate about the importance of various aspects of a home. An important variable not included is the location, which every real estate agent knows is crucial to pricing a house. This is ameliorated by the fact that all these houses are in the same general area. However, knowing more specific information about where they are located would almost certainly help the model. The price found from this model is to be used as a starting point for comparing a home with comparable homes in the area.

The model may be improved by re-expressing one or more of the predictors, especially *Age* and *Bathrooms*. We recommend caution in interpreting the coefficients across the entire range of these predictors.

15.4   Testing the Multiple Regression Model

There are several hypothesis tests in the multiple regression output, but all of them talk about the same thing. Each is concerned with whether the underlying model parameters (the slopes and intercept) are actually zero. The first of these hypotheses is one we skipped over for simple regression (for reasons that will be clear in a minute).

Now that we have more than one predictor, there's an overall test we should perform before we consider inference for the coefficients. We ask the global question: Is this multiple regression model any good at all? If home prices were set randomly or based on other factors than those we have as predictors, then the best estimate would just be the mean price.

To address the overall question, we'll test the null hypothesis that all the slope coefficients are zero:

$$H_0: \beta_1 = \cdots = \beta_k = 0 \text{ vs. } H_A: \text{at least one } \beta \neq 0.$$

We can test this hypothesis with an **F-test.** (It's the generalization of the *t*-test to more than one predictor.) The sampling distribution of the statistic is labeled with the letter F (in honor of Sir Ronald Fisher). The F-distribution has two degrees of freedom, k, the number of predictors, and $n - k - 1$. In the Guided Example, there were $k = 5$ predictors and $n = 1057$ homes, which means that the F-value of 321.8 has 5 and $1057 - 5 - 1 = 1051$ degrees of freedom. The regression output (page 502) shows that it has a P-value < 0.0001. The null hypothesis is that all the coefficients are 0, in which case the regression model predicts no better than the mean. The alternative is that at least one coefficient is non-zero. The test is one-sided—bigger F-values mean smaller P-values. If the null hypothesis were true, the F-statistic would have a value near 1. The F-statistic here is quite large, so we can easily reject the null hypothesis and conclude that the multiple regression model for predicting house prices with these five variables is better than just using the mean.[4]

Once we check the F-test and reject its null hypothesis—and, if we are being careful, *only* if we reject that hypothesis—we can move on to checking the test statistics for the individual coefficients. Those tests look like what we did for the slope of a simple regression in Chapter 14. For each coefficient, we test the null hypothesis that the slope is zero against the (two-sided) alternative that it isn't zero. The regression table gives a standard error for each coefficient and the ratio of

F-test for Simple Regression?

Why didn't we check the F-test for simple regression? In fact, we did. When you do a simple regression with statistics software, you'll see the F-statistic in the output. But for simple regression, it gives the same information as the *t*-test for the slope. It tests the null hypothesis that the slope coefficient is zero, and we already test that with the *t*-statistic for the slope. In fact, the square of that *t*-statistic is equal to the F-statistic for the simple regression, so it really is the identical test.

---

[4] To know how big the F-value has to be, we need a table of F-values. There are F tables in the back of the book, and most regression tables include a P-value for the F-statistic.

the estimated coefficient to its standard error. If the assumptions and conditions are met (and now we need the Nearly Normal Condition or a large sample), these ratios follow a Student's t-distribution:

$$t_{n-k-1} = \frac{b_j - 0}{SE(b_j)}.$$

Where did the degrees of freedom $n - k - 1$ come from? We have a rule of thumb that works here. The degrees of freedom value is the number of data values minus the number of estimated coefficients, minus 1 for the intercept. For the house price regression on five predictors, that's $n - 5 - 1$. Almost every regression report includes both the t-statistics and their corresponding P-values.

We can build a confidence interval in the usual way, with an estimate plus or minus a margin of error. As always, the margin of error is the product of the standard error and a critical value. Here the critical value comes from the t-distribution on $n - k - 1$ degrees of freedom, and the standard errors are in the regression table. So a confidence interval for each slope β_j is:

$$b_j \pm t^*_{n-k-1} \times SE(b_j).$$

The tricky parts of these tests are that the standard errors of the coefficients now require harder calculations (so we leave it to technology), and the meaning of a coefficient, as we have seen, depends on all the other predictors in the multiple regression model.

That last point is important. If we fail to reject the null hypothesis for a multiple regression coefficient, it does *not* mean that the corresponding predictor variable has no linear relationship to y. It means that the corresponding predictor contributes nothing to modeling y *after allowing for all the other predictors*.

The multiple regression model looks so simple and straightforward. It *looks* like each β_j tells us the effect of its associated predictor, x_j, on the response variable, y. But that is not true. This is, without a doubt, the most common error that people make with multiple regression. In fact:

- The coefficient β_j in a multiple regression can be quite different from zero even when it is possible there is no simple linear relationship between y and x_j.
- It is even possible that the multiple regression slope changes sign when a new variable enters the regression. We saw this for the *Price* on *Bedrooms* real estate example when *Living Area* was added to the regression.

So we'll say it once more: The coefficient of x_j in a multiple regression depends as much on the *other* predictors as it does on x_j. Failing to interpret coefficients properly is the most common error in working with regression models.

FOR EXAMPLE   Testing a multiple regression model

Continuing the example (p 498):

Leiram tries another model, adding the variable *Age* to see if that improves the model:

```
Response Variable: Respond Amount
R² = 91.50%; Adjusted R² = 91.23%
s = 18.179 with 149 − 5 = 144 degrees of freedom
```

Variable	Coeff	SE(Coeff)	t-ratio	P-value
Intercept	114.91	9.244	12.439	<0.0001
Income	0.00091	0.00012	7.619	<0.0001
Total Amount Spent	0.154	0.00855	18.007	<0.0001
Number of Purchases	-14.79	0.719	-20.570	<0.0001
Age	-0.1264	0.2144	-0.5898	0.5563

QUESTION   Has the variable *Age* improved the model? Would you leave the term in? Comment.

ANSWER   Of course, we would like to see the residual plots, but given this output it appears that although the R^2 value has increased from 0.9148 to 0.9150, the *t*-ratio for *Age* is only -0.5898 with a P-value of 0.5563. This indicates that there is no evidence to suggest that the slope for *Age* is different from 0. We cannot reject that null hypothesis. There is no reason to leave *Age* in this model.

15.5   Adjusted R^2 and the F-statistic

In Chapter 14, for simple linear regression, we interpreted R^2 as the variation in y accounted for by the model. The same interpretation holds for multiple regression, where now the model contains more than one predictor variable. The R^2 value tells us how much (as a fraction or percentage) of the variation in y is accounted for by the model with all the predictor variables included.

There are some relationships among the standard error of the residuals, s_e, the F-ratio, and R^2 that are useful for understanding how to assess the value of the multiple regression model. To start, we can write the standard error of the residuals as:

$$s_e = \sqrt{\frac{SSE}{n - k - 1}},$$

where $SSE = \sum e^2$ is called the **Sum of Squared Residuals** (or errors—the E is for error). A larger SSE (and thus s_e) means that the residuals are more variable and that our predictions will be correspondingly less precise.

We can look at the total variation of the response variable, y, which is called the **Total Sum of Squares** and is denoted SST: $SST = \sum (y - \bar{y})^2$. For any regression model, we have no control over SST, but we'd like SSE to be as small as we can make it by finding predictor variables that account for as much of that variation as possible. In fact, we can write an equation that relates the total variation SST to SSE:

$$SST = SSR + SSE,$$

where $SSR = \sum (\hat{y} - \bar{y})^2$ is called the **Regression Sum of Squares** because it comes from the predictor variables and tells how much of the total variation in the response is due to the regression model. For a model to account for a large portion of the variability in y, SSR should be large and SSE should be small. In fact, R^2 is just the ratio of SSR to SST:

$$R^2 = \frac{SSR}{SST} = 1 - \frac{SSE}{SST}.$$

When the SSE is nearly 0, the R^2 value will be close to 1.

In Chapter 14, we saw that for the relationship between two quantitative variables, testing the standard null hypothesis about the correlation coefficient,

Mean Squares

Whenever a sum of squares is divided by its degrees of freedom, the result is called a mean square. For example, the Mean Square for Error, which you may see written as MSE, is found as $SSE/(n - k - 1)$. It estimates the variance of the errors.

Similarly $SST/(n - 1)$ divides the total sum of square by *its* degrees of freedom. That is sometimes called the Mean Square for Total and denoted MST. We've seen this one before; the MST is just the variance of *y*.

And SSR/k is the Mean Square for Regression.

$H_0: \rho = 0$, was equivalent to testing the standard null hypothesis about the slope, $H_0: \beta_1 = 0$. A similar result holds here for multiple regression. Testing the overall hypothesis tested by the F-statistic, $H_0: \beta_1 = \beta_2 = \cdots = \beta_k = 0$, is equivalent to testing whether the true multiple regression R^2 is zero. In fact, the F-statistic for testing that all the slopes are zero can be found as:

$$F = \frac{R^2/k}{(1 - R^2)/(n - k - 1)} = \frac{\dfrac{SSR}{SST}\dfrac{1}{k}}{\dfrac{SSE}{SST}\dfrac{1}{n - k - 1}} = \frac{SSR/k}{SSE/(n - k - 1)} = \frac{MSR}{MSE}.$$

In other words, using an F-test to see whether any of the true coefficients is different from 0 is equivalent to testing whether the R^2 value is different from zero. A rejection of either hypothesis says that at least one of the predictors accounts for enough variation in *y* to distinguish it from noise. Unfortunately, the test doesn't say which slope is responsible. You must look at individual *t*-tests on the slopes to determine that. Because removing one predictor variable from the regression equation can change any number of slope coefficients, it is not straightforward to determine the right subset of predictors to use.

R^2 and Adjusted R^2

Adding a predictor variable to a multiple regression equation does not always increase the amount of variation accounted for by the model, but it can never reduce it. Adding new predictor variables will always keep the R^2 value the same or increase it. It can never decrease it. But, even if the R^2 value grows, that doesn't mean that the resulting model is a better model or that it has greater predictive ability. If you have a model with *k* predictors (all of which have statistically significant coefficients at some α level) and want to see if including a new variable, x_{k+1}, is warranted, you could fit the model with all $k + 1$ variables and simply test the slope of the added variable with a *t*-test of the slope.

This method can test whether the most recently added variable adds significantly to the model, but choosing the "best" subset of predictors is not necessarily straightforward. The trade-off between a small (parsimonious) model and one that fits the data well is one of the great challenges of any serious model-building effort. Various statistics have been proposed to provide guidance for this search, and one of the most common is called adjusted R^2. **Adjusted** R^2 imposes a "penalty" for each new term that's added to the model in an attempt to make models of different sizes (numbers of predictors) comparable. It differs from R^2 because it can shrink when a predictor is added to the regression model or grow when a predictor is removed if the predictor in question doesn't contribute usefully to the model. It can even be negative.

For a multiple regression with *k* predictor variables and *n* cases, it is defined as

$$R^2_{adj} = 1 - (1 - R^2)\frac{n - 1}{n - k - 1} = 1 - \frac{SSE/(n - k - 1)}{SST/(n - 1)}.$$

In the Guided Example, the regression of *Price* on *Bedrooms, Bathrooms, Living Area, Fireplaces,* and *Age* resulted in an R^2 of 0.6049. All the coefficients had P-values well below 0.05. The adjusted R^2 value for this model is 0.6030. Adding the variable *Lot Size* to the model gives the following regression model:

	Coeff	SE(Coeff)	t-ratio	P-value
Intercept	15360.011	7334.804	2.094	0.03649
Living Area	73.388	4.043	18.154	<0.00001
Bedrooms	−6096.387	2757.736	−2.211	0.02728
Bathrooms	18824.069	3676.582	5.120	<0.00001
Fireplaces	9226.356	3191.788	2.891	0.00392
Age	−152.615	48.224	−3.165	0.00160
Lot Size	847.764	1989.112	0.426	0.67005

```
Residual standard error: 48440 on 1041 degrees of freedom
Multiple R-squared: 0.6081, Adjusted R-squared: 0.6059
F-statistic: 269.3 on 6 and 1041 DF, P-value: <0.0001
```

The most striking feature of this regression table as compared to the one in the Guided Example on page 502, is that although most of the coefficients have changed very little, the coefficient of *Lot Size* is far from significant, with a P-value of 0.670. Yet, the adjusted R^2 value is actually higher than for the previous model. This is why we warn against putting too much faith in this statistic. Especially for large samples, the adjusted R^2 does not always adjust downward enough to make sensible model choices. The other problem with comparing these two models is that 9 homes had missing values for *Lot Size*, which means that we're not comparing the models on exactly the same data set. When we matched the two models on the smaller data set, the adjusted R^2 value actually did "make the right decision" but just barely—0.6059 versus 0.6060 for the model without *Lot Size*. One might expect a larger difference considering we added a variable whose *t*-ratio is much less than 1.

The lesson to be learned here is that there is no "correct" set of predictors to use for any real business decision problem, and finding a reasonable model is a process that takes a combination of science, art, business knowledge, and common sense. Look at the adjusted R^2 value for any multiple regression model you fit, but be sure to think about all the other reasons for including or not including any given predictor variable.

FOR EXAMPLE R^2 and adjusted R^2

QUESTION The model for *Respond Amount* that included *Age* as a predictor (see For Example on pages 504 and 505) had an adjusted R^2 value of 0.9123 (or 91.23%). The original model (see For Example on page 493) had an adjusted R^2 value of 0.9131. Explain how that is consistent with the decision we made to drop *Age* from the model.

ANSWER The *t*-ratio indicated that the slope for *Age* was not significantly different from 0, and thus that we should drop *Age* from the model. Although the R^2 value for the model that included *Age* was slightly higher, the adjusted R^2 value was slightly lower. Adjusted R^2 adds a penalty for each added term that allows us to compare models of different sizes. Because the adjusted R^2 value for the smaller model (without *Age*) is higher than the larger model (with *Age*), it indicates that the inclusion of *Age* is not justified, which is consistent with the recommendation from the *t*-ratio.

---

*15.6   The Logistic Regression Model

Business decisions often depend on whether something will happen or not. Will my customers leave my wireless service at the end of their subscription? How likely is it that my customers will respond to the offer I just mailed? The response variable is either Yes or No—a dichotomous response. By definition, that's a categorical response, so we can't use linear regression methods to predict it.

If we coded this categorical response variable by giving the Yes values the value 1 and the No values the value 0, we could "pretend" that it's quantitative and try to fit a linear model. Let's imagine that we're trying to model whether someone will respond to an offer based on how much they spent in the last year at our company. The regression of *Purchase (1 = Yes: 0 = No)* on *Spending* might look like Figure 15.6.

The model shows that as past spending increases, something having to do with responding to the offer increases. Look at *Spending* near $500. Nearly all the customers who spend that much responded to the offer, while for those customers who spent less than $100 last year, almost none responded. The line seems to show the proportion of those responding for different values of *Spending*. Looking at the plot, what would you predict the proportion of responders to be for those that spent about $225? You might say that the proportion was about 0.50.

Figure 15.6 A linear model of *Respond to Offer* shows that as *Spending* increases, the proportion of customers responding to the offer increases as well. Unfortunately, the line predicts values outside the interval (0,1), but those are the only sensible values for proportions or probabilities.

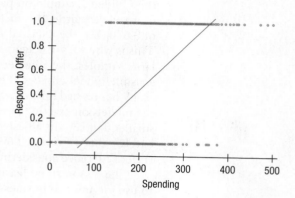

What's wrong with this interpretation? We know proportions (or probabilities) must lie between 0 and 1. But the linear model has no such restrictions. The line crosses 1.0 at about $350 in *Spending* and goes below 0 at about $75 in *Spending*. However, by transforming the probability, we can make the model behave better and get more sensible predictions for all values of *Spending*. We could just cut off all values greater than 1 at 1 and all values less than 0 at 0. That would work fine in the middle, but we know that things like customer behavior don't change that abruptly. So we'd prefer a model that curves at the ends to approach 0 and 1 gently. That's likely to be a better model for what really happens. A simple function is all we need. There are several that can do the job, but one common model is the logistic regression model. Here is a plot of the logistic regression model predictions for the same data.

Figure 15.7 A logistic regression model of *Respond to Offer* on *Spending* predicts values between 0 and 1 for the proportion of customers who *Respond* for all values of *Spending*.

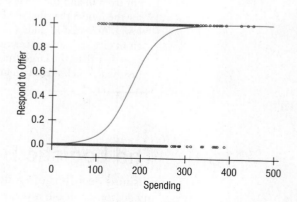

For many values of *Spending*, especially those near the mean value of *Spending*, the predictions of the probability of responding are similar for the linear and logistic models, but the logistic transformation approaches the limits at 0 and 1 smoothly. A logistic regression is an example of a nonlinear regression model. The computer finds the coefficients by solving a system of equations that in some ways mimic the least squares calculations we did for linear regression. The **logistic regression** model looks like this:

$$\ln\left(\frac{p}{1-p}\right) = \beta_0 + \beta_1 x_1 + \cdots + \beta_k x_k.$$

where ln( ) denotes the natural logarithm and p is the probability of a "success."

In other words, it's not the probabilities themselves that are modeled by a (multiple) regression model, but a transformation (the logistic function) of the probabilities. When the probabilities are transformed back and plotted, we get an S-shaped curve (when plotted against each predictor), as seen in Figure 15.7.

As with multiple regression, computer output for a logistic regression model provides estimates for each coefficient, standard errors, and a test for whether the coefficient is zero. Unlike multiple regression, the coefficients for each variable are tested with a Chi-square statistic[5] rather than a *t*-statistic, and the overall R^2 is no longer available. (Some computer programs use a *z*-statistic to test individual coefficients. The tests are equivalent.) Because the probabilities are not a linear function of the predictors, it is more difficult to interpret the coefficients. The fitted logistic regression equation can be written:

$$\ln\left(\frac{\hat{p}}{1-\hat{p}}\right) = b_0 + b_1 x_1 + \cdots + b_k x_k.$$

Some researchers try to interpret the logistic regression equation directly by realizing that the expression on the left of the logistic regression equation, $\ln\left(\frac{\hat{p}}{1-\hat{p}}\right)$, can be thought of as the predicted log (base e) odds of the outcome. This is sometimes denoted $\text{logit}(\hat{p})$. The higher the log odds, the higher the probability. Negative log odds indicate that the probability is less than 0.5 because then $\frac{p}{1-p}$ will be less than 1 and thus have a negative logarithm. Positive log odds indicate a probability greater than 0.5. If the coefficient of a particular predictor is positive, it means that higher values of it are associated with higher log odds and thus a higher probability of the response. So, the *direction* is interpretable in the same way as a multiple regression coefficient. But an increase of one unit of the predictor *increases* the predicted *log odds* by an amount equal to the coefficient of that predictor, after allowing for the effects of all the other predictors. It does not increase the predicted *probability* by that amount.

The transformation back to probabilities is straightforward, but nonlinear. Once you have fit the logistic regression equation

$$\ln\left(\frac{\hat{p}}{1-\hat{p}}\right) = \text{logit}(\hat{p}) = b_0 + b_1 x_1 + \cdots + b_k x_k,$$

you can find the individual probability estimates from the equation:

$$\hat{p} = \frac{1}{1 + e^{-(b_0 + b_1 x_1 + \cdots + b_k x_k)}} = \frac{e^{(b_0 + b_1 x_1 + \cdots + b_k x_k)}}{1 + e^{(b_0 + b_1 x_1 + \cdots + b_k x_k)}}.$$

Log Odds

Racetrack enthusiasts know that when p is a probability, $\frac{p}{1-p}$ is the *odds* in favor of a success. For example, when the probability of success, $p = 1/3$, we'd get the ratio $\frac{1/3}{2/3} = \frac{1}{2}$. We'd say that the odds in favor of success are 1:2 (or we'd probably say the odds *against* it are 2:1). Logistic regression models the *logarithm* of the odds as a linear function of x. In fact, nobody really thinks in terms of the log of the odds ratio. But it's the combination of that ratio and the logarithm that gets us the nice S-curved shape. What is important is that we can work backward from a log odds ratio to get the probability—which is often easier to think about.

---

[5] Yes, the same sampling distribution model as we used in Chapter 13.

The assumptions and conditions for fitting a logistic regression are similar to multiple regression. We still need the independence assumption and randomization condition. We no longer need the linearity or equal spread condition. However, it's a good idea to plot the response variable against each predictor variable to make sure that there are no outliers that could unduly influence the model. A customer who both spent $10,000 and responded to the offer could possibly change the shape of the curve shown in Figure 15.7. However, residual analysis for logistic regression models is beyond the scope of this book.

FOR EXAMPLE   Logistic regression

Leiram wants to build a model to predict *whether* a customer will respond to the new catalog offer. The variable *Respond* indicates 1 for responders and 0 for nonresponders. (Data in **Market Response**). After trying several models, she settles on the following logistic regression model:

Response Variable: Respond

Variable	Coeff	SE(Coeff)	z-value	P-value
Intercept	−3.298	0.4054	−8.136	<0.0001
Age	0.08365	0.012226	6.836	<0.0001
Days Since Last Purchase	0.00364	0.000831	4.378	<0.0001

QUESTIONS   Interpret the model. Would you keep both predictor variables in the model? Why? Who would be more likely to respond, a 20-year-old customer who has recently purchased another item, or a 50-year-old customer who has not purchased in a year?

ANSWERS   The model says:

$$\text{logit}(\hat{p}) = -3.298 + 0.08365\ Age + 0.00364\ Days\ Since\ Last\ Purchase,$$

which means that the probability of responding goes up with both *Age* and *Days Since Last Purchase*. Both terms have large z-values and correspondingly low P-values, which indicate that they contribute substantially to the model. We should keep both terms. Because both terms are positively associated with increasing (log) odds of purchasing, both an older person and one who has *not* purchased recently are more likely to respond. A 50-year-old who has not purchased recently would be more likely to respond than the 20-year-old who has.

GUIDED EXAMPLE   Time on Market

A real estate agent used information on 1115 houses, such as we used in our multiple regression Guided Example. She wants to predict whether a house sold in the first 3 months it was on the market based on other variables. (Data in **Time on Market**.) The variables available include:

Sold	1 = Yes—the house sold within the first 3 months it was listed; 0 = No, it did not sell within 3 months
Price	The price of the house as sold in 2002
Living Area	The size of the living area of the house in square feet
Bedrooms	The number of bedrooms
Bathrooms	The number of bathrooms (a half bath is a toilet and sink only)
Age	Age of the house in years
Fireplaces	Number of fireplaces in the house

PLAN	**Setup** State the objective of the study. Identify the variables. **Model** Think about the assumptions and check the conditions.

We want to build a model to predict whether a house will sell within the first 3 months it's on the market based on *Price* ($), *Living Area* (sq ft), *Bedrooms* (#), *Bathrooms* (#), *Fireplaces* (#), and *Age* (years). Notice that now *Price* is a predictor variable in the regression.

Outlier Condition

To fit a logistic regression model, we check that there are no outliers in the predictors that may unduly influence the model.

Here are scatterplots of *Sold* (1 = Yes; 0 = No) against each predictor. (Plots of *Sold* against the variables *Bathrooms*, *Bedrooms*, and *Fireplaces* are uninformative because these predictors are discrete.)

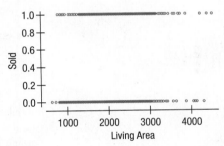

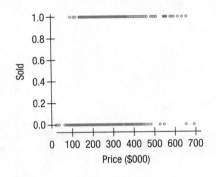

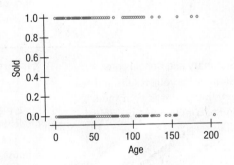

✓ **Outlier Condition.** There do not seem to be any outliers in the predictors. Of course, there can't be any in the response variable; it is only 0 or 1.

✓ **Independence Assumption.** We can regard the house prices as being independent of one another since they come from a fairly large geographic area.

✓ **Randomization Condition.** These 1115 houses are a random sample from a larger collection of houses.

We can fit a logistic regression predicting *Sold* from the six predictor variables.

(continued)

DO

Mechanics We always fit logistic regression models with computer software. An output table like this one isn't exactly what any of the major packages produce, but it is enough like all of them to look familiar.

Here is the computer output for the logistic regression using all six predictors.

	Coeff	SE(Coeff)	z-value	P-value
Intercept	−3.222e+00	3.826e−01	−8.422	<0.0001
Living Area	−1.444e−03	2.518e−04	−5.734	<0.0001
Age	4.900e−03	2.823e−03	1.736	0.082609
Price	1.693e−05	1.444e−06	11.719	<0.0001
Bedrooms	4.805e−01	1.366e−01	3.517	0.000436
Bathrooms	−1.813e−01	1.829e−01	−0.991	0.321493
Fireplaces	−1.253e−01	1.633e−01	−0.767	0.442885

The estimated equation is:

$$logit(\hat{p}) = -3.22 - 0.00144 \text{ Living Area}$$
$$+ 0.0049 \text{ Age} + 0.0000169 \text{ Price}$$
$$+ 0.481 \text{ Bedrooms} - 0.181 \text{ Bathrooms}$$
$$- 0.125 \text{ Fireplaces}$$

Three of the P-values are quite small, two are large (*Bathrooms* and *Fireplaces*), and one is marginal (*Age*). After examining several alternatives, we chose the following model.

	Coeff	SE(Coeff)	z-value	P-value
Intercept	−3.351e+00	3.601e−01	−9.305	<0.0001
Living Area	−1.574e−03	2.342e−04	−6.719	<0.0001
Age	6.106e−03	2.668e−03	2.289	0.022102
Price	1.672e−05	1.428e−06	11.704	<0.0001
Bedrooms	4.631e−01	1.354e−01	3.421	0.000623

The estimated logit equation is:

$$logit(\hat{p}) = -3.351 - 0.00157 \text{ Living Area}$$
$$+ 0.00611 \text{ Age} + 0.0000167 \text{ Price}$$
$$+ 0.463 \text{ Bedrooms}$$

While interpretation is difficult, it appears that among houses of a given size (and age and bedrooms), higher-priced homes may have a higher chance of selling within 3 months. For a house of given size and price and age, having more bedrooms may be associated with a greater chance of selling within 3 months.

REPORT

Summary and Conclusions
Summarize your results and state any limitations of your model in the context of your original objectives.

MEMO

Re: Logistic regression analysis of selling

A logistic regression model of *Sold* on various predictors was fit, and a model based on *Living Area*, *Bedrooms*, *Price*, and *Age* found that these four predictors were statistically significant in predicting the probability that a house will sell within 3 months. More thorough analysis to understand the meaning of the coefficients is needed, but each of these variables appears to be an important predictor of whether the house will sell quickly.

However, knowing more specific information about other characteristics of a house and where it is located would almost certainly help the model.

WHAT CAN GO WRONG?

Interpreting Coefficients

- **Don't claim to "hold everything else constant" for a single individual.** It's often meaningless to say that a regression coefficient says what we expect to happen if all variables but one were held constant for an individual and the predictor in question changed. While it's mathematically correct, it often just doesn't make any sense. For example, in a regression of salary on years of experience, years of education, and age, subjects can't gain a year of experience or get another year of education without getting a year older. Instead, we *can* think about all those who satisfy given criteria on some predictors and ask about the conditional relationship between y and one x for those individuals.

- **Don't interpret regression causally.** Regressions are usually applied to observational data. Without deliberately assigned treatments, randomization, and control, we can't draw conclusions about causes and effects. We can never be certain that there are no variables lurking in the background, causing everything we've seen. Don't interpret b_1, the coefficient of x_1 in the multiple regression, by saying: "If we were to change an individual's x_1 by 1 unit (holding the other x's constant), it would change his y by b_1 units." We have no way of knowing what applying a change to an individual would do.

- **Be cautious about interpreting a regression model as predictive.** Yes, we do call the x's predictors, and you can certainly plug in values for each of the x's and find a corresponding *predicted value*, $\hat{y}$. But the term "prediction" suggests extrapolation into the future or beyond the data, and we know that we can get into trouble when we use models to estimate $\hat{y}$ values for x's not in the range of the data. Be careful not to extrapolate very far from the span of your data. In simple regression, it was easy to tell when you extrapolated. With many predictor variables, it's often harder to know when you are outside the bounds of your original data.[6] We usually think of fitting models to the data more as modeling than as prediction, so that's often a more appropriate term.

- **Be careful when interpreting the signs of coefficients in a multiple regression.** Sometimes our primary interest in a predictor is whether it has a positive or negative association with y. As we have seen, though, the sign of the coefficient also depends on the other predictors in the model. Don't look at the sign in isolation and conclude that "the direction of the relationship is positive (or negative)." Just like the value of the coefficient, the sign is about the relationship after allowing for the linear effects of the other predictors. The sign of a variable can change depending on which other predictors are in or out of the model. For example, in the regression model for house prices, we saw the coefficient of *Bedrooms* change sign when *Living Area* was added to the model as a predictor. It isn't correct to say either that houses with more bedrooms sell for more on average or that they sell for less. The truth is more subtle and requires that we understand the multiple regression model.

- **If a coefficient's *t*-statistic is not significant, don't interpret it at all.** You can't be sure that the value of the corresponding parameter in the underlying regression model isn't really zero.

---

[6] With several predictors we can wander beyond the data because of the *combination* of values even when individual values are not extraordinary. For example, houses with 1 bathroom and houses with 5 bedrooms can both be found in the real estate records, but a single house with 5 bedrooms and only 1 bathroom would be quite unusual. The model we found is not appropriate for predicting the price of such an extraordinary house.

WHAT ELSE CAN GO WRONG?

- **Don't fit a linear regression to data that aren't straight.** This is the most fundamental regression assumption. If the relationship between the *x*'s and *y* isn't approximately linear, there's no sense in fitting a linear model to it. What we mean by "linear" is a model of the form we have been writing for the regression. When we have two predictors, this is the equation of a plane, which is linear in the sense of being flat in all directions. With more predictors, the geometry is harder to visualize, but the simple structure of the model is consistent; the predicted values change consistently with equal size changes in any predictor.

  Usually we're satisfied when plots of *y* against each of the *x*'s are straight enough. We'll also check a scatterplot of the residuals against the predicted values for signs of nonlinearity.

- **Watch out for changing variance in the residuals.** The estimate of the error standard deviation shows up in all the inference formulas. But that estimate assumes that the error standard deviation is the same throughout the extent of the *x*'s so that we can combine all the residuals when we estimate it. If s_e changes with any *x*, these estimates won't make sense. The most common check is a plot of the residuals against the predicted values. You can also check plots of residuals against several of the predictors. If they show a thickening and especially if they also show a bend, then consider re-expressing *y*.

- **Make sure the errors are nearly Normal.** All of our inferences require that, unless the sample size is large, the true errors be modeled well by a Normal model. Check the histogram and Normal probability plot of the residuals to see whether this assumption looks reasonable.

- **Watch out for high-influence points and outliers.** We always have to be on the lookout for a few points that have undue influence on our model, and regression is certainly no exception.

ETHICS IN ACTION

Alpine Medical Systems, Inc. is a large provider of medical equipment and supplies to hospitals, doctors, clinics, and other health care professionals. Alpine's VP of Marketing and Sales, Kenneth Jadik, asked one of the company's analysts, Nicole Haly, to develop a model that could be used to predict the performance of the company's sales force.

Based on data collected over the past year, as well as records kept by Human Resources, she considered five potential independent variables: (1) gender, (2) starting base salary, (3) years of sales experience, (4) personality test score, and (5) high school grade point average. The dependent variable (sales performance) is measured as the sales dollars generated per quarter.

In discussing the results with Nicole, Kenneth asks to see the full regression model with all five independent variables included. Kenneth notes that a *t*-test for

the coefficient of gender shows no significant effect on sales performance and recommends that it be eliminated from the model. Nicole reminds him of the company's history of offering lower starting base salaries to women, recently corrected under court order. If instead, starting base salary is removed from the model, gender is statistically significant, and its coefficient indicates that women on the sales force outperform men (taking into account the other variables). Kenneth argues that because gender is not significant when all predictors are included, it is the variable that should be omitted.

- Identify the ethical dilemma in this scenario.

- What are the undesirable consequences?

- Propose an ethical solution that considers the welfare of all stakeholders.

WHAT HAVE WE LEARNED?

Learning Objectives

Know how to perform a multiple regression, using the technology of your choice.

- Technologies differ, but most produce similar-looking tables to hold the regression results. Know how to find the values you need in the output generated by the technology you are using.

Understand how to interpret a multiple regression model.

- The meaning of a multiple regression coefficient depends on the other variables in the model. In particular, it is the relationship of y to the associated x after removing the linear effects of the other x's.

Be sure to check the Assumptions and Conditions before interpreting a multiple regression model.

- The **Linearity Assumption** asserts that the form of the multiple regression model is appropriate. We check it by examining scatterplots. If the plots appear to be linear, we can fit a multiple regression model.
- The **Independence Assumption** requires that the errors made by the model in fitting the data be mutually independent. Data that arise from random samples or randomized experiments usually satisfy this assumption.
- The **Equal Variance Assumption** states that the variability around the multiple regression model should be the same everywhere. We usually check the **Equal Spread Condition** by plotting the residuals against the predicted values. This assumption is needed so that we can pool the residuals to estimate their standard deviation, which we will need for inferences about the regression coefficients.
- The **Normality Assumption** says that the model's errors should follow a Normal model. We check the **Nearly Normal Condition** with a histogram or normal probability plot of the residuals. We need this assumption to use Student's-t models for inference, but for larger sample sizes, it is less important.

Know how to state and test hypotheses about the multiple regression coefficients.

- The standard hypothesis test for each coefficient is

$$H_0: \beta_j = 0 \text{ vs.}$$
$$H_A: \beta_j \neq 0$$

- We test these hypotheses by referring the test statistic

$$\frac{b_j - 0}{SE(b_j)}$$

to the Student's t-distribution on $n - k - 1$ degrees of freedom, where k is the number of coefficients estimated in the multiple regression.

Interpret other associated statistics generated by a multiple regression.

- R^2 is the fraction of the variation in y accounted for by the multiple regression model.
- Adjusted R^2 attempts to adjust for the number of coefficients estimated.
- The F-statistic tests the overall hypothesis that the regression model is of no more value than simply modeling y with its mean.
- The standard deviation of the residuals,

$$s_e = \sqrt{\frac{\sum e^2}{n - k - 1}}$$

provides an idea of how precisely the regression model fits the data.

Terms	
Adjusted R^2	An adjustment to the R^2 statistic that attempts to allow for the number of predictors in the model. It is sometimes used when comparing regression models with different numbers of predictors: $$R^2_{adj} = 1 - (1 - R^2)\frac{n-1}{n-k-1} = 1 - \frac{SSE/(n-k-1)}{SST/(n-1)}.$$
F-test	The F-test is used to test the null hypothesis that the overall regression is no improvement over just modeling y with its mean: $$H_0: \beta_1 = \cdots = \beta_k = 0 \ vs \ H_A: \text{at least one } \beta \neq 0$$ If this null hypothesis is not rejected, then you should not proceed to test the individual coefficients.
*Logistic regression	A regression model that models a binary (0/1) response variable based on quantitative predictor variables.
Multiple regression	A linear regression with two or more predictors whose coefficients are found by least squares. When the distinction is needed, a least squares linear regression with a single predictor is called a *simple regression*. The multiple regression model is: $y = \beta_0 + \beta_1 x_1 + \cdots + \beta_k x_k + \varepsilon$.
Regression Sum of Squares, SSR	A measure of the total variation in the response variable due to the model. $SSR = \sum (\hat{y} - \bar{y})^2$.
Sum of Squared Errors or Residuals, SSE	A measure of the variation in the residuals. $SSE = \sum (y - \hat{y})^2$.
Total Sum of Squares, SST	A measure of the variation in the response variable. $SST = \sum (y - \bar{y})^2$. Note that $\frac{SST}{n-1} = Var(y)$.
t-ratios for the coefficients	The t-ratios for the coefficients can be used to test the null hypotheses that the true value of each coefficient is zero against the alternative that it is not. The t-distribution is also used in the construction of confidence intervals for each slope coefficient.

TECHNOLOGY HELP: Regression Analysis

All statistics packages make a table of results for a regression. The table for multiple regression looks very similar to the table for simple regression.

Most packages offer to plot residuals against predicted values. Some will also plot residuals against the x's. With some packages, you must request plots of the residuals when you request the regression. Others let you find the regression first and then analyze the residuals afterward. Either way, your analysis is not complete if you don't check the residuals with a histogram or Normal probability plot and a scatterplot of the residuals against the x's or the predicted values. In most packages, the commands to obtain residuals and residual plots are the same for single and multiple regression.

Multiple regressions are always found with a computer or programmable calculator. Before computers were available, a full multiple regression analysis could take months or even years of work.

EXCEL

- Select **Data Analysis** from the **Analysis Group** on the Data Tab.
- Select **Regression** from the **Analysis Tools** list.
- Click the **OK** button.
- Enter the data range holding the y-variable in the box labeled **Y-range**.
- Enter the range of cells holding the x-variables in the box labeled **X-range**.
- Select the **New Worksheet Ply** option.
- Select **Residuals** options. Click the **OK** button.

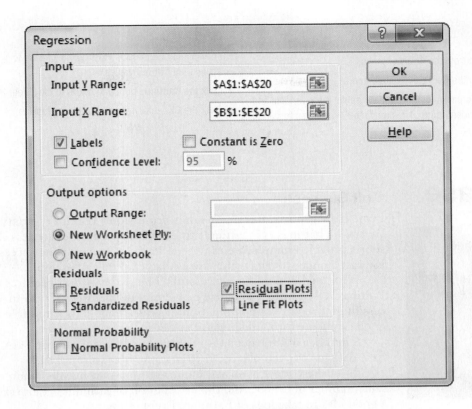

Comments

Although the Excel Data Analysis dialog offers a Normal probability plot of the residuals, the data analysis add-in does not make a correct probability plot, so don't use this option.

XLSTAT

- In **Modeling Data** menu, choose **Linear Regression**.
- Enter *y*-variable and *x*-variable cell ranges.
- Specify desired statistics in Outputs and Charts tabs, respectively.

Comments

For both Excel and XLStat, the Y and X cell ranges do not need to be in the same rows of the spreadsheet, although they must cover the same number of cells. It is a good idea to arrange your data in parallel columns as in a data table. The X-variables must be in adjacent columns. No cells in the data range may hold nonnumeric values or be left blank.

JMP

- From the **Analyze** menu, select **Fit Model**.
- Put the response variable in the **Y** dialog box and the predictor variables in the **Add** dialog box under **Construct Model Effects**.
- Click on **Run Model**.

Comments

JMP chooses a regression analysis when the response variable is "Continuous." The predictors can be any combination of quantitative or categorical. If you get a different analysis, check the variable types.

MINITAB

- Choose **Regression** from the **Stat** menu.
- Choose **Regression ...** from the **Regression** submenu.
- In the Regression dialog, assign the Y-variable to the Response box and assign the X-variables to the Predictors box.
- Click the **Graphs** button.
- In the Regression-Graphs dialog, select **Standardized residuals**, and check **Normal plot of residuals** and **Residuals versus fits**.
- Click the **OK** button to return to the Regression dialog.
- Click the **OK** button to compute the regression.

R

Suppose the response variable *y* and predictor variables $x_1, \ldots, x_k$ are in a data frame called mydata. To fit a multiple regression of *y* on x_1 and x_2:

- mylm = lm(y ~ x_1 + x_2, data = mydata)
- summary(mylm) # gives the details of the fit, including the ANOVA table
- plot(mylm) # gives a variety of plots

  To fit the model with *all* the predictors in the data frame,

- mylm = lm(y ~., data = mydata) # The period means use all other variables

Comments

To get confidence or prediction intervals use:

- predict(mylm, interval = "confidence")

  or

- predict(mylm, interval = "prediction")

SPSS

- Choose **Regression** from the **Analyze** menu.
- Choose **Linear** from the **Regression** submenu.
- When the Linear Regression dialog appears, select the Y-variable and move it to the dependent target. Then move the X-variables to the independent target.

- Click the **Plots** button.
- In the Linear Regression Plots dialog, choose to plot the *SRESIDs against the *ZPRED values.
- Click the **Continue** button to return to the Linear Regression dialog.
- Click the **OK** button to compute the regression.

Brief **Case**

Golf Success

Professional sports, like many other professions, require a variety of skills for success. That makes it difficult to evaluate and predict success. Fortunately, sports provide examples we can use to learn about modeling success because of the vast amount of data which are available. Here's an example.

What makes a golfer successful? The game of golf requires many skills. Putting well or hitting long drives will not, by themselves, lead to success. Success in golf requires a combination of skills. That makes multiple regression a good candidate for modeling golf achievement.

A number of Internet sites post statistics for the current PGA players. We have data for 181 top players of 2013 in the file **Golfers 2013**.

All of these players earned money on the tour, but they didn't all play the same number of events. And the distribution of earnings is quite skewed. So it's a good idea to take logs of Earnings/Event as the response variable.

The variables in the data file include:

Log$/E	The logarithm of earnings per event
Greens in Regulation	Greens in Regulation. Percentage of holes played in which the ball is on the green with two or more strokes left for par.
Putt Average	Average number of putts per hole in which the green was reached in regulation.
Save pct	Each time a golfer hits a bunker by the side of a green but needs only one or two additional shots to reach the hole, he is credited with a save. This is the percentage of opportunities for saves that are realized.
Yds/Drive	Average Drive Distance (yards). Measured as averages over pairs of drives in opposite directions (to account for wind).
Driving Acc	Drive Accuracy. Percent of drives landing on the fairway.

Investigate these data. Find a regression model to predict golfers' success (measured in log earnings per event). Write a report presenting your model including an assessment of its limitations. Note: Although you may consider several intermediate models, a good report is about the model you think best, not necessarily about all the models you tried along the way while searching for it.

EXERCISES

SECTION 15.1

1. A house in the upstate New York area from which the chapter data was drawn has 2 bedrooms and 1000 square feet of living area. Using the multiple regression model found in the chapter,

$$\widehat{Price} = 20{,}986.09 - 7483.10\ Bedrooms + 93.84\ Living\ Area.$$

a) Find the price that this model estimates.
b) The house just sold for $135,000. Find the residual corresponding to this house.
c) What does that residual say about this transaction?

2. A candy maker surveyed chocolate bars available in a local supermarket and found the following least squares regression model:

$$\widehat{Calories} = 28.4 + 11.37 Fat(g) + 2.91\ Sugar(g).$$

a) The hand-crafted chocolate she makes has 15g of fat and 20g of sugar. How many calories does the model predict for a serving?
b) In fact, a laboratory test shows that her candy has 227 calories per serving. Find the residual corresponding to this candy. (Be sure to include the units.)
c) What does that residual say about her candy?

SECTION 15.2

3. What can predict how much a motion picture will make? We have data on a number of recent releases that includes the *USGross* (in $M), the *Budget* ($M), the *Run Time* (minutes), and the average number of *Stars* awarded by reviewers. The first several entries in the data table look like this:

Movie	USGross ($M)	Budget ($M)	Run Time (minutes)	Stars
White Noise	56.094	30	101	2
Coach Carter	67.264	45	136	3
Elektra	24.409	65	100	2
Racing Stripes	49.772	30	110	3
Assault on Precinct 13	20.040	30	109	3
Are We There Yet?	82.674	20	94	2
Alone in the Dark	5.178	20	96	1.5
Indigo	51.100	25	105	3.5

We want a regression model to predict *USGross*. Parts of the regression output computed in Excel look like this:

```
Dependent variable is: USGross($)
R squared = 47.4% R squared (adjusted) = 46.0%
s = 46.41 with 120 - 4 = 116 degrees of freedom

Variable     Coefficient  SE(Coeff)  t-ratio  P-value
Intercept    -22.9898     25.70      -0.895   0.3729
Budget($)      1.13442     0.1297     8.75    <0.0001
Stars         24.9724      5.884      4.24    <0.0001
Run Time      -0.403296    0.2513    -1.60    0.1113
```

a) Write the multiple regression equation.
b) What is the interpretation of the coefficient of *Budget* in this regression model?

4. A middle manager at an entertainment company, upon seeing the analysis of Exercise 3, concludes that the longer you make a movie, the less money it will make. He argues that his company's films should all be cut by 30 minutes to improve their gross. Explain the flaw in his interpretation of this model.

SECTION 15.3

5. For the movies examined in Exercise 3, here is a scatterplot of *USGross* vs. *Budget*:

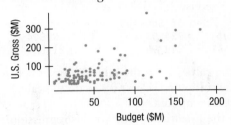

What (if anything) does this scatterplot tell us about the following Assumptions and Conditions for the regression?

a) Linearity condition
b) Equal Spread condition
c) Normality assumption

6. For the movies regression, here is a histogram of the residuals. What does it tell us about these Assumptions and Conditions?

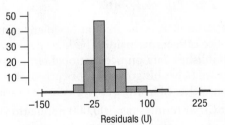

a) Linearity condition
b) Nearly Normal condition
c) Equal Spread condition

SECTION 15.4

7. In the regression output for the movies of Exercise 3,

a) What is the null hypothesis tested for the coefficient of *Stars* in this table?
b) What is the *t*-statistic corresponding to this test?
c) What is the P-value corresponding to this *t*-statistic?
d) Complete the hypothesis test. Do you reject the null hypothesis?

8. a) What is the null hypothesis tested for the coefficient of *Run Time* in the regression of Exercise 3?

b) What is the *t*-statistic corresponding to this test?
c) Why is this *t*-statistic negative?
d) What is the P-value corresponding to this *t*-statistic?
e) Complete the hypothesis test. Do you reject the null hypothesis?

SECTION 15.5

9. In the regression model of Exercise 3,

a) What is the R^2 for this regression? What does it mean?
b) Why is the "Adjusted R Square" in the table different from the "R Square"?

10. Here is another part of the regression output for the movies in Exercise 3:

Source	Sum of Squares	df	Mean Square	F-ratio
Regression	224995	3	74998.4	34.8
Residual	249799	116	2153.44	

a) Using the values from the table, show how the value of R^2 could be computed. Don't try to do the calculation, just show what is computed.
b) What is the F-statistic value for this regression?
c) What null hypothesis can you test with it?
d) Would you reject that null hypothesis?

CHAPTER EXERCISES

The next 12 exercises consist of two sets of 6 (one even-numbered, one odd-numbered). Each set guides you through a multiple regression analysis. We suggest that you do all 6 exercises in a set. Remember that the answers to the odd-numbered exercises can be found in the back of the book.

11. Police salaries 2013. Is the amount of violent crime related to what police officers are paid? The U.S. Bureau of Labor Statistics publishes data on occupational employment and wage estimates (www.bls.gov/oes/) for each of the 50 states. Here are data released from 2011 to 2013. The variables are:

  Violent Crime (crimes per 100,000 population)
  Police Officer Wage (mean $/hr)
  Graduation Rate (%)

One natural question to ask of these data is how police officer wages are related to violent crime across these states.

First, here are plots and background information.

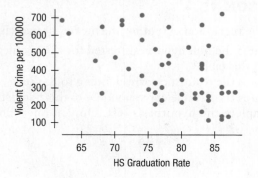

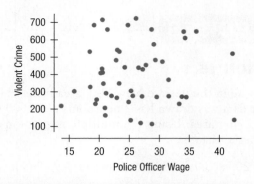

	Correlations		
	Violent Crime	**Graduation Rate**	**Police Officer Wage**
Violent Crime	1.000		
Graduation Rate	−0.473	1.000	
Police Officer Wage	0.051	−0.473	1.000

a) Name and check (to the extent possible) the regression assumptions and their corresponding conditions.
b) If we found a regression to predict *Violent Crime* just from *Police Officer Wage*, what would the R^2 of that regression be?

12. Ticket prices. On a typical night in New York, about 25,000 people attend a Broadway show, paying an average price of more than $75 per ticket. *Variety* (www.variety.com), a news weekly that reports on the entertainment industry, publishes statistics about the Broadway show business. The data file on the disk holds data about shows on Broadway for most weeks of a recent three-year period. (A few weeks are missing data.) The following variables are available for each week:

  Receipts ($ million)
  Paid Attendance (thousands)
  # Shows
  Average Ticket Price ($)

Viewing this as a business, we'd like to model *Receipts* in terms of the other variables.

First, here are plots and background information.

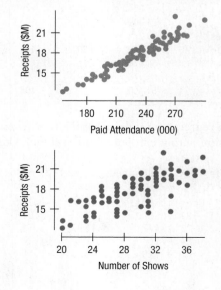

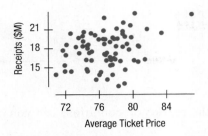

Average Ticket Price

Correlations				
	Receipts	**Paid Attendance**	**# Shows**	**Average Ticket Price**
Receipts ($M)	1.000			
Paid Attendance	0.961	1.000		
# Shows	0.745	0.640	1.000	
Average Ticket Price	0.258	0.331	−0.160	1.000

a) Name and check (to the extent possible) the regression assumptions and their corresponding conditions.
b) If we found a regression to predict *Receipts* only from *Paid Attendance*, what would the R^2 of that regression be?

T 13. Police salaries, part 2. Here's a multiple regression model for the variables considered in Exercise 11.

```
Dependent variable is: Violent Crime
R squared = 22.41%  R squared (adjusted) = 19.10%
s = 156.9 with 47 degrees of freedom
```

Source	Sum of Squares	df	Mean Square	F-ratio
Regression	334246	2	167123	6.786
Residual	1157508	47	24628	

Variable	Coeff	SE(Coeff)	t-ratio	P-value
Intercept	1370.22	292.180	4.690	<0.0001
Police Officer Wage	0.7947	3.598	0.221	0.8262
Graduation Rate	−12.641	3.451	−3.663	0.00063

a) Write the regression model.
b) What does the coefficient of *Police Officer Wage* mean in the context of this regression model?
c) In a state in which the average police officer wage is $20/hour and the high school graduation rate is 70%, what does this model estimate the violent crime rate would be?
d) Is this likely to be a good prediction? Why do you think that?

T 14. Ticket prices, part 2. Here's a multiple regression model for the variables considered in Exercise 12:[7]

```
Dependent variable is: Receipts($M)
R squared = 99.9%  R squared (adjusted) = 99.9%
s = 0.0931 with 74 degrees of freedom
```

Source	Sum of Squares	df	Mean Square	F-ratio
Regression	484.789	3	161.596	18634
Residual	0.641736	74	0.008672	

Variable	Coeff	SE(Coeff)	t-ratio	P-value
Intercept	−18.320	0.3127	−58.6	<0.0001
Paid Attendance	0.076	0.0006	126.7	<0.0001
# Shows	0.0070	0.0044	1.59	0.116
Average Ticket Price	0.24	0.0039	61.5	<0.0001

a) Write the regression model.
b) What does the coefficient of *Paid Attendance* mean in this regression? Does that make sense?
c) In a week in which the paid attendance was 200,000 customers attending 30 shows at an average ticket price of $70, what would you estimate the receipts would be?
d) Is this likely to be a good prediction? Why do you think that?

T 15. Police salaries, part 3. Using the regression table in Exercise 13, answer the following questions.

a) How was the *t*-ratio of 0.221 found for *Police Officer Wage*? (Show what is computed using numbers from the table.)
b) How many states are used in this model. How do you know?
c) The *t*-ratio for *Graduation Rate* is negative. What does that mean?

T 16. Ticket prices, part 3. Using the regression table in Exercise 14, answer the following questions.

a) How was the *t*-ratio of 126.7 found for *Paid Attendance*? (Show what is computed using numbers found in the table.)
b) How many weeks are included in this regression? How can you tell?
c) The *t*-ratio for the intercept is negative. What does that mean?

T 17. Police salaries, part 4. Consider the coefficient of *Police Officer Wage* in the regression table of Exercise 13.

a) State the standard null and alternative hypotheses for the true coefficient of *Police Officer Wage*.
b) Test the null hypothesis (at $\alpha = 0.05$) and state your conclusion.

T 18. Ticket prices, part 4. Consider the coefficient of *# Shows* in the regression table of Exercise 14.

a) State the standard null and alternative hypotheses for the true coefficient of *# Shows*.
b) Test the null hypothesis (at $\alpha = 0.05$) and state your conclusion.
c) A Broadway investor challenges your analysis. He points out that the scatterplot of *Receipts* vs. *# Shows* in Exercise 12 shows a strong linear relationship and claims that your result in part a can't be correct. Explain to him why this is not a contradiction.

T 19. Police salaries, part 5. A Senate aide accepts your analysis in Exercise 17 but claims that it demonstrates that if the state pays police more, it will actually *increase* the rate

---

[7] Some values are rounded to simplify the exercises. If you recompute the analysis with your statistics software you may see slightly different numbers.

of violent crime. Explain why this interpretation is not a valid use of this regression model. Offer some alternative explanations.

T 20. Ticket prices, part 5. The investor in Exercise 18 now accepts your analysis but claims that it demonstrates that it doesn't matter how many shows are playing on Broadway; receipts will be essentially the same. Explain why this interpretation is not a valid use of this regression model. Be specific.

T 21. Police salaries, part 6. Here are some plots of residuals for the regression of Exercise 13.

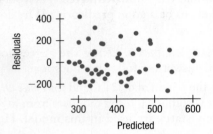

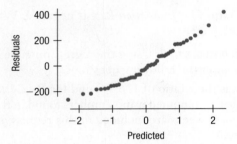

Which of the regression conditions can you check with these plots?

Do you find that those conditions are met?

T 22. Ticket prices, part 6. Here are some plots of residuals for the regression of Exercise 14.

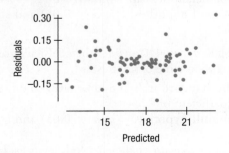

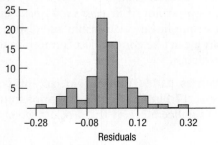

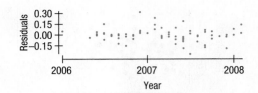

Which of the regression conditions can you check with these plots?

Do you find that those conditions are met?

23. Real estate prices. A regression was performed to predict selling *Price* of houses in dollars from their *Area* in square feet, *Lotsize* in square feet, and *Age* in years. The R^2 is 92%. The equation from this regression is given here.

$$\widehat{Price} = 169{,}328 + 35.3\,Area + 0.718\,Lotsize - 6543\,Age$$

One of the following interpretations is correct. Which is it? Explain what's wrong with the others.

a) Each year a house ages, it is worth $6543 less.
b) Every extra square foot of area is associated with an additional $35.50 in average price, for houses with a given lot size and age.
c) Every additional dollar in price means lot size increases 0.718 square feet.
d) This model fits 92% of the data points exactly.

24. Wine prices. Many factors affect the price of wine, including such qualitative characteristics as the variety of grape, location of winery, and label. Researchers developed a regression model considering two quantitative variables: the tasting score of the wine and the age of the wine (in years) when released to market. They found the following regression equation, with an R^2 of 65%, to predict the price (in dollars) of a bottle of wine.

$$\widehat{Price} = 6.25 + 1.22\,Tasting\ Score + 0.55\,Age$$

One of the following interpretations is correct. Which is it? Explain what's wrong with the others.

a) Each year a bottle of wine ages, its price increases about $.55.
b) This model fits 65% of the points exactly.
c) For a unit increase in tasting score, the price of a bottle of wine increases about $1.22.
d) After allowing for the age of a bottle of wine, a wine with a one unit higher tasting score can be expected to cost about $1.22 more.

25. Appliance sales. A household appliance manufacturer wants to analyze the relationship between total sales and the company's three primary means of advertising (television, magazines, and radio). All values were in millions of dollars. They found the following regression equation.

$$\widehat{Sales} = 250 + 6.75\,TV + 3.5\,Radio + 2.3\,Magazine$$

One of the following interpretations is correct. Which is it? Explain what's wrong with the others.

a) If they did no advertising, their income would be $250 million.

b) Every million dollars spent on radio makes sales increase $3.5 million, all other things being equal.

c) Every million dollars spent on magazines increases TV spending $2.3 million.

d) Sales increase on average about $6.75 million for each million spent on TV, after allowing for the effects of the other kinds of advertising.

26. Wine prices, part 2. Here are some more interpretations of the regression model to predict the price of wine developed in Exercise 24. One of these interpretations is correct. Which is it? Explain what is wrong with the others.

a) The minimum price for a bottle of wine that has not aged is $6.25.

b) The price for a bottle of wine increases on average about $.55 for each year it ages, after allowing for the effects of tasting score.

c) Each year a bottle of wine ages, its tasting score increases by 1.22.

d) Each dollar increase in the price of wine increases its tasting score by 1.22.

27. Cost of pollution. What is the financial impact of pollution abatement on small firms? The U.S. government's Small Business Administration studied this and reported the following model.

$$\overline{Pollution\ abatement/employee} = -2.494 - 0.431$$
$$\ln(Number\ of\ Employees) + 0.698\ \ln(Sales)$$

Pollution abatement is in dollars per employee.

a) The coefficient of ln(*Number of Employees*) is negative. What does that mean in the context of this model? What does it mean that the coefficient of ln(*Sales*) is positive?

b) The model uses the (natural) logarithms of the two predictors. What does the use of this transformation say about their effects on pollution abatement costs?

T 28. OECD economic regulations. A study by the U.S. Small Business Administration used historical data to model the GDP per capita of 24 of the countries in the Organization for Economic Co-operation and Development (OECD) (Crain, M. W., *The Impact of Regulatory Costs on Small Firms*, available at www.sba.gov/advocacy/7540/49291). One analysis estimated the effect on GDP of economic regulations, using an index of the degree of OECD economic regulation and other variables. They found the following regression model.

$$\overline{GDP/Capita}\,(1998–2002) = 10487 - 1343\ OECD$$
$$Economic\ Regulation\ Index + 1.078\ GDP/Capita\,(1988)$$
$$- 69.99\ Ethno\text{-}linguistic\ Diversity\ Index$$
$$+ 44.71\ Trade\ as\ share\ of\ GDP\ (1998–2002)$$
$$- 58.4\ Primary\ Education\,(\%\,Eligible\ Population)$$

All *t*-statistics on the individual coefficients have P-values < 0.05, except the coefficient of *Primary Education*.

a) The researchers hoped to show that more regulation leads to lower GDP/Capita. Does the coefficient of the OECD Economic Regulation Index demonstrate that? Explain.

b) The F-statistic for this model is 129.61 (5, 17 df). What do you conclude about the model?

c) If *GDP/Capita(1988)* is removed as a predictor, then the F-statistic drops to 0.694 and none of the *t*-statistics is significant (all P-values > 0.22). Reconsider your interpretation in part a.

29. Home prices. Many variables have an impact on determining the price of a house. A few of these are size of the house (square feet), lot size, and number of bathrooms. Information for a random sample of homes for sale in the Statesboro, Georgia, area was obtained from the Internet. Regression output modeling the asking price with square footage and number of bathrooms gave the following result.

```
Dependent Variable is: Asking Price
s = 67013   R-Sq = 71.1%   R-Sq(adj) = 64.6%
```

Predictor	Coeff	SE(Coeff)	t-ratio	P-value
Intercept	−152037	85619	−1.78	0.110
Baths	9530	40826	0.23	0.821
Area	139.87	46.67	3.00	0.015

```
Analysis of Variance
```

Source	DF	SS	MS	F	P-value
Regression	2	99303550067	49651775033	11.06	0.004
Residual	9	40416679100	4490742122		
Total	11	1.39720E+11			

a) Write the regression equation.

b) How much of the variation in home asking prices is accounted for by the model?

c) Explain in context what the coefficient of Area means.

d) The owner of a construction firm, upon seeing this model, objects because the model says that the number of bathrooms has no effect on the price of the home. He says that when *he* adds another bathroom, it increases the value. Is it true that the number of bathrooms is unrelated to house price? (*Hint:* Do you think bigger houses have more bathrooms?)

30. Home prices, part 2. Here are some diagnostic plots for the home prices data from Exercise 29. These were generated by a computer package and may look different from the plots generated by the packages you use. (In particular, note that the axes of the Normal probability plot are swapped relative to the plots we've made in the text. We only care about the pattern of this plot, so it shouldn't affect your interpretation.) Examine these plots and discuss whether the assumptions and conditions for the multiple regression seem reasonable.

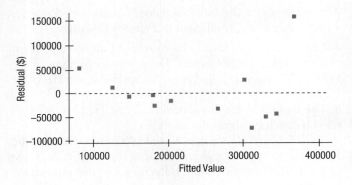

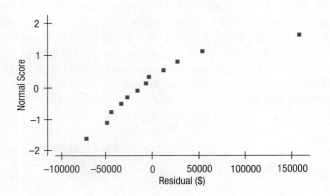

Normal Probability Plot

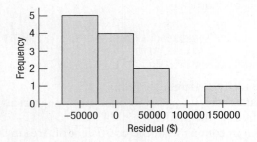

Assume that the residual plots show no violations of the conditions for using a linear regression model.

a) What is the regression equation?

b) From this model, what is the predicted salary (in thousands of dollars) of a secretary with 10 years (120 months) of experience, 9th grade education (9 years of education), 50 on the standardized test, 60 wpm typing speed, and the ability to take 30 wpm dictation?

c) Test whether the coefficient for words per minute of typing speed (X4) is significantly different from zero at $\alpha = 0.05$.

d) How might this model be improved?

e) A correlation of age with salary finds $r = 0.682$, and the scatterplot shows a moderately strong positive linear association. However, if X6 = Age is added to the multiple regression, the estimated coefficient of age turns out to be $b_6 = -0.154$. Explain some possible causes for this apparent change of direction in the relationship between age and salary.

32. Wal-Mart revenue. Here's a regression of monthly revenue of Wal-Mart Corp, relating that revenue to the Total U.S. Retail Sales, the Personal Consumption Index, and the Consumer Price Index.

```
Dependent variable is: Wal-Mart_Revenue
R squared = 66.7%  R squared (adjusted) = 63.8%
s = 2.327 with 39 - 4 = 35 degrees of freedom
```

Source	Sum of Squares	df	Mean Square	F-ratio
Regression	378.749	3	126.250	23.3
Residual	189.474	35	5.41354	

Variable	Coeff	SE(Coeff)	t-ratio	P-value
Intercept	87.0089	33.60	2.59	0.0139
Retail Sales	0.000103	0.000015	6.67	<0.0001
Persnl Consmp	0.00001108	0.000004	2.52	0.0165
CPI	−0.344795	0.1203	−2.87	0.0070

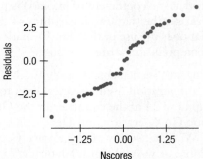

a) Write the regression model.

b) Interpret the coefficient of the Consumer Price Index (CPI). Does it surprise you that the sign of this coefficient is negative? Explain.

c) Test the standard null hypothesis for the coefficient of CPI and state your conclusions.

31. Secretary performance. The AFL-CIO has undertaken a study of 30 secretaries' yearly salaries (in thousands of dollars). The organization wants to predict salaries from several other variables. The variables to be considered potential predictors of salary are:

X1 = months of service

X2 = years of education

X3 = score on standardized test

X4 = words per minute (wpm) typing speed

X5 = ability to take dictation in words per minute

A multiple regression model with all five variables was run on a computer package, resulting in the following output.

Variable	Coeff	Std. Error	t-value
Intercept	9.788	0.377	25.960
X1	0.110	0.019	5.178
X2	0.053	0.038	1.369
X3	0.071	0.064	1.119
X4	0.004	0.0307	0.013
X5	0.065	0.038	1.734

s = 0.430   R-sq = 0.863

T 33. Gross domestic product. The gross domestic product (GDP) is an important measure of the overall economic strength of a country. GDP per capita makes comparisons between different size countries more meaningful. A researcher looking at GDP, fit the following model based on an educational variable, *Primary School Completion Rate (%)*, and finds:

```
Dependent variable is: GDP per Capita
R squared = 3.44%
s = 15945.46 with 96 − 1 = 95 df
```

Term	Estimate	Std Error	t-Ratio	P-value
Intercept	1935.5693	5987.938	0.320	0.7472
Primary Completion Rate	122.3288	66.8131	1.830	0.0703

a) Explain to the researcher why, on the basis of the regression summary, she might want to consider other predictor variables in the model.

b) Explain why you are not surprised that the sign of the slope is positive.

The researcher adds two variables to the regression and finds:

```
Dependent variable is: GDP per Capita
R squared = 80.00%
s = 7327.65 with 96 − 4 = 92 df
```

Term	Estimate	SE(Coeff)	t-Ratio	P-value
Intercept	2775.98251	2803.32606	0.99	0.3247
Cell phones/ 100 people	92.84968	37.38697	2.48	0.0148
Internet users per 100 people (2004)	480.49061	54.04020	8.89	<0.0001
Primary completion rate	−63.28454	32.25614	−1.96	0.0528

c) Explain how the slope of *Primary completion rate* can now be negative.

T 34. Lobster industry 2012. The lobster industry is an important one in Maine, with annual landings worth about $300,000,000 and employment consequences that extend throughout the state's economy. Here's a multiple regression to predict the *LogValue* from other variables published by the Maine Department of Marine Resources (maine.gov/dmr). The predictors are number of *Traps* (millions), number of licensed *Fishers*, and *Pounds/Trap* during the years 1957 to 2012.

```
Dependent variable is: LogValue
R squared = 97.5% R squared (adjusted) = 97.4%
s = 0.0801 with 56 − 4 = 52 degrees of freedom
```

Source	Sum of Squares	df	Mean Square	F-ratio
Regression	13.2861	3	4.4287	690.108
Residual	0.33370	52	0.00642	

Variable	Coeff	SE(Coeff)	t-ratio	P-value
Intercept	0.85601	0.08766	9.77	<0.0001
Traps(M)	0.56347	0.01245	45.26	<0.0001
Fishers	−0.0000444	0.00000953	−4.67	<0.0001
Pounds/Trap	0.00381	0.00145	2.62	0.0114

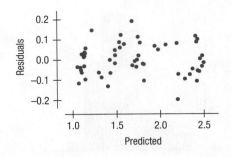

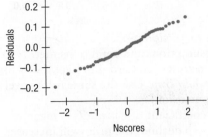

a) Write the regression model.

b) Are the assumptions and conditions met?

c) Interpret the coefficient of *Fishers*. Would you expect that restricting the number of lobstering licenses to even fewer fishers would increase the value of the harvest?

d) State and test the standard null hypothesis for the coefficient of *Pounds/Trap*. Scientists claim that this is an important predictor of the harvest. Do you agree?

T 35. Lobster industry 2012, revisited. Here's a multiple regression to predict the *Price* from the number of *Traps* (millions), the number of *Fishers*, and *Pounds/Trap* during the years 1957 to 2012.

```
Dependent variable is: Price/lb
R squared = 94.2% R squared (adjusted) = 93.8%
s = 0.2850 with 56 − 4 = 52 degrees of freedom
```

Source	Sum of Squares	df	Mean Square	F-ratio
Regression	68.3949	3	22.7983	280.70
Residual	4.2234	52	0.0812	

Variable	Coefficient	SE(Coeff)	t-ratio	P-value
Intercept	1.0942	0.3118	3.51	0.0009
Traps(M)	1.2357	0.0443	27.90	<0.0001
Fishers	−0.000149	0.0000339	−4.40	<0.0001
Pounds/Trap	−0.0180	0.00517	−3.47	0.0011

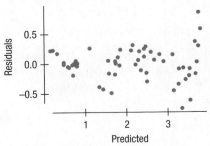

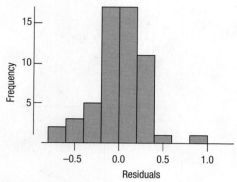

a) Write the regression model.
b) Are the assumptions and conditions met?
c) State and test the standard null hypothesis for the coefficient of *Pounds/Trap*. Use the standard α-level of .05 and state your conclusion.
d) Does the coefficient of *Pounds/Trap* mean that when the pounds per trap declines the price will increase?

T 36. HDI. In 1990, the United Nations created a single statistic, the Human Development Index or HDI, to summarize the health, education, and economic status of countries. Using data from 96 countries, here is a multiple regression model trying to predict *HDI*.

```
Dependent variable is: HDI
R squared = 98.23%
s = 0.024 with 96 − 8 = 88 df
```

Term	Estimate	Std Error	t-Ratio	P-value
Intercept	0.09010	0.04208	2.14	0.0351
Exp yrs school	0.01376	0.00176	7.79	<0.0001
Life exp	0.00333	0.00063	5.32	<0.0001
Maternal mortality ratio	−0.000120	0.0000254	−4.74	<0.0001
Mean yrs school	0.01686	0.00166	10.19	<0.0001
Urban%	0.000745	0.000181	4.13	<0.0001
GDP/cap	0.000000802	0.000000263	3.05	0.003
Cell phones/ 100 people	0.000460	0.000141	3.26	0.0016

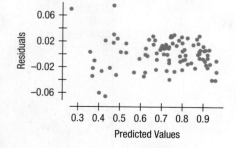

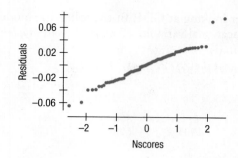

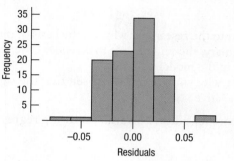

a) Write the regression model.
b) Are the assumptions and conditions met?
c) State and test the standard null hypothesis for the coefficient of *Expected Years of Schooling*. Use the standard-level of $\alpha = 0.05$ and state your conclusion.
d) What effects do your observation in response to part b have on your test in part c?

T 37. Wal-Mart revenue, part 2. Wal-Mart is the second largest retailer in the world. The data file on the disk holds monthly data on Wal-Mart's revenue, along with several possibly related economic variables.

a) Using computer software, find the regression equation predicting Wal-Mart revenues from the *Retail Index*, the Consumer Price index (*CPI*), and *Personal Consumption*.
b) Does it seem that Wal-Mart's revenue is closely related to the general state of the economy?

T 38. Wal-Mart revenue, part 3. Consider the model you fit in Exercise 37 to predict Wal-Mart's revenue from the Retail Index, CPI, and Personal Consumption index.

a) Plot the residuals against the predicted values and comment on what you see.
b) Identify and remove the four cases corresponding to December revenue and find the regression with December results removed.
c) Does it seem that Wal-Mart's revenue is closely related to the general state of the economy?

T *39. Clinical trials. An important challenge in clinical trials is patients who drop out before the trial is completed. This can cost pharmaceutical companies millions of dollars because patients who have received a tested treatment for months must be combined with those who received it for a much shorter time. Can we predict who will drop out of a study early? We have data for 428 patients from a clinical trial of depression. We have data on their *Age*, and their *Hamilton Rating Depression Scale* (*HRDS*) and whether or not they completed the study (*Drop1* = *Yes; 0* = *No*).

Here is the output from a logistic regression model of *Drop* on *HRDS* and *Age*.

Term	Estimate	Std Error	z	P-Value
Intercept	−0.441972938	0.488270354	−0.9055	0.3654
AGE	−0.037904831	0.011511729	−3.292	0.001
HDRS	0.046817607	0.015903137	2.944	0.0032

a) Write out the estimated regression equation.
b) What is the predicted log odds (logit) of the probability that a 30-year-old patient with an HDRS score of 30 will drop out of the study?
c) What is the predicted dropout probability of that patient?
d) What is the predicted log odds (logit) of the probability that a 60-year-old patient with an HDRS score of 8 will drop out of the study?
e) What is the associated predicted probability?

*40. Cost of higher education. Are there fundamental differences between liberal arts colleges and universities? In this case, we have information on the top 25 liberal arts colleges and the top 25 universities in the United States. We will consider the type of school as our response variable and will use the percent of students who were in the top 10% of their high school class and the amount of money spent per student by the college or university as our explanatory variables. The output from this logistic regression is given here.

Logistic Regression Table

Predictor	Coeff	SE(Coeff)	z	P
Intercept	−13.1461	3.98629	−3.30	0.001
Top 10%	0.0845469	0.0396345	2.13	0.033
$/Student	0.0002594	0.0000860	3.02	0.003

a) Write out the estimated regression equation.
b) Is percent of students in the top 10% of their high school class statistically significant in predicting whether or not the school is a university? Explain.
c) Is the amount of money spent per student statistically significant in predicting whether or not the school is a university? Explain.

T 41. Motorcycles. More than one million motorcycles are sold annually (www.webbikeworld.com). Off-road motorcycles (often called "dirt bikes") are a market segment (about 18%) that is highly specialized and offers great variation in features. This makes it a good segment to study to learn about which features account for the cost (manufacturer's suggested retail price, MSRP) of a dirt bike. Researchers collected data on 2005 model dirt bikes (lib.stat.cmu.edu/datasets/dirtbike_aug.csv). Their original goal was to study market differentiation among brands (*The Dirt on Bikes: An Illustration of CART Models for Brand Differentiation*, Jiang

Lu, Joseph B. Kadane, and Peter Boatwright, server1 .tepper.cmu.edu/gsiadoc/WP/2006-E57.pdf), but we can use these to predict msrp from other variables.

Here are scatterplots of three potential predictors, *Wheelbase (in)*, *Displacement (cu in)*, and *Bore (in)*.

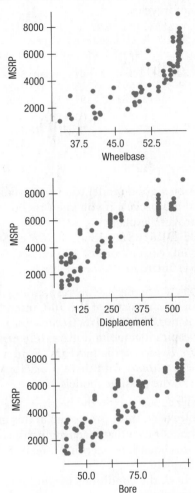

Comment on the appropriateness of using these variables as predictors on the basis of the scatterplots.

T 42. Motorcycles, part 2. In Exercise 41, we saw data on off-road motorcycles and examined scatterplots. Review those scatterplots. Here's a regression of *MSRP* on both *Displacement* and *Bore*. Both of the predictors are measures of the size of the engine. The displacement is the total volume of air and fuel mixture that an engine can draw in during one cycle. The bore is the diameter of the cylinders.

Dependent variable is: MSRP
R squared = 77.0% R squared (adjusted) = 76.5%
s = 979.8 with 98 − 3 = 95 degrees of freedom

Variable	Coeff	SE(Coeff)	t-ratio	P-value
Intercept	318.352	1002	0.318	0.7515
Bore	41.1650	25.37	1.62	0.1080
Displacement	6.57069	3.232	2.03	0.0449

a) State and test the standard null hypothesis for the coefficient of *Bore*.

b) Both of these predictors seem to be linearly related to *MSRP*. Explain what your result in part a means.

⊤ 43. Motorcycles, part 3. Here's another model for the *MSRP* of off-road motorcycles.

```
Dependent variable is: MSRP
R squared = 90.9% R squared (adjusted) = 90.6%
s = 617.8 with 95 - 4 = 91 degrees of freedom
```

Source	Sum of Squares	df	Mean Square	F-ratio
Regression	346795061	3	115598354	303
Residual	34733372	91	381685	

Variable	Coeff	SE(Coeff)	t-ratio	P-value
Intercept	-2682.38	371.9	-7.21	<0.0001
Bore	86.5217	5.450	15.9	<0.0001
Clearance	237.731	30.94	7.68	<0.0001
Engine strokes	-455.897	89.88	-5.07	<0.0001

a) Would this be a good model to use to predict the price of an off-road motorcycle if you knew its bore, clearance, and engine strokes? Explain.

b) The Suzuki DR650SE had an *MSRP* of $4999 and a 4-stroke engine, with a bore of 100 inches. Can you use this model to estimate its *Clearance*? Explain.

⊤ 44. Demographics. The dataset corresponding to this exercise holds various measures of the 50 United States. The *Murder* rate is per 100,000, *HS Graduation* rate is in %, *Income* is per capita income in dollars, *Illiteracy* rate is per 1000, and *Life Expectancy* is in years. Find a regression model for *Life Expectancy* with three predictor variables by trying all four of the possible models.

a) Which model appears to do the best?

b) Would you leave all three predictors in this model?

c) Does this model mean that by changing the levels of the predictors in this equation, we could affect life expectancy in that state? Explain.

d) Be sure to check the conditions for multiple regression. What do you conclude?

⊤ 45. Burger King nutrition. Like many fast-food restaurant chains, Burger King (BK) provides data on the nutrition content of its menu items on its website. Here's a multiple regression predicting calories for Burger King foods from *Protein* content (g), *Total Fat* (g), *Carbohydrate* (g), and *Sodium* (mg) per serving.

```
Dependent variable is: Calories
R-squared = 100.0% R-squared [adjusted] = 100.0%
s = 3.140 with 31 - 5 = 26 degrees of freedom
```

Source	Sum of Squares	df	Mean Square	F-ratio
Regression	1419311	4	354828	35994
Residual	256.307	26	9.85796	

Variable	Coeff	SE(Coeff)	t-ratio	P-value
Intercept	6.53412	2.425	2.690	0.0122
Protein	3.83855	0.0859	44.7	<0.0001
Total fat	9.14121	0.0779	117	<0.0001
Carbs	3.94033	0.0338	117	<0.0001
Na/Serv.	-0.69155	0.2970	-2.33	0.0279

a) Do you think this model would do a good job of predicting calories for a new BK menu item? Why or why not?

b) The mean of *Calories* is 455.5 with a standard deviation of 217.5. Discuss what the value of *s* in the regression means about how well the model fits the data.

c) Does the R^2 value of 100.0% mean that the residuals are all actually equal to zero?

⊤ 46. Health expenditures. Can the amount of money that a country spends on health (as % of GDP) be predicted by other economic indicators? Here's a regression predicting *Expenditures on Public Health (as % of GDP)* from *Expected Years of Schooling* and *Internet Users (per 100 people)*:

```
Dependent variable is: Expenditures on Public Health
R squared = 55.17%
s = 1.710 with 96 - 3 = 93
```

Term	Estimate	Std Error	t-Ratio	P-value
Intercept	0.19941	0.95009	0.21	0.8342
Expected Years of Schooling (of children) (years)	0.23244	0.08259	2.81	0.006
Internet Users Per 100 People (2004)	0.05142	0.01031	4.99	<0.0001

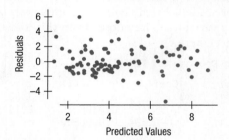

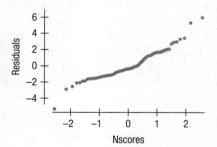

a) Write the regression model.

b) Are the assumptions and conditions met?

c) State and test the standard null hypothesis for the coefficient of *Expected Years of Schooling*. Use the standard α-level of $\alpha = 0.05$ and state your conclusion.

d) Does the coefficient of *Internet Users/100 people* mean that when Internet use increases, the expenditures on public health will increase as well?

JUST CHECKING ANSWERS

1 58.4% of the variation in *Percent Body Fat* can be accounted for by the multiple regression model using *Height*, *Age*, and *Weight* as predictors.

2 For a given *Height* and *Weight*, an increase of one year in *Age* is associated with an increase of 0.137% in *Body Fat* on average.

3 The multiple regression coefficient is interpreted for *given* values of the other variables. That is, for people of the *same Weight* and *Age*, those who are one inch taller on average have 1.2764% less *Body Fat*.

4 Histograms are used to examine the shapes of distributions of individual variables. We check especially for multiple modes, outliers, and skewness. They are also used to check the shape of the distribution of the residuals for the Nearly Normal Condition.

5 Scatterplots are used to check the Linearity Condition in plots of *y* vs. any of the *x*'s. They are used to check plots of the residuals or Studentized residuals against the predicted values, against any of the predictors, or against Time to check for patterns.

6 The Normal model is needed only when we use inference; it isn't needed for computing a regression model. We check the Nearly Normal Condition on the residuals.

16

Introduction to Data Mining

Paralyzed Veterans of America

The Paralyzed Veterans of America (PVA) is a philanthropic and service organization chartered by the U.S. government to serve the needs of U.S. veterans who suffer from spinal cord injury or disease. Since 1946, PVA has raised money to support a variety of activities, including advocacy for veterans' health care, research and education in spinal cord injury and disease, and support for veterans' benefits and rights.

PVA raises money in a variety of ways, but the majority of their fund-raising comes from direct mail campaigns, in which they send free address labels or greeting cards to potential donors on their mailing list and request a donation for receiving these gifts.

PVA sends out regular solicitations to a list of more than four million donors. In 2012, PVA received about $150 million in donations, but the effort cost them over $33 million in postage, administrative, and gift expenses. Most of that money was spent on mailings to people who never responded. In fact, in any one solicitation, groups like PVA are lucky to get responses from more than a few percent of the people they contact. Response rates for commercial companies such as large credit card banks are so low that they often measure responses not in percentage points, but in basis points—hundredths of a percent. If the PVA could avoid mailing to just half of the people who won't respond, they could save $17 million a year and produce only half as much wasted paper. Can statistical methods help them decide who should get what mail?

16.1   The Big Data Revolution

You learned about Big Data in Chapter 1, and you hear about it on the news almost every day. The technological advances of the past few decades have enabled us to capture information virtually everywhere and automatically. When you visit a social networking site, data about you, your likes, your photos, your friends, and your habits are uploaded to servers, waiting for analysis. Your smartphone sends information about the calls you make, the texts you send, and even your location to central servers. Whether this information is being used in a prudent way by the government is the subject of recent news and citizen concerns,[1] but the collection of data by business, science, and government is exploding and will continue to explode into the foreseeable future. And these data will be used to solve problems using techniques that are based on the methods you've learned to master in this course.

Terms like *data mining*, *data science*, and *analytics* are really just describing the application of techniques that are built on the methods that you've learned in this course to large data sets. In Chapter 1, we defined **data mining** as the process that uses a variety of data analysis tools to discover patterns and relationships in data to help build useful models and make predictions. The more general term **business analytics** (or sometimes simply *analytics*) describes any use of statistical analysis to drive business decisions from data, whether the purpose is predictive or simply descriptive. **Big data** is a term for data sets so large and complex that it becomes difficult to use traditional methods to capture, store, visualize, and analyze them. But the definition of how "big" the database should be to qualify keeps changing with advancing computer speeds and storage capabilities.

What can companies hope to learn from all these data, and how do they analyze them? With information about the transactions you make, the websites you visit, and the social media you interact with, retailers can customize offers to you and predict when and how to deliver those offers to maximize the chance that you will purchase. Advertisers can place their message in specific channels to maximize the impact for the group of consumers they are targeting.

The Obama presidential campaign used analytics rather than conventional wisdom to determine when and where to place ads during the fall of 2012. As part of their sophisticated targeting operations, they bought "detailed data on TV viewing by millions of cable subscribers, showing which channels they were watching, sometimes on a second-by-second basis."[2] The campaign then "used a third-party company to match viewing data to its own internal list of voters and poll responses," helping to make the most out of their advertising budget during the final weeks of the 2012 campaign. The *New York Times* reported that the system "allowed Mr. Obama's team to direct advertising with a previously unheard-of level of efficiency,"[3] with some analysts crediting this data mining with Obama's victory.

Most of the models used in analytics and data mining are based on the regression ideas that you've studied. But because data mining has benefited from work in other fields as well, it has a much richer set of tools than those in this book. The skills you've acquired by learning to fit multiple regression models diagnose them, and assess their limitations already prepare you to start entering the world of big data.

---

[1] "Momentum Builds Against N.S.A. Surveillance," *New York Times*, July 28, 2013.
[2] "Obama Campaign Took Unorthodox Approach to Ad Buying," *Washington Post*, November 14, 2012.
[3] "Secret of the Obama Victory? Rerun Watchers, for One Thing," *New York Times*, November 12, 2012.

16.2   Direct Marketing

Direct mail, often called "junk mail," generates about four million tons of paper waste per year. Direct e-mail, often called "spam," accounts for 95% of all e-mails according to the European Network and Information Security Agency (ENISA). Companies use direct mail and e-mail because, compared to other options, it is inexpensive and effective. To make it more effective, companies want to identify people who are most likely to respond. In other words, the companies would like to *target* their promotions and offers.

To help them decide how likely a customer might be to respond to a particular offer, a company might build a model to estimate the probability. The data used to build the model is usually a combination of data on customers that the company has collected and other data that it has bought. Data mining techniques can help companies direct their mailings to people more likely to respond to the solicitations. For example, analysts might use past customer behavior and other demographic information and build a model (or models) to predict who is most likely to respond. Techniques like multiple linear and logistic regression play an important role in these analyses, but if the number of variables is too large, researchers might use more sophisticated methods to reduce the number of predictor variables to a manageable size first.

Companies and philanthropic organizations collect an incredible amount of information about their customers. You've contributed to this data gathering yourself, perhaps without even knowing it. Every time you purchase something with a credit card, use a loyalty card at a supermarket, order by phone or on the Internet, or call an 800 number, your transaction is recorded. For a credit card company, this **transactional data** might contain hundreds of entries a year for each customer. For a philanthropic organization, the data might also include a history of when solicitations were sent, whether the donor responded to each solicitation, and how much they contributed. For a credit card bank, the transactional data would include every purchase the customer made. Every time a transaction is attempted, information is sent back to the transactional database to check that the card is valid, has not been reported stolen, and that the credit limit of the customer has not been exceeded—that's why there's a small delay before your transaction is approved every time you charge something. Although companies collect transactional data to facilitate the transactions themselves, they recognize that these data hold a wealth of information about their business. The challenge is to find ways to extract—to mine—that information.

In addition to transactional data, companies often have separate databases containing information about customers and about products (inventory, price, and shipping costs, for example). The databases can be linked to each other in a relational database. The properties of relational databases and some simple examples were discussed in Chapter 1.

The variables in the customer database are of two types: individual and regional. The individual variables are typically first gathered when the customer opens an account, registers on a website, or fills out a warranty card and are specific to that customer. They might include **demographic variables** such as a customer's age, income, and number of children. The company can add to these variables others that arise from the customer's interactions with the company, including some that may summarize variables in the transactional database. For example, the total amount spent each month might be a variable in the customer database that is updated from the customer's individual purchases, stored in the transactional database. The company might also purchase additional demographic data. Some demographic data are based on ZIP codes and can be obtained from agencies like the U.S. Census Bureau. These data can provide information on the average income, education, home value, and ethnic composition of the neighborhood in which a customer lives, but are not customer-specific.

Customer-specific data can also be purchased from a variety of commercial organizations. For example, a credit card company may want to send out an offer of free flight insurance to customers who travel frequently. To help know which customers those are, the company may buy information about customers' magazine subscriptions to identify customers who subscribe to travel or leisure magazines. The sharing and selling of individual information is controversial and raises privacy concerns, especially when the purchased data involve health records and personal information. In fact, concerns over the sharing of health data in the United States led to a set of strict guidelines known as HIPAA (Health Insurance Portability and Accountability Act) regulations. Restrictions on the collection and sharing of information about customers vary widely from country to country.

The PVA customer database[4] is a typical mix of variables. It contains 481 variables on each donor. There are 479 potential predictor variables and two responses: *TARGET_B*, a 0/1 variable which indicates whether the donor contributed to the most recent campaign and *TARGET_D* which gives the dollar amount of the contribution. Table 16.1 shows the first 18 records for a subset of the 481 variables found in the PVA data. We can guess what some of the variables mean from their names, but others are more mysterious.

ODATEDW	OSOURCE	TCODE	STATE	ZIP	DOB	RFA_2A	AGE	OWN	INC	SEX	WEALTH	AVGGIFT	TARGET_B	TARGET_D
9401	L16	2	GA	30738	6501	F	33	U	5	F	2	11.66667	0	0
9001	L01	1	MI	49028	2201	F	76	H	1	M	2	8.777778	0	0
8601	DNA	1	TN	37079	0	E		U	1	M		8.619048	1	10
8601	AMB	1	WI	53719	3902	G	59			M		16.27273	0	0
8601	EPL	2	TX	79925	1705	E	81	H	2	F	6	10.15789	0	0
8701	LIS	1	IN	46771	0	F				M		8.871333	0	0
9201	GRI	1	IL	60016	1807	F	79	H	4	M	6	13.8	0	0
9401	HOS	0	KS	67218	5001	G	48	U	7	F	7	18.33333	0	0
8901	DUR	0	MI	48304	1402	F	84	H	7	M	9	12.90909	0	0
8601	AMB	0	FL	34746	1412	F	83	H	2	F	3	9.090909	0	0
9501	CWR	2	LA	70582	0	D		U	5	F		5.8	0	0
9501	ARG	0	MI	48312-	4401	E	54			F		8	0	0
8601	ASC	0	TX	75644	2401	G	74	U	7	F		13.20833	0	0
9501	DNA	28	CA	90059	2001	E	78	H	1	F		10	0	0
9201	SYN	0	FL	33167	1906	F	79	U	2	M	3	10.09091	0	0
9401	MBC	2	MO	63084	3201	F	66	H	5	F		10	0	0
9401	HHH	28	WI	54235	0	F		H	7	F		20	0	0
9101	L02	28	AL	36108	4006	F	58	H	5	F		10.66667	0	0

Table 16.1 Part of the customer records from the PVA data set. Shown here are 15 of the 481 variables and 18 of the nearly 100,000 customer records used in the 1998 KDD (Knowledge Discovery and Data Mining) data mining competition.[5] The actual PVA customer database contains several million customer records.

---

[4] The PVA made some of their data available for the 1998 Knowledge Discovery and Data Mining (KDD) contest. The object of the contest was to build a model to predict which donors should receive the next solicitation based on demographic and past giving information. The results were presented at the KDD conference (www.kdnuggets.com). The variables discussed in this chapter are some of the ones that PVA made available. We have used various subsets of the PVA data set in previous chapters, sometimes changing the variable names for clarity. The entire data set can be found at the University of California at Irvine Data Repository: kdd.ics.uci.edu/databases/kddcup98/kddcup98.html

[5] The KDD Cup is the leading Data Mining competition in the world and is organized by the SIGKDD interest group of the ACM (Association for Computing Machinery).

Information about the variables, including their definitions, how they are collected, the date of collection, etc. is collectively called **metadata**. The variables in the PVA data set shown on the previous page are typical of the types of data found in the customer records of many companies. We shouldn't be surprised to find out that *AGE* is the donor's age measured in years and that *ZIP* is the donor's postal ZIP code. But without the metadata, it would be hard to know that *TCODE* is the code used before the title on the address label (0 = blank, 1 = Mr., 2 = Mrs., 28 = Ms., and so on) or that *RFA_2A* is a summary of past giving.

About 10% of the PVA variables describe past giving behavior collected by the PVA itself. More than half of the variables are regional data based on ZIP codes, most likely purchased from the Census Bureau. The rest of the variables are donor-specific, either gathered by the PVA or purchased from other organizations.

Sometimes disparate databases are gathered or merged in a central repository called a **data warehouse**. Maintaining the data warehouse is a huge job, and companies spend millions of dollars each year for the software, hardware, and technical personnel to do just that. Once a company has invested in creating and maintaining a data warehouse, it's only natural that they would want to get as much value as they can from it. For example, at PVA, analysts might use the data to build models to predict who will respond to a direct mail campaign; a credit card company, similarly, might want to predict who is most likely to accept an offer for a new credit card or services.

16.3   The Goals of Data Mining

The purpose of data mining is to extract useful information hidden in these large databases. With a database as large as a typical data warehouse, that search can be like looking for a needle in a haystack. How can analysts hope to find what they're looking for? They may start their search by using a sequence of queries to deduce facts about customer behavior, asking specific questions based on the data. Guided by their knowledge of the specific business, they may try a query-driven approach, asking a series of specific questions to deduce patterns. Such an approach is **online analytical processing** or **OLAP**. Analysts in sales, marketing, budgeting, inventory, and finance often use OLAP to answer specific questions involving many variables. An OLAP question for the PVA might be: "How many customers under the age of 65 with incomes between $40,000 and $60,000 in the Western region who have not donated in the previous two years gave more than $25 to the most recent solicitation?" Although OLAP is efficient in answering such multivariable queries, it is question-specific. OLAP produces answers for specific queries, typically as tables, but does not build a predictive model, and so it is not appropriate to generalize using OLAP.

In contrast to an OLAP query, the outcome of a data mining analysis is a **predictive model**—a model that uses predictor variables to predict a response. For a quantitative response variable (as in linear regression), the model will predict the *value* of the response, whereas for a categorical response, the model estimates the probability that the response variable takes on a certain value (as in logistic regression). Both linear and logistic multiple regressions are common tools of a data miner. Like all statistical models and unlike query-based methods such as OLAP, data mining generalizes to other similar situations through its predictive model. For example, an analyst using an OLAP query might find that customers in a certain age group responded to a recent product promotion. But without building a model, the analyst can't understand the relationship between the customer's age and the success of the promotion, and thus may be unable to predict how the product will do with a larger set of customers. The goal of a data mining project is to increase business understanding and knowledge by building a model to answer a specific set of questions raised at the beginning of the project.

Data mining is similar to traditional statistical analysis in that it involves exploratory data analysis and modeling. However, there are several aspects of data mining that distinguish it from more traditional statistical analysis. Although there is no consensus on exactly what data mining is and how it differs from statistics, some of the most important differences include:

- **The size of the databases.** Although no particular size is required for an analysis to be considered data mining, an analysis involving only a few hundred cases or only a handful of variables would not usually be considered data mining.
- **The exploratory nature of data mining.** Unlike a statistical analysis that might test a hypothesis or produce a confidence interval, the outcome of a data mining effort is typically a model used for prediction. Usually, the data miner is not interested in the values of the parameters of a specific model or in testing hypotheses.
- **The data are "happenstance."** In contrast to data arising from a designed experiment or survey, the typical data on which data mining is performed have not been collected in a systematic way. Thus, warnings not to confuse association with causation are especially pertinent for data mining. Moreover, the sheer number of variables involved makes any search for relationships among variables prone to Type I errors.
- **The results of a data mining effort are "actionable."** For business applications, there should be a consensus of what the problem of interest is and how the model will help solve the problem. There should be an action plan in place for a variety of possible outcomes of the model. Exploring large databases out of curiosity or just to see what they contain is not likely to be productive.
- **The modeling choices are automatic.** Typically the data miner will try several different types of models to see what each has to say, but doesn't want to spend a lot of time choosing which variables to include in the model or making the kinds of choices one might make in a more traditional statistical analysis. Unlike the analyst who wants to understand the terms in a stepwise regression, the data miner is more concerned with the predictive ability of the model. As long as the resulting model can help make decisions about who should receive the next offer, who is best suited for an online coupon, or who is most likely to switch cable providers next month, the data miner is likely to be satisfied.

16.4   Data Mining Myths

Data mining software usually contains a variety of exploratory and model-building tools and a graphical user interface designed to guide the user through the data mining process. Some people buy data mining software hoping that the tools will find information in their databases and write reports that spew knowledge with little or no effort or input from the user. Software vendors often capitalize on this hope by exaggerating the capabilities and the automatic nature of data mining to increase sales of their software. Data mining can often *assist* the analyst to find meaningful patterns and help predict future customer behavior, but the more the analyst knows about his or her business, the more likely he or she will be successful using data mining. Data mining is not a magic wand that can overcome poor data quality or collection. A product may have tools for detecting outliers and may be able to impute (assign a value for) missing data values, but all the issues you learned about good statistical analysis are still relevant for data mining.

Here are some of the more common myths about what data mining can do.

- **Myth 1:** Find answers to unasked questions

  Even though data mining can build a model to help answer a specific question, it does not have the ability to answer questions that haven't been asked. In fact, formulating a precise question is a key first step in any data mining project.

- **Myth 2:** Automatically monitor a database for interesting patterns

  Data mining techniques build predictive models and can answer queries, but do not find interesting patterns on their own any more than regression models do.

- **Myth 3:** Eliminate the need to understand the business

  In fact, the more an analyst understands his or her business, the more effective the data mining effort will be.

- **Myth 4:** Eliminate the need to collect good data

  Good data is as important for a data mining model as for any other statistical model that you've encountered. While some data mining software contains tools to help with missing data and data transformations, there is no substitute for quality data.

- **Myth 5:** Eliminate the need for good data analysis skills

  The better the data analysis skills of the miner—the skills you've learned in every chapter of this book—the better the analysis will be when using data mining tools. Data mining tools are more powerful and flexible than a statistical tool like regression but are similar in the way they work and in how they're implemented.

16.5   Successful Data Mining

The size of a typical data warehouse makes any analysis challenging. The ability to store data is growing faster than the ability to use it effectively. Commercial data warehouses often contain terabytes (TB)—more than 1,000,000,000,000 (1 trillion) bytes—of data (one TB is equivalent to about 260,000 digitized songs), and warehouses containing petabytes (PB—one PB = 1000 TB) are now common. The tracking database of the United Parcel Service (UPS) is estimated to be on the order of 16 TB, or roughly the digital size of all the books in the U.S. Library of Congress. According to *Wired* magazine, about 20 TB of photos are uploaded to Facebook every month. All the U.S. Census data from 1790 to 2000 would take about 600 TB. However, it's estimated that the servers at *Google* process a petabyte of data every 72 minutes.[6] Data miners hope to uncover some important strategic information lying hidden within these massive collections of data.

To have a successful data mining outcome, the first step is to have a well-defined business problem. With 500 variables, there are over 100,000 possible two-way relationships between pairs of variables. The number of pairs that will be related just by chance is likely to be large. And it's human nature to find many of these relationships interesting and even to posit plausible reasons why two variables might be associated. Some of these variables may appear to provide a useful predictive model, when in fact they do not. A well-defined business objective can help you avoid going down a lot of blind paths.

As in painting a house, much of the effort of a data mining project is in the preparation, cleaning, and checking of the data. It is estimated that for any data mining project, 65% to 90% of the time spent is spent in such **data preparation**. Data preparation involves investigating missing values, correcting wrong and inconsistent entries, reconciling data definitions, and possibly creating new variables from the original ones. The data may need to be extracted from several databases and combined. Errors must be corrected or eliminated, and outliers must be identified.

So, a successful data mining effort often requires a substantial amount of time devoted to basic preparation of the data before any modeling is performed. It will have a clear objective, agreed to by a team of people who will share the work and the responsibility of the data mining. It will have an action plan once the results of

---

[6] *Wired*, Issue 16.07, June 2008.

the data mining are known, whether those results are what the team expected or desired. Finally, the data mining should be accompanied by as much knowledge about both the data and the business question as possible. Blind searches for patterns in large databases are rarely fruitful and will waste valuable analytic resources.

FOR EXAMPLE   Data preparation

Data mining analyses often deal with large numbers of variables. For the PVA data, there are 479 potential predictors for modeling whether a donor will donate to the next campaign. Each of these variables may present issues that should be dealt with before proceeding to a full analysis. But usually a smaller number are selected for examination. Here's one example.

QUESTION Looking just at the variable *AGE*, what corrections or adjustments would be appropriate so that it can be a more effective predictor in a regression-based model?

ANSWER Let's look at a histogram and boxplot for the variable *AGE* for all 94,649 records of the PVA data set.

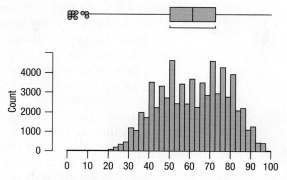

	Quantiles	
100.0%	maximum	98.000
99.5%		95.000
97.5%		90.000
90.0%		83.000
75.0%	quartile	75.000
50.0%	median	62.000
25.0%	quartile	48.000
10.0%		39.000
2.5%		31.000
0.5%		26.000
0.0%	minimum	1.000
Mean		61.598019
Std Dev		16.666865
Std Err Mean		0.0624749
upper 95% Mean		61.720469
lower 95% Mean		61.475568
N		71170

Figure 16.1   Graphical output and statistical summaries of the variable *AGE* help to identify questionable values.

There is a group of cases with ages below 20 that should immediately draw our attention. Are people that young likely to be donors? A closer examination reveals that there are even a handful of cases whose ages are below 15, of whom 17 are younger than 5 years old and 9 are 1 year old. There are also 23,479 missing values for the variable *AGE*. Our domain knowledge (what we know about the world and likely donors to such an organization) tells us that values below a certain age (probably below 19, but certainly below 10) or above a certain age (100? 110?) are almost certainly wrong, so we can see whether any of these values can be corrected from other records, or we can remove them from the analysis.[7]

It is harder to know whether the data in other variables are correct. And, while checking one or two variables to see if their values make sense is practical, examining bar charts, histograms, and boxplots for 479 variables would be a daunting task.

16.6 Data Mining Problems

Data mining can address different types of problems, some of which we have already encountered. When the goal is to predict a quantitative response variable, the problem is generically called a **regression problem**, regardless of whether linear regression is used (or even considered) as one of the models. When the response variable is categorical, the problem is referred to as a **classification problem** because the model will either guess the most likely category for each or assign a probability to each class for that code. For example, for the classification problem of predicting *whether or not* a particular donor will give to the next campaign, a model will produce either the most probable class (*donate or not*) or the probabilities of each of these. Predicting the *amount* of money a donor will donate is a regression problem.

Both of the previous problems are referred to as **supervised problems**. In a supervised setting, we are given a set of data for which we know the response. That is, for the PVA data, we *know* for at least one group of donors the values of the responses *TARGET-B* and *TARGET-D*. We know whether they gave to the last campaign and how much they donated. The data miner would construct a model based on a portion of the original data, called the **training set**. In order to assess how well the model will work in the future on data that the model hasn't encountered, the modeler uses the original data set that was withheld from the model building and then tests the predictions of the model on these data. This second data set is called the **test set**.

By contrast, there are problems for which there is no particular response variable. In these **unsupervised problems**, the goal may be to build clusters of cases with similar attributes. For example, a company may want to cluster their customers into groups with similar buying behaviors and tastes. Such analyses resemble segmentation analyses performed by marketing analysts. In this case, there is no response variable. All the predictors (or a subset of them) are used to build clusters based on an index that measures the similarity between the customers. Many different algorithms are available to find clusters.

16.7 Data Mining Algorithms

The methods used for the problems discussed in the previous section are often referred to as **algorithms**, a term that describes a sequence of steps with a specific purpose. You may hear a method (even one like linear regression) referred to as a model, an algorithm, a tool, or generically as a method. The terms seem to be

---

[7] Further analysis revealed that age was calculated from date of birth by a computer algorithm, but only the last two digits of the date of birth year were used. So, some people (perhaps no longer living) who were in the database had ages 100 years too young.

used interchangeably. This section touches on only a few of the models used in data mining. Some of the most common methods used for prediction are decision trees and neural networks (discussed below), support vector machines, belief nets, the regression methods discussed in Chapters 14 and 15, and random forests. It's a lively research field and new algorithms are appearing all the time.

Tree Models

In data mining, the term **decision tree** is used to describe a *predictive model*. These tree models use *data* to select predictor variables that give predictions of the response. The models are entirely driven by the data with no input from the user.

The way a tree model works is fairly simple. To illustrate the process, imagine that we want to predict whether someone will default on a mortgage. To build the tree, we use past data from a group of customers on whom we have similar information and for whom we know whether they defaulted or not. For this simplified example, let's assume we'll predict it only from the following variables:

- Age (years)
- Household Income ($)
- Years on the Job (years)
- Debt ($)
- Homeowner (Yes/No)

The tree model tries to find predictors that can distinguish those people who will default on their mortgages from those who won't. To do that, it first examines *every* potential predictor variable and every possible way of *splitting* that variable into two groups. For example, it looks at *Age*, and for every value of *Age*, it calculates the default rate of those *above* that value and *below* that value. Keeping track of the difference in default rates for the two groups defined by every possible predictor variable split at every possible way, it chooses that pair of predictors and *split points* that produce the greatest difference in default rates.[8] The tree algorithm decides which variables to split and where to split them by trying (essentially) all combinations of variables and split points until it finds the best splits. For a categorical predictor, the model considers every possible way of putting the categories into two groups. There are several criteria used by different algorithms to define "best," but they all have in common some attempt to find a way of splitting the cases so that the resulting two groups will have the largest possible difference in default rates. After the algorithm finds the first split, it continues searching again on the two resulting groups, finding the next best variable and split point (possibly using the same variable as the previous split again). It continues in this way until one of several criteria is met (for example, number of customers too small, not a large enough difference in rates is found) and then stops at what are called **terminal nodes**, where the model produces a prediction. The predictions at the terminal nodes are simply the average (if the response is quantitative) or the proportions of each category (for a classification problem) of the cases at that node.

The tree for our hypothetical mortgage example is shown in Figure 16.2. To understand the tree, start at the top and imagine being presented with a new customer. The first question the tree asks is "Is the *Household Income* more than $40,000?" If the answer is yes, move down to the right. For these cases, *Debt* is the next variable split, this time at $10,000. If this customer's *Debt* exceeded $10,000 move down and to the right again. The tree estimates that customers like this (both *Household Income* > $40,000 and *Debt* > $10,000) defaulted at a rate of 5% (0.05). Just to the left are customers whose *Household Income* was > $40,000 but whose

---

[8] This is a slight simplification. There are actually several possible variants on the criteria for splitting. Readers who want to know the details can read more advanced books on data mining or decision trees.

Debt was ≤ $10,000. They defaulted at a rate of 1%. For customers with incomes less than $40,000 (the left branch of the first split), the next variable split was not *Debt*, but *Job*. For those who have been at their job more than 5 years, the default rate was 6%, but was 11% for those with less time at their present job. The modeler may, at this point, label the outcomes as *risk* categories, for example by calling 1% *Very low*, 5 and 6% *Moderate*, and 11% *High*.

Figure 16.2   Part of a tree model on a hypothetical examination of mortgage default. The tree selects *Household Income* as the most important variable to split on and selects $40,000 as the cut point. For those customers whose income is more than $40,000, *Debt* is the next most important variable, while for those whose income is less than $40,000, how long they've held their current job is more important.

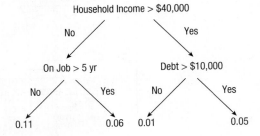

Tree models are very easy to implement and, in principle, are easy to interpret. They have the advantage of showing their logic clearly and are easy to explain to people who don't have a deep background in statistics. Even when they aren't used as a final model in a data mining project, they can be very useful for selecting a smaller subset of variables on which to do further analysis.

Unlike many other data mining algorithms, a tree can handle a great number of potential predictor variables, both categorical and quantitative. This makes it a great model for starting a data mining project. The modeler can then choose whether to use this model as a final model for predicting the response, or to use the variables suggested by the tree as inputs to other models that are more computationally intensive and less able to handle very large numbers of predictors.

FOR EXAMPLE   Decision tree

Decision trees are calculated by data mining software. Several different algorithms may be used, but the important thing to know is how to understand what the resulting decision tree says about the data. Figure 16.3 shows a tree model found for the PVA data using *TARGET_B* as a response variable. *TARGET_B* is "YES" if the donor contributed to the most recent solicitation and "NO" if not.

QUESTION   What can we learn about the PVA data from this decision tree?

ANSWER   The goal of the decision tree is to characterize those potential donors who responded to the most recent solicitation with a donation. The PVA tree model starts at the root node with all 94,649 customers (listed under Count in the top box of the figure).

The first branching of the tree is according to the variable *RFA_4*, a variable that summarizes past giving. The algorithm identifies this as the best single predictor of *TARGET_B*. So, the first split occurs by splitting the levels of that variable into two groups, with 29,032 customers split to the left and the remaining 65,617 customers to the right. You can tell that because *RFA_4* is listed first in both boxes of the first split. The variable *RFA_4* comprises codes that contain information on how recently the last gift solicitation was received, how frequently the donor has given, and how large the last gift was. (Examples of these codes are A3C, S4B, etc., seen in parentheses

(continued)

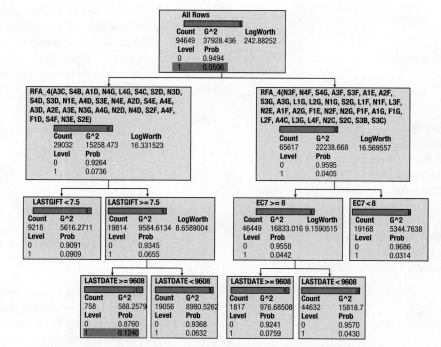

Figure 16.3   Part of a tree model run on the PVA data using JMP® software.

after the variable name in the output display.) The percentages next to the levels 0 and 1 indicate the proportion of those who donated or not (*TARGET_B 1 or 0*, respectively) in each node. So although only 7.36% of those in the left node contributed, that was almost twice the 4.05% in the right node who contributed.

The tree continues to split each subpopulation. Notice that the variable chosen to split the left branch of the tree (*LASTGIFT*) is not the same as the variable chosen to split the right branch (*EC7*). The terminal nodes are found in the bottom row. The terminal node at the bottom left of the figure shows that 12.4% of the 758 potential donors in this small subgroup donated, a substantial improvement over the average of all potential donors (5.06%) shown in the root node at the top. Even at this stage of the analysis, if the PVA directed their solicitations to names on their list that satisfied the conditions found in the path through the tree to this node, they might reduce their costs and increase their yield substantially.

It's interesting to see that of the 479 potential predictor variables, 3 of the top 4 variables selected by the decision tree algorithm involve past giving history rather than demographics. *RFA_4* is a summary of past giving, *LASTGIFT* measures the amount of the last gift, and *LASTDATE* reports how recently the donor made the last gift. Only *EC7* is a demographic variable, measuring the percentage of residents in the donor's ZIP code that have at least a bachelor's degree. By letting the tree grow further, we could grow the list of potential predictors to 20 or 30, a much larger number than shown here, but still a manageable subset of the original 479 variables.

Neural Networks

Another popular data mining tool is the multilayer perceptron or (artificial) **neural network**. This algorithm sounds more impressive than it really is. Although it was inspired by models that tried to mimic the function of the brain, it's really just an automatic, flexible, nonlinear regression tool. That is, it models a single response variable with a function of some number of predictor variables but, unlike multiple regression, it constructs a more complex function for the relationship. This can have the advantage of fitting the data better but has the disadvantage that complex

RFA_2A	2.31%
RFA_2	1.64%
RFA_2F	1.27%
LASTDATE	1.18%
INCOME	1.03%
PEPSTRFL	1.00%
ADATE_4	0.83%
ADATE_3	0.74%
RDATE_7	0.73%
LIFESRC	0.73%
HVP1	0.69%
RDATE_6	0.61%
DMA	0.55%
GENDER	0.54%
RDATE_3	0.51%

Table 16.2 The top 15 variables ranked in importance by the neural network node in Clementine®. The importance percentages indicate relative importance only and are not meant to be interpreted in an absolute sense.

functions are harder to interpret and understand. Indeed, modelers often don't even look at the models themselves. There are certainly things to be learned by studying how neural networks build models, but, for our purposes, we'll view them as "black box" models—models that don't produce an equation or graphical display that we can examine, but simply predict the response without much information about *how* it's doing it.

Even though they are black boxes, neural network algorithms do leave some clues. A listing of the most important variables is a common output feature, and some neural network software even provide plots of the predicted relationships between the response and the most important variables as measured by how much the predictions change when the variable is deleted. Table 16.2 shows a list of the top 15 variables in the PVA problem ranked by "importance" from a neural network output in a data mining software package called Clementine®.

Many other algorithms are used by data miners, and new ones are being developed all the time. Typically, a data miner will build several different models and then test them before selecting one to use, or the data miner might decide to form a "committee" of models, combining the output from several models, much in the way a CEO might take input from a group of advisors in order to make a decision. For a classification problem, the final prediction for a case might be the class that's predicted most often by the models in the committee. For a regression problem, the prediction might simply be the average prediction of all the models in the committee. Averaging models provides protection against choosing the "wrong" model, but usually at a cost of not being able to interpret the resulting predictions. Finding the best ways to combine many different models is an active and exciting area of current research in data mining.

16.8 The Data Mining Process

Data mining projects require a number of different skills. For that reason, a data mining project should be a team effort. No one person is likely to have the business knowledge, computer and database management skills, software expertise, and statistical training needed for all the steps in the process. Because data mining projects tend to be complex, it's useful to map out the steps for a successful project. A group of data mining experts have shared their combined expertise in a project called the Cross Industry Standard Process for Data Mining (CRISP-DM). A CRISP-DM schematic of the data mining cycle appears in Figure 16.4.

Figure 16.4 A picture of the CRISP-DM data mining process.

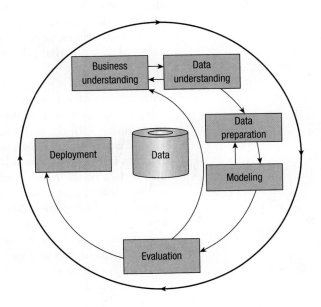

In this schematic, the process starts with the *Business understanding* phase. This is where the problem to be addressed is carefully articulated. It is best to have a specific problem before starting. A goal of understanding how to best manage customers sounds good but is not precise enough for a data mining project. A better, more specific question might be to understand which customers are most likely to switch cell phone providers in the next three months. It's important to have all the members of the data mining team involved at this stage, and the team should be representative of all parts of the business that may be affected by the resulting business decisions. If key constituents aren't represented, the model might not answer the right question. It can't be overemphasized how important it is to have consensus on a precise, correctly formulated problem to be addressed before continuing with the data mining effort.

The phase of *Data understanding* is central to the entire data mining project. If you want to know which customers are likely to switch providers, you have to understand your data and have data that can support such an exploration. For example, you'll need a sample that contains both customers who have recently switched and those who have been loyal customers for some time. There should also be variables in the database that could reasonably explain or predict the behavior. The choice of those variables has to rely on the business knowledge of the team. At the beginning of the project, it's best to include all the variables that might be useful, but keep in mind that having *too* many variables can make the model selection phase more difficult. It is crucial to understand the data warehouse, what it contains, and what its limitations are at this stage.

Once the variables have been selected and the response variable(s) agreed upon, it's time to begin the *Data preparation* phase for modeling. As mentioned earlier, this can be a time-consuming part of the process and is likely to be a team effort. Investigating missing values, correcting wrong and inconsistent entries, reconciling data definitions, and merging data sources are all challenging issues. Some of these can be handled automatically, while others require painstaking detailed analysis. The team will have to decide how much effort is reasonable to make the data set as complete and reliable as possible, given the time and resource constraints of the project.

Once the data have been prepared, the analysts begin the *Modeling* phase by exploring and developing models. If the number of variables is very large, a preliminary model (such as a tree) might be considered in order to narrow the candidate predictor variables down to a reasonable number. If the number of predictors is small enough, modelers can use traditional graphical analysis (histograms, bar charts) for each variable to start and then investigate the relationship between each predictor and the response with bivariate graphs (scatterplots, boxplots, or segmented bar charts, depending on the variable types). The more knowledge of the data and variables that goes into the model, the higher the chance of success for the project.

The analysts should now have several models that fit the response variable with differing levels of accuracy in the training set. Once the analyst has several models that seem reasonable (based on domain knowledge and performance on the training set), the *Evaluation* phase can begin. Looking at the structure of the models and deciding what predictors are important for each model should give the analysts information about how the predictors can predict the response variable of interest. In the *Evaluation* phase, the candidate models are tested against the test set, and various criteria are used to judge the models. For example, for a regression problem with a quantitative response, the sum of squares of the residuals from predicting the response on the test set might be compared. For a (two-level) classification problem, the two types of errors that arise (predicting YES when the true response is NO and vice versa) are weighed. Different costs on the two types of errors might be warranted, and the total cost of misclassification should reflect that. At this point, the business question that motivated the project should be revisited. Does the model help to answer the question? If not, it may be necessary to go back to one of the previous steps to investigate why that happened.

If the model (or average of several models) seems to give insight into the business problem, then it's time for the *Deployment* phase. Usually, that means using the model to predict an outcome on a larger data set than the original, or on more recent data than were used to build the model. The reason that so many phases of the CRISP-DM diagram have arrows both to and from them is that this process is not a straightforward movement through the phases—it is an iterative and interconnected process. Knowledge gained at any phase may trigger a reexamination of an earlier phase. Even the "final" phase, deployment, is not really final. In many business situations, the environment changes rapidly, so models can become stale quickly. Although a data mining project is complex and involves the efforts of many people, mapping out the different phases can help assure that the project is as successful as possible.

16.9 Summary

There are many similarities between the modeling process of data mining and the basic approach to modeling that you've been learning throughout this book. What makes data mining different really has to do with the large number of algorithms and types of models available to the data miner and with the size and complexity of the data sets. But most of what makes a data mining project successful would make any statistical analysis successful. The same principles of understanding and exploring variables and their relationships are key to both processes. A famous statistician, Jerry Friedman, was once asked if there was a difference between statistics and data mining. Before answering, he asked if the questioner wanted the long answer or the short answer. Because the response was "the short answer," Jerry said simply, "No." We never got to hear the long answer, but we suspect it may have contained some of the differences discussed in this chapter.

Having a good set of statistical and data analysis tools is a great beginning to becoming a successful data miner. Being willing to learn new techniques, whether they come from statistics, computer science, machine learning, or other disciplines, is essential. Learning to work with other people whose skills complement yours will not only make the task more pleasant, but also is key to a successful data mining project. The need to understand the information contained in large databases will increase in the years ahead, so it will continue to be important to know about this rapidly growing field.

WHAT CAN GO WRONG?

- **Be sure that the question to be answered is specific.** Make sure that the business question to be addressed is specific enough so that a model can help to answer it. A goal as vague as "improving the business" is not likely to lead to a successful data mining project.

- **Be sure that the data have the potential to answer the question.** Check the variables to see whether a model can reasonably be built to predict the response. For example, if you want to know which type of customers are going to a particular Web page, make sure that data are being collected that link the Web page to the customer visiting the site.

- **Beware of overfitting to the data.** Because data mining tools are powerful and because the data sets used to train them are usually large, it is easy to think you're fitting the data well. Make sure you validate the model on a test set—a data set not used to fit the model.

- **Make sure that the data are ready to use in the data mining model.** Typically, data warehouses contain data from several different sources. It's important to confirm that variables with the same name actually measure the same thing in two different databases. Missing values, incorrect entries, and different time scales are all challenges to be overcome before the data can be used in the model building phase.

- **Don't try it alone.** Data mining projects require a variety of skills and a lot of work. Assembling the right team of people to carry out the effort is crucial.

ETHICS IN ACTION

With U.S. consumers becoming more environmentally conscious, there has been an explosion of eco-friendly products on the market. One notable entry has been the gas-electric hybrid car.

A large nonprofit environmental group would like to target customers likely to purchase hybrid cars in the future with a message expressing the urgency to do so sooner rather than later. They understand that direct mailings are very effective in this regard, but staying true to their environmental concerns, they want to avoid doing a nontargeted mass mailing.

The executive team met to discuss the possibility of using data mining to help identify their target audience. The initial discussion revolved around data sources. Although they have several databases on demographics and

transactional information for consumers who have purchased green products and donated to organizations that promote sustainability, someone suggested that they get data on contributions to political parties. After all, there was a Green party, and Democrats tend to be more concerned about environmental issues than Republicans. Another member of the team was genuinely surprised that this was even possible. She wondered how ethical it is to use information about individuals that they may assume is being kept confidential.

- Identify the ethical dilemma in this scenario.

- What are the undesirable consequences?

- Propose an ethical solution that considers the welfare of all stakeholders.

WHAT HAVE WE LEARNED?

Learning Objectives   **Understand the uses and value of data mining in business.**

Recognize data mining approaches.

- OLAP approaches answer specific questions about data in a database that may involve many variables.
- Predictive model approaches build complex models that attempt to predict a response.

Be aware that data preparation is essential to sound data mining.

- Data must be cleaned of errors and impossible or implausible values.
- Missing values may be a serious issue.
- Data preparation may take the majority of time and effort devoted to a data mining project.

Be able to articulate the goal of a data mining project and understand that data mining efforts differ according to the kind of dependent variable.

- When the goal is to predict a quantitative response, the problem is called a regression problem.
- When the goal is to classify individuals according to a categorical variable, the problem is called a classification problem.

Be aware of the variety of models and algorithms that are used to build predictive models.

- Tree models build interpretable models that select predictor variables and split points to best predict the response in either a regression or classification problem.
- Neural networks build black box models that use all the predictor variables for both regression and classification problems.
- Many other data mining algorithms are used as well, and new competitors are continually being developed.

Understand that successful data mining is a team effort that requires a variety of skills.

- Defining the problem and objective of interest is crucial for success.
- The CRISP data mining process provides a good outline for a data mining project.

Terms

Algorithm	A set of instructions used for calculation and data processing. The algorithm specifies how the model is built from the data.
Big data	The collection and analysis of data sets so large and complex that traditional methods typically brought to bear on the problem would be overwhelmed.
Business analytics (or analytics or predictive analytics)	The process of using statistical analysis and modeling to make predictions and drive business decisions.
Classification problem	A prediction problem that involves a categorical response variable. (See also regression problem.)
Data mining	A process that uses a variety of data analysis tools to discover patterns and relationships in data that are useful for making predictions.
Data preparation	The process of cleaning data and checking its accuracy prior to modeling. Data preparation includes investigating missing values, correcting wrong and inconsistent entries, and reconciling data definitions.
Data warehouse	A digital repository for several large databases.
Decision tree (data mining version)	A model that predicts either a categorical or quantitative response in which the branches represent variable splits and the terminal nodes provide the predicted value.
Demographic variable	A variable containing information about a customer's personal characteristics or the characteristics of the region in which the customer lives. Commonly used demographics include age, income, race, and education.
Metadata	Information about the data including when and where the data were collected.
Neural network	A model based on analogies with human brain processing that uses combinations of predictor variables and nonlinear regression to predict either a categorical or quantitative response variable.
Online analytical processing (OLAP)	An approach for providing answers to queries that typically involve many variables simultaneously.
Predictive model	A model that provides predictions for the response variable.
Regression problem	A prediction problem that has a quantitative response variable. (See classification problem.)
Supervised problem	A classification or regression problem in which the analyst is given a set of data for which the response is known to use to build a model.
Terminal node	The final leaves of a decision tree where the predictions for the response variable are found.
Test set	The data set used in a supervised classification or regression problem *not used* to build the predictive model. The test set is withheld from the model building stage of the process, and the model's predictions on the test set are used to evaluate the model performance.
Training set	The data set used in a supervised classification or regression problem to build the predictive model.
Transactional data	Data describing an event involving a transaction, usually an exchange of money for goods or services.
Unsupervised problem	A problem, unlike classification or regression problems, where there is no response variable. The goal of an unsupervised problem is typically to put similar cases together into homogeneous groups or clusters.

Case **Study**

House Prices

What determines how much a house is worth? Of course, the first three rules of real estate are Location, Location and Location but in the same region, house appraisals typically compare a house to other similar houses in the area to determine a fair price. Can we find a regression model that will do that based only on attributes of the house?

A realtor gathered information on 1728 houses in a community in the Northeast U.S.

The data are in **Real_Estate_Case_Study**.

The variables are:

> Price (dollars)
>
> Lot size (acres)
>
> Age (years)
>
> Land value (dollars)
>
> Living area (square feet)
>
> Pct College (% of residents in the area with a college education—a demographic fact about the neighborhood.)
>
> Bedrooms (count)
>
> Fireplaces (count)
>
> Bathrooms (count: Half baths (without a tub or shower) are counted as half.)
>
> Rooms (count)

The realtor wants to find a model to predict the value of a house. She is interested in identifying "good buys"—houses with asking prices lower that might be expected from their attributes. She also hopes to learn what attributes of a house contribute to its market value.

Build a model to predict the price of a house. These are real data on houses, taken from public records, so they may contain errors or unusual values. Here are some things to consider:

1. Examine displays to see whether the conditions of your model are satisfied.
2. Are there any extraordinary values? Look at the extreme values of the variables to see if they seem reasonable. (A large collection of real data often has odd or incorrect values.) If you find values that are not plausible or that seem so unusual that they might represent uncharacteristic houses, you may choose to set them aside, but be sure to explain what you have done.
3. Are the data internally consistent? For example, how many rooms are in these houses? Simple calculations with the three variables counting rooms can point out anomalies.
4. Build a regression model with the data you find trustworthy.
5. Write a report describing your model and explaining what you have learned about house pricing. Be sure to discuss the R^2 (or the adjusted R^2), and the standard deviation of the residuals.

Answers

CHAPTER 1

SECTION EXERCISE ANSWERS

1. a) Each row represents a different house. It is a case.
   b) There are 7 variables including the house identifier.

3. a) House_ID is an identifier (categorical, not ordinal). Neighborhood is categorical (nominal). Mail_ZIP is categorical (nominal – ordinal in a sense, but only on a national level). Acres is quantitative (units – acres). Yr_Built is quantitative (units – year). Full_Market_Value is quantitative (units – dollars). Size is quantitative (units – square feet).
   b) These data are cross-sectional. All variables were measured at about the same time.

5. It is not clear if the data were obtained from a survey. They are certainly not from an experiment. Most likely they are just a collection of recent sales. We don't know if those sales are representative of all sales, so we should be cautious in drawing conclusions from these data about the housing market in general.

CHAPTER EXERCISE ANSWERS

7. Answers will vary.

9. *Who*—college students; *What*—2025 opinion and whether they'd purchase an electric vehicle; *When*—current; *Where*—your location; *Why*—Automobile manufacturer wants college student opinions; *How*—survey; *Variables*—There are 2 categorical variables. "Whether they'd purchase" is ordinal. The data are cross-sectional.

11. Answers will vary.

13. *Who*—MBA applicants at school in Northeast United States; *What*—sex, age, whether or not they accepted, whether or not they attended, and the reasons; *When*—not specified; *Where*—School in the northeastern United States; *Why*—The researchers wanted to investigate any patterns in female acceptance and attendance; *How*—Data obtained internally from admissions office; *Variables*—There are 5 variables. Sex, whether or not they accepted, whether or not they attended, and the reasons are all categorical variables; only age (years) of applicant is quantitative. The data are a cross section.

15. *Who*—experiment volunteers; *What*—herbal cold remedy or sugar solution, and cold severity; *When*—not specified; *Where*—major pharmaceutical firm; *Why*—Scientists were testing the efficacy of a herbal compound on the severity of the common cold; *How*—The scientists set up an experiment; *Variables*—There

are 2 variables. Type of treatment (herbal or sugar solution) is categorical, and can be considered either quantitative or categorical and ordered. *Concerns*—The severity of a cold seems subjective and difficult to quantify. Also, the scientists may feel pressure to report negative findings about the herbal product. The data are cross-sectional and from a designed experiment.

17. *Who*—vineyards; *What*—size of vineyard (possibly in acres), number of years in existence, state, varieties of grapes grown, average case price (probably in dollars), gross sales (probably in dollars), and percent profit; *When*—not specified; *Where*—not specified; *Why*—Business analysts hoped to provide information that would be helpful to producers of U.S. wines; *How*—not specified; *Variables*—There are 5 quantitative variables and 2 categorical variables. Size of vineyard, number of years in existence, average case price, gross sales, and percent profit are quantitative variables. State and variety of grapes grown are categorical variables. The data are a cross section collected in a designed survey.

19. *Who*—every model of automobile in the United States; *What*—vehicle manufacturer, vehicle type, weight (probably in pounds), horsepower (in horsepower), and gas mileage (in miles per gallon) for city and highway driving; *When*—This information is collected currently; *Where*—United States; *Why*—The Environmental Protection Agency uses the information to track fuel economy of vehicles; *How*—The data is collected from the manufacturer of each model; *Variables*—There are 6 variables. City mileage, highway mileage, weight, and horsepower are quantitative variables. Manufacturer and type of car are categorical variables. The data are a cross section.

21. *Who*—restaurants; *What*—% of customers liking restaurant, average meal cost ($), food rating (1–30), decor rating (1–30), service rating (1–30). *When*—current; *Where*—United States; *Why*—service to provide information for consumers; *How*—not specified; *Variables*—There are 5 variables. % liking and average cost are quantitative. Ratings (food, decor, and service) are ordered categories. The data are a cross section.

23. *Who*—students in an MBA statistics class; *What*—total personal investment in stock market ($), number of different stocks held, total invested in mutual funds ($), and name of each mutual fund; *When*—not specified; *Where*—School in the northeastern United States; *Why*—The information was collected for use in classroom illustrations; *How*—An online survey was conducted. Presumably, participation was required for all members of the class; *Variables*—There are 4 variables.

Name of mutual fund is a categorical variable. Number of stocks held, total amount invested in market ($), and in mutual funds ($) are quantitative variables. The data are a cross section.

25. *Who*—Indy 500 races; *What*—year, winner, chassis, engine, time (hours), speed (mph), and car #. *When*—1911–2014; *Where*—Indianapolis, Indiana; *Why*—It is interesting to examine the trends in Indy 500 races; *How*—Official statistics are kept for the race every year; *Variables*—There are 7 variables. Winner, chassis, engine, and car # are categorical variables. Year, time, and speed are quantitative variables. The quantitative variables can be viewed as time series.

27. Each row should be a single mortgage loan. Columns hold the borrower's name (which identifies the rows) and amount.

29. Each row is a week. Columns hold week number (to identify the row), sales prediction, sales, and difference.

31. Cross-sectional.

33. Time series.

CHAPTER 2
SECTION EXERCISE ANSWERS

1. a) Frequency table:

None	AA	BA	MA	PhD
164	42	225	52	29

b) Relative frequency table (divide each number by 512 and multiply by 100):

None	AA	BA	MA	PhD
32.03%	8.20%	43.95%	10.16%	5.66%

3. a)
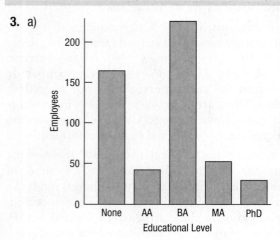

b)
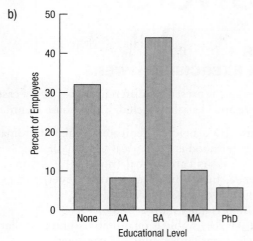

c)
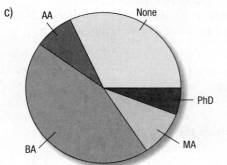

5. a) The vast majority of employees have either no college degree or a bachelor's degree (32% and 44%, respectively). About 10% have master's degrees, 8% have associate's degrees, and nearly 6% have PhDs.

b) I would not be comfortable generalizing this to any other division or company. These data were collected only from my division. Other companies might have vastly different educational distributions.

7. a)

	Totals
<1 Year	95
1–5 Years	205
More Than 5 Years	212

b) Yes

None	AA	BA	MA	PhD
164	42	225	52	29

9. a)

Percent	None	AA	BA	MA	PhD
<1 Year	6.1	7.1	22.2	38.5	41.4
1–5 Years	25.6	21.4	49.8	51.9	51.7
More Than 5 Years	68.3	71.4	28.0	9.6	6.9

b) No. The distributions look quite different. More than $^2/_3$ of those with no college degree have been with the company longer than 5 years, but almost none of the PhDs (less than 7%) have been there that long.

c)

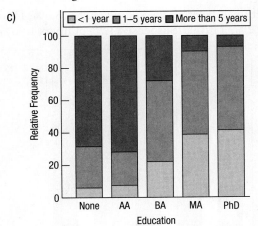

d) It's possible to see it in the table, but the stacked bar chart makes the differences much clearer.

e) A mosaic plot would display the different counts for each degree type. Areas of the plot representing each cell would then reflect the cell counts accurately.

CHAPTER EXERCISE ANSWERS

11. Answers will vary.

13. Answers will vary.

15. a) Yes, the categories divide a whole.
   b) Coca-Cola.

17. a) The pie chart does a better job of showing portions of a whole.
   b) There is no bar for "Other."

19. a) Yes, it is reasonable to assume that heart and respiratory disease caused approximately 38% of U.S. deaths in this year, since there is no possibility for overlap. Each person could only have one cause of death.
   b) Since the percentages listed add up to 73.7%, other causes must account for 26.3% of U.S. deaths.
   c) A bar graph or pie chart would be appropriate if a category for Other with 26.3% were added.

21. Cisco systems continues to dominate the market for desktop conferencing. Citrix and Microsoft are battling for second place. A pie chart or bar chart would be appropriate.

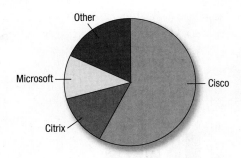

23. a) They don't total 100%. Others must have refused to answer or didn't know.
   b)

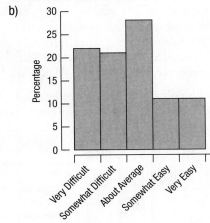

   c) Only if the "other" category were added.
   d) Answers will vary. Nearly half (43%) of business owners said that it would be somewhat or very difficult to obtain credit. Only 22% said it would be somewhat or very easy. Of the remaining, 28% said it would be about average and 7% didn't answer.

25. The bar chart shows that grounding is the most frequent cause of oil spillage for these 459 spills, and allows the reader to rank the other types as well. If being able to differentiate between close counts is required, use the bar chart. The pie chart is also acceptable as a display, but it's difficult to tell whether, for example, a greater percentage of spills is caused by grounding or collisions. To showcase the causes of oil spills as a fraction of all 459 spills, use the pie chart.

27. a) 31%
   b) Because the bars do not start at zero, it looks like India's percentage is about 6 times as big, but in fact, it's not even twice as big.
   c) Start the percentages at 0% on the vertical axis, not 40%.

d)

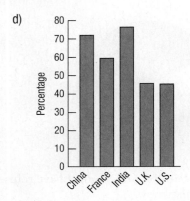

e) The percentage of people who say that wealth is important to them is highest in China and India (around 70%), followed by France (around 60%) and then the United States and the United Kingdom where the percentage was only about 45%.

29. a) They must be column percentages because the sums are greater than 100% across the rows and all the columns add to 100%.

b)

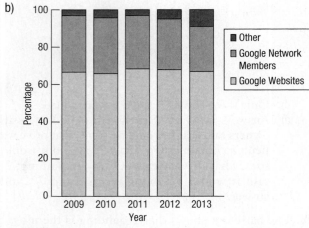

c) The main source of revenue for Google is from their own websites, which fluctuates around 67% during this period. The second largest source is from other network websites which decreased from 30% in 2009 to 24% in 2013. Licensing and other revenue was 3% in 2009, but increased to 9% by 2013.

31. a) 45.1%
    b) 34.9%
    c) 5.3%
    d) 59.8%
    e) 41.3%
    f) 65.8%
    g) Companies that reported a positive change on October 24 were more likely to report a negative change for the year than companies who reported a negative change on October 24.

33. a) 11.5%
    b) 45.5%

c) 18.2%
d) 2009 was 11.2%; 2010 was 8.2%; change = 3.0% fewer in 2010.

35. a) 3.4% G, 23.7% PG, 44.1% PG-13, 28.8% R or NC-17
    b) Action/Adventure 25.5%, Comedy 18.6%, Drama 40.2%, Thriller/Suspense 13.7%, Others 1.96%
    c)

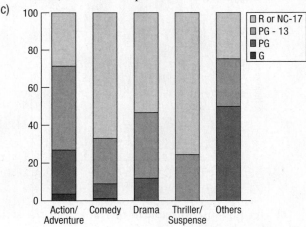

d) *Genre* and *Rating* are not independent. Thriller/Horror movies are all PG-13 or R/NC-17, but 27% of Action/Adventure movies were either G or PG in 2014. Comedy is over 67% R or NC-17 in this year, while Action/Adventure was only 28.8%.

37. a) 62.7%
    b) 62.8%
    c) 62.5%
    d) 23.9% from Asia, 1.9% Europe, 7.8% Latin America, 3.7% Middle East, and 62.7% North America.
    e) The column percentages are given in the table.

	MBA Program		
	Two-Yr	Evening	Total
Asia	18.90	31.73	**23.88**
Europe	3.05	0.00	**1.87**
Latin America	12.20	0.96	**7.84**
Middle East	3.05	4.81	**3.73**
North America	62.80	62.50	**62.69**
Total	**100.00**	**100.00**	**100.00**

f) No. The distributions appear to be different. For example, the percentage from Latin America among those in Two-Year programs is nearly 12.2% while for those in Evening programs it is less than 1%.

39. a) 4%
    b) 5%
    c) 4.5%
    d) 56.0%
    e) 71%

f) Here are row percentages:

	G	PG	PG-13	R/NC17	Total
2010–2014	3%	27%	56%	14%	100.0%
2005–2009	5%	24%	59%	12%	100.0%

The distributions are quite similar, although there were a few more PG and R/NC17 films in the more recent 5 years than the previous.

41. The study by the University of Texas Southwestern Medical Center provides evidence of an association between having a tattoo and contracting hepatitis C. Around 33% of the subjects who were tattooed in a commercial parlor had hepatitis C, compared with 13% of those tattooed elsewhere, and only 3.5% of those with no tattoo. If having a tattoo and having hepatitis C were independent, we would have expected these percentages to be roughly the same.

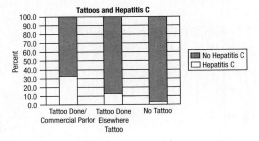

43. a) 66%
   b) It is higher.
   c) No, because we're not given counts or totals.
   d) Young (18- to 34-year-old) women appear to consider being professionally successful more important in their lives than do young men. But older respondents showed no difference by sex.

45. a)

	Caucasian	Hispanic	African-American	Other
Population	66.0%	16.0%	12.0%	6.0%
Moviegoers	63.0%	19.0%	12.0%	6.0%
Tickets	56.0%	26.0%	11.0%	7.0%

   b) The distributions of moviegoers are quite similar to the population as a whole, but Hispanics seem to buy proportionally more tickets and Caucasians fewer. Hispanics appear to go to the movies more often, on average, than Caucasians.

47. a) 14.25%
   b) 58.7%
   c) Younger women are more likely than older women to say that professional success is important to them.

49. a) The marginal totals have been added to the table:

		Hospital Size		
		Large	Small	Total
Procedure	**Major Surgery**	120 of 800	10 of 50	130 of 850
	Minor Surgery	10 of 200	20 of 250	30 of 450
	Total	130 of 1000	30 of 300	160 of 1300

160 of 1300, or about 12.3% of the patients had a delayed discharge.
   b) Major surgery patients were delayed 15.3% of the time.
       Minor surgery patients were delayed 6.7% of the time.
   c) Large hospital had a delay rate of 13%.
       Small hospital had a delay rate of 10%.
       The small hospital has the lower overall rate of delayed discharge.
   d) Large hospital: Major surgery 15% and Minor surgery 5%.
       Small hospital: Major surgery 20% and minor surgery 8%.
       Even though the small hospital had the lower overall rate of delayed discharge, the large hospital had a lower rate of delayed discharge for each type of surgery.
   e) Yes. While the overall rate of delayed discharge is lower for the small hospital, the large hospital did better with *both* major surgery and minor surgery.
   f) The small hospital performs a higher percentage of minor surgeries than major surgeries. 250 of 300 surgeries at the small hospital were minor (83%). Only 200 of the large hospital's 1000 surgeries were minor (20%). Minor surgery had a lower delay rate than major surgery (6.7 to 15.3%), so the small hospital's overall rate was artificially inflated. The larger hospital is the better hospital when comparing discharge delay rates.

51. a) 1284 applicants were admitted out of a total of 3014 applicants. 1284/3014 = 42.6%
   b) 1022 of 2165 (47.2%) of males were admitted. 262 of 849 (30.9%) of females were admitted.

		Males Accepted (of Applicants)	Females Accepted (of Applicants)	Total
Program	**1**	511 of 825	89 of 108	600 of 933
	2	352 of 560	17 of 25	369 of 585
	3	137 of 407	132 of 375	269 of 782
	4	22 of 373	24 of 341	46 of 714
	Total	1022 of 2165	262 of 849	1284 of 3014

   c) Since there are four comparisons to make, the following table organizes the percentages of males and females accepted in each program. Females are accepted at a higher rate in every program.

Program	Males	Females
1	61.9%	82.4%
2	62.9%	68.0%
3	33.7%	35.2%
4	5.9%	7.0%

d) The comparison of acceptance rate within each program is most valid. The overall percentage is an unfair average. It fails to take the different numbers of applicants and different acceptance rates of each program. Women tended to apply to the programs in which gaining acceptance was difficult for everyone. This is an example of Simpson's Paradox.

CHAPTER 3
SECTION EXERCISE ANSWERS

1. a)

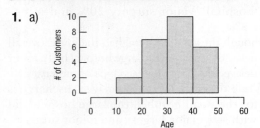

b)

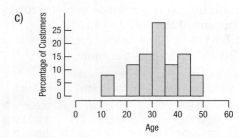

c)

d) 1|14
   2|0225699
   3|0002224558
   4|244488

3. a) Unimodal
   b) Around 35 years old

c) Fairly symmetric
d) No outliers.

5. a) About the same. The distribution is fairly symmetric.
   b) 31.84 years
   c) 32 years

7. a) Q1 = 26; Q3 = 38 (Answers may vary slightly.)
   b) Q1 = 25.5; Q3 = 40
   c) IQR = 12 years
   d) SD = 9.84 years

9. a) The distribution is skewed to the right. There are a few negative values. The range is about $6000. Center around $2000; IQR around $2000.
   b) The mean will be larger because the distribution is right skewed.
   c) Because of the skewness, the median is a better summary.

11. a) 11 has a z-score of -2.12; 48 has a z-score of $+1.64$.
   b) The min, 11, is more extreme.
   c) 61.36 years old

13. a)

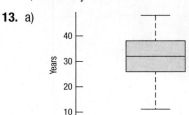

   b) No.
   c) 38 + 1.5*12 = 56 years old

15. a) Skewed to the right, since the mean is much greater than the median.
   b) Yes, at least one high outlier, since 250 is far greater than Q3 + 1.5 IQRs.
   c) We don't know how far the high whisker should go because we don't know the largest value inside the fence, or where other possible outliers might be.

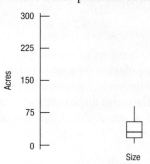

17. The ages of the women are generally higher than those of the men by about 10 years. As the boxplot shows, more than 3/4 of the women are older than all the men.

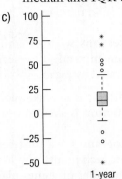

19. Sales in Location #1 were higher than sales in Location #2 in nearly every week. The company might want to compare other stores in locations like these to see if this phenomenon holds true for other locations.

21. The upper outlier limit is
$123 + 1.5*(123 - 44.9) = 240.15$.
The lower outlier limit is
$44.9 - 1.5*(123 - 44.9) < 0$. Yes, the maximum value is an outlier. We should look at a boxplot to know how to proceed.

23. a) Yes.
b) No—data are from a single time point.
c) No—response is "time" but measured at only one time point.
d) Yes.

25. A logarithmic transformation might make the distribution of revenues more symmetric.

CHAPTER EXERCISE ANSWERS

27. Answers will vary.

29. a) It is symmetric, with a center between 14.5 and 15 inches. *Size* varies from 12.5 inches to 17.5 inches and has no outliers.
b) The bin from 14.5 to 15 and the bin from 15 to 15.5
c) Men in the bin from 14.5 to 15 would need a 15.5-inch shirt. Those in the 15 to 15.5 bin would need a 16-inch shirt. So, 15.5 and 16 inches.

31. a) The distribution is roughly unimodal and symmetric with most values between 0 and 30% and many extreme values, both negative and positive. The median return seems to be about 15%. More than half of the funds returned between 0 and 20%.
b) It is a little hard to tell, but it appears that most funds may have done better than 11.5%. The median appears to be closer to 15%.

33. a)

Min.	1st Qu.	Median	3rd Qu.	Max.
−49.1	8.5	13.2	22.2	78.6

b) Median is 13.2%; IQR is 13.7%. Better to report median and IQR because of all the outliers.

c)

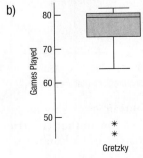

d) Now we can see that half of the ETFs earned 13.2% or more, which is higher than the 11.5% for the S&P 500.

35. Answers may vary slightly, depending on the stem used. The skewness of the distribution, the gap and the high outlier are all evident.

0	1111111222223333333444444
0	55666678
1	03
1	5
2	
2	5

Key: 0 | 8 = 80 acres

37. a) Wayne Gretzky—Games played per season

8	000000122
7	8899
7	0344
6	
6	4
5	
5	
4	58
4	

Key: 7 | 8 = 78 games

(Stem and leaf could also be displayed with smaller numbers on top.)

b)

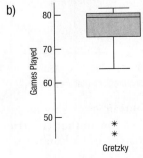

c) The distribution of the number of games played per season by Wayne Gretzky is skewed to the low end and has low outliers. The median is 79, and the range is 37 games.

d) There are two outlier seasons with 45 and 48 games. He may have been injured. The season with 64 games is also separated by a gap.

39. a) The median because the distribution is skewed.

b) Lower, because the distribution is skewed toward the low end.

c) That display is not a histogram. It's a time series plot using bars to represent each point. The histogram should split up the number of games played into bins, not display the number of games played over time.

41. a) Descriptive Statistics: Price ($)

Minimum	Q1	Median	Q3	Maximum
2.21	2.51	2.61	2.72	3.05

b) Range = max − min = 3.05 − 2.21 = $0.84;
IQR = Q3 − Q1 = 2.72 − 2.51 = $0.21

c)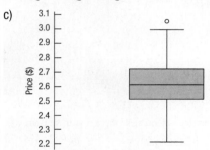

d) Symmetric with one high outlier. The mean is $2.62, with a standard deviation of $0.156.

e) There is one unusually high price that is greater than $3.00 per frozen pizza.

43. As we can see from the histogram and boxplot,

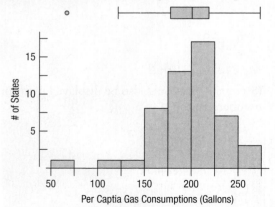

the main part of the distribution is roughly unimodal and symmetric, but there is a low outlier (District of Columbia). Without the District of Columbia (which

is not a state), the median consumption is 201.59 gal/capita. Fifty percent of the states have consumption between 178.5 and 218.6 gal/capita.

45. a) 1611 yards

b) Between Quartile 1 = 5585.75 yards, and Quartile 3 = 6131 yards.

c) The distribution of golf course lengths appears roughly symmetric, so the mean and the standard deviation are appropriate.

d) The distribution of the lengths of all the golf courses in Vermont is roughly unimodal and symmetric. The mean length of the golf courses is approximately 5900 yards and the standard deviation is 386.6 yards.

47. a) A boxplot is shown. A histogram would also be appropriate.

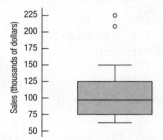

b) Descriptive Statistics: Sales ($) (Different statistics software may yield slightly different results.)

Variable	N	Mean	SE Mean	StDev	Minimum
Sales ($)	18	107845	11069	46962	62006

Q1	Median	Q3	Maximum
73422.5	95975	112330.0	224504

The mean sale is $107,845, and the median is $95,975. The mean is higher because the outliers pull it up.

c) The median because the distribution has outliers.

d) The standard deviation of the distribution is $46,962 and the IQR is $38,907.50. (Answers may vary slightly due to different quartile algorithms.)

e) The IQR because the outliers inflate the standard deviation.

f) The mean would decrease. The standard deviation would decrease. The median and IQR would be less affected.

49. A histogram shows that the distribution is unimodal and skewed to the left. There do not appear to be any outliers. The median failure rate for these 17 models is 16.2%. The middle 50% of the models have failure rates between 10.87% and 21.2%. The best rate is 3.17% for the 60GB Video model, and the worst is the 40GB Click Wheel at 29.85%.

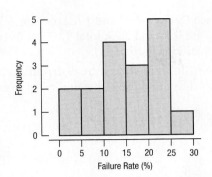

51. a) Gas prices generally increased over the nine-year period except for the dip in 2009. The distribution of prices in 2003 appears symmetric with a relatively small spread. Median prices rose in 2006, fell slightly by 2009 and then increased to new highs in 2012. There are low outliers in 2009 and 2012, but they appear to be quite near the lower fences.

b) Gas prices were most stable in 2003. That distribution has the smallest IQR. Prices in 2006 and 2009 were least stable.

53. a) There is one wine in Seneca more expensive than any other.

b) Both Cayuga and Seneca produce cheaper wines than Keuka, but Seneca has several high outliers.

c) Keuka Lake.

d) Cayuga Lake vineyards and Seneca Lake vineyards have approximately the same average price of about $15 a bottle, while a typical Keuka Lake vineyard has a price near $22. Keuka Lake vineyards have consistently high prices between about $18 and $26 a bottle. Cayuga Lake vineyards have prices from $13 to $19, and Seneca Lake vineyards have highly variable prices from about $12 to over $30.

55. a) The median speed is the speed at which 50% of the winning horses ran slower. Find 50% on the left, move straight over to the graph and down to a speed of about 36 mph.

b) Q1 = 34.5 mph; Q3 = 36.5 mph

c) Range = 7 mph
IQR = 2 mph

d)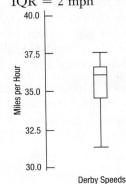

e) The distribution of winning speeds in the Kentucky Derby is skewed to the left. The lowest winning speed is just under 31 mph, and the fastest speed is about 37.5 mph. The median speed is approximately 36 mph, and 75% of winning speeds are above 34.5 mph. Only a few percent of winners have had speeds below 33 mph.

57. a) Class 3.

b) Class 3.

c) Class 3 because it is the most highly skewed. Median is higher.

d) Class 1.

e) Probably Class 1. But without the actual scores, it is impossible to calculate the exact IQRs.

59. There is an extreme outlier for the slow-speed drilling. One hole was drilled almost an inch away from the center of the target! If that distance is correct, the engineers at the computer production plant should investigate the slow-speed drilling process closely. It may be plagued by extreme, intermittent inaccuracy. The outlier in the slow-speed drilling process is so extreme that no graphical display can display the distribution in a meaningful way while including that outlier. That distance should be removed before looking at a plot of the drilling distances.

With the outlier removed, we can see that the slow drilling process is more accurate. The greatest distance from the target for the slow drilling process, 0.000098 inches, is still more accurate than the smallest distance for the fast drilling process, 0.000100 inches.

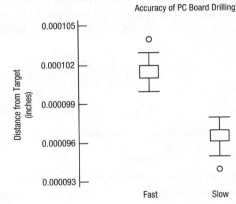

61. a) The mean of 54.41 is meaningless. These are categorical values.

b) Typically, the mean and standard deviation are influenced by outliers and skewness.

c) No. Summary statistics are only appropriate for quantitative data.

63. Although the numbers are small, the Trading-Leveraged Equity funds (with one exception) seemed to have performed the best, followed by the Health funds. The Large Growth funds were the most consistent performers. The Technology and "Other" funds were the worst performers.

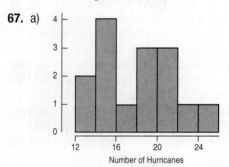

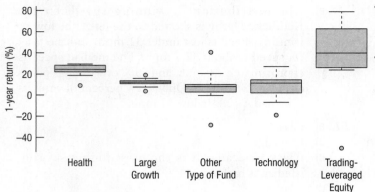

65. a) Even though MLS ID numbers are categorical identifiers, they are assigned sequentially, so this graph has some information. Most of the houses listed long ago have sold and are no longer listed.

b) A histogram is generally not an appropriate display for categorical data.

67. a)

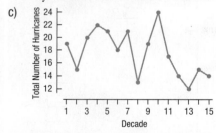

b) The distribution is fairly uniform considering the small numbers in each bin. There do not appear to be any decades that would be considered to be outliers.

c)

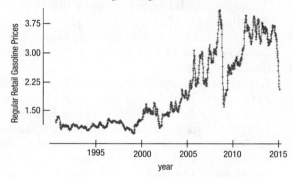

d) This graph does not support the claim that the number of hurricanes has increased in recent decades.

69. What is the *x*-axis? If it is time, what are the units? Months? Years? Decades? How is "productivity" measured?

71. The house that sells for \$400,000 has a *z*-score of $(400,000 - 167,900)/77,158 = 3.01$, but the house with 4000 sq. ft. of living space has a *z*-score of $(4000 - 1819)/663 = 3.29$. So it's even more unusual.

73. U.S. *z*-scores are -0.04 and 1.63, total $= 1.59$. Ireland *z*-scores are 0.25 and 2.77, total 3.02. So Ireland "wins" the consumption battle.

75. a) The histogram shows the distribution is strongly skewed to the right with a mode just over \$1.

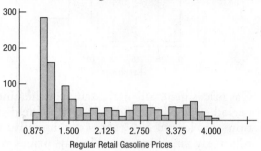

b) Prices were relatively stable until the late 1990s when they started increasing. They increased steadily from 2002 until 2008 (the financial crisis) when they dropped precipitously before increasing steadily again until mid 2011. After a period of fluctuation, they dropped sharply again at the end of 2014 and beginning of 2015.

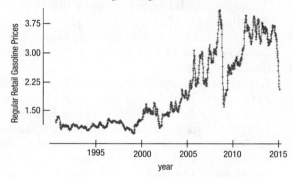

c) The time series plot is more informative because the prices have changed so much over time.

77. a) The distribution has a mode around 5.5 and possible modes at 8 and 9.5.

b) The unemployment rate grew rapidly between 2007 and 2010 and has been declining since then.

c) The time series plot. There is very little about the distribution that is interesting. The story about unemployment is told by the time series.

d) Unemployment decreased from 2004 to 2007 from just over 6% to just under 5%. It then increased rapidly until 2010 when it reached nearly 10%. Since then it has decreased steadily.

79. a) The distribution is skewed. That makes it difficult to estimate anything meaningful from the graph.

b) Transform these data using either square roots or logs.

CHAPTER 4
SECTION EXERCISE ANSWERS

1. a)

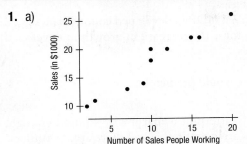

b) Positive.
c) Linear.
d) Strong.
e) No.

3. a) Years of experience.
b) Salary.
c) Salary.

5. a) True.
b) False. It will not change the correlation.
c) False. Correlation has no units.

7. Correlation does not demonstrate causation. The analyst's argument is that sales staff cause sales. However, the data may reflect the store hiring more people as sales increase, so any causation would run the other way.

9. a) False. The line usually touches none of the points. We minimize the sum of the squared errors.
b) True.
c) False. It is the sum of the squares of all the residuals that is minimized.

11. a) $2 \times 0.965 = 1.93$ SDs
b) $17.6 + 1.93 \times 5.34 = 27.906$ or $27,906
c) 0.965 SDs below the mean
d) $12,447

13. a) $b_1 = 0.914$ if found by hand. $b_1 = 0.913$ if found by technology. (Difference is due to rounding error.)
b) It means that an additional 0.914 ($1000) or $914 of sales is associated with each additional sales person working.
c) $b_0 = 8.10$
d) It would mean that, on average, we expect sales of 8.10 ($1000) or $8100 with 0 sales people working. Doesn't really make sense in this context.
e) $\widehat{Sales} = 8.10 + 0.914$ *Number of Sales People Working*
f) $24.55 ($1000) or $24,550. (24,540 if using the technology solution.)
g) 0.45 ($1000) or $450. ($460 with technology.)
h) Underestimated.

15. The winners may be suffering from regression to the mean. Perhaps they weren't really better than other rookie executives, but just happened to have a lucky year.

17. a) Thousands of dollars.
b) 2.77 (the largest residual in magnitude).
c) 0.07 (the smallest residual in magnitude).

19. $R^2 = 93.12\%$. About 93% of the variance in *Sales* can be accounted for by the regression of *Sales* on *Number of Sales Workers*.

21. 16, 16, 36, 49, 49, 64, 100. They are skewed to the high end.

CHAPTER EXERCISE ANSWERS

23. a) Number of text messages: explanatory; cost: response. To predict cost from number of text messages. Positive direction. Linear shape. Possibly an outlier for contracts with fixed cost for texting.
b) Fuel efficiency: explanatory; sales volume: response. To predict sales from fuel efficiency. There may be no association between mpg and sales volume. Environmentalists hope that a higher mpg will encourage higher sales, which would be a positive association. We have no information about the shape of the relationship.
c) Neither variable is explanatory. Both are responses to the lurking variable of temperature.
d) Price: explanatory variable; demand: response variable. To predict demand from price. Negative direction. Linear shape in a narrow range, but curved over a larger range of prices.

25. a) None.
b) 3 and 4
c) 2, 3, and 4
d) 2 and 4
e) 3 and 1

27. a)

Number of Broken Pieces
per Batch (24 batches)

b) Unimodal, skewed to the right. The skewness.
c) The positive, somewhat linear relation between batch number and broken pieces.

29. a) 0.006
b) 0.777
c) −0.923
d) −0.487

31. a) *Price.*
   b) *Sales.*
   c) Sales decrease by 24,369.49 pounds for every additional dollar charged.
   d) It is just a base value. It means nothing because stores won't set their price to $0.
   e) 56,572.32 pounds
   f) 3427.69 pounds

33. a) *Salary.*
   b) *Wins.*
   c) On average, teams who spend $1M more in salary win, 0.219 games more.
   d) Number of wins predicted for a team that spends $0 on salaries. This is not meaningful here.
   e) 2.19 games more
   f) Slightly worse. They are expected to win 9.96 games.
   g) −1.96 games
   h) Not very useful. The residual standard deviation says that we cannot predict better than about +/−5 games. For a prediction of 8 games won, that's 3 to 13 games. All but 2 teams that year were already in that range.

35. About 47,084 pounds

37. "Packaging" isn't a variable. At best, it is a category. There's no basis for computing a correlation.

39. The model is meaningless because the variable Region is not quantitative. The slope makes no sense because Region has no units. The boxplot comparisons are useful but the regression is meaningless.

41. a) There is a strong negative linear association between Carbon Footprint and Highway mpg.
   b) Quantitative variables. The hybrid cars appear to be essentially different from the others, so it is probably not appropriate to report a correlation for all the data; we have two different kinds of vehicles combined together.
   c) $r = -0.931$. Removing the hybrids leaves a more consistent collection of cars, which are more linearly associated.

43. a) Positive association.
   b) Plot is not linear, violating the linearity condition. There may be an outlier at 17 rooms.

45. a) The variables are both quantitative (with units % of GDP), the plot is reasonably straight, but there are a couple of outliers that influence the fit (especially 2009). The spread is roughly constant (although the spread is large). We should be cautious in interpreting the model too strictly.
   b) About 31.6% of the variation in the growth rates of developing countries is accounted for by the growth rates of developed countries.
   c) Years 1970–2011

47. a) $\overline{Growth\ (Developing\ Countries)} = 3.38 + 0.468$
   $Growth\ (Developed\ Countries)$

   b) The predicted growth of developing countries in years of 0 growth in developed countries. Yes, this makes sense.
   c) On average, GDP in developed countries increased 0.468% for a 1% increase in growth in developed countries.
   d) 5.25%
   e) More; we would predict 4.62%.
   f) 1.47%

49. a) Yes, the scatterplot is marginally straight enough, variables are quantitative and there are no outliers.
   b) Teams that score more runs generally have higher attendance but the relationship is only moderately strong, with a correlation of 0.233.
   c) There is a weak positive association, but correlation doesn't imply causation.

51. a) The predicted value of the money *Flow* if the *Return* was 0%.
   b) An increase of 1% in mutual fund return was associated with an increase of $771 million in money flowing into mutual funds.
   c) $9747 million
   d) −$4747 million; overestimated.

53. a) Model seems appropriate. Residual plot looks fine.
   b) Model not appropriate. Relationship is nonlinear.
   c) Model not appropriate. Spread is increasing.

55. There are two outliers that inflate the R^2 value and affect the slope and intercept. Without those two points, the R^2 drops from 79% to about 31%. The analyst should set aside those two customers and refit the model.

57. 0.03. For every $1000 increase in ad expenditure, sales are most likely to increase by $30,000.

59. a) R^2 is an indication of the strength of the model, not the appropriateness of the model.
   b) The agent should have said, "The model predicts that annual sales will be $10 million when $1.5 million is spent on advertising."

61. a) Quantitative variable condition: Both variables are quantitative (*GPA* and *Starting Salary*).
   b) Linearity condition: Examine a scatterplot of *Starting Salary* by *GPA*.
   c) Outlier condition: Examine the scatterplot.
   d) Equal spread condition: Plot the regression residuals versus predicted values.

63. a)

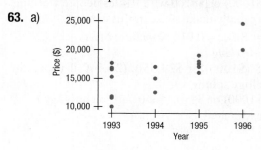

b) There is a strong, positive, association between *Price* and *Year* of used BMW 850CSi's with some upward curvature.

c) Yes, but with caution because of the curvature.

d) 0.757

e) 57.3% of the variability in *Price* of a used BMW 850 can be accounted for by the *Year* the car was made.

f) The relationship is not perfect. Other factors, such as options, condition, and mileage, may account for some of the variability in price.

65. a) The association between cost of living in 2013 and 2007 is very weak. There are no outliers. The scatterplot indicates that the linear model is appropriate but not likely to reveal much.

b) 7.0% of the variability in cost of living in 2013 can be explained by variability in cost of living in 2007.

c) -0.26

d) $\overline{Index\ 2013} = 191.6 - 0.683\,\overline{Index\ 2007}$

e) Moscow was predicted to have an index of $191.6 - .683(134.4) = 99.80$, but the 2013 index was only 81.58, so the residual is $81.58 - 99.80 = -18.22\%$.

67. a) 0.578

b) CO_2 levels account for 33.4% of the variation in mean temperature.

c) $\overline{Mean\ Temperature} = 15.3066 + 0.004\,CO_2$

d) The predicted mean temperature has been increasing at an average rate of $0.004°C$/ppm of CO_2.

e) One *could* say that with no CO_2 in the atmosphere, there would be a temperature of $15.3066°C$, but this is extrapolation to a nonsensical point.

f) No.

g) Predicted $16.7626°C$.

CHAPTER 5
SECTION EXERCISE ANSWERS

1. a) Independent (unless a large group of one gender comes to the ATM machine together).

b) Independent. The last digit of one student's SS number provides no information about another.

c) Not independent. How you perform on one test provides information about other tests.

3. a) Won't work, but won't hurt. Each number drawn is equally likely and independent of the others, so this set of numbers is just as likely as any other in the next drawing.

b) Won't work, but won't hurt. Each number drawn is equally likely and independent of the others and of previous drawings, so the previous winners are just as likely as any other in the next drawing.

5. a) 0.40

b) 0.60

c) $0.60^2 = 0.36$

d) $0.60 + 0.60 - (0.36) = 0.84$ or $1 - (0.4^2) = 0.84$

7. a) $755/1200 = 0.629$

b) Marginal.

c) $210/1200 = 0.175$

d) Joint.

9. a) $210/500 = 0.42$

b) $210/755 = 0.278$

c) $415/500 = 0.83$

11. a) $100\% - 30\% = 70\%$

b) Joint.

c)

		Online Banking		
		Yes	No	
Age	Under 50	0.25	0.15	0.40
	50 or Older	0.05	0.55	0.60
		0.30	0.70	1.00

d) $0.25/0.40 = 0.625$

e) No, because the conditional probability of banking online for those under 50 is 0.625. The probability of banking online is 0.30 which is not the same.

13. a)

b) 3.7%

c) 0.405

15. a) White: 0.68, Black: 0.11, Hispanic/Latino/Other: 0.21; White Male: 0.54, White Female: 0.46; Black Male: 0.52, Black Female: 0.48; Hispanic/Latino/Other Male: 0.58, Hispanic/Latino/Other Female: 0.42.

b) 0.0528

c) 0.46

d) 0.6893

CHAPTER EXERCISE ANSWERS

17. a) Individual outcomes can't be predicted, although in the long run the relative frequencies may be known (and for a roulette wheel should be equal).

b) This is likely a personal probability expressing his degree of belief that there will be a new high.

19. a) There is no such thing as the "law of averages." The overall probability of an airplane crash does not change due to recent crashes.

b) There is no such thing as the "law of averages." The overall probability of an airplane crash does not change due to a period in which there were no crashes.

21. a) It would be foolish to insure your neighbor's house for $300. Although you would probably simply collect $300, there is a chance you could end up paying much more than $300. That risk is not worth the $300.

b) The insurance company insures many people. The overwhelming majority of customers pay and never have a claim. The few customers who do have a claim are offset by the many who simply send their premiums without a claim. The relative risk to the insurance company is low.

23. a) Yes.
b) Yes.
c) No, probabilities sum to more than 1.
d) Yes.
e) No, sum isn't 1 and one value is negative.

25. 0.078

27. The events are disjoint. Use the addition rule.
a) 0.72
b) 0.89
c) 0.28

29. a) 0.5184
b) 0.0784
c) 0.4816

31. a) The repair needs for the two cars must be independent of one another.

b) This may not be reasonable. An owner may treat the two cars similarly, taking good (or poor) care of both. This may decrease (or increase) the likelihood that each needs to be repaired.

33. a) 0.68
b) 0.32
c) 0.04

35. a) The events are disjoint (an M&M can't be two colors at once), so use the addition rule where applicable.
  i) 0.30
  ii) 0.30

  iii) 0.90
  iv) 0

b) The events are independent (picking out one M&M doesn't affect the outcome of the next pick), so use the multiplication rule.
  i) 0.027
  ii) 0.128
  iii) 0.512
  iv) 0.271

37. a) Disjoint.
b) Independent.
c) No. Once you know that one of a pair of disjoint events has occurred, the other one cannot occur, so its probability has become zero.

39. a) 0.125
b) 0.125
c) 0.875
d) Independence.

41. a) 0.0225
b) 0.092
c) 0.00008
d) 0.556

43. a) Your thinking is correct. There are 47 cards left in the deck, 26 black and only 21 red.

b) This is not an example of the Law of Large Numbers. The card draws are not independent.

45. a) 0.39
b) 0.82

47. a) 0.642
b) 0.035
c) $1 - (410 + 815)/1270$

49. a)

b) 0.16
c) 0.667

51. a) $P(\text{Tweet and} <30) = P(\text{Tweet}) \times P(<30 \mid \text{Tweet}) = 0.15 * 0.5 = 0.075$

b) $P(\text{tweet} \mid <30) = 0.23$
$P(\text{tweet and} <30)/P(<30) = 0.075/0.23 = 0.326$

53.
a) 0.10
b) 0.17
c) 0.026
d) 0.24

55.
a) 0.1062
b) 0.4469
c) 0.0785

57.
a) 0.11
b) 0.27
c) 0.407
d) 0.344

59. No. 28.8% of men with OK blood pressure have high cholesterol, but 40.7% of men with high blood pressure have high cholesterol.

61.
a) 0.026
b) 0.84
c) 0.20
d) 0.053
e) 0.554
f) $P(\text{more time} \mid <25 \text{ years}) = 0.158$. $P(\text{more time}) = 0.10$. These probabilities are not equal, so the events are not independent.

63.
a) 0.47
b) 0.266
c) $P(\text{pool}) = 0.21$, $P(\text{pool} \mid \text{garage}) = 0.266$. Having a garage and a pool are not independent events.
d) $P(\text{pool and garage}) = 0.17$. Having a garage and a pool are not disjoint events.

65.
a) 0.71
b) $P(<500\text{K} \mid 2 \text{ Br}) = 0$. $P(<500\text{K}) = 0.71$. Since not equal, they are not independent.

67.
a) 15.4%
b) 11.4%
c) 73.9%
d) 18.5%

*69. a)

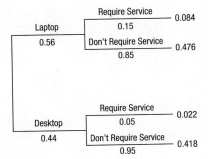

b) 10.6%
c) 0.792

CHAPTER 6
SECTION EXERCISE ANSWERS

1.
a) Discrete.
b) Yes/no.

3. 0.7

5. 0.781

7.
a) $19
b) $7

9.
a) $\mu = 30; \sigma = 6$
b) $\mu = 26; \sigma = 5$
c) $\mu = 30; \sigma = 5.39$
d) $\mu = -10; \sigma = 5.39$

11.
a) 110 and 12.
b) 450 and 40.
c) 190 and 14.422.
d) X and Y are independent for the SD calculation, but not necessarily for the sum.

13.
a) Yes. Outcomes are independent with probability $p = 1/6$. The outcomes are {getting a 6} and {not getting a 6}.
b) No. More than two outcomes are possible.
c) No. The chance of a woman (or man) changes depending on who has already been picked.
d) Yes, assuming responses (and cheating) are independent among the students.

15. Yes.

17. Geometric; $p = 0.287$

19.
a) 0.5488
b) 0.4512

CHAPTER EXERCISE ANSWERS

21.
a) $1, 2, \ldots, n$
b) Discrete.

23.
a) 1.7
b) 0.9

25.
a) 2.25 lights
b) 1.26 lights

27.
a) No, the probability he wins the second changes depending on whether he won the first.
b) 0.42
c) 0.08
d)

X	0	1	2
$P(X = x)$	0.42	0.50	0.08

e) $E(X) = 0.66$ tournaments; $\sigma = 0.62$ tournaments.

29.
a) No, the probability of one battery being dead will depend on the state of the other one since there are only 10 batteries.

b)

Number Good	0	1	2
P(number good)	$\left(\frac{3}{10}\right)\left(\frac{2}{9}\right) = \frac{6}{90}$	$\left(\frac{3}{10}\right)\left(\frac{7}{9}\right) + \left(\frac{7}{10}\right)\left(\frac{3}{9}\right) = \frac{42}{90}$	$\left(\frac{7}{10}\right)\left(\frac{6}{9}\right) = \frac{42}{90}$

c) $\mu = 1.4$ batteries
d) $\sigma = 0.61$ batteries

31. $\mu = E$ (total wait time) $= 74.0$ seconds

$\sigma = SD$ (total wait time) ≈ 20.57 seconds
(Answers to standard deviation may vary slightly due to rounding of the standard deviation of the number of red lights each day.) The standard deviation may be calculated only if the stoplights are independent of each other. This seems reasonable.

33. a) $\mu = 13.6, \sigma = 2.55$
b) Assuming the hours are independent of each other.
c) A typical 8-hour day will have about 11 to 16 repair calls.
d) 19 or more repair calls would be a lot! That's more than two standard deviations above average.

35. a) $B =$ number basic; $D =$ number deluxe; Net Profit$= 120B + 150D - 200$
b) \$928.00
c) \$187.45
d) Mean—no; SD—yes (sales are independent).

37. a) \$50
b) \$100

39. a) Let $X_i =$ price of ith Hulk figure sold; $Y_i =$ price of ith Iron Man figure sold; Insertion Fee $=$ \$0.55; $T =$ Closing Fee $= 0.0875(X_1 + X_2 + \cdots + X_{19} + Y_1 + \cdots + Y_{13})$; Net Income $= (X_1 + X_2 + \cdots + X_{19} + Y_1 + \cdots + Y_{13}) - 32(0.55) - 0.0875(X_1 + X_2 + \cdots + X_{19} + Y_1 + \cdots + Y_{13})$
b) $\mu = E$ (net income) $=$ \$313.24
c) $\sigma = SD$ (net income) $=$ \$6.625
d) Yes, to compute the standard deviation.

41. a) No, these are not Bernoulli trials. The possible outcomes are 1, 2, 3, 4, 5, and 6. There are more than two possible outcomes.
b) Yes, these may be considered Bernoulli trials. There are only two possible outcomes: Type A and not Type A. Assuming the 120 donors are representative of the population, the probability of having Type A blood is 43%. The trials are not independent because the population is finite, but the 120 donors represent less than 10% of all possible donors.
c) No, these are not Bernoulli trials. The probability of choosing a man changes after each promotion and the 10% condition is violated.
d) No, these are not Bernoulli trials. We are sampling without replacement, so the trials are not independent. Samples without replacement may be considered Bernoulli trials if the sample size is less than 10% of the population, but 500 is more than 10% of 3000.
e) Yes, these may be considered Bernoulli trials. There are only two possible outcomes: sealed properly and not sealed properly. The probability that a package is unsealed is constant at about 10%, as long as the packages checked are a representative sample of all.

43. a) 0.0819
b) 0.0064
c) 0.16
d) 0.992

45. $E(X) = 14.29$, so 15 patients

47. a) 0.078 pixels
b) 0.280 pixels
c) 0.374
d) 0.012

49. a) 0.293
b) 0.3597
c) 0.347

51. a) 0.0745
b) 0.502
c) 0.211
d) 0.0166
e) 0.0179
f) 0.9987

53. a) A uniform; all numbers should be equally likely to be selected.
b) 0.5
c) 0.001

55. a) The Poisson model.
b) 0.9502
c) 0.0025

57. a) 0.65
b) 0.75
c) 7.69 picks

59. a) $\mu = 10.44, \sigma = 1.16$
b) 0.812
c) 0.475
d) 0.00193
e) 0.998

CHAPTER 7
SECTION EXERCISE ANSWERS

1. In economics she scored 1.25 standard deviations above the mean. On the math exam she scored 1.50 standard deviations above the mean, so she did "better" on the math exam.

3. You scored 2.2 standard deviations above the mean.

5. a) According to the 68–95–99.7 Rule, only 5% of the distribution is beyond 2 standard deviations from the mean, so only 2.5% is more than 2 standard deviations above the mean. So less than 3% of the distribution is above a z-score of 2.20. You qualify.
   b) You need to assume that the distribution is unimodal and symmetric for the 68–95–99.7 Rule to apply.

7. a)

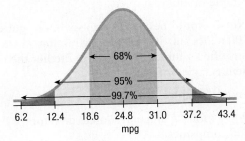

   b) 18.6 to 31.0 mpg
   c) 16%
   d) 13.5%
   e) less than 12.4 mpg

9. a) 6.68%
   b) 98.78%
   c) 71.63%
   d) 61.71%

11. a) 0.842
    b) −0.675
    c) −1.881
    d) −1.645 to 1.645

13. Yes. The histogram is unimodal and symmetric and the Normal probability plot is straight.

15. a) $\mu = E(\text{miles remaining}) = 164$ miles
       $\sigma = SD(\text{miles remaining}) \approx 19.80$ miles
    b) 0.580

17. a) 0.141 (0.175 with continuity correction)
    b) Answers may vary. That's a fairly high proportion, but the decision depends on the relative costs of not selling seats and bumping passengers.

19. a) A uniform; all numbers should be equally likely to be selected.
    b) 0.02
    c) 0.10

CHAPTER EXERCISE ANSWERS

21. a) 16%
    b) 50%
    c) 95%
    d) 0.15%

23. a) 6.2%
    b) 8.0%
    c) 2.6%
    d) $4.4\% < x < 8.0\%$

25. a) 50%
    b) 16%
    c) 2.5%
    d) More than 1.298 is more unusual.

27. a) $x > 1.233$
    b) $x < 1.190$
    c) $1.104 < x < 1.276$
    d) $x < 1.104$

29. a) 36.9% (using technology).
    b) 78.1%
    c) 99.8%
    d) 0.003%

31. a) $x > 8.51\%$
    b) $x < 4.69\%$
    c) $5.26\% < x < 7.14\%$
    d) $x > 4.69\%$

33. a) 5.74%
    b) 10.39%
    c) 2.43%

35. a) 79.58
    b) 18.50
    c) 95.79
    d) −2.79

37. $z_{SAT} = 1.30$; $z_{ACT} = 2$. The ACT score is the better score because it is farther above the mean in standard deviation units than the SAT score.

39. Any Job Satisfaction score more than 2 standard deviations below the mean or less than $100 - 2(12) = 76$ might be considered unusually low. We would expect to find someone with a Job Satisfaction score less than $100 - 3(12) = 64$ very rarely.

41. a) About 16%
    b) One standard deviation below the mean is −1.27 hours, which is impossible.
    c) Because the standard deviation is larger than the mean, the distribution is strongly skewed to the right, not symmetric.

43. a)

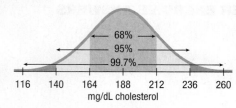

68%
95%
99.7%

116   140   164   188   212   236   260
mg/dL cholesterol

b)   30.85%
c)   17.00%
d)   IQR = Q3 − Q1 = 32.38
e)   Above 212.87 mg/dL

45. a)   To know about their consistency and how long they might last. Standard deviation measures variability, which translates to consistency in everyday use. A type of battery with a small standard deviation would be more likely to have life spans close to their mean life span than a type of battery with a larger standard deviation.
b)   The second company's batteries have a higher mean life span, but a larger standard deviation, so they have more variability. The decision is not clear-cut. The first company's batteries are not likely to fail in less than 21 months, but that wouldn't be surprising for the second company. But the second company's batteries could easily last longer than 39 months—a span very unlikely for the first company.

47. CEOs can have between 0 and maybe 40 (or possibly 50) year's experience. A standard deviation of 1/2 year is impossible because many CEOs would be 10 or 20 SDs away from the mean, whatever it is. An SD of 16 years would mean that 2 SDs on either side of the mean is plus or minus 32, for a range of 64 years. That's too high. So, the SD must be 6 years.

49. a)   1 oz
b)   0.5 oz
c)   0.023
d) $\mu = 4$ oz, $\sigma = 0.5$ oz
e)   0.159
f) $\mu = 12.3$ oz, $\sigma = 0.54$ oz

51. a)   12.2 oz
b)   0.51 oz
c)   0.058

53. a) $\mu = 37.6$ min, $\sigma = 3.7$ min
b)   No, 30 min is more than 2 SDs below the mean.

55. a) $\mu = 1920, \sigma = 48.99; P(T > 2000) = 0.051$
b) $\mu = \$220, \sigma = 11.09$; No—$300 is more than 7 SDs above the mean.
c) $P(D - \frac{1}{2}C > 0) \approx 0.26$

57. 0.077 or 0.094 using continuity correction

59. 0.025 or 0.0296 using continuity correction

61. a)   0.5
b)   0.25

CHAPTER 8
SECTION EXERCISE ANSWERS

1. a)   False. Sampling error cannot be avoided, even with unbiased samples.
b)   True.
c)   True.
d)   False. Randomization will match the characteristics in a way that is unbiased. We can't possibly think of all the characteristics that might be important or match our sample to the population on all of them.

3. a)   Organic farmers in the northeast United States.
b)   NOFA membership listing.
c)   Proportion who think global climate change is affecting crop yield.
d)   Simple random sample.

5. a)   No. It would be nearly impossible to get exactly 500 males and 500 females by random chance.
b)   Stratified sample, stratified by whether the respondent is male or female.

7. Systematic sample.

9. a)   Population—Human resources directors of Fortune 500 companies.
b)   Parameter—Proportion who don't feel surveys intruded on their work day.
c)   Sampling Frame—List of HR directors at Fortune 500 companies.
d)   Sample—23% who responded.
e)   Method—Attempted census (nonrandom).
f)   Bias—Hard to generalize because who responds is related to the question itself (nonresponse bias).

11. a)   Organic farmers in the Northeast.
b)   The members who attended the recent symposium.
c)   The sampling frame is not necessarily representative of the entire group of farmers. Those who attended the symposium may have different opinions from those who didn't. His sample isn't random and may be biased toward those most interested in the topic. Finally, the script is biased and may lead to an estimate of a higher proportion who think government should be doing more to fight global warming than is true within the population.

13. a)   Answers will vary. Question 1 seems appropriate. Question 2 predisposes the participant to agree that $50 is a reasonable price, and does not seem appropriate.
b)   Question 1 is the more neutrally worded. Question 2 is biased in its wording.

15. a)   True.
b)   False. Often parts of the population are not sampled.

c) False. Measurement error refers to inaccurate responses. Sampling error refers to sample-to-sample variability.

d) True.

17. a) This is a multistage design, with a cluster sample at the first stage and a simple random sample for each cluster.

b) If any of the three churches you pick at random is not representative of all churches then you'll introduce sampling error by the choice of that church.

CHAPTER EXERCISE ANSWERS

19. a) Voluntary response.

b) We have no confidence at all in estimates from such studies.

21. a) The population of interest is all adults in the United States.

b) The sampling frame is U.S. adults with telephones. Because they use computers to generate random numbers from all possible numbers this seems representative.

23. a) Population—U.S. teens.

b) Parameter—Proportion who have access to computers and proportion who generally access the Internet by cell phone.

c) Sampling Frame—not specified.

d) Sample—802 teens and parents.

e) Method—Not specified. Probably a random digit dialed survey.

f) Bias—If teens were interviewed in front of their parents, it might have biased their responses.

25. a) Population—Adults.

b) Parameter—Proportion who think drinking and driving is a serious problem.

c) Sampling Frame—Bar patrons.

d) Sample—Every 10th person leaving the bar.

e) Method—Systematic sampling.

f) Bias—Those interviewed had just left a bar. They probably think drinking and driving is less of a problem than do adults in general.

27. a) Population—Soil around a former waste dump.

b) Parameter—Concentrations of toxic chemicals.

c) Sampling Frame—Accessible soil around the dump.

d) Sample—16 soil samples.

e) Method—Not clear.

f) Bias—Don't know if soil samples were randomly chosen. If not, may be biased toward more or less polluted soil.

29. a) Population—Snack food bags.

b) Parameter—Weight of bags, proportion passing inspection.

c) Sampling Frame—All bags produced each day.

d) Sample—10 randomly selected cases, 1 bag from each case for inspection.

e) Method—Multistage sampling.

f) Bias—Should be unbiased.

31. Bias. Only people watching the news will respond, and their preference may differ from that of other voters. The sampling method may systematically produce samples that don't represent the population of interest.

33. a) Voluntary response. Only those who both see the ad *and* feel strongly enough will respond.

b) Cluster sampling. One town may not be typical of all.

c) Attempted census. Will have nonresponse bias.

d) Stratified sampling with follow-up. Should be unbiased.

35. a) This is a systematic sample.

b) It is likely to be representative of those waiting for the roller coaster. Indeed, it may do quite well if those at the front of the line respond differently (after their long wait) than those at the back of the line.

c) The sampling frame is patrons willing to wait for the roller coaster on that day at that time. It should be representative of the people in line, but not of all people at the amusement park.

37. Only those who think it worth the wait are likely to be in line. Those who don't like roller coasters are unlikely to be in the sampling frame, so the poll won't get a fair picture of whether park patrons overall would favor still more roller coasters.

39. a) Biased toward yes because of "pollute." "Should companies be responsible for any costs of environmental cleanup?"

b) Biased toward no because of "enforce" and "strict." "Should companies have dress codes?"

41. a) Not everyone has an equal chance. People with unlisted numbers, people without phones, and those at work cannot be reached. There may be a slight bias against people from large households.

b) Generate random numbers and call at random times.

c) Under the original plan, those families in which one person stays home are more likely to be included. Under the second plan, many more are included. People without phones are still excluded.

d) It improves the chance of selected households being included.

e) This takes care of phone numbers. Time of day may be an issue. People without phones are still excluded.

43. a) Answers will vary.
  b) The amount of change you typically carry. Parameter is the true mean amount of change. Population is the amount on each day around noon.
  c) Population is now the amount of change carried by your friends. The average estimates the mean of these amounts.
  d) Possibly for your class. Probably not for larger groups. Your friends are likely to have similar needs for change during the day.

45. a) Assign numbers 001 to 120 to each order. Use random numbers to select 10 transactions to examine.
  b) Sample proportionately within each type. (Do a stratified random sample.)

47. a) Select three cases at random; then select one jar randomly from each case.
  b) Use random numbers to choose three cases from numbers 61 through 80; then use random numbers between 1 and 12 to select the jar from each case.
  c) No. Multistage sampling.

49. a) Depends on the Yellow Pages listings used. If from regular (line) listings, this is fair if all doctors are listed. If from ads, probably not, as those doctors may not be typical.
  b) Not appropriate. This cluster sample will probably contain listings for only one or two business types.

CHAPTER 9
SECTION EXERCISE ANSWERS

1. a) Normal.
  b) 0.36
  c) They wouldn't change. The shape is still approximately Normal and the mean is still the true proportion.

3. a) 0.0339
  b) 0.5
  c) 0.842 (0.843 using a rounded answer from part a)
  d) 0.01
  e) 0.039

5. Yes. Assuming the survey is random, they should be independent. We don't know the true proportion, so we can't check np and nq, but we have observed 10 successes, and 30 failures which is sufficient.

7. a) 0.0357
  b) 400

9. a) p is the proportion of all backpacks entering the stadium which contain alcoholic beverages; $\hat{p}$ is the proportion in the sample $\hat{p} = 17/130 = 13.08\%$. Yes. This seems to be a random sample.
  b) p is the proportion of all visitors to the website who approve of recent bossnapping. $\hat{p}$ is the proportion in the sample $\hat{p} = 49.2\%$. No. This is a volunteer sample and may be biased.
  c) This question is about the mean weight, not a proportion. The methods of this chapter are not appropriate.

11. a) 0.35
  b) 0.034
  c) (0.283, 0.417)

13. a) False. Doesn't make sense. Workers are not proportions.
  b) True.
  c) False. Our best guess is 0.48 not 0.95.
  d) False. Our best guess is 0.48, but we're not sure that's correct.
  e) False. The statement should be about the true proportion, not future samples.

15. a) Narrower. (0.295, 0.405) (using 1.645 standard errors on each side)
  b) Narrower. (0.296, 0.404)
  c) Wider. (0.263, 0.437) (using 2.576 standard errors on each side).
  d) 4 times as large: 800 students.

17. a) About 2401 (using 1.96 standard errors)
  b) About 4148 (using 2.576 standard errors)
  c) About 385 (using 1.96 standard errors)

19. a) 141
  b) 318
  c) 564

CHAPTER EXERCISE ANSWERS

21. All the histograms are centered near 0.05. As n gets larger, the histograms approach the Normal shape, and the variability in the sample proportions decreases.

23. a)

n	Observed mean	Theoretical mean	Observed SD	Theoretical SD
20	0.0497	0.05	0.0479	0.0487
50	0.0516	0.05	0.0309	0.0308
100	0.0497	0.05	0.0215	0.0218
200	0.0501	0.05	0.0152	0.0154

  b) They are all quite close to what we expect from the theory.

c) The histogram is unimodal and symmetric for $n = 200$.
d) The success/failure condition says that np and nq should both be at least 10, which is not satisfied until $n = 200$ for $p = 0.05$. The theory predicted my choice.

25. a) Symmetric, because probabilities of success and failure are equal.
b) 0.5
c) Standard deviation would be 0.125
d) $np = 8 < 10$

27. a) About 68% should have proportions between 0.4 and 0.6, about 95% between 0.3 and 0.7, and about 99.7% between 0.2 and 0.8.
b) $np = 12.5, nq = 12.5$; both are ≥ 10.
c)

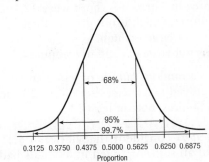

$np = nq = 32$; both are ≥ 10. Samples are random, but stock movements might not be independent.
d) Becomes narrower (less spread around 0.5).

29. a) This is a fairly unusual result: about 2.28 SDs above the mean.
b) The probability of that is about 0.012. So, in a class of 100 it is certainly reasonable that one person would do this well.

31. a)

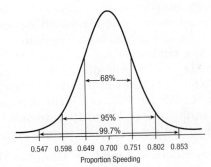

b) Both $np = 56$ and $nq = 24 \geq 10$. Drivers *may* be independent of each other, but if flow of traffic is very fast, they may not be. Or weather conditions may affect all drivers. In these cases they may get more or fewer speeders than they expect.

33. a) Assume that these children are typical of the population. They can be considered a random sample that represents fewer than 10% of all children. We expect 20.4 nearsighted and 149.6 not; both are at least 10.
b)

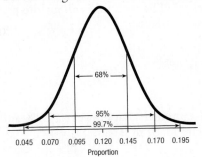

c) Probably between 12 and 29.

35. a) $\mu = 7\%, \sigma = 1.8\%$
b) Assume that clients pay independently of each other, that we have a random sample of all possible clients, and that these represent less than 10% of all possible clients. $np = 14$ and $nq = 186$ are both at least 10.
c) 0.048

37.

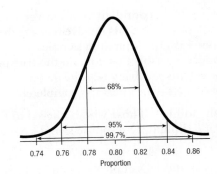

These are not random samples, and not all colleges may be typical (representative). $np = 320, nq = 80$ are large enough.

39. Yes; if their students were typical, a retention rate of $551/603 = 91.4\%$ would be 7 standard deviations above the expected rate of 80%.

41. 0.212. Reasonable that those polled are independent of each other and represent less than 10% of all potential voters. We assume the sample was selected at random. Success/Failure Condition met: $np = 208, nq = 192$. Both ≥ 10.

43. 0.088 using $N(0.08, 0.022)$ model.

45. a) Not correct. This implies certainty.
b) Not correct. Different samples will give different results. Most likely, none of the samples will have *exactly* 88% on-time orders.
c) Not correct. A confidence interval says something about the unknown population proportion, not the sample proportion in different samples.

d) Not correct. In this sample, we *know* that 88% arrived on time.

e) Not correct. The interval is about the parameter, not about the days.

47. a) False.
  b) True.
  c) True.
  d) False.

49. We are 90% confident that between 29.9% and 47.0% of U.S. cars are made in Japan.

51. a) 0.025 or 2.5%.
  b) The pollsters are 90% confident that the true proportion of adults who do not use e-mail is within 2.5% of the estimated 38%.
  c) A 99% confidence interval requires a larger margin of error. In order to increase confidence, the interval must be wider.
  d) 0.039 or 3.9%.
  e) Smaller margins of error will give us less confidence in the interval.

53. a) (12.7%, 18.6%)
  b) We are 95% confident that between 12.7% and 18.6% of all accidents involve teenage drivers.
  c) About 95% of all random samples of size 582 will produce intervals that contain the true proportion of accidents involving teenage drivers.
  d) Contradicts—the interval is completely below 20%.

55. Probably nothing. Those who bothered to fill out the survey are a voluntary response sample, which may be biased.

57. This was a random sample of less than 10% of all Internet users; there were $703 \times 0.18 = 127$ successes and 576 failures, both at least 10. We are 95% confident that between 15.2% and 20.8% of Internet users have downloaded music from a site that was not authorized. (Answer could be 15.2% to 20.9% if $n = 127$ is used instead of 0.18.)

59. a) $385/550 = 0.70$; 70% of U.S. chemical companies in the sample are certified.
  b) This was a random sample, but we don't know if it is less than 10% of all U.S. chemical companies; there were $550(0.70) = 385$ successes and 165 failures, both at least 10. We are 95% confident that between 66.2% and 73.8% of the chemical companies in the United States are certified. It appears that the proportion of companies certified in the United States is less than in Canada.

61. a) There may be response bias based on the wording of the question.
  b) (45.5%, 51.5%)

c) The margin of error based on the pooled sample is smaller, since the sample size is larger.

63. a) This was a random sample of less than 10% of all English children; there were $2700(0.20) = 540$ successes and 2160 failures, both at least 10 (18.2%, 21.8%).
  b) We are 98% confident that between 18.2% and 21.8% of English children are deficient in vitamin D.
  c) About 98% of random samples of size 2700 will produce confidence intervals that contain the true proportion of English children that are deficient in vitamin D.
  d) No. The interval says nothing about causation.

65. a) This was a random sample of less than 10% of all companies in Vermont; there were 12 successes and 0 failures, which is not greater than 10, so the sample is not large enough.
  b) Sample is not large enough to compute CI.

67. a) This was a random sample of less than 10% of all self-employed taxpayers; there were 20 successes and 206 failures, both at least 10.
  b) (5.1%, 12.6%)
  c) We are 95% confident that between 5.1% and 12.6% of all self-employed individuals had their tax returns audited in the past year.
  d) If we were to select repeated samples of 226 individuals, we'd expect about 95% of the confidence intervals we created to contain the true proportion of all self-employed individuals who were audited.

69. a) The 95% confidence interval for the true proportion of all 18- to 29-year-olds who believe the United States is ready for a woman president will be about twice as wide as the confidence interval for the true proportion of all U.S. adults, since it is based on a sample about one-fourth as large. (Assuming approximately equal proportions.)
  b) This was a random sample of less than 10% of all U.S. 15- to 29-year-old adults; there were $250 \times 0.62 = 155$ successes and 95 failures, both at least 10. We are 95% confident that between 56.0% and 68.0% of 18- to 29-year-olds believe the United States is ready for a woman president.

71. a) This was a random sample of less than 10% of all Internet users; there were $703(0.64) = 450$ successes and 253 failures, both at least 10. We are 90% confident that between 61.0% and 67.0% of Internet users would still buy a CD.
  b) In order to cut the margin of error in half, they must sample 4 times as many users; $4 \times 703 = 2812$ users.

73. 1801

75. 384 total, using $p = 0.15$

77. Since $z^* \approx 1.634$, which is close to 1.645, the pollsters were probably using 90% confidence.

79. This was a random sample of less than 10% of all customers; there were 67 successes and 433 failures, both at least 10. From the data set, $\hat{p} = 67/500 = 0.134$. We are 95% confident that the true proportion of customers who spend \$1000 per month or more is between 10.4% and 16.4%.

CHAPTER 10
SECTION EXERCISE ANSWERS

1. a) Let p = probability of winning on the slot machine. $H_0: p = 0.01$ vs. $H_A: p \neq 0.01$.
   b) Let p = proportion of patients cured by the new drug. $H_0: p = 0.3$ vs. $H_A: p \neq 0.3$.
   c) Let p = proportion of clients now using the website. $H_0: p = 0.4$ vs. $H_A: p \neq 0.4$.

3. a) False. A high P-value shows that the data are consistent with the null hypothesis, but provides no evidence for rejecting the null hypothesis.
   b) False. It results in rejecting the null hypothesis, but does not prove that it is false.
   c) False. A high P-value shows that the data are consistent with the null hypothesis but does not prove that the null hypothesis is true.
   d) False. Whether a P-value provides enough evidence to reject the null hypothesis depends on the risk of a type I error that one is willing to assume (the α level).

5. a) $SD(\hat{p}) = \sqrt{\dfrac{p_0 q_0}{n}} = \sqrt{\dfrac{(0.35)(0.65)}{300}} = 0.028$
   b) $z = \dfrac{\hat{p} - p_0}{SD(\hat{p})} = \dfrac{0.46 - 0.35}{0.028} = 3.93$
   c) Yes, that's an unusually large z-value.

7. a) $H_0: p = 0.40$ vs. $H_A: p \neq 0.40$. Two-sided.
   b) $H_0: p = 0.42$ vs. $H_A: p > 0.42$. One-sided.
   c) $H_0: p = 0.50$ vs. $H_A: p > 0.50$. One-sided.

9. a) True.
   b) False. The alpha level is set independently and does not depend on the sample size.
   c) False. The P-value would have to be less than 0.01 to reject the null hypothesis.
   d) False. It simply means we do not have enough evidence at that alpha level to reject the null hypothesis.

11. a) $z = \pm 1.96$
   b) $z = 1.645$
   c) $z = 2.33$; n is not relevant for critical values for z.

13. a) (0.196, 0.304)
   b) No, because 0.20 is a plausible value.
   c) The SE is based on $\hat{p}$, so
   $$SE(\hat{p}) = \sqrt{\dfrac{\hat{p}\hat{q}}{n}} = \sqrt{\dfrac{(0.25)(0.75)}{250}} = 0.0274.$$
   The SD is based on the hypothesized value 0.20, so
   $$SD(\hat{p}) = \sqrt{\dfrac{p_0 q_0}{n}} = \sqrt{\dfrac{(0.20)(0.80)}{250}} = 0.0253.$$
   d) The SE since it is sample based.

15. a) Type I error. The actual value is not greater than 0.3 but they rejected the null hypothesis.
   b) No error. The actual value is 0.50, which was not rejected.
   c) Type II error. The null hypothesis was not rejected, but it was false. The true relief rate was greater than 0.25.

CHAPTER EXERCISE ANSWERS

17. a) Let p be the percentage of products delivered on time. $H_0: p = 0.90$ vs. $H_A: p > 0.90$.
   b) Let p be the proportion of houses taking more than 3 months to sell. $H_0: p = 0.50$ vs. $H_A: p > 0.50$.
   c) Let p be the error rate. $H_0: p = 0.02$ vs. $H_A: p < 0.02$.

19. Statement d is correct.

21. If the rate of seat-belt usage after the campaign is the same as the rate of seat-belt usage before the campaign, there is a 17% chance of observing a rate of seat-belt usage after the campaign this large or larger in a sample of the same size by natural sampling variation alone.

23. Statement e is correct.

25. No, we can say only that there is a 27% chance of seeing the observed effectiveness just from natural sampling if $p = 0.7$. There is no *evidence* that the new formula is more effective, but we can't conclude that they are equally effective.

27. a) 0.186 using the normal model; 0.252 using exact probabilities.
   b) It seems reasonable to think there really may have been half of each. We would expect to get 12 or more reds out of 20 more than 15% of the time, so there's no real evidence that the company's claim is not true. The two sided P-value is greater than 0.30.

29. a) Not a random sample; there may be a nonresponse bias. Not clear that it is less than 10% of population; more than 10 successes and failures; CI is (28.7%, 35.3%). Be cautious in generalizing from this sample.
   b) Since 35% is in the interval, there is not strong evidence that fewer than 35% of all CEOs have

earned their MBA, but I would be cautious in concluding anything from this study.

c) $\alpha = 0.01$; it's an upper-tail test based on a 98% confidence interval.

31. a) Conditions are satisfied: random sample; less than 10% of population; more than 10 successes and failures; (0.4997, 0.5609); we are 95% confident that the true proportion of U.S. investors who believe the price of energy is hurting the U.S. investment climate is between 50.0% and 56.1%.

b) Yes, since 62% is not within the confidence interval, there is evidence that these reports are significantly different from 62%.

c) $\alpha = 0.05$; it's a two-tail test based on a 95% confidence interval.

33. a) Less likely

b) Alpha levels must be chosen *before* examining the data. Otherwise the alpha level could always be selected to reject the null hypothesis.

35. 1. Use p, not $\hat{p}$, in hypotheses.

2. The question is about *failing* to meet the goal. H_A should be $p < 0.96$.

3. Did not check $nq = (200)(0.04) = 8$. Since $nq < 10$, the Success/Failure condition is violated. Didn't check the 10% condition.

4. $\hat{p} = \dfrac{188}{200} = 0.94$; $SD(\hat{p}) = \sqrt{\dfrac{pq}{n}} = \sqrt{\dfrac{(0.96)(0.04)}{200}} \approx 0.014$ The student used $\hat{p}$ and $\hat{q}$.

5. z is incorrect; should be $z = \dfrac{0.94 - 0.96}{0.014} \approx -1.43$.

6. $P = P(z < -1.43) = 0.076$

7. There is only weak evidence that the new system has failed to meet the goal.

37. a) One-tailed. The company isn't worried about a lawsuit if "too many" minorities are hired.

b) Deciding the company is discriminating when it is not.

c) Deciding the company is not discriminating when it is.

d) The probability of correctly detecting actual discrimination.

e) Increases power.

f) Higher, since n is larger.

39. a) Let p = the proportion of children with genetic abnormalities. $H_0\colon p = 0.05$ vs. $H_A\colon p > 0.05$.

b) SRS (not clear from information provided); $384 < 10\%$ of all children; $np = (384)(0.05) = 19.2 > 10$ and $nq = (384)(0.95) = 364.8 > 10$.

c) $z = 6.28$, $P < 0.0001$

d) If 5% of children have genetic abnormalities, the chance of observing 46 children with genetic abnormalities in a random sample of 384 children is essentially 0.

e) Reject H_0. There is strong evidence that more than 5% of children have genetic abnormalities.

f) We don't know that environmental chemicals cause genetic abnormalities, only that the rate is higher now than in the past.

41. a) Let p = the proportion of Americans concerned about global warming in 2012. $H_0\colon p = 0.51$ vs. $H_A\colon p > 0.51$.

b) It is reasonable to assume that Gallup uses random sampling. $1024 < 10\%$ of all Americans; $np = (1024) \times (0.51) => 10$ and $nq = (1024) \times (0.49) > 10$.

c) $z = 2.561$, $P = 0.0052$

d) Reject H_0 at $\alpha = 0.05$. There is evidence to suggest that the proportion of Americans concerned about global warming has increased from 2011 to 2012.

e) This result is statistically significant at $\alpha = 0.05$, but it's not clear that this is a trend that will continue.

43. a) SRS (not clear from information provided); $1000 < 10\%$ of all workers; $n\hat{p} = 520 > 10$ and $n\hat{q} = 480 > 10$; (0.489, 0.551); we are 95% confident that between 48.9% and 55.1% of workers have invested in individual retirement accounts.

b) Let p = the proportion of workers who have invested. $H_0\colon p = 0.44$ vs. $H_A\colon p \neq 0.44$; since 44% is not in the 95% confidence interval, we reject H_0 at $\alpha = 0.05$. There is strong evidence that the proportion of workers who have invested in individual retirement accounts was not 44%. In fact, our sample indicates an increase in the percentage of adults who invest in individual retirement accounts.

45. Let p = the proportion of cars with faulty emissions. $H_0\colon p = 0.20$ vs. $H_A\colon p > 0.20$; two conditions are not satisfied: $22 > 10\%$ of the population of 150 cars and $np = (22)(0.20) = 4.4 < 10$. It's not a good idea to proceed with a hypothesis test.

47. Let p = the proportion of readers interested in an online edition. $H_0\colon p = 0.25$ vs. $H_A\colon p > 0.25$; SRS; $500 < 10\%$ of all potential subscribers; $np = (500)(0.25) = 125 > 10$ and $nq = (500)(0.75) = 375 > 10$; $z = 1.24$, $P = 0.1076$. Since the P-value is high, we fail to reject H_0. There is insufficient evidence to suggest that the proportion of interested readers is greater than 25%. The magazine should not publish the online edition.

49. Let p = the proportion of female executives. $H_0\colon p = 0.40$ vs. $H_A\colon p < 0.40$; data are for all executives in this company and may not be able to be generalized to all companies; $np = (43)(0.40) = 17.2 > 10$ and $nq = (43)(0.60) = 25.8 > 10$; $z = -1.31$, $P = 0.0955$. Since the P-value is high, we fail to reject H_0.

There is insufficient evidence to suggest proportion of female executives is any different from the overall proportion of 40% female employees at the company.

51. Let p = the proportion of first-time home buyers with income $<\$40{,}000$. $H_0: p = 0.15$ vs. $H_A: p > 0.15$; assume that the home buyers are independent of each other (i.e., not related); $100 < 10\%$ of all home buyers; $np = (100)(0.15) = 15 > 10$ and $nq = (100)(0.85) = 85 > 10$; $z = 2.80$, $P = 0.0026$. Since the P-value is low, we reject H_0. There is strong evidence that the proportion of home buyers with income $<\$40{,}000$ is greater than 15%. The 90% confidence interval is $(0.179, 0.321)$ which does not contain 0.15.

53. H_0: These MBA students are exposed to unethical practices at a similar rate to others in the program $(p = 0.30)$.
H_A: These students are exposed to unethical practices at a different rate than other students $(p \neq 0.30)$.
There is no reason to believe that students' rates would influence others; the professor considers this class typical of other classes; $120 < 10\%$ of all students in the MBA program; 27% of $120 = 32.4$—use 32 graduates; $np = 36 > 10$ and $nq = 84 > 10$; $z = -0.717$, $P = 0.4732$. Since the P-value is > 0.05, we fail to reject the null hypothesis. There is little evidence that the rate at which these students are exposed to unethical business practices is different from that reported in the study.

55. a) $z = -0.729$
b) -0.729 is between $(-3.29, 3.29)$ if we assume a two-sided 0.1% significance level.
c) We conclude that the percent of U.S. investors concerned about a politically divided government has not changed significantly between the two surveys.

57. a) The regulators decide that the shop is not meeting standards when it actually is.
b) The regulators certify the shop when it is not meeting the standards.
c) Type I.
d) Type II.

59. a) The probability of detecting that the shop is not meeting standards when they are not.
b) 40 cars produces higher power because n is larger
c) 10%; more chance to reject H_0.
d) A lot; larger problems are easier to detect.

61. a) One-tailed; we are testing to see if a decrease in the dropout rate is associated with the software.
b) H_0: The dropout rate does not change following the use of the software $(p = 0.13)$.
H_A: The dropout rate decreases following the use of the software $(p < 0.13)$.

c) The professor buys the software but the dropout rate has not actually decreased.
d) The professor doesn't buy the software and the dropout rate has actually decreased.
e) The probability of buying the software when the dropout rate has actually decreased.

63. a) H_0: The dropout rate does not change following the use of the software $(p = 0.13)$.
H_A: The dropout rate decreases following the use of the software $(p < 0.13)$.
One student's decision about dropping out should not influence another's decision; this year's class of 203 students is probably representative of all statistics students; $203 < 10\%$ of all students; $np = (203)(0.13) = 26.39 > 10$ and $nq = (203)(0.87) = 176.61 > 10$; $z = -3.21$, $P = 0.0007$. Since the P-value is very low, we reject H_0. There is strong evidence that the dropout rate has dropped since use of the software program was implemented. As long as the professor feels confident that this class of statistics students is representative of all potential students, then he should buy the program.
b) The chance of observing 11 or fewer dropouts in a class of 203 is only 0.07% if the dropout rate is really 13%.

CHAPTER 11
SECTION EXERCISE ANSWERS

1. a) Prices cannot be less than 0, but there is nothing to prevent some from being expensive, so they are likely to be skewed to the high end.
b) It should resemble the population distribution and be skewed to the right.
c) Nearly Normal. The Central Limit Theorem tells us this.

3. a) Normal.
b) 215 mg/dL
c) 4.63 mg/dL
d) Only c would change, to 3.0 mg/dL.

5. a) One would expect many small fish and a few large ones.
b) We don't know the exact distribution, but we know it is skewed, so it can't be Normal.
c) Probably not. With a skewed distribution, a sample of size 5 is not large enough to use the Central Limit Theorem.
d) The standard deviation is 0.30.
e) 4.5 pounds is more than 3 standard deviations above the mean, so by the 68–95–99.7 rule, it would happen less than 0.15% of the time.

7. a) 1.968 years
  b) 0.984 years (half as large)

9. a) 24
  b) 99

11. a) 2.064
  b) 1.984

13. a) (27.78, 35.90) years
  b) 4.06 years
  c) (27.92, 35.76) years. Slightly narrower.

15. Independence: The data were from a random survey and should be independent.
Randomization: the data were selected randomly.
10% Condition: These customers are fewer than 10% of the customer population.
Nearly Normal: The histogram is unimodal and symmetric, which is sufficient.

17. a) $H_0: \mu = 25$
  b) Two-sided; $H_A: \mu \neq 25$
  c) $t_{24} = 3.476$
  d) P-value < 0.002 or P-value $= 0.00195$
  e) Reject the null. There is strong evidence the mean age is not 25.

19. a) Four times as big, or $n = 100$.
  b) 100 times as big, or $n = 2500$.

CHAPTER EXERCISE ANSWERS

21. a) 1.74
  b) 2.37

23. As the variability of a sample increases, the width of a 95% confidence interval increases, assuming that the sample size remains the same.

25. a) ($4.382, $4.598)
  b) ($4.400, $4.580)
  c) ($4.415, $4.565)

27. a) Not correct. A confidence interval is for the mean weight gain of the population of all cows. It says nothing about individual cows.
  b) Not correct. A confidence interval is for the mean weight gain of the population of all cows, not individual cows.
  c) Not correct. We don't need a confidence interval about the average weight gain for cows in this study. We are certain that the mean weight gain of the cows in this study is 56 pounds.
  d) Not correct. This statement implies that the average weight gain varies. It doesn't.
  e) Not correct. This statement implies that there is something special about our interval, when this interval is actually one of many that could have been

generated, depending on the cows that were chosen for the sample.

29. The assumptions and conditions for a t-interval are not met. With a sample size of only 20, the distribution is too skewed. There is also a large outlier that is pulling the mean higher.

31. a) The data are a random sample of all weekdays; the distribution is unimodal and symmetric with no outliers.
  b) ($122.20, $129.80)
  c) We are 90% confident that the interval $122.20 to $129.80 contains the true mean daily income of the parking garage.
  d) 90% of all random samples of size 44 will produce intervals that contain the true mean daily income of the parking garage.
  e) $128 is a plausible value.

33. a) We can be more confident that our interval contains the mean parking revenue.
  b) Wider (and less precise) interval.
  c) By collecting a larger sample, they could create a more precise interval without sacrificing confidence.

35. a) (2230.4, 2469.6)
  b) The assumptions and conditions that must be satisfied are:
    1 Independence: probably OK.
    2 Nearly Normal condition: can't tell.
    3 Sample size of 51 is large enough.
  c) We are 95% confident the interval $2230.4 to $2469.6 contains the true mean increase in sales tax revenue.
    Examples of what the interval *does not* mean: The mean increase in sales tax revenue is $2350 95% of the time. 95% of all increases in sales tax revenue increases will be between $2230.4 and $2469.6. There's 95% confidence the next small retailer will have an increase in sales tax revenue between $2230.4 and $2469.6.

37. a) Given no time trend, the monthly on-time departure rates should be independent. However, if there is a time trend, the inference may not be correct. Though not a random sample, these months should be representative. The histogram looks unimodal, but slightly left-skewed; not a concern with a sample this large.
  b) (80.16%, 81.14%)
  c) We can be 90% confident that the interval from 80.16 to 81.14 holds the true mean monthly percentage of on-time flight departures.
    However, this does not imply that any particular month will have its on-time departure percentage in this interval.

39. a) The histogram of the lab fees shows 2 extreme outliers, so with the outliers included, the conditions for inference are violated.

b) With the outliers left in, the 95% confidence interval is (44.9, 81.6) minutes. If we remove the two extreme outliers, it is (54.6, 68.4).

In either case, we would be reluctant to conclude that the mean is above 55 minutes. The sample size is small and the presence of two large outliers advises us to be cautious about conclusions from this sample.

41. a) The assumptions and conditions that must be satisfied are:
The data come from a nearly normal distribution. The air samples were selected randomly, and there is no bias present in the sample.

b) The histogram of air samples is not nearly normal, but the sample size is large, so inference is OK.

43. a) $14.90 - 11.6$ or ± 3.3 miles per hour

b) We can find the needed sample size from: $1.96 \times 8/2 = \sqrt{n}$. $(7.84)^2 = n = 61.46$ use 62.

45. a) Interval: $653 to $707

b) The confidence interval suggests the mean audit cost is now greater than $650.

47. a) The timeplot shows no pattern, so it seems that the measurements are independent. Although this is not a random sample, an entire year is measured, so it is likely that we have representative values. We certainly have fewer than 10% of all possible wind readings. The histogram appears nearly normal.

b) A 95% confidence interval for the true mean speed is (7.795, 8.243) mph. Because there are many plausible values below 8 mph, we cannot be confident that the true mean is at least 8 mph. We would not recommend that the turbine be placed here.

49. The 1% P-value means that, if the mean monthly sales due to online purchases has not changed, there is a 1% chance (or 1 out of every 100 samples) that the resulting mean sales would occur assuming the historical mean for sales. This is rare and considered to be a significant value.

51. a) $H_0: \mu = 23.3$ years; $H_A: \mu > 23.3$ years

b) The randomization condition: The 40 online shoppers were selected randomly. Nearly normal condition: the distribution of the sample should be examined to check for serious skewness or outliers but the sample of 40 shoppers is large enough that it should be safe to proceed.

c) P-value = 0.145

d) If the mean age of shoppers remains at 23.3 years, there is a 14.5% chance of getting a sample mean of 24.2 years or older simply from natural sampling variation.

e) There is not enough evidence to suggest that the mean age of online shoppers has increased from the mean of 23.3 years.

53. a) Random sample. Nearly Normal Condition is reasonable by examining a Normal probability plot. The histogram is roughly unimodal (although somewhat uniform) and symmetric with no outliers. (Show your plot.)

b) (114.9, 148.8) calories

c) The mean number of calories in a serving of vanilla yogurt is between 115 and 149, with 95% confidence. We conclude that the diet guide's claim of 120 calories is toward the lower end, but is reasonable.

CHAPTER 12
SECTION EXERCISE ANSWERS

1. a) $\bar{y}_1 = 58.0$ years; $\bar{y}_2 = 47.0$ years
b) 11 years
c) $s_1^2 = 50$; $s_2^2 = 54$
d) $s_1 = 7.07$ years; $s_2 = 7.35$ years
e) 4.16 years

3. a) 9.6 years
b) 1.91 years
c) 5.03

5. a) 2.64
b) 9.985
c) 5
d) 0.0247
e) 0.0459
f) For either method of calculating df, there is reasonably strong evidence to suggest that we reject the null hypothesis and conclude that the mean age of houses is not the same in the two neighborhoods.

7. a) <0.001
b) <0.001
c) In both cases, reject H_0. There is very strong evidence to suggest that there is a difference in the mean ages of houses in the two neighborhoods.

9. a) (1.72, 20.28) years
b) No.
c) It suggests that we should reject H_0 since 0 is not a plausible value for the true mean difference.

11. a) (5.79, 13.41) years
b) The sample sizes are larger.
c) No.
d) It suggests that we should reject H_0 since 0 is not a plausible value for the true mean difference.

13. a) $t = 2.642$; P-value $= 0.0246$. Reject H_0 and conclude that there is sufficient evidence to suggest that the true means are different.

b) (1.72, 20.28) years

c) Not really. Because the standard deviations are fairly close, the two methods will result in essentially the same confidence intervals and hypothesis tests.

15. a) $t = 5.01$; P-value < 0.001. Reject H_0. There is strong evidence to reject the hypothesis that the means are equal.

b) (5.77, 13.43) years

c) Very close to previous results because the standard deviations are fairly close. In such a case, the two methods will result in essentially the same confidence intervals and hypothesis tests.

17. a) Paired. Each pair consists of a volunteer using each chair.

b) Not paired. The samples are random within each neighborhood.

c) Paired. Each pair consists of an hour in which the productivity of the two workers is compared.

19. a) Yes, each pair is a store in which the customers with and without the program are compared.

b) 3 customers

c) 4.52 customers

d) 1.43 customers

e) 2.098

f) 9

g) One-sided. They want to know if traffic increased.

h) 0.0327

i) We can reject the null hypothesis. There is evidence of an increase in traffic.

21. (0.379, 5.621) customers

CHAPTER EXERCISE ANSWERS

23. The P-value is too high to reject H_0 at any reasonable α-level.

25. a) 2.927 points

b) Larger.

c) We are 95% confident that the mean score for the CPMP math students will be between 5.573 and 11.427 points higher on this assessment than the mean score of the traditional students.

d) Since the entire interval is above 0, there is strong evidence that students who learn with CPMP will have higher mean scores in applied algebra than those in traditional programs.

27. a) $H_0: \mu_C - \mu_T = 0$; $H_A: \mu_C - \mu_T \neq 0$

b) If the mean scores for the CPMP and traditional students are really equal, there is less than a 1 in

10,000 chance of seeing a difference as large or larger than the observed difference of 9.4 points just from natural sampling variation.

c) There is strong evidence that the CPMP students have a different mean score than the traditional students. The evidence suggests that the CPMP students have a lower mean score.

29. $H_0: \mu_C - \mu_T = 0$; $H_A: \mu_C - \mu_T \neq 0$
$P = 0.1602$; fail to reject H_0. There is no evidence that the CPMP students have a different mean score on the word problems test than the traditional students.

31. a) (1.36, 4.64); df $= 33.1$

b) Since the CI does not contain 0, there is evidence that Route A is faster on average.

33. a) $H_0: \mu_C - \mu_A = 0$; $H_A: \mu_C - \mu_A \neq 0$

b) Independent groups assumption: The percentage of sugar in the children's cereals is unrelated to the percentage of sugar in adult cereals. Randomization condition: It is reasonable to assume that the cereals are representative of all children's cereals and adult cereals, in regard to sugar content. Nearly Normal condition: The histogram of adult cereal sugar content is skewed to the right, but the sample sizes are reasonably large. The Central Limit Theorem allows us to proceed.

c) (32.49, 40.80)%

d) Since the 95% confidence interval does not contain 0, we can conclude that the mean sugar content for the two cereals is significantly different at the 5% level of significance.

35. a) $H_0: \mu_{\text{Taxable}} - \mu_{\text{Municipal}} = 0$
$H_A: \mu_{\text{Taxable}} - \mu_{\text{Municipal}} \neq 0$

b) Distributions are symmetric. Funds are independent. Outliers in boxplot not severe enough to cause concern.

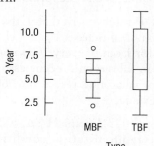

c) Difference Between Means $= 1.055$
t-Statistic $= 1.020$ w/16.54 df
$P = 0.3225$

d) There does not appear to be a significant difference between these types of funds.

37. $H_0: \mu_N - \mu_C = 0$; $H_A: \mu_N - \mu_C > 0$. Independent groups assumption: Student scores in one group should not have an impact on the scores of students in the other group. Randomization condition: Students were randomly assigned to classes. Nearly Normal condition: The histograms of the scores are unimodal and symmetric. P = 0.017; reject H_0. There is evidence that the students taught using the new activities have a higher mean score on the reading comprehension test than the students taught using traditional methods.

39. a) The hypotheses are: $H_0: \mu_{1995–2004} - \mu_{2005–2014} = 0$; $H_A: \mu_{1995–2004} - \mu_{2005–2014} \neq 0$.
   b) Both time periods have values that show as possible outliers in boxplots, but neither one seems particularly extreme. The two time periods are plausibly independent.
   c) Performing a two-sample t-test, $t = -0.88$ with 13 df, $P = 0.397$. Fail to reject H_0.

41. a) If the mean memory scores for people taking ginkgo biloba and people not taking it are the same, there is a 93.74% chance of seeing a difference in mean memory score this large or larger simply from natural sampling variability.
   b) Since the P-value is so high, there is no evidence that the mean memory test score for ginkgo biloba users is higher than the mean memory test score for non-users.
   c) Type II.

43. a) Males: (18.67, 20.11) pegs; females: (16.95, 18.87) pegs
   b) It may appear to suggest that there is no difference in the mean number of pegs placed by males and females, but a two-sample t-interval should be constructed to assess the difference in mean number of pegs placed.
   c) (0.29, 2.67) pegs
   d) We are 95% confident that the mean number of pegs placed by males is between 0.29 and 2.67 pegs higher than the mean number of pegs placed by females.
   e) Two-sample t-interval.
   f) If you attempt to use two confidence intervals to assess a difference in means, you are actually adding standard deviations. But it's the variances that add, not the standard deviations. The two-sample difference of means procedure takes this into account.

45. a) $H_0: \mu_N - \mu_S = 0; H_A: \mu_N - \mu_S \neq 0$
   $t = 6.47$, df = 53.49, P < 0.001
   Since the P-value is low, we reject H_0. There is strong evidence that the mean mortality rate is different for towns north and south of Derby. There is evidence that the mortality rate north of Derby is higher.
   b) The possible outlier slightly raised the mean and inflated the variance for North Derby. The t-statistic is large and the outlier not particularly

extreme, so omitting that point would likely make little difference in our conclusion.

47. a) Paired data assumption: The data are before and after job satisfaction rating for the same workers. Randomization condition: The workers were randomly selected to participate. Nearly Normal condition: A histogram of differences between before and after job satisfaction ratings is roughly unimodal and symmetric.
   b) $H_0: \mu_d = 0$; $H_A: \mu_d > 0$; $t = 3.60$;   df = 9; P-value = 0.0029; reject H_0. There is evidence that the mean job satisfaction rating has increased since the implementation of the exercise program.

49. Independent groups assumption: Assume that orders in June are independent of orders in August. Independence assumption: Orders were a random sample. Nearly normal condition: Hard to check with small sample, but no outliers. $H_0: \mu_J - \mu_A = 0; H_A: \mu_J - \mu_A > 0$; $t = -1.17; P = 0.136$; fail to reject H_0. Thus, although the mean delivery time during August is higher, the difference in delivery time from June is not significant. A larger sample may produce a different statistical result, but the size of the difference may not be important from a business standpoint.

51. a) We are 95% confident that the mean number of brand names remembered by viewers of shows with violent content will be between 1.6 and 0.6 lower than the mean number of brand names remembered by viewers of shows with neutral content.
   b) If they want viewers to remember their brand names, they should consider advertising on shows with neutral content, as opposed to shows with violent content.

53. a) She might attempt to conclude that the mean number of brand names recalled is greater after 24 hours.
   b) The groups are not independent. They are the same people, asked at two different time periods.
   c) A person with high recall right after the show might tend to have high recall 24 hours later as well. Also, the first interview may have helped the people to remember the brand names for a longer period of time than they would have otherwise.
   d) Randomly assign half of the group watching that type of content to be interviewed immediately after watching, and assign the other half to be interviewed 24 hours later.

55. a) The differences that were observed between the group of students with Internet access and those without were too great to be attributed to natural sampling variation.
   b) Type I.

c) No. There may be many other factors.
d) It might be used to market computer services to parents.

57. a) 8759 pounds
    b) Independent groups assumption: Sales in different seasons should be independent. Randomization condition: Not a random sample of weeks, but it is of stores. Nearly Normal condition: Can't verify, but we will proceed cautiously. We are 95% confident that the interval 3630.54 to 13,887.39 pounds contains the true difference in mean sales between winter and summer.
    c) Weather and sporting events may impact pizza sales.

59. Because runners are assigned to heats at random, we expect there to be no statistically significant difference between the heats.

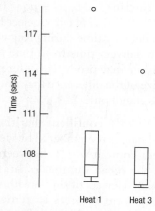

Boxplots show only a small difference in medians. But both heats had a high outlier time, making a formal test problematic.

61. $H_0: \mu_S - \mu_R = 0$; $H_A: \mu_S - \mu_R > 0$. Assuming the conditions are satisfied, it is appropriate to model the sampling distribution of the difference in means with a Student's t-model, with 7.03 degrees of freedom (from the approximation formula). $t = 4.57$; $P = 0.0013$; reject H_0. There is strong evidence that the mean ball velocity for Stinger tees is higher than the mean velocity for regular tees.

63. a) $H_0: \mu_M - \mu_R = 0$; $H_A: \mu_M - \mu_R > 0$. Independent groups assumption: The groups are not related in regards to memory score. Randomization condition: Subjects were randomly assigned to groups. Nearly Normal condition: We don't have the actual data. We will assume that the distributions of the populations of memory test scores are Normal. $t = -0.70$; df $= 45.88$; $P = 0.7563$; fail to reject H_0. There is no evidence that the mean number of objects remembered by those who listen to Mozart is higher than the mean number of objects remembered by those who listen to rap music.

b) We are 90% confident that the mean number of objects remembered by those who listen to Mozart is between 0.19 and 5.35 objects lower than the mean of those who listened to no music.

65. a) Paired data assumption: These data are paired as they are the 3- and 5-year returns of the same mutual funds. Randomization condition: random sample of funds. Nearly Normal condition: Histogram of differences is slightly skewed right, but we will proceed with caution.
    b) $H_0: \mu_{5yr} - \mu_{3yr} = 0$; $H_A: \mu_{5yr} - \mu_{3yr} > 0$
    c) In fact the mean paired difference is -9.58, which is the opposite direction from the alternative hypothesis, so the P-value will be higher than 0.50. (The actual one-sided P-value is nearly 1.0.)
    d) A confidence interval between -10.01% and -9.21% has a 95% chance of capturing the true mean difference. No evidence that mean 5 year returns are higher than mean 3 year returns.

67. a) Independent groups assumption: The prices in the two towns are not related. Randomization condition: Each sample was a random sample of prices. Nearly Normal condition: Both histograms are reasonably unimodal and symmetric with no outliers, so will use the two-sample t-test.
    b) $H_0: \mu_1 = \mu_2$; $H_A: \mu_1 \neq \mu_2$
    c) $t = -0.58$; $P = 0.567$; fail to reject H_0.
    d) We conclude that the mean price of homes in these two towns is not significantly different.

69. a) Independent groups assumption: The home runs hit in different leagues are independent. Randomization condition: Not a random sample, but we will assume it's representative. Nearly Normal condition: Both histograms are reasonably unimodal and symmetric with no outliers.
    b) $H_0: \mu_{AL} - \mu_{NL} = 0$; $H_A: \mu_{AL} - \mu_{NL} > 0$
    c) $t = 0.629$; $P = 0.268$
    d) We cannot conclude that the mean number of home runs is higher, on average, in the American League. We fail to reject the null hypothesis.
    e) $t = 0.641$; df $= 28$; $P = 0.263$. Same conclusion as in part d.

71. Adding variances requires that the variables be independent. These price quotes are for the same cars, so they are paired. Drivers quoted high insurance premiums by the local company will be likely to get a high rate from the online company, too.

73. a) The histogram—we care about differences in price.
    b) Insurance cost is based on risk, so drivers are likely to see similar quotes from each company, making the differences relatively smaller.
    c) The price quotes are paired; they were for a random sample of fewer than 10% of the agent's

customers; the histogram of differences looks approximately Normal.

75. H_0: $\mu_{Local-Online} = 0$; H_A: $\mu_{Local-Online} > 0$; $t = 0.826$ with 9 df. With a P-value of 0.215, we cannot reject the null hypothesis. These data don't provide evidence that online premiums are lower, on average.

77. a) H_0: $\mu_d = 0$; H_A: $\mu_d \neq 0$
b) $t_{144} = 2.406$; 2-sided $P = 0.017$
We are able to reject the null hypothesis (with a P-value of 0.017) and conclude that mean number of keystrokes per hour has changed.
c) 95% CI for mean keystrokes per hour $22.7 \pm t_{0.025, 144} s / \sqrt{n} = (4.05, 41.35)$

79. a) Paired data assumption: The data are paired by type of exercise machine. Randomization condition: Assume that the men and women participating are representative of all men and women in terms of number of minutes of exercise required to burn 200 calories. Nearly Normal condition: The histogram of differences between women's and men's times is roughly unimodal and symmetric. We are 95% confident that women take an average of 4.8 to 15.2 minutes longer to burn 200 calories than men when exercising at a light exertion rate.
b) Nearly Normal condition: There is no reason to think that this histogram does not represent differences drawn from a Normal population. We are 95% confident that women exercising with light exertion take an average of 4.9 to 20.4 minutes longer to burn 200 calories than women exercising with hard exertion.
c) Since these data are averages, we expect the individual times to be more variable. Our standard error would be larger, resulting in a larger margin of error.

81. a) Randomization condition: These stops are probably representative of all such stops for this type of car, but not for all cars. Nearly Normal condition: A histogram of the stopping distances is roughly unimodal and symmetric. We are 95% confident that the mean dry pavement stopping distance for this type of car is between 133.6 and 145.2 feet.
b) Independent groups assumption: The wet pavement stops and dry pavement stops were made under different conditions and not paired in any way. Randomization condition: These stops are probably representative of all such stops for this type of car, but not for all cars. Nearly Normal condition: The histogram of wet pavement stopping distances is more uniform than unimodal, but no outliers. We are 95% confident that the mean stopping distance on wet pavement is between 50.4 and 75.6 feet longer than the mean stopping distance on dry pavement.

83. a) The data are paired. They are for the same airlines in two different years and our concern is about whether the involuntary DB rate has changed. We would measure improvement for each airline.
b) A paired *t*-test finds a mean of paired differences (2013–2014) of 0.25. The corresponding *t* statistic is 1.978 with 13df. That gives a P-value of 0.0695, which is not small enough to conclude that there is a real difference.

CHAPTER 13

SECTION EXERCISE ANSWERS

1. a) (30, 30, 30, 30), 30 for each season
b) 1.933
c) 3

3. a) 3
b) No, it's smaller than the mean.
c) It would say that there is no evidence to reject the null hypothesis that births are distributed uniformly across the seasons.
d) 7.815
e) Do not reject the null hypothesis. As in part c, there is no evidence to suggest that births are not distributed uniformly across the seasons.

5. a) $(-0.913, 0.913, 0.365, -0.365)$
b) No, they are quite small for z-values.
c) Because we did not reject the null hypothesis, we shouldn't expect any of the standardized residuals to be large.

7. a) The age distributions of customers at the two branches are the same.
b) Chi-square test of homogeneity.
c)

	Age			
	Less Than 30	**30–55**	**56 or Older**	**Total**
In-Town Branch	25	45	30	**100**
Mall Branch	25	45	30	**100**
Total	**50**	**90**	**60**	**200**

d) 9.778
e) 2
f) 5.991
g) Reject H_0 and conclude that the age distributions at the two branches are not the same.

9. a) Men 62%, women 71%
b) Difference = 9%
c) SE = 0.022
d) (0.046, 0.133) or 4.6% to 13.3%

11. a)

	May-2010	Aug-2011	Aug-2012	Total
18–29	261.663	277.540	303.798	843
30–49	333.674	353.921	387.405	1075
50–64	261.663	277.540	303.798	843
Total	**857**	**909**	**995**	**2761**

b) 1.861

c) 4

d) We cannot reject the null hypothesis, the growth in use of social networking over these three years appears to be independent of age groups.

CHAPTER EXERCISE ANSWERS

13. a) Chi-square test of independence; one sample, two variables. We want to see if the variable *Account type* is independent of the variable *Trade type*.

b) Some other statistical test; the variable *Account size* is quantitative, not counts.

c) Chi-square test of homogeneity; we have two samples and one variable, *Courses*. We want to see if the distribution of *Courses* is the same for the two groups.

15. a) 10

b) Goodness-of-fit.

c) H_0: The die is fair. (All faces have $p = 1/6$.)
H_A: The die is not fair. (Some faces are more or less likely to come up than others.)

d) Count data; rolls are random and independent of each other; expected frequencies are all greater than 5.

e) 5

f) $\chi^2 = 5.600$; $P = 0.3471$

g) Since $P = 0.3471$ is high, fail to reject H_0. There is not enough evidence to conclude that the die is unfair.

17. a) Weights are quantitative, not counts.

b) Count the number of each type of nut, assuming the company's percentages are based on counts rather than weights (which is not clear).

19. a) Goodness-of-fit.

b) Count data; assume the lottery mechanism uses randomization and guarantees independence; expected frequencies are all greater than 5.

c) H_0: Likelihood of drawing each numeral is equal.
H_A: Likelihood of drawing each numeral is *not* equal.

d) $\chi^2 = 6.46$; df $= 9$; $P = 0.693$; fail to reject H_0.

e) The P-value says that if the drawings were in fact fair, an observed chi-square value of 6.46 or higher would occur about 69% of the time. This is not unusual at all, so we won't reject the null hypothesis that the values are uniformly distributed. The variation that we observed seems typical of that expected if the digits were drawn equally likely.

21. a) 40.3%

b) 8.15%

c) 62.0%

d) 286.67

e) Chi-square test of independence. H_0: Survival was independent of status on the ship. H_A: Survival was not independent of the status.

f) 3

g) We reject the null hypothesis. Survival depended on status. We can see, for example, that first-class passengers were more likely to survive than passengers of any other class.

23. a) Independence.

b) H_0: College choice is independent of birth order. H_A: There is an association between college choice and birth order.

c) Count data; not a random sample of students, but assume that it is representative; expected counts are low for both the Social Science and Professional Colleges for both third and fourth or higher birth order. We'll keep an eye on these when we calculate the standardized residuals.

d) 9

e) With a P-value this low, we reject the null hypothesis. There is some evidence of an association between birth order and college enrollment.

f) None are particularly large, but 4 cells have expected counts less than 5, so we should be cautious in making conclusions. Perhaps we should group 3 with 4 or more and redo the analysis. Unfortunately, 3 of the 4 largest standardized residuals are in cells with expected counts less than 5. We should be very wary of drawing conclusions from this test.

25. a) Chi-square test for homogeneity.

b) Count data; assume random assignment to treatments (although not stated); expected counts are all greater than 5.

c) H_0: The proportion of infection is the same for each group.
H_A: The proportion of infection is different among the groups.

d) $\chi^2 = 7.776$; df $= 2$; $P = 0.02$; reject H_0

e) Since the P-value is low, we reject the null hypothesis. There is strong evidence of difference in the proportion of urinary tract infections for cranberry juice drinkers, lactobacillus drinkers, and women that drink neither of the two beverages.

f) The standardized residuals are:

	Cranberry	Lactobacillus	Control
Infection	−1.87276	1.191759	0.681005
No infection	1.245505	−0.79259	−0.45291

The significant difference appears to be primarily due to the success of cranberry juice.

27. a) Homogeneity.
   b) H_0: The numbers of families spending each amount on eating out did not change between 2008 and 2010.
   c) Yes.
   d) We can reject the null hypothesis at the 0.05 level. There appears to be a difference in the distribution of expenses for eating out.
   e) In 2010, more families opted for spending less than $50/week eating out and fewer chose to spend more than that. There was a noticeable drop in large (more than $150/week) expenditures.

29. a) $P = 0.3766$. With a P-value this high, we fail to reject. There is not enough evidence to conclude that either men or women are more likely to make online purchases of books.
   b) Type II.
   c) $(−4.09\%, 10.86\%)$

31. a) $P < 0.001$. There is strong evidence that the proportions of the two groups are not equal.
   b) $(0.096, 0.210)$ (proportion of old − proportion of young)

33. a) The standard error of the difference in proportions is 0.0094, but the difference in proportions is 0.009, which is about one standard error. That is not a statistically significant difference.
   b) A 90% CI is 0.009 ± 0.0155.

35. This is a test of homogeneity. The null hypothesis in either case is that there was no change over the three years. Chi-square = 24.33 with 8 df P = 0.0020, so the null hypothesis is rejected.

37. a) Chi-square test for homogeneity.
   b) Count data; adult respondents were surveyed randomly; expected counts are all greater than 5.
   c) H_0: The distribution of attitudes about the importance of financial success was the same for men and women 18 to 34 years old.
   H_A: The distribution of attitudes was not the same.
   d) $\chi^2 = 17.43$; df = 3; $P = 0.0006$
   e) The P-value is small so we reject the null hypothesis and conclude that men and women in this age group have different attitudes about the importance of this type of success.

39. a) A test of independence would be appropriate.
   b) The data are counts; no cells are too small.
   c) $\chi^2 = 14.43$ with 2 df, P = 0.0007
   d)

	No	Yes
Yes	−1.22856	1.53931
No	2.01107	−2.51975
Retired	0.250431	0.313775

   e) Those who are employed are more likely to own stock than either those who are not employed or those who are retired.

41. a)

	Men	Women
Excellent	6.667	5.333
Good	12.778	10.222
Average	12.222	9.778
Below Average	8.333	6.667

   Count data; assume that these executives are representative of all executives that have ever completed the program; expected counts are all greater than 5.
   b) Decreased from 4 to 3.
   c) $\chi^2 = 9.306$; $P = 0.0255$. Since the P-value is low, we reject the null hypothesis. There is evidence that the distributions of responses about the value of the program for men and women executives are different.

43. H_0: There is no association between race and the section of the apartment complex in which people live.
   H_A: There is an association between race and the section of the apartment complex in which people live.
   Count data; assume that the recently rented apartments are representative of all apartments in the complex; expected counts are all greater than 5.
   $\chi^2 = 14.058$; df = 1; $P < 0.001$.
   Since the P-value is low, we reject the null hypothesis. There is strong evidence of an association between race and the section of the apartment complex in which people live. An examination of the components shows us that blacks are less likely to rent in Section A (component = 6.2215) and that blacks are more likely to rent in Section B (component = 5.0517).

45. $\hat{p}_B − \hat{p}_A = 0.206$
   95% CI = $(0.107, 0.306)$

47. a) Chi-square test for independence.
   b) Count data; assume that the sample was taken randomly; expected counts are all greater than 5.
   c) H_0: Outsourcing is independent of industry sector.

H_A: There is an association between outsourcing and industry sector.

d) $\chi^2 = 2815.968$; df $= 9$; P-value is essentially 0.

e) Since the P-value is so low, we reject the null hypothesis. There is strong evidence of an association between outsourcing and industry sector.

49. a) Chi-square test for homogeneity. (Could be independence if the categories are considered exhaustive.)

b) Count data; assume that the sample was taken randomly; expected counts are all greater than 5.

c) H_0: The distribution of employee job satisfaction level attained is the same for different management styles.

H_A: The distribution of employee job satisfaction level attained is different for different management styles.

d) $\chi^2 = 178.453$; df $= 12$; P-value is essentially 0.

e) Since the P-value is so low, we reject the null hypothesis. There is strong evidence that the distribution of employee job satisfaction level attained is different across management styles. Generally, exploitative authoritarian management is more likely to have lower levels of employee job satisfaction than consultative or participative styles.

51. Assumptions and conditions for test of independence satisfied. H_0: Reading online journals or blogs is independent of generation. H_A: There is an association between reading online journals or blogs and generation. $\chi^2 = 48.408$; df $= 8$; $P < 0.001$. We reject the null hypothesis and conclude that reading online journals or blogs is not independent of generational age.

53. Chi-square test of homogeneity (unless these two types of firms are considered the only two types in which case it's a test of independence).

Count data; assume that the sample was random; expected counts are all greater than 5.

H_0: Systems used have same distribution for both types of industry.

H_A: Distributions of type of system differs in the two industries.

$\chi^2 = 157.256$; df $= 3$; P-value is essentially 0.

Since the P-value is low, we can reject the null hypothesis and conclude that the type of ERP system used differs across industry type. Those in manufacturing appear to use more of the inventory management and ROI systems.

55. Chi-square test of independence.

Count data; we assume that this time period is representative; expected counts are less than 5 in three cells, but very close.

H_0: GDP % change is independent of Region of the United States.

H_A: GDP % change is not independent of Region of the United States.

$\chi^2 = 19.0724$; df $= 3$; $P < 0.001$.

The P-value is low so we reject the null hypothesis and conclude that GDP % change and Region are not independent. The result seems clear enough even though the expected counts of 3 cells were slightly below 5.

CHAPTER 14
SECTION EXERCISE ANSWERS

1. a) 9.8, 8.5, 5.9, 2.0, 8.5, 11.1

b) $-0.8, -0.5, -1.9, 1.0, 1.5, 0.9$

c) 1.46

3. $s_x = 12.5167, s_e = 1.46, n = 6, SE(b_1) = 0.052$

5. a) 5.837

b) 3

c) 0.01

d) Yes, the P-value is smaller than 0.05.

7. a) Linearity: the plot appears linear.

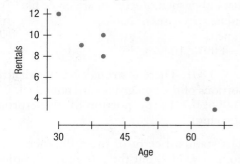

b) Independence: It is likely that the customers are independent. It is likely that errors made by the regression will be independent.

c) Equal spread: the plot appears to have a consistent spread.

d) Normality: We can't tell from what we know.

9. a) 19.06 ($1000)

b) $19.06 \pm 1.12 = (17.94, 20.19)$ in $1000's

c) $19.06 \pm 3.59 = (15.47, 22.65)$ in $1000's

11. a) 95% of potential customers in the community from households that have annual incomes of $80,000 can be expected to spend between $35.60 and $55.60 eating out each week. Pricing an individual meal in this range might attract that demographic.

b) We can be 95% confident that the mean amount spent weekly by individuals whose household incomes are $80,000 is between $40.60 and $50.60. The mean is a summary for this demographic.

c) We can estimate means more accurately than we can estimate values for individuals. That's why the confidence interval for the mean is narrower than the corresponding interval for an individual.

13. a) 464.8 ($000)
b) $464.8 \pm 97.52 = (366.2, 563.4)$ in $1000
c) No. Extrapolating what their sales might be with 500 employees based on data with between 2 and 20 employees working is likely to be very wrong.

CHAPTER EXERCISE ANSWERS

15. a)

Scatterplot of Minutes vs. Age

b) This scatterplot appears to have curvature at both ends of the age distribution so a linear regression may not be completely appropriate.
c) The regression equation is
$$\widehat{Minutes} = 750 - 11.5\, Age.$$
d) The residual plots are:

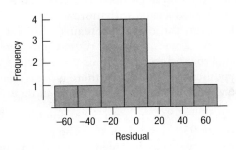

Histogram (response is Minutes)

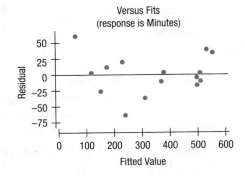

Versus Fits (response is Minutes)

The nearly normal condition is satisfied. There may be some curvature to the residual plot.

17. a) $\widehat{Budget} = -63.9981 + 1.03\, Run\, Time.$ The model suggests that movies cost about $1,030,000 per minute to make.

b) A negative starting value makes no sense, so it is best to interpret it as just a starting value.
c) Amounts by which movie costs differ from predictions made by this model vary, with a standard deviation of about $33 million.
d) 0.15$M/min
e) If we constructed other models based on different samples of movies, we'd expect the slopes of the regression lines to vary, with a standard deviation of about $150,000 per minute.

19. a) The scatterplot looks straight enough, the residuals look random and roughly normal, and the residuals don't display any clear change in variability although there may be some increasing spread.
b) I'm 95% confident that the cost of making longer movies increases at a rate of between 0.41 and 1.01 million dollars per minute. (CI is 0.41 to 1.02 using raw data.)

21. a) H_0: There is no linear relationship between calcium concentration in water and mortality rates for males. $(\beta_1 = 0)$ H_A: There is a linear relationship between calcium concentration in water and mortality rates for males. $(\beta_1 \neq 0)$
b) $t = -6.65, P < 0.0001$; reject the null hypothesis. There is strong evidence of a linear relationship between calcium concentration and mortality. Towns with higher calcium concentrations tend to have lower mortality rates.
c) For 95% confidence, use $t^*_{59} \approx 2.001$, or estimate from the table $t^*_{50} \approx 2.009$; $(-4.20, -2.26)$.
d) We are 95% confident that the average mortality rate decreases by between 2.26 and 4.20 deaths per 100,000 for each additional part per million of calcium in drinking water.

23. a)

Variable	Coefficient	SE(Coeff)	t-ratio	P-value
Intercept	452.496	12.05	37.6	≤ 0.0001
Year	−0.1869	0.0020	−31.1	≤ 0.0001

The LFPR for men has declined by about 18.7% per year during this period.
b) The residuals show no strong pattern that would call our assumptions into question. However, these data are measured over time, so they are probably not mutually independent.

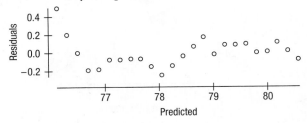

c) The null hypothesis that the slope is 0 is rejected. $P < 0.0001$. We can be confident that the apparent decline in the LFPR is not a random fluctuation.

d) $R^2 = 97.7\%$ of the variation is accounted for by this model.

25. a)

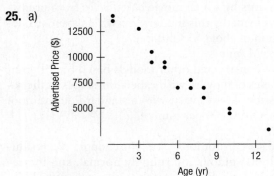

b) Yes, the plot seems linear.

c) $\widehat{Advertised\ Price} = 14286 - 959 \times Age$

d)

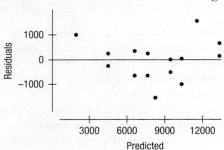

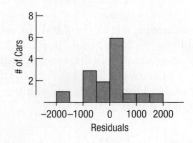

The residual plot shows some possible curvature. Inference may not be valid here, but we will proceed (with caution).

27. Based on these data, we are 95% confident that a used car's *Price* decreases between $819.50 and $1098.50 per year.

29. a)

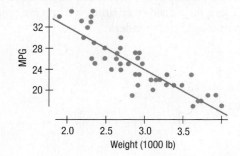

$\widehat{mpg} = 48.7393 - 8.21362(weight)$

b) Yes, the conditions seem satisfied. Histogram of residuals is unimodal and symmetric; residual plot looks okay, but some "thickening" of the plot with increasing values. There may be one possible outlier.

c) H_0: There is no linear relationship between the weight of a car and its fuel efficiency. $(\beta_1 = 0)$
H_A: There is a linear relationship between the weight of a car and its mileage. $(\beta_1 \neq 0)$

d) $t = -12.2$, df $= 48$, $P < 0.0001$; reject the null hypothesis. There is strong evidence of a linear relationship between weight of a car and its mileage. Cars that weigh more tend to have lower gas mileage.

31. a) H_0: There is no linear relationship between SAT Verbal and Math scores. $(\beta_1 = 0)$
H_A: There is a linear relationship between SAT Verbal and Math scores. $(\beta_1 \neq 0)$

b) Assumptions seem reasonable, since conditions are satisfied. Residual plot shows no patterns (one outlier); histogram is unimodal and roughly symmetric.

c) $t = 11.9$, df $= 160$, $P < 0.0001$; reject the null hypothesis. There is strong evidence of a linear relationship between SAT Verbal and Math scores. Students with higher SAT Verbal scores tend to have higher SAT Math scores.

33. a) $H_0: \beta_1 = 0$ vs. $H_A: \beta_1 \neq 0$

b) The P-value of 0.0076 shows strong evidence to reject the null hypothesis. It seems that higher salaries are associated with more wins.

c) However, there are two teams with low salaries and only 2 wins. Without those teams, the P-value is 0.396, and would lead us not to reject the null hypothesis. It seems that there is little relationship between salary and wins excluding those two teams.

35. a) $(-9.57, -6.86)$ mpg per 1000 pounds

b) We are 95% confident that the mean mileage of cars decreases by between 6.86 and 9.57 miles per gallon for each additional 1000 pounds of weight.

37. a) H_0: No linear relationship between *Population density* and *Ozone*, $\beta_1 = 0$.
H_A: $\beta_1 \neq 0$. $t = -2.29$; $P = 0.035$. With a P-value less than 0.05, we reject H_0.

b) Not particularly. Population density accounts for only 23.5% of the variation in ozone. However, s is only .009 ppm.

39. a) A 90% confidence interval for the slope (using a t^* value on 17 df) is $(-5.055, -0.685)$.

b) The mean *Ozone* level for cities with a population density of 0.002 million people/square km is 0.0665 ppm. A 90% confidence interval for that mean is $(0.0641, 0.0689)$.

41. a) 69 tablets.

b) Yes. The scatterplot is roughly linear with lots of scatter; plot of residuals vs. predicted values shows no overt patterns; Normal probability plot of residuals is reasonably straight.

c) H_0: No linear relationship between *Battery Life* and *Screen Brightness*, $\beta_1 = 0$. H_A: *Battery Life* is associated with *Screen Brightness*, $\beta \neq 0$. $t = 1.842$; P-value $= 0.070$. There is a slight *positive* association between *Screen Brightness* and *Battery Life*, but the P-value is too large to conclude that it did not occur by chance.

d) Not particularly. $R^2 = 4.82\%$ and $s = 1.946$ hours. Since the range of battery life is only about 9 hours, an s of 1.946 is quite large.

e) *Hours* $= 6.21 + 0.0049$ *Screen Brightness*

f) $(-0.00041, 0.0101)$ hours per cd/m^2 units

g) *Battery life* increases, on average, between -0.00041 and 0.0101 *hours* per one cd/m^2 units, with 95% confidence.

43. a) H_0: $\beta_1 = 0$ (There is no linear relationship between the 2011 index and the 2010 index.) vs. H_A: $\beta_1 \neq 0$ (There is a linear relationship.)

b) The P-value is very small, so we can reject the null hypothesis and conclude that expensive cities continue to be expensive.

c) 84% of the variation in the cost of living index in 2011 can be accounted for by the index in 2010.

d) Yes, the slope is positive and statistically significant.

45. a) The plot has a cluster of expensive cities in the upper right and a cluster of inexpensive cities in the lower left with a large gap between them.

b) The R^2 is inflated by the large separation of the two groups.

47. a)

b) Yes, there appears to be a linear association

c)
```
Dependent variable is: Attend/G
  R-squared = 59.3%  R-squared (adjusted) = 55.9%
  s = 6025 with 14 − 2 = 12 degrees of freedom
Variable     Coeff     SE(Coeff)   t-Ratio   P-Value
Intercept   −120463      35750      −3.37     0.0056
PitchAge     5305.56      1269       4.18     0.0013
```

Teams with pitchers who are a year older tend to have about 5305 more fans on average at each game.

49. a) We are 95% confident that the mean fuel efficiency of cars that weigh 2500 pounds is between 27.34 and 29.07 miles per gallon.

b) We are 95% confident that a car weighing 3450 pounds will have fuel efficiency between 15.44 and 25.36 miles per gallon.

51. a) $(-4.99, 5.29)$

b) Yes, it will probably be accurate, but not very useful since most of the pitchers fall in to this range.

53. a) We'd like to know if there is a linear association between *Price* and *Highway MPG* in cars. We have data on 2012 model-year cars giving their highway mpg and retail price. H_0: $\beta_1 = 0$ (no linear relationship between *Price* and *Highway MPG*); H_A: $\beta_1 \neq 0$.

b) The scatterplot shows that the Straight Enough Condition is reasonably satisfied. The residual plots show no particular pattern, so we can continue.

c) We find a relatively weak negative linear relationship between efficiency and price. An increase in price of $1000 is associated, on average, with a decrease of 0.42 highway mpg, but the R^2 of 10.3% suggests that the model accounts for relatively little of the variation in Energy use.

55. a) The 95% prediction interval shows the interval of uncertainty for a single country's predicted male unemployment rate, given its female unemployment rate.

b) The 95% confidence interval shows the interval of uncertainty for the mean male unemployment rate given a sample of female unemployment rates. Because this is an interval for an average, the variation or uncertainty is less, so the interval is narrower.

c) The two outliers are Mexico and Turkey, whose female rate is much lower than the male rate. There is no obvious justification for removing them, but the model would be stronger without them.

57. a) The 95% prediction interval shows the interval of uncertainty for the predicted energy use in 2010 based on energy use in 2006 for a single country.

b) The 95% confidence interval shows the interval of uncertainty for the mean energy use in 2010 based on the same energy use in 2006 for a sample of countries. Because this is an interval for an average, the variation or uncertainty is less, so the interval is narrower.

c) The outlier is Iceland. It does impact the regression. After removing it, the slope is lowered to 0.922 (SE 0.06) and the intercept becomes 5.744 (SE 9.06). The R^2 is lowered to 88.6% from 91.3%. The standard deviation of the residuals is now 13.10, which is much smaller than the original

22.53. This will make the intervals narrower and our predictions more precise.

59. a) $\widehat{Jan} = 120.73 + 0.6995\ Dec.$ We are told this is an SRS. One cardholder's spending should not affect another's. These are quantitative data with no apparent bend in the scatterplot. The residual plot shows some increased spread for larger values of January charges. A histogram of residuals is unimodal and slightly skewed to the right with several high outliers. We will proceed cautiously.

b) $1519.73

c) ($1330.24, $1709.24)

d) ($290.76, $669.76)

e) The residuals show increasing spread, so the confidence intervals may not be valid. I would be skeptical of interpreting them too literally.

CHAPTER 15
SECTION EXERCISE ANSWERS

1. a) $99,859.89

b) $35,140.11

c) The house sold for about $35,100 more than our estimate.

3. a) $\widehat{USGross} = -22.9898 + 1.13442\ Budget + 24.9724\ Stars - 0.403296\ RunTime$

b) After allowing for the effects of *RunTime* and *Stars*, each million dollars spent making a film yields about 1.13 million dollars in gross revenue.

5. a) Linearity: The plot is reasonably linear with no bends.

b) Equal spread: The plot fails the Equal Spread condition. It is much more spread out to the right than on the left.

c) Normality: A scatterplot of two of the variables doesn't tell us anything about the distribution of the residuals.

7. a) $H_0:\beta_{Stars} = 0$

b) $t = 4.24$

c) $P \leq 0.0001$

d) Reject the null hypothesis and conclude that the coefficient of *Stars* is not zero.

9. a) $R^2 = 0.474$ or 47.4%

About 47.4% of the variation in *USGross* is accounted for by the least squares regression on *Budget*, *RunTime*, and *Stars*.

b) Adjusted R^2 accounts for the number of predictors, so it differs from R^2, which does not.

CHAPTER EXERCISE ANSWERS

11. a) Linearity: The scatterplots show little pattern, but are not nonlinear.
Independence: States are not a random sample, but they may be independent of each other.
Equal variance: The scatterplots do not appear to have a changing spread.
Normality: To check the Nearly Normal condition, we'll need to look at the residuals; we can't check it with these plots.

b) 0.26%

13. a) $\widehat{Violent\ Crime} = 1370.22 + 0.795\ Police\ Officer\ Wage - 12.641\ Graduation\ Rate$

b) After allowing for the effects of graduation rate (or, alternatively, among states with similar graduation rates), states with higher police officer wages have more crime at the rate of 0.7953 crimes per 100,000 for each dollar per hour of average wage.

c) 501.244 crimes per 100,000

d) Not very good; R^2 is only 22.4% and $s = 156.9$.

15. a) $0.221 = \dfrac{0.7947}{3.598}$

b) 50 states. There are 47 degrees of freedom and that's equal to $n - k - 1$. With two predictors, $50 - 2 - 1 = 47$.

c) The *t*-ratio is negative because the coefficient is negative.

17. a) $H_0:\beta_{Officer\ Wage} = 0$ vs. $H_A:\beta_{Officer\ Wage} \neq 0$

b) $P = 0.8262$; that's not small enough to reject the null hypothesis at $\alpha = .05$ and conclude that the coefficient is discernibly different from zero.

19. This is a causal interpretation, which is not supported by regression. For example, among states with high graduation rates, it may be that those with higher violent crime rates choose (or are obliged) to spend more to hire police officers, or that states with higher costs of living must pay more to attract qualified police officers but also have higher crime rates. Moreover, the coefficient of *Police Wage* cannot be distinguished from 0.

21. Constant variance condition: may be met by the residuals vs. predicted plot (but some may judge otherwise).
Nearly Normal condition: met by the Normal probability plot

23. a) Doesn't mention other predictors; suggests direct relationship

b) Correct

c) Can't predict x from y

d) Incorrect interpretation of R^2

25. a) Extrapolates far from the data

b) Suggests a perfect relationship

c) Can't predict one explanatory variable from another

d) Correct

27. a) The sign of the coefficient for ln(*Number of Employees*) is negative. This means that for businesses that have the same amount of sales, those with more employees spend less per employee on pollution abatement on average. The sign of the coefficient for ln(*Sales*) is positive. This means that for businesses with the same number of employees, those with larger sales spend more on pollution abatement on average.

b) The logarithms mean that the effects become less severe (in dollar terms) as companies get larger either in sales or in number of employees.

29. a) $\widehat{Price} = -152,037 + 9530\,Baths + 139.87\,Area$

b) $R^2 = 71.1\%$

c) For houses with the same number of bathrooms, each square foot of area is associated with an increase of $139.87 in the price of the house, on average.

d) The regression model says that for houses of the same size, those with more bathrooms are not priced higher. It says nothing about what would happen if a bathroom were added to a house. That would be a predictive interpretation, which is not supported by regression.

31. a) The regression equation is: $\widehat{Salary} = 9.788 + 0.110\,Service + 0.053\,Education + 0.071\,Test\,Score + 0.004\,Typing\,wpm + 0.065\,Dictation\,wpm$

b) $Salary = 29.205$ or $29,205

c) The *t*-value is 0.013 with 24 df. P-value = 0.9897 which is not significant at $\alpha = 0.05$.

d) Take out the explanatory variable for typing speed since it is not significant.

e) *Age* is likely to be collinear with several of the other predictors already in the model. For example, secretaries with longer terms of *Service* will probably also be older.

33. a) This model explains less than 4% of the variation in *GDP per Capita*. The P-value is not particularly low.

b) Because more education is generally associated with a higher standard of living, it is not surprising that the simple association between *Primary Completion Rate* and GDP is positive

c) The coefficient now is measuring the association between *GDP/Capita* and *Primary Completion Rate* after accounting for the two other predictors.

35. a) $\widehat{Price} = 1.094 + 1.236\,Traps\,(M) - 0.000149\,Fishers - 0.0180\,Pounds/Trap$

b) The residuals show greater spread on the right and their distribution shows a possible outlier on the high end. We might wonder whether values from year to year are mutually independent. We should interpret the model with caution.

c) $H_0: \beta_{Pounds/Trap} = 0$ vs. $H_A: \beta_{Pounds/Trap} \neq 0$; the P-value is 0.0011. It appears that *Pounds/Trap* does contribute to the model.

d) No. We can't draw causal conclusions from a regression. A change in *Pounds/Trap* would likely affect other variables in the model.

37. a) R squared = 66.7% R squared (adjusted) = 63.8%
s = 2.327 with 39 − 4 = 35 degrees of freedom

Variable	Coeff	SE(Coeff)	t-ratio	P-value
Intercept	87.0089	33.60	2.59	0.0139
CPI	−0.344795	0.1203	−2.87	0.0070
Personal Consumption	1.10842e−5	4.403e−6	2.52	0.0165
Retail Sales	1.03152e−4	1.545e−5	6.67	<0.0001

$\widehat{Revenue} = 87.0 - 0.344\,CPI + 0.000011\,Personal\,Consumption + 0.0001\,Retail\,Sales$

b) R^2 is 66.7%, and all *t*-ratios are significant. It looks like these variables can account for much of the variation in Wal-Mart revenue.

39. a) $\widehat{Logit(Drop)} = -0.4419 - 0.0379\,Age + 0.0468\,HDRS$

b) −0.1749

c) 0.4564

d) −2.342

e) 0.0877

41. *Displacement* and *Bore* would be good predictors. Relationship with *Wheelbase* isn't linear.

43. a) Yes, R^2 of 90.9% says that most of the variability of *MSRP* is accounted for by this model.

b) No, a regression model may not be inverted in this way.

45. a) Yes, R^2 is very large.

b) The value of *s*, 3.140 calories, is very small compared with the initial standard deviation of *Calories*. This indicates that the model fits the data quite well, leaving very little variation unaccounted for.

c) A true value of 100% would indicate zero residuals, but with real data such as these, it is likely that the computed value of 100% is rounded up from a slightly lower value.

Tables and Selected Formulas

Two-tail probability		0.20	0.10	0.05	0.02	0.01	
One-tail probability		0.10	0.05	0.025	0.01	0.005	
Table T	df						df
Values of t_α	1	3.078	6.314	12.706	31.821	63.657	1
	2	1.886	2.920	4.303	6.965	9.925	2
	3	1.638	2.353	3.182	4.541	5.841	3
	4	1.533	2.132	2.776	3.747	4.604	4
	5	1.476	2.015	2.571	3.365	4.032	5
	6	1.440	1.943	2.447	3.143	3.707	6
	7	1.415	1.895	2.365	2.998	3.499	7
	8	1.397	1.860	2.306	2.896	3.355	8
	9	1.383	1.833	2.262	2.821	3.250	9
	10	1.372	1.812	2.228	2.764	3.169	10
	11	1.363	1.796	2.201	2.718	3.106	11
	12	1.356	1.782	2.179	2.681	3.055	12
	13	1.350	1.771	2.160	2.650	3.012	13
	14	1.345	1.761	2.145	2.624	2.977	14
	15	1.341	1.753	2.131	2.602	2.947	15
	16	1.337	1.746	2.120	2.583	2.921	16
	17	1.333	1.740	2.110	2.567	2.898	17
	18	1.330	1.734	2.101	2.552	2.878	18
	19	1.328	1.729	2.093	2.539	2.861	19
	20	1.325	1.725	2.086	2.528	2.845	20
	21	1.323	1.721	2.080	2.518	2.831	21
	22	1.321	1.717	2.074	2.508	2.819	22
	23	1.319	1.714	2.069	2.500	2.807	23
	24	1.318	1.711	2.064	2.492	2.797	24
	25	1.316	1.708	2.060	2.485	2.787	25
	26	1.315	1.706	2.056	2.479	2.779	26
	27	1.314	1.703	2.052	2.473	2.771	27
	28	1.313	1.701	2.048	2.467	2.763	28
	29	1.311	1.699	2.045	2.462	2.756	29
	30	1.310	1.697	2.042	2.457	2.750	30
	32	1.309	1.694	2.037	2.449	2.738	32
	35	1.306	1.690	2.030	2.438	2.725	35
	40	1.303	1.684	2.021	2.423	2.704	40
	45	1.301	1.679	2.014	2.412	2.690	45
	50	1.299	1.676	2.009	2.403	2.678	50
	60	1.296	1.671	2.000	2.390	2.660	60
	75	1.293	1.665	1.992	2.377	2.643	75
	100	1.290	1.660	1.984	2.364	2.626	100
	120	1.289	1.658	1.980	2.358	2.617	120
	140	1.288	1.656	1.977	2.353	2.611	140
	180	1.286	1.653	1.973	2.347	2.603	180
	250	1.285	1.651	1.969	2.341	2.596	250
	400	1.284	1.649	1.966	2.336	2.588	400
	1000	1.282	1.646	1.962	2.330	2.581	1000
	∞	1.282	1.645	1.960	2.326	2.576	∞
Confidence levels		80%	90%	95%	98%	99%	

Two tails

One tail

Right-tail probability		0.10	0.05	0.025	0.01	0.005
Table X	df					
Values of χ_α^2	1	2.706	3.841	5.024	6.635	7.879
	2	4.605	5.991	7.378	9.210	10.597
	3	6.251	7.815	9.348	11.345	12.838
	4	7.779	9.488	11.143	13.277	14.860
	5	9.236	11.070	12.833	15.086	16.750
	6	10.645	12.592	14.449	16.812	18.548
	7	12.017	14.067	16.013	18.475	20.278
	8	13.362	15.507	17.535	20.090	21.955
	9	14.684	16.919	19.023	21.666	23.589
	10	15.987	18.307	20.483	23.209	25.188
	11	17.275	19.675	21.920	24.725	26.757
	12	18.549	21.026	23.337	26.217	28.300
	13	19.812	22.362	24.736	27.688	29.819
	14	21.064	23.685	26.119	29.141	31.319
	15	22.307	24.996	27.488	30.578	32.801
	16	23.542	26.296	28.845	32.000	34.267
	17	24.769	27.587	30.191	33.409	35.718
	18	25.989	28.869	31.526	34.805	37.156
	19	27.204	30.143	32.852	36.191	38.582
	20	28.412	31.410	34.170	37.566	39.997
	21	29.615	32.671	35.479	38.932	41.401
	22	30.813	33.924	36.781	40.290	42.796
	23	32.007	35.172	38.076	41.638	44.181
	24	33.196	36.415	39.364	42.980	45.559
	25	34.382	37.653	40.647	44.314	46.928
	26	35.563	38.885	41.923	45.642	48.290
	27	36.741	40.113	43.195	46.963	49.645
	28	37.916	41.337	44.461	48.278	50.994
	29	39.087	42.557	45.722	59.588	52.336
	30	40.256	43.773	46.979	50.892	53.672
	40	51.805	55.759	59.342	63.691	66.767
	50	63.167	67.505	71.420	76.154	79.490
	60	74.397	79.082	83.298	88.381	91.955
	70	85.527	90.531	95.023	100.424	104.213
	80	96.578	101.879	106.628	112.328	116.320
	90	107.565	113.145	118.135	124.115	128.296
	100	118.499	124.343	129.563	135.811	140.177

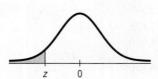

Table Z	SECOND DECIMAL PLACE IN Z										
Areas under the standard Normal curve	0.09	0.08	0.07	0.06	0.05	0.04	0.03	0.02	0.01	0.00	z
										0.0000[†]	−3.9
	0.0001	0.0001	0.0001	0.0001	0.0001	0.0001	0.0001	0.0001	0.0001	0.0001	−3.8
	0.0001	0.0001	0.0001	0.0001	0.0001	0.0001	0.0001	0.0001	0.0001	0.0001	−3.7
	0.0001	0.0001	0.0001	0.0001	0.0001	0.0001	0.0001	0.0001	0.0002	0.0002	−3.6
	0.0002	0.0002	0.0002	0.0002	0.0002	0.0002	0.0002	0.0002	0.0002	0.0002	−3.5
	0.0002	0.0003	0.0003	0.0003	0.0003	0.0003	0.0003	0.0003	0.0003	0.0003	−3.4
	0.0003	0.0004	0.0004	0.0004	0.0004	0.0004	0.0004	0.0005	0.0005	0.0005	−3.3
	0.0005	0.0005	0.0005	0.0006	0.0006	0.0006	0.0006	0.0006	0.0007	0.0007	−3.2
	0.0007	0.0007	0.0008	0.0008	0.0008	0.0008	0.0009	0.0009	0.0009	0.0010	−3.1
	0.0010	0.0010	0.0011	0.0011	0.0011	0.0012	0.0012	0.0013	0.0013	0.0013	−3.0
	0.0014	0.0014	0.0015	0.0015	0.0016	0.0016	0.0017	0.0018	0.0018	0.0019	−2.9
	0.0019	0.0020	0.0021	0.0021	0.0022	0.0023	0.0023	0.0024	0.0025	0.0026	−2.8
	0.0026	0.0027	0.0028	0.0029	0.0030	0.0031	0.0032	0.0033	0.0034	0.0035	−2.7
	0.0036	0.0037	0.0038	0.0039	0.0040	0.0041	0.0043	0.0044	0.0045	0.0047	−2.6
	0.0048	0.0049	0.0051	0.0052	0.0054	0.0055	0.0057	0.0059	0.0060	0.0062	−2.5
	0.0064	0.0066	0.0068	0.0069	0.0071	0.0073	0.0075	0.0078	0.0080	0.0082	−2.4
	0.0084	0.0087	0.0089	0.0091	0.0094	0.0096	0.0099	0.0102	0.0104	0.0107	−2.3
	0.0110	0.0113	0.0116	0.0119	0.0122	0.0125	0.0129	0.0132	0.0136	0.0139	−2.2
	0.0143	0.0146	0.0150	0.0154	0.0158	0.0162	0.0166	0.0170	0.0174	0.0179	−2.1
	0.0183	0.0188	0.0192	0.0197	0.0202	0.0207	0.0212	0.0217	0.0222	0.0228	−2.0
	0.0233	0.0239	0.0244	0.0250	0.0256	0.0262	0.0268	0.0274	0.0281	0.0287	−1.9
	0.0294	0.0301	0.0307	0.0314	0.0322	0.0329	0.0336	0.0344	0.0351	0.0359	−1.8
	0.0367	0.0375	0.0384	0.0392	0.0401	0.0409	0.0418	0.0427	0.0436	0.0446	−1.7
	0.0455	0.0465	0.0475	0.0485	0.0495	0.0505	0.0516	0.0526	0.0537	0.0548	−1.6
	0.0559	0.0571	0.0582	0.0594	0.0606	0.0618	0.0630	0.0643	0.0655	0.0668	−1.5
	0.0681	0.0694	0.0708	0.0721	0.0735	0.0749	0.0764	0.0778	0.0793	0.0808	−1.4
	0.0823	0.0838	0.0853	0.0869	0.0885	0.0901	0.0918	0.0934	0.0951	0.0968	−1.3
	0.0985	0.1003	0.1020	0.1038	0.1056	0.1075	0.1093	0.1112	0.1131	0.1151	−1.2
	0.1170	0.1190	0.1210	0.1230	0.1251	0.1271	0.1292	0.1314	0.1335	0.1357	−1.1
	0.1379	0.1401	0.1423	0.1446	0.1469	0.1492	0.1515	0.1539	0.1562	0.1587	−1.0
	0.1611	0.1635	0.1660	0.1685	0.1711	0.1736	0.1762	0.1788	0.1814	0.1841	−0.9
	0.1867	0.1894	0.1922	0.1949	0.1977	0.2005	0.2033	0.2061	0.2090	0.2119	−0.8
	0.2148	0.2177	0.2206	0.2236	0.2266	0.2296	0.2327	0.2358	0.2389	0.2420	−0.7
	0.2451	0.2483	0.2514	0.2546	0.2578	0.2611	0.2643	0.2676	0.2709	0.2743	−0.6
	0.2776	0.2810	0.2843	0.2877	0.2912	0.2946	0.2981	0.3015	0.3050	0.3085	−0.5
	0.3121	0.3156	0.3192	0.3228	0.3264	0.3300	0.3336	0.3372	0.3409	0.3446	−0.4
	0.3483	0.3520	0.3557	0.3594	0.3632	0.3669	0.3707	0.3745	0.3783	0.3821	−0.3
	0.3859	0.3897	0.3936	0.3974	0.4013	0.4052	0.4090	0.4129	0.4168	0.4207	−0.2
	0.4247	0.4286	0.4325	0.4364	0.4404	0.4443	0.4483	0.4522	0.4562	0.4602	−0.1
	0.4641	0.4681	0.4721	0.4761	0.4801	0.4840	0.4880	0.4920	0.4960	0.5000	−0.0

[†] For $z \leq -3.90$ the areas are 0.0000 to four decimal places.

Table Z (cont.)

Areas under the standard Normal curve

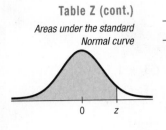

z	0.00	0.01	0.02	0.03	0.04	0.05	0.06	0.07	0.08	0.09
0.0	0.5000	0.5040	0.5080	0.5120	0.5160	0.5199	0.5239	0.5279	0.5319	0.5359
0.1	0.5398	0.5438	0.5478	0.5517	0.5557	0.5596	0.5636	0.5675	0.5714	0.5753
0.2	0.5793	0.5832	0.5871	0.5910	0.5948	0.5987	0.6026	0.6064	0.6103	0.6141
0.3	0.6179	0.6217	0.6255	0.6293	0.6331	0.6368	0.6406	0.6443	0.6480	0.6517
0.4	0.6554	0.6591	0.6628	0.6664	0.6700	0.6736	0.6772	0.6808	0.6844	0.6879
0.5	0.6915	0.6950	0.6985	0.7019	0.7054	0.7088	0.7123	0.7157	0.7190	0.7224
0.6	0.7257	0.7291	0.7324	0.7357	0.7389	0.7422	0.7454	0.7486	0.7517	0.7549
0.7	0.7580	0.7611	0.7642	0.7673	0.7704	0.7734	0.7764	0.7794	0.7823	0.7852
0.8	0.7881	0.7910	0.7939	0.7967	0.7995	0.8023	0.8051	0.8078	0.8106	0.8133
0.9	0.8159	0.8186	0.8212	0.8238	0.8264	0.8289	0.8315	0.8340	0.8365	0.8389
1.0	0.8413	0.8438	0.8461	0.8485	0.8508	0.8531	0.8554	0.8577	0.8599	0.8621
1.1	0.8643	0.8665	0.8686	0.8708	0.8729	0.8749	0.8770	0.8790	0.8810	0.8830
1.2	0.8849	0.8869	0.8888	0.8907	0.8925	0.8944	0.8962	0.8980	0.8997	0.9015
1.3	0.9032	0.9049	0.9066	0.9082	0.9099	0.9115	0.9131	0.9147	0.9162	0.9177
1.4	0.9192	0.9207	0.9222	0.9236	0.9251	0.9265	0.9279	0.9292	0.9306	0.9319
1.5	0.9332	0.9345	0.9357	0.9370	0.9382	0.9394	0.9406	0.9418	0.9429	0.9441
1.6	0.9452	0.9463	0.9474	0.9484	0.9495	0.9505	0.9515	0.9525	0.9535	0.9545
1.7	0.9554	0.9564	0.9573	0.9582	0.9591	0.9599	0.9608	0.9616	0.9625	0.9633
1.8	0.9641	0.9649	0.9656	0.9664	0.9671	0.9678	0.9686	0.9693	0.9699	0.9706
1.9	0.9713	0.9719	0.9726	0.9732	0.9738	0.9744	0.9750	0.9756	0.9761	0.9767
2.0	0.9772	0.9778	0.9783	0.9788	0.9793	0.9798	0.9803	0.9808	0.9812	0.9817
2.1	0.9821	0.9826	0.9830	0.9834	0.9838	0.9842	0.9846	0.9850	0.9854	0.9857
2.2	0.9861	0.9864	0.9868	0.9871	0.9875	0.9878	0.9881	0.9884	0.9887	0.9890
2.3	0.9893	0.9896	0.9898	0.9901	0.9904	0.9906	0.9909	0.9911	0.9913	0.9916
2.4	0.9918	0.9920	0.9922	0.9925	0.9927	0.9929	0.9931	0.9932	0.9934	0.9936
2.5	0.9938	0.9940	0.9941	0.9943	0.9945	0.9946	0.9948	0.9949	0.9951	0.9952
2.6	0.9953	0.9955	0.9956	0.9957	0.9959	0.9960	0.9961	0.9962	0.9963	0.9964
2.7	0.9965	0.9966	0.9967	0.9968	0.9969	0.9970	0.9971	0.9972	0.9973	0.9974
2.8	0.9974	0.9975	0.9976	0.9977	0.9977	0.9978	0.9979	0.9979	0.9980	0.9981
2.9	0.9981	0.9982	0.9982	0.9983	0.9984	0.9984	0.9985	0.9985	0.9986	0.9986
3.0	0.9987	0.9987	0.9987	0.9988	0.9988	0.9989	0.9989	0.9989	0.9990	0.9990
3.1	0.9990	0.9991	0.9991	0.9991	0.9992	0.9992	0.9992	0.9992	0.9993	0.9993
3.2	0.9993	0.9993	0.9994	0.9994	0.9994	0.9994	0.9994	0.9995	0.9995	0.9995
3.3	0.9995	0.9995	0.9995	0.9996	0.9996	0.9996	0.9996	0.9996	0.9996	0.9997
3.4	0.9997	0.9997	0.9997	0.9997	0.9997	0.9997	0.9997	0.9997	0.9997	0.9998
3.5	0.9998	0.9998	0.9998	0.9998	0.9998	0.9998	0.9998	0.9998	0.9998	0.9998
3.6	0.9998	0.9998	0.9999	0.9999	0.9999	0.9999	0.9999	0.9999	0.9999	0.9999
3.7	0.9999	0.9999	0.9999	0.9999	0.9999	0.9999	0.9999	0.9999	0.9999	0.9999
3.8	0.9999	0.9999	0.9999	0.9999	0.9999	0.9999	0.9999	0.9999	0.9999	0.9999
3.9	1.0000[†]									

SECOND DECIMAL PLACE IN Z

† For $z \geq 3.90$, the areas are 1.0000 to four decimal places.

Selected Formulas

$Range = Max - Min$

$IQR = Q3 - Q1$

Outlier Rule-of-Thumb: $y < Q1 - 1.5 \times IQR$ or $y > Q3 + 1.5 \times IQR$

$$\bar{y} = \frac{\sum y}{n}$$

$$s = \sqrt{\frac{\sum (y - \bar{y})^2}{n - 1}}$$

$$z = \frac{y - \mu}{\sigma} \text{ (model based)} \qquad z = \frac{y - \bar{y}}{s} \text{ (data based)}$$

$$r = \frac{\sum z_x z_y}{n - 1}$$

$$\hat{y} = b_0 + b_1 x \qquad \text{where } b_1 = r\frac{s_y}{s_x} \text{ and } b_0 = \bar{y} - b_1\bar{x}$$

$$P(\mathbf{A}) = 1 - P(\mathbf{A}^C)$$

$$P(\mathbf{A} \text{ or } \mathbf{B}) = P(\mathbf{A}) + P(\mathbf{B}) - P(\mathbf{A} \text{ and } \mathbf{B})$$

$$P(\mathbf{A} \text{ and } \mathbf{B}) = P(\mathbf{A}) \times P(\mathbf{B}|\mathbf{A})$$

$$P(\mathbf{B}|\mathbf{A}) = \frac{P(\mathbf{A} \text{ and } \mathbf{B})}{P(\mathbf{A})}$$

If $\mathbf{A}$ and $\mathbf{B}$ are independent, $P(\mathbf{B}|\mathbf{A}) = P(\mathbf{B})$

$$E(X) = \mu = \sum xP(x) \qquad\quad Var(X) = \sigma^2 = \sum (x - \mu)^2 P(x)$$

$$E(X \pm c) = E(X) \pm c \qquad\quad Var(X \pm c) = Var(X)$$

$$E(aX) = aE(X) \qquad\qquad\quad Var(aX) = a^2 Var(X)$$

$$E(X \pm Y) = E(X) \pm E(Y) \quad Var(X \pm Y) = Var(X) + Var(Y)$$
$$\text{if } X \text{ and } Y \text{ are independent}$$

Geometric: $\qquad P(x) = q^{x-1}p \qquad\quad \mu = \frac{1}{p} \qquad \sigma = \sqrt{\frac{q}{p^2}}$

Binomial: $\qquad P(x) = {_n}C_x p^x q^{n-x} \quad \mu = np \qquad \sigma = \sqrt{npq}$

$\hat{p} = \frac{x}{n} \qquad \mu(\hat{p}) = p \qquad SD(\hat{p}) = \sqrt{\frac{pq}{n}}$

Poisson probability model for successes: Poisson (λ)

λ = mean number of successes.

X = number of successes.

$$P(X = x) = \frac{e^{-\lambda}\lambda^x}{x!}$$

Expected value: $\qquad E(X) = \lambda$

Standard deviation: $\qquad SD(X) = \sqrt{\lambda}$

Sampling distribution of $\bar{y}$:

(CLT) As n grows, the sampling distribution approaches the Normal model with

$$\mu(\bar{y}) = \mu_y \qquad SD(\bar{y}) = \frac{\sigma}{\sqrt{n}}$$

Inference:

Confidence interval for parameter = ***statistic*** $\pm$ ***critical value*** $\times$ ***SE***(***statistic***)

Test statistic $= \dfrac{statistic - parameter}{SD(statistic)}$

Parameter	Statistic	SD (statistic)	SE (statistic)
p	$\hat{p}$	$\sqrt{\dfrac{pq}{n}}$	$\sqrt{\dfrac{\hat{p}\hat{q}}{n}}$
μ	$\bar{y}$	$\dfrac{\sigma}{\sqrt{n}}$	$\dfrac{s}{\sqrt{n}}$
$\mu_1 - \mu_2$	$\bar{y}_1 - \bar{y}_2$	$\sqrt{\dfrac{\sigma_1^2}{n_1} + \dfrac{\sigma_2^2}{n_2}}$	$\sqrt{\dfrac{s_1^2}{n_1} + \dfrac{s_2^2}{n_2}}$
μ_d	$\bar{d}$	$\dfrac{\sigma_d}{\sqrt{n}}$	$\dfrac{s_d}{\sqrt{n}}$
σ_ε	$s_e = \sqrt{\dfrac{\sum(y - \hat{y})^2}{n - 2}}$	(divide by $n - k - 1$ in multiple regression)	
β_1	b_1	(in simple regression)	$\dfrac{s_e}{s_x\sqrt{n - 1}}$
μ_ν	$\hat{y}_\nu$	(in simple regression)	$\sqrt{SE^2(b_1) \times (x_\nu - \bar{x})^2 + \dfrac{s_e^2}{n}}$
y_ν	$\hat{y}_\nu$	(in simple regression)	$\sqrt{SE^2(b_1) \times (x_\nu - \bar{x})^2 + \dfrac{s_e^2}{n} + s_e^2}$

Pooling: For testing difference between proportions: $\hat{p}_{pooled} = \dfrac{y_1 + y_2}{n_1 + n_2}$

For testing difference between means: $s_p = \sqrt{\dfrac{(n_1 - 1)s_1^2 + (n_2 - 1)s_2^2}{n_1 + n_2 - 2}}$

Substitute these pooled estimates in the respective SE formulas for both groups when assumptions and conditions are met.

Chi-square: $\chi^2 = \sum \dfrac{(Obs - Exp)^2}{Exp}$

Assumptions for Inference	And the Conditions That Support or Override Them

Proportions (z)

- **One sample**
  1. Individuals are independent.
  2. Sample is sufficiently large.

    1. SRS and $n < 10\%$ of the population.
    2. Successes and failures each ≥ 10.

Means (t)

- **One Sample** (df $= n - 1$)
  1. Individuals are independent.
  2. Population has a Normal model.

    1. SRS and $n < 10\%$ of the population.
    2. Histogram is unimodal and symmetric.*

- **Matched pairs** (df $= n - 1$)
  1. Data are matched.
  2. Individuals are independent.
  3. Population of differences is Normal.

    1. (Think about the design.)
    2. SRS and $n < 10\%$ OR random allocation.
    3. Histogram of differences is unimodal and symmetric.*

- **Two independent samples** (df from technology)
  1. Groups are independent.
  2. Data in each group are independent.
  3. Both populations are Normal.

    1. (Think about the design.)
    2. SRSs and $n < 10\%$ OR random allocation.
    3. Both histograms are unimodal and symmetric.*

Distributions/Association (χ^2)

- **Goodness of fit** (df $=$ # of cells $- 1$; one variable, one sample compared with population model)
  1. Data are counts.
  2. Data in sample are independent.
  3. Sample is sufficiently large.

    1. (Are they?)
    2. SRS and $n < 10\%$ of the population.
    3. All expected counts ≥ 5.

- **Homogeneity** [df $= (r - 1)(c - 1)$; many groups compared on one variable]
  1. Data are counts.
  2. Data in groups are independent.
  3. Groups are sufficiently large.

    1. (Are they?)
    2. SRSs and $n < 10\%$ OR random allocation.
    3. All expected counts ≥ 5.

- **Independence** [df $= (r - 1)(c - 1)$; sample from one population classified on two variables]
  1. Data are counts.
  2. Data are independent.
  3. Sample is sufficiently large.

    1. (Are they?)
    2. SRSs and $n < 10\%$ of the population.
    3. All expected counts ≥ 5.

Regression with k predictors (t, df $= n - k - 1$)

- **Association** of each quantitative predictor with the response variable
  1. Form of relationship is linear.

  2. Errors are independent.

  3. Variability of errors is constant.

  4. Errors follow a Normal model.

    1. Scatterplots of y against each x are straight enough. Scatterplot of residuals against predicted values shows no special structure.
    2. No apparent pattern in plot of residuals against predicted values.
    3. Plot of residuals against predicted values has constant spread, doesn't "thicken."
    4. Histogram of residuals is approximately unimodal and symmetric, or Normal probability plot is reasonably straight.*

* Less critical as n increases

Quick Guide to Inference

Plan			Do				Report
Inference about?	**One group or two?**	**Procedure**	**Model**	**Parameter**	**Estimate**	**SE**	**Chapter**
Proportions	One sample	1-Proportion z-Interval	z	p	$\hat{p}$	$\sqrt{\dfrac{\hat{p}\hat{q}}{n}}$	9
		1-Proportion z-Test				$\sqrt{\dfrac{p_0 q_0}{n}}$	10
Means	One sample	t-Interval t-Test	t df $= n - 1$	μ	$\bar{y}$	$\dfrac{s}{\sqrt{n}}$	11
	Two independent groups	2-Sample t-Test 2-Sample t-Interval	t df from technology	$\mu_1 - \mu_2$	$\bar{y}_1 - \bar{y}_2$	$\sqrt{\dfrac{s_1^2}{n_1} + \dfrac{s_2^2}{n_2}}$	13
	Matched pairs	Paired t-Test Paired t-Interval	t df $= n - 1$	μ_d	$\bar{d}$	$\dfrac{s_d}{\sqrt{n}}$	13
Distributions (one categorical variable)	One Sample	Goodness-of-Fit	χ^2 df $= cells - 1$			$\displaystyle\sum \frac{(Obs - Exp)^2}{Exp}$	14
	Many independent groups	Homogeneity χ^2 Test	χ^2 df $= (r-1)(c-1)$				
Independence (two categorical variables)	One sample	Independence χ^2 Test					
Association (two quantitative variables)	One sample	Linear Regression t-Test or Confidence Interval for β	t df $= n - 2$	β_1	b_1	$\dfrac{s_e}{s_x \sqrt{n-1}}$ (compute with technology)	15, 16
		*Confidence Interval for μ_ν		μ_ν	$\hat{y}_\nu$	$\sqrt{SE^2(b_1) \times (x_\nu - \bar{x})^2 + \dfrac{s_e^2}{n}}$	
		*Prediction Interval for y_ν		y_ν	$\hat{y}_\nu$	$\sqrt{SE^2(b_1) \times (x_\nu - \bar{x})^2 + \dfrac{s_e^2}{n} + s_e^2}$	
Association (one quantitative and two or more categorical variables)	One sample	Multiple Regression t-test or Confidence interval for each β_j	t df $= n - (k+1)$	β_j	b_j	(from technology)	17, 18
		F test for regression model	F df $= k$ and $n - (k+1)$			MST/MSE	18
Association (one quantitative and two or more categorical variables)	Two or more	ANOVA	F df $= k - 1$ and $N - k$			MST/MSE	20

Photo Acknowledgments

Title page
Page iii, Yvan Dubé/Getty Images

Meet the Authors
Page VI, Courtesy of Norean Radke Sharpe, Courtesy of Richard D. De Veaux, Courtesy of Paul F. Velleman

Chapter 1
Page 1, Anton Balazh/Shutterstock; Page 6, Margo Harrison/Shutterstock; **Page 13**, Pearson Education, Inc.; **Page 14**, Lightpoet/Fotolia

Chapter 2
Page 22, Courtesy of Keen Shoes/Photo © Ben Moon, Copyright © 2012 CereusData LLC. All rights reserved; **Page 31**, Adisa/Shutterstock; **Page 39**, Feng Yu/Shutterstock, Dorling Kindersley Ltd.;

Chapter 3
Page 49, Richard Drew/AP Images; **Page 62**, Lindasj22/Shutterstock; **Page 64**, Pressmaster/Shutterstock; **Page 66**, Africa Studio/Fotolia; **Page 76**, Ithaca Times; **Page 82**, Iriana Shiyan/Shutterstock

Chapter 4
Page 97, David Parker/Alamy; **Page 101**, Library of Congress Prints and Photographs Division [LC-USZ62-61365]; **Page 104**, Epsicons/Shutterstock; **Page 106**, Crystal Kirk/Shutterstock; **Page 112**, Pearson Education, Inc., Frank Richards/Pearson Education, Inc.; **Page 118**, Andy Dean Photography/Shutterstock; **Page 124**, Ufulum/Shutterstock; **Page 129**, Michael Shake/Fotolia **Page 142**, Zuma Press, Inc/Alamy

Chapter 5
Page 145, Laitr Keiows/Fotolia; **Page 148**, Frank Richards/Pearson Education, Inc.; **Page 149**, Matthew Mawson/Alamy, PhotoEdit; **Page 154**, Avigo Photos/Fotolia; **Page 160**, Iriana Shiyan/Shutterstock; **Page 161**, Kim Steele/Getty Images; **Page 168**, Yuri Arcurs/Shutterstock

Chapter 6
Page 179, Kroomjai/Shutterstock; **Page 184**, Steve Design/Shutterstock; **Page 188**, Universal Uclick; **Page 189**, Photos 12/Alamy; **Page 192**, Keith Brofsky/Photodisc/Getty Images; **Page 200**, Elena Elisseeva/Shutterstock

Chapter 7
Page 207, Gary/Fotolia; **Page 214**, David Buffington/Getty Images; **Page 219**, Walter Hodges/Getty Images; **Page 228**, Getty Images

Chapter 8
Page 237, Bettmann/Corbis; **Page 239**, Michael Lamotte/Cole Group/Getty Images; **Page 244**, Digital Vision/Getty Images; **Page 245**, Tatiana Popova/Shutterstock; **Page 248**, Universal Uclick; **Page 252**, The Wizard of Id © 2001 John L. Hart/Distributed by Creators Syndicate. Reprinted with permission. All rights reserved; **Page 256**, Goodluz/Shutterstock

Chapter 9
Page 265, Bertys30/Fotolia; **Page 270**, Galina Barskaya/Shutterstock; **Page 277**, Universal Uclick; **Page 278**, Ittel/Sipa/Newscom; **Page 282**, Michael Blann/Photodisc/Getty Images; **Page 288**, Worldpics/Shutterstock; **Page 298**, Ji Pruitt/Shutterstock

Chapter 10
Page 299, Slavko Sereda/Shutterstock; **Page 303**, Image Source Plus/Alamy; **Page 308**, Richard Paul Kane/Shutterstock; **Page 310**, A. Barrington Brown/Science Source; **Page 315**, Tischenko Irina/Shutterstock; **Page 326**, Luoman/Getty Images

Chapter 11
Page 335, Tom Norring/Danita Delimont/Alamy; **Page 338**, Oldtime/Alamy; **Page 343**, Courtesy of the International Statistical Institute; **Page 349**, Sapsiwai/Shutterstock; **Page 352**, Jim Lopes/Shutterstock; **Page 361**, Scott Leigh/Getty Images

Chapter 12
Page 371, Goodshoot/Getty Images; **Page 373**, Peter Cade/Photodisc/Getty Images; **Page 376**, Ruslan Kudrin/Shutterstock, Alexey Stiop/Shutterstock, Michaeljung/Shutterstock, Iko/Fotolia, Bestweb/Shutterstock, Jorg Hackemann/Shutterstock, Jason Stitt/Shutterstock, Inga Dudkina/Dreamstime LLC, Plamen Petrov/Dreamstime LLC, Ron Zmiri/Dreamstime LLC; **Page 380**, Turtix/Shutterstock; **Page 382**, Lenaer/Shutterstock; **Page 383**, Monkeybusinessimages/iStock/Getty Images; **Page 390**, Ariel Skelley/Blend Images/Getty Images; **Page 400**, EricVega/E+/Getty Images

Chapter 13

Page 377, Jupiterimages/Stockbyte/Getty Images; **Page 417**, Ken Graff/iStock/Getty Images; **Page 422**, Tischenko Irina/Shutterstock; **Page 432**, Minerva Studio/ Fotolia; **Page 441**, Jakub Jirsák/Fotolia; **Page 453**, WavebreakmediaMicro/Fotolia

Chapter 14

Page 455, Courtesy of Nambé; **Page 462**, Courtesy of Nambé; **Page 471**, Peter Teller/Photodisc/Getty Images; **Page 474**, Larry Crowe/AP Images; **Page 475**, Pearson Education, Inc.

Chapter 15

Page 489, David Hughes/Shutterstock; **Page 493**, Monkey Business/Fotolia; **Page 499**, Rob Marmion/ Shutterstock; **Page 509**, Yamini Chao/Digital Vision/ Getty Images; **Page 510**, Brian A Jackson/Shutterstock; **Page 518**, Djtaylor/Fotolia

Chapter 16

Page 531, Phattman/Fotolia; **Page 536**, Plush Studios/ The Agency Collection/Getty Images; **Page 548**, Alexmisu/Shutterstock

Subject Index

Note: Page numbers in **boldface** indicate chapter-level topics; FE indicates For Example; and n indicates a footnote.

Note: Page numbers in **boldface** indicate chapter-level topics; FE indicates For Example; and n indicates a footnote.

Note: Page numbers in **boldface** indicate chapter-level topics; FE indicates For Example; and n indicates a footnote.

Note: Page numbers in **boldface** indicate chapter-level topics; FE indicates For Example; and n indicates a footnote.

Note: Page numbers in **boldface** indicate chapter-level topics; FE indicates For Example; and n indicates a footnote.

Note: Page numbers in **boldface** indicate chapter-level topics; FE indicates For Example; and n indicates a footnote.

Note: Page numbers in **boldface** indicate chapter-level topics; FE indicates For Example; and n indicates a footnote.

Note: Page numbers in **boldface** indicate chapter-level topics; FE indicates For Example; and n indicates a footnote.

Note: Page numbers in **boldface** indicate chapter-level topics; FE indicates For Example; and n indicates a footnote.

Assumptions for Inference	And the Conditions That Support or Override Them

Proportions (z)

- **One sample**
  1. Individuals are independent.
  2. Sample is sufficiently large.

  1. SRS and $n < 10\%$ of the population.
  2. Successes and failures each ≥ 10.

Means (t)

- **One Sample** (df $= n - 1$)
  1. Individuals are independent.
  2. Population has a Normal model.

  1. SRS and $n < 10\%$ of the population.
  2. Histogram is unimodal and symmetric.*

- **Matched pairs** (df $= n - 1$)
  1. Data are matched.
  2. Individuals are independent.
  3. Population of differences is Normal.

  1. (Think about the design.)
  2. SRS and $n < 10\%$ OR random allocation.
  3. Histogram of differences is unimodal and symmetric.*

- **Two independent samples** (df from technology)
  1. Groups are independent.
  2. Data in each group are independent.
  3. Both populations are Normal.

  1. (Think about the design.)
  2. SRSs and $n < 10\%$ OR random allocation.
  3. Both histograms are unimodal and symmetric.*

Distributions/Association (χ^2)

- **Goodness of fit** (df $= \#$ of cells $- 1$; one variable, one sample compared with population model)
  1. Data are counts.
  2. Data in sample are independent.
  3. Sample is sufficiently large.

  1. (Are they?)
  2. SRS and $n < 10\%$ of the population.
  3. All expected counts ≥ 5.

- **Homogeneity** [df $= (r - 1)(c - 1)$; many groups compared on one variable]
  1. Data are counts.
  2. Data in groups are independent.
  3. Groups are sufficiently large.

  1. (Are they?)
  2. SRSs and $n < 10\%$ OR random allocation.
  3. All expected counts ≥ 5.

- **Independence** [df $= (r - 1)(c - 1)$; sample from one population classified on two variables]
  1. Data are counts.
  2. Data are independent.
  3. Sample is sufficiently large.

  1. (Are they?)
  2. SRSs and $n < 10\%$ of the population.
  3. All expected counts ≥ 5.

Regression with k predictors (t, df $= n - k - 1$)

- **Association** of each quantitative predictor with the response variable
  1. Form of relationship is linear.

  1. Scatterplots of y against each x are straight enough. Scatterplot of residuals against predicted values shows no special structure.

  2. Errors are independent.
  3. Variability of errors is constant.

  2. No apparent pattern in plot of residuals against predicted values.
  3. Plot of residuals against predicted values has constant spread, doesn't "thicken."

  4. Errors follow a Normal model.

  4. Histogram of residuals is approximately unimodal and symmetric, or Normal probability plot is reasonably straight.*

Analysis of Variance (F, df dependent on number of factors and number of levels in each)

- **Equality** of the mean response across levels of categorical predictors
  1. Additive Model (if there are 2 factors with no interaction term).

  1. Interaction plot shows parallel lines (otherwise include an interaction term if possible).

  2. Independent errors.
  3. Equal variance across treatment levels.

  2. Randomized experiment or other suitable randomization.
  3. Plot of residuals against predicted values has constant spread. Boxplots (partial boxplots for 2 factors) show similar spreads.

  4. Errors follow a Normal model.

  4. Histogram of residuals is unimodal and symmetric, or Normal probability plot is reasonably straight.

* Less critical as n increases